Asymmetric Synthesis of Drugs and Natural Products

Asymmetric Synthesis of Drugs and Natural Products

Ahindra Nag

CRC Press
Taylor & Francis Group
Boca Raton London New York

CRC Press is an imprint of the
Taylor & Francis Group, an **informa** business

CRC Press
Taylor & Francis Group
6000 Broken Sound Parkway NW, Suite 300
Boca Raton, FL 33487-2742

© 2018 by Taylor & Francis Group, LLC
CRC Press is an imprint of Taylor & Francis Group, an Informa business

No claim to original U.S. Government works

International Standard Book Number-13: 978-1-138-03361-0 (Hardback)

This book contains information obtained from authentic and highly regarded sources. Reasonable efforts have been made to publish reliable data and information, but the author and publisher cannot assume responsibility for the validity of all materials or the consequences of their use. The authors and publishers have attempted to trace the copyright holders of all material reproduced in this publication and apologize to copyright holders if permission to publish in this form has not been obtained. If any copyright material has not been acknowledged please write and let us know so we may rectify in any future reprint.

Except as permitted under U.S. Copyright Law, no part of this book may be reprinted, reproduced, transmitted, or utilized in any form by any electronic, mechanical, or other means, now known or hereafter invented, including photocopying, microfilming, and recording, or in any information storage or retrieval system, without written permission from the publishers.

For permission to photocopy or use material electronically from this work, please access www.copyright.com (http://www.copyright.com/) or contact the Copyright Clearance Center, Inc. (CCC), 222 Rosewood Drive, Danvers, MA 01923, 978-750-8400. CCC is a not-for-profit organization that provides licenses and registration for a variety of users. For organizations that have been granted a photocopy license by the CCC, a separate system of payment has been arranged.

Trademark Notice: Product or corporate names may be trademarks or registered trademarks, and are used only for identification and explanation without intent to infringe.

Library of Congress Cataloging-in-Publication Data

Names: Nag, Ahindra, editor.
Title: Asymmetric synthesis of drugs and natural products / editor, Ahindra Nag.
Description: Boca Raton : CRC Press, 2018. | Includes bibliographical references.
Identifiers: LCCN 2017039701| ISBN 9781138033610 (hardback : alk. paper) | ISBN 9781315302317 (ebook)
Subjects: LCSH: Pharmaceutical chemistry. | Asymmetric synthesis. | Asymmetry (Chemistry) | Drugs--Synthesis. | Organic compounds--Synthesis.
Classification: LCC RS403 .A88 2018 | DDC 615.1/9--dc23
LC record available at https://lccn.loc.gov/2017039701

Visit the Taylor & Francis Web site at
http://www.taylorandfrancis.com

and the CRC Press Web site at
http://www.crcpress.com

Printed and bound in Great Britain by
TJ International Ltd, Padstow, Cornwall

Contents

Preface vii
Contributors ix

Chapter 1 Basic Stereochemical Approaches to Natural Products and Drugs 1

Ahindra Nag

Chapter 2 Diastereoselective Addition of Organometallic Reagents to Chiral Carbonyl
Compounds 29

Ivana Gergelitsová and Jan Veselý

Chapter 3 Enantiomerically Pure Compounds by Enantioselective Synthetic Chiral Metal
Complexes 75

Frady G. Adly and Ashraf Ghanem

Chapter 4 Chirality Organization of Peptides and π-Conjugated Polyanilines 133

Toshiyuki Moriuchi, Satoshi D. Ohmura, and Toshikazu Hirao

Chapter 5 Diastereoselective Syntheses of 3,4,5-Trihydroxypiperidine Iminosugars 163

Adam McCluskey and Michela I. Simone

Chapter 6 Use of Specific New Artificial or Semisynthetic Biocatalysts
for Synthesis of Regio- and Enantioselective Compounds 177

Marco Filice, Oscar Romero, and Jose M. Palomo

Chapter 7 Bioactive Natural Products and Their Structure–Activity Relationships Studies 191

Athar Ata and Hadeel Alhazmi

Chapter 8 Asymmetric Biocatalysis in Organic Synthesis of Natural Products 205

*Renata Kołodziejska, Aleksandra Karczmarska-Wódzka,
and Agnieszka Tafelska-Kaczmarek*

Chapter 9 Asymmetric Synthesis of Biaryls and Axially Chiral Natural Products 271

Renata Kołodziejska and Agnieszka Tafelska-Kaczmarek

Chapter 10 Palladium-Catalyzed Asymmetric Transformations of Natural Products
and Drug Molecules 293

*Mariette M. Pereira, Carolina S. Vinagreiro, Fábio M.S. Rodrigues,
and Rui M.B. Carrilho*

Chapter 11 Enantioselective Organocatalysis from Concepts to Applications
in the Synthesis of Natural Products and Pharmaceuticals 325

Marta Meazza and Ramon Rios

Chapter 12 Chiral Building Blocks for Drugs Synthesis *via* Biotransformations 345

Pilar Hoyos, Vittorio Pace, and Andrés R. Alcántara

Chapter 13 Chiral Medicines 449

Apurba Bhattacharya and Rakeshwar Bandichhor

Chapter 14 Drug Delivery Systems 461

Ahindra Nag

Problems to Be Solved 475

Index 485

Preface

In recent years, there has been considerable interest in the chemical and pharmaceutical industries for enantiomers of natural products and drugs due to considerable advances in the synthesis, analysis, separation of chiral molecules, and less complex and more selective pharmacodynamics profile of chiral drugs therapy for more effectiveness and safety. All activities are initiated to produce intermediates and bulk active chiral drugs for the global pharmaceutical market, which is predicted to grow to $800 billion by the year 2020.

There are several books available in asymmetric synthesis of drugs, but the subject material in most of them is present in a diffused or highly specialized form, which makes it very difficult for students to go through the various textbooks, journals, and pharmacopeias. The major objective of writing this book was to cater to the needs of undergraduate and postgraduate students in a lucid, condensed, and cohesive form.

Chapter 1 discusses the fundamental knowledge and instrumental analysis of chiral compounds. The other 12 chapters discuss methods of asymmetric synthesis of natural products and drugs, which follows the syllabus of various reputed universities so that students can use this book as a textbook. It will also be helpful to researchers and scientists in the research and development field. Bioactivity of natural products and their structure–activity relationship studies have been discussed in Chapter 9. The last chapter is on drug delivery systems, which will be helpful for understanding the goal, metabolism, and safety consideration of chiral drugs.

I express my indebtedness and gratitude to all the professors who have contributed chapters in the book. I also acknowledge my indebtedness to my wife Jayita Nag and my sons Aritra and Anindya Nag for their encouragement and sustained cooperation. I am also thankful to my research scholars Himadri Maity, Bipasha Halder, Suvendhu Halder, and Rahul Guin for assistance in the completion of the book. The cooperation of publishers, Renu Upadhya, Sikha Garg, and Taylor & Francis Group, USA, in bringing out the book is very much appreciated.

It is our hope that this book will meet the demand of all those who study the subject either in a course or through research application. I gladly invite constructive suggestions from fellow professors, students, scientists, and researchers for further improvement of the text.

Ahindra Nag

Contributors

Frady G. Adly
Department of Biomedical Science
University of Canberra
Canberra, Australian Capital Territory, Australia

Andrés R. Alcántara
Organic & Pharmaceutical Chemistry Department
Complutense University of Madrid
Madrid, Spain

Hadeel Alhazmi
Department of Chemistry
The University of Winnipeg
Winnipeg, Manitoba, Canada

Athar Ata
Department of Chemistry
The University of Winnipeg
Winnipeg, Manitoba, Canada

Rakeshwar Bandichhor
Center of Excellence, Research & Development, Integrated Product Development
Dr. Reddy's Laboratories Ltd.
Andhra Pradesh, India

Apurba Bhattacharya
Department of Chemistry
Texas A&M Kingsville
Kingsville, Texas

Rui M.B. Carrilho
Department of Chemistry
University of Coimbra
Coimbra, Portugal

Marco Filice
Department of Biocatalysis
Institute of Catalysis
Madrid, Spain

Ashraf Ghanem
Department of Biomedical Science
University of Canberra
Canberra, Australian Capital Territory, Australia

Ivana Gergelitsová
Department of Organic Chemistry
Charles University
Prague, Czech Republic

Toshikazu Hirao
The Institute of Scientific and Industrial Research
Osaka University
Osaka, Japan

Pilar Hoyos
Organic & Pharmaceutical Chemistry Department
Complutense University of Madrid
Madrid, Spain

Aleksandra Karczmarska-Wódzka
Department of Pharmacology and Therapy
Nicolaus Copernicus University in Toruń, Collegium Medicum in Bydgoszcz
Bydgoszcz, Poland

Renata Kołodziejska
Department of Biochemistry
Nicolaus Copernicus University in Toruń, Collegium Medicum in Bydgoszcz
Bydgoszcz, Poland

Adam McCluskey
Discipline of Chemistry
University of Newcastle
Callaghan, New South Wales, Australia

Marta Meazza
School of Chemistry
University of Southampton
Southampton, United Kingdom

Toshiyuki Moriuchi
Department of Applied Chemistry
Graduate School of Engineering
Osaka University
Osaka, Japan

Ahindra Nag
Department of Chemistry
Indian Institute of Technology
Kharagpur, India

Satoshi D. Ohmura
Department of Science and Technology
Tokushima University
Tokushima, Japan

Vittorio Pace
Department of Pharmaceutical Chemistry
University of Vienna
Vienna, Austria

Jose M. Palomo
Department of Biocatalysis
Institute of Catalysis
Madrid, Spain

Mariette M. Pereira
Department of Chemistry
University of Coimbra
Coimbra, Portugal

Ramon Rios
School of Chemistry
University of Southampton
Southampton, United Kingdom

Fábio M.S. Rodrigues
Department of Chemistry
University of Coimbra
Coimbra, Portugal

Oscar Romero
Department of Biocatalysis
Institute of Catalysis
Madrid, Spain

Michela I. Simone
Discipline of Chemistry
Priority Research Centre for Chemical Biology
  & Clinical Pharmacology
University of Newcastle
Callaghan, New South Wales, Australia

Agnieszka Tafelska-Kaczmarek
Department of Chemistry
Nicolaus Copernicus University in Toruń
Toruń, Poland

Jan Veselý
Department of Organic Chemistry
Charles University
Prague, Czech Republic

Carolina S. Vinagreiro
Department of Chemistry
University of Coimbra
Coimbra, Portugal

1 Basic Stereochemical Approaches to Natural Products and Drugs

Ahindra Nag

CONTENTS

1.1　Basic Concept of Chirality 1
1.2　Meso Compounds 3
1.3　Tautomerism and Valance Tautomerism 4
1.4　Conformation 5
1.5　Fischer Projection and Absolute Configuration 9
1.6　Chiral Resolution 9
　　　1.6.1　Crystal Picking 10
　　　1.6.2　Chemical Separation 10
　　　1.6.3　Biochemical Separation 11
　　　1.6.4　Chromatographic Separation 12
1.7　Application of Enantiomers in Drugs and Natural Products 14
Problems 17
Answers 20
References 26

1.1　BASIC CONCEPT OF CHIRALITY

An asymmetric carbon atom (known as the **stereogenic center**) is attached to four different groups[1] termed a **chiral** (pronounced as kiral). The word chiral derives from the Greek word *cheira* meaning hand, which is closely related to optical activity. For a molecule to have chirality, it must not possess a plane, a center, or a fourfold alternating axis of symmetry. Molecules which are mirror images of each other are termed **enantiomers** (from the Greek *entatios* meaning opposite) and need chiral recognition to be separated. Enantiomers react[1-6] at different rates with other chiral compounds and may have different solubilities in the presence of an optically active solvent. They may display different absorption spectra under circulatory polarized light. Enantiomers may have different optical rotations, which could be either (+), that is, **dextrorotatory** (clockwise), or (−), that is, **levorotatory** (anticlockwise), and can be determined by a polarimeter. The optical purity of a mixture of enantiomers is given by

$$\% \text{ Optical purity of sample} = 100 * \frac{\text{Specific rotation of sample}}{\text{Specific rotation of a pure enantiomer}}$$

$$\text{Specific rotation} \left[\alpha\right]_D = \alpha_{obs}/cl$$

where
　　α_{obs} is the experimentally observed rotation
　　c is the concentration in g/mL
　　l is the path length of the cell used, expressed in dm (=10 cm)

Enantiomeric excess[7] is one of the indicators of the success of an asymmetric synthesis. The enantiomeric excess (ee$_p$) of a product is determined by chromatographic methods using the following formula:

$$ee_p = \frac{[d]-[l]}{[d]+[l]}$$

where [d] and [l] are the concentrations of the mixture of the products.

Enantiomeric ratio[7] (**E**) is calculated from the extent of conversion (c) and the enantiomeric excess (ee$_p$) of the product using the following equation:

$$E = \frac{\ln\left[1-c(1+ee_p)\right]}{\ln\left[1-c(1-ee_p)\right]}$$

The stereoisomers of a compound that are not mirror images of each other are termed **diastereomers** (Figure 1.1). They have different physical characteristics, such as refractive indices, specific rotations, melting and boiling points, crystalline structures, and different solubilities in different solvents. They possess different chemical properties when reacted with chiral and achiral reagents and different rates of reaction due to different values of ΔH. **Diastereoselectivity**[2] is the preference for the formation of one or more diastereomers over the other in an organic reaction. **Diastereomeric excess (De)** is defined as the proportion of major diastereomer (A) minus that of minor diastereomer (B):

$$De = 100(A-B)/A+B$$

Many natural products such as carbohydrates, proteins, terpenoids, alkaloids, and steroids have two or more stereogenic centers (Figure 1.2a through c).

Diastereomers are divided into two subclasses[2]: (1) **epimers** and (2) **anomers**.

1. *Epimers*: The compounds (Figure 1.3) that differ in geometry at one of the several stereocenters are called epimers, such as D-glucose and D-mannose, inositol and epi-inositol, lipoxin and epilipoxin, and so on. Inositol is a sugar alcohol, and its sweetness is half of sucrose, whereas lipoxin is a bioactive autacoid metabolite made up of various cells. Doxorubicin and epirubicin are two epimers that are used as drugs to treat cancer. **Epimerization** is a chemical process where an epimer is transformed into its chiral counterpart.
2. *Anomers*: The structure of anomeric compounds (Figure 1.4) differs only in an anomeric carbon such as α- or β-glucose. Hence, epimers differ only at the chiral center but not in an anomeric carbon. **Anomerization** is the process of converting one anomer to another.

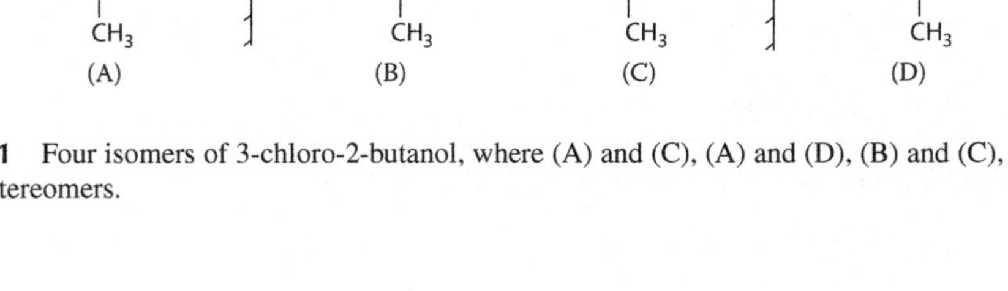

FIGURE 1.1 Four isomers of 3-chloro-2-butanol, where (A) and (C), (A) and (D), (B) and (C), and (B) and (D) are diastereomers.

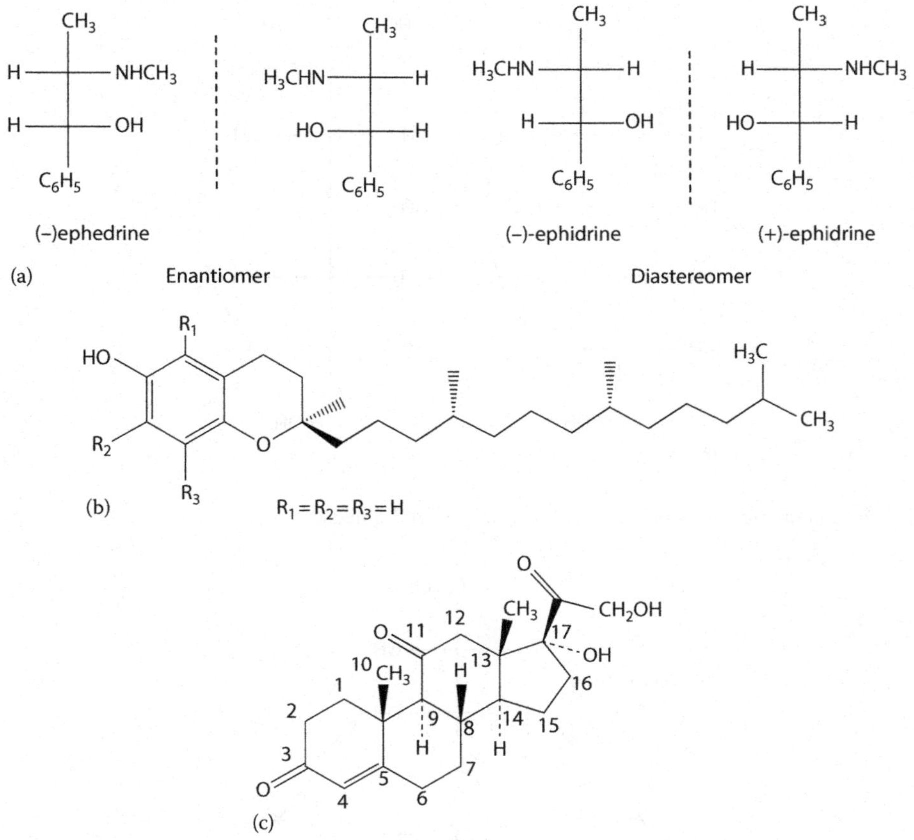

Basic Stereochemical Approaches to Natural Products and Drugs

FIGURE 1.2 (a) Ephedrine alkaloids (two stereogenic centers). (b) Vitamin E (three chiral centers). (c) Cortisone has six chiral centers (C-8, C-9, C-10, C-13, C-14, C-17).

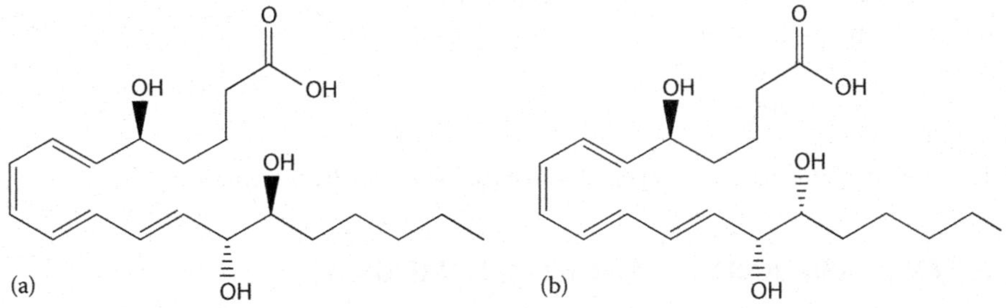

FIGURE 1.3 Epimer compounds: (a) lipoxin and (b) epilipoxin.

1.2 MESO COMPOUNDS

A meso compound (Figure 1.5) is defined as a 1:1 mixture of enantiomers. They have no observable optical rotation and cancel each other out, that is, a meso compound is a molecule with multiple stereocenters, which is superimposable on its mirror image. This particular property leads to specific qualities that meso compounds do not share with most other stereoisomers. In a meso compound, two substituents are common: for example, in 2,4-pentanediol, the second and fourth carbon atoms are stereocenters and all four substituents are common.

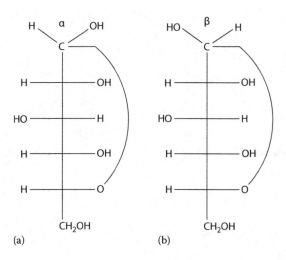

FIGURE 1.4 Anomeric compounds: (a) α-glucose and (b) β-glucose.

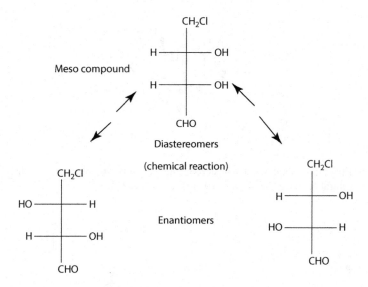

FIGURE 1.5 Relation of a racemic compound with enantiomer and diastereomer.

1.3 TAUTOMERISM AND VALANCE TAUTOMERISM

Tautomerism is a special type of functional isomerism in which the isomers are in dynamic equilibrium with each other: for example, ethyl acetoacetate (Figure 1.6), which is an equilibrium mixture of two forms (keto and enol). The difference between resonance and tautomerism is that resonance-contributing forms have no real existence, whereas in tautomerism, the two forms are in rapid equilibrium and they are isolable in their pure state. The tautomers differ from each other in the relative position of electron distribution, but in all resonance forms, the position of atomic nuclei remains unchanged and the actual structure of the entity remains one and the same.

Valence tautomerism is defined as the change in interatomic distances through the formation of new bonds by redistribution of valence electrons with a molecule: for example, cyclooctatetraene and bicyclooctatriene (Figure 1.7) show a pair of valence tautomers.

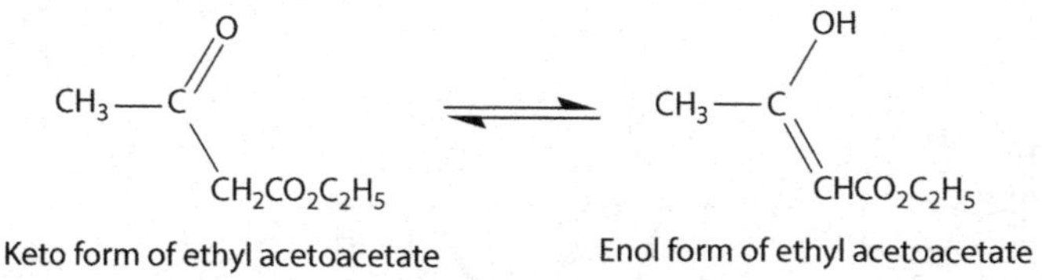

Basic Stereochemical Approaches to Natural Products and Drugs

FIGURE 1.6 Tautomerism of ethyl acetoacetate.

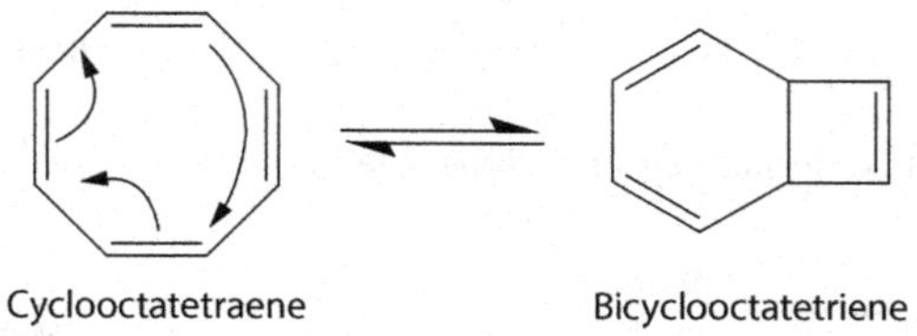

FIGURE 1.7 Valence tautomerism.

1.4 CONFORMATION

Arrangements of a molecule that can be obtained by rotation of carbon–carbon single bonds are called **conformations**, and a set of stereoisomers that differ in conformation, that is, in torsion angles and potential energy minima, are called conformational isomers (Figure 1.8). This information is supplied by **conformational formulas** such as *gauche* and *anti,* where *gauche* is defined as synonymous with a synclinal alignment of groups attached to adjacent atoms and *anti* is defined as a conformer whose two bulky groups have a torsion angle of about 180° (as two CH_3 groups of butane).

The **chair conformation** is the most stable state as it is formed with less torsional strain and the bond angle is nearly 109.5° whereas in the **boat conformation**, the hydrogens at C1 and C4 are close together, in which Van der Waal's interaction is known as **flagpole interaction**. The angle strain and flagpole interaction make the boat conformation less stable than the chair formation. Bonds that are almost parallel to the C_3 axis in the chair conformation of cyclohexane are called **axial bonds**, and the bonds that are directed away from the core of the cyclohexane molecule and are also in the approximate plane of the chair conformation of the same molecule are called **equatorial bonds** (Figure 1.9). There may likely be a difference of free energy in the ground state for equatorial and axial isomers than in the transition state, and an axial isomer with a higher ground-state energy will react at a faster rate, which is defined as **steric assistance** or **steric acceleration**. Stereoisomerism resulting from restricted rotation around single bonds because of a high rotational barrier is termed **atropisomerism**[8]: for example, tetra-substituted biphenyls at the ortho position.

When two atoms or groups attached to adjacent atoms and the torsion angle of three bonds are zero, it is said to be an **eclipsed conformation**. The conformation of groups attached to two adjacent atoms is said to be **staggered** if such groups are far away from a possible eclipsed arrangement with

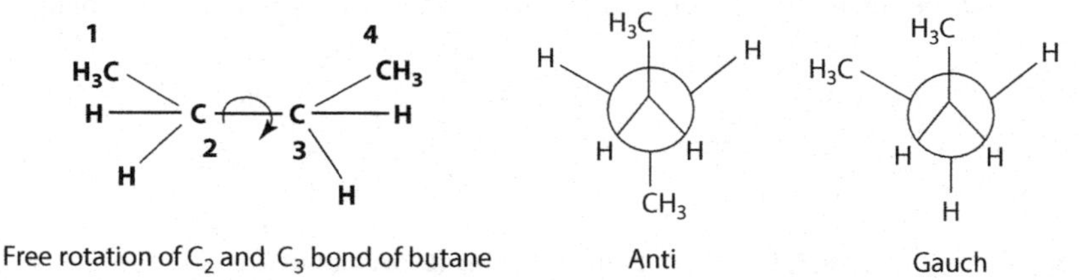

FIGURE 1.8 Conformation of butane.

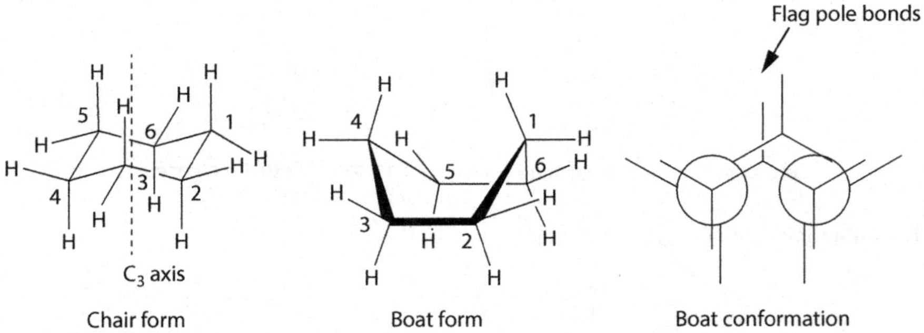

FIGURE 1.9 Chair and boat conformations of cyclohexane.

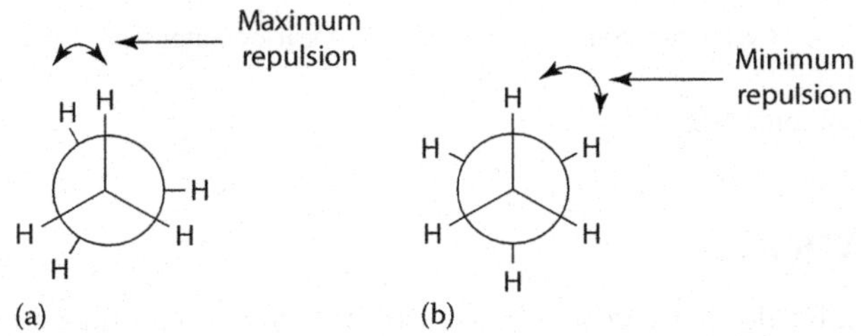

FIGURE 1.10 (a) Eclipsed (least stable) and (b) staggered (more stable) conformations.

minimum torsion angles. The eclipsed conformation is less stable than the staggered conformation because the hydrogen atoms and the bonding pair of electrons are as close to each other as possible, whereas in the staggered form they are at a maximum distance, which causes minimum repulsion (Figure 1.10).

In the *cis* and *trans* isomers (Figure 1.11), there are different substituents: where the higher-priority groups are on opposite sides of the bond, it is the (**E**)-isomer or configuration (German: *Entegegen*, meaning opposite side), and where the higher-priority groups are on the same side, it is the (**Z**)-isomer or configuration (German: *Zusammen*, meaning same side).

Cis and *trans* isomers can be interconverted by radical formation and free rotation around a C–C bond, which causes inversion of the configuration. This process is called **stereomutation** (Figure 1.12).

When the single bond between the two centers is free to rotate, *cis* and *trans* isomers are invalid. Two widely accepted prefixes used to distinguish diastereomers on sp³-hybridized bonds in an open-chain molecule are *syn* and *anti*. **Anti-elimination** takes place in a conformation which is staggered around the carbon–carbon bond, and the bond from the other carbon to the hydrogen is *anti*-coplanea. The dihedral angle between these bonds is 180°. In *anti*-elimination, the base attacks the β-hydrogen on the opposite side of the leaving group. Weakly ionizing solvents promote *anti*-elimination when the leaving group is charged. **Syn-elimination** takes place under a forcing condition, and the leaving groups cannot achieve *anti*-periplanarity (Figure 1.13). Here, departing groups are coplanar and eclipsed.

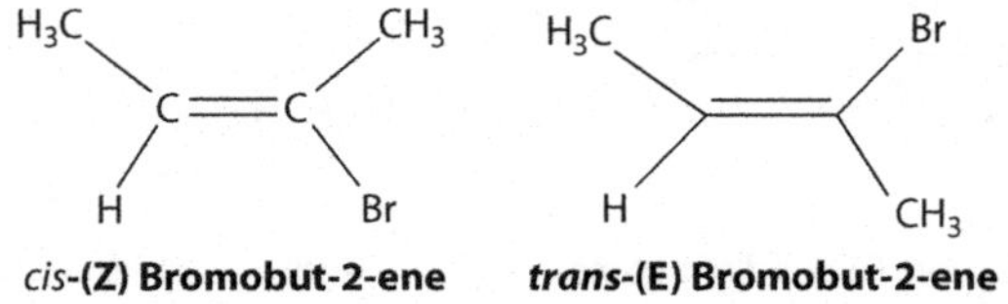

FIGURE 1.11 Geometrical isomerism of bromobut-2-ene.

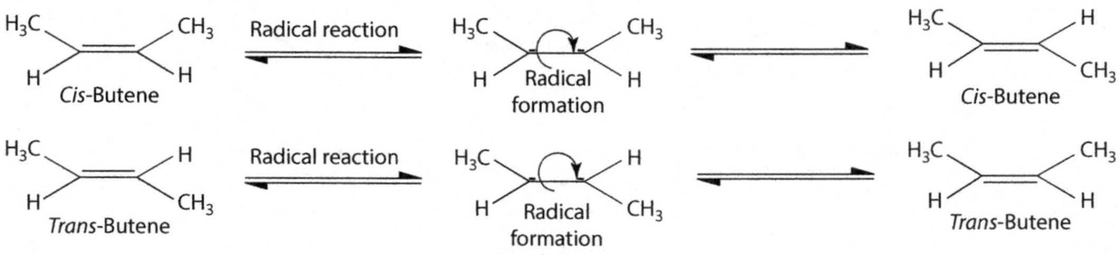

Basic Stereochemical Approaches to Natural Products and Drugs

FIGURE 1.12 Stereomutation process. (From Mehata, B. and Mahata, M., *Organic Chemistry,* Prientice-Hall of India, New Delhi, India, 2014. With permission.)

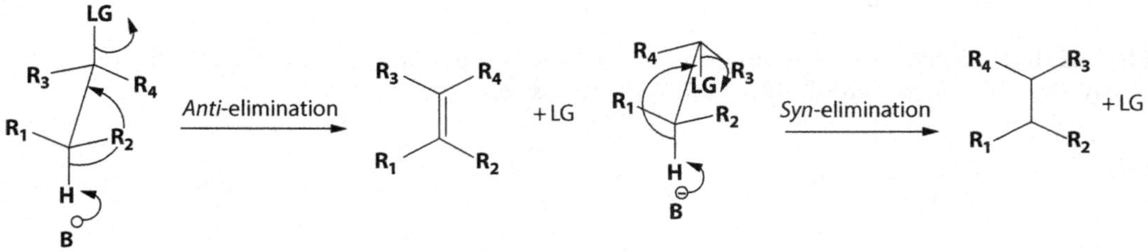

FIGURE 1.13 *Anti-* and *syn-*elimination reaction. LG, leaving group; B, base.

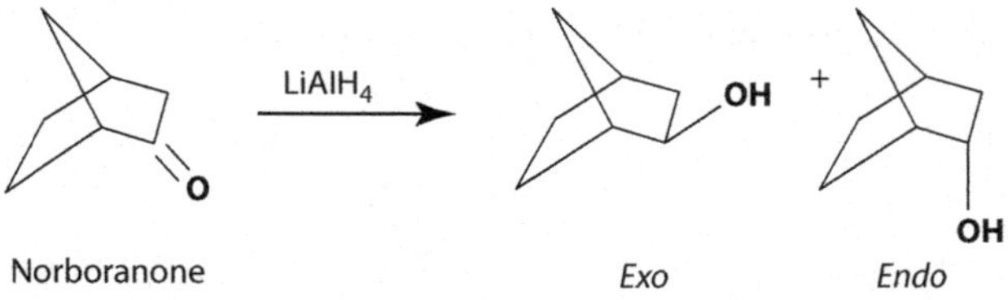

FIGURE 1.14 *Exo* and *Endo* compounds.

Weakly ionizing solvents promote *syn*-elimination when the leaving group is uncharged. In *syn*-elimination, the base attacks the β-hydrogen on the same side as the leaving group.

When a substitution is on the left side or on the same side on top of the bridge of a bicyclic compound, it is called **exo** (Greek *exo*, meaning out), and when the substitute is on the opposite side of the top of the bridge, it is called **endo** (Greek *endo*, meaning within) (Figure 1.14).

Optically pure α-fluoro-α-methyl-β-hydroxyl esters of erythro[8] or threo configuration were prepared by Kitazume et al. where (*S*)-α-fluoro-α-methyl-β-keto ester was transformed into erythro-β-hydroxy ester with >99% selectivity, whereas *R*-enantiomer was reduced to threo-β-hydroxy ester with >98% selectivity (Figure 1.15).

A line of electrons gets least priority and is ranked below the hydrogen (according to sequence rules) as oxime and azo, designated as *E* and *Z* isomers (Figure 1.16).

The rate of chemical reactions involving bond making (conformational transformations), which correlates the stereochemistry of starting materials and products in terms of transition states and intermediates, is defined as **dynamic stereochemistry**. Generally, in a **stereospecific** reaction, the stereochemically different starting materials lead to stereochemically different products: for example, two 2-butene diastereomers will produce two different stereomeric products (Figure 1.17).

But **stereoselective reaction** yields only one set of stereoisomers predominately as the formation of two isomeric *trans*-stilbene takes place only after dehydrochlorination of 1,2-diphenyl-1-chloroethane (Figure 1.18).

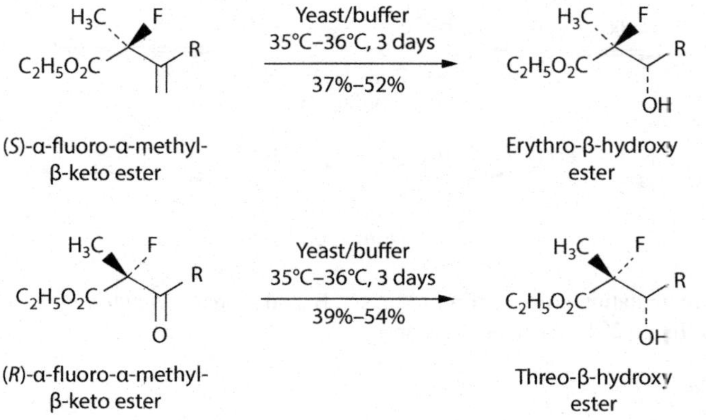

FIGURE 1.15 Synthesis of optically pure α-fluoro-α-methyl-β-hydroxyl esters. (From Kitazume, T. and Kobayashi, T., *Synthesis*, 1987(2), 187, 1987. With permission.)

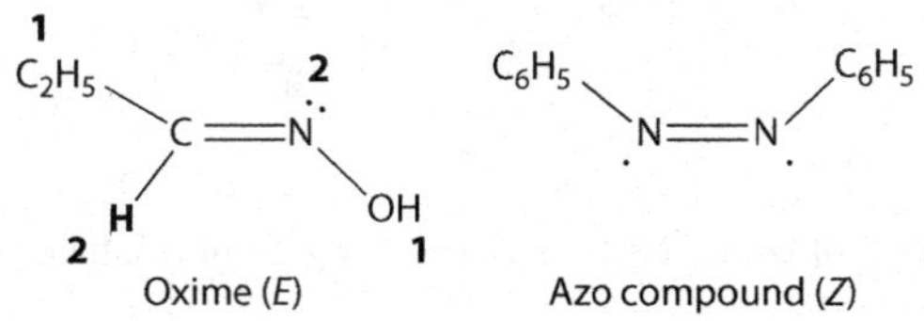

FIGURE 1.16 Geometrical isomerism of oxime (*E*) and azo compound (*Z*).

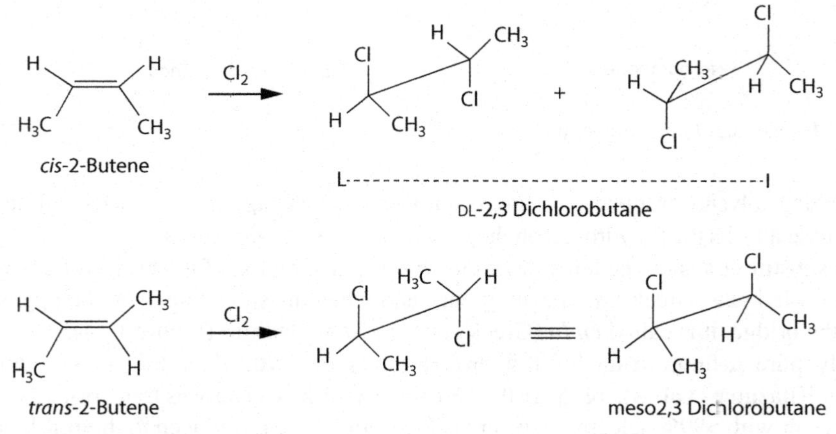

FIGURE 1.17 Stereospecific reactions.

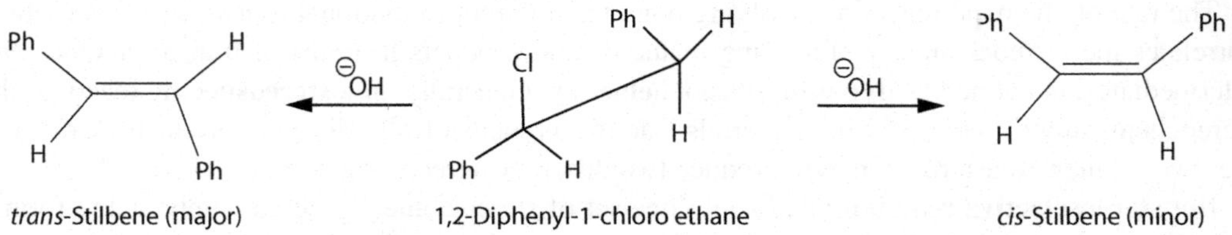

FIGURE 1.18 Stereoselective reactions.

1.5 FISCHER PROJECTION AND ABSOLUTE CONFIGURATION

Fischer projection is a planar representation of a three-dimensional structure (configuration) (Figure 1.19) where vertical lines represent bonds going away from the observer and horizontal lines represent bonds coming toward the observer (out of the plane). Fischer projection is very useful in organic chemistry, especially for monosaccharide and amino acid, and can also be used for differentiating enantiomers of chiral molecules. But Fischer projection structures cannot be rotated by 90° or 270° in the plane of the page because the orientation of bonds relative to one another can change, converting a molecule into its enantiomers, but a rotation by 180° is allowed.

Absolute configuration proposed by Cohn, Ingold, and Prelog is a special arrangement of four ligands attached to a chiral center (Figure 1.20). The directly attached atom having the highest atomic number has the highest priority, and the other ligand atoms follow in order of decreasing atomic numbers. The ligand of lowest atomic number is directed away from the viewer and is behind the plane of the paper. Multiple-bond atoms should be counted as having the same number as single-bond atoms. Similarly, *cis* precedes *trans*, Z precedes E, and R precedes S. If the sequence of the molecule from the highest to the lowest priority proceeds clockwise, the molecule is specified as R (rectus = right), and if its sequence proceeds counterclockwise, the molecule is designated as S (sinster = left). An enantiomerically pure substance that has either (+) *R* or (−) *S* configuration is termed **homochiral**, whereas isomeric molecules of opposite chirality, that is, where one is *R* and the other is *S*, are called **heterochiral**. Other models have been discussed in Chapter 2.

1.6 CHIRAL RESOLUTION

The separation of a racemic mixture into two optically active forms (+ or −) is known as **chiral resolution**.[13] This system is called **enantiomeric enrichment**, which is a process to increase the percentage of an enantiomer of enantiomerically impure compounds so that the enantiomeric excess gets close to 100%. As diastereomeric compounds have different physical and chemical properties, they can be separated by chiral resolution into their enantiomers. One disadvantage of the chiral resolution method compared to direct asymmetric synthesis is that in one of the enantiomers, only 50% of the desired enantiomer is obtained. **Kinetic resolution** is based on the principal for partial or complete resolution of a racemate, which depends on the transition state of either enantiomer in a chemical reaction with a chiral nonracemic reagent, catalyst, or enzyme, and the reaction is not allowed to proceed to completion. There are different methods for chiral resolution such as (1) **crystal picking**, (2) **chemical separation**, (3) **biochemical separation**, and (4) **chromatographic method**.

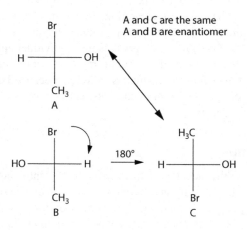

FIGURE 1.19 Fischer projection structures.

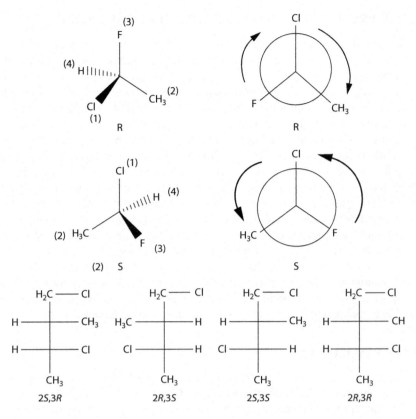

FIGURE 1.20 Absolute configuration of *R* proposed by Cohn, Ingold, and Prelog on the basis of priority substituents. (From Mehata, B. and Mahata, M., *Organic Chemistry*, Prientice-Hall of India, New Delhi, India, 2014. With permission.)

1.6.1 CRYSTAL PICKING

Louis Pasteur first discovered[1] this method in 1853 by the formation of diastereomeric salts and their fractional crystallization. At least 5%–10% of all racemates are known to crystallize as mixtures of enantiopure crystals, which are called **conglomerates**. This method is optimized by a resolving agent and its molar ratio, effect of temperature, and pH. The application of this method is laborious and time-consuming. This is the physical separation of crystals of (+) or (−) by hand picking or through a lens from a racemic mixture, which may produce crystals that are enantiomorphic. This manual method of separation of conglomerate crystals into individual enantiomers is called **triage**. Otherwise, in a supersaturated solution of a racemic mixture, a pure crystal of an enantiomer is added to produce a preferential enantiomer, which is called **preferential crystallization** by inoculation. Addition of a (−) seed of hydrobenzoin in ethanolic solution of (±) hydrobenzoin will produce a (−) crystal with 97% optical purity.

1.6.2 CHEMICAL SEPARATION

In this method, enantiomers of the racemic mixtures are converted into diastereomers through salt formation, where the salts differ in solubility in different solvents and can be separated by fractional crystallization and chromatographic separation, which are preferred for covalent diastereomers. Organic bases such as cinchonidine, brucine, cinchonine, ephedrine, quinine, and strychnine are used. The addition of optically active base (−) strychnine in a racemic mixture of lactic acid will form salts of

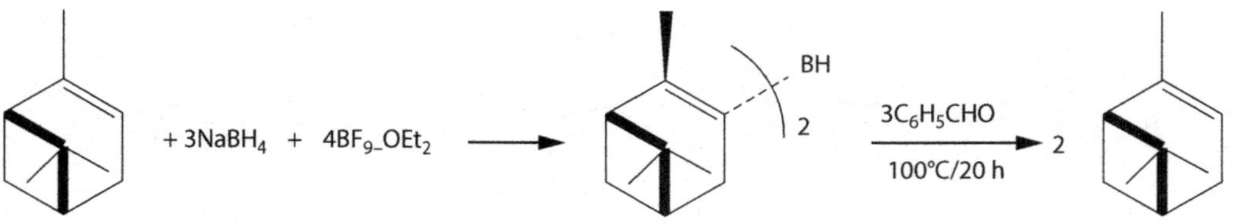

FIGURE 1.21 Resolution of α-pinene.

(−) strychnine (+) lactate and (−) strychnine (−) lactate. Treatment with diluted mineral acid removes active base (−) strychnine and leaves separate isomers of (+) lactic acid and (−) lactic acid. For resolution of acids, synthetic amines are used, especially α-phenyl amine. The resolved acids in turn can be used for the resolution of racemic amines. Kozma et al.[9] separated racemic mixtures by the distillation method. They suggested that if the racemic compound reacted with half an equivalent amount of resolving agent, the enantiomeric mixture that remained after crystallization could be separated by distillation. Fogassy et al.[10] suggested that in the resolution of racemic ibuprofen with (R)-phenyl ethyl amine, the free enantiomer (S)-ibuprofen can be separated from the salt ((R)-ibuprofen) to produce (R)-phenyl ethyl amine by extraction with a supercritical fluid (carbon dioxide) (Figure 1.29). The optical purity (91.3%) of the enantiomer (+) α-pinene was increased to 98.9% by hydroboration of α-pinene with sodium borohydride and boron trifluoride etherate to get diisopinocamphenyl borane (Ipc $_2$BH).[11] In this process, 15% excess α-pinene was used to stop the isomer of diisopinocamphenyl borane, and filtrate solution after treatment with benzaldehyde at 100°C for 20 h will produce (+) α-pinene that is 98.9% optically pure and Ipc $_2$BH in an almost pure form (Figure 1.21).

1.6.3 BIOCHEMICAL SEPARATION

When certain microorganisms (bacteria, yeast, mold fungi) are added to a solution of a racemic mixture, they cause the decomposition of one of the optically active forms and assimilate on one of the enantiomers more rapidly than the others.[12] *Penicillium glaucum* decomposes[13] (+) tartaric acid more rapidly than (−) isomer, and (+) isomer can be easily obtained from the residue after treatment with the mold. Singh et al.[14] observed that *Aspergillus fumigatus* L-*amino acid oxidase* (L-aao) had the ability to cause the resolution of racemic mixtures of DL-amino acids, DL-alanine, DL-phenylalanine, DL-tyrosine, and DL-aspartic acid. A chiral column, Crownpak CR+ was used for the analysis of the amino acids. The enzyme was able to cause the resolution of the three DL-amino acids, resulting in the production of optically pure D-alanine (100% resolution), D-phenylalanine (80.2%), and D-tyrosine (84.1%), respectively. The resolution of a racemic mixture[15] of amino acid can be converted to D-amino acid and L-amino acid by an enzyme "acylase I" extracted from hog kidney. The mixture consists of hydrolyzed substances of L-amino acid and unhydrolyzed substances of D-amino acid. Basak et al.[16] synthesized anti-inflammatory pure chiral drugs. The methyl esters of ibuprofen, naproxen, or flurbiprofen were reduced with sodium borohydride in methanol, and the mixture was extracted with ethyl acetate. The combined organic layers were dried over anhydrous sodium sulfate and then evaporated. The product was purified by silica gel chromatography using hexane/ethyl acetate at 3:1 ratio. The alcohol was then acetylated with acetic anhydride, triethylamine, and a catalytic amount of DMAP using dichloromethane as a solvent. Furthermore, the acetyl derivative was hydrolyzed with the help of the enzyme (PPL) in a phosphate buffer (pH 8.0) and acetone. The oxidation of the (+) alcohols was carried out with PDC using DMF as a solvent. After 72 h, the mixture was partitioned between ethyl acetate and aqueous NaHCO$_3$. The aqueous layer was adjusted to pH 2 and then re-extracted with ethyl acetate. The organic layer was dried with Na$_2$SO$_4$, filtered, and evaporated to leave the acid as a white solid (yield ~50%) having identical ^{1}H NMR with the authentic racemic sample (Figure 1.22).

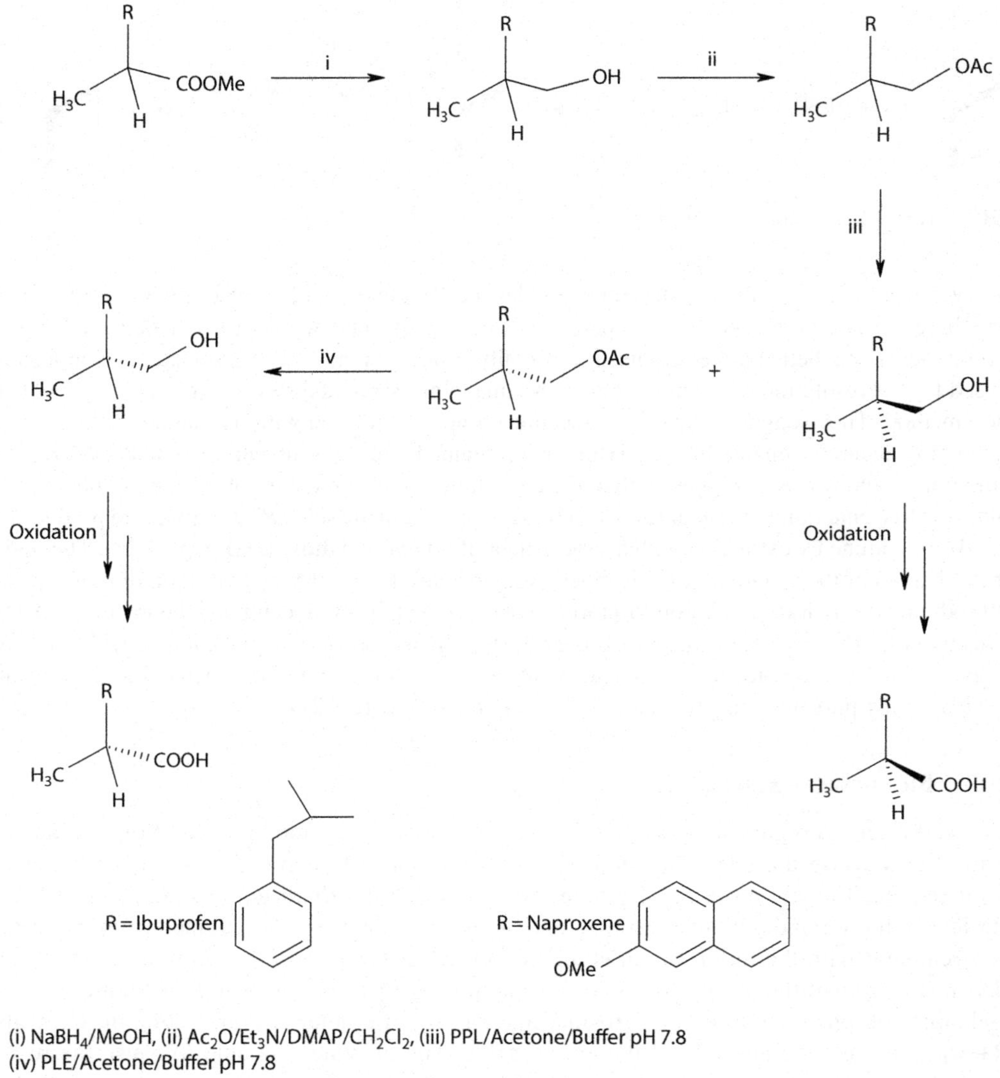

(i) NaBH$_4$/MeOH, (ii) Ac$_2$O/Et$_3$N/DMAP/CH$_2$Cl$_2$, (iii) PPL/Acetone/Buffer pH 7.8
(iv) PLE/Acetone/Buffer pH 7.8

FIGURE 1.22 Enantioselective hydrolysis of ester group to produce anti-inflammatory drugs. (From Nag, A. and Dey, B., *Computer-Added Drug Design and Delivery Systems*, McGraw Hill Publishers, New York, 2011. With permission.)

1.6.4 Chromatographic Separation

Separation of the chiral molecules by the chromatographic method can be difficult because of their identical physical properties.

UV spectra is a helpful instrument to distinguish *cis–trans* isomers such as *cis-* and *trans-*stilbene; in both cases, the absorption is significantly different because *trans*-stilbene is rigidly planar, whereas in *cis*-stilbene, one of the phenyl rings swivels to 43° out of the plane to avoid steric repulsion of orthohydrogens (Figure 1.23). Thus, there is a difference in λ_{max} for *cis* (205 nm) and *trans* (280 nm), and the barrier of inversion in *cis–trans* is 42.8 kcal mol.

Chiral compounds can be solved by choosing two different types of selectors, which can distinguish between a chiral additive in the mobile phase and one in the stationary phase. Of these approaches, chiral stationary phases are more commonly used for separation of enantiomers.

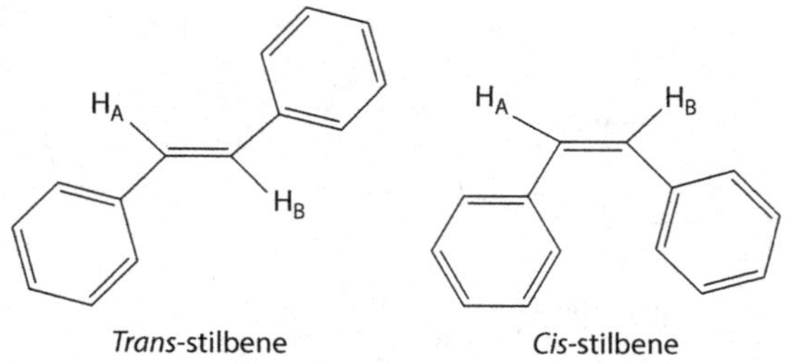

FIGURE 1.23 *Cis-* and *trans*-stilbene.

Generally, the separation of a racemic mixture is used in the following techniques: (1) formation of a diastereomeric mixture from the racemic mixture by a chiral resolving solvent, followed by separation by an achiral adsorbate or modified by a chiral reagent and (2) achiral adsorbent using the chiral eluting solvent as the mobile phase or direct resolution using chiral stationary phase, such as protein-based, polymer-based carbohydrates, brush-type phases, or cyclodextrins and chirobiotic phases. Racemic olefins, biphenyls, ketones, and carboxylic acid and its salts are separated by a microcrystalline triacetyl cellulose column.[17]

Chromatographic techniques such as thin-layer chromatography (TLC), gas chromatography (GC), supercritical fluid chromatography (SFC), and high-performance liquid chromatography (HPLC) have been used for enantiomer separation. TLC techniques are a developing branch of separation and quantization of drugs both in pharmaceutical dosage forms and in biological materials. Resolution of racemic thioridazine obtained from Thioril tablets (Cipla Ltd., Goa, India) into its enantiomers has been achieved by HPLC using a beta-cyclodextrin (CD)-bonded stationary phase. Thioridazine was isolated from commercial formulations and was purified using preparative TLC by Bhusan et al.[18]

Chiral resolution of quinidine and quinine was performed by Somberg et al.[19] using a Spectra-Physics HPLC instrument and UV variable wavelength detector set at 254 nm. The chromatographic column was a prepacked 25 mm × 4.6 mm ID Cyclobond I (5 μm) operated with a methanol/0.014 M sodium perchlorate (75:25 v/v) mobile phase at a flow rate of 0.2 mL/min. Quantitative analysis of optically active *d*- and *l*-enantiomers of menthol esters was carried out[7] on a Hitachi GC model G-300 by Nag et al. Supelco's Sp 2330 fused silica capillary column (30 mm × 0.32 mm i.d.) operating isothermally at 170°C was used to separate and identify the methyl esters synthesized by enzymatic reaction. The injector and detector temperatures were 250°C and 260°C respectively. Nitrogen was used as the carrier gas at a flow rate of 40 cm²/s. The amount of each enantiomer was estimated by peak area recorded and intergraded by the computer. A chiral liquid chromatographic method was developed by Nirogi et al.[20] for the enantiomeric resolution of linezolid, (*S*)(–)-*N*-[[-3-[3-fluoro-4-(4-morpholinyl)phenyl]-2-oxo-5-oxazolidinyl] methyl] acetamide, an antibiotic in bulk drugs. The enantiomers of linezolid were resolved on a Chiralcel OJ-RH column using a mobile phase system containing 150 mM di-sodium hydrogen phosphate buffer (pH 4.5)/acetonitrile (86:14, v/v). The percentage recovery of (*R*)-enantiomer ranged from 98.9 to 102.9 in bulk drug samples of linezolid.

Recently, determination of absolute configuration[2] by NMR spectroscopy has become very popular because the NMR instrument is available in most laboratories and only a small amount of sample is needed for the analysis. It can be conducted in a solution, which is applicable to both solid and liquid samples. NMR chemical shifts of both enantiomeric derivatives are assigned as $\Delta\delta$ values ($\Delta\delta = \delta_s - \delta_R$) to determine absolute configuration. Mosher's NMR method[21] (1973) is useful for the determination of absolute configuration[20] in which a mixture of enantiomers is converted into a mixture of diastereomers by bonding them to another chemical, which is itself a chiral derivatizing agent (CDA). H¹ NMR of diastereomeric derivatives has different chemical shifts, and this helps the assignment of absolute configuration of molecules. The conditions for CDA are that it must not racemize under derivatization or analysis. Its attachment should be

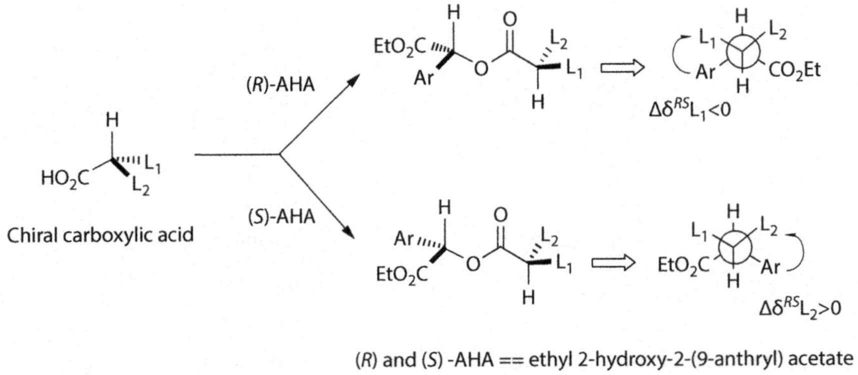

(R) and (S) -AHA == ethyl 2-hydroxy-2-(9-anthryl) acetate

FIGURE 1.24 Identification of (R) and (S)–9 AHA ester by NMR. (From Nag, A. and Dey, B., *Computer-Added Drug Design and Delivery Systems*, McGraw Hill Publishers, New York, 2011. With permission.)

mild enough so that the substrate does not racemize either. If analysis is completed by HPLC, the CDA must contain a chromophore to enhance detectability. If analysis is completed by NMR, the CDA should have a functional group that gives a singlet in the resultant NMR spectrum, where the singlet must be remote from other peaks. The chiral agent may be α-methoxy-α-(trifluoromethyl) phenylacetic acid, mandelic acid, or O-methyl mandelate. This method can be used in primary, secondary, and tertiary alcohols and amines, diols, carboxylic acids, and sulfoxides.[21] This method requires derivatization with both R and S reagents to produce two diastereomers, but a problem may arise during the change in chemical shift due to the magnetic anisotropy of the derivatizing agent. This could happen when the signals in the NMR spectra of the diastereoisomeric derivatives are too close to be distinguished (small $\Delta\delta^{RS}$). Then it requires a change to a more powerful reagent, which might be a costly solution in some cases. Manuel Seco et al.[22] suggested that valuable alternatives to increase the $\Delta\delta^{RS}$ values include acquiring the NMR spectra at lower temperatures with treatment of one enantiomer with R and S chiral derivatizing reagents. The difference in temperature changes will induce conformer changes, which will make a difference in the chemical shifts. This principal is utilized by scientists[23] for chiral acid and (R)- and (S)-ethyl 2-hydroxy-2-(9-anthryl) acetate at low temperature. Here, the sign of the $\Delta\delta^{RS}$ values of the esters of 9-AHA (5) is an indicator of the spatial location of L_1/L_2 relative to the aryl group (Figure 1.24).

1.7 APPLICATION OF ENANTIOMERS IN DRUGS AND NATURAL PRODUCTS

Currently, in the pharmaceutical market, the majority of newly introduced drugs are chiral. It is expected that nearly 95% of the pharmaceutical drugs will be chiral by 2020.[30] Pharmaceutical companies are using chirality as a tool to increase the span of their patented blockbuster drugs, and the global market will reach $5.1 billion by the year 2017. Different enantiomers may have the same chemical formula, but they differ widely in their biological properties. This is due to their three-dimensional structure; one form may be more suitable for specific interactions with other biological molecules such as receptor or enzymes. Drug–receptor interactions are stereoselective, that is, minor changes in the structure of a drug molecule can produce major changes in its pharmacological properties (**stereopharmacology**).

The advantages of single-enantiomer products are that they are less complex and more selective. They simplify interpretation of their basic pharmacology.[24] They have potential for an improved therapeutic index and can reduce potential for complex drug interaction and also adverse effects. They have longer or shorter duration of action and more appropriate dosing frequency. They decrease interindividual variability.

Basic Stereochemical Approaches to Natural Products and Drugs

Single-enantiomer products are a major concern in the modern pharmaceutical industry because the body, being amazingly chiral-selective, can metabolize each enantiomer through a separate pathway to produce different pharmacological activities as one isomer may produce the desired therapeutic activities, while the other may be inactive or produce unwanted effects. **Pharmacokinetic differences**[16] as a result of stereoisomerism can cause better absorption as L-methotrexate than D-methotrexate. After the application of a racemic mixture of thalidomide by a pharmaceutical company (West Germany in 1957) as a sedative for pregnant women, thousands of babies were born with missing or abnormal arms, hands, legs, or feet. It was banned in several countries in 1961. Later scientists discovered that its therapeutic activity resided exclusively in the R-(+)-enantiomer as a sedative but in the S-(+)-enantiomer as a teratogen[25] (a substance affecting the development of the fetus and causing structural or functional disability) (Figure 1.25).

In 1992, the U.S. Food and Drug Administration[26] issued a guideline specifying that chiral drugs could be brought to the market only for their therapeutic use as active isomers and that each enantiomer of the drug should be studied separately for its pharmacological and metabolic pathways.

D-Glucose is consumed by the human body, whereas other enantiomers, such as L-glucose, are biologically inactive, although L-glucose has led to the production and patenting of non-nutritive sweetness.[16] The use of racemic (±) 3(3′,4′-dihydroxyphenyl)-alanine (DOPA) causes nausea, vomiting, and granulocytopenia, whereas the use of enzyme dopamine decarboxylase produces (S)-(−) DOPA 3 (3,4-dihydroxyphenyl)-alanine (Figure 1.26) and reduces the adverse effects of granulocytopenia.[26]

L-(+) ascorbic acid (vitamin C) is more effective than D-(−) ascorbic acid in the human system.[16] Similarly L-penicillamine causes more toxicity and weight loss than D-penicillamine when used in the treatment for Wilson's disease[24] (Figure 1.27).

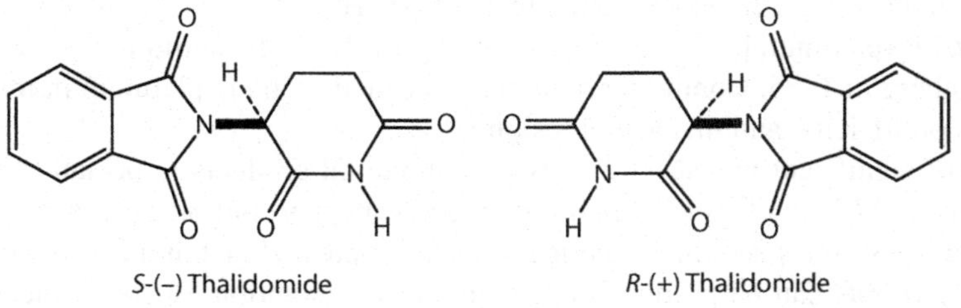

FIGURE 1.25 (S) and (R)-thalidomide.

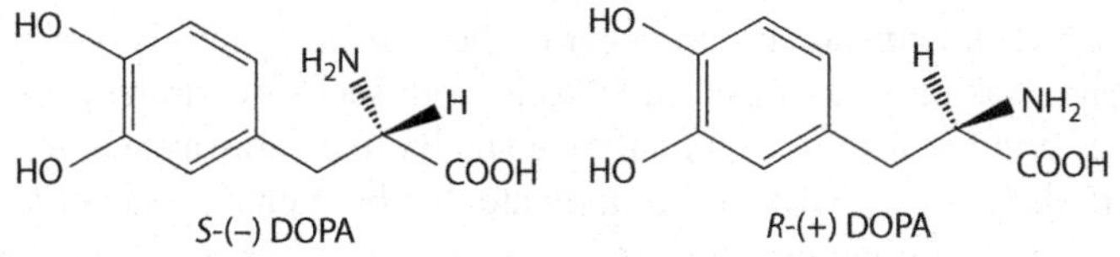

FIGURE 1.26 (S) and (R)–DOPA.

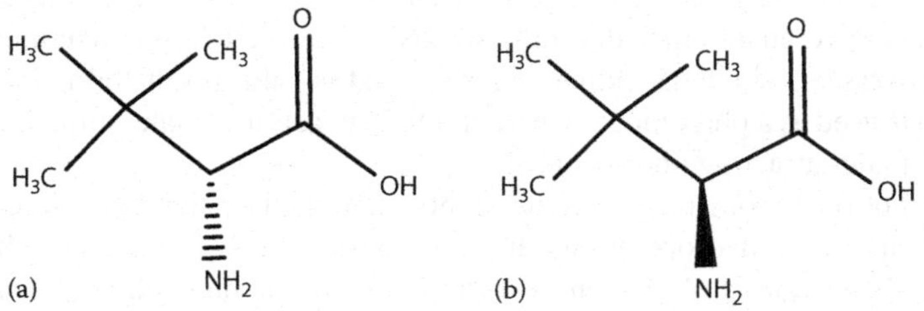

FIGURE 1.27 (a) L- and (b) D-penicillamine.

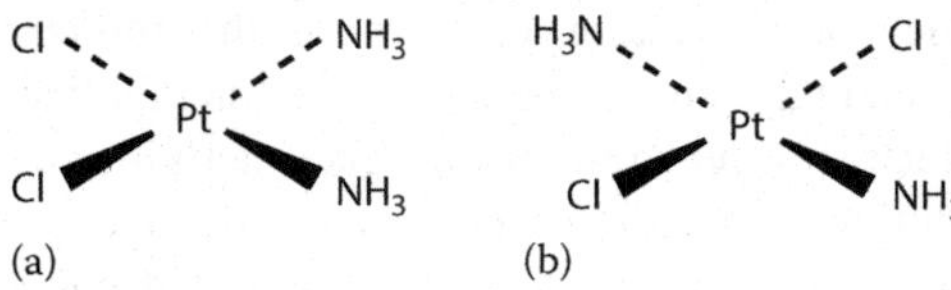

FIGURE 1.28 (a) *Cis-* and (b) *trans-*platinum.

Propranolol[24] exists in two different enantiomers, and isomers of propranolol have been compared for their β-blocking capacity (β-adrenergic receptor antagonists). *R*-(+)-propranolol has less than one-hundredth the potency of *S*-(−)-propranolol, but *R*-(+)-propranolol is used as a contraceptive. In the case of the well-known painkiller ibuprofen, the (*S*)-enantiomer has the desired pharmacological activity, while the (*R*)-enantiomer is totally inactive, but *R*-ketoprofen[27] undergoes less than 10% chiral inversion in humans, which reduces dose requirement in comparison to racemate and also the potential for gastric ulcers. *S*-omeprazole and *S*-pantoprazole are found to be more effective proton pump inhibitors than their corresponding racemates. *R*-salbutamol[27] is a more commonly used bronchodilator for its $β_2$ agonistic activity than *S*-salbutamol because the dose of *R*-salbutamol fits the three-dimensional confirmation of the $β_2$ adrenoreceptor.

In the areas of organometallic chemistry, chirality is an important factor as the restricted rotation around the carbon–carbon double bond is responsible for the *cis* and *trans* configurations. In the *cis* configuration, functional groups are on the same side of the carbon chain, whereas in the *trans* configuration, functional groups are on the opposite side of the carbon chain. The difference between the two is that the *cis* isomer is a polar molecule whereas the *trans* isomer is nonpolar. The two isomers of *cis-* and *trans-*platinum are square planar (Pt(NH$_3$)$_2$Cl) where *cis*-platinum has antitumor activity while *trans-*platinum has no such activity (Figure 1.28). Its antitumor properties were only found by Rosenberg in 1965. Common problems associated with *cis*-platinum in the clinic include nephrotoxicity, ototoxicity, and myelosuppresion.[28]

Chirality is of significant industrial importance in natural products, especially food taste, odor, and agrochemicals. *l*-Menthol[7] is used in the industry as peppermint oil, in candy, in beverages, in tobacco, in local anesthetics, and in cosmetic products, whereas *d*-menthol has an undesirable taste. Optically pure alcohols can be produced either in aqueous solution by stereoselective hydrolysis of the corresponding racemic esters or in organic solvents by esterification of the corresponding racemic alcohols. Dextrorotatory asparagine[3] (HOOC-CH(NH$_2$) CH$_2$CONH$_2$) has a sweet taste, whereas levorotatory asparagine is tasteless. Similarly, (*S*)-(+)-leucine[5] has a bitter taste whereas *R*-(−)-leucine[4] tastes sweet. **Carvone** is a member of the terpenoids family, which is found naturally in many essential oils. It is most abundant in the oils from seeds of caraway and is responsible for the flavor of caraway and spearmint. *S*-(+)-carvone smells like caraway whereas *R*(−)-carvone has a spearmint odor.[3] In the body, *in vivo* studies indicate that both enantiomers of carvone are mainly metabolized into dihydroxycaravonic acid.[29] **Limonene** is a colorless liquid classified as a cyclic terpene generally used in common cosmetic products, food manufacturing, and some medicines; its *R*-(+)-isomer possesses a strong smell of orange while the *S*-(+)-isomer has a lemon odor.[3]

Chiral compounds are used as pesticides; for example, the (*R*)-(+)-enantiomer of the herbicide dichlorprop is the active enantiomer that kills weeds, while the (*S*)-(−)-enantiomer is inactive as a herbicide. It possesses a single asymmetric carbon and is sold in salt form. 1,7-Dioxaspiro[5.5] undecane (olean) is used as a pheromone, where *R*(+)-olean acts as a male-attracting pheromone and *S*(−)-olean as a female-attracting pheromone.[30]

Recently, one-pot operations have become an effective method that can avoid several purifications, save time, and minimize operations that are considered to be green method. Kobayashi[31] recently reported the synthesis of (*R*)- and (*S*)-rolipram with a single chiral center with four flow reactors in a single flow, all using heterogeneous catalysts. (−)-Oseltamivir phosphate is one of the most effective drugs for the treatment of influenza and was synthesized in a one-pot system by

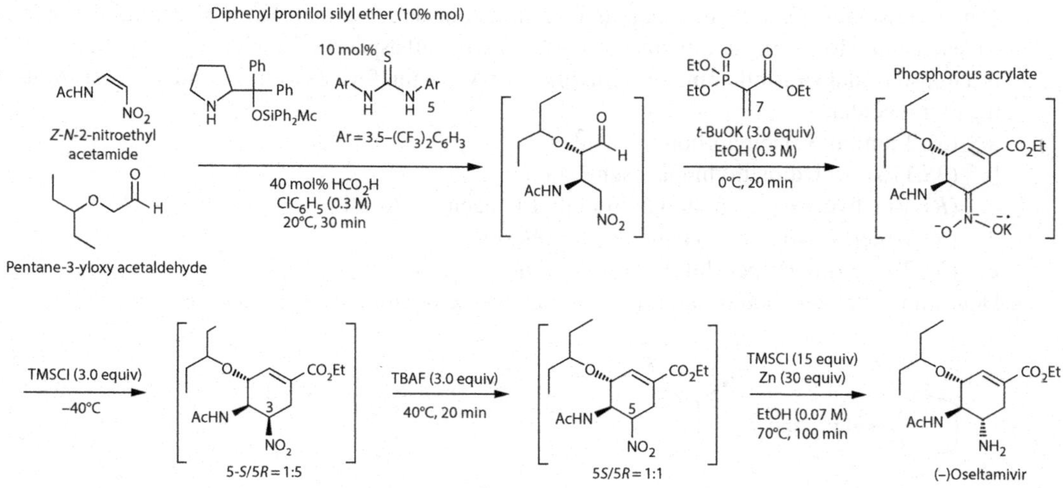

Basic Stereochemical Approaches to Natural Products and Drugs

FIGURE 1.29 Synthesis of (−)-oseltamivir. (From Hayashi, Y. and Ogasawara, S., *Org. Lett.*, 18(14), 3426, 2016. With permission.)

Hayshi et al.[32] The reaction was completed within 60 min by adding the reagents sequentially without evaporation or solvent swap. A continuous-flow synthesis of (−)-oseltamivir composed of five flow units was accomplished (Figure 1.29). In each unit, the following reactions were conducted efficiently: (1) a diphenylprolinol silyl ether–mediated Michael reaction; (2) a domino reaction of Michael and intermolecular Horner–Wadsworth–Emmons reactions; (3) protonation; (4) epimerization; and (5) reduction of a nitro group to an amine following the mechanism shown in Figure 1.29.

Thus, in the last two decades, the practice of new asymmetric synthesis of industrially chiral drugs and natural products has fascinated chemists to develop such auxiliaries, reagents, and catalysts that can incorporate stereoselectivity in reactions to make the targeted product.

PROBLEMS

1. Alanine has the structure shown here. Is it an L- or D-amino acid? How many dipeptides can be formed from alanine?

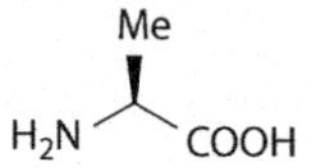

2. Explain the structural features of α-helix. Why is a peptide bond stronger and shorter than a C–N single bond? (There is no restriction about this bond.)
3. Draw the structures of a parallel β-sheet and a antiparallel β-sheet.
4. Draw the structures of *P*- and *M*-helical conformations of the 1,1′-disubstituted ferrocenes.
5. How can you detect free radicals? Give an example of redox radical reaction.
6. Which methyl group is responsible for chirality in camphor? Generally methyl group is used as chiral to investigate the enzymatic reaction. Explain why one racemic form camphor is known.
7. Which is more stable, 1,3,5-heptatriene or 1,3,6-heptatriene? What will be the product in the following reaction?

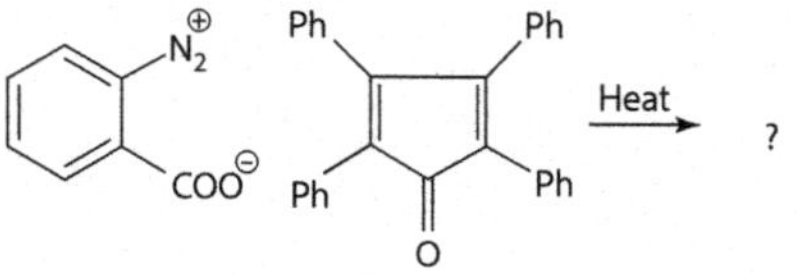

8. Name the smallest alkene molecule that exhibits diastereomerism. Draw structural formulas for each of the following compounds, clearly showing all aspects of stereochemistry. Naproxen is a nonsteroidal anti-inflammatory drug that relieves pain, fever, swelling, and stiffness. Mark the chiral center.
   a. (S)-2-amino-3-phenylpropionic acid
   b. (3S,11S)-3,11-dimethylnonacosan-2-one
   c. (R)-4-(1-hydroxy-2-(methylamino)ethyl)benzene-1,2-diol
   d. (R)-1-methyl-4-(1-methylethenyl)cyclohexene
   e. (R)-2-(2,6-dioxopiperidin-3-yl)isoindoline-1,3-dione
9. Determine whether each of the following pairs are enantiomers or diastereomers:
   a.
   b.
   c.
10. Indicate the relationship (identical, enantiomers, or diastereomers) between the following pairs of compounds:
    a.
    b.
    c.

d.

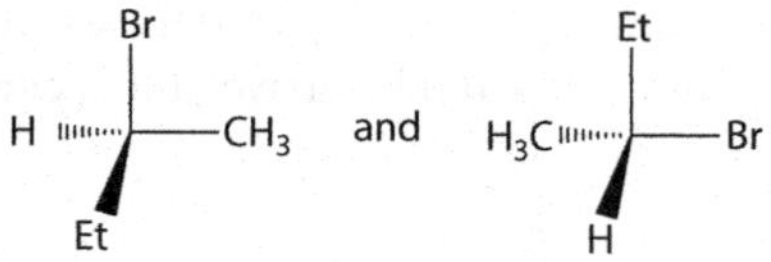

Br   Et

H⋯⫶⫶—CH₃   and   H₃C⋯⫶⫶—Br

Et   H

e. [norbornanone structures] and [norbornanone structure with O]

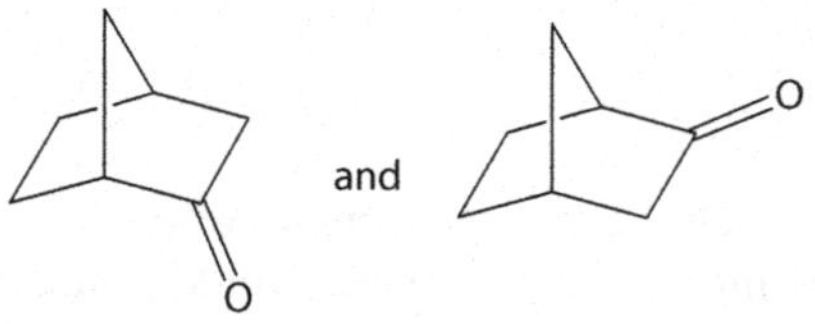

f. [cyclohexane with sulfoxide and t-butyl substituents] and [cyclohexane with sulfoxide and t-butyl substituents]

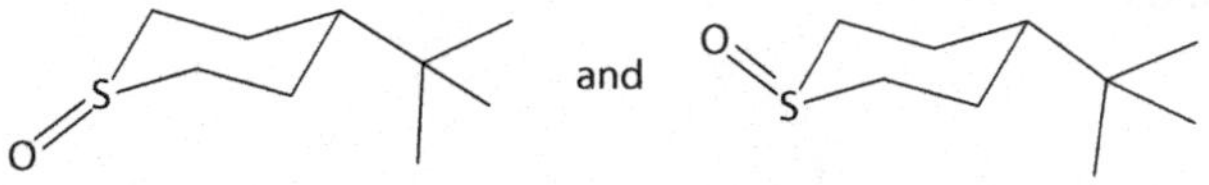

11. What is reductive elimination reaction? Hydrogen bromide is added to 3, 3-dimethyl-1-butene to give a mixture of diastereomeric bromide in equal amounts. Write the mechanism of the reaction. How will you obtain monobromo substitution of alkane treatment with bromine?
12. Why do alkyl boranes undergo isomerization? Give the product from hydroboration oxidation of 1-methylcyclopentane.
13. An allene is a compound in which one carbon atom has double bonds with each of its two adjunct carbon centers. Why are substituted allenes chiral?
14. What is homomorphic? What will be the products of alkali treatment of *cis-* and *trans-*4-bromo cyclohexanol on alkali treatment?
15. Indicate how to resolve racemic 1-phenylethylamines using the method of reversible conversion into diastereomers.

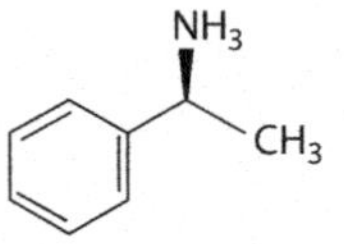

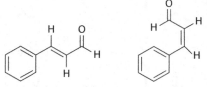

16. Though alkylamine such as *N*-methylethaneamine is achiral, N-substituted heterocyclic amines such as *N*-chloro or *N*-methoxyaziridine derivatives can be resolved enantiomers. Discuss the phenomenon based on their energy barriers.
17. How will you differentiate between two isomers, $HOCH_2C_6H_5CH_3$ and $C_6H_5CH_2CH_2OH$? The following compounds are daistereomers of cinnamaldehyde.
    Will the two hydrogens of central carbon–carbon double bonds in *cis-* and *trans-*stilbenes are different in ^1H NMR?
18. What is an atropisomer? Show an example and explain it.
19. What are enantiotopic protons? How would you differentiate between them using the ^1H NMR spectroscopy?

Trans-cinnamaldehyde   *Cis*-cinnamaldehyde

20 Asymmetric Synthesis of Drugs and Natural Products

20. For the following compound, a graduate student assumed β-orientation for the C-1/OH group based on the ¹H-¹H NMR coupling data. Please explain how did he arrive at this conclusion.

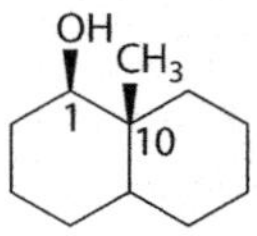

21. For the following compound, which NMR spectroscopy experiment would be helpful in determining the strereochemistry of the following compound? How would you differentiate between these two compounds?

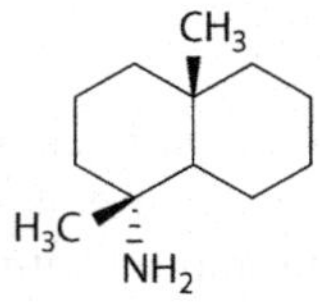

22. a. Reaction between this aldehyde and ketone on base gives a compound A with the ¹H NMR spectrum: $-\delta_H$(p.p.m): -1.10(9H,s), 1.17(9H,s), 6.4(1H,d,J = 15), and 7.0(1H,d J = 15) and 7.0(1H, d, J = 15). What is its structure with stereochemistry?
   b. When this compound reacts with HBr, it gives compound B with this NMR spectrum: δH(p.p.m): -1.08(9H,s), 1.13(9H,s), 2.71(1H, dd, J 1.9, 1.77), 3.25(2H, dd, J 10.0, 17.7), and 4.38(1H, dd, J 1.9, 10.0).
   c. Suggest a structure, assign the spectrum, and give a mechanism for the formation of B.

ANSWERS

1. L-amino acid
   There are four dipeptides that can be formed from alanine: Ala. Ala; alanylalanine; Ala. Gly; glycyalanine.
2. α-Helix has a right-handed helical conformation with a pitch of 5.4 Å, wherein each CO group is hydrogen bonded to the NH group of the fourth residue further toward the C-terminus. There are 3.6 amino acids per turn of the helix.
   C–N has considerable double bond character from delocalization of N's nonbonding e's to the O of C=O. The assembly is $\overset{\diagdown}{\underset{\underset{O}{\parallel}}{C}}-\overset{\diagup}{\underset{\diagdown}{N}}$.
3.

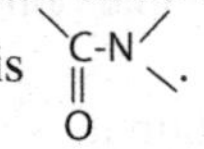

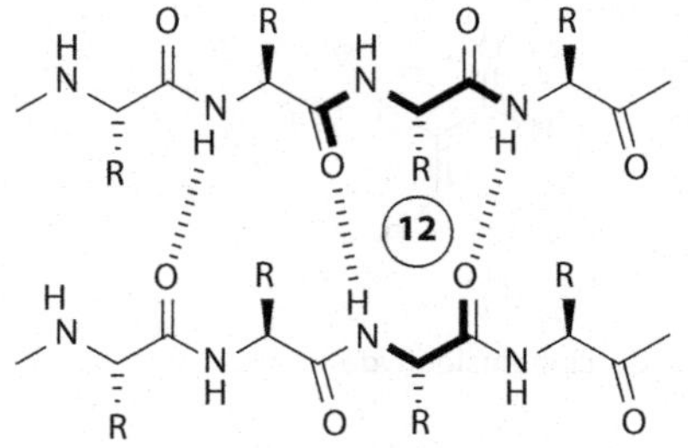

Parallel β-sheet 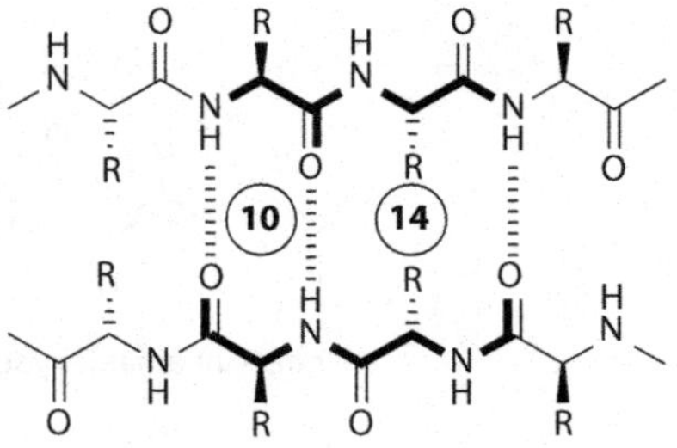   Antiparallel β-sheet

4.

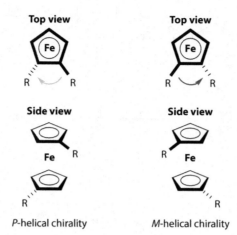

P-helical chirality   M-helical chirality

5. The reaction rate is studied with or without inhibitor such as hydroquinone, and the reaction rate is inhibited by the presence of inhibitor to detect free radicals reaction. Another method is electron spin resonance spectroscopic studies, which can detect the free radicals. In redox reactions one electron transfer reaction is employed to produce radicals such Kolbe electrolysis of carboxylic salts.

6. A C bonded to an H, a tritium, and a deuterium. Enzymes can distinguish between the different isotopes of hydrogen. Camphor has four stereoisomers as two racemic forms, but the bridge of camphor gives *cis*-form, and the structure impossibility of a *trans*-bridge eliminates a pair of enantiomers.

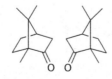

(*R*-left) and (*S*-right) Camphor

7. The number of conjugated double bonds increases the stability increase as 1,3,5-heptatriene ($CH_2=CH-CH=CH-CH=CH-CH=CH-CH_2$) has three conjugated double bonds, whereas 1,3,6-heptatriene ($CH_2=CH-CH=CH-CH_2-CH=CH_2$) has two conjugated double bonds.

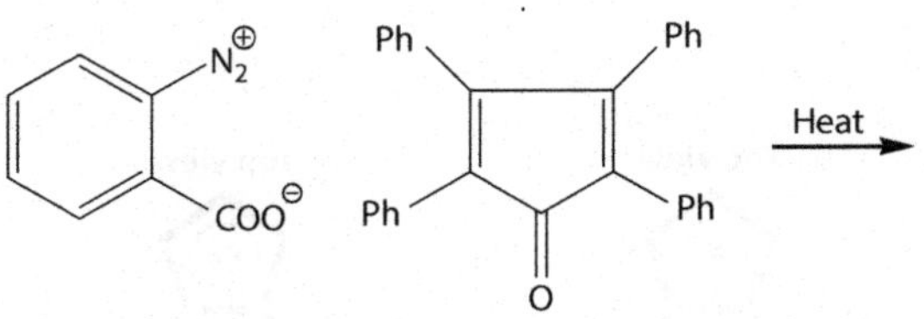

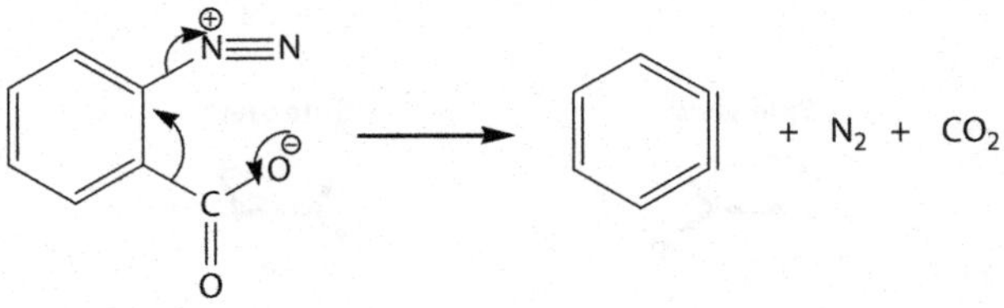

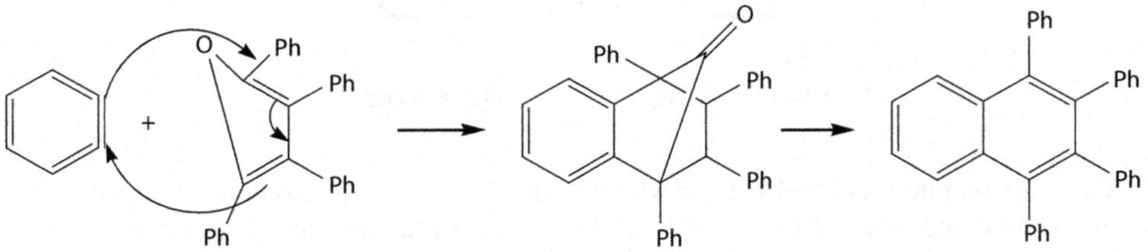

8. It is CHD = CHD.
   a. (S)-2-amino-3-phenylpropionic acid

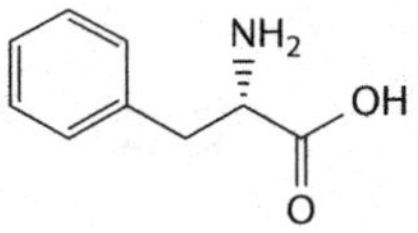

   b. (3S,11S)-3,11-dimethylnonacosan-2-one

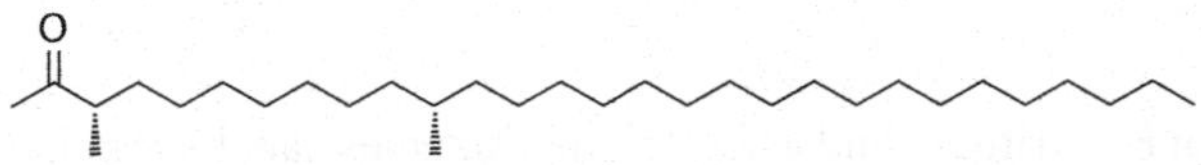

   c. (R)-4-(1-hydroxy-2-(methylamino)ethyl)benzene-1,2-diol

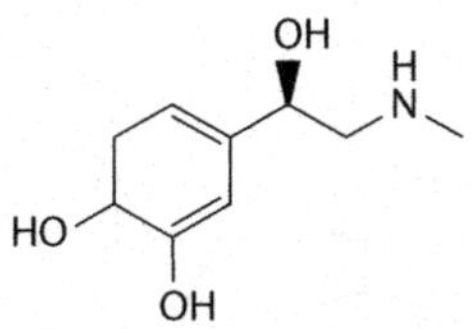

   d. (R)-1-methyl-4-(1-methylethenyl)cyclohexene

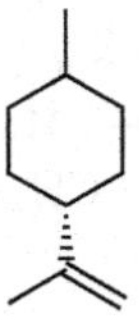

e. (R)-2-(2,6-dioxopiperidin-3-yl)isoindoline-1,3-dione

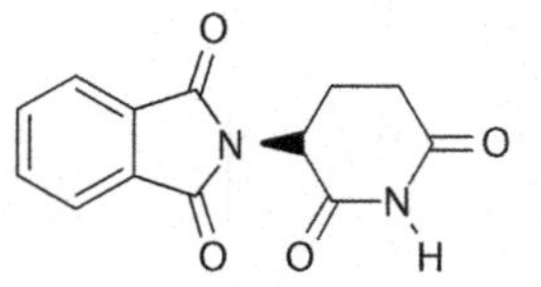

Chiral centre of naproxen is below

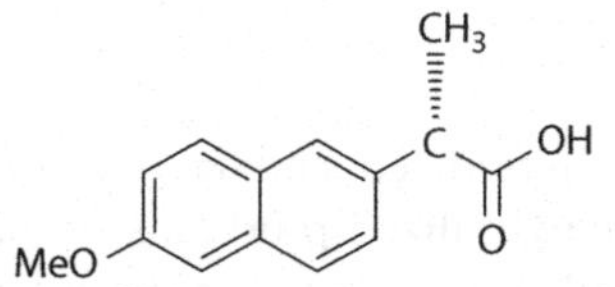

9. a. These are not enantiomers. The molecule on the left has *trans* substitutions, while the molecule on the right has *cis* substations. They are diastereomers.
   b. These are enantiomers. If we rotate the molecule on the left 180° to the right, we see that both molecules have a dash C=Cl bond and a wedge C–CH$_3$ bond creating a nonsuperimposable mirror image of the original molecule.
   c. These are not enantiomers. The molecule on the right can undergo a ring flip followed by a 129° rotation counterclockwise to make the mirror image of the molecule superimposable on the original; in other words, the two structures are identical. This is a *meso* compound.
10. (a) Diastereomers, (b) enantiomers, (c) homomers (identical), (d) enantiomers, (e) identical, (f) diastereomers.
11. Reductive elimination is the oxidation state of the metal center decreases while forming a new covalent bond between two ligands. Since oxidative addition and reductive elimination are reverse reactions, the same mechanisms apply for both processes, and the product equilibrium depends on the thermodynamics of both directions. The reaction of 3,3-dimethyl-1-buten with HBr as follows:

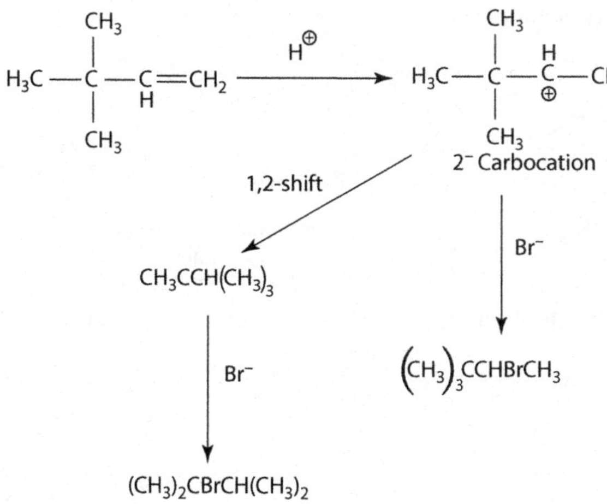

For monobromo substitution of bromine is added to the excess of alkane if alkane is added in bromine solution polysubstitution is formed due to localized concentration of bromine.

12. The reversible nature of hydroboration at higher temperature and the tendency of boron to become attached to the carbon at the end of chain where steric crowding is minimum is the cause of isomerization of alkylboranes. These isomerizations are useful in synthesis because they enable us to convert a more stable internal alkene to a less stable 1-alkene.

Hydroboration oxidation of 1-methylcyclopentane is as follows:

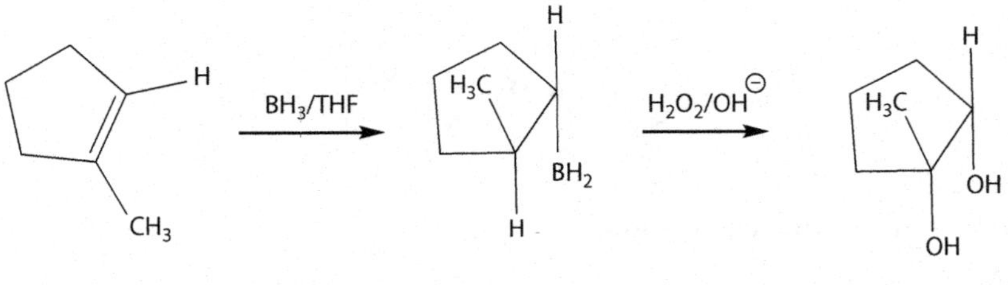

1-Methyl cyclopentene →(BH₃/THF)→ intermediate →(H₂O₂/OH⁻)→ trans-2-methyl cyclopentanol

13. The central carbon atom is sp hybridized, and the outer two carbons are sp² hybridized in the structure of allene. As two unhybridized p-orbitals on the central carbon atom are perpendicular each other, two molecular plains between the outer-central carbon atoms also must be perpendicular. Therefore, chirality could exist in allenes.

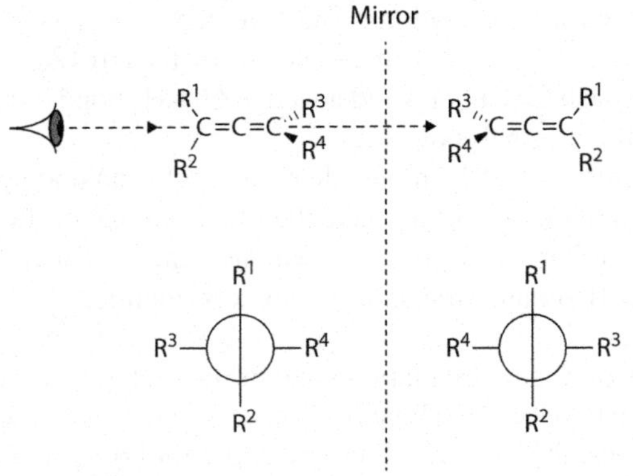

14. The ligands, which are structurally identical when detected from a chemical molecule, are called homomorphic.

Trans-4-bromo cyclohexanol will give epoxide product

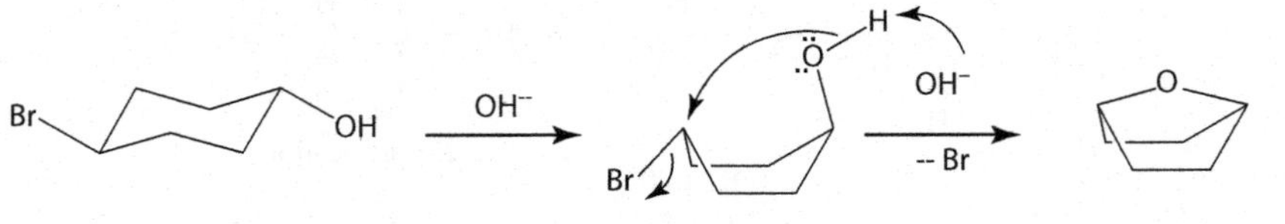

Trans-4 bromocyclohexanol → Intramolecular transition state → Epoxide product

Cis-4-bromo cyclohexanol on alkali treatment produces *cis*- and *trans*-cyclohexanol

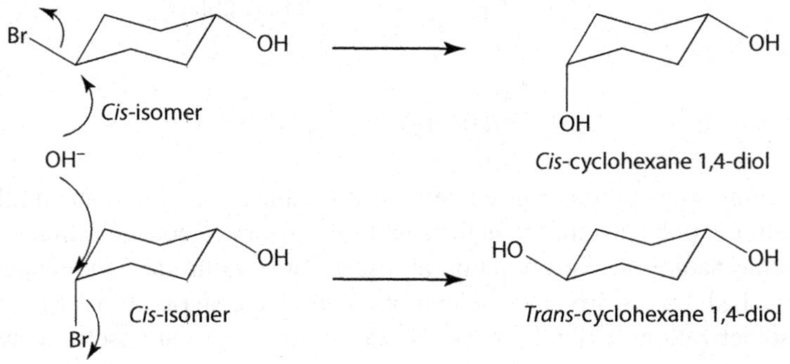

Cis-isomer → Cis-cyclohexane 1,4-diol

Cis-isomer → Trans-cyclohexane 1,4-diol

Basic Stereochemical Approaches to Natural Products and Drugs 25

15. For example, the racemic compounds are resolved by formation of diastereomeric salts. A warm solution of *rac*-1-phenylethylamine is added to a solution of enantiopure acid (e.g., tartaric acid, lactic acid). The solution is allowed to stand several times at ambient temperature, and the diastereomeric salt will be obtained as a precipitate. A mixture of the salt, organic solvent, and basic aqueous solution is shaken and the organic layer is separated. Thus, the obtained organic solution contains a chiral compound.

16. In *N*-methylethanamine, nitrogen inversion occurs rapidly and it is impossible to resolve the enantiomers. On the other hand, the inversion barrier of N-substituted heterocyclic amines becomes much higher by increasing the p-character of N–Cl or N–O bond with increasing electronegativity. Thereby, destabilization of the transition state might occur due to increase in s-character in the lone pair on the nitrogen atom.

17. The compound $HOCH_2C_6H_5CH_3$ has two singlets: one for CH_2 and other for CH_3. The four phenyl Hs will be a multiplet, whereas $C_6H_5CH_2CH_2OH$ five phenyl groups Hs multiplet and two triplet for the two methylene groups. In both cases a broad singlet for both H and OH. Yes, they are different in *cis*- and *trans*-stilbene.

18. Tropisomers are stereoisomers resulting from hindered rotation about a single bond where the barrier to rotate is high enough to separate each conformation. For example, 2,2'-bis(diphenylphosphino)-1,1'-binapthyl, which is known as BINAP, is one of the most famous ones.

(S)- BINAP   Hindered rotation   (R)- BIANP

19. Enantiotopic protons have the same chemical shift in the vast majority of situations. However, if they are placed in a chiral environment (e.g., a chiral solvent), they will have different chemical shifts.

    These two stereoisomers can be differentiated with the help of coupling constants.

    The protons of *cis* double bonds will exhibit coupling constants in the range of 9–12 Hz, while protons of *trans* double bond show coupling constants in the range of 15–18 Hz.

Trans-cinnamaldehyde   *Cis*-cinnamaldehyde

20. He has observed *trans*-diaxial ^1H-^1H couplings between two red protons (J = 9–12 Hz) and axial-equatorial couplings (J = 1–7 Hz) between H-1 and H-2, as shown in the following conformation of this compound.

21. NOESY spectrum will be helpful to determine the relative stereochemistry between two methyl groups as they will exhibit *cis*-relationship between them.

22.

Compound A is a *trans* isomer and compound B is one type of ABX system: AB are the diastereotopic CH_2 group (J_{AB} = 17.7) and H is the CHBr proton (J_{AX} = 10; J_{bx} = 1.9).

REFERENCES

1. E.L. Eliel; S.H. Wilen; and M.P. Doyle; *Basic Organic Stereochemistry*; Wiley Interscience, New York, 2001.
2. D. Nasipuri; *Stereochemistry of Organic Compounds: Principles and Applications*; New Age International Publishers, New Delhi, India, 2011.
3. R.S. Dhillion; I.P. Singh; and C. Baskar; *Stereiochemistry*; Narosa Publishers, New Delhi, India, 2016.
4. B. Mehata and M. Mahata; *Organic Chemistry*; Prientice-Hall of India, New Delhi, India, 2014.
5. I.L. Finner; *Organic Chemistry*, Vol. 1; Longman, London, U.K., 1974.
6. N. Tewari; *Advanced Organic Reaction Mechanism*; Books and Allied(P) Ltd, Kolkata, India, 2005.
7. D.-L. Wang; A. Nag; G.-C. Lee; and J.-F. Shaw; Factors affecting the resolution of *dl-menthol* by immobilized lipase-catalyzed esterification in organic solvents; *J. Agric. Food Chem.* 50, 262–265, 2002.
8. T. Kitazume and T. Kobayashi; Synthesis of optically active α-fluoro-α-methyl-β-hydroxyesters of *erythro* or *threo* configuration with Baker's yeast; *Synthesis* 1987(2), 187–188, 1987.
9. (a) D. Kozma; Z. Madarasz; C. Kassai; and E. Fogassy; Optical resolution of *N*-methylamphetamine via diastereoisomeric salt formation with 2*R*,3*R*-*O*,*O'*-di-*p*-toluoyltartaric acid; *Chirality* 11, 373, 1999. (b) C.C. Hung; Patent No. 212 667, 1999.
10. E. Fogassy; M. Ács; T. Szili; B. Simándi; and J. Sawinsky; Molecular chiral recognition in supercritical solvents; *Tetrahedron Lett.* 35, 257–260, 1994.

11. H.C. Brown and N.M. Yoon; Monoisopinocampheylborane—A new chiral hydroborating agent for relatively hindered (trisubstituted) olefins; *J. Am. Chem. Soc.* 99, 5514–5516, 1977.
12. R.O. Oktore; *Basic Separation Techniques in Biochemistry*; New Age International, New Delhi, India, 1998.
13. S. Singh; B.K. Gogoi; and R.L. Bezbarush; Racemic resolution of some DL-amino acids using *Aspergillus fumigatus* L-amino acid oxidase; *Curr. Microbiol.* 63, 94–99, 2011.
14. K. Michi and H. Tsuda; Enzymatic resolution of racemic amino acids; *Bull. Agric. Chem. Soc. Jpn.* 21(4), 235–238, 2014.
15. A. Nag and B. Dey; *Computer-Added Drug Design and Delivery Systems*; McGraw Hill Publishers, New York, 2011.
16. Y.-M. Liu; P. Gordon; S. Green; and J.V. Sweedler; Determination of salsolinol enantiomers by gas chromatography-mass spectrometry with cyclodextrin chiral columns; *Anal. Chim. Acta* 420(1), 81–88, 2000.
17. R. Bhushan and D. Gupta; HPLC resolution of thioridazine enantiomers from pharmaceutical dosage form using cyclodextrin-based chiral stationary phase; *J. Chromatogr. B: Anal. Technol. Biomed. Life Sci.* 837, 133–137, 2006.
18. J. Somberg and V. Ranade; *Optically active isomers of quinine and quinidine and their respective biological action, US 20030212098 AI Abstract*, 2003.
19. R. Nirogi; S. Kotal; R. Rajeswari Katta, S. Vennila; V. Kandikere, K. Mudigonda; and H.B. Vurimindi; Enantiomeric separation of linezolid by chiral reversed-phase liquid chromatography; *J. Chromatogr. Sci.* 46, 10, 2008.
20. J.A. Dale and H.S. Mosher; Nuclear magnetic resonance enantiomer reagents, configurational correlations via nuclear magnetic resonance chemical shifts of diastereomeric mandelate, *O*-methyl mendalate and α-methoxy-α-trifluoromethylphenyl acetate (MTPA) esters; *J. Am. Chem. Soc.* 95, 512–519, 1973.
21. M. Seco; E. Quinoa; and R. Rigueur; The assignment of absolute configuration by NMR; *Chem. Rev.* 104(1), 17–118, 2004.
22. M.J. Ferreiro; S.K. Latypov; E. Quiñoá; and R. Riguera; Assignment of the absolute configuration of r-chiral carboxylic acids by ¹H NMR spectroscopy; *J. Org. Chem.* 65, 2658–2666, 2000.
23. S.S. Jayakrishna and L.E. George; Chiral drugs as a matter of specialization in modern medicine; *Inter. J. Res. Pharma. Biomed. Soc.* 3(1), 3–5, 2012.
24. H.-J. Schmahl; H. Nau; and D. Neubert; The enantiomers of the teratogenic thalidomide analogue EM 12; *Arch. Toxicol.* 62(2), 200–204, 1988.
25. D. Carpenter; E. James; B.A. Zucker; and J. Avorn; Drug-review deadlines and safety problems; *New Engl. J. Med.* 358, 1354–1361, 2008.
26. C.M. Ramsay; J. Cowati; E. Flannery; C. McLachlan; and D.R. Taylor; Bronchoprotective and bronchodilator effects of single dose s of (S)-salbutamol, ®-salbutamol and racemic salbutamol in patient with bronchial asthma; *Eur. J. Clin. Pharmacol.* 55, 353–359, 1999.
27. E. Cvitkovic; J. Spaulding; V. Bethune; J. Martin; and W.F. Whitmore; Improvement of *cis*-dichlorodiammineplatinum (NSC 119875): Therapeutic index in an animal model; *Cancer* 39, 1357, 1977.
28. K.P.C. Vollhardt and N. Schore. *Organic Chemistry*, 5th edn.; Freeman, New York, 2007; J.H. Beatty; Limonene—A natural insecticide; *J. Chem. Educ.* 63(9), 768, 1986.
29. G. Haniotakis; W. Francke; K. Mori; H. Redlich; and V. Schurig; Sex-specific activity of (*R*)-(−)- and (*S*)-(+)-1,7-dioxaspiro[5.5]undecane, the major pheromone of *Dacus oleae*; *J. Chem. Ecol.* 12(6), 1556–1568, 1986.
30. T. Tsubogo; H. Oyamada; and S. Kobayashi; Multistep continuous-flow synthesis of (*R*)- and (*S*)-rolipram using heterogeneous catalyst; *Nature* 520, 329, 2015.
31. Y. Hayashi and S. Ogasawara; Time economical total synthesis of (−)-Oseltamivir; *Org. Lett.* 18(14), 3426–3429, 2016.

2 Diastereoselective Addition of Organometallic Reagents to Chiral Carbonyl Compounds

Ivana Gergelitsová and Jan Veselý

CONTENTS

2.1 Introduction 29
2.2 Models for Asymmetric Carbonyl Compound Addition 30
2.3 Models for 1,2-Asymmetric Carbonyl Addition 31
2.4 Models for 1,3-Asymmetric Carbonyl Addition 40
2.5 Addition of Achiral Reagents 43
2.6 Addition of Aldehyde 43
    2.6.1 Aldehydes Containing α-Oxygen Substituent 44
    2.6.2 Aldehydes Containing α-Nitrogen Substituent 47
    2.6.3 Aldehydes Containing Other Heteroatoms in α-Position 57
2.7 Addition of Chiral Ketones 63
    2.7.1 Acyclic α-Chiral Ketones 63
    2.7.2 Acyclic Chiral Ketones Containing Oxygen in α-Position 63
    2.7.3 Acyclic Chiral Ketones Containing Nitrogen in α-Position 66
    2.7.4 Acyclic Chiral Ketones Containing Other Heteroatoms in α-Position 67
    2.7.5 Cyclic Ketones 68
2.8 Conclusion 69
References 69

2.1 INTRODUCTION

Carbon–carbon bond-forming reactions constitute one of the key transformations enabling the construction of organic molecules. Nucleophilic addition of organometallic reagents to carbonyl-containing substrates represents one of the crucial approaches. The use of organometallic compounds in the area of organic synthesis has been known for more than one and a half century. Pioneering works of Edward Frankland,[1] followed by Reformatsky, and others have shown high potential of organozinc compounds. In the early twentieth century, the role of another metal, magnesium, was ascertained in Grignard reagents (Victor Grignard), and its use in organic synthesis was established. Soon after that, in the 1930s and 1940s, other organometallic compounds, such as organolithium (George Wittig and Karl Ziegler) and organocopper (Henry Gilman), were prepared, and their chemical reactivity was observed. The work of these pioneers outlined the principles of organometallic chemistry and made it an indispensable part of organic synthesis. Now, the chemistry of organoaluminum, organotitanium, organotin, organoboron, organoiron, organomanganese, organocerium, and other organolanthanide compounds is established and is still developing with the aim of achieving organic compounds with a high degree of selectivity. It is not feasible to include in this chapter an exhaustive list of all the contributions focused on stereoselective addition of organometallic reagents to carbonyl compounds. Thus, the most important stereochemical aspects and recent advances in this area are mentioned. The parts related to organoboron and organosilicon reagents due to different behavior of C–B/C–Si bond are excluded in this chapter. More information in this area is available in recent review articles.[2,3] Moreover, chemoselectivity in the addition of organometallic reagents to carbonyl substrates is not discussed in detail. Generally, the chemoselectivity of the additions can be seriously affected by undesired competing reactions, such as β-hydride elimination resulting in the reduction of carbonyl substrates or α-deprotonation/enolization resulting in aldol reactions.[4]

Addition reaction of organometallic reagents with prochiral carbonyl substrates (asymmetrical ketones or aldehydes) results in the formation of a new stereogenic center. The presence of another stereogenic center (one or more) in organometallic reagents or carbonyl compounds leads to diastereoisomers. Based on the character of the organometallic reagent and carbonyl substrate, we can classify these transformations into four categories:

1. Addition of achiral organometallics to achiral carbonyl substrate
2. Addition of achiral organometallics to chiral carbonyl substrate
3. Addition of chiral organometallics to achiral carbonyl substrate
4. Addition of chiral organometallics to chiral carbonyl substrate

In the last case, multiple asymmetric inductions can occur when there is a specific match–mismatch effect, affording products with good to high diastereoselectivity, whereas in the first case, the addition takes place without any asymmetric induction, such as the use of a chiral solvent or chiral ligand, to produce racemic alcohols. The second case has been chosen as the subject of this chapter, because it represents the traditional concept used throughout stereoselective organic synthesis. The main goal of this chapter is to give readers a general overview of the stereochemical outcome of the addition of organometallic reagents to chiral carbonyl compounds in connection with models used for their prediction.[5]

2.2 MODELS FOR ASYMMETRIC CARBONYL COMPOUND ADDITION

In 1894, Fisher formed the basis of asymmetric induction, as a result of his extensive work on saccharide synthesis, when he made the observation that further synthesis with asymmetric systems proceeds in an asymmetric manner.[6] Shortly thereafter, the first asymmetric synthesis of α-hydroxy acids **2** from the corresponding α-ketoesters **1** was achieved by McKenzie (Figure 2.1).[7] His work became a pillar for further investigations in the field of stereoselective nucleophilic additions.[8] The need arose to justify experimental observations in order to answer the question of how the stereogenic center, which is in close proximity to the prochiral reaction center, influences selectivity of the reaction. In the 1950s and 1960s, a number of models were developed in order to understand, predict, and control diastereoselectivity of the nucleophilic additions to carbonyl compounds with an adjacent stereocenter.[9–11] The topic was summarized in many reviews and textbooks,[12,13] but we report here only the main contributions to this field.

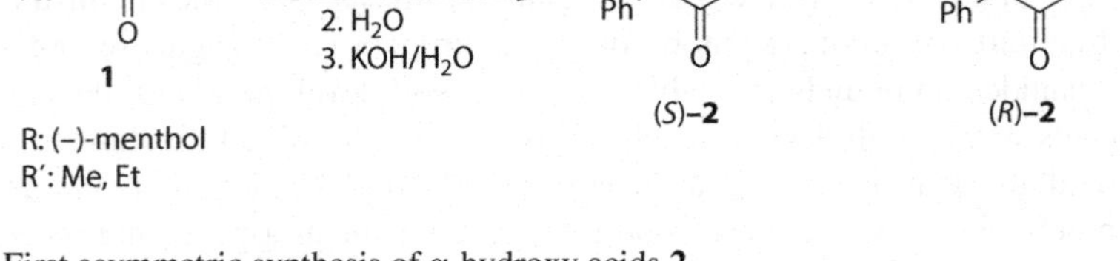

FIGURE 2.1 First asymmetric synthesis of α-hydroxy acids **2**.

2.3 MODELS FOR 1,2-ASYMMETRIC CARBONYL ADDITION

Cram's rules[14] became famous for their simplicity and represented the basis used in further extensions and modifications. The Cram model analyzes the stereochemical outcome of the additions of nucleophile to α-chiral carbonyl compounds.[15] The model considers steric factors of α-substituents and arranges them according to sizes L (large), M (medium), and S (small). The model suggests that the R_L group adopts a conformation *anti* to the carbonyl group due to steric reasons (carbonyl oxygen is assumed to coordinate with organometallic nucleophile, which makes it the bulkiest group in the molecule) and the carbonyl group is flanked by the less bulky substituents R_M and R_S. Nucleophile attacks the carbonyl group along the less-hindered trajectory at an angle of 90° (Figure 2.2).

A different model has to be considered if chelation between carbonyl oxygen and substituents on an adjacent stereocenter occurs.[16] The so-called Cram-chelation rule (Figure 2.3) assumes that the conformation where carbonyl oxygen eclipses the chelating substituent is locked by the present metal cation.[17] Nucleophile attacks from the side of the R_S substituent at a 90° angle.

While studying the addition of Grignard reagents and alkyl lithiums to β-chloroketones, Cornforth noticed that the chlorine substituent took the role of the R_L substituent, even though bulkier

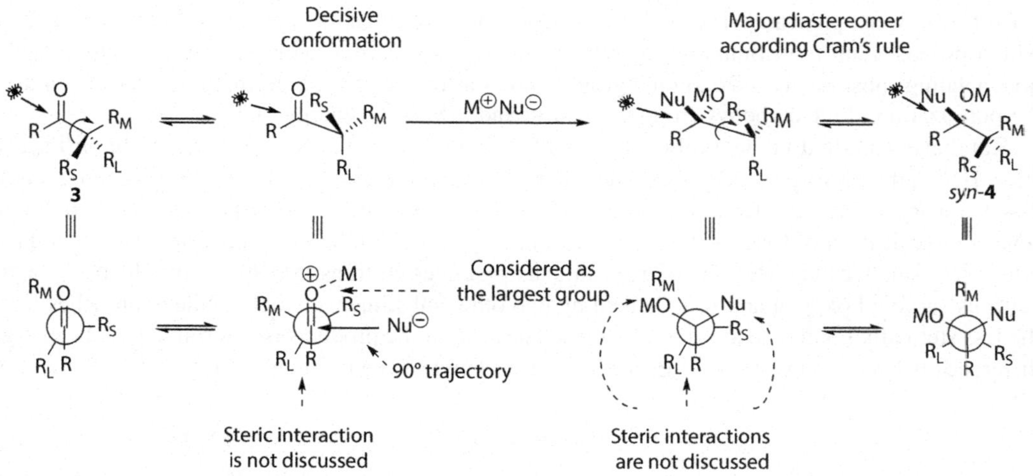

FIGURE 2.2 Cram's rule.

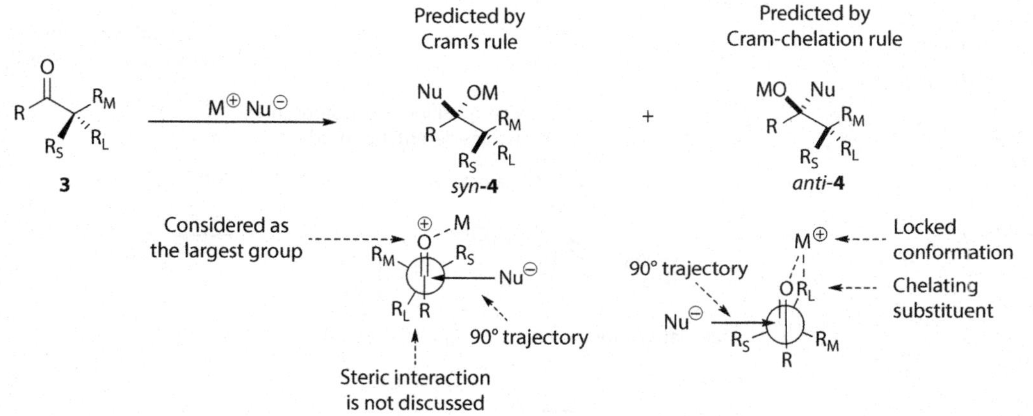

FIGURE 2.3 Cram's rule and Cram-chelation rule.

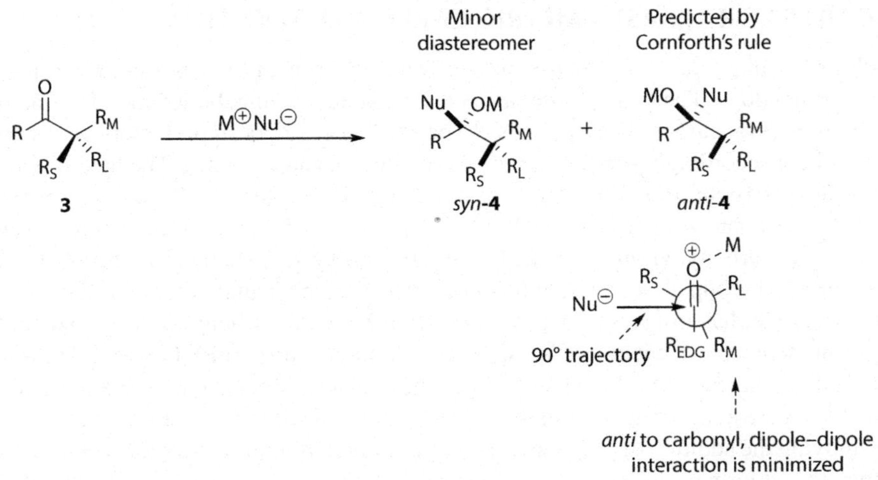

Figure 2.4

FIGURE 2.4 Cornforth's rule.

substituents were present.[18] This led to the Cornforth modification of Cram's rule (Figure 2.4), which proposes an *anti* orientation of carbon–chlorine (or carbon with any other electronegative nonchelating substituent) and carbonyl group dipoles as the favored conformation, because it makes the polarization of the carbonyl group the easiest and lowers the energy of TS.

Karabatsos studied nucleophilic additions to aldehydes, where R_L = Ph, R_M = Me or ^{i}Pr, and R_S = H,[19] and was surprised by the given selectivities (for R_M = Me, the ratio of the products is 2–4:1; for R_M = ^{i}Pr, the ratio of the products is 1–1.9:1; the ratio varies depending on used conditions).[20] Application of Cram's rule on the results would lead to the incorrect conclusion that ^{i}Pr is effectively smaller than Me. Therefore, Karabatsos suggested transition states, in which R_M or R_L eclipses carbonyl oxygen and R is flanked by two other substituents. Nucleophile approaches from the less sterically hindered side at a 90° angle. The ratio of the products is determined by the energy difference between interactions of R_M or R_L and carbonyl oxygen (Figure 2.5).

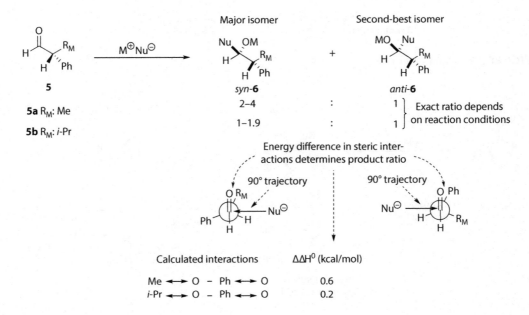

FIGURE 2.5 Karabatsos' rule.

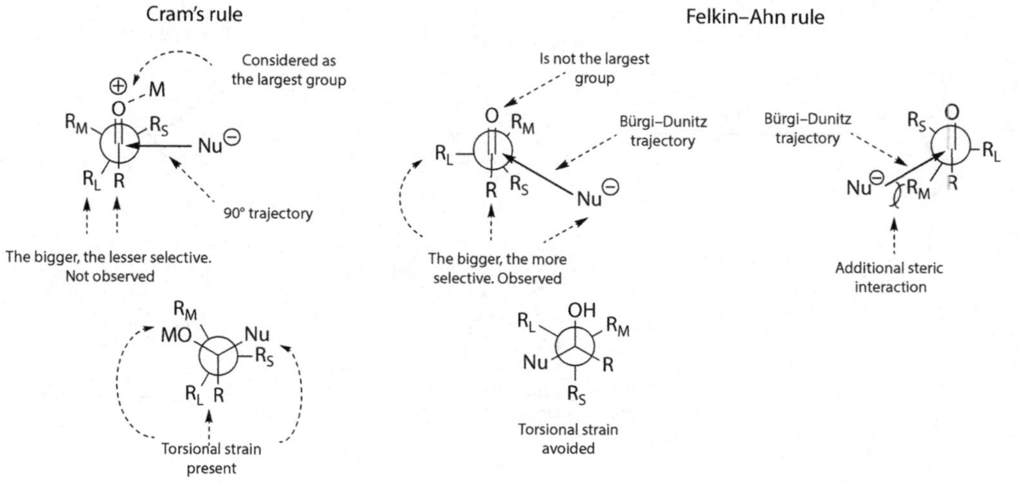

Diastereoselective Addition of Organometallic Reagents to Chiral Carbonyl Compounds 33

FIGURE 2.6 Cram's rule versus Felkin–Ahn rule.

Cram acyclic's rule failed to explain the observed trend that the bulkier the R_L, the more stereoselective is the reaction (exactly the opposite trend would be expected due to the strain between R_L and R substituents; see Figure 2.6), and to predict the stereoselective outcome of additions to cyclohexanones.[21] These observations led to the formulation of a new model, known as the Felkin– Ahn. Model, is based on contributions of Felkin, Ahn, and Eisenstein (Figure 2.6).[21–25] The model considers both steric and electronic effects of substituents. In contrast with previous models, it suggests a staggered conformation of the transition state, which avoids torsional (Pitzer) strain, which is quite high even when the degree of bonding is low. Moreover, it proposes that the size of carbonyl oxygen was overestimated and the main steric interactions involve R and incoming nucleophile. With these assumptions in mind, these authors proposed the conformation, in which R_L is the most electronegative substituent (substituent with the lowest lying σ*) oriented perpendicular to the plane of the carbonyl group and simultaneously antiparallel to the direction of nucleophilic attack (Figure 2.6). There are two such conformations: either R_S is next to R and R_M is next to carbonyl oxygen, or R_M is next to R and R_S is next to carbonyl oxygen. Calculations showed that the nucleophile attacks the carbonyl group following the Bürgi–Dunitz trajectory[26–28] at an approximate angle of 103° from the side of the R_S substituent, which privileged the conformer with R_S next to hydrogen and R_M next to carbonyl oxygen (Figure 2.6).

The perpendicular orientation of R_L or the most electronegative substituent is stabilized by electronic effects. First, it makes the separation between the electronegative group and the negatively charged nucleophile the farthest. Second, the perpendicular orientation enables the overlap of C-Lσ* and C = Oπ* orbitals. The combination of C-Lσ* and C = Oπ* results in lowering the energy of lowest unoccupied molecular orbital (LUMO). An antiperiplanar attack of nucleophile gives a more favorable overlap with combined orbitals than a synperiplanar attack does. Ahn proposed that the orbital effects might be more effective than steric ones, especially at longer distances.[24]

However, the conformation with R_M next to carbonyl oxygen has to be considered for relatively small R_M groups. If nucleophile is small, the conformation with R_L group next to carbonyl oxygen has to be taken into account. Conformers with R_L next to R are not considered in any case (Figure 2.7).[21] These conformations successfully explain the formation of minor diastereomers and show the power of the Felkin–Ahn model.

The model gives reasonable explanations for results of less stereoselective reactions as well. The *tert*-butyl ketones do not match the observed trend that the bigger the R, the higher the stereoselectivity.[21] The sharp drop of selectivity was observed while going from R = ^{i}Pr to R = ^{t}Bu. Any staggered

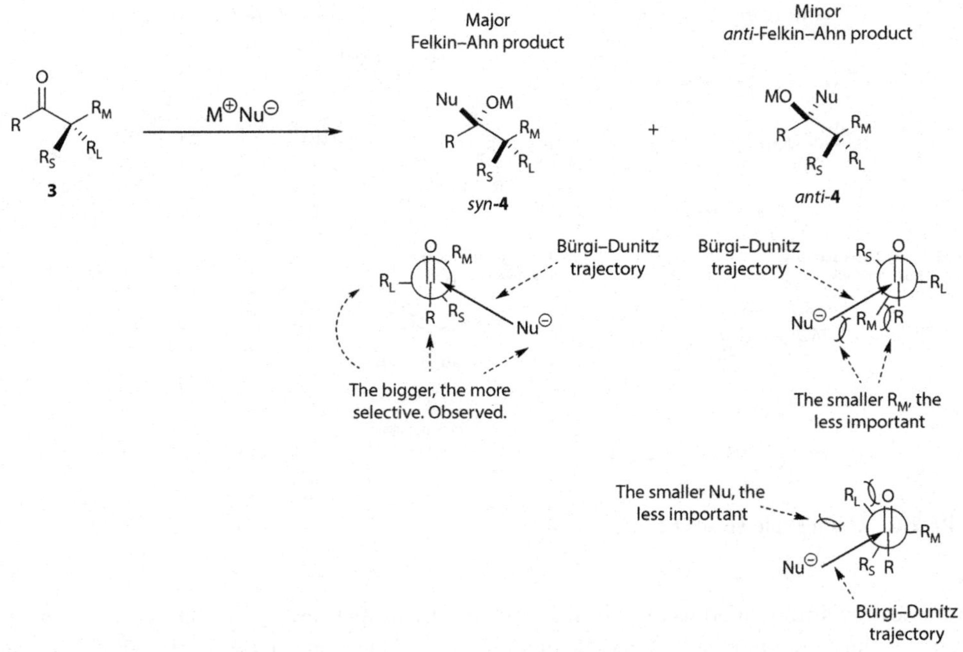

FIGURE 2.7 Felkin–Ahn rule describes the formation of major and minor diastereomers.

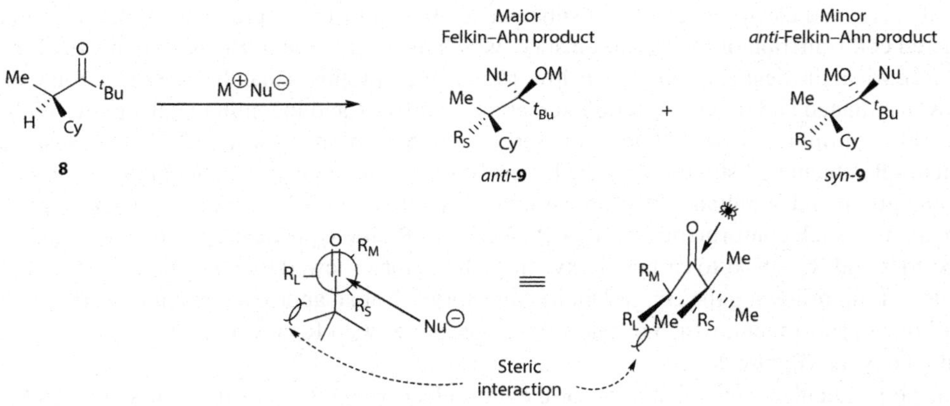

FIGURE 2.8 Felkin–Ahn model for *tert*-butyl ketones.

conformation of the transition state is strain free, due to the interaction between R_L and R = ^{t}Bu, which becomes almost as severe a strain as that between R_L and nucleophile (Figure 2.8).

There are two possible products of nucleophilic addition to conformationally locked cyclohexanones: axial and equatorial alcohols. It is generally accepted that bulky nucleophiles or hindered ketones (ketones with substitution in position 3 or 5) prefer equatorial attack to yield axial alcohol.[29] In contrast, axial attack is favored with small nucleophiles or unhindered ketones.[30,31] A large number of models were made in order to explain the stereochemical outcome of nucleophilic attack.[32] Among them, the Felkin–Ahn hypothesis is the most widely accepted (Figure 2.9).[22,33–35] Methylcyclohexane cannot adopt the Felkin–Ahn conformation, and therefore a nearly eclipsed transition state is considered. Selectivity depends on competition between torsional and steric strains. If the nucleophile is small, torsional strain dominates. On the other hand, with large nucleophiles, steric strain plays the main role.

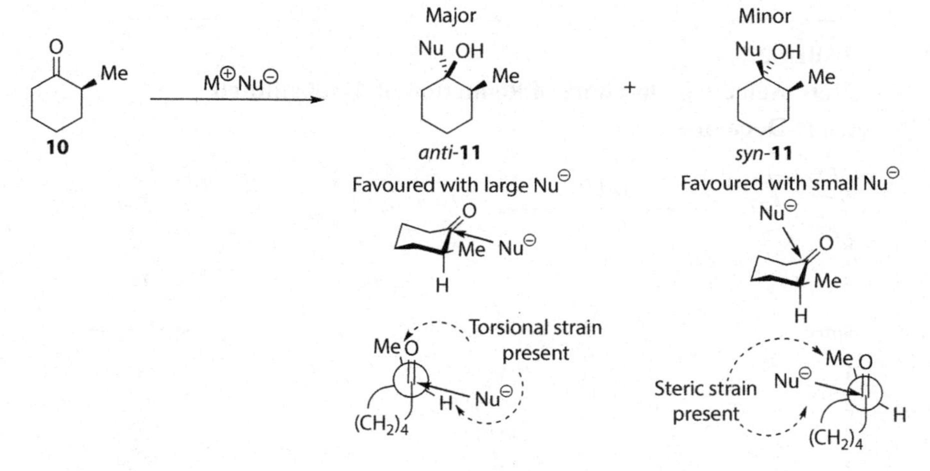

Diastereoselective Addition of Organometallic Reagents to Chiral Carbonyl Compounds

FIGURE 2.9 Felkin–Ahn rationalization for substituted cyclohexanones.

However, the Felkin–Ahn rule failed to explain the stereoselective outcome of nucleophilic additions to 5-substituted-2-adamantanones, 3-substituted cyclohexanones, 2,3-endo,endo-disubstituted-7-norbornanones, and 2,2-diarylcyclopentanones. Therefore, Cieplak used principles of microscopic reversibility and came up with a new model. He based his model on the stabilizing interaction of low-lying σ* of forming bonds by electron donation with σ of the perpendicular substituent. Nucleophile attacks the carbonyl group preferentially in the direction antiperiplanar to the best electron-donor vicinal bond. The model proposes that C—H is a better σ-donor than the C—C bond (the order is C—S > C—H > C—C > C—N > C—O; see Figure 2.10).

Although the Cieplak model successfully explains stereoselectivity of nucleophilic addition to 5-substituted-2-adamantanones,[36,37] 3-substituted cyclohexanones,[38] 2,3-endo,endo-disubstituted-7-norbornanones,[39–41] and 2,2-diarylcyclopentanones,[42] there was sharp discussion about the reliability of this model. First of all, the order of the electron-donor groups was disputed, mainly by Rozeboom and Houk who presented strong evidence of the opposite trend.[43–46] However, Laube and Addock published experimental results that supported the Cieplak order of donor ability.[38,47–52]

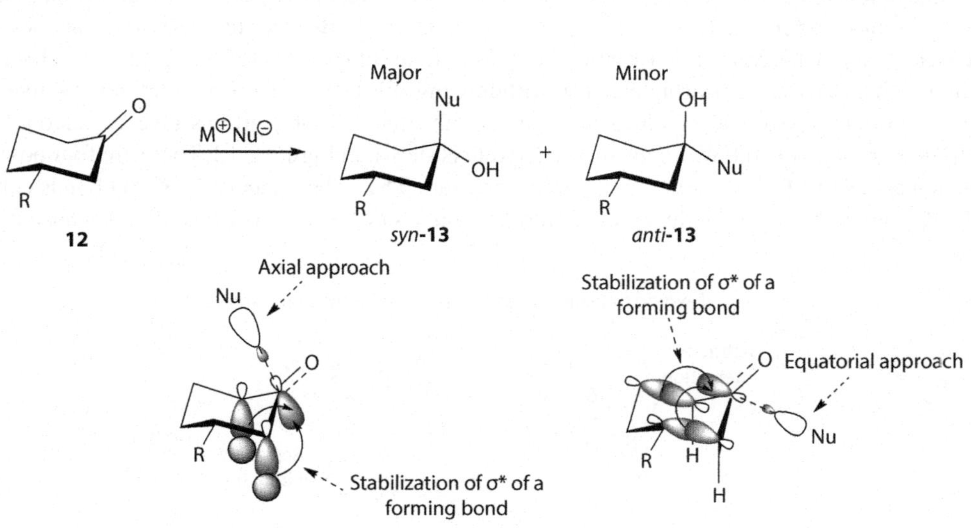

FIGURE 2.10 Cieplak model.

TABLE 2.1
Stereoselective Outcome of Reduction of 4-Substituted *trans*-Decalone

Entry	R	eq.15/ax.15
1	H	60/40
2	eq-OH	61/39
3	eq-OAc	71/29
4	eq-Cl	71/29
5	ax-OH	85/15
6	ax-OAc	83/17
7	ax-Cl	88/12

Second, the nature of the transition state was contradicted. At the end, the character of the proposed stabilizing interaction (electron donation into antibonding σ* of a forming bond) is considered as a bond-weakening process by Reetz.[53] Later on, Houk offered a new explanation for the stereochemical outcome of nucleophilic addition to 4-substituted *trans*-decalones based on the long-term electrostatic interaction between nucleophile and 4-substituent.[45,54] He noticed that the 4-axial substituent has considerable effect on stereoselectivity compared to the 4-equatorial one (Entries 2–4 versus Entries 5–7, Table 2.1). There is repulsive electrostatic interaction between nucleophile and 4-axial substituent in the transition state, which disfavors the equatorial attack and, hence, formation of axial alcohol (Figure 2.11). On the other hand, axial attack is favored due to an attractive electrostatic interaction that is present in the transition state. Calculations showed that the axial transition state is more stable than the equatorial one, in which torsional strain is present. Moreover, similar rationalization can be applied to explain the stereochemistry of nucleophilic additions to 3-cyclohexanones, 2,2-diarylcyclopentanones, and 1,2-endo, endo-disubstituted-7-norbornanones.[54]

Calculations confirmed that in the case of 1,2-endo,endo-disubstituted-7-norbornanones, the stereoselectivity is driven by electrostatic effects of remote substituents (Figure 2.12).[39,41,54] It can be concluded that derivatives bearing electron withdrawing substituent (EWG) preferentially give *anti* alcohol and derivatives bearing electron donating substituent (EDG) groups give *syn* alcohol. An illustrative example of an EWG-substituted derivate is shown in Figure 2.12. There are four possible transition states (**TS1–TS4**). Among them, the structure **TS1**, where aldehydic C=O bonds eclipse the C_1-C_2 bond and nucleophile, coming from the *endo* side, is the most favorable because of the

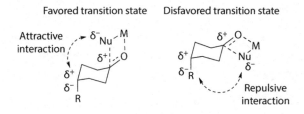

FIGURE 2.11 Long-term polar interactions involved in transition state of nucleophilic addition to 4-axial-substituted cyclohexanones.

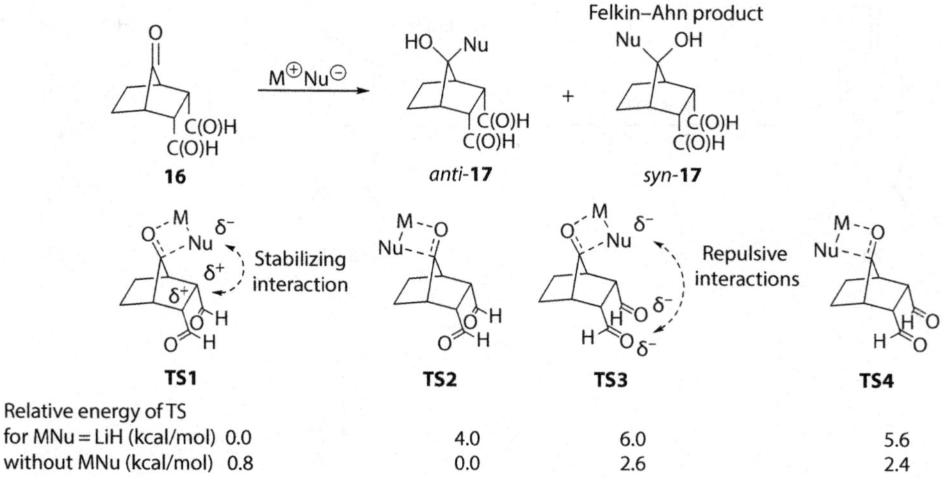

Diastereoselective Addition of Organometallic Reagents to Chiral Carbonyl Compounds 37

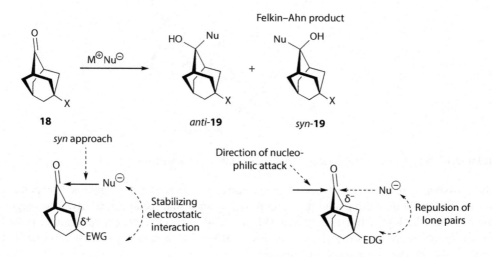

FIGURE 2.12 Nucleophilic addition to 1,2-endo,endo-dicarbaldehyde-7-norbornanones. Relative energies were calculated by MP2/6-31G*.

FIGURE 2.13 Nucleophilic addition to 5-substituted-2-adamantanones.

stabilizing interaction with positive carbon centers. This interaction is much weaker when nucleophile approaches from the *anti* side (**TS2**). There is a destabilizing interaction between nucleophile and carbonyl oxygens in transition state **TS3**.

The same approach can be used to determine stereoselectivity of nucleophilic addition to 5-substituted-2-adamantanones.[51] When the 5-substituent is the EWG group, *syn*-attack is favored due to the stabilizing interaction between EWG and nucleophile. Substitution by EDG leads to *anti*-attack due to repulsive interaction between lone pairs of EDG and nucleophile (Figure 2.13).

While working on the total synthesis of the antibiotic aranorosin, Wipf and Kim observed an interesting selectivity of nucleophilic addition to 4,4-disubstituted cyclohexanedienone (Scheme 2.1).[55,56] They realized that nucleophile comes preferentially from the side that is *anti* to the 4-oxygen substituent (α-attack, Scheme 2.2).

In this case, both the Felkin–Ahn (would predict β-attack) and the Cieplak models (would predict α-attack, however, without qualitative correlation between ratio of products and energy of σ of donor C–C bond) failed to explain the observed selectivities. Therefore, Kim and Wipf searched for

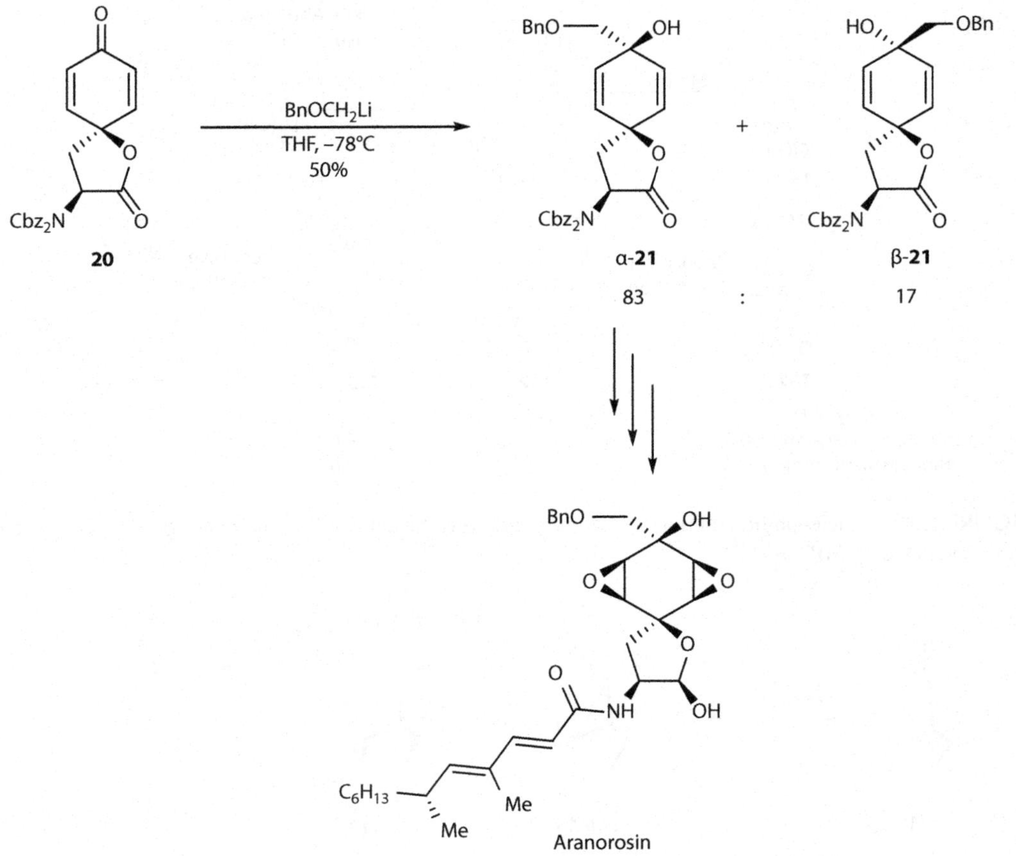

SCHEME 2.1 Nucleophilic addition step during total synthesis of aranorosin.

another possible explanation and showed linear correlation between the calculated perpendicular vector of the dipole moment μ^1 and the natural logarithm of the facial selectivity. The bigger the dipole moment, the higher the α-selectivity (Figure 2.14). This example proves another point that during nucleophilic additions to sterically unbiased carbonyl compounds, electrostatic control plays a more significant role than hyperconjugation.

Heatcock et al.[57] evaluated the Ahn–Eisenstein presumption that the substituent with the lowest energy of σ* is always the L substituent. They found out that electronic effect can be contra balanced with steric effect. They proposed equilibrium of four different conformations **C1–C4** (Figure 2.15). The preferred conformation depends on both electronic and steric properties of the substituent. By balancing these properties, they created the following order of substituents MeO, *t*-Bu, Ph, *i*-Pr, Et, Me, H. For small substituents, both the electronic and steric properties favor conformer **C1**. If there is a small difference in the electronic properties of substituents, steric plays a bigger role (favors conformer **C4**). They stated that their results cannot be rationalized by the Cieplak model, which would lead to the conformations shown in Figure 2.16.

Tomoda formulated a quantitative model known as the Exterior Frontier Orbital Extension (EFOE), which assumes the importance of ground-state conformational and electronic properties of unsaturated compounds. His model is based on the kinetic Salem–Klopman equation, which expresses the driving force of a chemical reaction by the summation of free independent terms[58,59]: the exchange repulsion term, which represents steric effects; the electrostatic interaction term, which plays a role especially in ionic reactions; and the donor–acceptor orbital interaction term. The EFOE model combines the steric term (evaluated by π-plane-divided accessible space [PDAS]) together

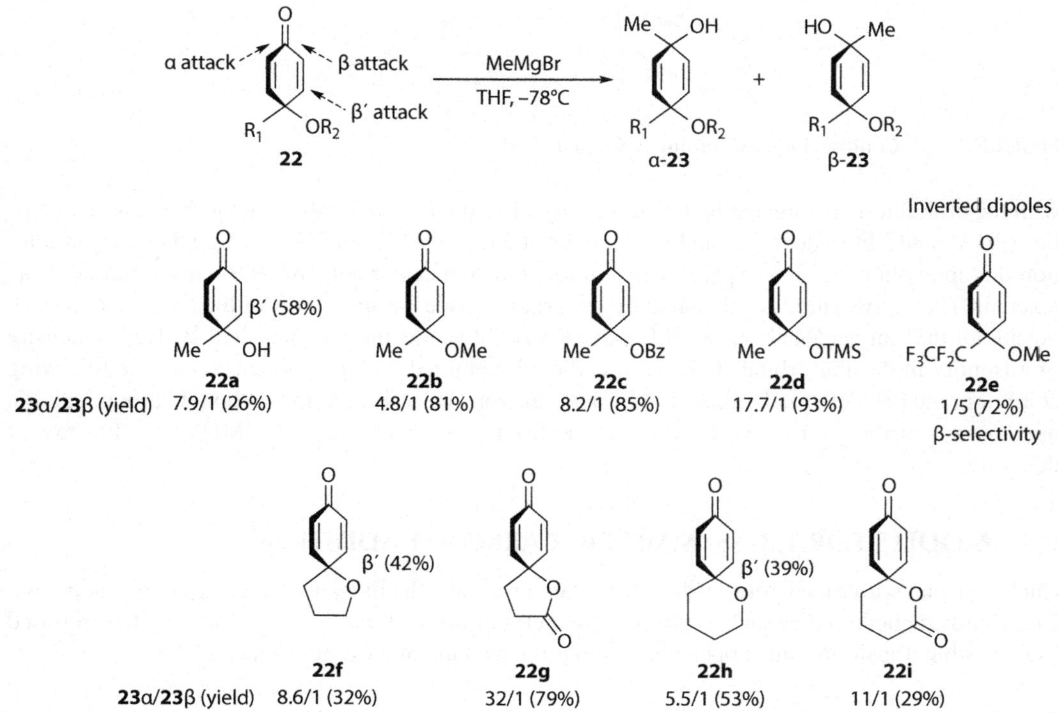

Diastereoselective Addition of Organometallic Reagents to Chiral Carbonyl Compounds 39

SCHEME 2.2 Selectivity of nucleophilic addition to 4,4-disubstituted cyclohexanedienone **22**.

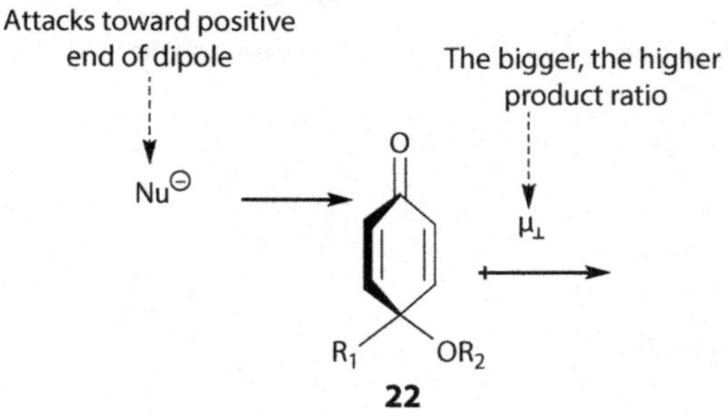

FIGURE 2.14 Kim and Wipf's explanation of α-selectivity.

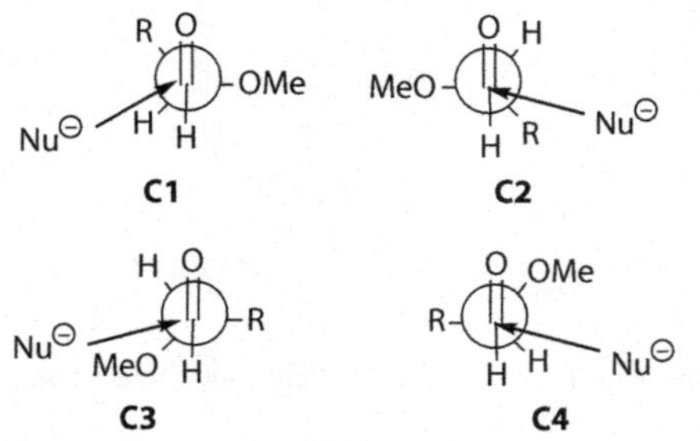

FIGURE 2.15 Equilibrium of four transition states suggested by Heathcock.

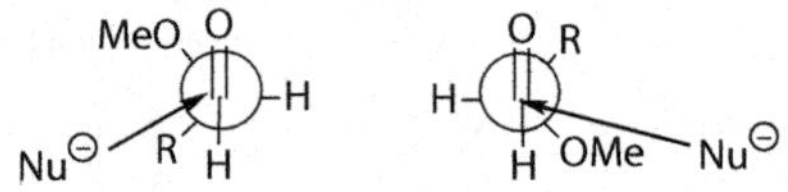

FIGURE 2.16 Conformations according to Cieplak model.

with the orbital term (evaluated by EFOE density), leaving out coulombic interactions. Both quantities (PDAS and EFOE density) can be calculated at once. The idea of PDAS is based on the assumption that the volume of outer space nearest to the reaction center contains steric information of the reactant. The PDAS value is calculated by integrating space within 2.65 Å of the molecular surface outside of the van der Waals radii. EFOE density is defined as the π-plane-divided electron density of a frontier molecular orbital (LUMO of carbonyl) summed over points that satisfy the following condition: the absolute total value of the wave functions belonging to the carbonyl carbon makes a maximum contribution to the total value of the frontier molecular orbital (FMO) wave function at the point.

2.4 MODELS FOR 1,3-ASYMMETRIC CARBONYL ADDITION

Chelation plays a crucial role in the case of controlling selectivity in 1,3-asymmetric induction. Cram studied nucleophiles such as Grignard reagents or organolithium compounds.[60,61] He proposed free possible transition states: open chain, polar open chain, and cyclic (Figure 2.17).

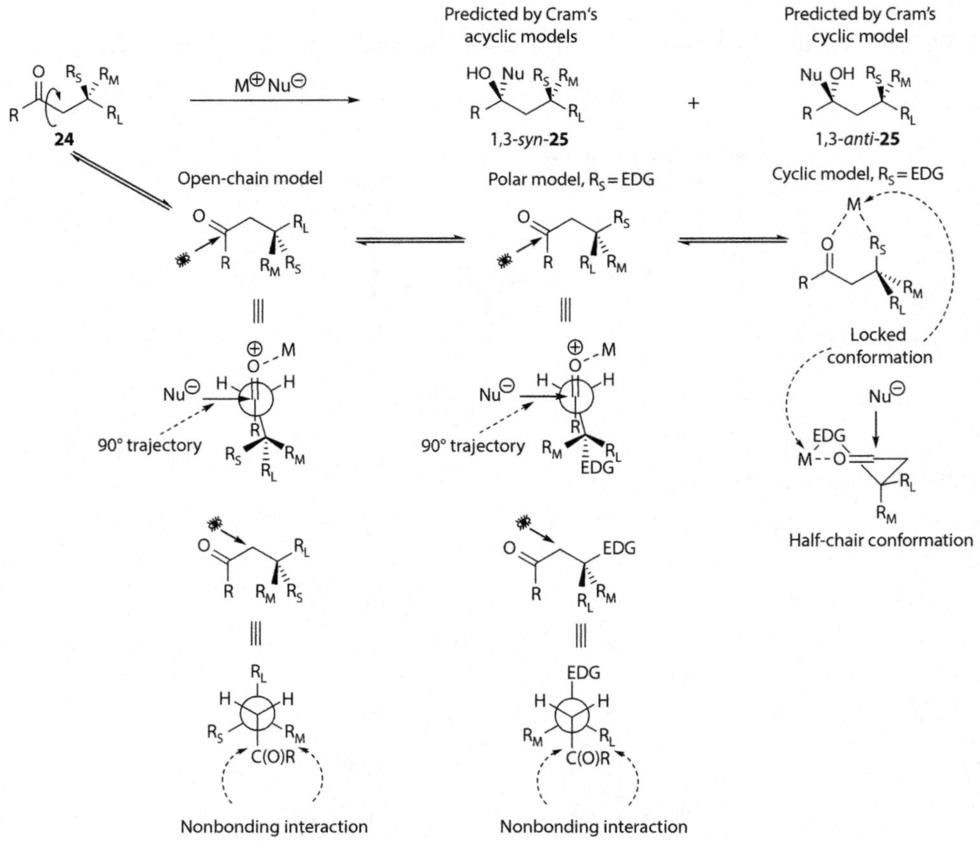

FIGURE 2.17 Transition states proposed by Cram for 1,3-asymmetric addition.

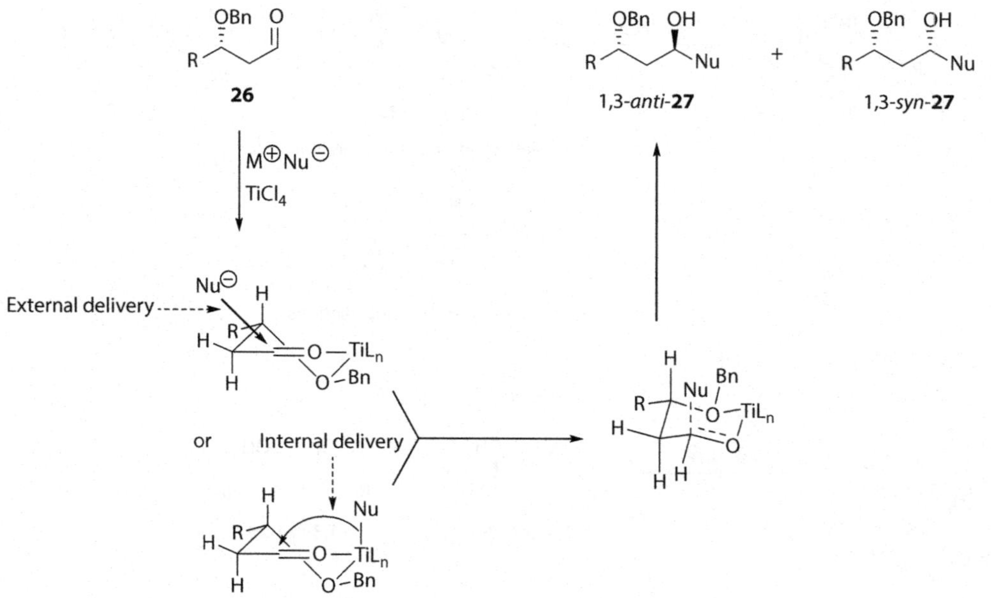

FIGURE 2.18 Cram–Reetz chelation model for 1,3-asymmetric induction.

Open-chain transition state is based on minimal steric interactions and it is similar to the Cram model for 1,2-asymmetric induction. It places R_L in β-position and carbonyl oxygen as far as possible. The polar open-chain model places two polar groups: EDG in β-position and carbonyl oxygen as far as possible. The cyclic model is based on important coordination between metal, EDG substituent, and carbonyl oxygen. The open-chain model was designed for 1,3-asymmetric induction of a substrate that does not contain coordinating (EDG) substituents. However, the degree of stereoinduction is low and strongly depends on temperature, nucleophile, and solvent, and the model does not fit observed data completely. Slightly better is the situation where a coordinating (EDG) substituent is present. In that case, the other two models are supposed to play a role. Among them, the cyclic model correlates most data. However, RMgX (Cram), RLi (Cram), and (R₂CuLi)[62] do not perform as well as in the case of 1,2-asymmetric induction.

Later on, Reetz played with the chelation concept.[63,64] He studied reactions of β-alkoxy derivatives with carbon nucleophiles and found out that $TiCl_4$ can perform chelation control and gives products with high diastereomeric ratio. Reetz supposed that the cyclic transition state is responsible for the stereochemical outcome (Figure 2.18), and it is therefore necessary to use such nucleophiles that do not destroy chelation. It is not clear whether the reacting species is the nucleophile alone or whether transmetalation between nucleophile and $TiCl_4$ creates $RTiCl_3$, which is responsible for intramolecular delivery of R.

The Jacques model[65] supposes staggered conformation of the transition state with *anti* orientation between $C_β$ and the forming bond, which is similar to the Felkin model. The decisive conformation is that where the interaction between the R substituent and substituents on $C_β$ is minimal (Figure 2.19).

Evans noticed that steric and electrostatic effects combine to influence the direction and degree of 1,3-induction. He made an open-chain model in order to explain 1,3-*anti* selectivity for nucleophilic additions to aldehydes bearing X = β-EDG under nonchelating conditions.[66–68] Like Felkin and Ahn, he proposed a staggered conformation **C1**, in which $C_β$ is oriented perpendicular to C=O contains minimal dipoles and nonbonding interactions (Figure 2.20). Conformations **C3** and **C6** suffer with destabilizing gauche interaction between $R_β$ and C=O. Conformations **C4** and **C5** become disfavored as the size of $R_β$ increases. There is a destabilizing dipolar interaction in conformations **C2** and **C5** for X = EDG.

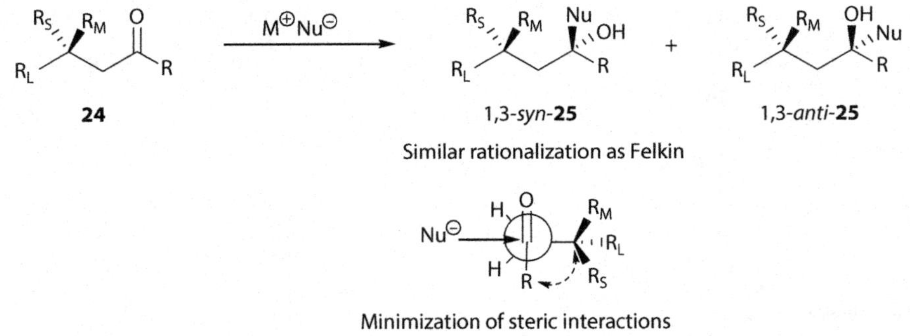

FIGURE 2.19 Jacques model.

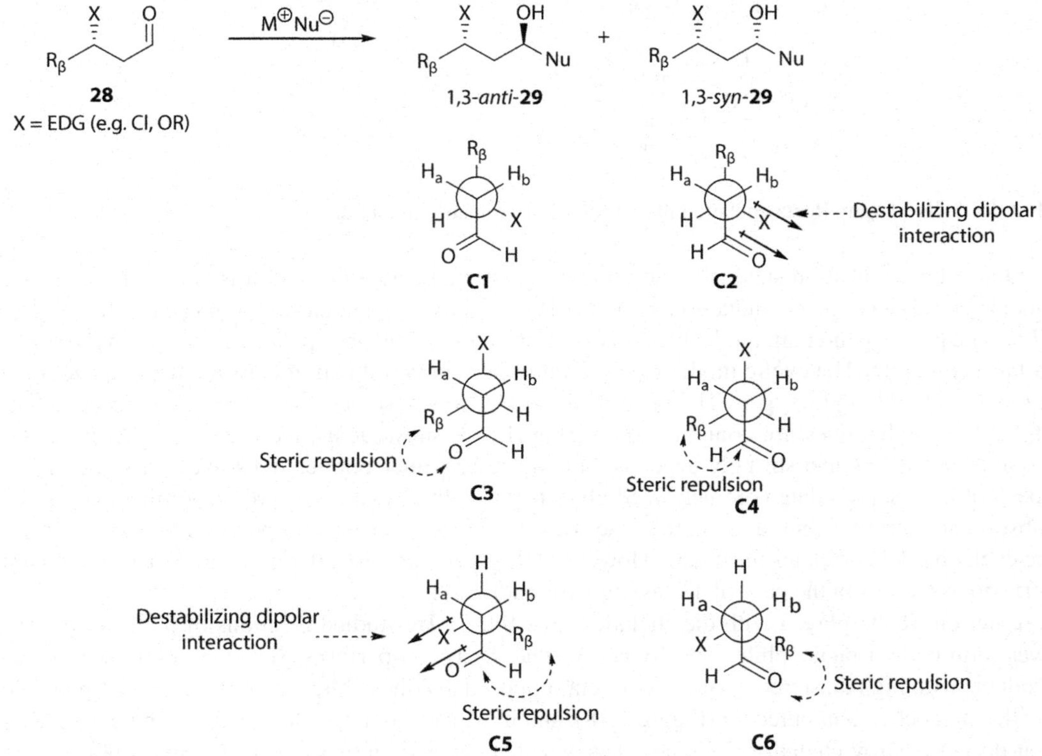

FIGURE 2.20 Evans model for 1,3-asymmetric induction.

The Felkin–Ahn rule and Cram-chelation rule proved to be reliable tools in determining the stereochemical outcome of nucleophilic additions to β-substituted carbonyl compounds in a qualitative sense. They have huge predictable power and are broadly used for this purpose. Felkin–Ahn product is a broadly accepted term for the major carbonyl addition product predicted by the Felkin–Ahn rule as well as by Cram, Karabatsos, and Cornforth. The minor product is often called the *anti*-Felkin–Ahn product or Cram-chelate product.

However, the degree of stereoselectivity can be affected by reaction conditions. Changes in solvent, temperature, and organometallic reagents among others influence the reaction mode. Modification of these conditions can enhance, diminish, or even reverse stereoselectivities. Chelation control and non–chelation control generally lead to the opposite diastereoselectivity.

There are open-chain and chelation-control models for predicting stereochemical outcomes of 1,3-asymmetric nucleophilic additions. Among them, especially models that include chelation control, such as the Reetz model, showed huge predictable power.

2.5 ADDITION OF ACHIRAL REAGENTS

Since McKenzie's work focused on asymmetric synthesis of α-hydroxy carboxylic acids from methyl benzoylformate,[7] the stereochemical outcome of nucleophilic addition of organometallic reagents to carbonyl compounds has been investigated. The diastereoselective addition of achiral organometallic reagents to chiral α-substituted carbonyl compounds has been the most thoroughly studied example of 1,2-addition, besides reduction of carbonyl compounds with metal hydride reagents and nitrile addition to carbonyls. As mentioned in the previous chapter, steric, electronic, and chelation properties of α-substituents in carbonyl compounds play a key role in the degree of stereoselection of the addition process. Diastereoselectivity of 1,2-addition is also affected by other important factors including solvent, temperature, additive, and specific features of substrate or organometallic reagent. In order to illustrate the effects of these modifications on stereodiscrimination, the most typical and recent examples of 1,2-additions of achiral organometallic compounds to chiral carbonyl compounds are described in this chapter. The following text is organized according to the nature of the carbonyl substrate and the presence of heteroatoms in it.

2.6 ADDITION OF ALDEHYDE

A typical example of addition of achiral organometallic reagents to a chiral aldehyde substrate is illustrated in Table 2.2. In general, additions of organometallic/alkylmetal reagents to ordinary chiral aldehydes **30** with no chelating ability should provide Felkin–Ahn diastereoisomer **31** preferentially (Table 2.2).

TABLE 2.2
Additions of Achiral Organometallic Reagents to α-Chiral Aldehydes

Entry	MNu	Additive	Yield	anti/syn	References
1	MeLi	—	91	80/20	[69,70]
2	MeLi	18-crown-6	90	88/12	[70]
3	MeLi	K-211[a]	86	90/10	[70]
4	BuLi	—	91	83/17	[69,70]
6	BuLi	18-crown-6	93	91/9	[70]
7	BuLi	15-crown-5	91	97/3	[70]
8	MeMgBr	—	64	72/28	[1,71,73]
9	MeMgOTs	—	76	92/8	[73]
10	MeMgOSO$_2$C$_6$H$_2$(Me)$_3$	—	63	94/6	[73]
11	EtMgBr	—	78	72/28	[71,72]
12	BuMgBr	—	89	87/13	[71,72]
13	PhMgBr	—	80	72/28	[73]
14	MeTiCl$_3$	—	n.p.	81/19	[74]

[a] Cryptofix K211 is 4,7,13,18-tetraoxa-1,10-diazabicyclo[8.5.5]eicosane.

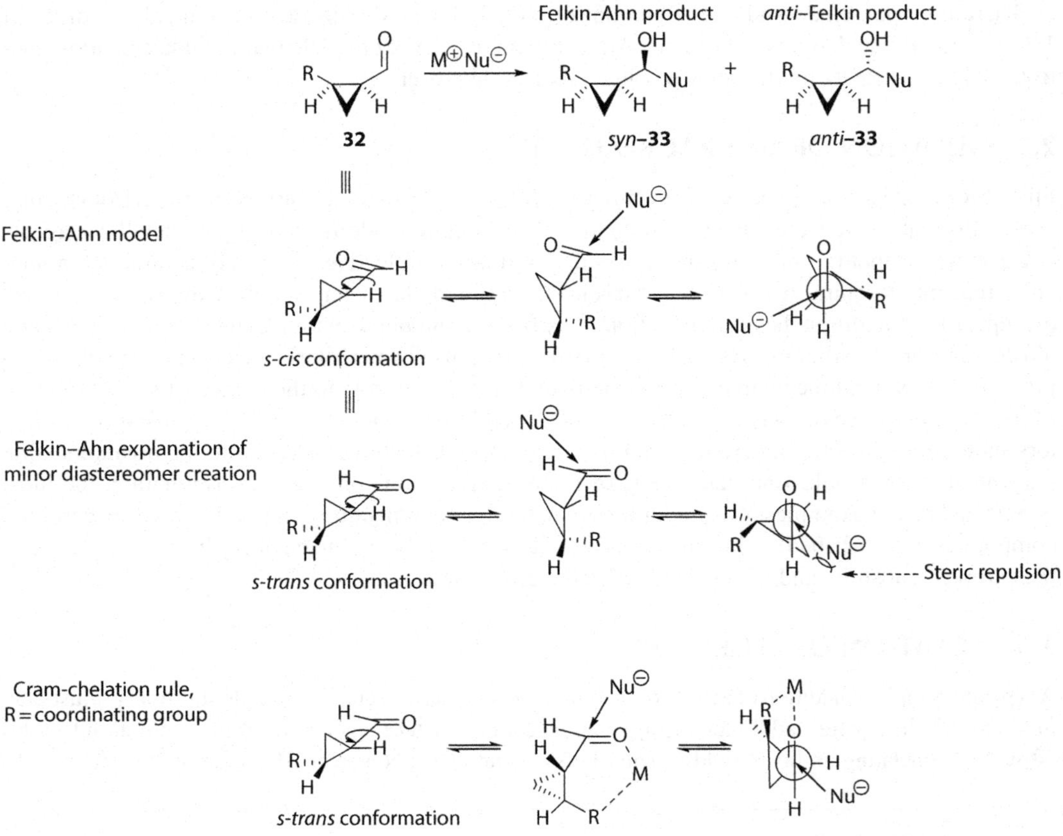

SCHEME 2.3 Ways of addition to chiral cyclopropyl carbaldehydes **32**.

In order to increase the Felkin–Ahn/*anti*-Felkin–Ahn selectivity in the addition process, it is convenient to use sterically more demanding alkylmetals, that is, the bulkiness of nucleophile (Nu⁻) (Entries 11–14) as well as the bulkiness of the metal as counterion (M⁺) (Entries 8 and 11–16) or metal ligands (Entries 8–10). In addition, Felkin–Ahn/*anti*-Felkin–Ahn selectivity can be enhanced by complexation of M⁺ with kryptofixes and crown ethers (Entries 2, 3, 6, and 7), especially when selective complexation based on the character of the additive can take place (Entry 7 versus Entry 6).

Not only acyclic α-branched but also cyclic chiral aldehydes can be used as substrates for additions with organometallic reagents. Addition reactions on chiral cyclopropyl carbaldehydes **32** proceed with high diastereoselectivities. Both the Felkin–Ahn adduct and the *anti*-Felkin–Ahn adduct can be formed predominantly depending on the substrate and nucleophile (Scheme 2.3). The observed selectivity can be explained either by the Felkin–Ahn rule when chelation can be ruled out or by the chelation model predicting the formation of the *anti*-Felkin–Ahn adduct (Table 2.3, Entries 1 and 2).[75]

2.6.1 Aldehydes Containing α-Oxygen Substituent

High diastereoselectivity in additions has been observed when aldehydes containing electronegative α-substituents were used. Especially the presence of the strongly chelating functional groups, α-alkoxy and α-amino, in substituents can significantly affect diastereofacial selectivity. When strongly chelating groups are present in the aldehyde substrate, the chelation-controlled transition state/model is favored in nonchelating solvents, affording preferentially *anti*-Felkin–Ahn adducts (Table 2.4).

TABLE 2.3
Addition to Chiral Cyclopropyl Carbaldehydes 34

Entry	R¹	R²	R³	R⁴	NuM	syn/anti	Reference
1	Ph	CONEt$_2$	H	H	MeMgBr	4/96	[75]
2	Ph	CONEt$_2$	H	H	EtMgBr	4/96	[75]

TABLE 2.4
Solvent Effect on Stereoselectivity of Nucleophilic Addition

Entry	R$_L$	MNu	Solvent	Yield (%)	anti/syn	Reference
1	CH$_2$CO$_2$Me	^{n}Bu$_2$CuLi	THF	21	62/38	[77]
2	CH$_2$CO$_2$Me	^{n}Bu$_2$CuLi	Et$_2$O	54	95/5	[77]
3	OBn	Ph≡—ZnBr	THF	75	81/19	[78]
4	OBn	Ph≡—ZnBr	Et$_2$O	95	95/5	[78]

Cainelli et al. reported an interesting study about the effect of temperature and mixture of solvents on diastereocontrol of the addition process.[76] They made an interesting insight into the solvent effect on stereoselectivity in addition, including α-silyloxy aldehydes 38 in ethereal solvents. They had shown that solute–solvent interactions can control a diastereoisomeric switch (Table 2.5, Entries 1–3, 7–12 versus Entries 4–6) from anti-Felkin–Ahn to Felkin–Ahn adducts based on the formation of two solvation clusters with different thermodynamic properties. Selected examples of solvent-controlled diastereofacial selectivity are summarized in Table 2.5.[76]

An interesting selectivity has been also observed in additions of organometallic reagents to α,β-epoxy carbaldehydes 40. Concretely, optically active α-trimethylsilyl-α,β-epoxy aldehydes afford the corresponding adducts with high diastereoselectivity (up to 26:1) upon treatment with organolithium or Grignard reagents (Table 2.6).[79] Noteworthy is that the presence of the silyl group is indispensable for getting high anti epoxy alcohol selectivity (Entries 1–7 versus Entries 8, 9).

When aldehydes containing a cyclic α-oxygen moiety, such as a dimer of acrolein 42, are used as substrates in additions with organometallic reagents, low diastereoselectivity of the corresponding adducts is observed (Table 2.7). It can be explained by the limited participation of pyran oxygen

TABLE 2.5
Effect of Solvent on Nucleophilic Addition of EtMgBr to Aldehyde 38

Entry	Solvent	T (°C)	anti/syn
1	THF	−79.5	78/22
2	THF	1.0	77/23
3	THF	20.5	75/25
4	Anisole	−30.4	39/61
5	Anisole	1.6	40/60
6	Anisole	21.0	42/58
7	PhOMe/THF 1:1	−59.1	78/22
8	PhOMe/THF 1:1	0.7	76/24
9	PhOMe/THF 1:1	20.1	74/26
10	PhOMe/4 equiv THF	−27.4	71/29
11	PhOMe/4 equiv THF	1	69/31
12	PhOMe/4 equiv THF	21.4	70/30

TABLE 2.6
Nucleophilic Addition to Chiral α-Trimethylsilyl-α,β-Epoxy Aldehydes 40

Entry	R^1	R^2	MNu	Yield (%)	syn/anti
1	^{n}Bu	SiMe$_3$	MeMgBr	89	12/88
2	^{n}Bu	SiMe$_3$	MeLi	89	16/84
3	^{n}Bu	SiMe$_3$	AllylMgBr	90	15/85
4	^{n}Am	SiMe$_3$	EtMgBr	86	3/97
5	^{n}Am	SiMe$_3$	^{i}PrMgBr	81	1/96
6	^{n}Am	SiMe$_3$	PhMgBr	97	11/89
7	^{n}Am	SiMe$_3$	^{n}Bu≡≡Li	93	9/91
8	^{n}Am	H	EtMgBr	Not reported	50/50
9	^{n}Am	H	^{n}Bu≡≡Li	Not reported	48/52

TABLE 2.7
Nucleophilic Addition to Dimer of Acrolein 42

Entry	MNu	syn/anti	References
1	EtLi	72/28	[80,81]
2	EtLi/TMEDA	80/20	[80,81]
3	EtCuLi	50/50	[82]
4	EtCuMgBr	12/88	[82]
5	Et$_2$Zn	15/85	[81]
6	Et$_2$TiCl$_2$	24/76	[82]

in the chelation process. Contrarily, a facile formation of chelates with organotin and organozinc reagents can significantly increase the *anti*-Felkin–Ahn/Felkin–Ahn adduct ratio (Entries 5, 6).

Diastereoselectivity of addition of organometallic reagents to chiral 2,3-dialkoxyaldehydes is highly dependent on the substrate as well as metallic source used. Under chelation conditions, when 1,2-chelation is favored over 1,3-chelation, 2,3-dialkoxy aldehydes **44** behave as 2-alkoxyaldehydes, affording 1,2-*anti*-Felkin–Ahn adducts. When 1,3-chelation can be taken into account, the reaction proceeds with 1,2-Felkin–Ahn selectivity for aldehyde substrates having *syn*-2,3-dialkoxy groups (Table 2.8).

Similar results in terms of diastereocontrol were observed by Michelet et al. in additions of a variety of organometallic reagents to β-C-glycoside aldehyde **46** (Table 2.9).[83] When organolithium reagents in the presence of hexamethylphosphoramide (HMPA) were taken into the reaction, Felkin–Ahn adducts were favored due to non–chelation control (Entry 1). As expected, the selectivity was reversed when the reaction was carried out with Grignard reagents in coordinating solvents, such as Et$_2$O or tehtrahydrofuran (THF) (Entries 2–4).

In comparison with previous examples, aldehydes containing both α- and β-oxygen substituents exhibit only mild preference in the formation of Felkin–Ahn adducts in reaction with organometallic reagents. The observed diastereoselectivity in addition of selected organometallic reagents to 2,3-isopropylideneglyceraldehyde (**49**) is summarized in Table 2.10.[84–88] The observed low selectivity can be explained by competitive chelation between α- and β-oxygen groups.

Highly substrate-dependent reaction diastereoselectivity was observed in reactions of 2-alkoxy-3-amino aldehydes with nucleophilic organometallic reagents. In general, rigid 5- and 6-membered cyclic aldehydes, such as **51** (Scheme 2.4)[89] and **53** (Table 2.11),[90] respectively, exhibited high degrees of reaction diastereoselectivity.

Excellent Felkin–Ahn selectivity was observed by Wee and Tang when 5-oxazolidinone carbaldehyde **55** was treated with organocerium nucleophiles (Scheme 2.5).[91,92] Other examples of high Felkin–Ahn selectivity in addition of organometallic reagents to 2-alkoxy-3-amino aldehydes were published by Shimizu and Overman and others.[93–95]

2.6.2 Aldehydes Containing α-Nitrogen Substituent

Suitably protected α-amino aldehydes are among the most widely used intermediates in organic synthesis. Among others they can be used for the formation of both diastereoisomeric *anti/syn*

TABLE 2.8
Addition of Organolithium Reagents to Chiral 2,3-Dialkoxyaldehydes 44

Entry	Aldehyde	RLi	Yield (%)	1,2-syn/1,2-anti	Reference
1	44a	Li—≡—(CH$_2$)$_4$OTBDPS	84	55/45	[97]
2	44b	Li—≡—Ph	91	53/47	[98]
3	44c	Li—≡—TES	44	20/80	[99]
4	44d	Li—≡—CO$_2$Et	69	46/54	[100]
5	44e	Li—≡—CO$_2$Et	79	99/1	[100]
6.	44f	BrMg—≡	99	60/40	[101,102]
7	44g	BrMg—≡—Ph	69	28/72	[103]

TABLE 2.9
Nucleophilic Addition to β-C-Glycoside Aldehyde 46

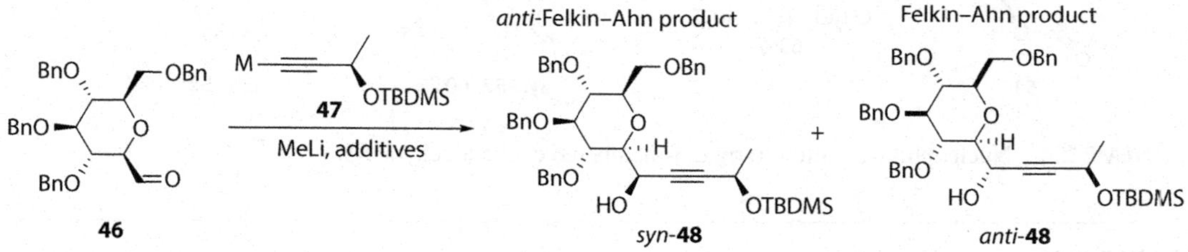

Entry	Additives	Solvent	T (°C)	Yield (%)	syn/anti
1	HMPA	—	−78°C → rt	50	40/60
2	MgBr$_2$, MgBr$_2$ (8 equiv.)	Et$_2$O	−30°C → rt	75	75/25
3	MgBr$_2$, CuI (8 equiv.)	THF/Me$_2$S	−78°C → rt	50	70/30
4	MgBr$_2$, ZnBr$_2$ (8 equiv.)	Et$_2$O	−78°C → rt	47	53/47
5	CeCl$_3$, CeCl$_2$ (8 equiv.)	THF	−78°C → rt	20	58/42
6	ZnBr$_2$, ZnBr$_2$ (8 equiv.)	Et$_2$O	0°C → rt	No reaction	—

TABLE 2.10
Nucleophilic Addition to 2,3-Isopropylideneglyceraldehyde (49)

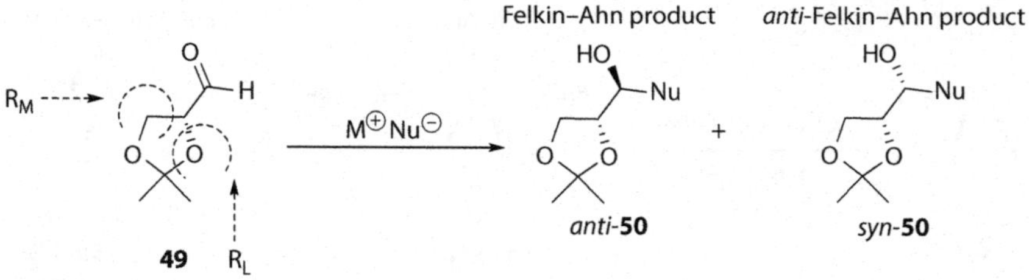

Entry	MNu	anti/syn	References
1	AllylMgBr	60/40	[84]
2	AllylTi(O^{i}Pr)$_3$	71/29	[84]
3	(Allyl)$_2$Zn	90/10	[85]
4	PhLi	48/52	[84]
5	PhMgBr	48/52	[84]
6	PhTi(O^{i}Pr)$_3$	9/91	[84]
7	MeLi	60/40	[84]
8	MeMgBr	67/33	[84]
9	^{n}BuLi	69/31	[84]
10	^{n}BuMgBr	75/25	[84]
11	^{t}BuMgBr	75/25	[86]
12	^{n}BuTi(O^{i}Pr)$_3$	90/10	[84]
13	≡—Li	50/50	[87,88]
14	═—Li	50/50	[87,88]

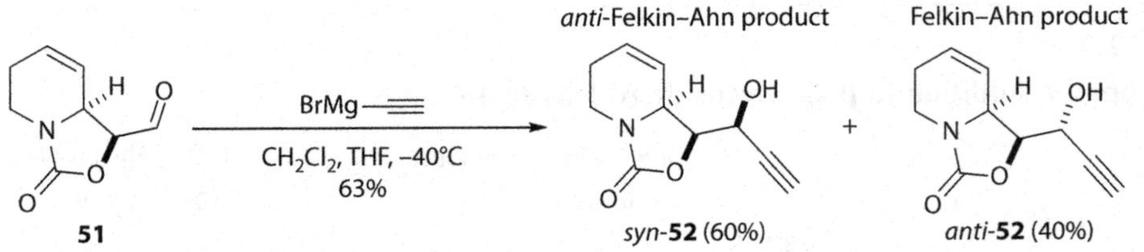

SCHEME 2.4 Nucleophilic addition to rigid 5-membered cyclic aldehyde **51**.

TABLE 2.11
Nucleophilic Additions to 6-Membered Cyclic Aldehydes 53

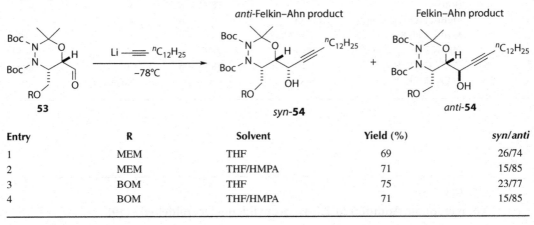

Entry	R	Solvent	Yield (%)	syn/anti
1	MEM	THF	69	26/74
2	MEM	THF/HMPA	71	15/85
3	BOM	THF	75	23/77
4	BOM	THF/HMPA	71	15/85

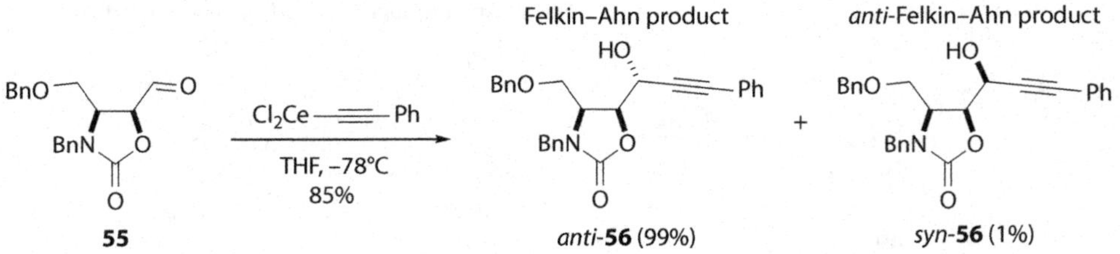

SCHEME 2.5 Highly diastereoselective addition of organocerium nucleophiles to aldehyde **55**.

β-amino alcohols, and they are versatile structural motifs occurring in a number of biologically active compounds.[96] When aldehydes contain an amine moiety in the α-position, high diastereoselection in 1,2-additions with organometallic reagents can be achieved by use of either nonchelating sterically demanding or strongly chelating α-nitrogen groups on aldehyde substrates. In general, bulky N-protecting groups favor the Felkin–Ahn transition state affording *anti*-amino alcohols predominantly, whereas strongly chelating α-nitrogen substituents generate *syn*-amino alcohols through a preferential formation of chelation-control transition state.[104]

A synthetically interesting N-protecting group that can be used for stereoselective formation of *anti*- and *syn*-β-amino alcohols is *N,N*-dibenzyl group (NBn$_2$). Moreover, the corresponding *N,N*-dibenzyl aldehydes **57** are readily accessible and stable toward racemization at room temperature. When nonchelating organolithium and Grignard reagents are used, Felkin–Ahn adducts (*anti*-amino alcohols) are formed with high selectivity (Entries 1, 2, 6, and 7, Table 2.12). Decreased diastereocontrol has been observed with highly reactive Grignard reagents (allylmagnesium chloride), cuprates, and organocerium reagents (Entries 3–5). Poor diastereocontrol has been observed only

TABLE 2.12
Nucleophilic Addition to β-Amino Alcohols 57 under Nonchelating Conditions

Entry	R	NR$_2$	MNu	Yield (%)	syn/anti	Reference
1	Me	NBn$_2$	MeMgI	87	5/95	[107]
2	Me	NBn$_2$	PhMgBr	85	3/97	[107]
3	Me	NBn$_2$	Me$_2$CuLi	80	25/75	[107]
4	Me	NBn$_2$	Me$_3$CeCl$_2$	70	10/90	[107]
5	Bn	NBn$_2$	AllylMgCl	82	28/72	[107]
6	Bn	NBn$_2$	MeMgI	85	8/92	[107]
7	Bn	NBn$_2$	PhMgBr	84	3/97	[107]
8	Bn	NHBoc	VinylMgBr	66	52/48	[106]
9	Me	NMe$_2$	PhMgBr	—	50/50	[108]
10	Bn	NMe$_2$	MeMgBr	—	35/65	[108]
11	Bn	NMe$_2$	PhMgBr	—	48/52	[108]

when N-substituents with medium steric bulk such as N-monosubstituted derivative or N,N-dimethyl derivatives are present in the carbonyl substrate (Entries 8–11).[105,106]

On the other hand, chelation of α-N,N-dibenzylamino aldehydes 59 with Ti^{4+} followed by addition of organozinc reagents affords the corresponding anti-Felkin–Ahn adducts (syn-amino alcohols) preferentially (Entries 1–3, Table 2.13). Similar results are obtained when MeTiCl$_3$ or Et$_2$Zn are used.

In order to study the degree of diastereocontrol in additions of organometallic reagents to α-amino aldehydes 61, a number of N-protecting groups have been investigated (Table 2.14). Among others, the N-trityl,[109] N-benzostabase[110] group exhibited high diastereocontrol for anti adducts (Entries 1–6) and the N-phenylfluorenyl,[111] N,N-di-ortho-methylbenzyl[108] group exhibited high diastereocontrol for syn adducts (Entries 7–9).

Aldehydes, which contain configurationally stable cyclic α-amine moieties, such as aziridine carbaldehydes, have been successfully applied for the preparation of 1,2-amino alcohols.[112–115] Moreover, products of stereoselective 1,2-additions, the corresponding aziridinyl carbinols, contain three consecutive stereocenters and represent valuable building blocks offering versatile access to structurally complex targets.[113–116] Since amine and aldehyde groups are not orthogonal to each other, an unprotected secondary amine cannot coexist with an aldehyde within the same molecule for a prolonged time, in which case the use of N-protecting groups is essential.

High anti-Felkin–Ahn selectivity using N-benzyl-protected cis-aziridine carbaldehydes was observed by Pedrosa et al. in the late 1990s.[118] In order to investigate the effect of N-protecting groups (Bn, Boc, Ts) as well as aziridine substitution and stereochemistry (cis, trans, 2,2,3-trisubstituted) on the selectivity of 1,2-addition with organometallic reagents, Jackson and Borhan performed a combined theoretical and experimental study summarized in Table 2.15.[119]

In general, N-benzyl-protected cis-aziridine (Entries 1–3, Table 2.16) and N-Boc-protected trans-aziridine carbaldehydes (Entries 3–5, Table 2.17) afford the corresponding anti-Felkin–Ahn adducts (syn-amino alcohols) with high to excellent diastereoselectivity (up to 99:1), whereas N-benzyl-protected trans-aziridines 65 (Entries 1, 2, Table 2.17) and N-Boc-protected cis-aziridines 63

TABLE 2.13
Nucleophilic Addition to β-Amino Alcohols under Chelating Conditions

Entry	R	Reagent	Yield (%)	syn/anti	Reference
1	Me	TiCl$_4$/(Me)$_2$Zn	82	94/6	[108]
2	Bn	TiCl$_4$/(Me)$_2$Zn	63	78/22	[108]
3	^{i}Pr	TiCl$_4$/(Me)$_2$Zn	65	65/35	[108]
4	Me	MeTiCl$_3$	82	94/6	[118]
5	Bn	MeTiCl$_3$	63	78/23	[107]
6	$_i$Pr	MeTiCl$_3$	65	65/35	[107]
7	Me	Et$_2$Zn	95	88/12	[118]
8	Bn	Et$_2$Zn	70	90/10	[118]
9	^{i}Bu	Et$_2$Zn	62	92/8	[118]
10	Ph	Et$_2$Zn	65	>99/1	[118]

TABLE 2.14
Influence of N-Protected Group on Nucleophilic Addition to α-Amino Aldehydes 61

Entry	R$_1$	R$_2$	R$_3$	MNu	Yield (%)	anti/syn	Reference
1				PhMgBr	88	>98/2	[109]
2				TMS—≡—Li	88	>98/2	[109]
3				^{n}BuLi	65	93/7	[109]
4	Bn			PhMgBr	75 (55)^a	85/15	[110]
5	Bn			PhLi	68 (53)	>95/5	[110]
6	Me			PhMgBr	73 (60)	>95/5	[110]
7	Bn			≡—MgBr	97	9/90	[111]
8	Me			≡—MgBr	93	31/69	[111]

^{a} Reduction of methylester with diisobutylaluminium hydride was conducted before nucleophilic addition, and total yield of crude compound is given. The yield of pure compound is reported within parentheses.

TABLE 2.15
Effect of N-Protecting Group and Substitution Pattern on Nucleophilic Addition to Aziridine Carbaldehydes

	N-Protecting Group		
Substitution Pattern	**Bn**	**Boc**	**Ts**
Cis	Excellent	Poor	Poor
Trans	Poor	Excellent	Moderate
2,2,3-Trisubstituted	—	Moderate	Excellent

TABLE 2.16
Nucleophilic Addition to *cis*-Aziridine Carbaldehydes 63

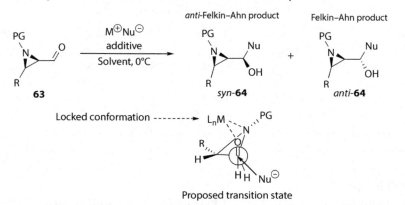

Entry	R	PG	MNu	Additive	Solvent	Yield (%)	syn/anti
1	Et	Bn	EtMgBr	—	Toluene	75	>99/1
2	Et	Bn	Et₂Zn	—	Et₂O	85	>99/1
3	Et	Bn	EtMgBr	TMEDA	Toluene	65	62/38
4	CH₃(CH₂)₅	Boc	HC₂MgBr	—	CH₂Cl₂	53	59/41
5	CH₃(CH₂)₅	Boc	HC₂MgBr	TMEDA	CH₂Cl₂	51	50/50
6	CH₃(CH₂)₅	Boc	PhMgBr	—	CH₂Cl₂	55	90/10

(Entries 4–6, Table 2.16) react with organometallic reagents in a low or almost unselective manner. Highly stereocontrolled additions of organometallic reagents occur with *N*-tosyl-protected 2,3-disubstituted aziridine-2-carbaldehydes **67** (Table 2.18) with preference of *anti*-Felkin–Ahn adducts (*syn*-amino alcohols, *syn/anti* ratio up to 99:1).

Authors concluded that the origin of the *anti*-Felkin–Ahn selectivity lies in substrate conformational preference, which is affected by the electronic nature of the N-protecting group, conformation of the aziridine, and *exo/endo* orientation of the aldehyde predominantly. In addition, they assume limited chelation-like control in these reactions as the presence of a complexing agent (e.g., TMEDA, MgBr₂·OEt₂) has almost no effect on Felkin–Ahn/*anti*-Felkin–Ahn (*anti/syn*) selectivity. From the previous discussion, we gather that *syn*-amino alcohols are readily available motifs using 1,2-addition of organometallic reagent to the corresponding substrates. On the other hand, *anti*-amino alcohols have to be accessed either by Mitsunobu inversion or by complementary hydride reduction of the related ketones.[120]

TABLE 2.17
Nucleophilic Addition to *trans*-Aziridine Carbaldehydes

Proposed transition state

Entry	R	PG	MNu	Additive	Solvent	Yield (%)	syn/anti[a]
1	Et	Bn	EtMgBr	—	Ether	75	55/45
2	Et	Bn	EtMgBr	TMEDA	Ether	85	55/45
3	Cy	Boc	MeMgBr	—	CH_2Cl_2	85	>99/1
4	Cy	Boc	MeMgBr	TMEDA	CH_2Cl_2	80	>99/1
5	Cy	Boc	MeMgBr	$MgBr_2 \cdot OEt_2$	CH_2Cl_2	78	>99/1

[a] Exists as 55:45 mixture of invertomers at N observable by NMR spectroscopy at 5°C.

TABLE 2.18
Nucleophilic Addition to *N*-Tosyl-Protected 2,2,3-Trisubstituted Aziridine Carbaldehydes 67

Entry	MNu	Yield (%)	syn/anti
1	≡—MgBr	92	>99/1
2	PhMgBr	85	>99/1

Similar results of *anti*-Felkin–Ahn selectivity in addition to Grignard reagents to *N*-Boc-protected *trans*-aziridine carbaldehydes were obtained by Righi et al. (Table 2.19), who developed stereoselective formation of *anti,syn*-1-bromo-2,3-amino alcohols **71** using one-pot organometallic addition/aziridine, thus starting a reaction sequence.[113,121] Besides aziridine-2-carbaldehydes with classical N-protecting groups (Bn, Boc, Ts), other N-protected aziridine carbaldehydes were also studied in stereocontrolled 1,2-additions (Table 2.19).[112,122]

Other interesting examples of nucleophilic additions of organometallic reagents to aldehydes containing nitrogen in α-position are depicted in Scheme 2.6. Excellent 1,2-*syn* diastereoselectivity was observed with the addition of organolithium and Grignard reagents to 3-substituted-4-oxoazetidine-2-carboxaldehydes **72** (Table 2.20, Scheme 2.7).[123–126]

TABLE 2.19
One-Pot Nucleophilic Addition and Consequent Ring-Opening Reaction

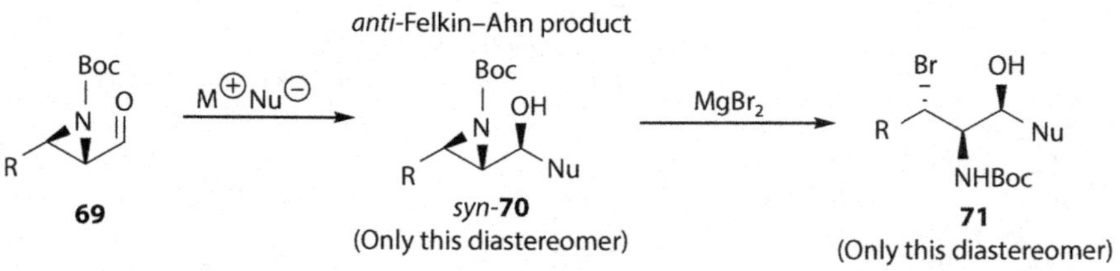

anti-Felkin–Ahn product

69 → **syn-70** (Only this diastereomer) → **71** (Only this diastereomer)

Entry	R	MNu	Yield of 71 (%)
1	^{n}Pr	MeMgBr	65
2	^{n}Pr	^{t}BuMgBr	60
3	^{n}Pr	VinylMgBr	55
4	Cy	MeMgBr	68
5	Cy	^{t}BuMgBr	65
6	Cy	VinylMgBr	52

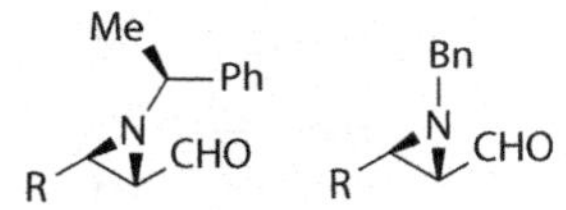

SCHEME 2.6 Other N-protected aziridine carbaldehydes studied in stereocontrolled additions.

Other aldehydes, which contain cyclic α-amine moiety, such as N-protected prolinals **75**, were also studied in additions with organometallic reagents. However, most of the N-protected prolinals **75** (*N*-Bn, *N*-Boc, *N*-Cbz) exhibit a low diastereocontrol in additions with organolithium,[127–129] Grignard,[130–132] or organozinc[133,134] and organotin[135] reagents. Interestingly, high *anti*-Felkin–Ahn diastereoselectivity (up to 99:1) in additions to *N*-Boc-protected prolinal **75** was observed only when phenolates with magnesium bromide as counterion were employed (Entries 1–3, Table 2.21). Opposite diastereoisomers (Felkin–Ahn control) were reached in addition of titanium phenolates to *N*-Boc-protected prolinal **75** (Entries 4, 5).[136]

A remarkable stability toward epimerization as well as an efficient stereocontrol is shown in *N*-tritylprolinal **77** in its reactions with a variety of organometallic nucleophiles. The corresponding proline-derived amino alcohols (*anti*-amino alcohols) were obtained with high Felkin–Ahn stereoselectivity (Entries 1–7, Table 2.22). The only exception was highly reactive allylMgBr (Entry 8).

One of the most widely used aldehydes containing nitrogen in α-position is a serinal derivative, Garner's aldehyde **79**.[137] The ease of availability of both enantiomers in large quantities and configurational and chemical stability makes this aldehyde a valuable synthetic intermediate in natural product synthesis.[138] Garner's aldehyde **79** has been widely used in 1,2-additions with a number of organometallic reagents (Table 2.23). Good to high Felkin–Ahn stereocontrol affording *anti*-amino alcohols **80** was observed in non-chelation-controlled processes with organolithium (Entries 1–8),[139–147] organorhodium (Entries 9, 10),[148,149] organonickel (Entry 11),[150] and Grignard reagents (Entries 12, 13),[145,151,152] especially in combination with complexing agents (e.g., HMPT) or Lewis acids (BF$_3$·Et$_2$O) precluding chelation. On the other hand, the use of organozinc reagents (Entries 14–17)[140,142,145,152] and organocuprates (Entries 18, 19)[93,153] favors reverse *anti*-Felkin–Ahn diastereocontrol, leading to *syn*-amino alcohols **80**.

In addition, diastereoselectivities also differ depending on the character of organometallic reagents and bulkiness of substituents at the hemiaminal carbon. Whereas alkenyl metals exhibit moderate diastereocontrol of addition, excellent selectivities were obtained with metal acetylides.[5a]

TABLE 2.20
Nucleophilic Addition to 4-Oxoazetidine-2-Carboxaldehydes 72

Felkin–Ahn + chelation-controlled product

anti-Felkin–Ahn product

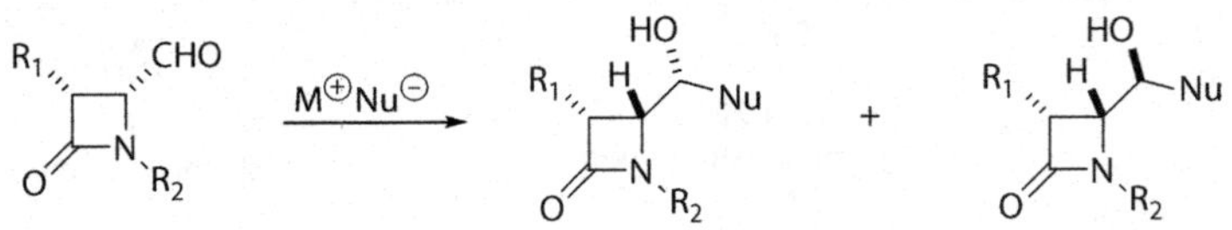

72 *anti*-**73** *syn*-**73**

Entry	R_1	R_2	MNu	Yield (%)	*anti/syn*
1	PhtN	Ar	Li—≡—Ph	69	67/33
2	CH_2O	Ar	Li—≡—Ph	55	83/17
3	PhtN	Ar	BrMg—≡—Ph	›68	100/0
4	CH_2O	Ar	BrMg—≡—Ph	›55	100/0
5	PhO	PMP	Li—≡—Ph	79	100/0
6	PhO	PMP	Li—≡—TMS	81	100/0
7	PhO	Allyl	Li—≡—Ph	66	100/0
8	PhO	Allyl	Li—≡—TMS	70	100/0
9	BnO	PMP	Li—≡—Ph	77	100/0
10	MeO	PMP	Li—≡—Ph	64	100/0
11	MeO	PMP	Li—≡—TMS	56	100/0
12	MeO	3-butenyl	Li—≡—Ph	54	70/30
13	MeO	3-butenyl	Li—≡—TMS	42	70/30

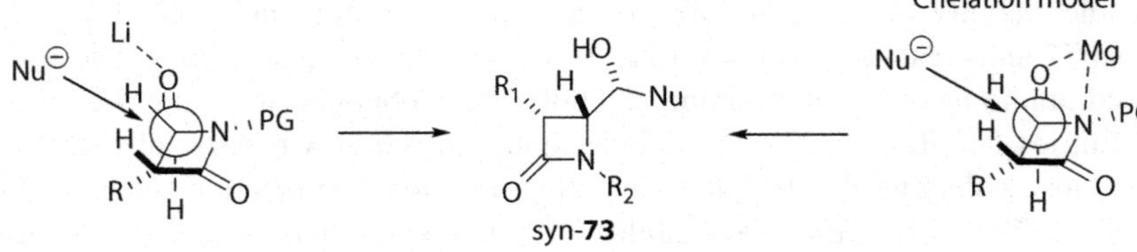

SCHEME 2.7 Felkin–Ahn and chelation transition states for nucleophilic addition to 4-oxoazetidine-2-carboxaldehydes **73**.

Until now, other aldehydes containing nitrogen in α-position and oxygen in β-position have been screened in additions with a variety of organometallic reagents, for example, organocerium compounds (Table 2.24).[91] In addition, this type of aldehydes became a basis for total synthesis of various tetrodotoxin derivatives[154–158] and other useful synthons.[159] For example, an excellent *anti*-Felkin–Ahn diastereoselectivity (up to 98 de) was observed in addition of organocerium reagents to *N*-benzyl-5-vinyl-oxazolidin-2-one-4-carbaldehyde **82**.[92] (For more details, see review of Plé and Haudrechy about alkynylation of chiral alkoxy-, amino-, and thio-substituted aldehydes.[5d])

TABLE 2.21
Nucleophilic Addition to *N*-Boc-Protected Prolinal 75

74 75 *syn*-**76** *anti*-**76**

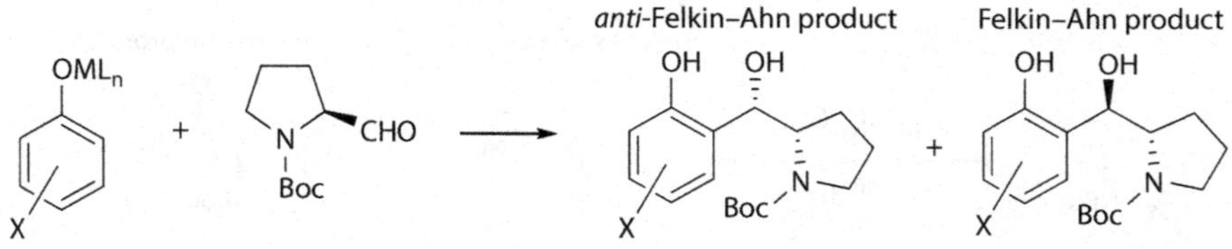

Entry	X	ML_n	Yield (%)	*syn/anti*
1	H	MgBr	65	›99/1
2	3-CH_3O	MgBr	70	99/1
3	3,4-OCH_2O	MgBr	75	97/3
4	3-CH_3O	$Ti(OCH(CH_3)_2)_3$	51	12/88
5	3,4-OCH_2O	$Ti(OCH(CH_3)_2)_3$	72	8/92

TABLE 2.22
Nucleophilic Addition to *N*-Tr-Protected Prolinal 77

77 *anti*-**78** *syn*-**78**

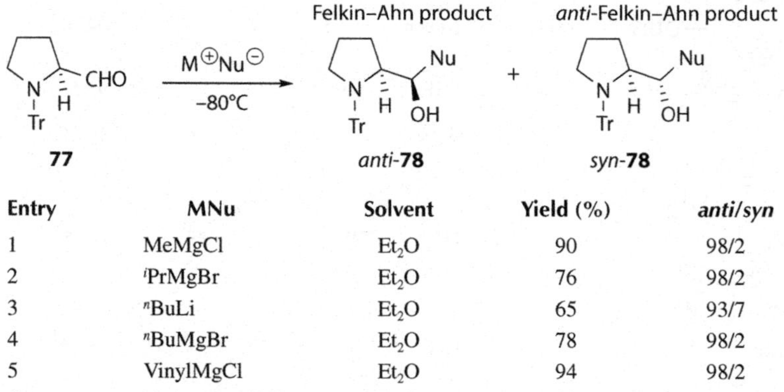

Entry	MNu	Solvent	Yield (%)	*anti/syn*
1	MeMgCl	Et_2O	90	98/2
2	^{i}PrMgBr	Et_2O	76	98/2
3	^{n}BuLi	Et_2O	65	93/7
4	^{n}BuMgBr	Et_2O	78	98/2
5	VinylMgCl	Et_2O	94	98/2
6	Li—≡—TMS	Et_2O	88	98/2
7	PhMgBr	Et_2O	90	98/2
8	AllylMgBr	Et_2O	85	63/37

Aldehydes containing nitrogen in α-position and sulfur in β-position (2-amino-3-thio-substituted aldehydes **84**) show similar diastereocontrol with the addition of organometallic reagents as the corresponding β-oxy analogs.[160–162] Selected examples are summarized in Table 2.25.

2.6.3 Aldehydes Containing Other Heteroatoms in α-Position

In comparison to substituted 2-amino and 2-oxy aldehydes, aldehydes with other heteroatoms in α-position have been less studied. High 1,2-Felkin–Ahn diastereocontrolled addition of organomagnesium and lithium reagents to 2-*tert*-butylsulfenyl aldehydes **86** has been observed by Enders et al. (Table 2.26).[163]

TABLE 2.23
Nucleophilic Addition to Garner Aldehyde 79

Entry	MNu	Additive	Solvent	T (°C)	Yield (%)	anti/syn	References
1	Li—≡—C₁₃H₂₇	—	THF	−23	83	89/11	[139]
2	Li—≡—C₁₃H₂₇	HMPT	THF	−78	71	>95/5	[140]
3	Li—≡—⟨OTBS⟩	—	THF	−78	80	94/6	[141]
4	Li—≡—⟨OTBS⟩	HMPT	Toluene	−78 → 0	85	95/5	[142]
5	Li—≡—CO₂Me	—	Et₂O	−78 → 0	62	8/92	[143]
6	Li—≡—CO₂Et	HMPT	THF	−78	75	93/7	[144]
7	Li⟨=⟩	—	THF	−78	—	83/17	[145]
8	Li—⟨=⟩—OTBS	DMPU	Toluene	−95	57	94/6	[146,147]
9	Rh—≡—Ph	—	THF	40	74	95/5	[148]
10	Rh—⟨=⟩—C₆H₁₃	—	aq DME	55	78	82/18	[149]
11	Ni—C(TMS)=CHMe	—	THF	Rt	78	>95/5	[150]
12	BrMg⟨=⟩	—	THF	−78	—	75/25	[145]
13	PhMgBr	—	THF	−78 → 0	—	83/17	[151]
14	ClMg—≡	ZnBr₂	THF/toluene	−78	—	50/50	[152]
15	ClZn⟨=⟩	—	Et₂O	−78 → r.t.	85	14/86	[145]
16	Li—≡—C₁₃H₂₇	ZnBr₂	Et₂O	−78 → r.t.	87	6/94	[140]
17	Li—≡—⟨OTBS⟩	ZnCl₂	Toluene/Et₂O	−78 → r.t.	65	9/91	[142]
18	BrMg—≡—Me	CuI	THF/Me₂S	−78	95	6/94	[153]
19	Li—⟨dithiane⟩	CuI (cat.) BF₃·Et₂O	THF	−50	70	>99/1	[93]

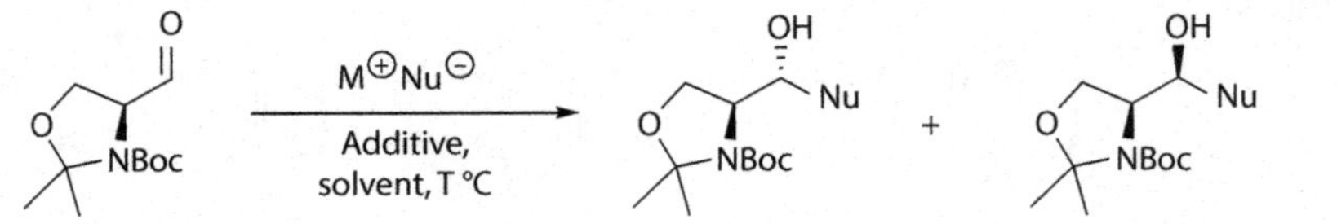

TABLE 2.24
Nucleophilic Addition to Aldehydes That Contain Both α-N and β-O Atoms 82

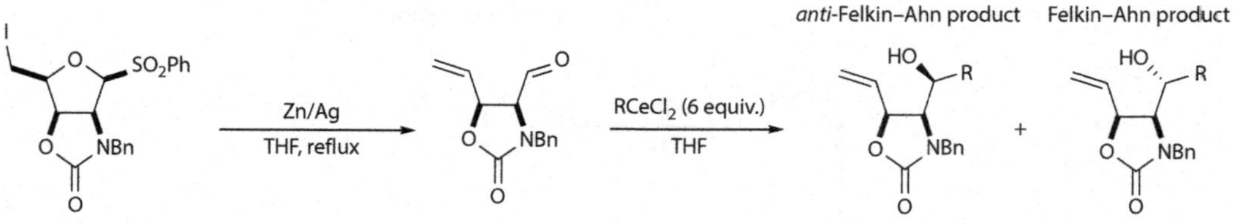

Entry	R	T (°C)/Time (h)	Yield (%)	syn/anti
1	MeLi	−78/5	85	95/5
2	MeLi	−78/5	35	89/11
3	VinylMgBr	−78/5	85	>99/1
4	VinylMgBr	−78/5	40	83/17
5	(OLi, O^{t}Bu enolate)	−78/5	92	86/1
6	Ph—≡—Li	−78/5	94	>99/1
7	2-FuranylLi	−78/5	85	>99/1
8	PhMgBr	−78/5; 0/1	52	88/12

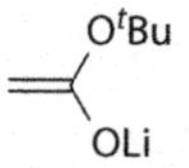

TABLE 2.25
Nucleophilic Addition to Aldehydes That Contain Both α-N and β-S Atoms

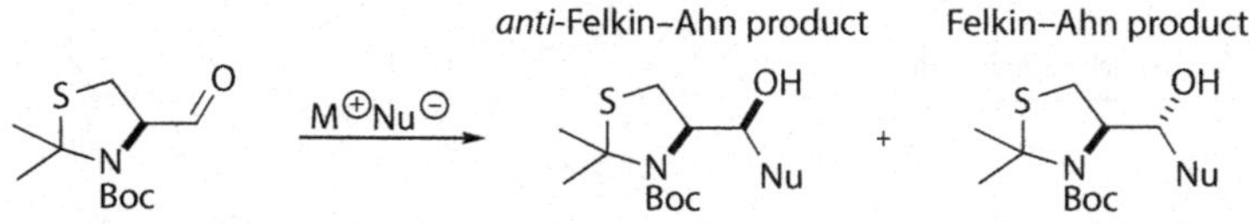

Entry	MNu	Solvent	Additive	Yield (%)	syn/anti	References
1	Li—≡—^{n}Bu	THF	—	88	40/60	[160]
2	Li—≡—^{n}Bu	THF	HMPA	71	17/83	[160]
3	BrZn—≡—^{n}Bu	Et$_2$O	—	48	95/5	[160]
4	ClZn—≡—^{n}Bu	Et$_2$O	—	86	>99/1	[160]
5	Li—≡—(CH$_2$)$_9$^{i}Pr	THF	HMPA	82	90/10	[161,162]

Good to high *anti*-Felkin–Ahn stereocontrol with the addition of organozinc reagents to the "thio" version of the Garner aldehyde **88** was observed by Fujisawa et al. affording *syn*-amino alcohols **89** (Entries 1 and 2, Table 2.27). The opposite diastereoselectivity was reached with organolithium compounds in the presence of a chelation additive, such as HMPA (Entries 3 and 4).[160] Besides aldehyde **88**, sulfone analog **90** has also been successfully used in diastereoselective additions with organometallic reagents, and extraordinary 1,2-Felkin–Ahn diastereoselectivity has been observed (Scheme 2.8).[161,162]

TABLE 2.26
Stereoselective Addition to Aldehyde 86

Entry	R$_1$	MNu	Yield (%)	anti/syn
1	Pr	BnMgBr	54	99/1
2	^{i}Pr	BnMgBr	68	>98/2
3	c-C$_6$H$_{11}$CH$_2$	BnMgBr	53	99/1
4	Bn	MeMgBr	84	91/9
5	Bn	PhMgBr	62	>97/3
6	Bn	Li—≡—Ph	62	>97/3
7	Bn	AllylMgBr	51	27/73

TABLE 2.27
Alkynyllithium and Alkynylzinc Addition to Aldehyde 88

Entry	M	Solvent	Additive	Yield (%)	syn/anti
1	ZnBr	Et$_2$O	—	48	95/5
2	ZnCl	Et$_2$O	—	86	>99/1
3	Li	THF	—	88	40/60
4	Li	THF	HMPA	71	17/83

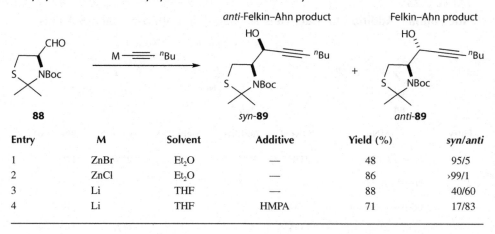

SCHEME 2.8 Nucleophilic addition to α-amino, α-sulfone-substituted aldehyde **90**.

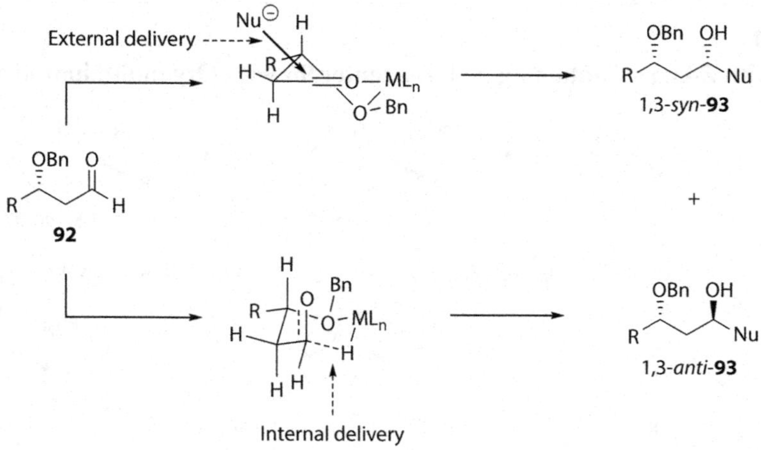

SCHEME 2.9 External and internal delivery of hydride in chelation-controlled nucleophilic additions to β-substituted alcohol.

TABLE 2.28
Chelation-Controlled 1,3-Induction—Reetz Model[64]

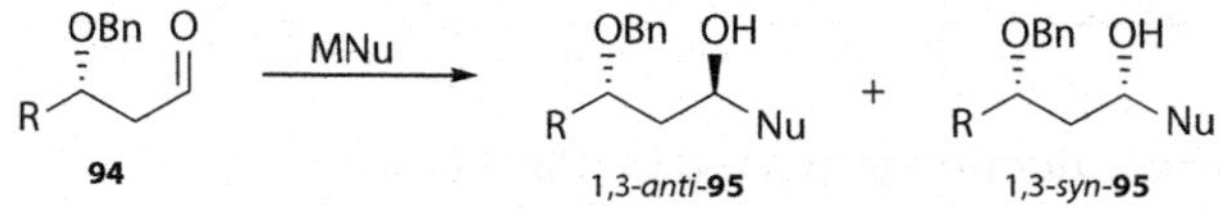

Entry	Reagent	R1	R2	1,3-anti/1,3-syn
1	MeTiCl$_3$	Me	Me	90/10
2	TiCl$_4$/AllylTMS	Me	Allyl	95/5
3	TiCl$_4$/ZnBu$_2$	Me	^{n}Bu	90/10
4	MeTiCl$_3$	^{n}Bu	Me	91/9
5	TiCl$_4$/AllylTMS	^{n}Bu	Allyl	95/5
6	TiCl$_4$/CH$_2$C(Me)CH$_2$TMS	Me	CH$_2$C(Me)CH$_2$	95/5
7	TiCl$_4$/CH$_2$C(Me)CH$_2$TMS	^{n}Bu	CH$_2$C(Me)CH$_2$	95/5

The mentioned examples of Felkin–Ahn/*anti*-Felkin–Ahn selectivity for additions including aldehydes containing a heteroatom in α-position do not comprise an exhaustive list, but they have been chosen to clearly demonstrate the fundamentals of these addition processes. In a broader sense, the presence of a stereogenic center in α-position has a substantial effect on the selectivity of the addition. Emplacement of the chiral center to β-position diminishes its influence on the addition process, especially when the substituents cannot participate in chelation. A chelate-controlled process can occur either through external addition of organometallic reagents to the reaction center chelated with chiral center or through internal attack of nucleophile, which is a part of the chelate itself (Scheme 2.9). The latter case occurs in additions of organometallics with a metal center having two or more free coordination sites and affords the corresponding 1,3-*anti* adducts (Table 2.28).

Moving of the stereogenic center further from the aldehyde moiety (to γ- or δ-position) reduces its influence on the addition process substantially,[64,164,165] affording the adducts with diastereoselectivity that is not synthetically useful. However, high diastereocontrolled addition processes have been observed in several structurally specific aldehydes. In 1991, Reetz et al. reported remote asymmetric induction (1,4-induction) in the addition of organolithium reagents to γ-dibenzylamino Z-enals **96**,[166]

TABLE 2.29
Reaction with Z-Enals Containing γ-Stereocenter 96 with Organolithium Reagents

Entry	R_1	R	Yield (%)	1,3-syn/1,3-anti
1	Me	Me	78	92/8
2	^{i}PrCH$_2$	Me	79	91/9
3	Bn	Me	80	92/8
4	Bn	^{n}Bu	80	87/13
5	Bn	Ph	84	91/9

TABLE 2.30
Remote Asymmetric Induction Reported by Castelano

Entry	R_1	RLi	Additive	Yield (%)	101/102
1	OMe	PhLi	—	67	95/5
2	OMe	MeLi	—	67	95/5
3	OMe	MeLi	HMPA	60	75/25
4	Et	MeLi	—	86	76/24
5	Et	MeLi	HMPA	99	76/24

affording the corresponding 1,2-adducts with high diastereoselectivities (Table 2.29). In addition, the corresponding E-isomer treated under the same conditions provided adducts with lower selectivity.

This observation can be rationalized by chelation with subsequent selective "intramolecular" delivery of the nucleophile to aldehyde. Alternatively, it can be explained by the selective attack from the sterically less-hindered side of enal, with the enal conformationally locked by 1,3-allylic strain.

Another example of remote asymmetric induction (1,5-induction) was reported by Castellano et al.[167] The addition of organolithium or organomagnesium reagents to densely functionalized trialdehyde **100** afforded the corresponding adduct **101** in good yield and high diastereoselectivity (Table 2.30). The selectivity can be rationalized by chelation and steric effects afforded by the crowded aromatic core in the intermediates (mono- and double adduct, Figure 2.21).

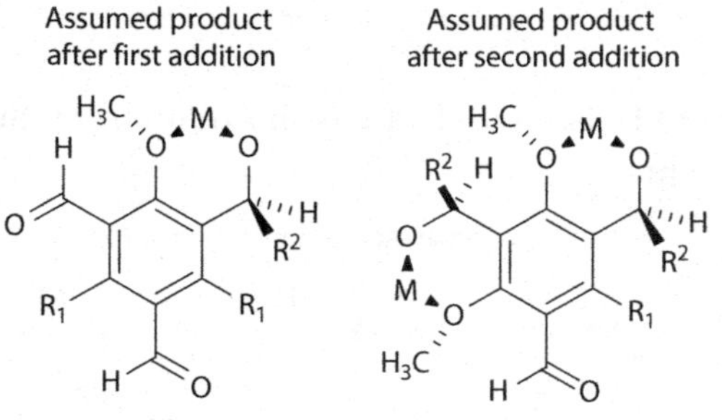

FIGURE 2.21 Structure of proposed intermediates **98** and **99**.

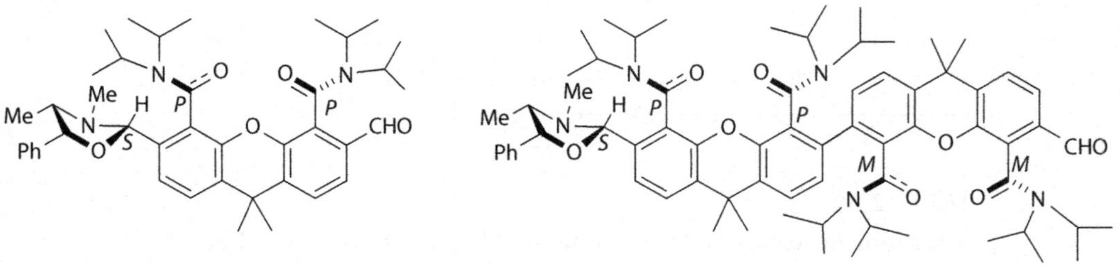

FIGURE 2.22 Xanthene-derived substrates used in asymmetric induction reported by Clayden.

Similar principles of remote asymmetric induction were also published by Clayden et al. based on the xanthene system (Figure 2.22).[168,169]

2.7 ADDITION OF CHIRAL KETONES

2.7.1 ACYCLIC α-CHIRAL KETONES

A diastereoselective addition of organometallic reagents to simple chiral ketones is generally more challenging than addition to aldehydes. This is because of lowered reactivity of ketones and also lowered stereoselection due to a smaller difference in steric demand between two alkyl substituents in ketones instead of alkyl and hydrogen in an aldehyde moiety. In general, the Felkin–Ahn model can be used in predicting the stereochemical outcome of addition to chiral ketones. The degree of stereoselection is affected by solvent, temperature, organometallic reagent, and the bulkiness of ketone substituents. Sterically demanding substituents (R) of ketone **105** favor Felkin–Ahn adduct formation over an *anti*-Felkin–Ahn product (Table 2.31). This can be explained by the enhanced influence of the L substituent in the Felkin–Ahn model when nucleophile is led to approach the carbonyl moiety from the side of the chiral center due to interaction with a bulky R substituent.[170,171]

2.7.2 ACYCLIC CHIRAL KETONES CONTAINING OXYGEN IN α-POSITION

In comparison to the previous section, the addition of organometallic reagents to acyclic chiral ketones containing oxygen in α-position is among the most thoroughly studied examples of chelation-controlled reactions. In general, additions of organometallic reagents to chiral α-alkoxy ketones proceed with a high degree of diastereoselectivity with a profound effect of the used solvent and metal (Table 2.32). Whereas organolithium reagents show low selectivity, the additions of more coordinating Grignard reagents provide *anti*-Felkin–Ahn adducts with excellent selectivity (Entries 1–4, 7–10).[172] Not only

TABLE 2.31
Dependence of Diastereoselectivity of Additions on the Bulkiness of R Substituent

Entry	L	M	R	anti/syn	Reference
1	Ph	Me	Et	86/14	[170]
2	Ph	Me	iPr	90/10	[170]
3	Ph	Me	Ph	87/13	[171]
4	Ph	Me	tBu	96/4	[171]

TABLE 2.32
Additions of Achiral Organometallic Reagents to α-Chiral Ketones

Entry	L	M	R	Solvent	MNu	anti/syn	References
1	C_7H_{15}	OMEM	Me	THF	BuMgBr	>99/1	[173]
2	C_7H_{15}	OMOM	Me	THF	BuMgBr	>99/1	[173]
3	C_7H_{15}	OBn	Me	THF	BuMgBr	>99/1	[173]
4	Me	OBn	Et	THF	MeMgCl	>99/1	[174]
5	Me	OBn	Et	THF	MeLi	60/40	[174]
6	Me	OBn	Et	THF	MeLi/TiCl$_4$	95/5	[174]
7	Me	OMe	Ph	THF	Me$_2$Mg	>99/1	[175,176]
8	C_7H_{15}	OMEM	Me	Pentane	BuMgBr	90/10	[173]
9	C_7H_{15}	OMEM	Me	DCM	BuMgBr	93/7	[173]
10	C_7H_{15}	OMEM	Me	Et$_2$O	BuMgBr	90/10	[173]
11	C_7H_{15}	OMEM	Me	Pentane	BuLi	67/33	[173]
12	C_7H_{15}	OMEM	Me	DCM	BuLi	75/25	[173]
13	C_7H_{15}	OMEM	Me	Et$_2$O	BuLi	50/50	[173]

acyclic chiral ketones with α-oxygen moiety but also cyclic α-alkoxy ketones afford a reaction with organometallic reagents with high degrees of selectivity (Scheme 2.10).[173]

The number of chiral auxiliaries required to control the addition process of organometallic reagents to a ketone moiety was also studied. Among others, auxiliaries containing well-chelating oxygen atoms included in a conformationally locked/rigid structural motif were introduced. Eliel introduced bicyclic 1,3-oxathiane moiety (derived from (+)-pulgenone) to the α-position

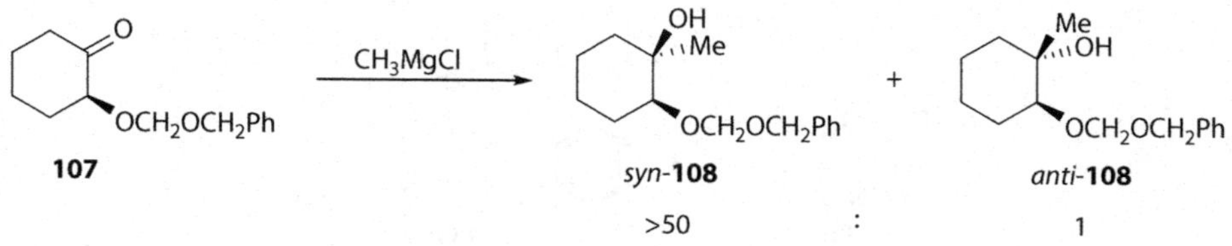

SCHEME 2.10 An example of a highly stereoselective addition to cyclic α-alkoxy ketone.

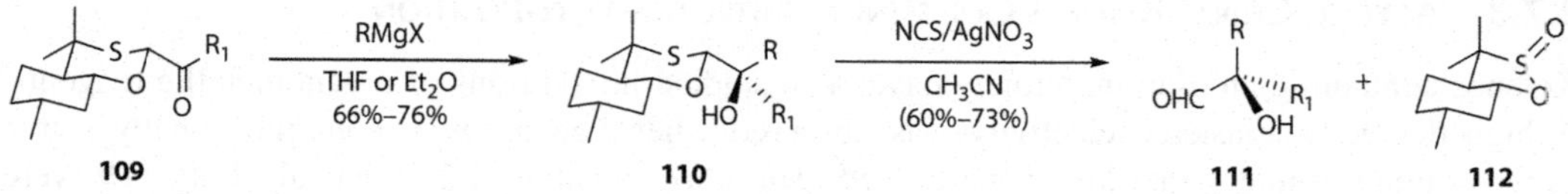

SCHEME 2.11 Additions to chiral bicyclic 1,3-oxathiane ketones introduced by Eliel.

of ketones.[177] Additions of Grignard reagents to such ketones **109** proceeded with a high degree of diastereoselectivity affording, after subsequent cleavage, chiral α-hydroxy aldehydes **111** (Scheme 2.11).[178,179] The observed high diastereocontrol of addition is supported by the chelation model, when coordination of magnesium atom to two hard oxygen atoms (instead of S and O) is involved first, followed by intramolecular addition of R from the less-hindered side. The reaction pathway is strongly influenced by the presence of other chelation groups, such as OBn, or less coordinating organometallic reagents (RLi).[180] Later on, other scaffolds related to the Eliel system were successfully used in stereoselective additions.[117,181–184]

The dioxolane moiety has also been successfully used in diastereoselective additions as a chiral auxiliary. In 1997, Myles et al. observed a high level of diastereoselectivity in additions of organomagnesium reagents to α-keto acetals containing a diphenyl dioxolane chiral auxiliary (Scheme 2.12).[185]

Other chiral auxiliaries were also introduced and applied in additions of organometallic reagents to ketones, such as tetrahydropyranyl group or binol-ester systems (Figure 2.23), affording the corresponding adducts with a high degree of diastereoselectivity.[186–190]

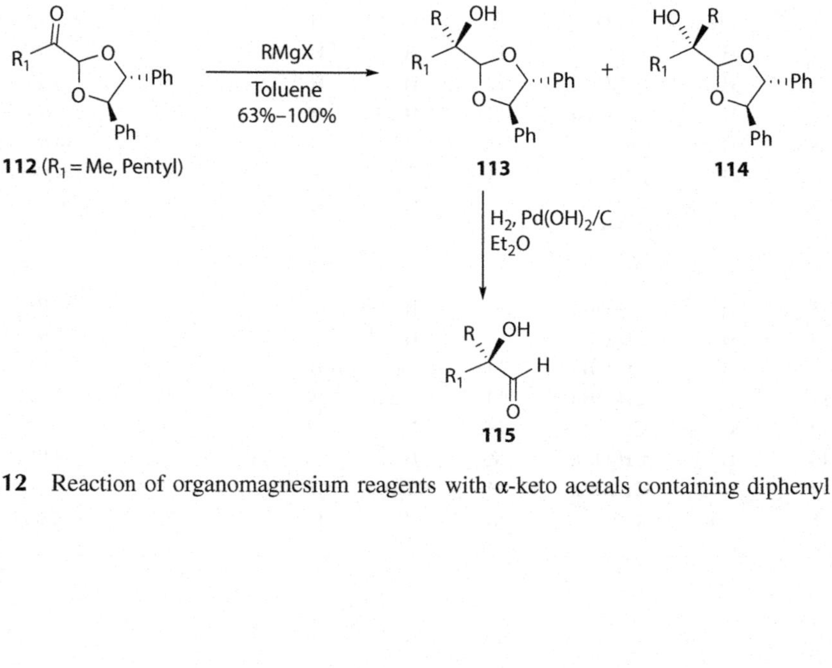

SCHEME 2.12 Reaction of organomagnesium reagents with α-keto acetals containing diphenyl dioxolane auxiliary.

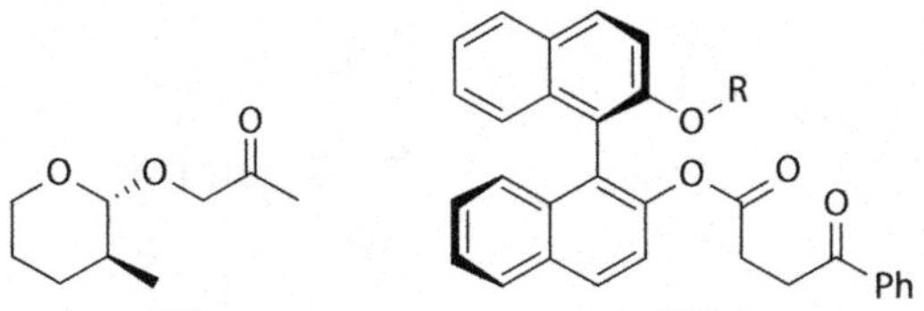

66   Asymmetric Synthesis of Drugs and Natural Products

FIGURE 2.23 Examples of ketones with chiral auxiliaries used in stereoselective additions.

2.7.3 ACYCLIC CHIRAL KETONES CONTAINING NITROGEN IN α-POSITION

Ketones containing an α-amine moiety have been studied in additions of organometallic reagents. A high degree of diastereoselectivity was observed when ketones with configurationally stable cyclic α-amine moiety (aziridine) were used. The corresponding 1,2-amino alcohols **119** were obtained in the reaction of **118** with Grignard reagents with excellent diastereoselectivity (up to 97:3, Table 2.33).[115] Not only N-protected aziridine ketones **118** but also ketones with unprotected aziridine scaffold **120** were used in diastereoselective nucleophilic additions (Table 2.34).[191] Remarkably high diastereoselectivity can be rationalized with bidentate chelation of Mg^{2+} by carbonyl oxygen and aziridine amide, formed after initial deprotonation.

In comparison to the number of chiral auxiliaries containing oxygen (see Section 2.7.2) among N-containing chiral auxiliaries used in diastereoselective additions, only proline derivatives have

TABLE 2.33
Diastereoselective Additions to Ketones with Aziridine Moiety

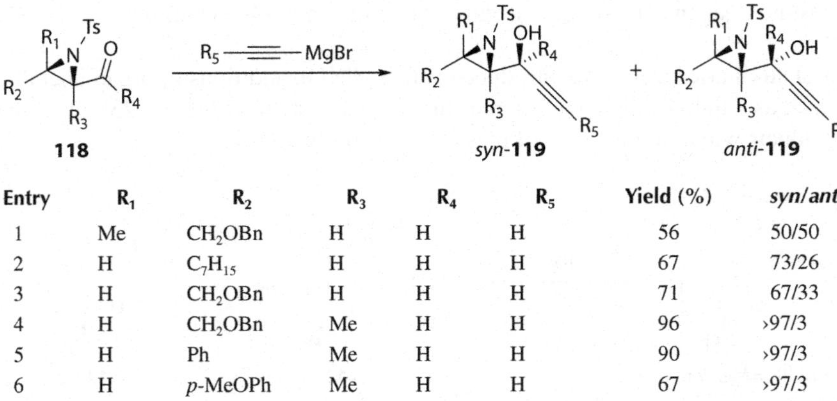

Entry	R₁	R₂	R₃	R₄	R₅	Yield (%)	syn/anti
1	Me	CH₂OBn	H	H	H	56	50/50
2	H	C₇H₁₅	H	H	H	67	73/26
3	H	CH₂OBn	H	H	H	71	67/33
4	H	CH₂OBn	Me	H	H	96	>97/3
5	H	Ph	Me	H	H	90	>97/3
6	H	p-MeOPh	Me	H	H	67	>97/3
7	H	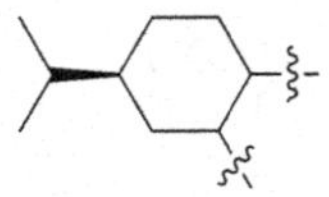 (menthyl)		H	H	93	>97/3
8	H	CH₂OBn	Me	H	Me	92	>97/3
9	H	CH₂OBn	Me	H	Ph	91	>97/3
10	H	CH₂OBn	Me	Me	H	89	>97/3
11	H	CH₂OBn	Me	Me	Ph	84	>97/3
12	H	CH₂OBn	Me	H	TMS	90	>97/3
13	H	CH₂OBn	Me	H	TIPS	98	>97/3
14	H	CH₂OBn	Me	C₂H	H	80	—
15	ⁿBu	Me	Me	H	H	73	55/45

TABLE 2.34
Addition of Grignard Reagents to Ketone 120

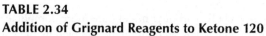

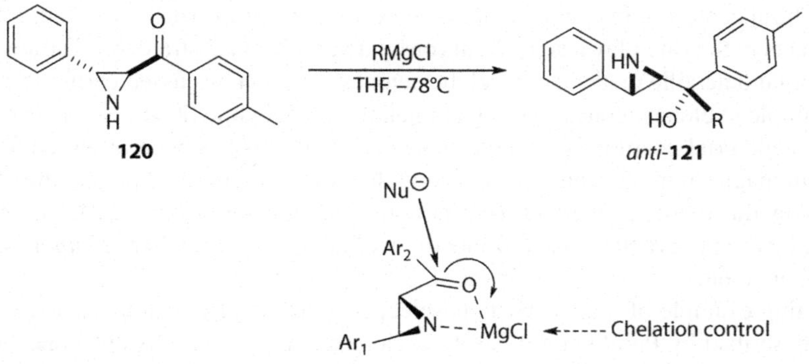

Entry	R	Yield (%)	*anti*/*syn*
1	Me	74	Only *anti*
2	Et	78	Only *anti*
3	Ph	56	Only *anti*

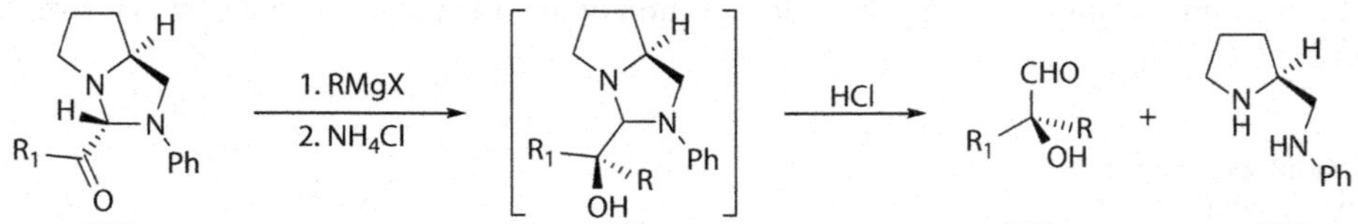

122 123 124 125

SCHEME 2.13 Diastereoselective addition of organomagnesium reagents to chiral aminal **122**.

received significant attention. Chiral aminal **122** obtained readily from diamine and glyoxal derivative was introduced by Mukaiyama in 1978 and exhibited a high degree of diastereoselectivity in additions of organomagnesium reagents (Scheme 2.13).[192–194]

2.7.4 Acyclic Chiral Ketones Containing Other Heteroatoms in α-Position

In comparison to the previous subsections, ketones with other heteroatoms in the α-position were much less studied in additions of organometallic reagents affording the corresponding products with low diastereoselectivity. One example of nearly non–stereoselective addition of organolithium compounds to ketone having a sulfur atom in α-position was described by Overman et al. in 2000 (Scheme 2.14).[195]

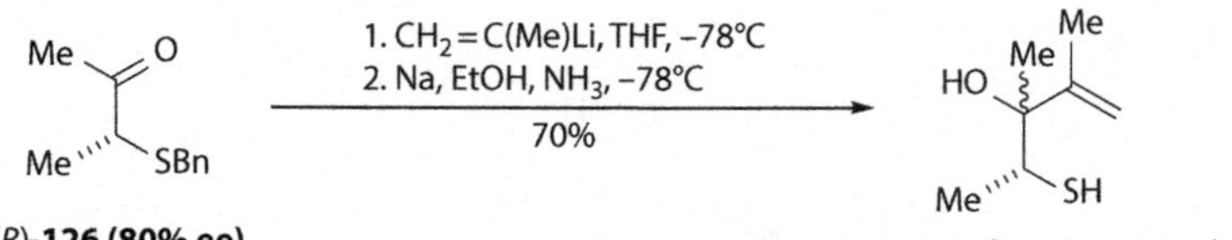

(*R*)-**126 (80% ee)** (4*R*)-**127** (dr = 3/2, 80% ee)

SCHEME 2.14 Addition reported by Overman.

2.7.5 CYCLIC KETONES

Additions of organometallic reagents to a cyclic carbonyl moiety are more complicated than in the case of acyclic ketones, where diastereoselectivity is predominantly affected by the configuration of the adjacent chiral center(s). In cyclic ketones, conformations of the ring, substitution patterns, and electronic effects in the ring play a significant role and vary from substrate to substrate. A model for additions of organometallic reagents to cyclohexanones has been discussed earlier (Section 2.2). In the case of simple cyclohexanones, the organometallic nucleophile attack on the carbonyl from a less-hindered equatorial position is favored (Scheme 2.15). This has been observed for organolithium and organomagnesium reagents, except acetylides, which prefer to approach the carbonyl from an axial position, due to torsional effects (for more information, see Section 2.2). Unfortunately, this model cannot be used to predict diastereoselectivity of additions to smaller or larger cyclic ketones, such as cyclopentanones and cycloheptanones.[196]

An interesting example of highly diastereoselective addition of organolithium reagent to cyclic ketones was described by Pulido et al.[197] Both cyclohexanones and cyclopentanones having (Z)-β-stannylvinyl group at C-3 afforded the corresponding adducts with a high degree of *syn* selectivity with alkylmagnesium and alkyllithium reagents. The direct effect of the stannate moiety is depicted in Scheme 2.16.[198]

Another example of diastereoselective additions of Grignard reagents to cyclic ketones, 4,4-disubstituted cyclohexadienone **20**, was announced by Wipf et al. (see Section 2.2, Scheme 2.2).[55] The predominant formation of the adduct **α-21** can be explained due to the absence of steric hindrance by dipol–dipol interactions between reagent and substrate. Other cyclic ketones that exhibit an interesting selectivity with the addition of an organometallic reagent are discussed in Section 2.2.[199,200]

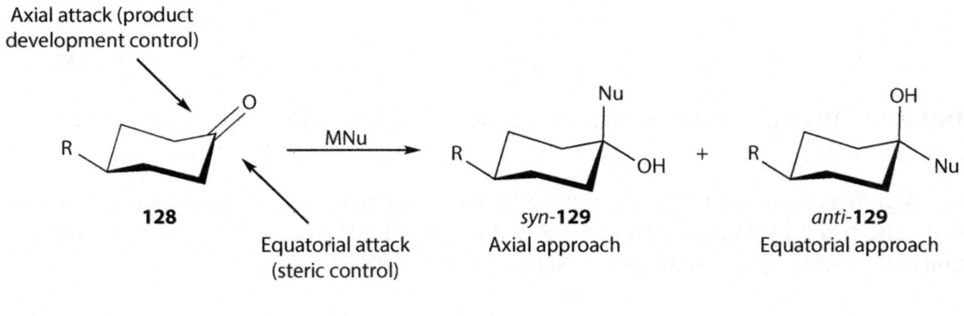

SCHEME 2.15 General scheme of additions to cyclohexanones.

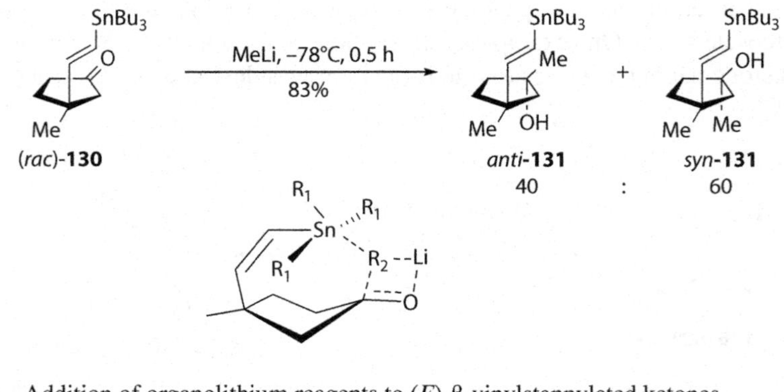

SCHEME 2.16 Addition of organolithium reagents to (*E*)-β-vinylstannylated ketones.

2.8 CONCLUSION

Diastereoselective additions of organometallic reagents to chiral carbonyl compounds have been used for the synthesis of a variety of natural, biologically active, and structurally interesting compounds. As illustrated in this chapter, high degrees of selectivity were achieved in additions of organolithium, organomagnesium, organotin, organocopper, organocerium reagents, and other metalloids to carbonyl-containing compounds. Although more than 60 years have gone by since Cram and coworkers outlined the origins of diastereoselectivity in the additions of nucleophiles to chiral carbonyl compounds, Cram's rule together with the Felkin–Ahn model are still the most commonly used tools for the elucidation and prediction of asymmetric induction in 1,2-additions. In the case of controlling selectivity in 1,3-asymmetric induction, the Reetz and Evans models describe sufficiently clearly the stereochemical outcome of such additions. Nevertheless, these fundamental models cannot cover all the aspects and factors affecting stereoselection; thus, further modifications and extensions have been developed.

Diastereoselectivity of 1,2-addition is affected not only by the stereochemistry of the carbonyl substrate and the characteristics of the organometallic reagent but also by many other factors including R solvent, temperature, additive, and specific features of the substrate or organometallic reagent. In order to illustrate the effects of the mentioned modifications on stereodiscrimination, the most typical and recent examples of 1,2-additions of achiral organometallic compounds to chiral carbonyl compounds were selected. By bringing all the most important models together with selected examples of 1,2-additions, we hope that this text can assist to further develop the area of diastereoselective additions to carbonyl-containing molecules.

REFERENCES

1. Frankland, E. *Ann. Chem. Pharm.* 1849, *71*, 177.
2. For selected reviews about chemistry of organosilicon compounds, see: (a) Somfai, P.; Seashore-Ludlow, B. Organosilicon reagents: Vinyl, alkynyl, arylsilanes. In *Comprehensive Organic Synthesis II*; Knochel, P.; Molander, G. A., Eds. Elsevier: Bremen, Germany, 2014, Vol. 1; pp. 27–48; (b) Yus, M.; González-Gómez, J. C.; Foubelo, F. *Chem. Rev.* 2013, *113*, 5595; (c) Yus, M.; Gonzales-Gomez, J. C.; Foubelo, F. *Chem. Rev.* 2011, *111*, 7774.
3. For selected reviews about organoboron chemistry, see: (a) Huo, H.-X., Duvall, J. R., Huanga, M.-Y.; Hong, R. *Org. Chem. Front.* 2014, *1*, 303; (b) Ramachandran, P. V.; Gagare, P. D.; Nicponski, D. R. Allylborons. In *Comprehensive Organic Synthesis II*; Knochel, P.; Molander, G. A., Eds. Elsevier: Bremen, Germany, 2014, Vol. 2; pp. 1–371; (c) Hall, D. G.; Lachance, H. Allylboration of carbonyl compounds. In *Organic Reactions*; Denmark, S. E., Ed. Wiley: Hoboken, NJ, 2008, Vol. 8; pp. 1–574; (d) Elford, T. G.; Hall, D. G. *Synthesis* 2010, *6*, 893; (e) Chemler, S. R.; Roush allylboronation. In *Name Reaction for Homologation*; Li, J., Corey, E. J., Eds. John Wiley & Sons: Hoboken, New Jersey, 2009, Vol. 2; pp. 613–640.
4. Yamazaki, S.; Yamabe, S. *J. Org. Chem.* 2002, *67*, 9346.
5. For selected recent reviews and chapters about stereocontrolled additions of organometals to compounds containing carbonyl functional group, see: (a) Huryn, D. M. Carbanions of alkali and alkaline-Earth cations: (II) Selectivity of carbonyl addition reactions. In *Comprehensive Organic Synthesis II*; Knochel, P.; Molander, G. A., Eds. Elsevier: Bremen, Germany, 2014, Vol. 1; pp. 27–48; (b) Gawley, R. E.; Aubé, J. *Principles of Asymmetric Synthesis*. Elsevier: Oxford, U.K., 2012, pp. 179–244; (c) O'Brien, A. G. *Tetrahedron* 2011, *67*, 9639; (d) Guillarme, S.; Plé, K.; Banchet, A.; Liard, A.; Haudrechy, A. *Chem. Rev.* 2006, *106*, 2355; (e) Mengel, A.; Reisner, O. *Chem. Rev.* 1999, *99*, 1191, and references therein.
6. Fisher, E. *Ber. Dtsch. Chem. Ges.* 1894, *27*, 3189.
7. McKenzie, A. *J. Chem. Soc.* 1904, *85*, 1249.
8. Curtin, D. Y.; Harris, E. E.; Meislich, E. K. *J. Am. Chem. Soc.* 1952, *74*, 2901.
9. Prelog, V. *Helv. Chim. Acta* 1953, *36*, 308.
10. Prelog, V. *Bull. Soc. Chim. Fr.* 1956, *23*, 967.
11. Cram, D. J.; Abd Elhafez, F. A. *J. Am. Chem. Soc.* 1952, *74*, 3210.
12. Mengel, A.; Reiser, O. *Chem. Rev.* 1999, *99*, 1191.
13. Gung, B. W. *Tetrahedron* 1996, *52*, 5263.

14. Cram, D. J.; Wilson, D. R. *J. Am. Chem. Soc.* 1963, *85*, 1245.
15. Cram, D. J.; Elhafez, F. A. A. *J. Am. Chem. Soc.* 1952, *74*, 5828.
16. Winstein, S.; Holmes, N. J. *J. Am. Chem. Soc.* 1955, *77*, 5562.
17. Cram, D. J.; Kopecky, K. R. *J. Am. Chem. Soc.* 1959, *81*, 2748.
18. Cornforth, J. W.; Cornforth, R. H.; Mathew, K. K. *J. Chem. Soc.* 1959, 112.
19. Karabatsos, G. J. *J. Am. Chem. Soc.* 1967, *89*, 1367.
20. Cram, D. J.; Abd Elhafez, F. A. *J. Am. Chem. Soc.* 1952, *74*, 5828; Cram, D. J.; Abd Elhafez, F. A. *J. Am. Chem. Soc.* 1954, *76*, 22.
21. Chérest, M.; Felkin, H.; Prudent, N. *Tetrahedron Lett.* 1968, *9*, 2199.
22. Ahn, N. T.; Eisenstein, O. *Nouv. J. Chim.* 1977, *1*, 61.
23. Ahn, N. T.; Eisenstein, O.; Lefour, J.-M.; Dâu, M.-E. *J. Am. Chem. Soc.* 1973, *95*, 6146.
24. Ahn, N. T.; Eisenstein, O. *Tetrahedron Lett.* 1976, *17*, 155.
25. Ahn, N. T. *Top. Curr. Chem.* 1980, *88*, 146.
26. Bürgi, H. B.; Dunitz, J. D.; Shefter, E. *J. Am. Chem. Soc.* 1973, *95*, 5065.
27. Bürgi, H. B.; Dunitz, J. D.; Lehn, J. M.; Wipff, G. *Tetrahedron* 1974, *30*, 1561.
28. Bürgi, H. B.; Dunitz, J. D. *Acc. Chem. Res.* 1983, *16*, 153.
29. Dauben, W. C.; Fonken, G. J.; Noyce, D. S. *J. Am Chem. Soc.* 1956, *78*, 2579.
30. Barton, D. H. R.; *J. Chem. Soc.* 1953, 1027.
31. Boone, J. R.; Ashby, E. C. Reduction of cyclic and bicyclic ketones by complex metal hydrides. In *Topics in Stereochemistry*; Allinger, N. L., Eliel, E. L., Eds. John Wiley & Sons: New York, Vol. 11, p. 53.
32. For explanation of stereochemical outcome of nucleophilic attack, see: (a) Liotta, C. L. *Tetrahedron Lett.* 1975, *519*, 523; (b) Liotta, C. L.; Burgess, L. M.; Eberhardt, W. H. *J. Am. Chem. Soc.* 1984, *106*, 4849; (c) Ashby, E. C.; Boone, J. R.; *J. Org. Chem.* 1976, *41*, 2890; (d) Giddings, M. R.; Hudec, J. *Can. J. Chem.* 1981, *59*, 459; (e) Wipke, W. T.; Gund, P. *J. Am. Chem. Soc.* 1976, *98*, 8107; (f) Perlburger, J. C.; Muller, P. *J. Am. Chem. Soc.* 1977, *99*, 6316; (g) Rei, M.-H. *J. Org. Chem.* 1979, *44*, 2760; (h) Rei, M. H. *J. Org. Chem.* 1983, *48*, 5386.
33. Cherest, M.; Felkin, H. *Tetrahedron Lett.* 1968, 2205; 1971, 383.
34. Cherest, M.; Felkin, H.; Frajerman, C. *Tetrahedron Lett.* 1971, 379.
35. Anh, N. T. *Fortschr. Chem. Forschung.* 1980, *88*, 145.
36. Cheung, C.-K.; Tseng, L.-T.; Lin, M. H.; Srivastava, S.; le Noble, W. J. *J. Am. Chem. Soc.* 1986, *108*, 1598.
37. Li, H.; le Noble, W. J. *Tetrahedron Lett.* 1990, *31*, 4391.
38. Cieplak, A. S.; Tait, B.; Johnson, C. R. *J. Am. Chem. Soc.* 1989, *111*, 8447.
39. Mehta, G.; Khan, F. A. *J. Am. Chem. Soc.* 1990, *112*, 6140.
40. Mehta, G.; Khan, F. A.; Ganguly, B.; Chandrasekhar, J. *J. Chem. Soc. Perkin Trans.* 1994, *2*, 2275.
41. Ganguly, B.; Chandrasekhar, J.; Khan, F. A.; Mehta, G. *J. Org. Chem.* 1993, *58*, 1734.
42. Halterman, R. L.; McEvoy, M. A. *J. Am. Chem. Soc.* 1990, *112*, 6690.
43. Rozeboom, M. D.; Houk, K. N. *J. Am. Chem. Soc.* 1982, *104*, 1189.
44. Wu, Y.-D.; Houk, K. N. *J. Am. Chem. Soc.* 1987, *109*, 908.
45. Wu, Y.-D.; Tucker, J. A.; Houk, K. N. *J. Am. Chem. Soc.* 1991, *113*, 5018.
46. Li, H.; Mehta, G.; Padma, S.; le Noble, W. J. *J. Org. Chem.* 1991, *56*, 2006.
47. Laube, T.; Ha, T.-K. *J. Am. Chem. Soc.* 1988, *110*, 5511.
48. Laube, T.; Stilz, H. U. *J. Am. Chem. Soc.* 1987, *109*, 5876.
49. Adcock, W.; Abeywickrema, A. N. *J. Org. Chem.* 1982, *47*, 2957.
50. Olah, G. A.; Forsyth, D. A. *J. Am. Chem. Soc.* 1977, *97*, 3137.
51. Adcock, W.; Cotton, J.; Trout, N. A. *J. Org. Chem.* 1994, *59*, 1867.
52. Gung, B. W.; Yanik, M. *J. Org. Chem.* 1996, *61*, 947.
53. Frenkin, G.; Kohler, K. F.; Reetz, M. T. *Angew. Chem. Int. Ed. Engl.* 1991, *30*, 1146.
54. Paddon-Row, M. N.; Wu, Y.-D.; Houk, K. N. *J. Am. Chem. Soc.* 1992, *114*, 10638.
55. Wipf, P.; Kim, Y. *J. Am. Chem. Soc.* 1994, *116*, 11678.
56. Wipf, P.; Jung, J.-K. *Chem. Rev.* 1999, *99*, 1469.
57. Lodge, E. P.; Heatcock, C. H. *J. Am. Chem. Soc.* 1987, *109*, 3353.
58. Klopman, G. *J. Am. Chem. Soc.* 1968, *90*, 223.
59. Salem, L. *J. Am. Chem. Soc.* 1968, *90*, 543.
60. Leitereg, T. J.; Cram, D. J. *J. Am. Chem. Soc.* 1968, *90*, 4011.
61. Leitereg, T. J.; Cram, D. J. *J. Am. Chem. Soc.* 1968, *90*, 4019.
62. Still, W. C.; Schneider, J. A. *Tetrahedron Lett.* 1980, *21*, 1035.
63. Reetz, M. T.; Jung, A. *J. Am. Chem. Soc.* 1983, *105*, 4833.

64. Reetz, M. T. *Angew. Chem. Int. Ed. Engl.* 1984, *23*, 556.
65. Brienne, M.-J.; Ouannes, C.; Jacques, J. *Bull. Soc. Chim. Fr.* 1968, *35*, 1036.
66. Evans, D. A.; Duffy, J. L.; Dart, M. J. *Tetrahedron Lett.* 1994, *35*, 8537.
67. Evans, D. A.; Duffy, J. L.; Dart, M. J. *Tetrahedron Lett.* 1994, *35*, 8541.
68. Evans, D. A.; Dart, M. J.; Duffy, J. L.; Yang, M. G.; Livingston, A. B. *J. Am. Chem. Soc.* 1995, *117*, 661.
69. Nakada, M.; Urano, Y.; Kobayashi, S.; Ohno, M. *J. Am. Chem. Soc.* 1988, *110*, 4826.
70. Yamamoto, Y.; Maruyama, K. *J. Am. Chem. Soc.* 1985, *107*, 6411.
71. Maruoka, K.; Itoh, T.; Yamamoto, H. *J. Am. Chem. Soc.* 1985, *107*, 4573.
72. Maruoka, K.; Itoh, T.; Sakurai, M.; Nonoshita, K.; Yamamoto, H. *J. Am. Chem. Soc.* 1988, *110*, 3588.
73. Reetz, M. T.; Harmat, N.; Mahrwald, R. *Angew. Chem., Int. Ed. Engl.* 1992, *31*, 342.
74. Reetz, M. T.; Steinbach, R.; Westermann, J.; Peter, R.; Wenderoth, B. *Chem. Ber.* 1985, *118*, 1441.
75. Ono, S.; Shuto, S.; Matsuda, A. *Tetrahedron Lett.* 1996, *37*, 221.
76. Cainelli, G.; Giacomini, D.; Galletti, P.; Orioli, P. *Eur. J. Org. Chem.* 2001, *2001*, 4509.
77. Janowitz, A.; Kunz, T.; Handke, G.; Reissig, H.-U. *Synlett*, 1989, *1989*, 24.
78. Mead, K. T. *Tetrahedron Lett.* 1987, *28*, 1019.
79. Takeda, Y.; Matsumoto, T.; Sato, F. *J. Org. Chem.* 1986, *51*, 4728.
80. Singh, S. M.; Oehlschlager, A. C. *Can. J. Chem.* 1988, *66*, 209.
81. Bhupathy, M.; Cohen, T. *Tetrahedron Lett.* 1985, *26*, 2619.
82. Cohen, T.; Bhupathy, M. *Tetrahedron Lett.* 1983, *24*, 4163.
83. Michelet, V.; Adiey, K.; Tanier, S.; Dujardin, G.; Genêt, J.-P. *Eur. J. Org. Chem.* 2003, *2003*, 2947.
84. Mulzer, J.; Angermann, A. *Tetrahedron Lett.* 1983, *24*, 2843.
85. Mulzer, J.; Kappert, M.; Huttner, G.; Jibril, I. *Angew. Chem., Int. Ed. Engl.* 1984, *23*, 704.
86. Mulzer, J.; Buettelmann, B.; Muench, W. *Liebigs Ann. Chem.* 1988, *1988*, 445.
87. Mulzer, J.; Greifenberg, S.; Beckstett, A.; Gottwald, M. *Liebigs Ann. Chem.* 1992, *11*, 1131.
88. Mulzer, J.; Funk, G. *Synthesis* 1995, *1995*, 101.
89. Guanti, G.; Banfi, L.; Narisano, E. *Tetrahedron Lett.* 1989, *30*, 5507.
90. Cossy, J.; Pévet, I.; Meyer, C. *Synlett* 2000, *2000*, 122.
91. Wee, A. G. H.; Tang, F. *Can. J. Chem.* 1998, *76*, 1070.
92. Wee, A. G. H.; Tang, F. *Tetrahedron Lett.* 1996, *37*, 6677.
93. Shimizu, M.; Wakioka, I.; Fujisawa, T. *Tetrahedron Lett.* 1997, *38*, 6027.
94. Caderas, C.; Lett, R.; Overman, L. E.; Rabinowitz, M. H.; Robinson, L. A.; Sharp, M. J.; Zablocki, J. *J. Am. Chem. Soc.* 1996, *118*, 9073.
95. Overman, L. E.; Robinson, L. A.; Zablocki, J. *J. Am. Chem. Soc.* 1992, *114*, 368.
96. Bergmeier, S. C. *Tetrahedron* 2000, *56*, 2561.
97. Yu, J.; Lai, J.-Y.; Ye, J.; Balu, N.; Reddy, L. M.; Duan, W.; Fogel, E. R.; Capdevila, J. H.; Falck, J. R. *Tetrahedron Lett.* 2002, *43*, 3939.
98. Kang, S. H.; Kim, W. J. *Tetrahedron Lett.* 1989, *30*, 5915.
99. Mukaiyama, T.; Suzuki, K.; Yamada, T. *Chem. Lett.* 1982, *11*, 929.
100. Su, Y.-L.; Yang, C.-S.; Teng, S.-J.; Zhao, G.; Ding, Y. *Tetrahedron* 2001, *57*, 2147.
101. Horton, D.; Hugues, J. B.; Tronchet, J. M. *J. Chem. Commun.* 1965, *1965*, 481.
102. Horton, D.; Tronchet, J. M. *J. Carbohydr. Res.* 1966, *2*, 315.
103. Ogura, H.; Ogiwara, M.; Itoh, T.; Takahashi, H. *Chem. Pharm. Bull.* 1973, *21*, 2051.
104. Hili, R.; Baktharaman, S.; Yudin, A. K. *Eur. J. Org. Chem.* 2008, *2008*, 5201.
105. Holladay, M. W.; Rich, D. H.; *Tetrahedron Lett.* 1983, *24*, 4401.
106. Hanson, G. H.; Lindberg, T. *J. Org. Chem.* 1985, *50*, 5399.
107. Reetz, M. T.; Drewes, M. W.; Schmitz, A. *Angew. Chem., Int. Ed. Engl.* 1987, *26*, 1141.
108. Reetz, M. T. *Angew. Chem., Int. Ed. Engl.* 1991, *30*, 1531.
109. Bejjani, J.; Chemla, F.; Audouin, M. *J. Org. Chem.* 2003, *68*, 9747.
110. Bonar-Law, R. P.; Davis, A. P.; Dorgan, B. J.; Reetz, M. T.; Wehrsig, A. *Tetrahedron Lett.* 1990, *31*, 6725.
111. Lee, B. W.; Lee, J. H.; Jang, K. C.; Kang, J. E.; Kim, J. H.; Park, K.-M.; Park, K. H. *Tetrahedron Lett.* 2003, *44*, 5905.
112. Hwang, G.; Chung, J.-H.; Lee, W. K. *J. Org. Chem.* 1996, *61*, 6183.
113. Righi, G.; Pietrantonio, S.; Bonini, C. *Tetrahedron* 2001, *57*, 10039.
114. Andrés, J. M.; de Elena, N.; Pérez-Encabo, A. *Tetrahedron* 1999, *55*, 14137.
115. Schomaker, J. M.; Geiser, A. R.; Huang, R.; Borhan, B. *J. Am. Chem. Soc.* 2007, *129*, 3794.
116. Righi, G.; Ciambrone, S.; *Tetrahedron Lett.* 2004, *45*, 2103.
117. Bailey, W. F.; Reed, D. P.; Clark, D. R.; Kapur, G. N. *Org. Lett.* 2001, *3*, 1865.
118. Andrés, J. M.; Barrio, R.; Martínez, M. A.; Pedrosa, R.; Pérez-Encabo, A. *J. Org. Chem.* 1996, *61*, 4210.

119. Kulshrestha, A.; Schomaker, J. M.; Holmes, D.; Staples, R. J.; Jackson, J. E.; Borhan, B. *Chem. Eur. J.* 2011, *17*, 12326.
120. Pierre, J. L.; Handel, H.; Baret, P. *Tetrahedron* 1974, *30*, 3213.
121. Righi, G.; Ronconi, S.; Bonini, C. *Eur. J. Org. Chem.* 2002, *2002*, 1573.
122. Park, C. S.; Choi, H. G.; Lee, H.; Lee, W. K.; Ha, H.-J. *Tetrahedron: Asymmetry* 2000, *11*, 3283.
123. Ren, X.-F.; Konaklieva, M. I.; Shi, H.; Dickey, S.; Lim, D. V.; Gonzalez, J.; Turos, E. *J. Org. Chem.* 1998, *63*, 8898.
124. Ren, X.-F.; Konaklieva, M. I.; Turos, E. *J. Org. Chem.* 1995, *60*, 4980.
125. Alcaide, B.; Almendros, P.; Alonso, J. M. *J. Org. Chem.* 2004, *69*, 993.
126. Alcaide, B.; Polanco, C.; Sierra, M. A. *J. Org. Chem.* 1998, *63*, 6786.
127. Hanson, G. J.; Baran, J. S.; Lindberg, T. *Tetrahedron Lett.* 1986, *27*, 3577.
128. Koskinen, A. M. P.; Paul, J. M. *Tetrahedron Lett.* 1992, *33*, 6853.
129. Tsutsumi, S.; Okonogi, T.; Shibahara, S.; Ohuchi, S.; Hatsuchiba, E.; Patchett, A. A.; Christensen, B. G. *J. Med. Chem.* 1994, *37*, 3492.
130. Reed, P. E.; Katzenellenbogen, J. A. *J. Org. Chem.* 1991, *56*, 2624.
131. Ito, H.; Ikeuchi, Y.; Taguchi, T.; Hanzawa, Y. *J. Am. Chem. Soc.* 1994, *116*, 5469.
132. Barrett, A. G. M.; Damiani, F. *J. Org. Chem.* 1999, *64*, 1410.
133. Andrés, J. M.; Pedrosa, R.; Pérez, A.; Pérez-Encabo, A. *Tetrahedron* 2001, *57*, 8521.
134. Harris, B. D.; Bhat, K. L.; Joullié, M. M. *Heterocycles* 1986, *24*, 1045.
135. Vara Prasad, J. V. N.; Rich, D. H. *Tetrahedron Lett.* 1990, *31*, 1803.
136. Bigi, F.; Casnati, G. Sartori, G.; Araldi, G.; Bocelli, G. *Tetrahedron Lett.* 1989, *30*, 1121.
137. Garner, P. *Tetrahedron Lett.* 1984, *25*, 5855.
138. Passiniemi, M.; Koskinen, A. M. P.; Beilstein J. *Org. Chem.* 2013, *9*, 2641.
139. Lemke, A.; Büschleb, M.; Ducho, C. *Tetrahedron* 2010, *66*, 208.
140. Herold, P. *Helv. Chim. Acta* 1988, *71*, 354.
141. Karjalainen, O. K.; Koskinen, A. M. P. *Org. Biomol. Chem.* 2011, *9*, 1231.
142. Gruza, H.; Kiciak, K.; Krasiński, A.; Jurczak, J. *Tetrahedron: Asymmetry* 1997, *8*, 2627.
143. Hanessian, S.; Yang, G.; Rondeau, J.-M.; Neumann, U.; Betschart, C.; Tinelnot-Blomley, M. *J. Med. Chem.* 2006, *49*, 4544.
144. Garner, P.; Park, J. M. *J. Org. Chem.* 1990, *55*, 3772.
145. Coleman, R. S.; Carpenter, A. J. *Tetrahedron Lett.* 1992, *33*, 1697.
146. Passiniemi, M.; Koskinen, A. M. P. *Tetrahedron Lett.* 2008, *49*, 980.
147. Passiniemi, M.; Koskinen, A. M. P. *Org. Biomol. Chem.* 2011, *9*, 1774.
148. Dhondi, P. K.; Carberry, P.; Choi, L. B.; Chisholm, J. D. *J. Org. Chem.* 2007, *72*, 9590.
149. Fürstner, A.; Krause, H. *Adv. Synth. Catal.* 2001, *343*, 343.
150. Sa-ei, K.; Montgomery, J. *Tetrahedron* 2009, *65*, 6707.
151. Williams, L.; Zhang, Z.; Shao, F.; Carroll, P. J.; Joullié, M. M. *Tetrahedron* 1995, *52*, 11673.
152. Reginato, G.; Mordini, A.; Tenti, A.; Valacchi, M.; Broguiere, J. *Tetrahedron: Asymmetry* 2008, *19*, 2882.
153. Zhang, X.; van der Donk, W. A. *J. Am. Chem. Soc.* 2007, *129*, 2212.
154. Asai, M.; Nishikawa, T.; Ohyabu, N.; Yamamoto, N.; Isobe, M. *Tetrahedron* 2001, *57*, 4543.
155. Nishikawa, T.; Asai, M.; Ohyabu, N.; Yamamoto, N.; Isobe, M. *Angew. Chem., Int. Ed. Engl.* 1999, *38*, 3081.
156. Nishikawa, T.; Urabe, D.; Yoshida, K.; Iwabuchi, T.; Asai, M.; Isobe, M. *Org. Lett.* 2002, *4*, 2679.
157. Nishikawa, T.; Urabe, D.; Yoshida, K.; Iwabuchi, T.; Asai, M.; Isobe, M. *Chem. Eur. J.* 2004, *10*, 452.
158. Nishikawa, T.; Asai, M.; Isobe, M. *J. Am. Chem. Soc.* 2002, *124*, 7847.
159. Tsujimoto, T.; Nishikawa, T.; Urabe, D.; Isobe, M. *Synlett* 2005, *2005*, 433.
160. Fujisawa, T.; Nagai, M.; Koike, Y.; Shimizu, M. *J. Org. Chem.* 1994, *59*, 5865.
161. Takikawa, H.; Muto, S.; Nozawa, D.; Kayo, A.; Mori, K. *Tetrahedron Lett.* 1998, *39*, 6931.
162. Takikawa, H.; Nozawa, D.; Kayo, A.; Muto, S.-E.; Mori, K. *J. Chem. Soc., Perkin Trans. 1*, 1999, *1999*, 2467.
163. Enders, D.; Piva, O.; Burkamp, F. *Tetrahedron* 1996, *52*, 2893.
164. Ramón, D. J.; Yus, M. Alkylation of carbonyl and amino groups. In *Science of Synthesis*; Molander, G., Ed. Georg Thieme Verlag KG: Stuttgart, Germany, 2011, Vol. 2, Chapter 2.7; pp. 349–359.
165. Nógrádi, M. *Stereoselective Synthesis: A Practical Approach*, 2nd edn. VCH: New York, 1995; pp. 136–154.
166. Reetz, M. T.; Wang, F.; Harms, K. *J. Chem. Soc., Chem. Commun.* 1991, *1991*, 1309.
167. Lampkins, A. J.; Abdul-Rahim, O.; Castellano, R. K. *J. Org. Chem.* 2006, *71*, 5815.
168. Clayden, J.; Lund, A.; Vallverdú, L.; Helliwell, M. *Nature* 2004, *431*, 966.
169. (a) Clayden, J. *Chem. Soc. Rev.* 2009, *38*, 817. (b) Clayden, J.; Vassiliou, N. *Org. Biomol. Chem.* 2006, *4*, 2667.

170. Alvarez-Ibarra, C.; Arjona, O.; Perez-Ossorio, R.; Perez-Rubalcaba, A.; Quiroga, M. L.; Santesmases, M. J. *J. Chem. Soc.,Perkin Trans. 2* 1983, *1983*, 1645.
171. Karabatsos, G. J. *Tetrahedron Lett.* 1972, *52*, 5289.
172. For recent examples, see: (a) Vitellozzia, L.; McAllistera, G. D.; Genski, T.; Taylor, R. J. K. *Synthesis* 2016, *48*, 48; (b) Schnabel, C.; Hiersemann, M. *Org. Lett.* 2009, *11*, 2555; (c) Noguchi, Y.; Hirose, T.; Furuya, Y.; Ishiyama, A.; Otoguro, K.; Ōmura, S.; Sunazuka, T. *Tetrahedron Lett.* 2012, *53*, 1802; (d) Heinrich, C. F.; Miesch, M.; Miesch, L. *Org. Biomol. Chem.* 2015, *13*, 2153.
173. Still, W. C.; McDonald, J. H.. *Tetrahedron Lett.* 1980, *21*, 1031.
174. Reetz, M. T.; Huellmann, M. *J. Chem. Soc., Chem. Commun.* 1986, *21*, 1600.
175. Chen, X.; Hortelano, E. R.; Eliel, E. L.; Frye, S. V. *J. Am. Chem. Soc.* 1990, *112*, 6130.
176. Chen, X.; Hortelano, E. R.; Eliel, E. L.; Frye, S. V. *J. Am. Chem. Soc.* 1992, *114*, 1778.
177. Lynch, J. E.; Eliel, E. L. *J. Am. Chem. Soc.* 1984, *106*, 2943.
178. Frye, S. V.; Eliel, E. L. *J. Am. Chem. Soc.* 1988, *110*, 484.
179. Eliel, E. L.; Frye, S. V.; Hortelano, E. R.; Chen, X.; Bai, X. *Pure Appl. Chem.* 1991, *63*, 1591.
180. Bai, X.; Eliel, E. L. *J. Org. Chem.* 1992, *57*, 5166.
181. Martinez-Ramos, F.; Vargas-Díaz, M. E.; Chacón-García, L.; Tamariz, J.; Joseph-Nathan, P.; Zepeda, L. G. *Tetrahedron Asymmetry* 2001, *12*, 3095.
182. Chacón-Garcia, L.; Lagunas-Rivera, S.; Pérez-Estrada, S.; Vargas-Díaz, M. E.; Joseph-Nathan, P.; Tamariz, J.; Zepeda, L. G. *Tetrahedron Lett.* 2004, *45*, 2141.
183. Vargas-Díaz, M. E.; Chacón-Garcia, L.; Velázquez, P.; Tamariz, J.; Joseph-Nathan, P.; Zepeda, L. G. *Tetrahedron Asymmetry* 2003, *14*, 3225.
184. Vargas-Díaz, M. E.; Joseph-Nathan, P.; Tamariz, J.; Zepeda, L. G. *Org. Lett.* 2007, *9*, 13.
185. Akhoon, K. M.; Myles, D. C. *J. Org. Chem.* 1997, *62*, 6041.
186. Charette, A. B.; Benslimane, A. F.; Mellon, C. *Tetrahedron Lett.* 1995, *36*, 8557.
187. Tamai, Y.; Akiyama, M.; Okamura, A.; Miyano, S. *J. Chem. Soc. Chem. Commun.* 1992, *1992*, 687.
188. Tamai, Y.; Hattori, T.; Date, M.; Koike, S.; Kamikubo, Y.; Akiyama, M.; Seino, K.; Takayama, H.; Oyama, T.; Miyano, S. *J. Chem. Soc. Perkin Trans.* 1999, *1*, 1685.
189. Tamai, Y.; Hattori, T.; Date, M.; Takayama, H.; Kamikubo, Y.; Minato, Y.; Miyano, S. *J. Chem. Soc. Perkin Trans.* 1999, *1*, 1141.
190. Date, M.; Tamai, Y.; Hattori, T. J.; Takayama, H.; Kamikubo, Y.; Miyano, S. *Chem. Soc. Perkin Trans.* 2001, *1*, 645.
191. Yu, L.; Kokai, A.; Yudin, A. K. *J. Org. Chem.* 2007, *72*, 1737.
192. Mukaiyama, T.; Sakito, Y.; Asami, M. *Chem. Lett.* 1978, *7*, 1253.
193. Mukaiyama, T.; Sakito, Y.; Asami, M. *Chem. Lett.* 1979, *8*, 705.
194. Mukaiyama, T. *Tetrahedron* 1981, *37*, 4111.
195. Overman, L. E.; Ponce, A. M. *J. Am. Chem. Soc.* 2000, *122*, 8672.
196. Ashby, E. C.; Laemmle, J. T. *Chem. Rev.* 1975, *75*, 521.
197. Barbero, A.; Pulido, F. J.; Rincón, J. A.; Cuadrado, P.; Galisteo, D.; Martínez-García, H. *Angew. Chem. Int. Ed.* 2001, *40*, 2102.
198. Barbero, A.; Pulido, F. J.; Rincón, J. A. *J. Am. Chem. Soc.* 2003, *125*, 12049.
199. Dimitroff, M.; Fallis, A. G. *Tetrahedron Lett.* 1998, *39*, 2527.
200. Dimitroff, M.; Fallis, A. G. *Tetrahedron Lett.* 1998, *39*, 2531.

3 Enantiomerically Pure Compounds by Enantioselective Synthetic Chiral Metal Complexes

Frady G. Adly and Ashraf Ghanem

CONTENTS

3.1 Introduction 75
3.2 History 76
    3.2.1 Mechanism of Dirhodium(II)-Catalyzed Cyclopropanation Reactions 77
    3.2.2 Types of Dirhodium(II)-Carbenoid Intermediates 79
    3.2.3 Modes of Interaction between the Dirhodium(II) Complex and the Carbene 80
    3.2.4 Approach of the Alkene 80
3.3 Modifications in the Dirhodium(II) Frame 82
    3.3.1 Electronic Modifications 82
    3.3.2 Steric Modifications 83
3.4 Dirhodium(II) Carboxylate Complexes 83
    3.4.1 Conformations in Dirhodium(II) Carboxylate Complexes 83
    3.4.2 Dirhodium(II) Catalysts Derived from Proline-Based Ligands 84
3.5 Dirhodium(II) Catalysts Derived from Chiral N-Protected Amino Acid Ligands 92
    3.5.1 Homoleptic Complexes 92
    3.5.2 Heteroleptic Complexes 109
    3.5.3 Dirhodium(II) Catalysts Derived from Substituted Cyclopropane Carboxylate Ligands 110
3.6 Dirhodium(II) Carboxamidates 113
    3.6.1 Homoleptic Complexes 113
    3.6.2 Heteroleptic Complexes 121
3.7 Effects of Axial Ligands on Enantioselectivity 124
3.8 Conclusion 126
References 127

3.1 INTRODUCTION

Historically, two approaches have been reported for achieving enantiomerically enriched compounds. The first approach is through asymmetric chemical transformations involving chiral pools, chiral auxiliaries, chiral reagents, and/or enantioselective catalysis/biocatalysis.[1] The second approach is through the preparation of the racemic mixture and the separation of enantiomers by means of chiral separation.[2] These two pathways, however, have their own disadvantages. While the former requires a suitable precursor—auxiliary, reagent, or catalyst—the latter is generally time consuming and yields only up to 50% of the desired enantiomer.

Asymmetric catalysis is considered one of the ultimate solutions for getting access to enantiomerically enriched compounds at which, a metal complex carrying chiral ligands has its own merits to return many equivalents of the desired enantiomerically enriched chiral product. Due to the increasing number of available methodologies for producing enantiomerically enriched organic compounds, the scope of asymmetric catalysis has greatly expanded to include a broad range of chemical transformations.[3] Ideally, a practical asymmetric catalyst should provide high yield and selectivity (chemo-, diastereo-, and enantioselectivity) for a broad range of substrates in different reaction conditions, while being inexpensive and readily available in both enantiopure forms. A large number of complexes have been already reported and many of these complexes have been studied and used in asymmetric catalysis.[4–19]

Dirhodium(II) paddlewheel complexes are among the most attractive catalysts because of their high level of applicability, efficiency, and selectivity.[16,20–33] The ability of such complexes to effectively catalyze a variety of reactions has demonstrated their synthetic potential, particularly in the context of chiral catalysis. Further, dirhodium(II) catalysts with very high turnover numbers have been reported.[34,35] Therefore, the cost of rhodium was enormously overcast by the capability of very small amounts of the catalyst to return large quantities of value-added chemicals.

3.2 HISTORY

The story commenced in 1960 when Chernyaev and coworkers[36] uncovered a fruitful part of rhodium chemistry. They reported an air-stable green crystalline complex achieved by refluxing rhodium(III) chloride in formic acid. This complex was initially formulated as the rhodium(I) species, $H[Rh(O_2CH)_2 \cdot 0.5H_2O]$. However, it was quickly found that this compound lacks acid character. The product was subsequently identified as dirhodium(II) tetraformate monohydrate $[Rh_2(O_2CH)_4 \cdot H_2O]$ by x-ray diffraction.[37] This was the first example of a binuclear rhodium(II) carboxylate complex that possesses the now known "lantern" structure (Figure 3.1).

Later in the 1970s, dirhodium(II) tetraacetate, $Rh_2(OAc)_4$, which was prepared in the mid-1960s,[38,39] was reported as an exceptionally effective catalyst for a wide variety of catalytic transformations involving diazo compounds.[40,41] This discovery by Teyssie and coworkers holds a unique importance in the history of the dirhodium(II) catalysis field.[40–42] In fact, $Rh_2(OAc)_4$ and related complexes are not susceptible to redox transformations with diazo compounds, do not form π-complexes with olefins, and are resistant to ligand exchange under ordinary catalytic conditions. Besides, it was found that yields from $Rh_2(OAc)_4$-catalyzed cyclopropanations were not decreased even when catalyst to diazo substrate molar ratios as low as 0.0005 were used.[43] Also, due to the facile ligand exchange property of $Rh_2(OAc)_4$, this complex has become a versatile synthetic tool to access a variety of other dirhodium(II) complexes.[20,21] In addition to achiral ligands, chiral ligands were also successfully introduced, which opened the door for the design and preparation of a variety of chiral dirhodium(II) complexes.[20,21,33] Furthermore, the exploration of their catalytic activities promoted the rhodium-carbenoid chemistry to unprecedented levels of chemo-, regio-, and stereoselectivity.

Chiral dirhodium(II) catalysts were found to be exceptional for a wide range of chemical transformations, particularly in metal-nitrenoid and carbenoid chemistry.[16] These chemical

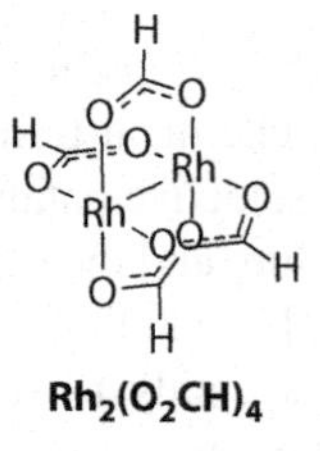

$Rh_2(O_2CH)_4$

FIGURE 3.1 Lantern structure of dirhodium(II) tetraformate, $(Rh_2(O_2CH)_4)$.

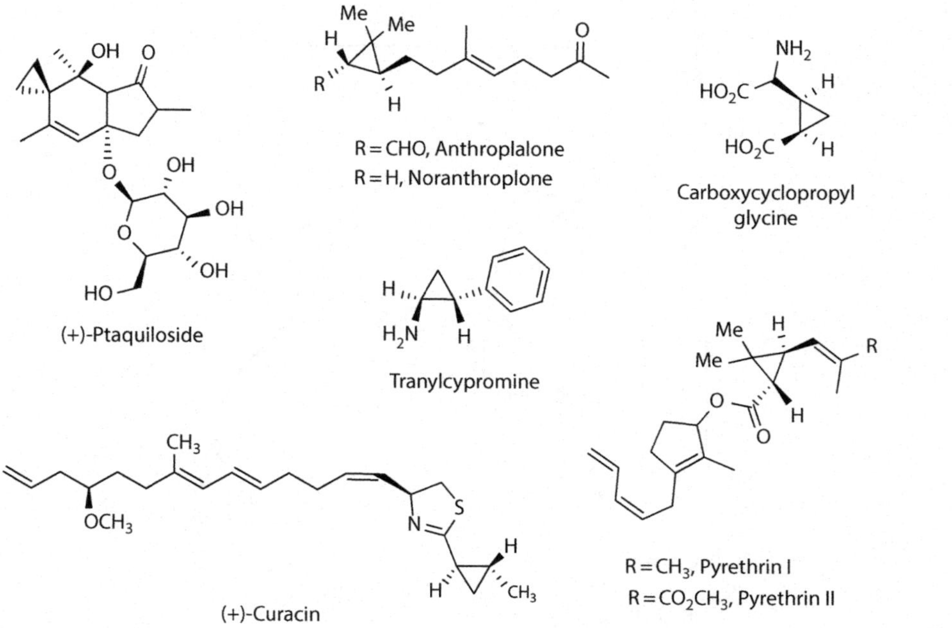

Enantiomerically Pure Compounds by Enantioselective Synthetic Chiral Metal Complexes 77

FIGURE 3.2 Model examples of cyclopropane containing naturally occurring and synthetic compounds.

transformations involve aziridinations,[44–46] C–H insertions,[23,47,48] ylide transformations,[8,10,11,49,50] Lewis acid–catalyzed processes,[11,51–55] cross-coupling reactions[56], and cyclopropanation and cyclopropenation reactions.[7,57–59]

This chapter focuses on the utilization of chiral dirhodium(II) paddlewheel complexes in the enantioselective synthesis of chiral three-membered carbocycles. The discussion will cover both inter- and intramolecular asymmetric cyclopropanation and cyclopropenation reactions. The synthesis of these particular structures was chosen as being very important entities in the field of organic and natural products chemistry. They function as a useful utility in organic synthesis that has the potential to give access to complex structures with a defined orientation of functional groups.[60–64] Also, even though it is a highly strained entity, the cyclopropyl group is a well-known structure motif in nature.[65,66] Naturally occurring and synthetic chiral cyclopropanes are endowed with a large spectrum of biological properties including enzyme inhibition, insecticidal, herbicidal, antibacterial, antifungal, antiviral, and antitumor activities (Figure 3.2).[66–68] For example, they have been found as structural moieties in cyclopropane-based peptidomimetics[69,70]; pyrethroid insecticides[71,72]; antipsychotic agents[73]; selective inhibitors for papain and cystein proteases[74]; the antidepressant and anxiolytic agent, tranylcypromine[75]; the antimitotic agent, (+)-curacin[76]; and the anticancer agent, (+)-ptaquiloside.[77] This is in addition to a number of best-selling pharmaceuticals containing the cyclopropyl group, which give the cyclopropane chemistry more economic importance.[78]

3.2.1 Mechanism of Dirhodium(II)-Catalyzed Cyclopropanation Reactions

The cyclopropyl group can be efficiently constructed through the reaction of a dirhodium(II)-carbenoid species (formed when a diazo compound reacts with a dirhodium(II) catalyst) and an alkene (Scheme 3.1a). The globally accepted mechanism for this reaction was originally proposed by Yates[79] in 1952 for copper-catalyzed diazo decomposition. This mechanism involves the initial complexation of the negatively polarized carbon of the diazo compound to the axial site of the Rh(II) catalyst, which is coordinatively unsaturated (Scheme 3.1b). Subsequent irreversible extrusion of N_2 from the intermediate **2** generates the Rh(II)-carbene complex **4**. In this mechanism, the extrusion

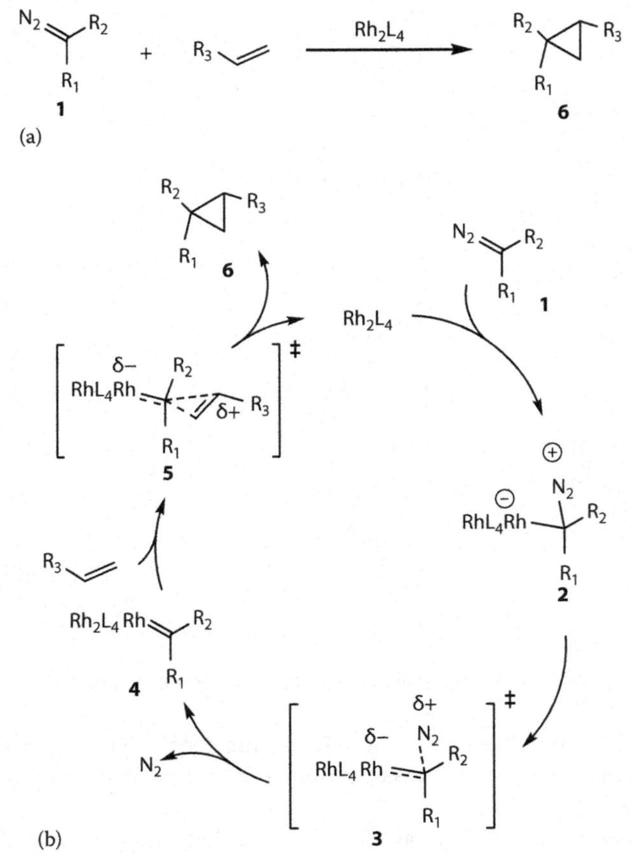

SCHEME 3.1 (a) Dirhodium(II)-catalyzed cyclopropanation reaction and (b) mechanistic pathway for dirhodium(II)-carbene intermediate formation and cyclopropanation.[80–87]

of N_2 is considered to be the rate-limiting step, and a number of kinetic studies have provided some support for this mechanism.[80–87] Computational as well as kinetic studies of dirhodium(II) carboxylate–catalyzed reactions indicated that carbene binding occurs only at one of the two rhodium-active sites at a time.[80,87,88]

Dirhodium(II)-carbenoid intermediate **4** can be represented by two valence bond structures, either as a double bond to the metal center (**7a**) or as a charge-separated structure (**7b**). Although representation of **4** as a charge-separated structure de-emphasizes the back-bonding stabilization from the rhodium atom (Figure 3.3),[10] it does emphasize the electrophilic nature of the carbenoid carbon.

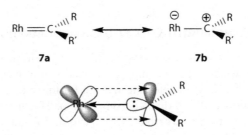

FIGURE 3.3 Carbenoid resonance structure and back-bonding from the metal atom.[10]

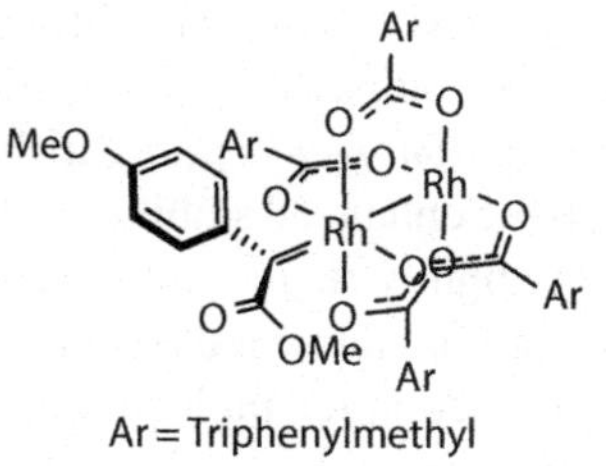

Enantiomerically Pure Compounds by Enantioselective Synthetic Chiral Metal Complexes   79

Ar = Triphenylmethyl
8

FIGURE 3.4 Metastable dirhodium(II)-carbene complex studied by Davies and coworkers.[89]

In fact, the formation of this intermediate in dirhodium(II)-catalyzed carbenoid transformations remained elusive until Davies and coworkers[89] provided a direct evidence for being a genuine dirhodium(II)-carbene complex. They reported the generation of the metastable dirhodium(II)-carbenoid intermediate **8** (Figure 3.4), which was stable in chloroform and kept at 0°C for a ~20 h period. The authors were able to characterize the Rh=C bond by vibrational and NMR spectroscopy, extended x-ray absorption, fine structure analysis, and quantum chemical calculations.[89]

3.2.2 Types of Dirhodium(II)-Carbenoid Intermediates

Electrophilicity of the metal-carbenoid intermediate is an important feature to control the reactivity; low electrophilicity can lead to less reactivity, while too much electrophilicity can cause side reaction. Electronic properties of the dirhodium(II)-carbenoid intermediates can be easily tuned by varying the substituents linked to the carbene carbon. Based on electron-withdrawing and electron-donating properties of these substituents, diazo substrates were classified into three groups: *monoacceptor*, *diacceptor*, and *donor–acceptor*.[17,30] The terms "*donor*" and "*acceptor*" refer to electron-donating or electron-withdrawing groups, respectively. Typical *acceptor* groups can be keto, cyano, trifluoromethyl, phosphonate, or sulfonate, while typical *donor* groups are vinyl, alkynyl, aryl, or heteroaryl (Figure 3.5). In general, an *acceptor* substituent on the diazo substrate will kinetically retard the formation of rhodium-carbenoid intermediate, but once formed, these intermediates are more electrophilic and hence show high reactivity, while being less stable and selective.[47] On the contrary, introducing a *donor* substituent will make the diazo substrate more reactive toward carbenoid generation. But once the carbenoid intermediate is formed, it tends to be less electrophilic and reactive, while being more stable and selective (Figure 3.5).[47]

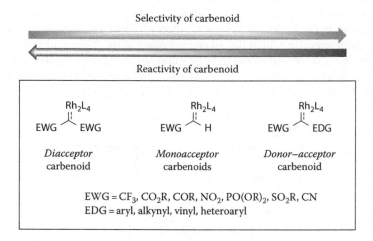

FIGURE 3.5 Classification of dirhodium(II) carbenoids[17,30] and relationship between their reactivity and selectivity.[47]

An emerging area in asymmetric synthesis involves the utilization of *donor–acceptor* carbenoids due to their enhanced levels of selectivity compared to other carbenoid classes. In a number of cases, the reactivity and stability observed with *donor–acceptor* carbenoids differ substantially from those *monoacceptor* and *diacceptor*. Due to the enhanced stability of this class of intermediates, they have potential capability toward a variety of highly regio- and stereoselective reactions, especially when combined with chiral dirhodium(II) carboxylate catalysts. Their unique selectivities have already facilitated in the efficient construction of complex molecular architectures.[90]

3.2.3 Modes of Interaction between the Dirhodium(II) Complex and the Carbene

The preferred mode of interaction between the dirhodium(II) complex and the carbene was already explained using MM2 followed by extended Hückel calculations through the interaction between dirhodium(II) tetraacetate and vinyl carbene.[91] The results revealed that the carbene favorably align staggered to the oxygen of the carboxylates rather than an eclipsed alignment (Figure 3.6). The preferred staggered orientation is not only due to steric hindrance but also required for metal back-bonding stabilization of the carbenoid as the d_{yz} and d_{xz} orbitals of the rhodium atoms are hybridized to generate two new orbitals that lie in this staggered positions.[92]

Later, Davies and coworkers[83] analyzed both methyl diazoacetate and methyl vinyldiazoacetate as models for *acceptor* and *donor–acceptor* systems, respectively, to justify the greater chemoselectivity displayed by the latter. However, in 2009, these models were reevaluated by the same research group[82] as the calculated small potential energy barrier wasn't consistent with a large amount of experimental data, which suggests that *donor–acceptor* are highly selective species. The authors recognized that the previously used calculations were not appropriately describing rhodium in the system. The new models revealed that an *acceptor* carbenoid (e.g., methyl diazoacetate) prefers the eclipsed conformation, while a *donor–acceptor* carbenoid (e.g., methyl α-phenyldiazoacetate) adopts the staggered conformation, as the more sterically bulky nature of the phenyl group does not permit the eclipsed conformation (Figure 3.7). The authors communicated that this observation might have major implications on the developed models to describe the enantioinduction of chiral dirhodium(II) complexes since the two carbenoids will orient themselves differently relative to the given chiral ligand environment.[82]

3.2.4 Approach of the Alkene

The final step in the dirhodium(II)-catalyzed cyclopropanation mechanism involves the approach of the alkene and generation of cyclopropane final product **6** (Scheme 3.1b). Bonge and Hansen[93] studied the mechanism of dirhodium(II)-catalyzed cyclopropanations with ethyl bromo-, chloro-, and iododiazoacetate through density functional theory (DFT) calculations. They found that, in addition to transition states in which the alkene approaches the carbenoid in an end-on manner, side-on trajectory states were also found to be of importance (Figure 3.8). The relative energies of the side-on trajectory transition states compared to the end-on trajectory transition states are shown to be affected by the alkene substrate, as well as the carbenoid substituents.

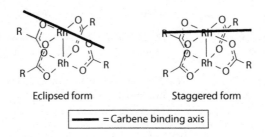

FIGURE 3.6 Alignment of carbene on dirhodium(II) complex.[26,91]

Enantiomerically Pure Compounds by Enantioselective Synthetic Chiral Metal Complexes

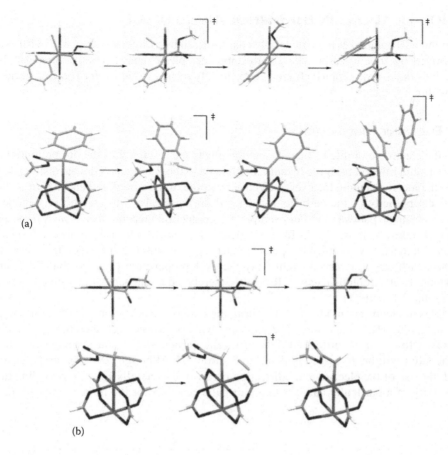

FIGURE 3.7 Calculated structures for (a) methyl α-phenyldiazoacetate as *donor–acceptor* system and (b) methyl diazoacetate as *acceptor* system, top and side views. (Reprinted with permission from Hansen, J., Autschbach, J., and Davies, H.M.L., *J. Org. Chem.*, 74, 6555, 2009. Copyright 2016.)

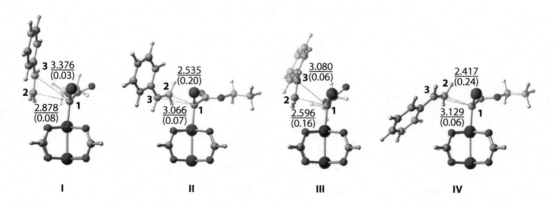

FIGURE 3.8 Transition states of $Rh_2(O_2CH)_4$-catalyzed cyclopropanation of styrene with ethyl bromodiazoacetate: I and III represent end-on trajectory transition states; II and IV represent side-on trajectory transition states. (Reprinted with permission from Bonge, H.T. and Hansen, T., *J. Org. Chem.*, 75, 2309, 2010. Copyright 2016.)

3.3 MODIFICATIONS IN THE DIRHODIUM(II) FRAME

All dirhodium(II) complexes are structurally characterized by a bimetallic core with a Rh–Rh single bond, bridged by four μ$_2$-carboxylate, carboxamidate, phosphonate, or other ligands. Reported modifications in the dirhodium(II) framework is mostly related to either electronic or steric modifications within its ligands.

3.3.1 Electronic Modifications

Altering the electronic profile of ligands mainly affects the reactivity of the dirhodium(II) catalyst as electronically different bridging ligands coordinated to the Rh–Rh axes donate distinct degrees of charge to the metal changing the overall electronic profile of the complex. This, in turn, will impact on the electrophilicity of the carbenoid generated during catalysis to significantly influence the reaction mechanistic pathway.[31,94] For example, the competition reaction between cyclopropanation and aryl C–H insertion illustrated in Table 3.1 clearly demonstrates the role of the catalyst electronic profile on reactivity.[94] A completely opposite reactivity was observed at which the more electron-rich catalyst, Rh$_2$(cap)$_4$, returned exclusively the cyclopropanation product (Table 3.1, entry 3), while the highly electrophilic catalyst, Rh$_2$(pfb)$_4$, merely led to the generation of the C–H insertion product (Table 3.1, entry 1).

Examples to electronic modifications within the complex scaffold included employing mixed valence dirhodium(II,III) species,[46,95] complexes with Bi–Rh heterobimetallic species,[96–99] and complexes with axially coordinated *N*-heterocyclic carbenes[100,101] and applying reaction additives. The latter will be discussed in detail in Section 3.7. Although several examples have been reported, the use of the electronic modification pathway is generally limited to the fine tuning of the selectivity of a particular catalyst in a particular reaction for the preparation of a particular product.[31]

TABLE 3.1
Model Example for the Effect of Catalyst Electrophilic Profile on Reaction Pathway[94]

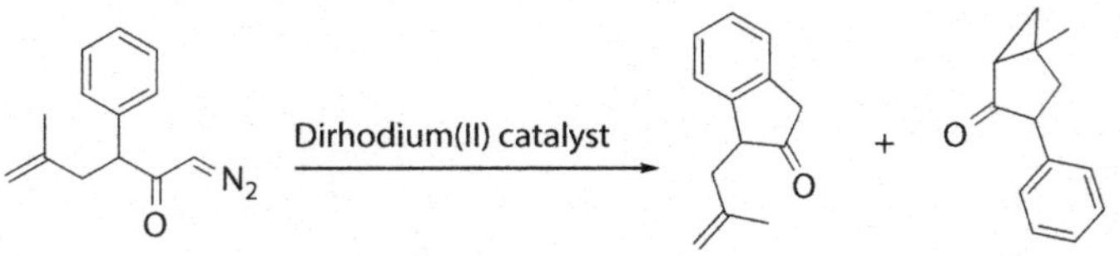

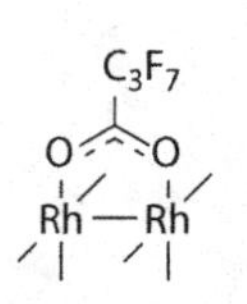

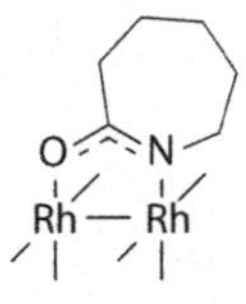

Entry	Catalyst	Yield (%)	a:b
1	Rh$_2$(pfb)$_4$	86	100:0
2	Rh$_2$(OAc)$_4$	92	52:48
3	Rh$_2$(cap)$_4$	75	0:100

TABLE 3.2
Model Example for the Effect of Catalyst Steric Profile on Reaction Chemoselectivity[102]

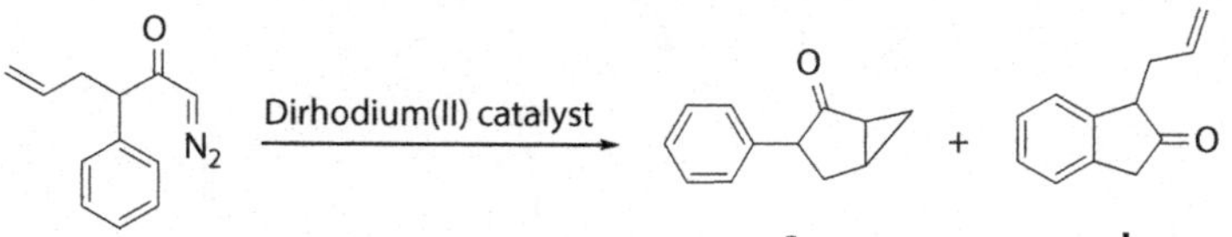

Entry	Catalyst	Yield (%)	c:d
1	Rh$_2$(OAc)$_4$	63	71:29
2	Rh$_2$(tpa)$_4$	83	0:100

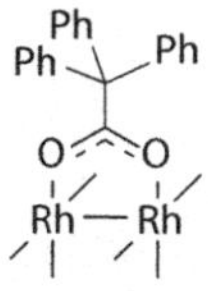

3.3.2 Steric Modifications

On the other hand, the importance of ligand sterics has been confirmed through multiple reports not only on chemo- and regioselectivity but also on the enantioselectivity of the catalyst.[31] In the example illustrated in Table 3.2, a completely opposite reactivity was observed when ligand sterics was altered from methyl in Rh$_2$(OAc)$_4$ to triphenylmethyl in Rh$_2$(tpa)$_4$.[102] Aryl C–H insertion was the exclusive pathway observed with Rh$_2$(tpa)$_4$ (Table 3.2, entry 2). This was opposite to cyclopropanation being the preferred mode of reactivity in competition with aryl C–H insertion in the Rh$_2$(OAc)$_4$-catalyzed reaction (Table 3.2, entry 1).

Using the ligand's steric profile for controlling, predicting, and justifying the observed selectivity of a dirhodium(II) catalyst is essentially the fundamental pathway employed in the field of dirhodium(II) development.[22,31] The importance related to the exploration and manipulation of ligand sterics in chiral dirhodium(II) development will be manifested within the upcoming sections.

3.4 DIRHODIUM(II) CARBOXYLATE COMPLEXES

3.4.1 Conformations in Dirhodium(II) Carboxylate Complexes

In general, conformation of chiral dirhodium(II) paddlewheel complexes is believed to be a critical factor in their chemistry, and this topic was previously reviewed by Hansen and Davies.[22] In most dirhodium(II) catalysts, the ligand-blocking groups would have a considerable conformational mobility. However, due to steric constraints, these blocking groups either adapt an "up (α)" or "down (β)" conformation (Figure 3.9). The blocking groups cannot lie in the periphery of the catalyst as it would bump into the adjacent ligand. Based on this and by considering the α- and β-arrangement for all four ligands, four possible conformations may be generated: α,α,α,α (C_4-symmetry), α,α,α,β (C_1-symmetry), α,α,β,β (C_2-symmetry), and α,β,α,β (D_2-symmetry)[22,26,91] (Figure 3.9).

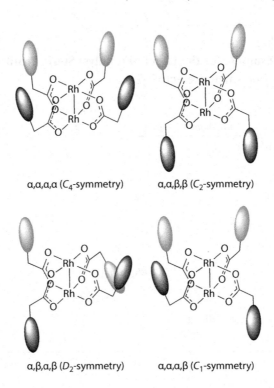

FIGURE 3.9 Models for different ligand arrangements (the sterically blocking groups around the rhodium-active sites are depicted as ovals).[22]

3.4.2 Dirhodium(II) Catalysts Derived from Proline-Based Ligands

Mckervey and coworkers[103–105] were the first to investigate N-protected proline derivatives as ligands for dirhodium(II) catalysts (**9**, Figure 3.10). The significance of this type of carboxylate ligands, however, was not acknowledged until Davies et al.[26,91] prepared a series of prolinate-based chiral dirhodium(II) catalysts (**10–20**, Figure 3.10) and found that the long aliphatic chain variant, Rh$_2$(S-DOSP)$_4$ (**11**), was an outstanding catalyst among the obtained series.

The high levels of asymmetric induction observed with Rh$_2$(S-DOSP)$_4$ have been postulated to emanate from its favored D_2-symmetrical arrangement of ligands in solution (the N-dodecylarylsulfonyl groups are aligned in an α,β,α,β arrangement) (Figure 3.11). This generates a complex with two identical rhodium-active sites with sufficient sterically overburden moieties that can limit the trajectories approaching the axial carbene ligand.[22,26,91]

The utility of Rh$_2$(S-DOSP)$_4$ was expanded by the discovery that it is an extraordinary catalyst for reactions involving *donor–acceptor*-substituted carbenoids.[34,91,106–110] For example, in the cyclopropanation reaction of styrene with styryldiazoacetate, Rh$_2$(S-DOSP)$_4$ was the optimum catalyst (Table 3.3). Even at −78°C, Rh$_2$(S-DOSP)$_4$ led to the generation of the cyclopropane product in 98% *ee*. This chemistry was effectively employed in the total synthesis of (+)-sertraline[111] and cyclopropyl amino acids.[91] The high asymmetric induction of Rh$_2$(S-DOSP)$_4$ is, however, limited to *donor–acceptor* carbenoid systems.[26]

Rh$_2$(S-DOSP)$_4$-catalyzed cyclopropanation of 1,1-diarylethylenes with methyl phenylacetate resulted in the formation of the corresponding cyclopropane products with high enantioselectivity (up to 99% *ee*) and modest diastereoselectivity (up to 80% *de*). This cyclopropanation reaction was successfully employed in the total synthesis of a cyclopropyl analogue of tamoxifen (Scheme 3.2).[110]

A solid-phase version of the same reaction was also reported, which included the cyclopropanation of a resin-bonded alkene with phenyldiazoacetate. The obtained stereoselectivities were almost identical to those observed for the solution-phase reactions (Scheme 3.3).[112]

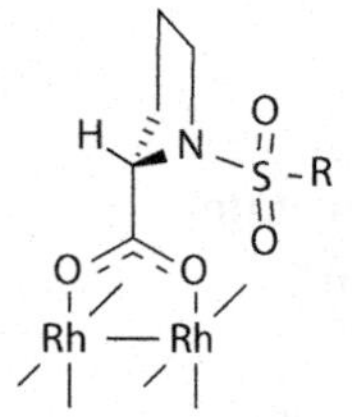

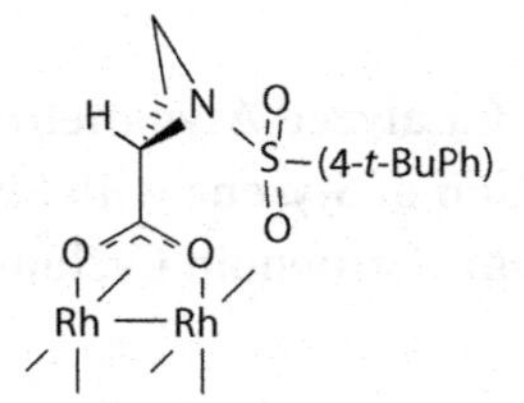

Enantiomerically Pure Compounds by Enantioselective Synthetic Chiral Metal Complexes

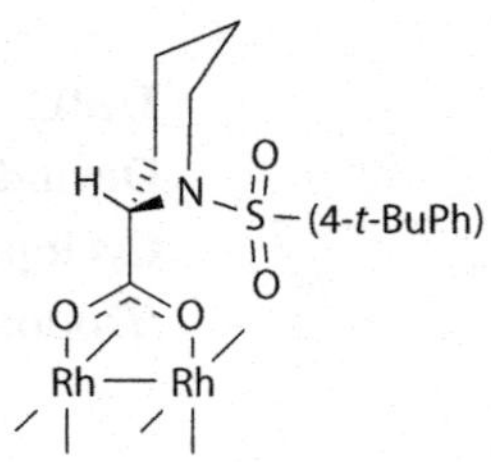

9 R = Ph, [**Rh₂(S-BSP)₄**]
10 R = 4-t-BuC₆H₄, [**Rh₂(S-TBSP)₄**]
11 R = 4-(C₁₂H₂₅)C₆H₄, [**Rh₂(S-DOSP)₄**]
12 R = 4-(MeO)C₆H₄
13 R = 4-(NO₂)C₆H₄
14 R = 3,5-(CF₃)C₆H₃
15 R = 2,4,6-(i-Pr)C₆H₂
16 R = i-Pr

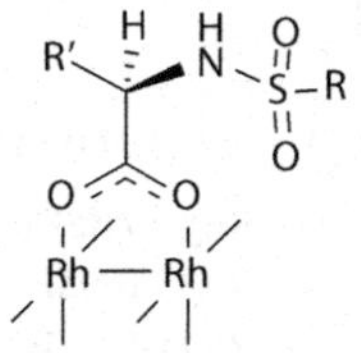

19 R = 4-t-BuC₆H₄, R' = i-Pr
20 R = 4-(Me)C₆H₄, R' = PhCH₂

FIGURE 3.10 Dirhodium(II) carboxylates derived from chiral prolinate ligands (Mckervey complex **9**[103] and Davies complexes **10–20**[91]).[20]

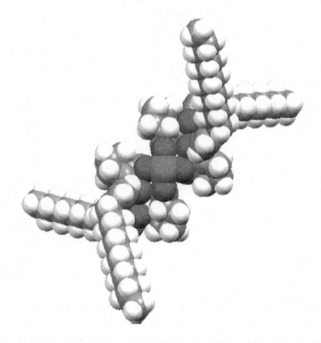

FIGURE 3.11 A 3D model for Rh₂(S-DOSP)₄ (top view). (Reprinted from Hansen, J. and Davies, H.M.L., *Coord. Chem. Rev.*, 252, 545, 2008. Copyright 2016, with permission from Elsevier.)

The extent of this chemistry was even more expanded when Rh₂(S-DOSP)₄-catalyzed decomposition of heteroaryldiazoacetates led to highly diastereoselective and enantioselective cyclopropanations (up to 97% *ee*). Heteroaryldiazoacetates carrying both electron-rich and electron-deficient heterocycles, including thiophene, furan, pyridine, indole, oxazole, isoxazole, and benzoxazole, were suitable for this transformation (Scheme 3.4).[108]

TABLE 3.3
Dirhodium(II)-Catalyzed Asymmetric Cyclopropanation of Styrene with Styryldiazoacetate (*Donor–Acceptor* Carbenoid Cyclopropanation)[26]

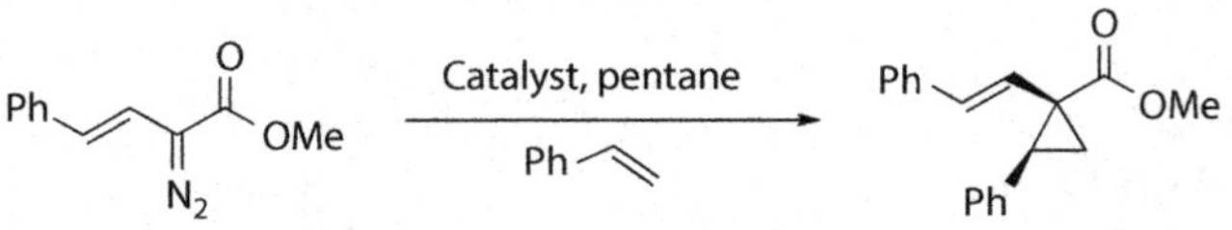

Entry	Catalyst	Temp. (°C)	ee (%)
1	Rh$_2$(*S*-TBSP)$_4$ (**10**)	25	90
2	Rh$_2$(*S*-DOSP)$_4$ (**11**)	25	92
3	Rh$_2$(*S*-DOSP)$_4$ (**11**)	−78	98

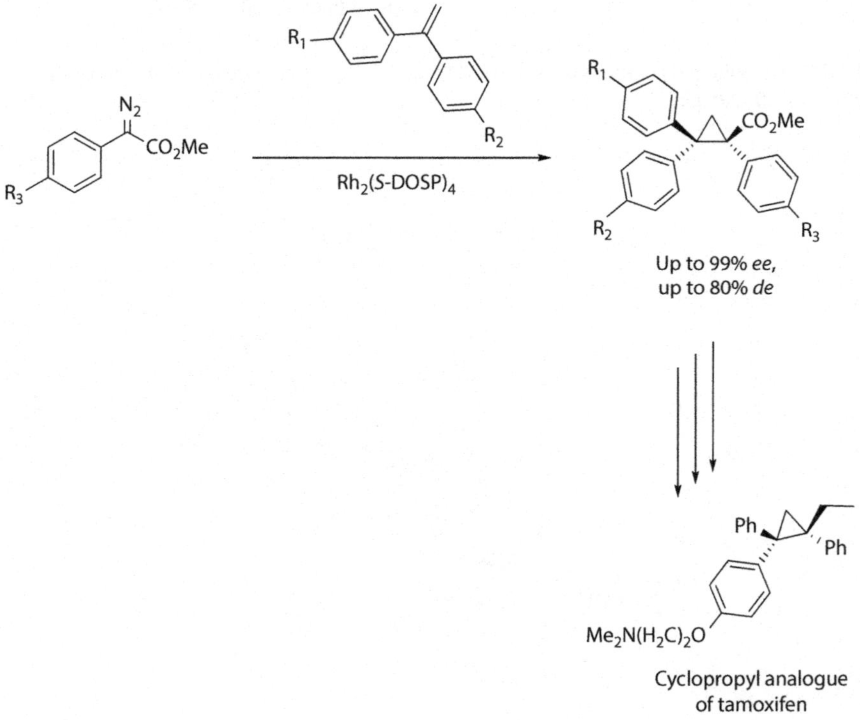

Up to 99% *ee*, up to 80% *de*

Cyclopropyl analogue of tamoxifen

SCHEME 3.2 Asymmetric total synthesis of cyclopropyl analogue of tamoxifen.[110]

This chemistry was even extended to include Rh$_2$(*S*-DOSP)$_4$-catalyzed decomposition of alkynyldiazoacetates. The reaction resulted in the formation of the corresponding alkynyl-substituted cyclopropanes in good-to-excellent enantioselectivity (Scheme 3.5).[113]

In 2006, the complementary nature of Rh(II)- and Pd(II)-catalyzed reactions was also featured. Aryldiazoacetates carrying reactive group for palladium(II) cross-coupling transformations (iodide, triflate, organoboron, and organostannane) can undergo effective enantioselective dirhodium(II)-catalyzed cyclopropanations with no interference from this extra functionality.[114]

It was also reported that Rh$_2$(*R*-DOSP)$_4$ (the enantiomer of **11**, Figure 3.10) can be utilized for the decomposition of aryldiazoacetates in the presence of pyrroles or furans to generate the corresponding

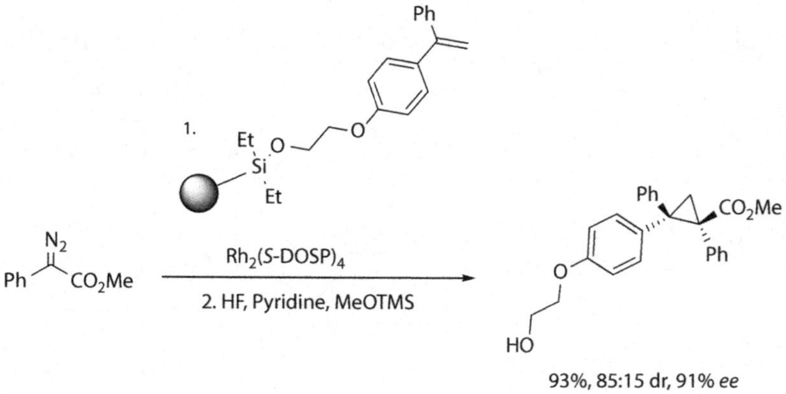
Ph
1.
Et
O
O
Si
Et
N2
Ph
CO2Me
Rh2(S-DOSP)4
2. HF, Pyridine, MeOTMS
Ph
CO2Me
Ph
O
HO
93%, 85:15 dr, 91% ee

SCHEME 3.3 Solid-phase cyclopropanation of a resin-bonded olefin with phenyldiazoacetate.[112]

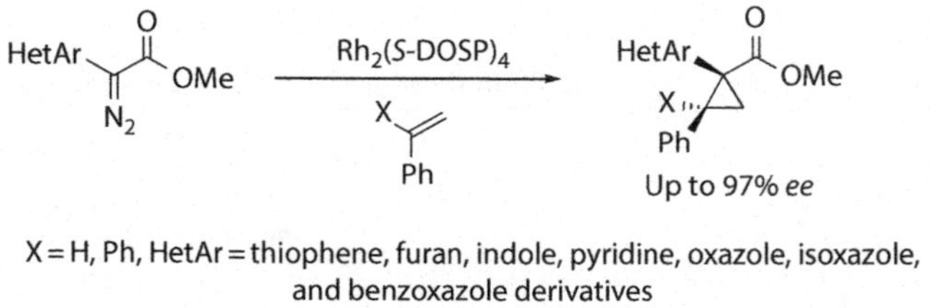
HetAr
O
OMe
N2
Rh2(S-DOSP)4
X
Ph
HetAr
O
OMe
X
Ph
Up to 97% ee
X = H, Ph, HetAr = thiophene, furan, indole, pyridine, oxazole, isoxazole,
and benzoxazole derivatives

SCHEME 3.4 Asymmetric cyclopropanations involving heteroaryldiazoacetates.[108]

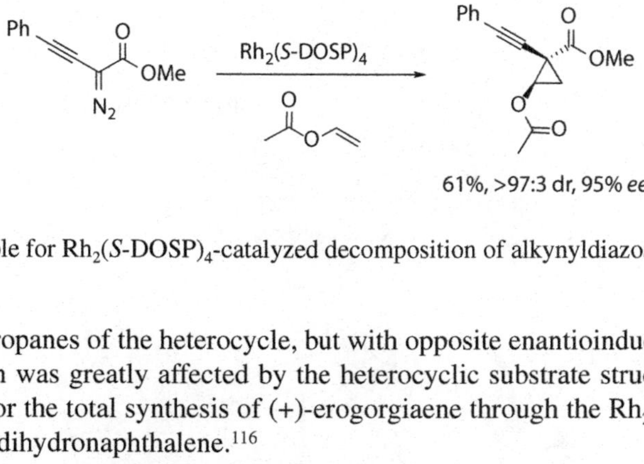
Ph
O
OMe
N2
Rh2(S-DOSP)4
O
O
Ph
O
OMe
O
O
61%, >97:3 dr, 95% ee

SCHEME 3.5 Example for Rh_2(*S*-DOSP)$_4$-catalyzed decomposition of alkynyldiazoacetates.[113]

mono- or bis-cyclopropanes of the heterocycle, but with opposite enantioinduction (Scheme 3.6).[115] The enantioinduction was greatly affected by the heterocyclic substrate structure. This methodology was employed for the total synthesis of (+)-erogorgiaene through the Rh_2(*S*-DOSP)$_4$-catalyzed cyclopropanation of dihydronaphthalene.[116]

In 2013, a study that provides guidelines for selecting the optimal chiral dirhodium(II) catalyst for cyclopropanation of substituted aryldiazoacetates was executed. The results indicated that Rh_2(*S*-DOSP)$_4$ would not result in high enantioselectivity with all aryldiazoacetates. Rh_2(*S*-DOSP)$_4$, however, was found to be the most effective catalyst for the broadest range of substituted methyl aryldiazoacetates.[109]

A general model has been suggested to account for the observed outcomes of dirhodium(II) prolinate-catalyzed asymmetric cyclopropanations (Figure 3.12). In this model, these catalysts are believed to adapt the D_2-symmetrical conformation. The *si*-face of the carbenoid is protected by a sulfonyl ligand behaving as a blocking group. The alkene will approach from the *re*-face over the electron-donating group (EDG), to generate the observed configuration.[26,91,108,110]

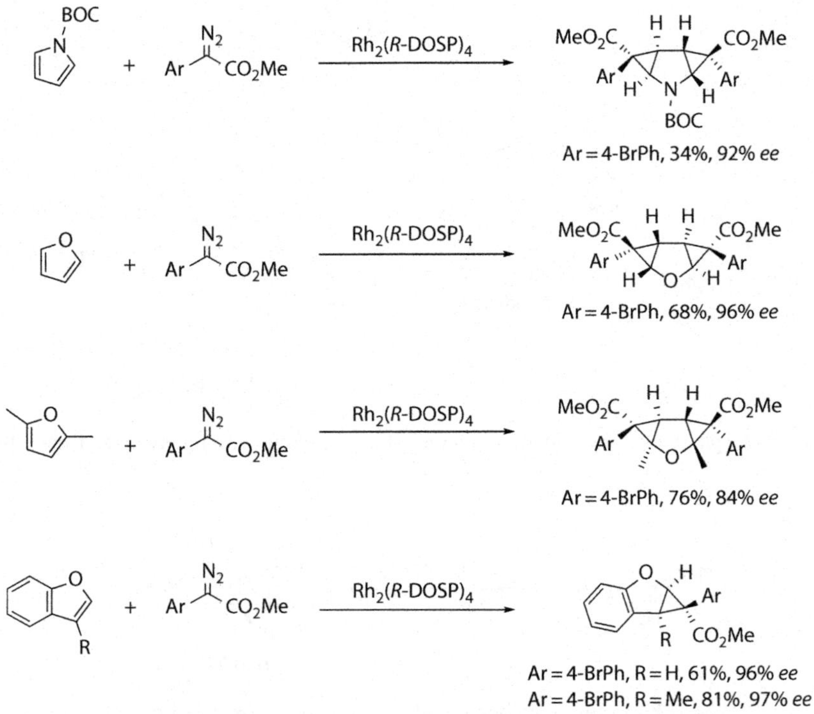

SCHEME 3.6 Rh$_2$(R-DOSP)$_4$-catalyzed decompositions of aryldiazoacetates in the presence of pyrroles or furans.[115]

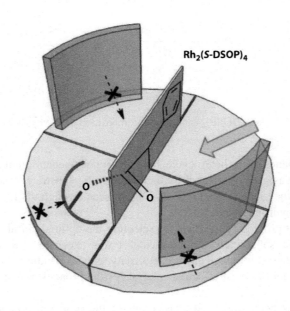

FIGURE 3.12 A model for the asymmetric induction by dirhodium(II) (S)-prolinate catalysts.[26,91,108,110] (Adapted from Davies, M.L.H. and Morton, D., *Chem. Soc. Rev.*, 40, 1857, 2011. With permission of the Royal Society of Chemistry.)

TABLE 3.4
Cyclopropanation of Allenes with *p*-Bromophenyldiazoacetate[117,118]

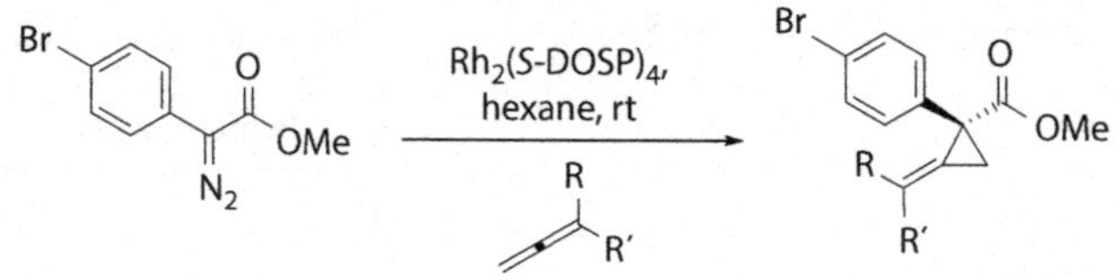

Entry	R	R'	Yield (%)	ee (%)
1	Ph	H	76	90
2	*p*-ClC$_6$H$_4$	H	61	84
3	C$_5$H$_{11}$	H	60	88
4	CH$_2$PhBn	H	54	>80
5	Ph	CH$_3$Me	33	86
6	CH$_3$Me	CH$_3$Me	30	90
7	(CH$_3$)$_3$SiTMS	CH$_3$Me	79	85

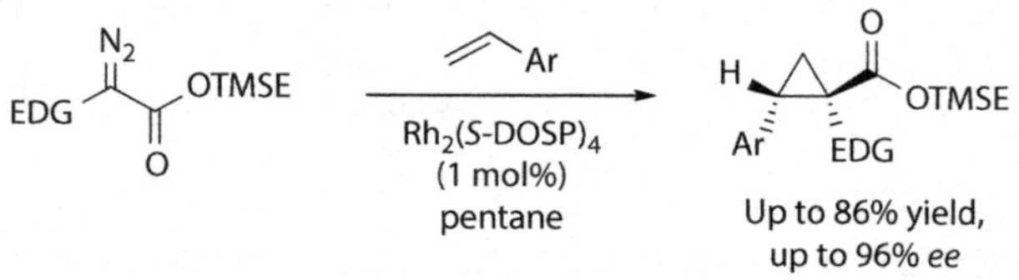

SCHEME 3.7 Chiral Rh$_2$(*S*-DOSP)$_4$-catalyzed cyclopropanation of TMSE aryldiazoesters/styryldiazoacetates and styrenes.[119]

Gregg et al.[117,118] explored the Rh$_2$(*S*-DOSP)$_4$-catalyzed enantioselective cyclopropanation of allenes with aryldiazoacetate esters. The reaction generated the corresponding alkylidene cyclopropane product in 80%–90% *ee* (Table 3.4).

Recently, highly functionalized cyclopropanecarboxylates were readily prepared by dirhodium(II)-catalyzed cyclopropanation of alkenes with trimethylsilylethyl (TMSE) aryldiazoacetates and styryldiazoacetates at which Rh$_2$(*S*-DOSP)$_4$ was the optimum catalyst for this particular transformation (Scheme 3.7).[119]

Rh$_2$(*S*-DOSP)$_4$ also proved to be an excellent catalyst in the enantioselective cyclopropenation of alkynes by aryldiazoacetates to afford cyclopropenes carrying quaternary stereocenters (Scheme 3.8a).[120] Furthermore, Rh$_2$(*S*-DOSP)$_4$-catalyzed cyclopropanation of alkylstyrenes with vinyldiazoacetates at −45°C proceeds smoothly with high levels of enantioinduction to the corresponding cyclopropene product containing quaternary carbon atoms. Further under vigorous conditions (DMB, reflux, 2 h), the vinylcyclopropene products will undergo a dirhodium(II)-catalyzed ring expansion reaction to corresponding cyclopentadienes (Scheme 3.8b).[121]

Computational data showed that the high level of enantioselectivity associated with this transformation is controlled by the orientation of the alkyne substrate while approaching the carbenoid through a relatively late transition state (Figure 3.13). This specific orientation occurs due to the presence of a hydrogen-bonding interaction between the alkyne hydrogen and a carboxylate ligand on the dirhodium(II) catalyst.[121]

To further proof the D_2-symmetry hypothesis, second-generation prolinate-based dirhodium(II) complexes were designed and prepared (Figure 3.14). This was achieved by the synthesis of bidentate ligands with two sulfonylprolinates connected together. As the ligands themselves are having a C_2-symmetry, higher D_2-symmetric catalysts were attainable. These new complexes possess a more rigid version of the complexes carrying the C_1-symmetric ligands (the arylsulfonyl groups

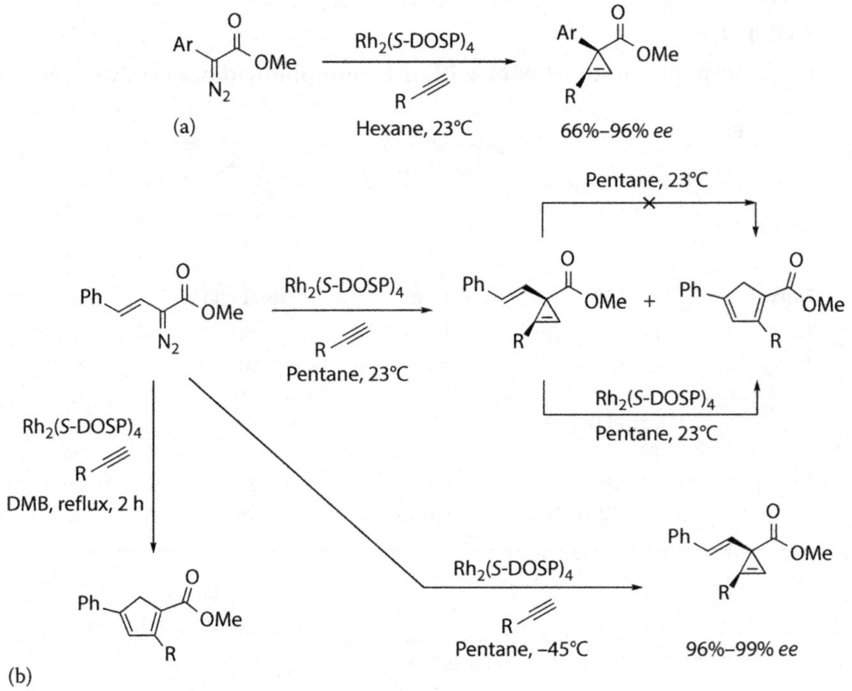

SCHEME 3.8 Rh$_2$(S-DOSP)$_4$-catalyzed synthesis of cyclopropenes. (a) Enantioselective cyclopropenation of alkynes by aryldiazoacetates and (b) cyclopropenation of alkylstyrenes with vinyldiazoacetates.[121]

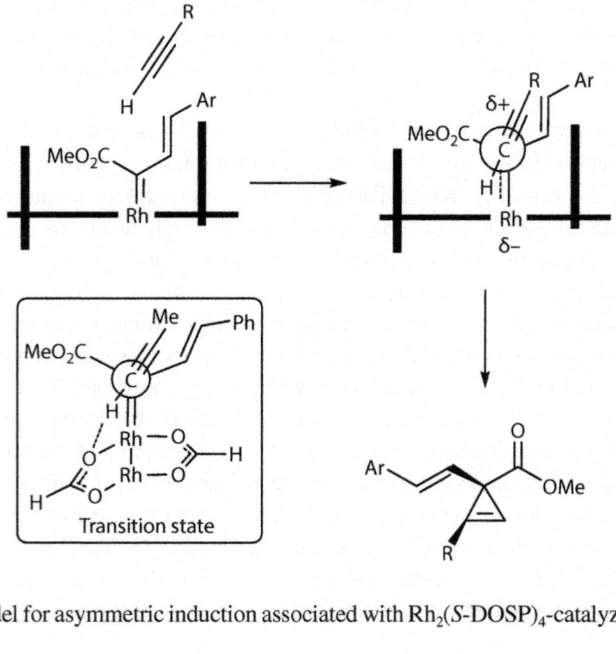

FIGURE 3.13 A model for asymmetric induction associated with Rh$_2$(S-DOSP)$_4$-catalyzed cyclopropenation.[121]

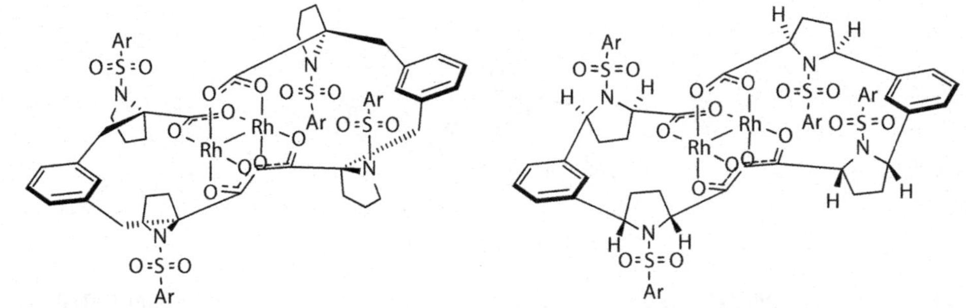

Ar = 4-t-BuC$_6$H$_4$

Ar = 4-t-BuC$_6$H$_4$, [Rh$_2$(S-biTBSP)$_2$]
Ar = 4-C$_{12}$H$_{25}$C$_6$H$_4$, [Rh$_2$(S-biDOSP)$_2$]
Ar = 2,4,6-tri-i-prC$_6$H$_4$, [Rh$_2$(S-biTISP)$_2$]

FIGURE 3.14 Second-generation prolinate-based complexes.[122,123]

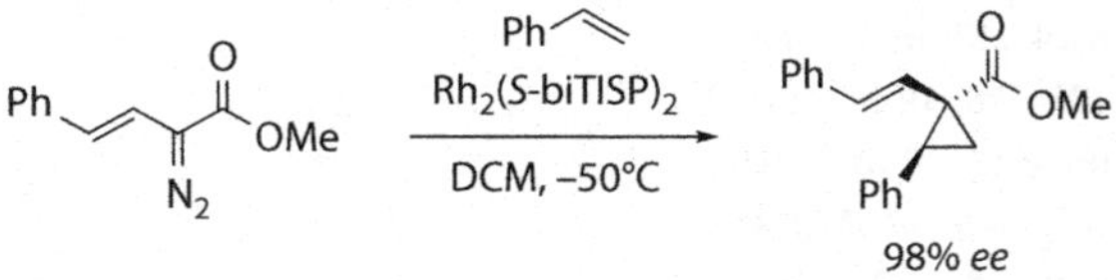

SCHEME 3.9 Example for Rh$_2$(S-biTISP)$_2$-catalyzed asymmetric cyclopropanation.[123]

are conformationally locked in the α,β,α,β arrangement around the rhodium-active center).[122,123] For these new catalysts, not only both catalyst faces are equivalent but also all staggered binding orientations of the axial substrate ligands are identical with respect to the approaching substrate.[22,123]

The successful demonstration that Rh$_2$(S-biTISP)$_2$ is an excellent catalyst in enantioselective cyclopropanation (Scheme 3.9) added more support to the proposed concept that the selectivity of Rh$_2$(S-DOSP)$_4$ is due to the D_2-symmetric arrangement of its ligands around the rhodium core.[123]

In contrast to intermolecular cyclopropanation, dirhodium(II) prolinates result in modest enantioselectivities for intramolecular cyclopropanation reactions with *donor–acceptor* carbenoids. For example, intramolecular Rh$_2$(S-DOSP)$_4$-catalyzed cyclopropanation of allyl vinyldiazoacetate generated the corresponding fused cyclopropyl lactone in 72% *ee* (Scheme 3.10).[124]

In 2015, Su et al.[125] reported a series of dirhodium(II) tetrakis[(4S)-3-(arylsulfonyl)oxazolidine-4-carboxylate], dirhodium tetrakis[(4S,5R)-5-methyl-3-(arylsulfonyl)oxazolidine-4-carboxylate], and dirhodium tetrakis[(4R)-3-(arylsulfonyl)thiazolidine-4-carboxylate 1,1-dioxide] complexes with different *para*-substituted arylsulfonyl groups derived from L-serine, L-threonine, and L-cysteine, respectively (Figure 3.15).

The prepared complexes were tested in the asymmetric aziridination and cyclopropanation reactions. Results indicated that the heterocyclic ring and the substituents on the arylsulfonyl group have critical effects on the degree of asymmetric induction. In general, a higher enantioselectivity was

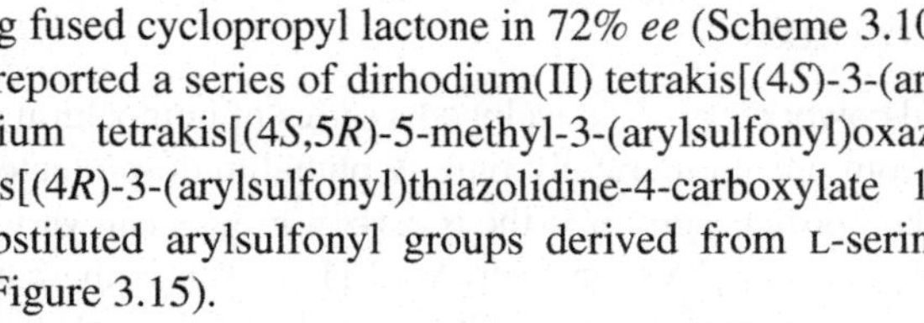

SCHEME 3.10 Example for intramolecular Rh$_2$(S-DOSP)$_4$-catalyzed cyclopropanation of allyl vinyldiazoacetate.[124]

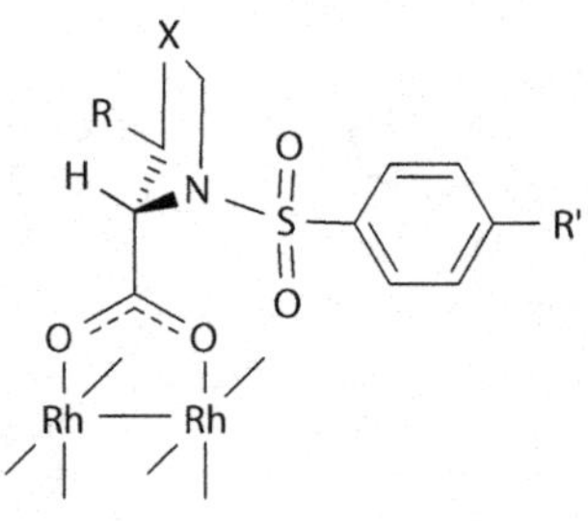

X = O, R = H, R' = NO₂, **[Rh₂(4S-NOSO)₄]**
X = O, R = H, R' = F, **[Rh₂(4S-FLSO)₄]**
X = O, R = H, R' = CF₃, **[Rh₂(4S-TFSO)₄]**
X = O, R = H, R' = Me, **[Rh₂(4S-MESO)₄]**
X = O, R = H, R' = t-Bu, **[Rh₂(4S-TBSO)₄]**
X = O, R = H, R' = MeO, **[Rh₂(4S-MOSO)₄]**
X = O, R = H, R' = n-C₁₂H₂₅, **[Rh₂(4S-DOSO)₄]**
X = O, R = Me, R' = NO₂, **[Rh₂(4S, 5R-MNOSO)₄]**
X = O, R = Me, R' = F, **[Rh₂(4S, 5R-MFLSO)₄]**
X = O, R = Me, R' = CF₃, **[Rh₂(4S, 5R-MTFSO)₄]**
X = O, R = Me, R' = Me, **[Rh₂(4S, 5R-MMESO)₄]**
X = O, R = Me, R' = t-Bu, **[Rh₂(4S, 5R-MTBSO)₄]**
X = O, R = Me, R' = MeO, **[Rh₂(4S, 5R-MMOSO)₄]**
X = O, R = Me, R' = n-C₁₂H₂₅, **[Rh₂(4S, 5R-MDOSO)₄]**

X = SO₂, R = H, R' = NO₂, **[Rh₂(4R-NOST)₄]**
X = SO₂, R = H, R' = F, **[Rh₂(4R-FLST)₄]**
X = SO₂, R = H, R' = CF₃, **[Rh₂(4R-TFST)₄]**
X = SO₂, R = H, R' = Me, **[Rh₂(4R-MEST)₄]**
X = SO₂, R = H, R' = t-Bu, **[Rh₂(4R-TBST)₄]**
X = SO₂, R = H, R' = MeO, **[Rh₂(4R-MOST)₄]**
X = SO₂, R = H, R' = n-C₁₂H₂₅, **[Rh₂(4R-DOST)₄]**

FIGURE 3.15 Structure of Su prolinate-based complexes.[125]

observed with oxazolidine-4-carboxylate-based catalysts compared to thiazolidine-4-carboxylate 1,1-dioxide-based catalysts. Among the prepared complexes, Rh₂(4S-DOSO)₄ and Rh₂(4S,5R-MNOSO)₄ resulted in the highest levels of enantioselectivity in aziridination (94% *ee*) and cyclopropanation (98% *ee*) of styrene, respectively.

3.5 DIRHODIUM(II) CATALYSTS DERIVED FROM CHIRAL N-PROTECTED AMINO ACID LIGANDS

3.5.1 Homoleptic Complexes

Hashimoto, Ikegami, and coworkers[126–133] developed a series of homochiral dirhodium(II) carboxylate complexes derived from enantiomerically pure *N*-phthalimido-protected L-amino acid ligands (**21–35**, Figure 3.16). The optimum group at the α-carbon groups can vary depending on the reaction,[22] but, in general, the *tert*-butyl variant, Rh₂(*S*-PTTL)₄ (**32**), is the catalyst with the broadest application in asymmetric cyclopropanations. Other catalysts were also introduced by vertically extending the length of the *N*-phthalimide moiety (**36–39**, Figure 3.16).[21,127]

For this particular family of catalysts, there are still ambiguities that surround the arrangement of ligands during catalysis in solution. These uncertainties led to doubts related to their mechanism of asymmetric discrimination during carbenoid transformations. As a consequence, a number of proposals have emerged trying to justify the enantioselection mechanisms during dirhodium(II)-catalyzed reactions. The first model was proposed by Hashimoto[128,139,140], and it was proposed after the x-ray crystal structure of Rh₂(*S*-PTPA)₄ was determined. Rh₂(*S*-PTPA)₄ was found to have two adjacent *N*-phthalimido rings oriented on the upper face of the complex, while the other two are oriented toward its lower face in solid state. In Hashimoto's model, it was proposed that dirhodium(II) carboxylates derived from *N*-phthalimido-protected amino acid ligands adapt preferentially the α,α,β,β conformation during catalysis (Figure 3.17).

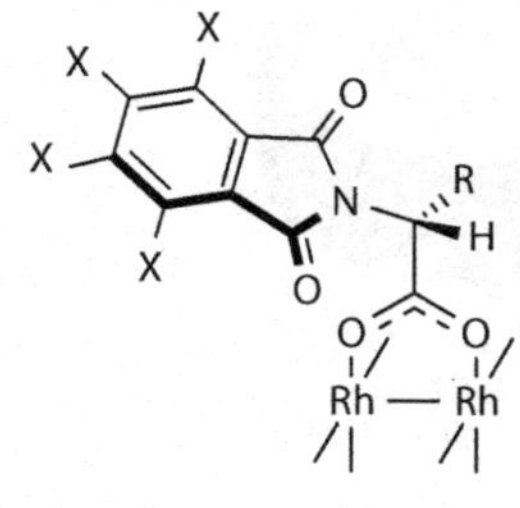

21 R = PhCH$_2$, X = H [**Rh$_2$(R-PTPA)$_4$**]
22 R = t-Bu, X = H [**Rh$_2$(R-PTTL)$_4$**]
23 R = t-Bu, X = Cl, [**Rh$_2$(R-TCPTTL)$_4$**]

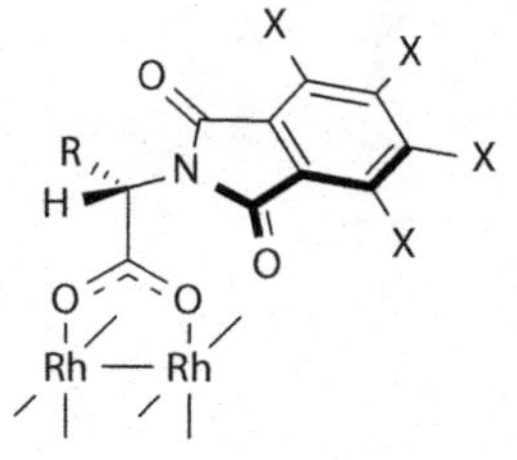

24 R = PhCH$_2$, X = H, [**Rh$_2$(S-PTPA)$_4$**]
25 R = PhCH$_2$, X = Cl, [**Rh$_2$(S-TCPTPA)$_4$**]
26 R = Me, X = H, [**Rh$_2$(S-PTA)$_4$**]
27 R = Me, X = Cl, [**Rh$_2$(S-TCPTA)$_4$**]
28 R = Et, X = H, [**Rh$_2$(S-PTTEA)$_4$**]
29 R = i-Pr, X = H, [**Rh$_2$(S-PTV)$_4$**]
30 R = i-Pr, X = Cl, [**Rh$_2$(S-TCPTV)$_4$**]
31 R = i-Pr, X = Br, [**Rh$_2$(S-TBPTV)$_4$**]
32 R = t-Bu, X = H, [**Rh$_2$(S-PTTL)$_4$**]
33 R = t-Bu, X = Cl, [**Rh$_2$(S-TCPTTL)$_4$**]
34 R = t-Bu, X = F, [**Rh$_2$(S-TFPTTL)$_4$**]
35 R = t-Bu, X = Br, [**Rh$_2$(S-TBPTTL)$_4$**]

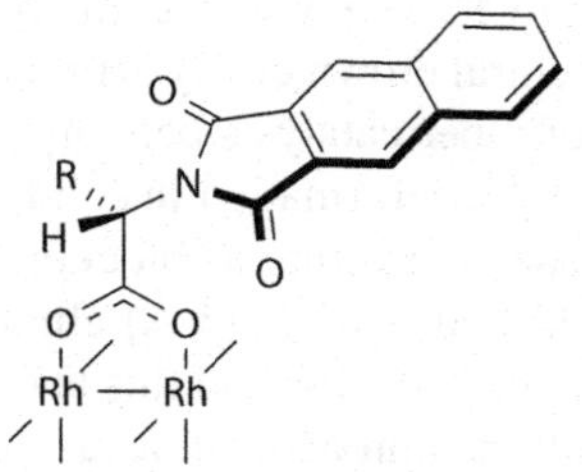

36 R = PhCH$_2$, [**Rh$_2$(S-BPTPA)$_4$**]
37 R = Me, [**Rh$_2$(S-BPTA)$_4$**]
38 R = i-Pr, [**Rh$_2$(S-BPTV)$_4$**]
39 R = t-Bu, [**Rh$_2$(S-BPTTL)$_4$**]

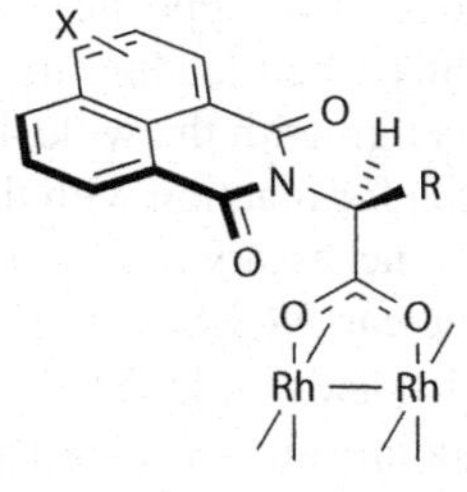

40 R = Me, X = H, [**Rh$_2$(S-NTV)$_4$**]
41 R = t-Bu, X = H, [**Rh$_2$(S-NTTL)$_4$**]
42 R = t-Bu, X = 4-Cl, [**Rh$_2$(S-4-Cl-NTTL)$_4$**]
43 R = t-Bu, X = 3-Cl, [**Rh$_2$(S-3-Cl-NTTL)$_4$**]
44 R = t-Bu, X = 4-Br, [**Rh$_2$(S-4-Br-NTTL)$_4$**]
45 R = t-Bu, X = 4-NO$_2$, [**Rh$_2$(S-4-NO$_2$-NTTL)$_4$**]
46 R = t-Bu, X = 3-NO$_2$, [**Rh$_2$(S-3-NO$_2$-NTTL)$_4$**]
47 R = PhCH$_2$, X = H, [**Rh$_2$(S-NTPA)$_4$**]
48 R = PhCH$_2$, X = 4-Cl, [**Rh$_2$(S-4-Cl-NTPA)$_4$**]
49 R = PhCH$_2$, X = 3-Cl, [**Rh$_2$(S-3-Cl-NTPA)$_4$**]
50 R = PhCH$_2$, X = 4-Br, [**Rh$_2$(S-4-Br-NTPA)$_4$**]
51 R = PhCH$_2$, X = 4-NO$_2$, [**Rh$_2$(S-4-NO$_2$-NTPA)$_4$**]
52 R = PhCH$_2$, X = 3-NO$_2$, [**Rh$_2$(S-3-NO$_2$-NTPA)$_4$**]

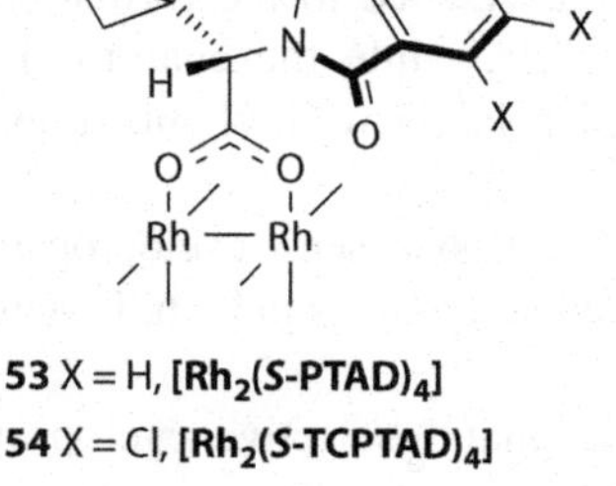

53 X = H, [**Rh$_2$(S-PTAD)$_4$**]
54 X = Cl, [**Rh$_2$(S-TCPTAD)$_4$**]

FIGURE 3.16 Structures of reported chiral dirhodium(II) carboxylates derived from chiral N-protected amino acid ligands (Hashimoto complexes **21–39**, Dauban complex **40**,[134] Müller and Ghanem complexes[135] **41–52**, and Davies complexes **53–54**[136–138]).[20]

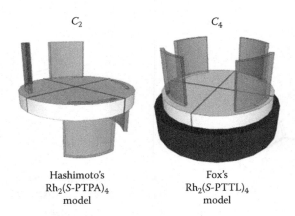

FIGURE 3.17 Models for the distinct ligand orientations used to rationalize the observed enantioselectivity of dirhodium(II) carboxylates derived from N-protected amino acid ligands.[147] (Reprinted with permission from Qin, C., Boyarskikh, V., Hansen, J.H., Hardcastle, K.I., Musaev, D.G., and Davies, H.M.L., *J. Am. Chem. Soc.*, 133, 19198, 2011. Copyright 2016.)

Fox subsequently proposed that these complexes adapt an "all-up" conformation during catalysis. Fox's proposal was based on the x-ray crystal structure of $Rh_2(S\text{-PTTL})_4$ with the four *N*-phthalimido groups oriented toward one face of the complex creating a "chiral crown cavity" (Figure 3.17).[141,142] Other research groups have also reported x-ray structures of other catalysts belonging to the same family, and all of these catalysts were having the same "all-up" conformation in solid state.[135,143–146] According to Fox's model, two opposite *N*-phthalimido groups are acting as blocking walls, while the other two are slightly tilted leading to a narrow (~11 Å) and wide (~15 Å) chiral cavity. The carbene is predicted to align with the wide dimension of the chiral cavity, and this leaves the *si*-face of the carbene accessible for reaction with the alkene *via* end-on approach, while, according to Fox, the *tert*-butyl groups are necessary to limit reactivity to only one of the catalyst faces (Figure 3.17).

Later, 2D Heteronuclear NOESY studies by Charette[144] and Duddeck,[146] independently, confirmed that $Rh_2(S\text{-PTTL})_4$ and similar catalysts have a mobile conformation in solution, which allows the existence of other conformers with at least one *N*-phthaloyl group flipped down. Also Gardiner and Ghanem[148] succeeded to grow crystals of $Rh_2(S\text{-PTPA})_4(EtOAc)_2$ and $Rh_2(S\text{-PTPA})_4(MeOH)(H_2O)$ from ethyl acetate and methanolic solutions of $Rh_2(S\text{-PTPA})_4$, respectively. The authors found that different conformations exist for each of these adducts at which all of them feature inequivalent axial coordination sites unlike the $Rh_2(S\text{-PTPA})_4$ adduct previously reported by Hashimoto. The bis(EtOAc) and MeOH/H₂O adducts exhibit α,α,α,β and α,α,α,α conformations, respectively.

Fox et al.[141] found that $Rh_2(S\text{-PTTL})_4$ is an excellent catalyst for intermolecular cyclopropanation reactions involving α-alkyldiazo compounds. The catalyst led to the formation of desired cyclopropanes with high diastereoselectivity and yield levels, while the enantioselectivity was highly dependent on the structure of the diazoester at which larger α-alkyl substituents afforded increasingly higher enantiomeric excess values (Table 3.5).

Awatta and Arai[149] further reported the $Rh_2(S\text{-PTTL})_4$-catalyzed asymmetric cyclopropanation of diazooxindole affording the corresponding spiro-cyclopropyloxindole products in high diastereoselectivity and moderate to good enantioselectivity (Table 3.6).

Likewise, Charette et al.[144] reported the first catalytic enantioselective cyclopropanation of alkenes with α-nitro diazoacetophenones as part of their *trans*-directing group investigations. $Rh_2(S\text{-TCPTTL})_4$ (**33**, Figure 3.16) proved to be the most suitable catalyst for this kind of asymmetric cyclopropanations. For example, the $Rh_2(S\text{-TCPTTL})_4$-catalyzed cyclopropanation of styrene with α-nitro-α-diazo-*p*-methoxyacetophenone gave the product in 93% *ee* (Table 3.7). The corresponding products obtained from this reaction were used as precursors for the synthesis of optically active *cis*-cyclopropane α-amino acids.

TABLE 3.5
Rh$_2$(S-PTTL)$_4$-Catalyzed Intermolecular Cyclopropanation Reactions of α-Alkyldiazo Compounds[141]

Entry	R	Yield (%)	dr (E:Z)	ee (%)
1	Me	95	91:9	3
2	Et	95	92:8	79
3	n-Pr	100	>95:5	94
4	n-Bu	96	>95:5	96
5	i-Bu	92	>95:5	99

TABLE 3.6
Rh$_2$(S-PTTL)$_4$-Catalyzed Asymmetric Cyclopropanation of Various Olefins with Diazooxindole[149]

Entry	R-Substituent	Yield (%)	dr	ee (%)
1	Ph	>99	98:2	66
2	4-ClC$_6$H$_4$	>99	97:3	65
3	3-ClC$_6$H$_4$	98	98:2	60
4	2-ClC$_6$H$_4$	92	96:4	66
5	4-FC$_6$H$_4$	>99	96:4	64
6	4-MeC$_6$H$_4$	>99	96:4	62
7	4-MeOC$_6$H$_4$MeOC$_6$H$_4$	>99	93:7	48
8	n-C$_3$H$_5$	65	88:12	74

This method was further extended to different *diacceptor* α-EWG-diazoacetophenones bearing an α-PMP-ketone group, as diastereo- and enantioselectivity control group. They were found to be effective carbene precursors for Rh$_2$(S-TCPTTL)$_4$-catalyzed highly stereoselective cyclopropanation of alkenes (Table 3.8).[143]

Rh$_2$(S-TBPTTL)$_4$ (35, Figure 3.16) was also reported as an exceptional catalyst for asymmetric cyclopropanations of 1-aryl-substituted and related conjugated alkenes with *tert*-butyl-α-diazopropionate.[150] High levels of enantioinduction (up to 93% *ee*), as well as virtually full *trans*-diastereoselectivity, were successfully achieved (Table 3.9). This protocol is considered the first example for a catalytic asymmetric cyclopropanation of alkenes with α-diazopropionates and partially complements with the above-discussed Fox cyclopropanation methodology.

TABLE 3.7
Rh$_2$(S-TCPTTL)$_4$-Catalyzed Cyclopropanation of Styrene with α-Nitro-α-Siazo-p-Methoxyacetophenone[144]

Entry	Catalyst	Yield (%)	dr	ee (%)
1	Rh$_2$(S-PTA)$_4$ (26)	55	53:47	22
2	Rh$_2$(S-PTPA)$_4$ (24)	56	55:45	43
3	Rh$_2$(S-PTV)$_4$ (29)	76	72:28	16
4	Rh$_2$(S-PTTL)$_4$ (32)	80	33:67	2
5	Rh$_2$(S-TCPTA)$_4$ (27)	81	97:3	92
6	Rh$_2$(S-TCPTPA)$_4$ (25)	72	94:6	91
7	Rh$_2$(S-TCPTV)$_4$ (30)	82	98:2	91
8	Rh$_2$(S-TCPTTL)$_4$ (33)	81	98:2	93
9	Rh$_2$(S-TBPTV)$_4$ (31)	89	96:4	80
10	Rh$_2$(S-TCPTTL)$_4$ (33)	70	99:1	92
11	Rh$_2$(S-TCPTTL)$_4$ (33)	80	98:2	93

TABLE 3.8
Rh$_2$(S-TCPTTL)$_4$-Catalyzed Cyclopropanation of Several α-EWG-Diazoacetophenones Bearing an α-p-Methoxyphenyl (PMP)-Ketone Group[143]

Entry	EWG	Temp. (°C)	Yield (%)	dr	ee (%)
1	NO$_2$	−50	81	98:2	93
2	CN	−35	98	95:5	84
3	CO$_2$Me	−40	60	99:1	88
4	Ph	−50	9	>95:5	98

Hashimoto et al.[145] expanded the application of Rh$_2$(S-TBPTTL)$_4$ catalyst to include asymmetric cyclopropenation of 1-alkynes with 2,4-dimethyl-3-pentyl-α-alkyl-α-diazoacetates. This transformation afforded high levels of enantioselectivity (up to 99% *ee*), as well as good to high chemoselectivities were achieved (Table 3.10). For example, the Rh$_2$(S-TBPTTL)$_4$-catalyzed cyclopropenation of phenylacetylene with 2,4-dimethyl-3-pentyl-α-diazopropionate afforded the corresponding cyclopropene product in up to 95% *ee*.

Doyle and coworkers[151] studied the Rh$_2$(S-PTTL)$_4$-catalyzed intramolecular cyclopropanation of enoldiazoacetamides. However, the obtained *donor–acceptor* cyclopropene derivatives underwent

TABLE 3.9
Rh$_2$(S-TBPTTL)$_4$-Catalyzed Cyclopropanation of Styrene with α-Diazopropionates[150]

Entry	R	Catalyst	Temp. (°C)	Yield (%)	dr (E:Z)	ee (%)
1	CH(i-Pr)$_2$	Rh$_2$(S-PTTL)$_4$ (32)	−60	89	94:6	47
2	CH(i-Pr)$_2$	Rh$_2$(S-TFPTTL)$_4$ (34)	−60	86	92:8	33
3	CH(i-Pr)$_2$	Rh$_2$(S-TCPTTL)$_4$ (33)	−60	89	95:5	69
4	CH(i-Pr)$_2$	Rh$_2$(S-TBPTTL)$_4$ (35)	−78	86	>99:1	86
5	Et	Rh$_2$(S-TBPTTL)$_4$ (35)	−60	80	91:9	35
6	t-Bu	Rh$_2$(S-TBPTTL)$_4$ (35)	−78	87	>99:1	92

TABLE 3.10
Dirhodium(II)-Catalyzed Cyclopropenation of Phenylacetylene with 2,4-Dimethyl-3-Pentyl-α-Diazopropionate[145]

Entry	Catalyst	Temp. (°C)	Yield (%)	ee (%)
1	Rh$_2$(S-PTTL)$_4$	23	86	46
2	Rh$_2$(S-PTV)$_4$	23	78	36
3	Rh$_2$(S-PTA)$_4$	23	91	38
4	Rh$_2$(S-TFPTTL)$_4$	23	80	35
5	Rh$_2$(S-TCPTTL)$_4$	23	89	72
6	Rh$_2$(S-TBPTTL)$_4$	23	94	85
7	Rh$_2$(S-TBPTTL)$_4$	−40	89	94
8	Rh$_2$(S-TBPTTL)$_4$	−60	90	95

an *in situ* interconversion to the corresponding β-lactam derivatives (Scheme 3.11). The β-lactam derivatives were formed in high yield, exclusive *cis*-diastereoselectivity, and high enantiocontrol.

In 2006, Davies et al.[136] suggested that the logical way for the further development of Rh$_2$(S-PTTL)$_4$ catalyst and analogues is to put a much sterically overburden hydrocarbon than the *tert*-butyl group connected to the α-carbon of the ligand. The authors used the highly enantioselective C–H functionalization chemistry that they developed to access the synthetic amino acid, L-adamantylglycine, in highly enantiomerically pure form[136] (Scheme 3.12). This was then used for the synthesis of Rh$_2$(S-PTAD)$_4$ catalyst (53, Figure 3.16).

Very recently, Adly, Ghanem, and coworkers[152] reported the x-ray crystal structure of Rh$_2$(S-PTAD)$_4$ complex. Rh$_2$(S-PTAD)$_4$ was observed to form a bis(EtOAc) adduct when crystalized from ethyl acetate/*n*-hexane with a full α,α,α,α conformation in solid state (Figure 3.18). This observation confirmed that there is still enough room for a Lewis basic ligand to coordinate to the "achiral" axial rhodium coordination site (the site shrouded by the adamantyl

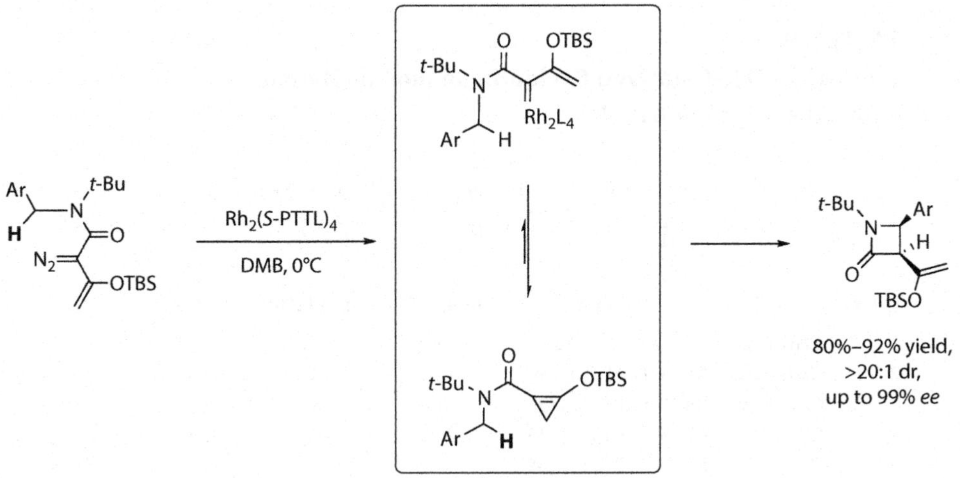

SCHEME 3.11 Enantioselective *cis*-β-lactam synthesis from enoldiazoacetamides and *donor–acceptor* cyclopropene derivatives.[151]

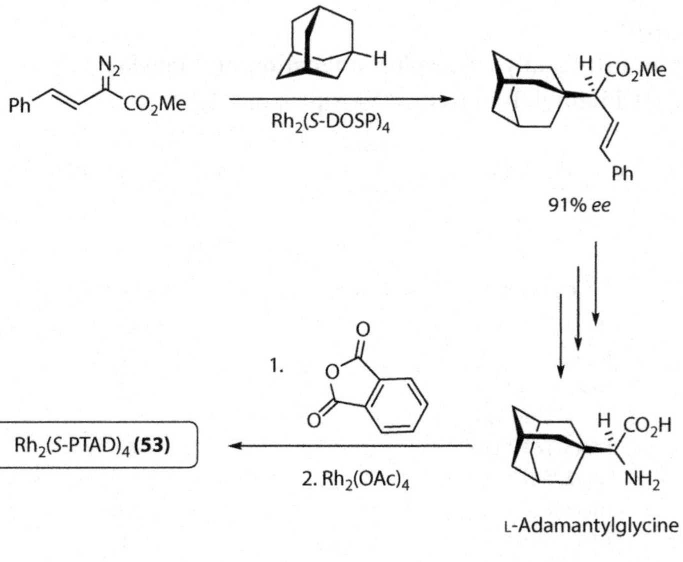

SCHEME 3.12 Asymmetric synthesis of Rh$_2$(*S*-PTAD)$_4$ catalyst (**53**).[136]

substituents) and provided a direct evidence that both Rh atoms can be still accessible by the diazo substrates even after the introduction of the more bulky adamantyl groups. This contradicted with the Fox model, which assumed that the bulky groups at the α-position would limit the reactivity to only one catalyst face.[141,142]

Rh$_2$(*S*-PTAD)$_4$ was utilized in the stereoselective synthesis of dimethyl 1,2-diphenylcyclopropylphosphonate. The results revealed that Rh$_2$(*S*-PTAD)$_4$ was very effective at which the reaction afforded the cyclopropylphosphonate product with high enantioselectivity levels (99% *ee*) compared to Rh$_2$(*S*-DOSP)$_4$ (34% *ee*), Rh$_2$(*S*-biTISP)$_2$ (88% *ee*), and Rh$_2$(*S*-PTTL)$_4$ (97% *ee*) (Table 3.11).[136]

Also, the Rh$_2$(*S*-PTAD)$_4$-catalyzed cyclopropanation of 1-aryl-2,2,2-trifluorodiazoethanes or α-diazo-2-phenylacetonitrile with electron-rich alkenes afforded the corresponding trifluoromethyl-substituted or nitrile-substituted cyclopropanes, respectively, with high diastereoselectivity and enantioselectivity levels (Tables 3.12 and 3.13).[137,153]

Enantiomerically Pure Compounds by Enantioselective Synthetic Chiral Metal Complexes

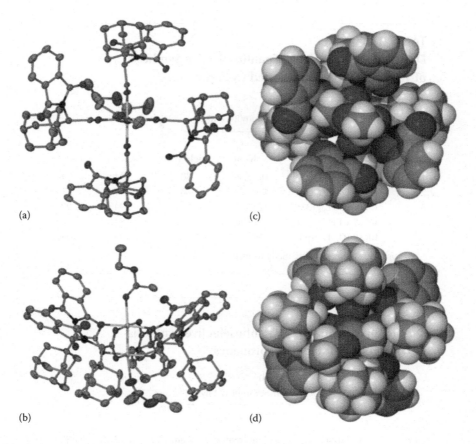

FIGURE 3.18 Molecular structure of bis(EtOAc) adduct of Rh$_2$(S-PTAD)$_4$; (a) viewed into the chiral crown cavity, (b) general view (all hydrogen atoms as well as a second similar molecule and lattice solvent are omitted for clarity). Space-filling representation viewed along the Rh–Rh axis, (c) into the chiral crown cavity, (d) onto the axial Rh coordination site shrouded by the adamantyl groups.[152] (Reprinted from Adly, F.G., Gardiner, M.G., and Ghanem, A., *Chem.-Eur. J.*, 22, 3447, 2016. Copyright 2016, with permission from John Wiley and Sons.)

TABLE 3.11
Enantioselective Synthesis of Dimethyl 1,2-Diphenylcyclopropylphosphonate[136]

Entry	Catalyst	Yield (%)	ee (%)
1	Rh$_2$(S-DOSP)$_4$ (**11**)	69	34
2	Rh$_2$(S-biTISP)$_2$	89	88
3	Rh$_2$(S-PTTL)$_4$ (**32**)	85	97
4	Rh$_2$(S-PTAD)$_4$ (**53**)	86	99

TABLE 3.12
Dirhodium(II)-Catalyzed Enantioselective Synthesis of Trifluoromethyl-Substituted Cyclopropanes[137]

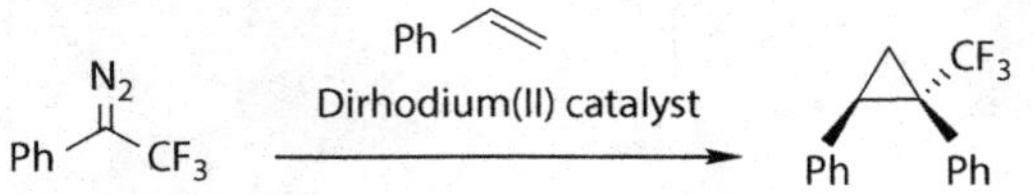

Entry	Catalyst	Solvent	Yield (%)	de (%)	ee (%)
1	Rh$_2$(S-DOSP)$_4$ (11)	Hexanes	80	94	40[a]
2	Rh$_2$(S-DOSP)$_4$ (11)	TFT	60	90	37[a]
3	Rh$_2$(S-PTTL)$_4$ (32)	TFT	95	>94	97
4	Rh$_2$(S-PTTL)$_4$ (32)	CH$_2$Cl$_2$	96	>94	86
5	Rh$_2$(S-PTAD)$_4$ (53)	TFT	94	>94	>98

[a] Opposite enantiomer preferentially formed.

TABLE 3.13
Dirhodium(II)-Catalyzed Enantioselective Synthesis of Nitrile-Substituted Cyclopropanes[153]

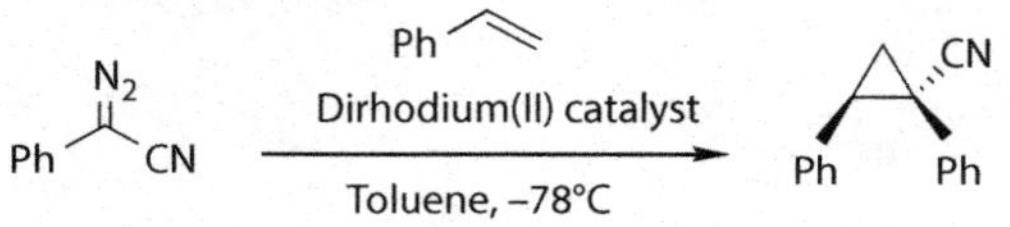

Entry	Catalyst	Loading (mol%)	Yield (%)	dr (E:Z)	ee (%)
1	Rh$_2$(S-DOSP)$_4$ (11)	2	85	95:5	34
2	Rh$_2$(S-PTTL)$_4$ (32)	2	84	96:4	90
3	Rh$_2$(S-PTAD)$_4$ (53)	2	86	97:3	90
4	Rh$_2$(S-PTAD)$_4$ (53)	1	80	97:3	85

Furthermore, the Rh$_2$(S-PTAD)$_4$-catalyzed reaction of a variety of α-aryl-α-diazoketones with activated olefins was reported to afford cyclopropyl ketones with high diastereoselectivity (up to >95:5 dr) and enantioselectivity (up to 98% ee) (Scheme 3.13).[154]

Rh$_2$(S-PTAD)$_4$ was also found to be the optimal catalyst for cyclopropanations involving *ortho*-substituted aryldiazoacetates. The catalyst provided high levels of enantioinduction; in particular, it was very effective with reactions involving 2-chlorophenyl aryldiazoacetate derivative (97% ee) (Scheme 3.14).[109]

It was also reported that aryldiazoacetates and vinyldiazoacetates are capable to undergo high enantioselective cyclopropanations with electron-deficient alkenes (Scheme 3.15). The best

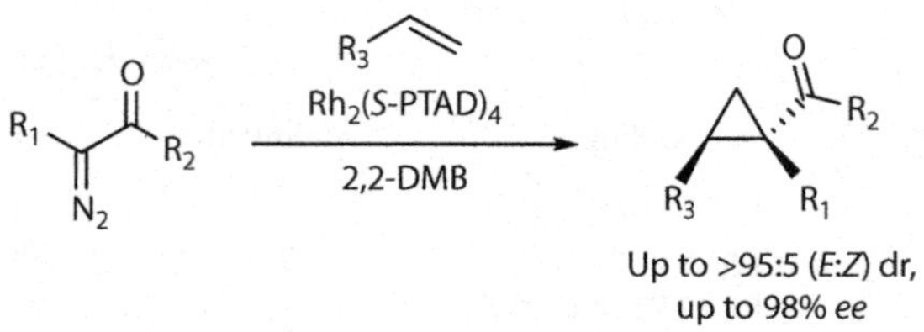

SCHEME 3.13 Rh$_2$(S-PTAD)$_4$-catalyzed cyclopropanation of diazo ketones.[154]

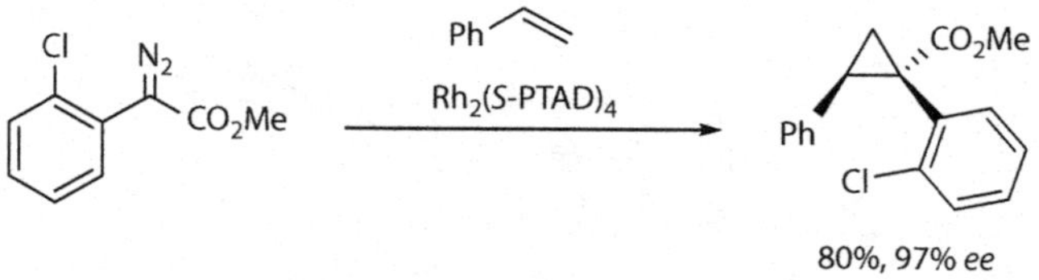

SCHEME 3.14 Rh$_2$(S-PTAD)$_4$-catalyzed enantioselective cyclopropanation of styrene with 2-chlorophenyl aryldiazoacetate derivative.[109]

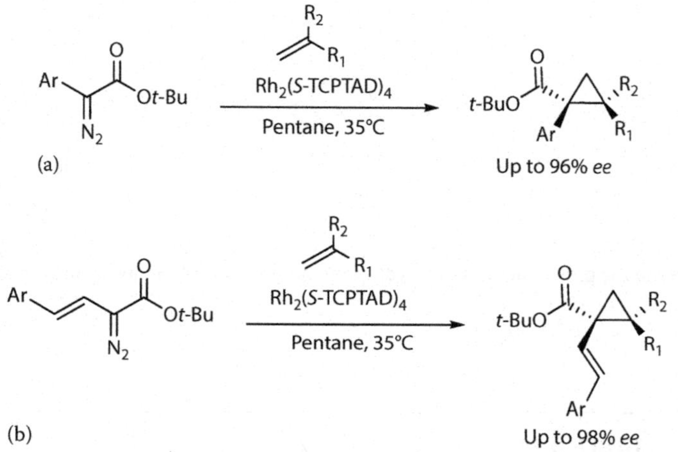

SCHEME 3.15 Rh$_2$(S-TCPTAD)$_4$-catalyzed reaction of aryldiazoacetates and vinyldiazoacetates with electron-deficient alkenes.[138]

catalyst for this high enantioselective transformation was the Rh$_2$(S-PTAD)$_4$ tetrachloro variant, Rh$_2$(S-TCPTAD)$_4$ (**54**, Figure 3.16).[138]

The transformation involves the initial formation of a prereaction complex between the carbene intermediates and the carbonyl group of the substrate, and the final reaction product type depends on the nature of the carbonyl group. Acrylates and acrylamides lead to the generation of cyclopropane products, while unsaturated aldehydes and ketones result in the generation of epoxide products (Scheme 3.16).[138] Computational data showed that, in all cases, the ylide is favorably generated. With acrylates and acrylamides, the ylide formation is reversible and, as a consequence, cyclopropane is eventually observed as a final reaction product. With aldehydes and ketones, however, ylide generation is not reversible because of its quick conversion into epoxide (Scheme 3.16).

Rh$_2$(S-PTAD)$_4$ was also effective for the enantioselective cyclopropenation of aryl alkynes with siloxyvinyldiazoacetate as a carbenoid precursor (Table 3.14).[155] By using such reaction, highly enantioenriched cyclopropenes carrying germinal acceptor groups can be successfully achieved after the removal of the silyl protecting group from the generated cyclopropene product.

Müller and coworkers[156–158] reported similar types of catalysts using N-1,8-naphthaloyl-L-amino acids (**41–52**, Figure 3.16). They also showed the suitability of Rh$_2$(S-NTTL)$_4$ catalyst (**41**, Figure 3.16) for Rh-catalyzed cyclopropanation of styrene with (silanyloxyvinyl)diazoacetates with exceptional diastereo- and enantioselectivity levels (Scheme 3.17a).[159] The scope of the catalyst was further extended to include cyclopropanation of dihydrofuran and dihydropyran (Scheme 3.17b and c).[160,161] When this chemistry was employed with ethyl diazo(triethylsilyl)acetate, it afforded the corresponding cyclopropane product in a good yield (69%), but with modest diastereoselectivity (64% *de*) and enantioselectivity (54% *ee*).[158]

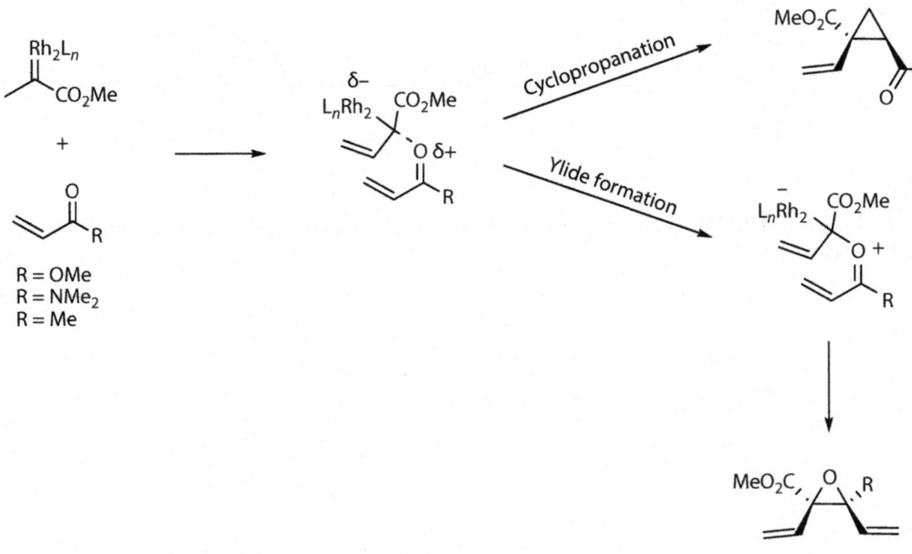

SCHEME 3.16 Schematic presentation of the cyclopropanation, ylide formation, and epoxidation pathways.[138]

TABLE 3.14
Reaction of Siloxyvinyldiazoacetate and Phenylacetylene Using Rh$_2$(S-PTAD)$_4$ Catalyst[155]

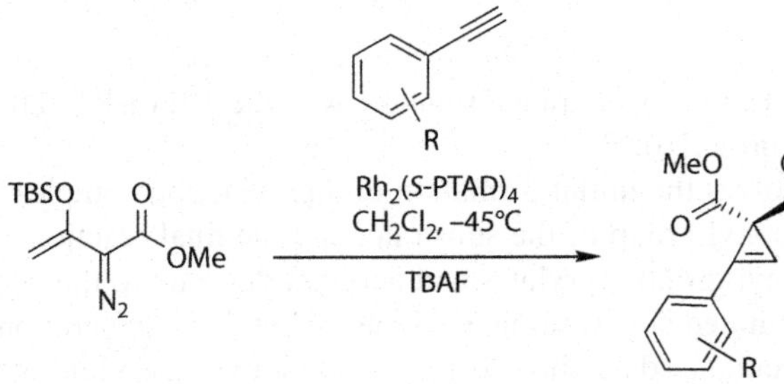

Entry	R	Yield (%)	ee (%)
1	H	83	98
2	*ortho*-Me	80	98
3	*para*-Me	86	93
4	*para*-Et	90	97
5	*para*-tBu	87	98
6	*para*-Br	92	95
7	*para*-Ph	88	98
8	*meta*-CF$_3$	94	93
9	*para*-Ethynyl	85	94
10	*meta*-Ethynyl	88	97
11	*para*-F, *meta*-Me	77	97
12	*ortho*-CH$_2$OTBS	94	99

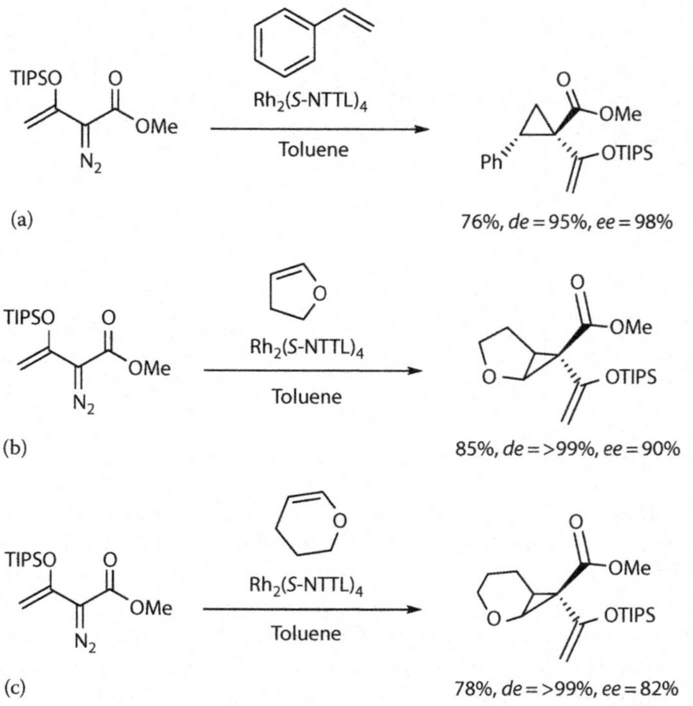

SCHEME 3.17 Rh$_2$(S-NTTL)$_4$-catalyzed cyclopropanations involving (silanyloxyvinyl)diazoacetates.[159–161]

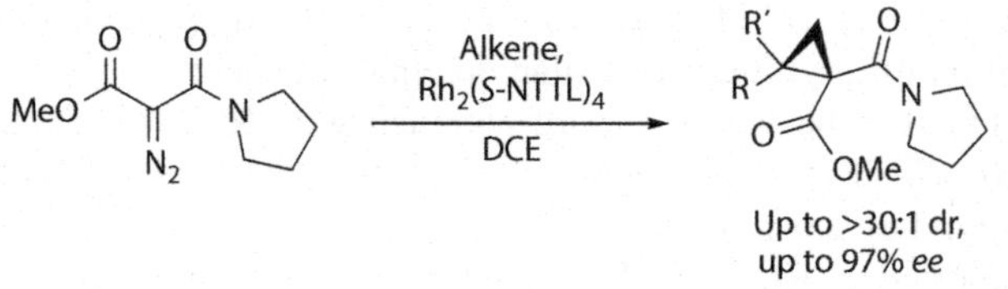

SCHEME 3.18 Rh$_2$(S-NTTL)$_4$-catalyzed enantioselective formation of 1,1-cyclopropane diesters (*trans*-directing group concept).[162–165]

Charette et al.[162–165] described the enantioselective formation of 1,1-cyclopropane diesters *via* Rh$_2$(S-NTTL)$_4$-catalyzed cyclopropanation of olefins (Scheme 3.18). They were the first to elaborate the concept of the *trans*-directing ability of amide groups in Rh(II)-catalyzed cyclopropanation reactions. This concept provided a solution for the stereoselective synthesis of 1,1-dicarboxycyclopropane derivatives.

The authors hypothesized that the in–out conformation for a carbene derived from malonates is operative. Placing one group in the same plane of the metal carbene liberates space for an alkene to approach and enhance the electrophilicity of the carbenoid.[162,163] The out-of-plane substituent can act as a *trans*-directing group and transition states I–IV (Figure 3.19a) would be plausible. Assuming that the use of a chiral catalyst would be effective at blocking the pro-(*S*)-face, they found that the four possible transition state structures would lead to a pair of enantiomers. These postulations may explain the low enantiocontrol obtained to date with Rh(II)-catalyzed cyclopropanation of malonates.

The strategy proposed by Charrette was to use a carbene that possesses two different groups with different *trans*-directing abilities, in combination with a catalyst that would be effective at blocking one of the two prochiral faces. Transition states VII and IX (Figure 3.19b) would not be accessible

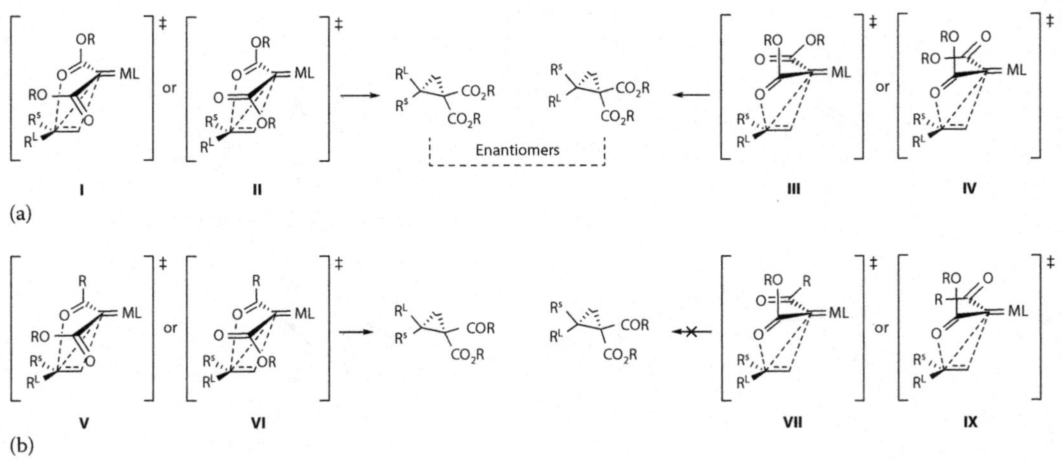

FIGURE 3.19 (a) Rh(II)-catalyzed cyclopropanation transition states of diazomalonates as carbene precursors; (b) the proposed strategy with carbenoids possessing two different groups with different *trans*-directing abilities.[162,163]

due to the greater *trans*-directing ability of the COR group. By applying this strategy and by using Rh$_2$(S-NTTL)$_4$ as a catalyst, the authors succeeded in obtaining the cyclopropane products with enantioselectivities up to 97% *ee* and diastereoselectivities up to >30:1 dr.[162,163]

The potential utility of this property was further illustrated in several functional group transformations and in the stereoselective synthesis of (S)-(+)-curcumene, (S)-(+)-nuciferal, (S)-(+)-nuciferol, (+)-erogorgiaene, (±)-xanthorrhizol, and (±)-2-hydroxycalamenene.[162]

The performance of Rh$_2$(S-NTTL)$_4$, Rh$_2$(S-PTTL)$_4$, and Rh$_2$(S-DOSP)$_4$ catalysts in intramolecular cyclopropanation of allyl 2-diazo-3-silanyloxybut-3-enoates was also examined by Müller et al.[161,166] (Table 3.15). The best results were obtained with Rh$_2$(S-PTTL)$_4$, where 89% *ee* was observed at −78°C. Rh$_2$(S-NTTL)$_4$ was slightly less selective, while Rh$_2$(S-DOSP)$_4$ was found to be not suitable with this kind of substrates.

Fokin et al.[167] reported a Rh$_2$(S-NTTL)$_4$-catalyzed asymmetric cyclopropanation that uses *N*-sulfonyl-1,2,3-triazoles as azavinyl carbene precursors. The azavinyl carbenes readily reacted with various olefins to afford the corresponding cyclopropane carboxaldehydes with very high

TABLE 3.15
Intramolecular Cyclopropanation Involving 1-Phenyl-1-Propenyl-2-Diazo-3-Silanyloxybut-3-Enoates[166]

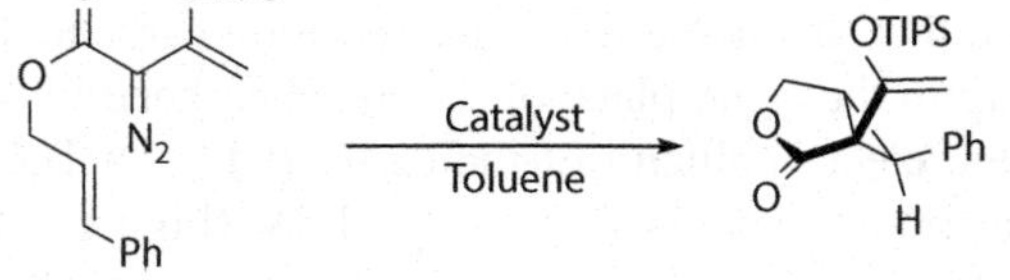

Entry	Catalyst	Temp. (°C)	Yield (%)	ee (%)
1	Rh$_2$(S-NTTL)$_4$ (**41**)	rt	77	73
2	Rh$_2$(S-PTTL)$_4$ (**32**)	rt	93	77
3	Rh$_2$(S-PTTL)$_4$ (**32**)	−78	66	89
4	Rh$_2$(S-DOSP)$_4$ (**11**)	rt	69	5

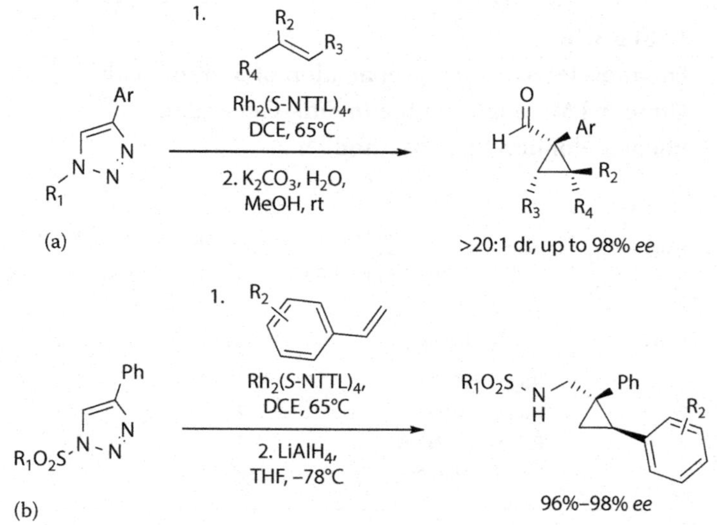

Enantiomerically Pure Compounds by Enantioselective Synthetic Chiral Metal Complexes 105

SCHEME 3.19 Enantioselective cyclopropanation involving 1,2,3-triazoles.[167]

enantioselectivity (Scheme 3.19a). The scope of the reaction with respect to the 1-sulfonyltriazole was investigated and results indicated that substrates possessing both electron-rich and electron-deficient aryl groups at C4 reacted smoothly to generate the desired cyclopropane products in excellent enantioselectivity (up to 98% ee). Moreover, the authors found that LiAlH$_4$ reduction of the imine product immediately after its synthesis is capable to provide an easy access to aminocyclopropanes in both good yields and excellent enantioselectivity (Scheme 3.19b).

In 2004, Müller, Ghanem, and coworkers[135,168] reported several Rh$_2$(S-NTTL)$_4$ analogues at which only one hydrogen on the heterocyclic tether is substituted generating ligands carrying "lower symmetry" N-protecting groups (Figure 3.20). The results revealed that Rh$_2$(S-4-Br-NTTL)$_4$-catalyzed cyclopropanation of styrene with dimethyl malonate proceeded with far improved levels of enantioselectivity (82% ee) compared to its parent, Rh$_2$(S-NTTL)$_4$ (37% ee) (Table 3.16).[168,169] The same catalyst was also effective for olefin cyclopropanation with Meldrum's acid giving 92% ee with styrene and 87% ee with 1-pentene.

The enhanced enantioselectivity levels observed with Rh$_2$(S-4-Br-NTTL)$_4$ were justified through the x-ray crystal structure of Rh$_2$(S-NTTL)$_4$ (Figure 3.21).[135] The x-ray revealed that

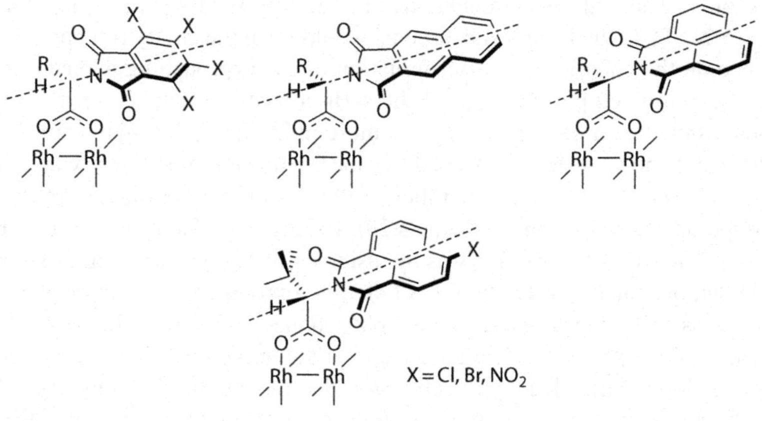

FIGURE 3.20 Ligands' backbone structure comparison.

TABLE 3.16
Enantioselective Cyclopropanation of Styrene with Dimethyl Malonate *via* the *In Situ*–Generated Phenyliodonium Ylide Method[168]

Entry	Catalyst	Yield (%)	ee (%)
1	Rh$_2$(S-NTTL)$_4$	72	37
2	Rh$_2$(S-4-Cl-NTTL)$_4$	77	66
3	Rh$_2$(S-4-Br-NTTL)$_4$	75	82
4	Rh$_2$(S-4-NO$_2$-NTTL)$_4$	60	66

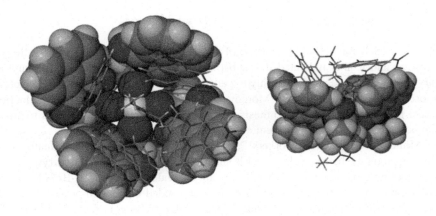

FIGURE 3.21 X-ray structure of Rh$_2$(S-NTTL)$_4$. (Reprinted from Ghanem, A., Gardiner, M.G., Williamson, R.M., and Müller, P., *Chem.-Eur. J.*, 16, 3291, 2010. Copyright 2016, with permission from John Wiley and Sons.)

N-1,8-naphthaloyl groups maintained the chiral nature of the crown cavity through a clockwise twist of these groups. The authors communicated that if Rh$_2$(S-4-Br-NTTL)$_4$ is retaining a similar structure as Rh$_2$(S-NTTL)$_4$, the bromo substituents would lie at the cavity rim and are likely to exert a strong influence on the enantiofacial discrimination of the approaching alkene (cavity rim steric impedance).[135] The improved performance of the 4-Br-substituted catalyst over the 4-Cl analogue was also justified as the larger halide would exert more influence at the cavity rim.

Very recently, Adly and Ghanem[170] explored this local reduction of symmetry approach further by reporting Rh$_2$(S-1,2-NTTL)$_4$ as a new member of the chiral dirhodium(II) family derived from C$_s$-symmetric N-protected *tert*-leucine (Scheme 3.20). The idea of reducing the local ligand's heterocyclic tether symmetry in Rh$_2$(S-1,2-NTTL)$_4$ was by fusing a ring at one side of the N-heterocyclic tether borders. Results demonstrated that Rh$_2$(S-1,2-NTTL)$_4$ is a promising backup catalyst for the cyclopropanation reactions involving *donor–acceptor* phosphonate carbenoids (Scheme 3.20); however, the reported results did not provide a clear advantage for the introduced "lower symmetry" approach.

The authors explored amended approaches for the reduction of symmetry of the ligand's N-heterocyclic tether at which they prepared four more complexes, namely, Rh$_2$(S-*tert*PTTL)$_4$, Rh$_2$(S-1-Ph-BPTTL)$_4$, Rh$_2$(S-BOTL)$_4$, and Rh$_2$(S-BHTL)$_4$ (Figure 3.22). Among the prepared

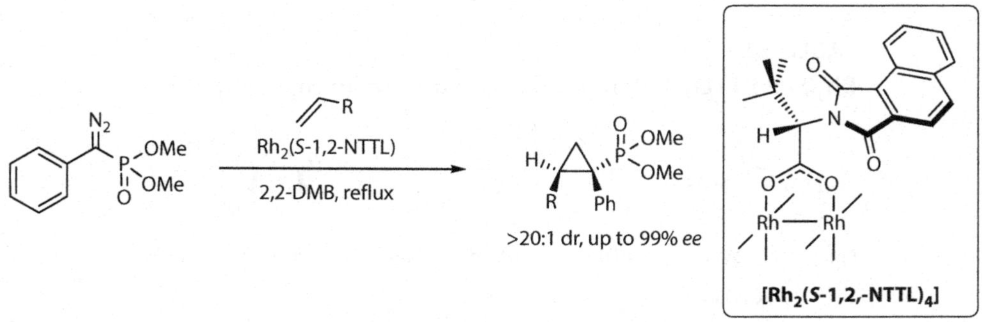

SCHEME 3.20 Rh$_2$(S-1,2-NTTL)$_4$-catalyzed asymmetric cyclopropanation of styrenes with dimethyl α-diazobenzylphosphonate.[170]

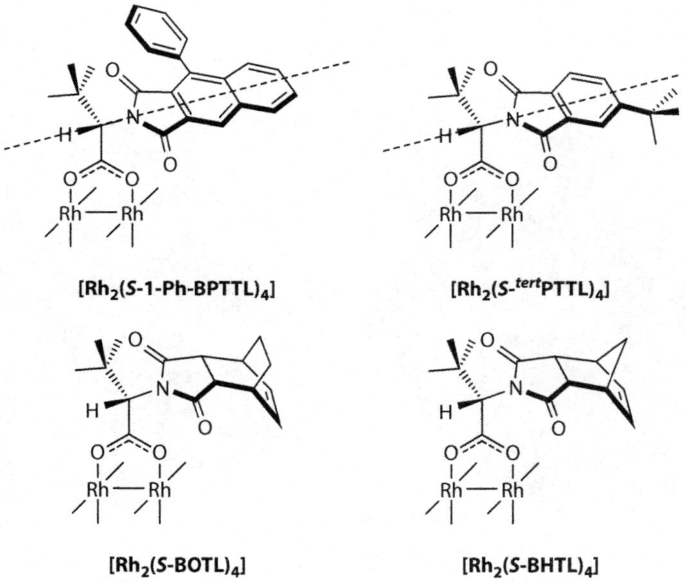

FIGURE 3.22 Structure of Adly and Ghanem complexes.[152]

complexes, Rh$_2$(S-^{tert}PTTL)$_4$ proved to be an exceptional catalyst with extraordinary enantioselectivity (up to 99% *ee*). Screening with a number of different *donor–acceptor* diazo systems revealed that, generally, Rh$_2$(S-^{tert}PTTL)$_4$ is a more enantioselective catalyst compared to Rh$_2$(S-PTTL)$_4$ and Rh$_2$(S-NTTL)$_4$, while having comparable enantioselectivity to Rh$_2$(S-PTAD)$_4$ (Table 3.17).

The authors explained the extraordinary enantioselectivity of Rh$_2$(S-^{tert}PTTL)$_4$ through its x-ray structure.[152] The x-ray structure revealed full chiral crown conformation featuring the *tert*-butyl substituent similarly disposed toward the "corner" of the square-shaped cavity (Figure 3.23). According to the authors, the square cavity of Rh$_2$(S-^{tert}PTTL)$_4$ contrasts with the rectangular cavity originally reported by Fox for Rh$_2$(S-PTTL)$_4$ (Figure 3.23).[141] The added substitution on the *N*-phthaloyl group in Rh$_2$(S-^{tert}PTTL)$_4$ nicely extended the width of each of cavity walls to the point that adjacent ligands are nearly at van der Waals contact (Figure 3.23). Furthermore, from the space-filling representations, comparisons of Rh$_2$(S-^{tert}PTTL)$_4$ and Rh$_2$(S-PTTL)$_4$ indicated that the extra *tert*-butyl substituents in Rh$_2$(S-^{tert}PTTL)$_4$ were introducing greater ligand conformational rigidity through the C$_\alpha$-CO$_2$ as well as N-C$_\alpha$ bond torsions of the ligands. According to the authors, these structural features are responsible for the observed enhanced enantioselectivity levels for Rh$_2$(S-^{tert}PTTL)$_4$ compared to Rh$_2$(S-PTTL)$_4$.

TABLE 3.17
Rh$_2$(S-^{tert}PTTL)$_4$-Catalyzed Asymmetric Cyclopropanations[152]

Entry	EWG	EDG	Solvent	Yield (%)	dr (E:Z)	ee (%)
1	PO(OMe)$_2$	Ph	2,2-DMB	92	>20:1	98
2	CF$_3$	Ph	2,2-DMB	99	>20:1	88
3	CO$_2$Me	p-OH-Ph	2,2-DMB	85	>18:1	78
4	CN	Ph	Toluene	81	25:1	82

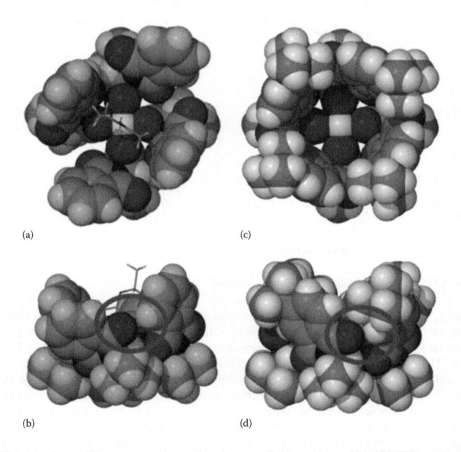

FIGURE 3.23 Space-filling structure comparison between EtOAc adduct of Rh$_2$(S-PTTL)$_4$ and bis(THF) adduct of Rh$_2$(S-^{tert}PTTL)$_4$: (a, c) top views of Rh$_2$(S-PTTL)$_4$ and Rh$_2$(S-^{tert}PTTL)$_4$, respectively; (b, d) side views of Rh$_2$(S-PTTL)$_4$ and Rh$_2$(S-^{tert}PTTL)$_4$, respectively. (Reprinted from Adly, F.G., Gardiner, M.G., and Ghanem, A., *Chem.-Eur. J.*, 22, 3447, 2016. Copyright 2016, with permission from John Wiley and Sons.)

3.5.2 Heteroleptic Complexes

In 2012, Fox reported the mixed ligated complex dirhodium(II) tris[N-phthaloyl-(S)-tert-leucinate] triphenylacetate, Rh$_2$(S-PTTL)$_3$(TPA) (Figure 3.24). The x-ray structure of the complex displayed all N-phthalimide groups on one face of the complex in structural similarity to Rh$_2$(S-PTTL)$_4$.[171]

Rh$_2$(S-PTTL)$_3$(TPA) returned the best enantioselectivity in the cyclopropanation reaction of styrene with ethyl α-diazobutanoate (88% ee) (Table 3.18). The scope of this catalyst was expanded to include other substrate classes (namely, aliphatic alkynes, silylacetylenes, and α-olefins) that were particularly challenging in intermolecular cyclopropanations with α-alkyl-α-diazoesters. Generally, Rh$_2$(S-PTTL)$_3$(TPA) was capable to catalyze enantioselective cyclopropanations with yields and enantioselectivities comparable and sometimes superior to Rh$_2$(S-PTTL)$_4$.[171]

In the same context, Charette et al.[172] reported the synthesis of several chiral heteroleptic dirhodium(II) tetracarboxylate catalysts. The major observation was that replacing one of the chlorinated ligands in Rh$_2$(S-TCPTV)$_4$ or Rh$_2$(S-TCPTTL)$_4$ with achiral nonchlorinated PTAiB ligand had a beneficial impact on their asymmetric induction (Table 3.19).

The x-ray structure of Rh$_2$(S-TCPTTL)$_3$(PTAiB) revealed that the achiral PTAiB points toward the opposite direction relative to the three N-phthalimido groups in the complex in solid state (Figure 3.25).

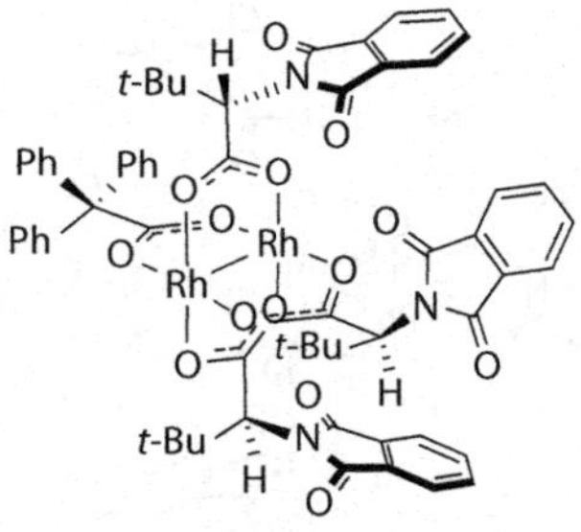

[Rh$_2$(S-PTTL)$_3$(TPA)]

FIGURE 3.24 Structure of Rh$_2$(S-PTTL)$_3$(TPA) reported by Fox.[171]

TABLE 3.18
Rh$_2$(S-PTTL)$_3$(TPA)-Catalyzed Cyclopropanation of Ethyl α-Diazobutanoate[171]

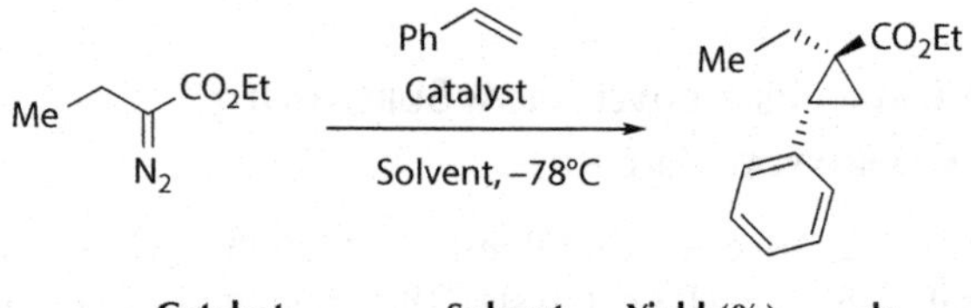

Entry	Catalyst	Solvent	Yield (%)	dr	ee (%)
1	Rh$_2$(S-PTTL)$_4$ (32)	Hexanes	95	92:8	79
2	Rh$_2$(S-BPTTL)$_4$ (39)	Toluene	80	88:12	73
3	Rh$_2$(S-NTTL)$_4$ (41)	Toluene	61	75:25	45
4	Rh$_2$(S-TBPTTL)$_4$ (35)	CH$_2$Cl$_2$	6	97:3	11
5	Rh$_2$(S-TBPTTL)$_4$ (35)	Toluene	10	88:12	16
6	Rh$_2$(S-TBPTTL)$_4$ (35)	Hexanes	23	84:16	16
7	Rh$_2$(S-PTTL)$_3$(TPA)	Toluene	66	95:5	81
8	Rh$_2$(S-PTTL)$_3$(TPA)	Hexanes	91	96:4	88

TABLE 3.19
Evaluation of Chiral Heteroleptic Rh$_2$(S-TCPTV)$_3$(PTAiB) and Rh$_2$(S-TCPTTL)$_3$(PTAiB) Complexes as Catalysts in Asymmetric Cyclopropanation[172]

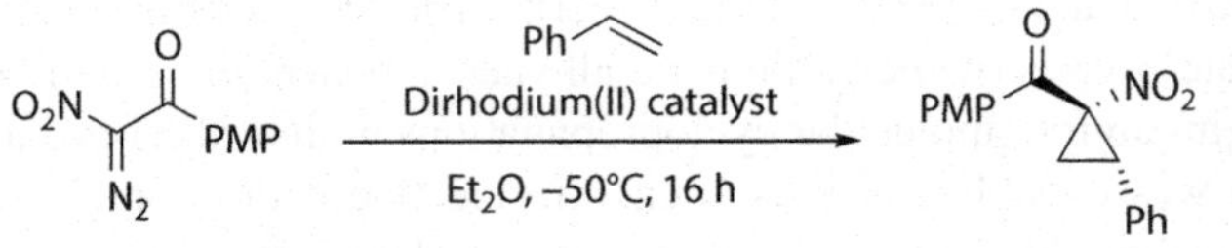

Entry	Catalyst	Yield (%)	dr	ee (%)
1	Rh$_2$(S-TCPTV)$_4$ (30)	82	98:2	91.1
2	Rh$_2$(S-TCPTTL)$_4$ (33)	81	98:2	92.9
3	Rh$_2$(S-TCPTV)$_3$(PTAiB)	76	93:7	95.0
4	Rh$_2$(S-TCPTTL)$_3$(PTAiB)	84	92:8	96.4

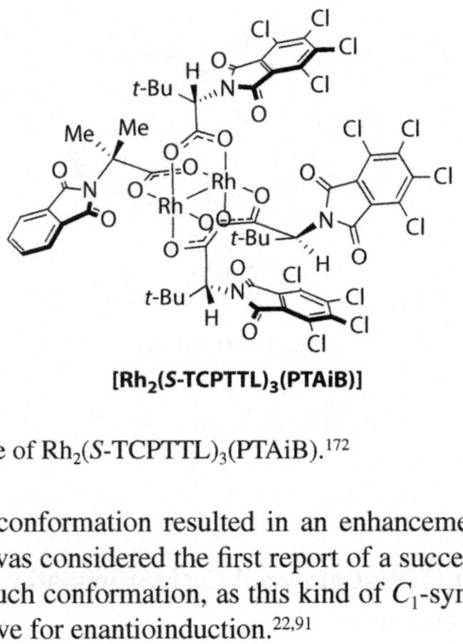

FIGURE 3.25 α,α,α,β-Structure of Rh$_2$(S-TCPTTL)$_3$(PTAiB).[172]

The reason why this α,α,α,β conformation resulted in an enhancement in enantioinduction is still ambiguous. Charette's report was considered the first report of a successful enantioselective transformation using a catalyst with such conformation, as this kind of C_1-symmetry conformation has long been overlooked as nonoperative for enantioinduction.[22,91]

3.5.3 DIRHODIUM(II) CATALYSTS DERIVED FROM SUBSTITUTED CYCLOPROPANE CARBOXYLATE LIGANDS

In 2011, Davies et al.[32,147,173] reported the usefulness of employing chiral cyclopropane carboxylic acids as ligands for chiral dirhodium(II) catalysts. They prepared dirhodium(II)tetrakis[(R)-1-(4-bromophenyl)-2,2-diphenylcyclopropane carboxylate], Rh$_2$(R-BrTPCP)$_4$ (Figure 3.26), as the first member of this class of chiral dirhodium(II) carboxylates. Very recently, the same group reported another set of catalysts that belongs to the same dirhodium(II) family (Figure 3.26) as being effective for enantioselective [3+2]-cycloaddition of nitrones with vinyldiazoacetates[174] and for catalyst-controlled site-selective and stereoselective C–H functionalization.[32]

The x-ray structure of Rh$_2$(R-p-BrTPCP)$_4$ showed that the cyclopropane carboxylate ligands are arranged in a D_2-symmetry around the rhodium core with two identical C_2-symmetric cavities similar to Rh$_2$(S-DOSP)$_4$ (Figure 3.27).[147] Also, the rectangular binding cavity of Rh$_2$(R-p-BrTPCP)$_4$ (8.5 × 10.5 Å) is significantly smaller than the Rh$_2$(S-PTTL)$_4$ cavity (12.8 × 14.1 Å).

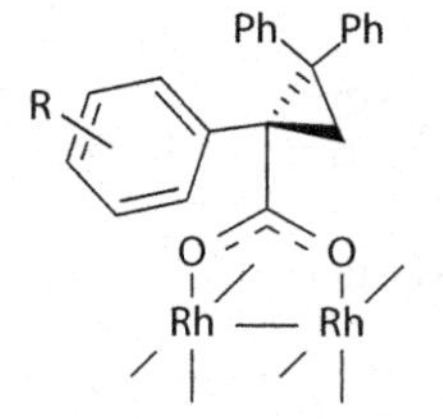

Enantiomerically Pure Compounds by Enantioselective Synthetic Chiral Metal Complexes   111

R = H, [Rh₂(R-TPCP)₄]
R = p-Ph, [Rh₂(R-p-PhTPCP)₄]
R = p-Br, [Rh₂(R-p-BrTPCP)₄]
R = p-ᵗBu, [Rh₂(R-p-ᵗBuTPCP)₄]
R = p-NO₂, [Rh₂(R-p-NO₂TPCP)₄]
R = p-CF₃, [Rh₂(R-p-CF₃TPCP)₄]
R = 3,5-diCF₃, [Rh₂(R-3,5-diCF₃TPCP)₄]
R = 3,5-diBr, [Rh₂(R-3,5-diBr-TPCP)₄]

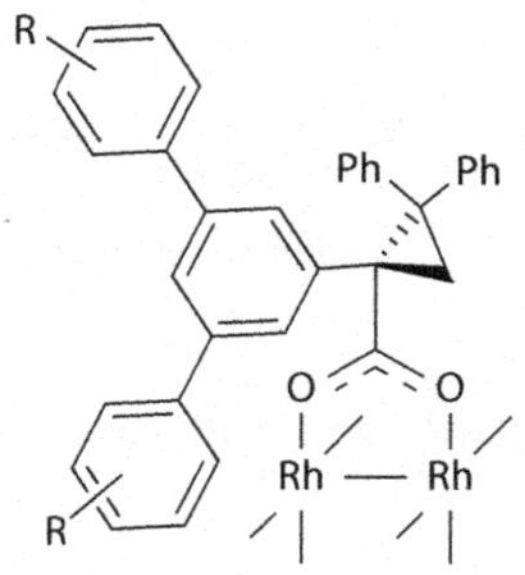

R = H, [Rh₂(R-3,5-diPhTPCP)₄]
R = 3,5-diCF₃, [Rh₂(R-3,5-di(3,5-diCF₃C₆H₃)TPCP)₄]
R = 3,5-diCH₃, [Rh₂(R-3,5-di(3,5-diCH₃C₆H₃)TPCP)₄]
R = 3,5-diPh, [Rh₂(R-3,5-di(3,5-diPhC₆H₃)TPCP)₄]
R = p-Ph, [Rh₂(R-3,5-di(p-diPhC₆H₄)TPCP)₄]
R = p-CF₃, [Rh₂(R-3,5-di(p-diCF₃C₆H₄)TPCP)₄]
R = p-ᵗBu, [Rh₂(R-3,5-di(p-diᵗBuC₆H₄)TPCP)₄]

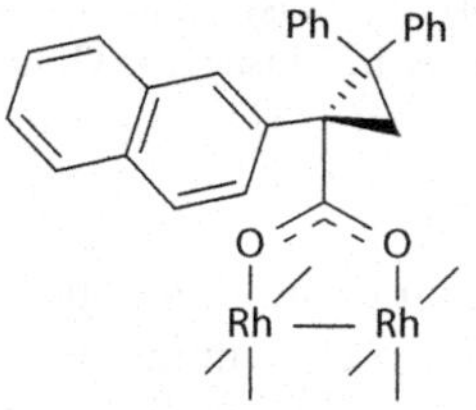

[Rh₂(R-NPCP)₄]

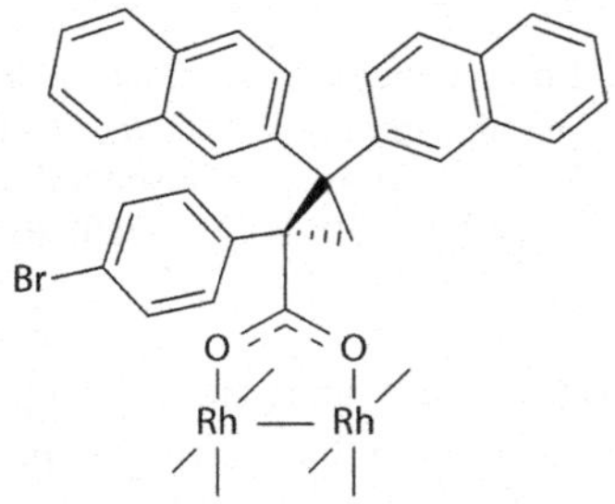

[Rh₂(S-BNPCP)₄]

FIGURE 3.26 Structures of reported chiral dirhodium(II) carboxylates derived from cyclopropane carboxylate ligands.[32,147,173,174]

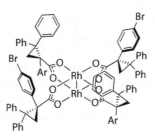

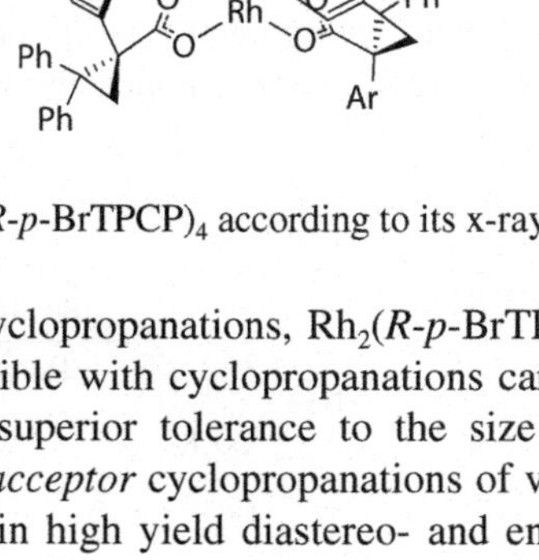

FIGURE 3.27 D_2-symmetry of Rh₂(R-p-BrTPCP)₄ according to its x-ray structure.[147]

In *donor–acceptor* carbenoid cyclopropanations, Rh₂(R-p-BrTPCP)₄ exhibited multiple advantages. In addition to being compatible with cyclopropanations carried out in methylene chloride, Rh₂(R-p-BrTPCP)₄ demonstrated superior tolerance to the size of the carbenoid ester group. Rh₂(R-BTPCP)₄-catalyzed *donor–acceptor* cyclopropanations of various alkenes afforded the corresponding cyclopropane product in high yield diastereo- and enantioselectivity (up to 97% *ee*) (Scheme 3.21).[147]

Density functional theory (DFT) calculations of the lowest-energy conformer revealed that it has two ligands in conrotatory-rotated fashion to reduce steric interactions with the axial carbene

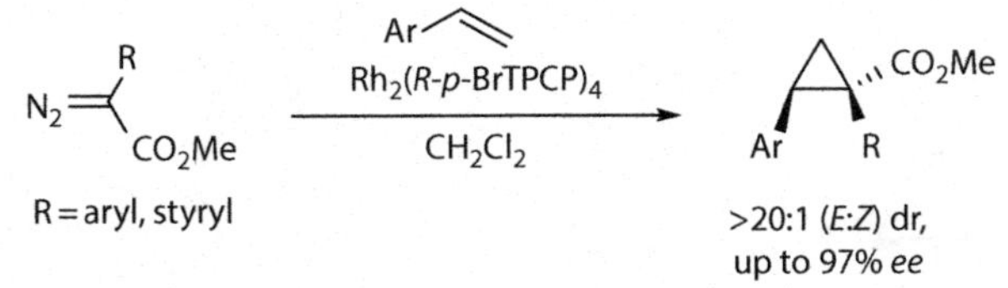

SCHEME 3.21 Rh$_2$(R-p-BrTPCP)$_4$-catalyzed stereoselective cyclopropanation.

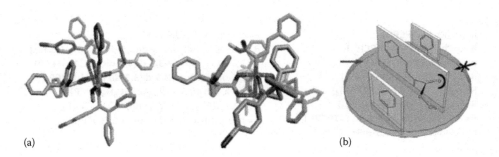

FIGURE 3.28 (a) Lowest-energy conformation of *s*-trans carbene, top view (left) and side view (right). (b) Predictive stereochemical model for Rh$_2$(R-p-BrTPCP)$_4$-catalyzed transformations. (Reprinted with permission from Qin, C., Boyarskikh, V., Hansen, J.H., Hardcastle, K.I., Musaev, D.G., and Davies, H.M.L., *J. Am. Chem. Soc.*, 133, 19198, 2011. Copyright 2016.)

ligand, while the other two ligands stayed in upward direction to minimize steric interactions with neighboring ligands (Figure 3.28a). This arrangement leads to a C_2-symmetric environment around the carbene-binding site incorporating two phenyl rings and two *p*-bromophenyl rings. The *donor* aryl/styryl group is blocked by one of the phenyl rings, while the other ring is located next to the *acceptor* ester group (Figure 3.28b).[147] As the ester group blocks the attack from its side (*re*-face), the approach of the alkene takes place over the *donor* group (*si*-face) to generate the cyclopropane product in the observed configuration.

The Rh$_2$(R-p-BrTPCP)$_4$-catalyzed synthesis of a variety of 2-arylbicyclo[1.1.0]butane carboxylate derivatives **55** under low catalyst loading (0.01 mol%) with high levels of enantioselectivity (70%–94% *ee*) was also reported, whereas the same reaction catalyzed by higher catalyst loading of Rh$_2$(tpa)$_4$ (1.0 mol%) afforded the cyclohexene carboxylate derivatives **56** (Scheme 3.22).[175]

Also recently, highly functionalized cyclopropanecarboxylates were readily prepared by Rh$_2$(R-p-PhTPCP)$_4$-catalyzed cyclopropanation of alkenes with trichloroethyl (TCE) aryldiazoacetates and styryldiazoacetates at which having such a labile protecting group on the ester, chiral triarylcyclopropane carboxylate ligands were conveniently prepared (Scheme 3.23).[119]

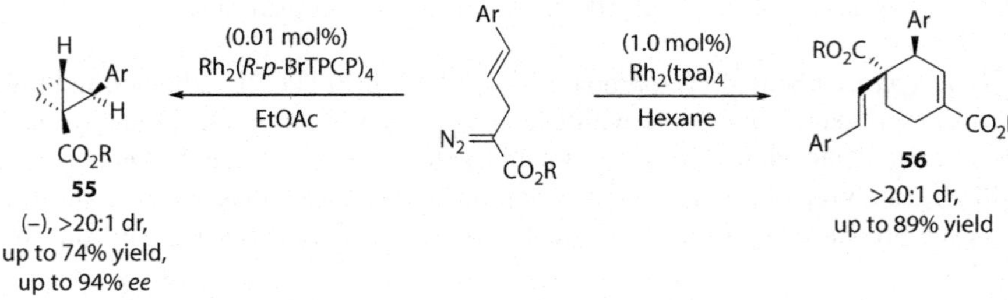

SCHEME 3.22 Divergent synthesis of **55** and **56**.[175]

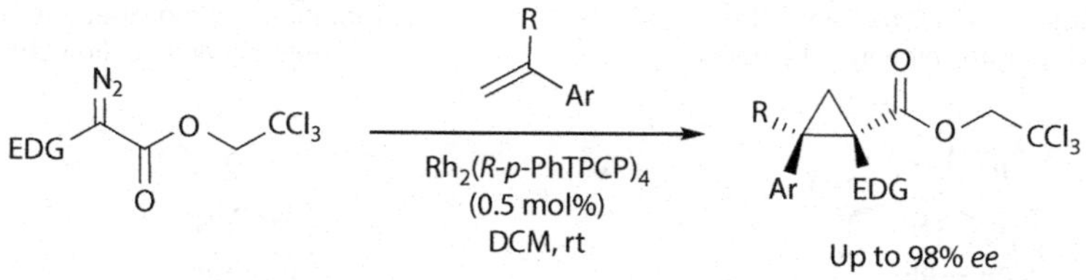

SCHEME 3.23 Chiral Rh$_2$(R-p-PhTPCP)$_4$-catalyzed cyclopropanation of TCE aryldiazoesters/styryldiazoacetates and styrenes.[119]

3.6 DIRHODIUM(II) CARBOXAMIDATES

3.6.1 Homoleptic Complexes

Although the first synthesis of dirhodium(II) carboxamidates occurred in the 1980s when dirhodium(II) tetraacetamidate was isolated from a melt of acetamide and Rh$_2$(OAc)$_4$,[176] chiral dirhodium(II) carboxamidates were developed by Doyle and coworkers.[10,177] Chiral carboxamidate ligands consist of bridging lactams derived from amino acids with an ester group connected to the carbon atom adjacent to the nitrogen (Figures 3.30 and 3.31). Early investigations in carboxamidate ligand design determined the usefulness of ester as the optimum stereodirecting group (introduces the (R)- or (S)-configuration to the ligand). The utilization of alkyl or aryl attachments resulted in significant deterioration in enantioselectivity.[178,179] In fact, the effect of the ester group size on catalyst enantioselectivity is substantial and can finely balance its steric factors.[180] Moreover, initial attempts led to the realization that acyclic amides were not generally suitable because ligand exchange requires access to the *cis* (E) amide form rather than the *trans* (Z) form (Figure 3.29).[177] Reported modifications of the lactam ring have already given rise to four carboxamidate ligand classes, namely, pyrrolidinates,[178] oxazolidinates,[181] imidazolidinates,[182] and azetidinates[183] (Figures 3.30 and 3.31). Again, they differ in reactivity and selectivity on the basis of their steric and/or electronic profiles.

Compared to dirhodium(II) carboxylates, dirhodium(II) carboxamides have a more complex paddlewheel structure. Due to the nature of the carboxamidate ligand, it bridges the Rh–Rh core *via* both an oxygen atom and a nitrogen atom. Because of the unsymmetrical bridging ligands, there are four possible geometrical isomers, namely, (2)-*cis*, (2)-*trans*, (1, 3), and (4, 0), based on the positions of nitrogens and oxygens connected to each rhodium (Figure 3.32).[28] Although examples for each geometrical isomer, except for the (2)-*trans* isomer, have been isolated and characterized,[28,188] monitoring the ligand exchange process with LC-MS showed that the (2)-*cis* geometry is the dominant isomer (>85%). Also, results indicated that all other isomers isomerize into this major isomer upon heating.[178,181,188]

The (2)-*cis* configuration is defined at which each rhodium atom is connected to two nitrogen atoms and two oxygen atoms in a *cis*-fashion. As a result, these complexes can only adapt C_2-symmetry due to this intrinsic ligand-binding preference.[22] The (2)-*cis* geometry was also constantly observed in the x-ray structures for different dirhodium(II) carboxamidate complexes, including

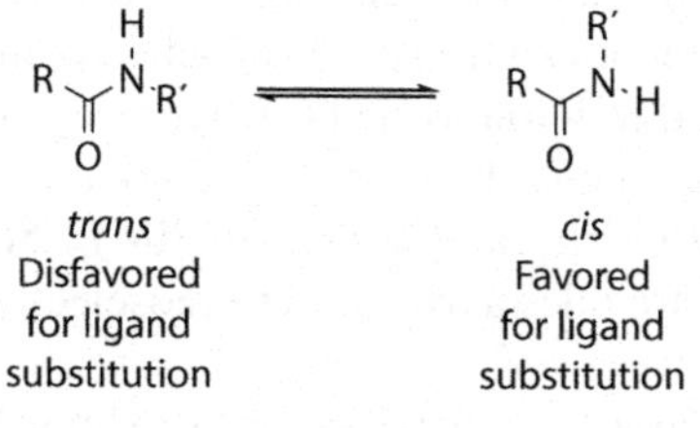

FIGURE 3.29 Favored and disfavored amide forms for ligand exchange.[177]

Dirhodium(II) carboxamidates derived from chiral pyrrolidinates

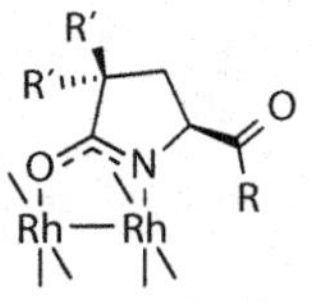

57 R = OMe, R' = H, [**Rh₂(5S-MEPY)₄**]
58 R = OCH₂CMe₃, R' = H, [**Rh₂(5S-NEPY)₄**]
59 R = O(CH₂)₁₇CH₃, R' = H, [**Rh₂(5S-ODPY)₄**]
60 R = NMe₂, R' = H, [**Rh₂(5S-DMAP)₄**]
61 R = OMe, R' = F, [**Rh₂(5S-dFMEPY)₄**]

Dirhodium(II) carboxamidates derived from chiral oxazolidinates

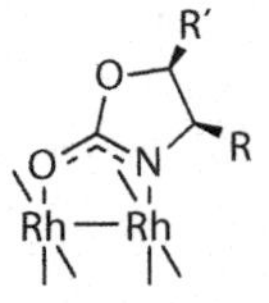

62 R = CO₂Me, R' = H, [**Rh₂(4S-MEOX)₄**]
63 R = H, R' = Me, [**Rh₂(4S-THREOX)₄**]
64 R = PhCH₂, R' = H, [**Rh₂(4S-BNOX)₄**]
65 R = i-Pr, R' = H, [**Rh₂(4S-IPOX)₄**]
66 R = Ph, R' = H, [**Rh₂(4S-PHOX)₄**]
67 R = Me, R' = Ph, [**Rh₂(4S-MPOX)₄**]

Dirhodium(II) carboxamidates derived from chiral azetidinates

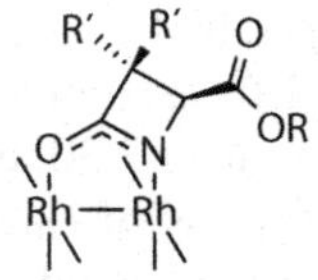

68 R = Me, R' = H, [**Rh₂(4S-MEAZ)₄**]
69 R = PhCH₂, R' = H, [**Rh₂(4S-BNAZ)₄**]
70 R = CH₂CMe₃, R' = H, [**Rh₂(4S-NEPAZ)₄**]
71 R = t-Bu, R' = H, [**Rh₂(4S-IBAZ)₄**]
72 R = ᶜC₆H₁₁, R' = H, [**Rh₂(4S-CHAZ)₄**]
73 R = ᶜC₆H₁₁, R' = F, [**Rh₂(4R-dFCHAZ)₄**]
74 R = i-Pr, R' = F, [**Rh₂(4R-dFIBAZ)₄**]
75 R = S/R-menthyl, R' = H, [**Rh₂(4S, S/R-MenthAZ)₄**]
76 R = 4-FC₆H₄CH₂, R' = H, [**Rh₂(4S-(4')-FBNAZ)₄**]

FIGURE 3.30 Structures of different classes of chiral dirhodium(II) carboxamidates (cont.).[20,28,184,185]

Rh₂(5R-MEPY)₄[178] (the enantiomer of **57**, Figure 3.30) and Rh₂(4S-MEOX)₄ (**62**, Figure 3.30). In these two complexes, the two ester groups (E) are oriented in counterclockwise fashion to effectively block one side of the axial carbene ligand during catalysis (Figure 3.33).[12,187]

The evolution of this (2)-*cis* geometry led to the development of imidazolidinone carboxylate-ligated catalysts carrying chiral *N*-acyl groups (**77–91**, Figure 3.31). The chiral *N*-acyl attachments of the imidazolidinone carboxylate catalysts were designed to potentially reinforce the inherent stereocontrol offered by the core system. As an example, the use of ligand diastereomers to prepare Rh₂(1'S,2'R,5'S,4S-MNACIM)₄ (**91**) and its diastereomer, Rh₂(1'R,2'S,5'R,4S-MNACIM)₄, showed significant difference in diastereo- and enantioselectivity, and the best diastereocontrol was achieved with **91** (Table 3.20).[189]

Further, viewing the diastereomers of Rh₂(MCPIM)₄ and Rh₂(BSPIM)₄ down the Rh–Rh axis illustrated that these catalysts are configured as shown in Figure 3.34a. Rh₂(4S,2'S,3'S-MCPIM)₄ (**85**) and Rh₂(4S,2'S-BSPIM)₄ (**88**) have the pendent ester and *N*-acyl side chains oriented in the same direction following a counterclockwise spiral (matched complexes) (Figure 3.34a). This orientation is particularly well suited to intramolecular reactions in which the active site for reaction is tethered to the dirhodium(II) axial coordination site. In contrast, the ester and *N*-acyl side chains are in configurational opposition in Rh₂(4S,2'R,3'R-MCPIM)₄ (**86**) and Rh₂(4S,2'R-BSPIM)₄ (**87**) (mismatched complexes), which provides a barrier to stereoselectivity enhancement in intramolecular transformations.[28,190]

In a different context, the electron-rich character of dirhodium(II) carboxamidate catalysts make them catalytically less active compared to dirhodium(II) carboxylates. They are very effective,

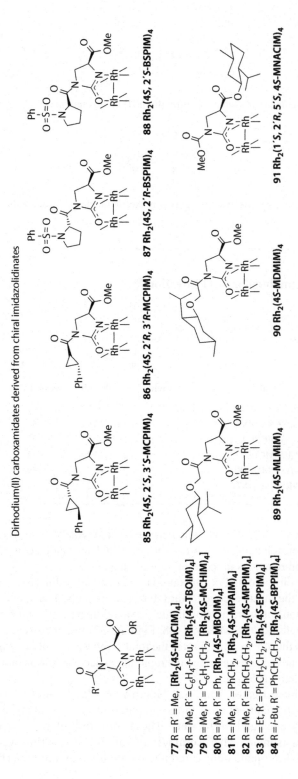

FIGURE 3.31 Structures of different classes of chiral dirhodium(II) carboxamidates.[20,28,186,187]

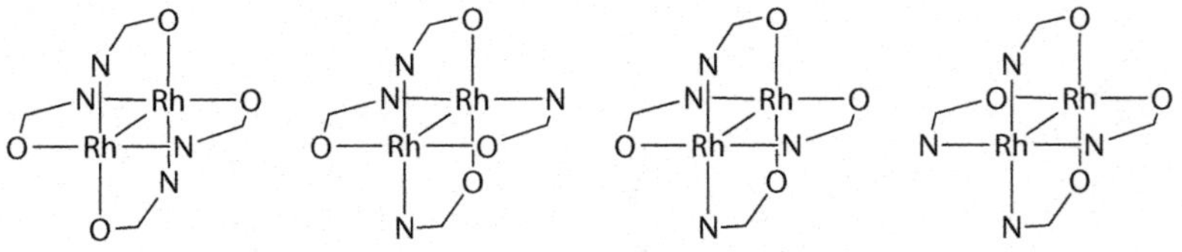

FIGURE 3.32 Possible geometrical isomers of dirhodium(II) carboxamidates.[28]

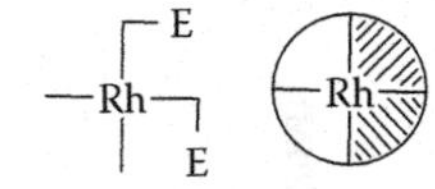

FIGURE 3.33 CO$_2$Me (E) entity is occupying two adjacent quadrants around the dirhodium core (top view).[12,187]

TABLE 3.20
Results for Intramolecular C–H Insertion[189]

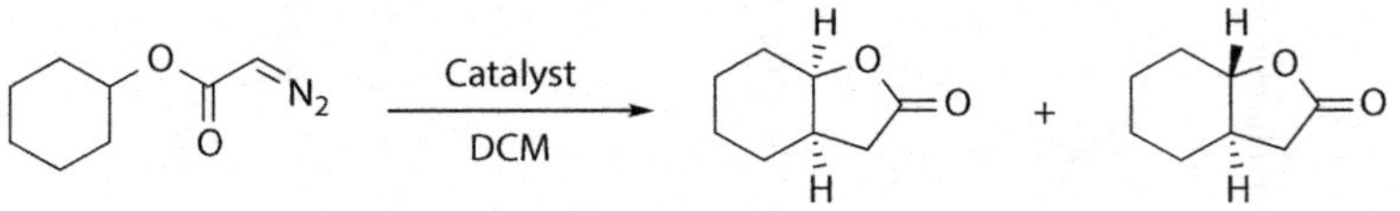

Entry	Catalyst	Yield (%)	dr (a:b)	ee of a (%)	ee of b (%)
1	Rh$_2$(1′S,2′R,5′S,4S-MNACIM)$_4$ (**91**)	80	100:0	95	—
2	Rh$_2$(1′R,2′S,5′R,4S-MNACIM)$_4$	71	79:21	84	68
3	Rh$_2$(4S,2′S,3′S-MCPIM)$_4$ (**85**)	78	99:1	97	—
4	Rh$_2$(4S,2′R,3′R-MCPIM)$_4$ (**86**)	63	80:20	72	13
5	Rh$_2$(4S,2′S-BSPIM)$_4$ (**88**)	88	97:3	>99	>99
6	Rh$_2$(4S,2′R-BSPIM)$_4$ (**87**)	89	98:2	74	33

however, in the decomposition of diazoacetate derivatives and widely employed for intramolecular cyclopropanation,[187] intermolecular cyclopropenation,[191] and intramolecular C–H insertion reactions,[28] and often result in reactions proceeding in more than 90% ee.

Among the different dirhodium(II) carboxamidate classes, azetidinate-based catalysts (**68–76**, Figure 3.32)[183,192,193] are the most reactive toward diazo compounds. They are compatible with diazoacetates that are unstable toward pyrrolidinate-, oxazolidinate-, and immidazolidinate-ligated catalysts.[183] As an example, Rh$_2$(4S-MEAZ)$_4$ (**68**, Figure 3.32) provided several advantages in the synthesis of **92**, which is the key intermediate in the total synthesis of milnacipran and its analogues (Scheme 3.24).[194] These advantages included the rapid diazo decomposition and affording the highest level of enantioselectivity (95% ee).

Furthermore, azetidinate-ligated Rh$_2$(4S,R-MenthAZ)$_4$ (**75**, Figure 3.30) offered excellent diastereocontrol and high enantiocontrol in the preparation of *cis*-cyclopropane products starting from substituted styrenes and diazoesters. These were in preference over their thermodynamically favored *trans* isomers. The usefulness of this transformation was illustrated in the total synthesis of the cyclopropane-configured phenylethylthiazoylthiourea (PETT) analogue **93** (Scheme 3.25).[193]

Enantiomerically Pure Compounds by Enantioselective Synthetic Chiral Metal Complexes 117

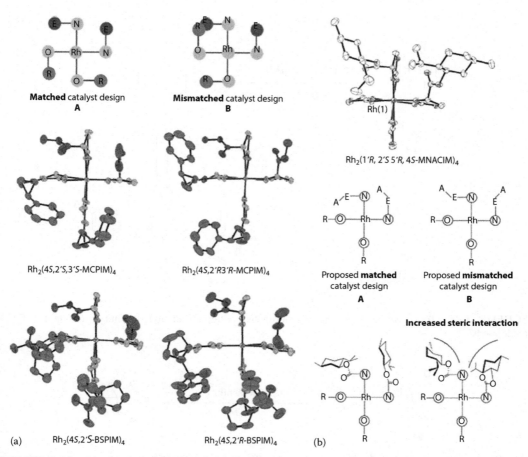

FIGURE 3.34 (a) Configurational differences between matched and mismatched Rh$_2$(MCPIM)$_4$ and Rh$_2$(BSPIM)$_4$ catalysts and (b) proposed matched and mismatched configurations of Rh$_2$(MNACIM)$_4$ catalyst. ([a]: Reprinted with permission from Doyle, M.P., Duffy, R., Ratnikov, M., and Zhou, L., *Chem. Rev.*, 110, 704, 2010. Copyright 2016. [b]: Reprinted from Doyle, M.P.; Morgan, J.P.; Colyer, J.T. *J. Organomet. Chem.*, 690, 5525, 2005. Copyright 2016, with permission from Elsevier.)

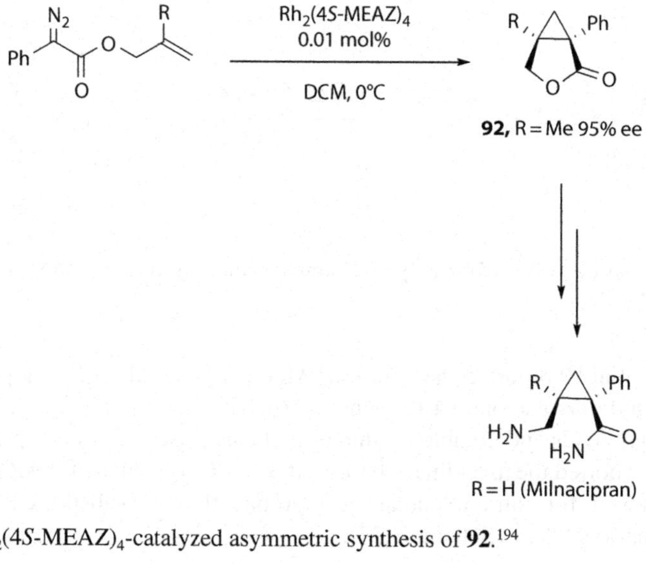

SCHEME 3.24 Rh$_2$(4S-MEAZ)$_4$-catalyzed asymmetric synthesis of **92**.[194]

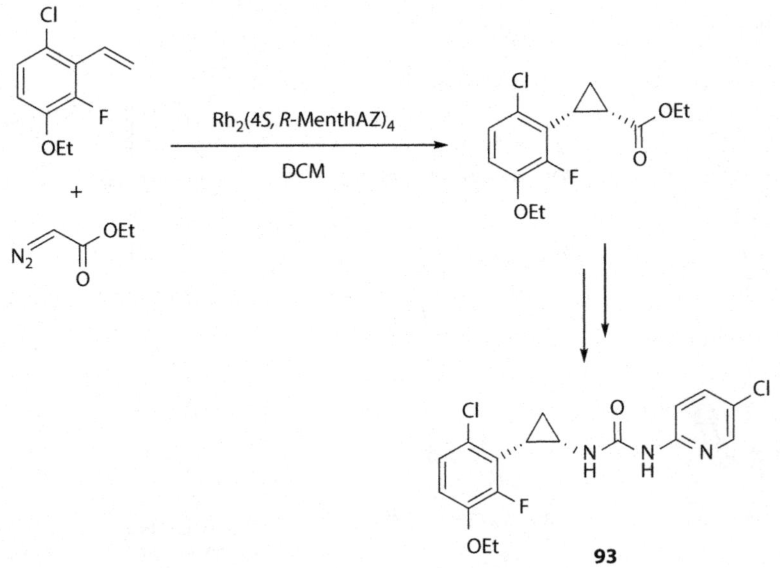

93

SCHEME 3.25 Total synthesis of the cyclopropane-configured phenylethylthiazoylthiourea (PETT) analogue **93**.[193]

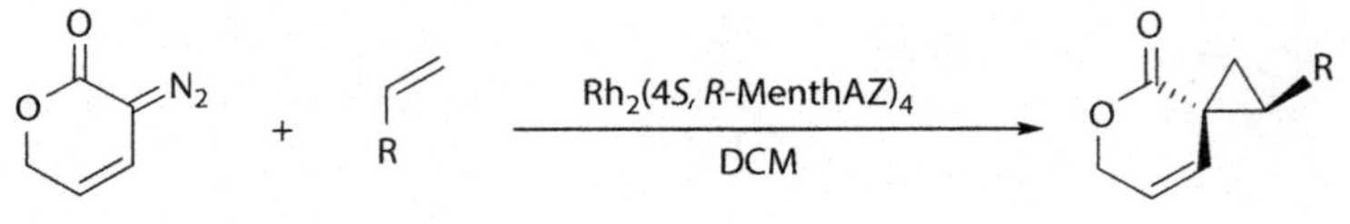

94

R = Ph — 74%, *de* = >90%, *ee* = 84%

R = (Me)C = CHPh — 81%, *de* = 78%, *ee* = 86%

R = CH = CHPh — 86%, *de* = 66%, *ee* = 73%

SCHEME 3.26 Rh_2(4*S*,*R*-MenthAZ)$_4$-catalyzed cyclopropanation of different olefins with vinyl diazolactone **94**.[195]

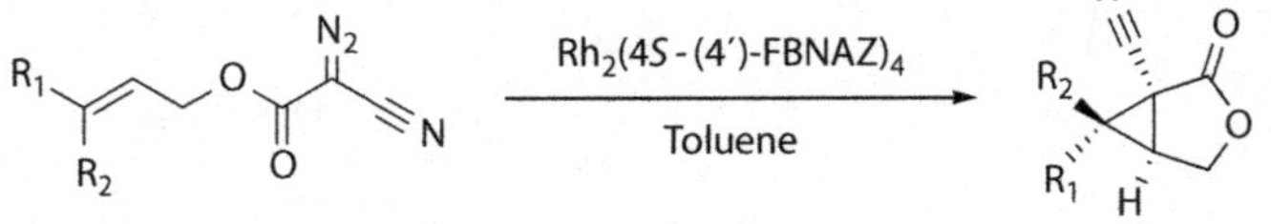

Up to 91% *ee*

SCHEME 3.27 Rh_2(4*S*-(4′)-FBNAZ)$_4$-catalyzed intramolecular cyclopropanation of substituted allylic α-cyano-α-diazoacetates.[185]

Similarly, Doyle et al.[195] reported the Rh_2(4*S*,*R*-MenthAZ)$_4$-catalyzed cyclopropanation of various alkenes with vinyl diazolactone **94** (Scheme 3.26). The corresponding cyclopropane products were obtained in high yields and notable diastereo- and enantioselectivities (up to 86% *ee*).

Charette et al.[185] reported the azetidinate-based catalyst, Rh_2(4*S*-(4′)-FBNAZ)$_4$ (**76**, Figure 3.30), as an effective catalyst for intramolecular cyclopropanation of substituted allylic α-cyano-α-diazoacetates with up to 91% *ee* (Scheme 3.27).

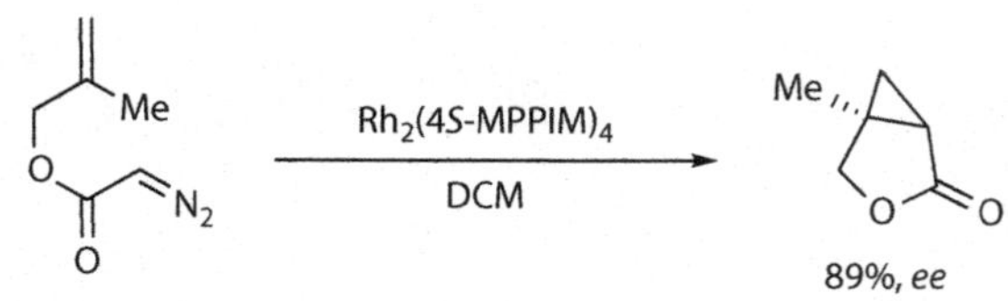

SCHEME 3.28 Rh$_2$(4S-MPPIM)$_4$-catalyzed intermolecular cyclopropanation of 2-methallyl diazoacetate.[196,197]

Doyle et al.[187] employed Rh$_2$(5S-MEPY)$_4$ (**57**, Figure 3.30) and its enantiomer, Rh$_2$(5R-MEPY)$_4$, in the intramolecular cyclopropanations involving several trisubstituted and *cis*-disubstituted allylic diazoacetates. Doyle succeeded in preparing the chiral cyclopropane-fused lactone products in good-to-excellent yields and exceptional enantioselectivity. When similar reactions involving the *trans*-disubstituted isomers were performed, however, the products were obtained with moderate enantioselectivity. The use of Rh$_2$(4S-MPPIM)$_4$ catalyst (**82**, Figure 3.31) resulted in 89%–96% *ee*,[196] emphasizing the importance of the steric bulk and positioning of the ligand on enantioselectivity control. As an example, the Rh$_2$(5S-MEPY)$_4$-catalyzed intermolecular cyclopropanation involving 2-methallyl diazoacetate provided the cyclopropane product with only 7% *ee*. Alternatively, when Rh$_2$(4S-MPPIM)$_4$ was employed in the same reaction, the cyclopropane product was obtained with 89% *ee* (Scheme 3.28).[196,197] In the same context, a study comparing Doyle's catalysts with Cu(I) and Ru(II) catalysts showed that dirhodium(II) carboxamidates are capable to generate far superior enantiomerically pure products.[197]

Generally, Doyle justified the observed absolute configuration and the high enantioselectivity level of dirhodium(II) carboxamidates in these intramolecular cyclopropanations through the transition state model illustrated in Figure 3.35.[187]

Likewise, Doyle expanded this intramolecular cyclopropanation chemistry to include cyclopropanation of allylic diazoacetamides to afford the corresponding cyclopropane-fused lactam product. As an example, the Rh$_2$(4S-MEOX)$_4$ (**62**, Figure 3.30)-catalyzed decomposition of *N*-allyl diazoacetamide provided the 3-azabicyclo[3.1.0]hex-2-one product in 98% *ee* (Scheme 3.29).[187]

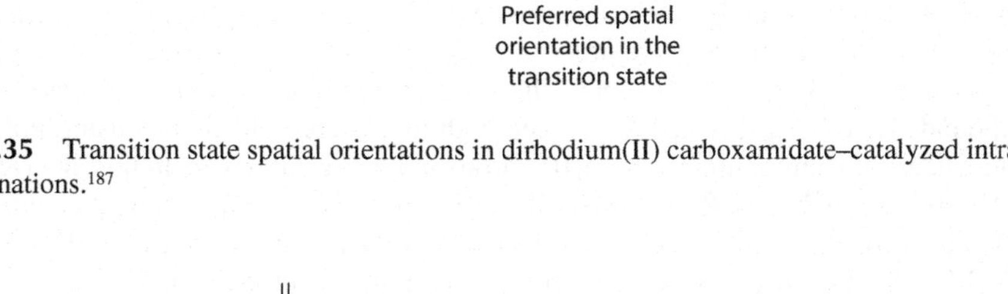

FIGURE 3.35 Transition state spatial orientations in dirhodium(II) carboxamidate–catalyzed intramolecular cyclopropanations.[187]

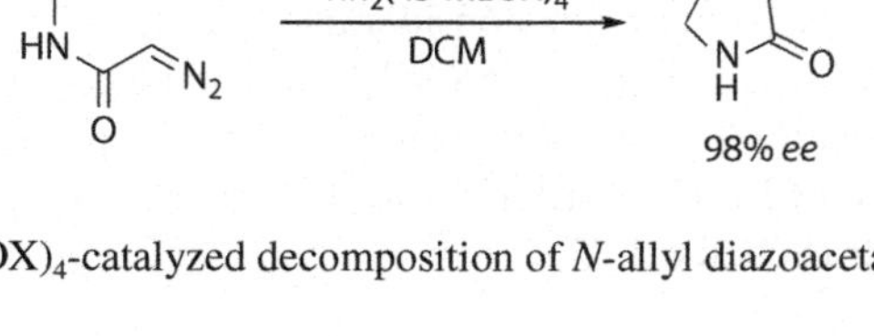

SCHEME 3.29 Rh$_2$(4S-MEOX)$_4$-catalyzed decomposition of *N*-allyl diazoacetamide.[187]

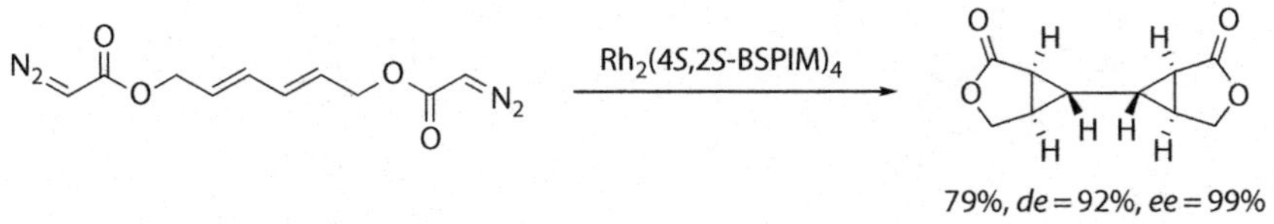

SCHEME 3.30 Rh₂(4S,2S-BSPIM)₄-catalyzed double intramolecular cyclopropanation.[198]

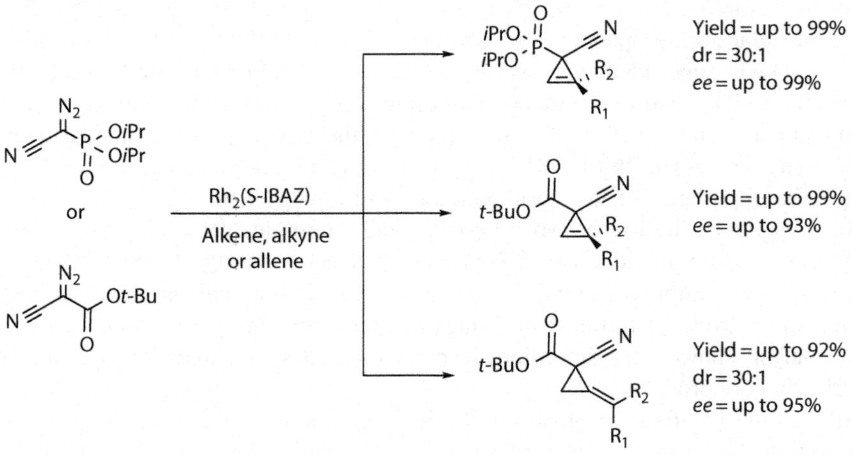

SCHEME 3.31 Rh₂(S-IBAZ)₄-catalyzed cyclopropan(en)ations.[199]

When using the allylic diazoacetamides as substrates, however, substitution of the extra hydrogen on the amide nitrogen was essential to get higher yield. This enhancement was returned to the formation of the reaction-favored conformer.

The Rh₂(4S,2S-BSPIM)₄-catalyzed cyclopropanation of bis-diazoacetate was also reported by Doyle et al. at which a sequence of two successive intramolecular cyclopropanations took place (Scheme 3.30).[198]

Recently, Charette reported a highly stereoselective Rh₂(S-IBAZ)₄-catalyzed cyclopropanation of alkenes and alkynes with *diacceptor* diazo compounds.[199] The *iso*steric profile of phosphonic and carboxylic acid derivatives permitted the alternative use of both α-cyano diazophosphonate and α-cyano diazocarboxylate esters in this chemistry, resulting in α-cycloprop(en)ylphosphonates and α-cyano cycloprop(en)ylcarboxylates, respectively, in high yields and stereoselectivities (Scheme 3.31). Charette expanded the scope of substrates to include substituted allenes and established the first catalytic enantioselective process to access *diacceptor* alkylidenecyclopropanes.

As we illustrated earlier in this chapter, the cyclopropanation of *donor–acceptor*-substituted diazo compounds have been reported to proceed with high stereoselectivities using dirhodium(II) carboxylate catalysts. Common dirhodium(II) carboxamidate catalysts including Rh₂(5S-MEPY)₄ (**57**), Rh₂(4S-MEOX)₄ (**62**), and Rh₂(4S-MBOIM)₄ (**80**), however, resulted in poor enantioselectivities with this class of diazo substrates.[186] An increase in enantiomeric excess of **95** was obtained upon using Rh₂(4S-TBOIM)₄ (**78**) leading to the cyclopropane product in 77% *ee* in methylene chloride as a reaction solvent (Table 3.21).[186] Further, the use of pentanes as a reaction solvent did not affect the catalyst enantioselectivity to a considerable extent.

Doyle and coworkers also reported the Rh₂(5S-MEPY)₄-catalyzed asymmetric synthesis of cyclopropenes with high enantiocontrol.[14,191] As an example, reacting propyldiethyl acetal with methyl diazoacetate in the presence of Rh₂(5S-MEPY)₄ afforded the desired cyclopropane product in 42% yield and more than 98% *ee* (Scheme 3.32).

TABLE 3.21
Dirhodium(II) Carboxamidate–Catalyzed Cyclopropanation of Styrene with Methyl Phenyldiazoacetate[186]

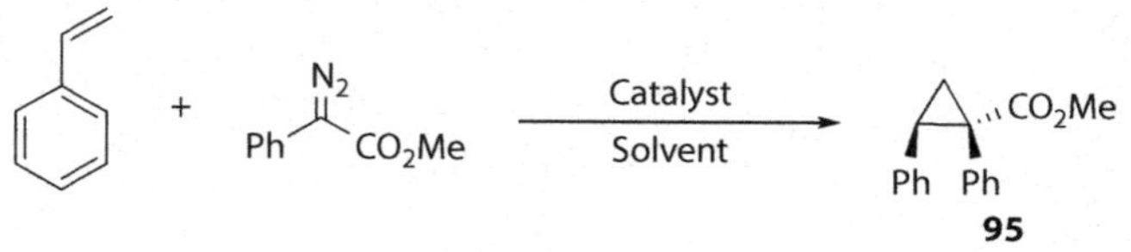

Entry	Catalyst	Solvent	Yield (%), (cis:trans)	ee (%)
1	Rh₂(5S-MEPY)₄ (**57**)	CH₂Cl₂	27 (97:3)	49
2	Rh₂(4S-MEOX)₄ (**62**)	CH₂Cl₂	57 (96:4)	41
3	Rh₂(4S-MBOIM)₄ (**80**)	CH₂Cl₂	73 (96:4)	48
4	Rh₂(4S-TBOIM)₄ (**78**)	CH₂Cl₂	63 (95:5)	77
5	Rh₂(4S-TBOIM)₄ (**78**)	Pentanes	69 (94:6)	75

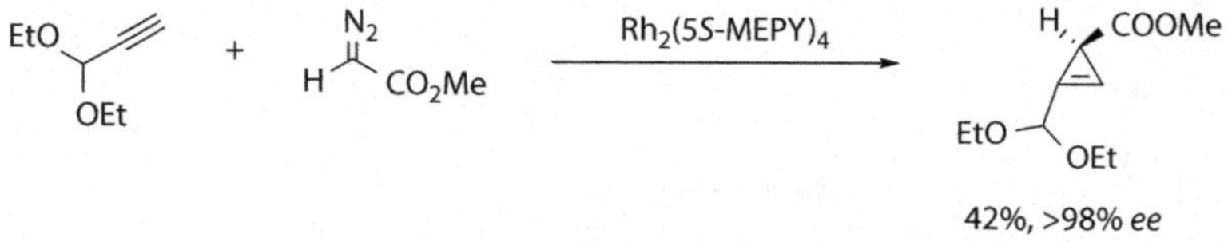

SCHEME 3.32 Rh₂(5S-MEPY)₄-catalyzed cyclopropenation of propyldiethyl acetal with methyl diazoacetate.

Intramolecular cyclopropanation has also been successfully reported with excellent levels of enantioselectivity.[200] Generally, Rh₂(4S-IBAZ)₄ was the best for both enantio- and chemoselective cyclopropenations (up to 97% ee) (Scheme 3.33). The propargyl analogue was a very good example that demonstrates the catalyst effect on selectivity (Scheme 3.33b and c). With the less reactive Rh₂(5S-MEPY)₄, it undergoes highly chemo- and enantioselective allylic cyclopropanation, while with the more reactive Rh₂(4S-IBAZ)₄, addition on the triple bond is favored.

In the same context, Poulter et al.[201] used Rh₂(5S-MEPY)₄ (**57**) in the synthesis of optically pure presqualene diphosphate at which the key step in this synthesis was the stereoselective intramolecular cyclopropanation of farnesyl diazoacetate, while Martin et al.[202] used Rh₂(5S-MEPY)₄ and its enantiomer, Rh₂(5R-MEPY)₄, to prepare conformationally restricted peptide isosteres and extended this work to the preparation of cyclopropane peptidomimetics as novel enkephalin analogues.

In 1996, the Hashimoto group reported the synthesis of a dirhodium(II) carboxamidate complex, Rh₂(S-PTPI)₄ (Figure 3.36), with 3-(S)-phthalimido-2-piperidinonate as chiral bridging ligands.[203,204] Later, the same research group reported its analogue, Rh₂(S-BPTPI)₄ (Figure 3.36), as a highly efficient Lewis acid catalyst for enantioselective hetero-Diels–Alder reactions.[54,55,205]

The use of Rh₂(S-PTPI)₄ saved a high order of enantioselectivity in the cyclopropanation reactions that involved styrenes, E-1-phenylbutadiene, and 1,1-disubstituted alkenes. The combinational use of 2,4-dimethyl-3-pentyl diazoacetate as a carbene source and ether as a reaction solvent was crucial for the success of this catalyst (Scheme 3.34).[204]

3.6.2 Heteroleptic Complexes

Corey et al.[206] communicated a new dirhodium(II) carboxamidate having (R,R)-4,5-diphenyl-N-triflylimidazolidinone (DTPI) as bridging ligands (Figure 3.37). The new catalyst was tested in enantioselective cyclopropanation of ethyl diazoacetate with terminal acetylenes. The authors described

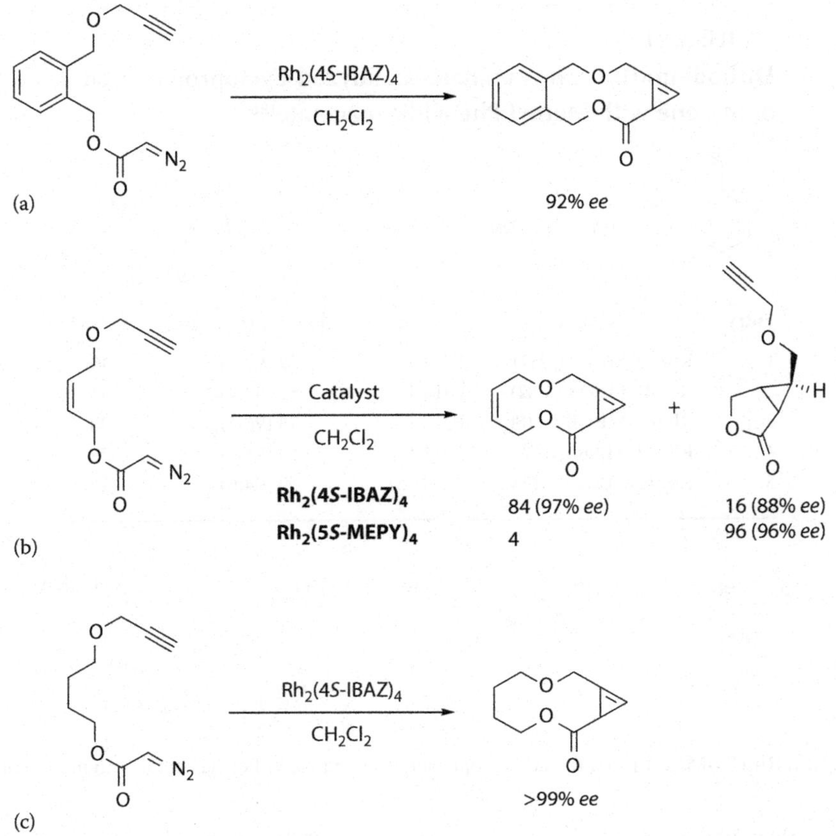

SCHEME 3.33 Rh$_2$(4S-IBAZ)$_4$-catalyzed intramolecular cyclopropenation.

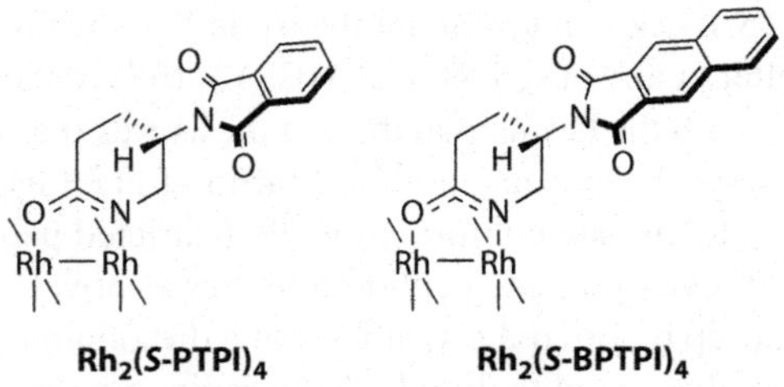

FIGURE 3.36 Structure of Hashimoto's Rh$_2$(S-PTPI)$_4$ and Rh$_2$(S-BPTPI)$_4$ catalysts.[54,55,203–205]

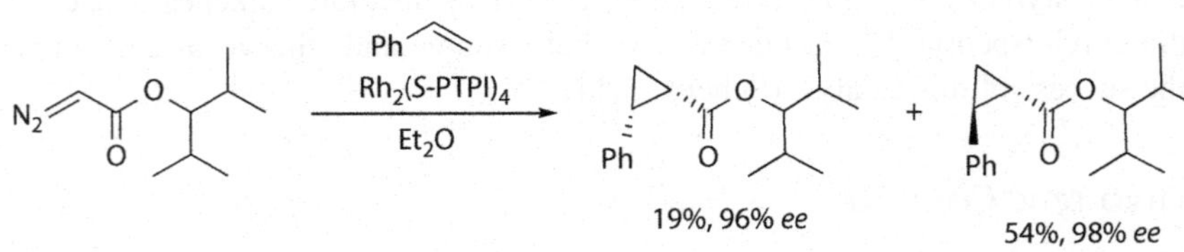

SCHEME 3.34 Example for Rh$_2$(S-PTPI)$_4$-catalyzed enantioselective cyclopropanation of styrene with 2,4-dimethyl-3-pentyl diazoacetate.[204]

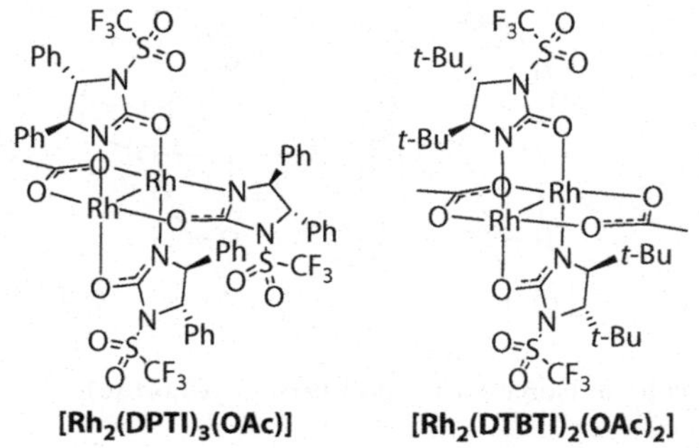

FIGURE 3.37 Structure of Rh$_2$(DPTI)$_3$(OAc) and Rh$_2$(DTBTI)$_2$(OAc)$_2$.[206,207]

TABLE 3.22
Rh$_2$(DTBTI)$_2$(OAc)$_2$-Catalyzed Cyclopropanation of Ethyl Diazoacetate and Terminal Alkynes

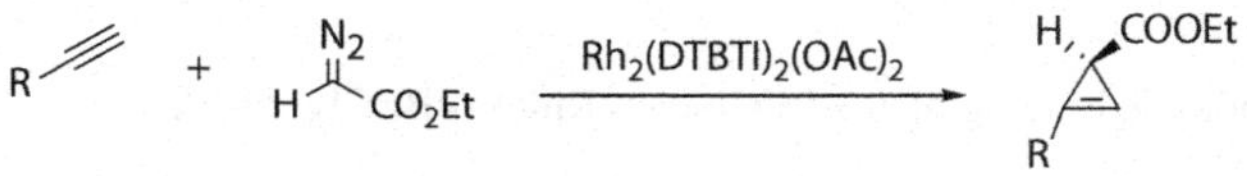

Entry	R	Temp. (°C)	Yield (%)	ee (%)
1	CH$_3$(CH$_2$)$_4$	23	87	89
2	CH$_3$(CH$_2$)$_4$	0	84	91
3	t-Bu	0	81	90
4	MeOCH$_2$	0	78	93

the catalyst to be outstanding in terms of enantioselectivity, yield, scope, and efficiency of catalyst recovery. Later, the same group reported the new C_2-symmetric complex, Rh$_2$(DTBTI)$_2$(OAc)$_2$, having only two *anti*-DTBTI ((*R,R*)-4,5-di-*tert*-butyl-*N*-triflylimidazolidinone) bridges (Figure 3.37).[207] This catalyst was highly effective in cyclopropenation reactions of a wide range of alkynes (Table 3.22). It was notable due to its robustness, as well as the easy synthesis of the associated chiral ligand.

Doyle et al.[208] reported Rh$_2$(1,6-BPGlyc)$_2$(OAc)$_2$ as a new member of the dirhodium(II) carboxamidate family. The bridging ligands were 1,6-bis-(*N*-benzyl)-diphenylglycoluril (1,6-BPGlyc) and acetate ligands (Figure 3.38).

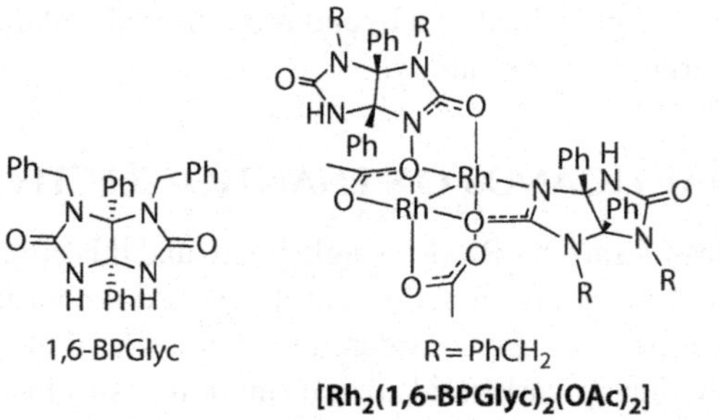

FIGURE 3.38 Structure of 1,6-BPGlyc and Rh$_2$(1,6-BPGlyc)$_2$(OAc)$_2$.[208]

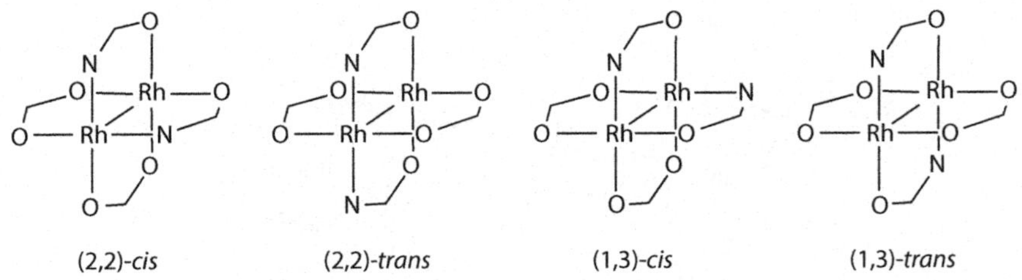

FIGURE 3.39 Possible geometrical isomers for $Rh_2(1,6\text{-BPGlyc})_2(OAc)_2$.[208]

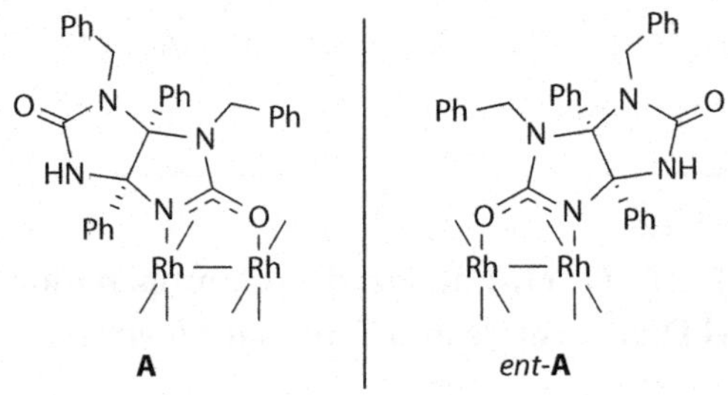

FIGURE 3.40 Enantiotopic binding of 1,6-BPGlyc on $Rh_2(1,6\text{-BPGlyc})_2(OAc)_2$.[208]

Despite the unusual steric profile of the ligand and the mixed substitution pattern of the new catalyst, there were only minor differences relative to previously reported dirhodium(II) carboxamidates.[208] As it contains only two carboxamide ligands, $Rh_2(1,6\text{-BPGlyc})_2(OAc)_2$ could have been formed in four possible isomers: (2)-*cis*, (2)-*trans*, (1, 3)-*cis*, and (1, 3)-*trans* (Figure 3.39). X-ray structure of the complex not only confirmed the C_2-symmetery of $Rh_2(1,6\text{-BPGlyc})_2(OAc)_2$ but also revealed the preference for the (1, 3)-*cis* isomer. The authors attributed this preference to the *trans*-effect-directing ligand substitution.

In addition to the selective ligand arrangement, the formation of $Rh_2(1,6\text{-BPGlyc})_2(OAc)_2$ is stereoselective. Glycoluril ligand is a *meso*-compound with two enantiotopic metal binding sites that would provide enantiomeric pairs upon substitution (**A**, *ent*-**A**) (Figure 3.40). Moreover, the (1, 3)-*cis* complexes are helically chiral (*M*, *P*) about the Rh–Rh bond axis due to the fused nature of the μ-NCO bridging ligands. The $Rh_2(1,6\text{-BPGlyc})_2(OAc)_2$ complex was found to be formed stereoselectively as a racemic mixture of the (A, A)-*P* and (*ent*-A, *ent*-A)-*P* diastereomers.[208]

The *cis*/*trans* selectivity of $Rh_2(1,6\text{-BPGlyc})_2(OAc)_2$ in the formation of **96** was closer to $Rh_2(5S\text{-MEPY})_4$ (**57**) than the electronically related $Rh_2(4S\text{-MPPIM})_4$ (**82**) (Table 3.23). The *cis*/*trans* selectivity of the formation of **97** was basically unchanged by $Rh_2(1,6\text{-BPGlyc})_2(OAc)_2$ in comparison to the results of both dirhodium(II) carboxylates and carboxamidates except that $Rh_2(1,6\text{-BPGlyc})_2(OAc)_2$ was more reactive than $Rh_2(5S\text{-MEPY})_4$. The overall conclusion by the authors was that the distinct features of glycoluril as a ligand only offered a platform for expanded diversity within the dirhodium(II) carboxamidate family.

3.7 EFFECTS OF AXIAL LIGANDS ON ENANTIOSELECTIVITY

The dirhodium core consists of a strong Rh–Rh single bond, and this core provides the dirhodium(II) complex with an excellent ability to form adducts at its two axial coordination sites. The two axial ligands are labile and, therefore, they are considered to be the sites that give the dirhodium(II) complexes its catalytic activity during carbenoid transformations. As illustrated earlier, the proposed mechanism considers that only one of the two coordination sites working as a carbene-binding site at a time throughout the catalytic cycle.[87,88]

TABLE 3.23
Cis/Trans Selectivity of Rh$_2$(1,6-BPGlyc)$_2$(OAc)$_2$ in the Formation of 96 and 97[208]

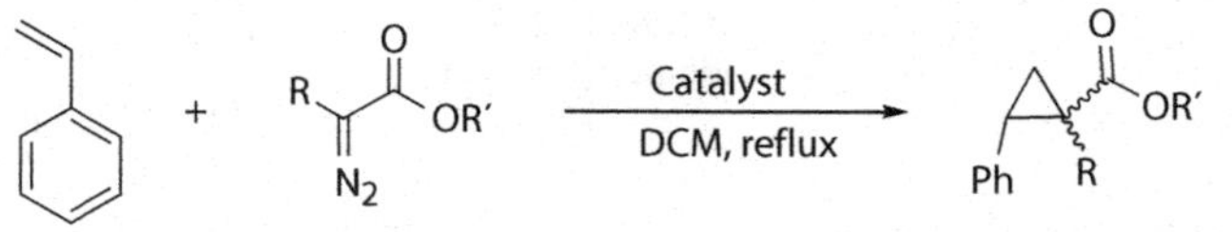

96 R = H, R' = Et
97 R = Ph, R' = Me

Entry	Catalyst	Yield (%), (Z:E) 96	97
1	Rh$_2$(4S-MPPIM)$_4$ (82)	64 (74:26)	66 (97:3)
2	Rh$_2$(5S-MEPY)$_4$ (57)	59 (46:54)	27 (97:3)
3	Rh$_2$(1,6-BPGlyc)$_2$(OAc)$_2$	43 (47:53)	55 (99:1)
4	Rh$_2$(OAc)$_4$	93 (38:62)	69 (98:2)

The two axial positions of the dirhodium(II) are often occupied by solvent molecules that have the ability to establish weaker bonds with the dirhodium core. As a consequence, the reaction solvent is able to critically affect the reaction outcome. Solvents with poor coordinative capabilities (e.g., DCM or nonpolar solvents) are most efficient for carbenoid transformations. However, solvents that coordinate into dirhodium(II) complexes (e.g., ACN or THF) can partially or totally inhibit the generation of the carbenoid.[27,28,87,88,209–212]

Kinetic studies revealed that axial ligands, such as ACN, inhibit this kind of transformations through a mixed kinetic inhibition mechanism. In this mechanism, the ligand can bind to both the free complex and the catalyst-substrate complex.[87,88] In 2000, the Jessop group studied the effect of solvent on the enantioselectivity of Rh$_2$(S-TBSP)$_4$-catalyzed asymmetric cyclopropanation of styrene with methyl α-phenyldiazoacetate (Scheme 3.35). They observed that the enantioselectivity is not only dependent on the coordinating ability of the solvent but also on its dielectric constant (the more polar, the lower *ee* value obtained).[211]

A few reports have emerged where the addition of Lewis base to the cyclopropanation reactions proved to be a useful and efficient method for tuning the properties of dirhodium(II) complexes.[27,35,143,164,165] Davies et al.[35] explored the addition of methyl benzoate to the

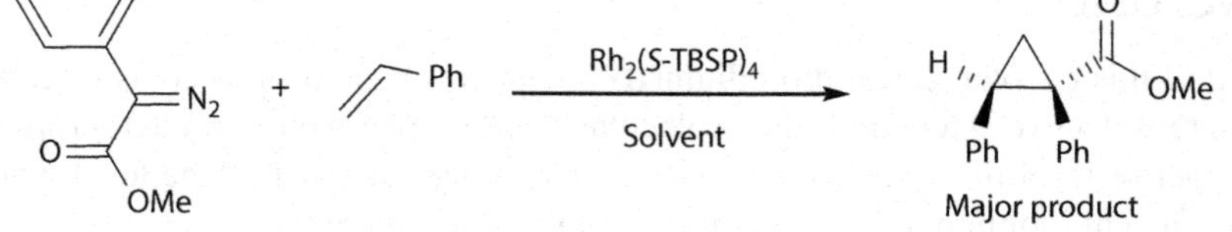

Solvent	ee
n-Hexane	90% ee
DCM	67% ee
THF	81% ee
ACN	73% ee
DMF	69% ee
scCO$_2$	84% ee

SCHEME 3.35 Solvent effect on the enantioselectivity of Rh$_2$(S-TBSP)$_4$-catalyzed cyclopropanation of styrene.[211]

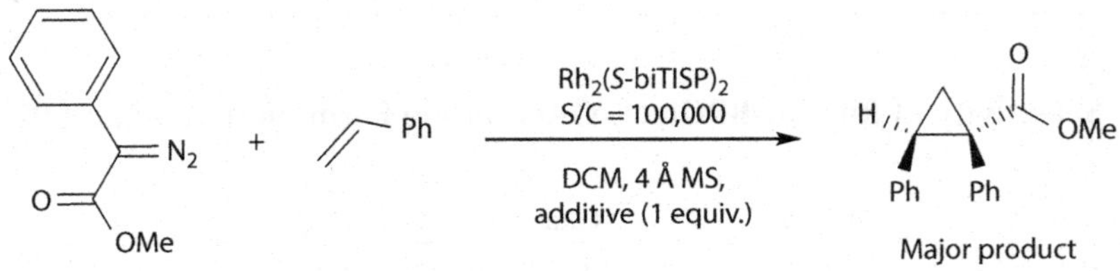

SCHEME 3.36 Additive effect on the activity of Rh$_2$(S-biTISP)$_2$ in cyclopropanation reaction of styrene with methyl α-phenyldiazoacetate.[35]

Additive	Result
None	42 h, 82%, 65% ee
PhCO$_2$Me	28 h, 85%, 83% ee
TMU	>36 h, 48%, 74% ee
DIPEA	>36 h, 5%
OP(Oct)$_3$	>36 h, 36%, 78% ee

Rh$_2$(S-biTISP)$_2$-catalyzed cyclopropanation reaction mixture. It did not only improve the enantioselectivity of the cyclopropanation but also allowed the utilization of very small amount of the catalyst (S/C = 100,000) with high efficiency (Scheme 3.36). At this stage, the authors were uncertain about the actual role of the additive; however, they believed that this might be because of the coordination of the methyl benzoate additive to the carbenoid or to the other rhodium center. Recently, in 2013, computational studies on dirhodium(II)-catalyzed cyclopropanations of electron-deficient alkenes carried out by the same research group gave a reasonable hypothesis. The authors concluded that the interaction between the carbenoid and the methyl benzoate carbonyl is able to protect the rhodium-carbene intermediates from self-destruction.[138]

Charette et al.[143,164,165] also found that TfNH$_2$ and DMAP can be used as additives to moderately improve the chiral induction in cyclopropanation reactions involving *diacceptor* diazo compounds. The additive's degree of success was highly dependent on the diazo substrate and reaction temperature. TfNH$_2$ and DMAP were shown to be optimal with Rh$_2$(S-NTTL)$_4$ and Rh$_2$(S-TCPTTL)$_4$, respectively. The authors believed that the system is quite complex as the coordination onto one of the reactive sites could not only modify the catalyst electronic properties but also can alter the spatial arrangement of the chiral bridging ligands.[143]

On the other hand, chiral dirhodium(II) carboxamidates have a rigid structure if compared to chiral dirhodium(II) carboxylates. To the best of my knowledge, it is not reported that any of the known dirhodium(II) carboxamidates exhibits solvent effects on stereocontrol.[12,28]

3.8 CONCLUSION

As illustrated in this chapter, chiral dirhodium(II) complexes have been used as effective catalysts for highly stereoselective inter- and intramolecular cyclopropanation and cyclopropenation reactions. This superior level of diastereo- and enantioselectivity has reached the level where they can act as a very efficient tool in building up complex molecular structures.

Despite the number of available highly efficient dirhodium(II) catalysts, it is evident that none can be considered as a "universal catalyst" that is able to afford high enantiomeric induction with all different classes of substrates and under different reaction conditions. But, the careful choice of a suitable catalyst for the desired reaction can afford the desired cyclopropane product with high levels of chemo-, diastereo-, and enantioselectivity.

Also, with the fast-advanced development in the fields of crystallography, computational modeling, and solution NMR spectroscopy, efforts may succeed in uncovering the mystery behind the conformation of dirhodium(II) complexes in solid state, in solution, and during catalysis. This, in turn, may assist in the design and development of novel, highly stereoselective chiral dirhodium(II) complexes.

REFERENCES

1. Lin, G.-Q.; You, Q.-D.; Cheng, J.-F. *Chiral Drugs: Chemistry and Biological Action*; John Wiley & Sons: Hoboken, NJ, 2011.
2. Younes, A. A.; Mangelings, D.; Van der Heyden, Y. *J. Chromatogr. A* 2012, *1269*, 154.
3. Zhang, X. *Chem. Eng. News Arch.* 2001, *79*, 142.
4. Ye, T.; McKervey, M. A. *Chem. Rev.* 1994, *94*, 1091.
5. Timmons, D. J.; Doyle, M. P. *J. Organomet. Chem.* 2001, *617–618*, 98.
6. Ren, T. *Coord. Chem. Rev.* 1998, *175*, 43.
7. Lebel, H.; Marcoux, J. F.; Molinaro, C.; Charette, A. B. *Chem. Rev.* 2003, *103*, 977.
8. Hodgson, D. M.; Stupple, P. A.; Pierard, F. Y. T. M.; Labande, A. H.; Johnstone, C. *Chem. Eur. J.* 2001, *7*, 4465.
9. Felthouse, T. R. *Prog. Inorg. Chem.* 1982, *29*, 73.
10. Doyle, M. P.; McKervey, M. A.; Ye, T. *Modern Catalytic Methods for Organic Synthesis with Diazo Compounds: From Cyclopropanes to Ylides*; Wiley: New York, 1998.
11. Doyle, M. P. *J. Org. Chem.* 2006, *71*, 9253.
12. Doyle, M. P. In *Modern Rhodium-Catalyzed Organic Reactions*; Evans, P. A. Ed.; Wiley-VCH Verlag GmbH & Co. KGaA: Weinheim, Germany, 2005; p. 341.
13. Doyle, M. P. *Aldrichim. Acta* 1996, *29*, 3.
14. Doyle, M. P. *Recl. Trav. Chim. Pays-Bas* 1991, *110*, 305.
15. Doyle, M. P. *Chem. Rev.* 1986, *86*, 919.
16. Davies, H. M. L.; Manning, J. R. *Nature* 2008, *451*, 417.
17. Davies, H. M. L.; Beckwith, R. E. J. *Chem. Rev.* 2003, *103*, 2861.
18. Davies, H. M. L. *Eur. J. Org. Chem.* 1999, *1999*, 2459.
19. Boyar, E. B.; Robinson, S. D. *Coord. Chem. Rev.* 1983, *50*, 109.
20. Adly, F. G.; Ghanem, A. *Chirality* 2014, *26*, 692.
21. El-Deftar, M.; Adly, F. G.; Gardiner, M. G.; Ghanem, A. *Curr. Org. Chem.* 2012, *16*, 1808.
22. Hansen, J.; Davies, H. M. L. *Coord. Chem. Rev.* 2008, *252*, 545.
23. Davies, H. M. L.; Morton, D. *Chem. Soc. Rev.* 2011, *40*, 1857.
24. Colacot, T. J. *Proc. Indian Acad. Sci. (J. Chem. Sci.)* 2000, *11*, 197.
25. Davies, H. M. L.; Bois, J. D.; Yu, J.-Q. *Chem. Soc. Rev.* 2011, *40*, 1855.
26. Davies, H. M. L. *Eur. J. Org. Chem.* 1999, *1999*, 2459.
27. Trindade, A. F.; Coelho, J. A. S.; Afonso, C. A. M.; Veiros, L. F.; Gois, P. M. P. *ACS Catal.* 2012, *2*, 370.
28. Doyle, M. P.; Duffy, R.; Ratnikov, M.; Zhou, L. *Chem. Rev.* 2010, *110*, 704.
29. Davies, H. M. L. *Curr. Org. Chem.* 1998, *2*, 463.
30. Davies, H. M. L.; Hedley, S. J. *Chem. Soc. Rev.* 2007, *36*, 1109.
31. Merlic, C. A.; Zechman, A. L. *Synthesis* 2003, *2003*, 1137.
32. Liao, K.; Negretti, S.; Musaev, D. G.; Bacsa, J.; Davies, H. M. L. *Nature* 2016, *533*, 230.
33. Deng, Y.; Qiu, H.; Srinivas, H. D.; Doyle, M. P. *Curr. Org. Chem.* 2016, *20*, 61.
34. Pelphrey, P.; Hansen, J.; Davies, H. M. L. *Chem. Sci.* 2010, *1*, 254.
35. Davies, H. M. L.; Venkataramani, C. *Org. Lett.* 2003, *5*, 1403.
36. Chernyaev, I. I.; Shenderetskaya, E. V.; Koryagina, A. A. *Russ. J. Inorg. Chem.* 1960, *5*, 559.
37. Chernyaev, I. I.; Shenderetskaya, E. V.; Maiorova, A. G.; Koryagina, A. A. *Russ. J. Inorg. Chem.* 1965, *10*, 537.
38. Nazarova, L. A.; Chernyaev, I. I.; Morozova, A. S. *Russ. J. Inorg. Chem.* 1965, *10*, 539.
39. Nazarova, L. A.; Chernyaev, I. I.; Morozova, A. S. *Russ. J. Inorg. Chem.* 1966, *11*, 2583.
40. Hubert, A. J.; Noels, A. F.; Anciaux, A. J.; Teyssie, P. *Synthesis* 1976, *9*, 600.
41. Paulisse, R.; Reimlinger, H.; Hayez, E.; Hubert, A. J.; Teyssie, P. *Tetrahedron Lett.* 1973, *14*, 2233.
42. Anciaux, A. J.; Demonceau, A.; Hubert, A. J.; Noels, A. F.; Petiniot, N.; Teyssie, P. *J. Chem. Soc., Chem. Commun.* 1980, *1980*, 765.
43. Doyle, M. P.; van Leusen, D.; Tamblyn, W. H. *Synthesis* 1981, *10*, 787.
44. Zhang, X.-J.; Yan, M.; Huang, D. *Org. Biomol. Chem.* 2009, *7*, 187.
45. Yamawaki, M.; Tanaka, M.; Abe, T.; Anada, M.; Hashimoto, S. *Heterocycles* 2007, *72*, 709.
46. Catino, A. J.; Nichols, J. M.; Forslund, R. E.; Doyle, M. P. *Org. Lett.* 2005, *7*, 2787.
47. Davies, H. M. L.; Denton, J. R. *Chem. Soc. Rev.* 2009, *38*, 3061.
48. Kubiak, R. W.; Mighion, J. D.; Wilkerson-Hill, S. M.; Alford, J. S.; Yoshidomi, T.; Davies, H. M. *Org. Lett.* 2016, *18*, 3118. DOI:10.1021/acs.orglett.6b01298.
49. Doyle, M. P.; Forbes, D. C.; Vasbinder, M. M.; Peterson, C. S. *J. Am. Chem. Soc.* 1998, *120*, 7653.

50. Doyle, M. P.; Hu, W. *Synlett* 2001, *2001*, 1364.
51. Doyle, M. P.; Phillips, I. M.; Hu, W. *J. Am. Chem. Soc.* 2001, *123*, 5366.
52. Doyle, M. P.; Valenzuela, M.; Huang, P. *Proc. Natl. Acad. Sci. USA* 2004, *101*, 5391.
53. Wang, Y.; Wolf, J.; Zavalij, P.; Doyle, M. P. *Angew. Chem., Int. Ed.* 2008, *47*, 1439.
54. Watanabe, N.; Shimada, N.; Anada, M.; Hashimoto, S. *Tetrahedron: Asymmetry* 2014, *25*, 63.
55. Anada, M.; Washio, T.; Shimada, N.; Kitagaki, S.; Nakajima, M.; Shiro, M.; Hashimoto, S. *Angew. Chem., Int. Ed.* 2004, *43*, 2665.
56. Hansen, J. H.; Parr, B. T.; Pelphrey, P.; Jin, Q.; Autschbach, J.; Davies, H. M. L. *Angew. Chem., Int. Ed.* 2011, *50*, 2544.
57. Doyle, M. P.; Protopopova, M. N. *Tetrahedron* 1998, *54*, 7919.
58. Davies, H. M. L.; Antoulinakis, E. G. *Org. React.* 2001, *57*, 1.
59. Tsuji, J. *Modern Rhodium-Catalyzed Organic Reactions*; Wiley-VCH: Weinheim, Germany, 2005.
60. Wong, H. N. C.; Hon, M.-Y.; Tse, C.-W.; Yip, Y.-C.; Tanko, J.; Hudlicky, T. *Chem. Rev.* 1989, *89*, 165.
61. Reissig, H.-U.; Zimmer, R. *Chem. Rev.* 2003, *103*, 1151.
62. Goldberg, A. F. G.; O'Connor, N. R.; Craig, R. A.; Stoltz, B. M. *Org. Lett.* 2002, *14*, 5314.
63. Hudlicky, T.; Reed, J. W. In *Comprehensive Organic Synthesis*; Trost, B. M.; Fleming, I. Eds.; Pergamon: New York, 1991; p. 899.
64. Rubin, M.; Rubina, M.; Gevorgyan, V. *Chem. Rev.* 2007, *107*, 3117.
65. Donaldson, W. A. *Tetrahedron* 2001, *57*, 8589.
66. Chen, D. Y.-K.; Pouwer, R. H.; Richard, J.-A. *Chem. Soc. Rev.* 2012, *41*, 4631.
67. Liu, H. W.; Walsh, C. T. In *The Chemistry of the Cyclopropyl Group*; Patai, S.; Rappoport, Z. Eds.; Wiley: Chichester, U.K., 1987; p. 959.
68. Salaun, J. *Top. Curr. Chem.* 2000, *207*, 1.
69. Gnad, F.; Reiser, O. *Chem. Rev.* 2003, *103*, 1603.
70. Reichelt, A.; Martin, S. F. *Acc. Chem. Res.* 2006, *39*, 433.
71. Staudinger, H.; Ruzicka, L. *Helv. Chim. Acta* 1924, *7*, 177.
72. Arlt, D.; Jautelat, M.; Lantzsch, R. *Angew. Chem., Int. Ed.* 1981, *20*, 703.
73. Zhang, X.; Hodgetts, K.; Rachwal, S.; Zhao, H.; Wasley, J. W. F.; Craven, K.; Brodbeck, R.; Kieltyka, A.; Hoffman, D.; Bacolod, M. D.; Girard, B.; Tran, J.; Thurkauf, A. *J. Med. Chem.* 2000, *43*, 3923.
74. Kumar, J. S. R.; Roy, S.; Datta, A. *Bioorg. Med. Chem. Lett.* 1999, *9*, 513.
75. Baldessarini, R. J. In *Goodman & Gilman's the Pharmacological Basis of Therapeutics*; Brunton, L. L.; Lazo, J. S.; Parker, K. L. Eds.; McGraw-Hill: New York, 2005.
76. Wipf, P.; Reeves, J. T.; Day, B. W. *Curr. Pharm. Des.* 2004, *10*, 1417.
77. Yamada, K.; Ojika, M.; Kigoshi, H. *Nat. Prod. Rep.* 2007, *24*, 798.
78. de Meijere, A.; Kozhushkov, S. I.; Fokin, A. A.; Emme, I.; Redlich, S.; Schreiner, P. R. *Pure Appl. Chem.* 2003, *75*, 549.
79. Yates, P. *J. Am. Chem. Soc.* 1952, *74*, 5376.
80. Nakamura, E.; Yoshikai, N.; Yamanaka, M. *J. Am. Chem. Soc.* 2002, *124*, 7181.
81. Doyle, M. P. *Acc. Chem. Res.* 1986, *19*, 348.
82. Hansen, J.; Autschbach, J.; Davies, H. M. L. *J. Org. Chem.* 2009, *74*, 6555.
83. Nowlan, D. T.; Gregg, T. M.; Davies, H. M. L.; Singleton, D. A. *J. Am. Chem. Soc.* 2003, *125*, 15902.
84. Sheehan, S. M.; Padwa, A.; Snyder, J. P. *Tetrahedron Lett.* 1998, *39*, 949.
85. Wong, F. M.; Wang, J.; Hengge, A. C.; Wu, W. *Org. Lett.* 2007, *9*, 1663.
86. Berry, J. F. *Dalton Trans.* 2012, *41*, 700.
87. Pirrung, M. C.; Liu, H.; Morehead, A. T. *J. Am. Chem. Soc.* 2002, *124*, 1014.
88. Pirrung, M. C.; Morehead, A. T. *J. Am. Chem. Soc.* 1996, *118*, 8162.
89. Kornecki, K. P.; Briones, J. F.; Boyarskikh, V.; Fullilove, F.; Autschbach, J.; Schrote, K. E.; Lancaster, K. M.; Davies, H. M. L.; Berry, J. F. *Science* 2013, *342*, 351.
90. Fraile, J. M.; García, J. I.; Gissibl, A.; Mayoral, J. A.; Pires, E.; Reiser, O.; Roldán, M.; Villalba, I. *Chem.-Eur. J.* 2007, *13*, 8830.
91. Davies, H. M. L.; Bruzinski, P. R.; Lake, D. H.; Kong, N.; Fall, M. J. *J. Am. Chem. Soc.* 1996, *118*, 6897.
92. Cotton, F. A.; Walton, R. A. *Multiple Bonds between Metal Atoms*; Clarendon Press: Oxford, U.K., 1993.
93. Bonge, H. T.; Hansen, T. *J. Org. Chem.* 2010, *75*, 2309.
94. Padwa, A.; Austin, D. J.; Hornbuckle, S. F.; Semones, M. A.; Doyle, M. P.; Protopopova, M. N. *J. Am. Chem. Soc.* 1992, *114*, 1874.
95. Catino, A. J.; Nichols, J. M.; Choi, H.; Gottipamula, S.; Doyle, M. P. *Org. Lett.* 2005, *7*, 5167.
96. Hansen, J.; Dikarev, E.; Autschbach, J.; Davies, H. M. L. *J. Org. Chem.* 2009, *74*, 6564.
97. Durivage, J. C.; Gruhn, N. E.; Li, B.; Dikarev, E. V.; Lichtenberger, D. L. *J. Cluster Sci.* 2008, *19*, 275.

98. Dikarev, E. V.; Gray, T. G.; Li, B. *Angew. Chem., Int. Ed.* 2005, *44*, 1721.
99. Dikarev, E. V.; Li, B.; Zhang, H. *J. Am. Chem. Soc.* 2006, *128*, 2814.
100. Trindade, A. F.; Gois, P. M. P.; Veiros, L. F.; Andre, V.; Duarte, M. T.; Afonso, C. A. M.; Caddick, S.; Cloke, F. G. N. *J. Org. Chem.* 2008, *73*, 4076.
101. Gois, P. M. P.; Trindade, A. F.; Veiros, L. F.; Andre, V.; Duarte, M. T.; Afonso, C. A. M.; Caddick, S.; Cloke, F. G. N. *Angew. Chem., Int. Ed.* 2007, *46*, 5750.
102. Hashimoto, S.; Watanabe, N.; Ikegami, S. *J. Chem. Soc., Chem. Commun.* 1992, *1992*, 1508.
103. Kennedy, M.; McKervey, M. A.; Maguire, A. R.; Roos, G. H. P. *J. Chem. Soc., Chem. Commun.* 1990, *1990*, 361.
104. Ye, T.; Garcia, F. C.; McKervey, M. A. *J. Chem. Soc., Perkin Trans. 1* 1995, *1995*, 1373.
105. Ye, T.; McKervey, M. A.; Brandes, B. D.; Doyle, M. P. *Tetrahedron Lett.* 1994, *35*, 7269.
106. Davies, H. M. L.; Bruzinski, P. R.; Fall, M. J. *Tetrahedron Lett.* 1996, *37*, 4133.
107. Davies, H. M. L.; Rusiniak, L. *Tetrahedron Lett.* 1998, *39*, 8811.
108. Davies, H. M. L.; Townsend, R. J. *J. Org. Chem.* 2001, *66*, 6595.
109. Chepiga, K. M.; Qin, C.; Alford, J. S.; Chennamadhavuni, S.; Gregg, T. M.; Olson, J. P.; Davies, H. M. L. *Tetrahedron* 2013, *69*, 5765.
110. Davies, H. M. L.; Nagashima, T.; Klino, J. L. *Org. Lett.* 2000, *2*, 823.
111. Corey, E. J.; Gant, T. J. *Tetrahedron Lett.* 1994, *35*, 5373.
112. Nagashima, T.; Davies, H. M. L. *J. Am. Chem. Soc.* 2001, *123*, 2695.
113. Davies, H. M. L.; Boebel, T. A. *Tetrahedron Lett.* 2000, *41*, 8189.
114. Ni, A.; France, J. E.; Davies, H. M. L. *J. Org. Chem.* 2006, *71*, 5594.
115. Hadley, S. J.; Ventura, D. L.; Dominiak, P. M.; Nygren, C. L.; Davies, H. M. L. *J. Org. Chem.* 2006, *71*, 5349.
116. Davies, H. M. L.; Dai, X.; Long, M. S. *J. Am. Chem. Soc.* 2006, *128*, 2485.
117. Gregg, T. M.; Farrugia, M. K.; Frost, J. R. *Org. Lett.* 2009, *11*, 4434.
118. Gregg, T. M.; Algera, R. F.; Frost, J. R.; Hassan, F.; Stewart, R. J. *Tetrahedron Lett.* 2010, *51*, 6429.
119. Negretti, S.; Cohen, C. M.; Chang, J. J.; Guptill, D. M.; Davies, H. M. L. *Tetrahedron* 2015, *71*, 7415.
120. Davies, H. M. L.; Lee, G. H. *Org. Lett.* 2004, *6*, 1233.
121. Briones, J. F.; Hansen, J.; Hardcastle, K. I.; Autschbach, J.; Davies, H. M. L. *J. Am. Chem. Soc.* 2010, *132*, 17211.
122. Davies, H. M. L.; Kong, N. *Tetrahedron Lett.* 1997, *38*, 4203.
123. Davies, H. M. L.; Panaro, S. A. *Tetrahedron Lett.* 1999, *40*, 5287.
124. Davies, H. M. L.; Doan, B. D. *J. Org. Chem.* 1999, *64*, 8501.
125. Kang, J.; Zhu, B.; Liu, J.; Wang, B.; Zhang, L.; Su, C.-Y. *Org. Chem. Front.* 2015, *2*, 890.
126. Takahashi, T.; Tsutsui, H.; Tamura, M.; Kitagaki, S.; Nakajima, M.; Hashimoto, S. *Chem. Commun.* 2001, *2001*, 1604.
127. Kitagaki, S.; Anada, M.; Kataoka, O.; Matsuno, K.; Umeda, C.; Watanabe, N.; Hashimoto, S. *J. Am. Chem. Soc.* 1999, *121*, 1417.
128. Tsutsui, H.; Matsuura, M.; Makino, K.; Nakamura, S.; Nakajima, M.; Kitagaki, S.; Hashimoto, S. *Isr. J. Chem.* 2001, *41*, 283.
129. Yamawaki, M.; Tsutsui, H.; Kitagaki, S.; Anada, M.; Hashimoto, S. *Tetrahedron Lett.* 2002, *43*, 9561.
130. Tsutsui, H.; Yamaguchi, Y.; Kitagaki, S.; Nakamura, S.; Anada, M.; Hashimoto, S. *Tetrahedron: Asymmetry* 2003, *14*, 817.
131. Minami, K.; Saito, H.; Tsutsui, H.; Nambu, H.; Anada, M.; Hashimoto, S. *Adv. Synth. Catal.* 2005, *347*, 1483.
132. Hashimoto, S.; Watanabe, N.; Ikegami, S. *Tetrahedron Lett.* 1990, *31*, 5173.
133. Tsutsui, H.; Abe, T.; Nakamura, S.; Anada, M.; Hashimoto, S. *Chem. Pharm. Bull.* 2005, *53*, 1366.
134. Collet, F.; Lescot, C.; Liang, C.; Dauban, P. *Dalton Trans.* 2010, *39*, 10401.
135. Ghanem, A.; Gardiner, M. G.; Williamson, R. M.; Müller, P. *Chem.-Eur. J.* 2010, *16*, 3291.
136. Reddy, R. P.; Lee, G. H.; Davies, H. M. L. *Org. Lett.* 2006, *8*, 3437.
137. Denton, J. R.; Sukumaran, D.; Davies, H. M. L. *Org. Lett.* 2007, *9*, 2625.
138. Wang, H.; Guptill, D. M.; Varela-Alvarez, A.; Musaev, D. G.; Davies, H. M. L. *Chem. Sci.* 2013, *4*, 2844.
139. Hashimoto, S.; Watanabe, N.; Sato, T.; Shiro, M.; Ikegami, S. *Tetrahedron Lett.* 1993, *34*, 5109.
140. Anada, M.; Kitagaki, S.; Hashimoto, S. *Heterocycles* 2000, *52*, 875.
141. DeAngelis, A.; Dmitrenko, O.; Yap, G. P. A.; Fox, J. M. *J. Am. Chem. Soc.* 2009, *131*, 7230.
142. DeAngelis, A.; Boruta, D. T.; Lubin, J.-B.; Plampin, J. N.; Yap, G. P. A.; Fox, J. M. *Chem. Commun.* 2010, *46*, 4541.
143. Lindsay, V. N. G.; Nicolas, C.; Charette, A. B. *J. Am. Chem. Soc.* 2011, *133*, 8972.

144. Lindsay, V. N. G.; Lin, W.; Charette, A. B. *J. Am. Chem. Soc.* 2009, *131*, 16383.
145. Goto, T.; Takada, K.; Shimada, N.; Nambu, H.; Anada, M.; Shiro, M.; Ando, K.; Hashimoto, S. *Angew. Chem., Int. Ed.* 2011, *50*, 6803.
146. Mattiza, J. T.; Fohrer, J. G. G.; Duddeck, H.; Gardiner, M. G.; Ghanem, A. *Org. Biomol. Chem.* 2011, *9*, 6542.
147. Qin, C.; Boyarskikh, V.; Hansen, J. H.; Hardcastle, K. I.; Musaev, D. G.; Davies, H. M. L. *J. Am. Chem. Soc.* 2011, *133*, 19198.
148. Gardiner, M. G.; Ghanem, A. Unpublished results 2010.
149. Awata, A.; Arai, T. *Synlett* 2013, *24*, 29.
150. Goto, T.; Takada, K.; Anada, M.; Ando, K.; Hashimoto, S. *Tetrahedron Lett.* 2011, *52*, 4200.
151. Xu, X.; Deng, Y.; Yim, D. N.; Zavalij, P. Y.; Doyle, M. P. *Chem. Sci.* 2015, *6*, 2196.
152. Adly, F. G.; Gardiner, M. G.; Ghanem, A. *Chem.-Eur. J.* 2016, *22*, 3447.
153. Denton, J. R.; Cheng, K.; Davies, H. M. L. *Chem. Commun.* 2008, *2008*, 1238.
154. Denton, J. R.; Davies, H. M. L. *Org. Lett.* 2009, *11*, 787.
155. Briones, J. F.; Davies, H. M. L. *Tetrahedron* 2011, *67*, 4313.
156. Müller, P.; Allenbach, Y.; Robert, E. *Tetrahedron: Asymmetry* 2003, *14*, 779.
157. Müller, P.; Ghanem, A. *Synlett* 2003, *12*, 1830.
158. Müller, P.; Lacrampe, F. *Helv. Chim. Acta* 2004, *87*, 2848.
159. Müller, P.; Bernardinelli, G.; Allenbach, Y. F.; Ferri, M.; Flack, H. D. *Org. Lett.* 2004, *6*, 1725.
160. Müller, P.; Bernardinelli, G.; Allenbach, Y.; Ferri, M.; Grass, S. *Synlett* 2005, *2005*, 1397.
161. Müller, P.; Allenbach, Y.; Chappellet, S.; Ghanem, A. *Synthesis* 2006, *10*, 1689.
162. Marcoux, D.; Goudreau, S. R.; Charette, A. B. *J. Org. Chem.* 2009, *74*, 8939.
163. Marcoux, D.; Charette, A. B. *Angew. Chem.* 2008, *47*, 10155.
164. Marcoux, D.; Lindsay, V. N. G.; Charette, A. B. *Chem. Commun.* 2010, *46*, 910.
165. Marcoux, D.; Azzi, S.; Charette, A. B. *J. Am. Chem. Soc.* 2009, *131*, 6970.
166. Müller, P.; Allenbach, Y.; Grass, S. *Tetrahedron: Asymmetry* 2005, *16*, 2007.
167. Chuprakov, S.; Kwok, S. W.; Zhang, L.; Lercher, L.; Fokin, V. V. *J. Am. Chem. Soc.* 2009, *131*, 18034.
168. Müller, P.; Ghanem, A. *Org. Lett.* 2004, *6*, 4347.
169. Ghanem, A.; Aboul-Enein, H. Y.; Müller, P. *Chirality* 2005, *17*, 44.
170. Adly, F. G.; Maddalena, J.; Ghanem, A. *Chirality* 2014, *26*, 764.
171. Boruta, D. T.; Dmitrenko, O.; Yap, G. P. A.; Fox, J. M. *Chem. Sci.* 2012, *3*, 1589.
172. Lindsay, V. N. G.; Charette, A. B. *ACS Catal.* 2012, *2*, 1221.
173. Davies, H. M. L.; Hansen, J.; Changming, Q. PCT/US2012/040608, 2012
174. Qin, C.; Davies, H. M. L. *J. Am. Chem. Soc.* 2013, *135*, 14516.
175. Qin, C.; Davies, H. M. L. *Org. Lett.* 2013, *15*, 310.
176. Dennis, A. M.; Korp, J. D.; Bernal, I.; Howard, R. A.; Bear, J. L. *Inorg. Chem.* 1983, *22*, 1522.
177. Doyle, M. P.; Brandes, B. D.; Kazala, A. P.; Pieters, R. J.; Jarstfer, M. B.; Watkins, L. M.; Eagle, C. T. *Tetrahedron Lett.* 1990, *31*, 6613.
178. Doyle, M. P.; Winchester, W. R.; Hoorn, J. A. A.; Lynch, V.; Simonsen, S. H.; Ghosh, R. *J. Am. Chem. Soc.* 1993, *115*, 9968.
179. Doyle, M. P.; Winchester, W. R.; Protopopova, M. N.; Müller, P.; Bernardinelli, G.; Ene, D. G.; Motallebi, S. *Helv. Chim. Acta* 1993, *76*, 2227.
180. Doyle, M. P.; Davies, S. B.; Hu, W. H. *Chem. Commun.* 2000, *10*, 867.
181. Doyle, M. P.; Dyatkin, A. B.; Protopopova, M. N.; Yang, C. I.; Miertschin, C. S.; Winchester, W. R.; Simonsen, S. H.; Lynch, V.; Ghosh, R. *Recl. Trav. Chim. Pays-Bas* 1995, *114*, 163.
182. Doyle, M. P.; Zhou, Q.-L.; Raab, C. E.; Roos, G. H. P.; Simonsen, S. H.; Lynch, V. *Inorg. Chem.* 1996, *35*, 6064.
183. Doyle, M. P.; Davies, S. B.; Hu, W. *Org. Lett.* 2000, *2*, 1145.
184. Doyle, M. P.; Hu, W.; Phillips, I. M.; Moody, C. J.; Pepper, A. G.; Slawin, A. G. *Adv. Synth. Catal.* 2001, *343*, 112.
185. Lin, W.; Charette, A. B. *Adv. Synth. Catal.* 2005, *347*, 1547.
186. Doyle, M. P.; Zhou, Q. L.; Charnsangavej, C.; Longoria, M. A.; McKervey, M. A.; Garcia, C. F. *Tetrahedron Lett.* 1996, *37*, 4129.
187. Doyle, M. P.; Austin, R. E.; Bailey, A. S.; Dwyer, M. P.; Dyatkin, A. B.; Kalinin, A. V.; Kwan, M. M. Y.; Liras, S.; Oalmann, C. J.; Pieters, R. J.; Protopopova, M. N.; Raab, C. E.; Roos, G. H. P.; Zhou, Q. L.; Stephen, F. M. *J. Am. Chem. Soc.* 1995, *117*, 5763.
188. Welch, C. J.; Tu, Q.; Wang, T.; Raab, C.; Wang, P.; Jia, X.; Bu, X.; Bykowski, D.; Hohenstaufen, B.; Doyle, M. P. *Adv. Synth. Catal.* 2006, *348*, 821.

189. Doyle, M. P.; Morgan, J. P.; Colyer, J. T. *J. Organomet. Chem.* 2005, *690*, 5525.
190. Doyle, M. P.; Morgan, J. P.; Fettinger, J. C.; Zavalij, P. Y.; Colyer, J. T.; Timmons, D. J.; Carducci, M. D. *J. Org. Chem.* 2005, *70*, 5291.
191. Doyle, M. P.; Protopopova, M.; Müller, P.; Ene, D.; Shapiro, E. A. *J. Am. Chem. Soc.* 1994, *116*, 8492.
192. Doyle, M. P.; Zhou, Q. L.; Simonsen, S. H.; Lynch, V. *Synlett* 1996, *1996*, 697.
193. Hu, W.; Timmons, D. J.; Doyle, M. P. *Org. Lett.* 2002, *4*, 901.
194. Doyle, M. P.; Hu, W. *Adv. Synth. Catal.* 2001, *343*, 299.
195. Bykowski, D.; Wu, K.-H.; Doyle, M. P. *J. Am. Chem. Soc.* 2006, *128*, 16038.
196. Doyle, M. P.; Zhou, Q. L.; Dyatkin, A. B.; Ruppar, D. A. *Tetrahedron Lett.* 1995, *36*, 7579.
197. Doyle, M. P.; Peterson, C. S.; Zhou, Q. L.; Nishiyama, H. *J. Chem. Soc., Chem. Commun.* 1997, *1997*, 211.
198. Doyle, M. P.; Wang, Y.; Ghorbani, P.; Bappert, E. *Org. Lett.* 2005, *7*, 5035.
199. Lindsay, V. N. G.; Fiset, D.; Gritsch, P. J.; Azzi, S.; Charette, A. B. *J. Am. Chem. Soc.* 2013, *135*, 1463.
200. Doyle, M. P.; Ene, D. G.; Peterson, C. S.; Lynch, V. *Angew. Chem., Int. Ed.* 1999, *38*, 700.
201. Rogers, D. H.; Yi, E. C.; Poulter, C. D. *J. Org. Chem.* 1995, *60*, 941.
202. Martin, S. F.; Dwyer, M. P.; Hartmann, B.; Knight, K. S. *J. Org. Chem.* 2000, *65*, 1305.
203. Watanabe, N.; Matsuda, H.; Kuribayashi, H.; Hashimoto, S. *Heterocycles* 1996, *42*, 537.
204. Kitagaki, S.; Matsuda, H.; Watanabe, N.; Hashimoto, S. *Synlett* 1997, *1997*, 1171.
205. Washio, T.; Nambu, H.; Anada, M.; Hashimoto, S. *Tetrahedron: Asymmetry* 2007, *18*, 2606.
206. Lou, Y.; Horikawa, M.; Kloster, R. A.; Hawryluk, N. A.; Corey, E. J. *J. Am. Chem. Soc.* 2004, *126*, 8916.
207. Lou, Y.; Remarchuk, T. P.; Corey, E. J. *J. Am. Chem. Soc.* 2005, *127*, 14223.
208. Nichols, J. M.; Liu, Y.; Zavalij, P.; Isaacs, L.; Doyle, M. P. *Inorg. Chim. Acta* 2008, *361*, 3309.
209. Doyle, M. P.; Forbes, D. C. *Chem. Rev.* 1998, *98*, 911.
210. Candeias, N. R.; Gois, P. M. P.; Afonso, C. A. M. *J. Org. Chem.* 2006, *71*, 5489.
211. Wynne, D. C.; Olmstead, M. M.; Jessop, P. G. *J. Am. Chem. Soc.* 2000, *122*, 7638.
212. Antos, J. M.; McFarland, J. M.; Iavarone, A. T.; Francis, M. B. *J. Am. Chem. Soc.* 2009, *131*, 6301.

4 Chirality Organization of Peptides and π-Conjugated Polyanilines

Toshiyuki Moriuchi, Satoshi D. Ohmura, and Toshikazu Hirao

CONTENTS

4.1 Introduction... 133
4.2 Chirality Organization of Peptides 135
    4.2.1 Chirality Organization of Peptides by Using Organic Molecular Scaffold 135
    4.2.2 Chirality Organization of Peptides by Using Organometallic Molecular Scaffold 138
4.3 Synthesis of Optically Active Polyanilines.. 143
    4.3.1 Polymerization of Anilines in the Presence of a Chiral Acid 143
    4.3.2 Doping of Emeraldine Bases with a Chiral Acid............................. 147
    4.3.3 Introduction of Chiral Groups into Polyanilines 148
4.4 Chiral Complexation of Emeraldine Bases with Chiral Complexes................ 154
4.5 Application of Optically Active Polyanilines 156
4.6 Conclusion 156
References.. 157

4.1 INTRODUCTION

Architectural control of molecular self-assembly is of great importance for the development of functional materials.[1–10] The utilization of self-assembling properties of amino acids, which possess chiral centers and hydrogen bonding sites, is considered to be a powerful tool for structurally defined molecular assemblies. Highly ordered molecular arrangements are observed in proteins to fulfill the specific functions as observed in enzymes, receptors, etc. Secondary structures of proteins such as α-helices, β-sheets, and β-turns are known to play an important role in the three-dimensional structures and biological activity of proteins.[11,12] Typical secondary structures are shown in Figure 4.1. α-Helix has a right-handed helical conformation wherein each CO group is hydrogen bonded to the NH group of the fourth residue further toward the C-terminus. In parallel β-sheets, a series of 12-membered hydrogen-bonded rings are organized, while an alternating series of 10- and 14-membered hydrogen-bonded rings are formed in antiparallel β-sheets. β-Turn forms a hydrogen bond between the CO group and the NH group of the third residue further toward the C-terminus to form a 10-membered hydrogen-bonded ring. A 7-membered hydrogen-bonded ring based on the formation of a hydrogen bond between the CO group and the NH group of the second residue further toward the C-terminus is formed in γ-turn. Considerable efforts have been devoted to design secondary structure mimics composed of short peptides to gain fundamental insight into the factors affecting the protein structure and stability. The utilization of molecular scaffolds is a potential strategy for the control of intramolecular interaction of peptides (Figure 4.2). A variety of molecular scaffolds have been employed to induce the β-sheet and turn structures of attached peptides.

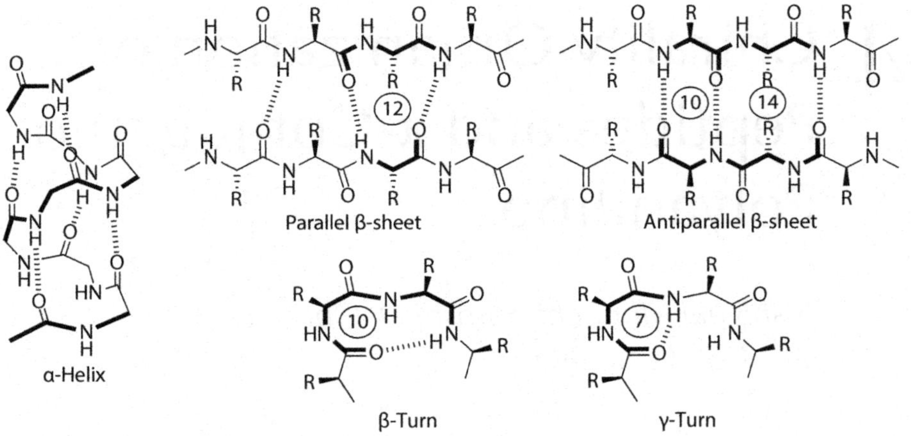

FIGURE 4.1 Secondary structures of proteins.

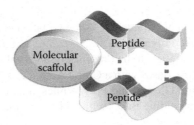

FIGURE 4.2 Chirality organization of peptides by using molecular scaffold.

On the other hand, π-conjugated polymers and oligomers have drawn much attention for the application of electrical materials depending on their redox properties.[13–17] Polyanilines (PAn's) are one of the promising conducting π-conjugated polymers with chemical stability. PAn has been extensively studied due to its unique electronic and redox properties as well as numerous potential uses in a wide range of applications in many fields. PAn's exist in three different discrete redox forms, which include the fully reduced leucoemeraldine, the semioxidized emeraldine, and the fully oxidized pernigraniline base forms as shown in Figure 4.3.[13] The redox properties have been demonstrated to be controlled by the introduction of an acceptor unit.[18,19] Another interesting function of PAn's is the coordination properties of two nitrogen atoms of the quinonediimine moiety. Two nitrogen atoms of the quinonediimine moiety of the emeraldine base form have been tested to be capable of participating in the complexation to afford novel conjugated polymer complexes.[20–22] The conjugated polymer complex can effectively serve as an oxidation catalyst.[23–29] Also, controlled complexation with redox-active quinonediimine derivatives has been demonstrated to afford the conjugated polymeric complexes, the conjugated trimetallic macrocycle, or the conjugated bimetallic complexes, depending on the coordination mode of the introduced metal complexes.[30–38] To induce further functionalization of PAn's, the structural control of PAn's has attracted a great deal of attention. Recently, optically active PAn's have attracted considerable interest because of their potential applications in chiral sensing, chiral recognition, and chiral separation. This chapter summarizes the molecular scaffolds for the creation of the β-sheet and turn structures of attached peptides (Figure 4.2) and the chirality organization of π-conjugated PAn's (Figure 4.3).

Chirality Organization of Peptides and π-Conjugated Polyanilines

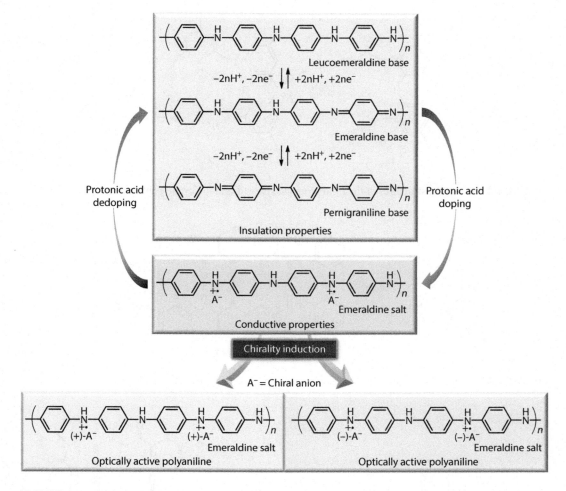

FIGURE 4.3 Redox behavior of PAn and chirality induction into the π-conjugated backbone of PAn.

4.2 CHIRALITY ORGANIZATION OF PEPTIDES

4.2.1 Chirality Organization of Peptides by Using Organic Molecular Scaffold

The rigid aromatic spacer **1** serves as an organic molecular scaffold to induce a hydrogen-bonding bridge in the cyclic peptide **2** consisting of the amino acid sequence Ile-Val-Gly (Figure 4.4). The NMR studies of **2** support the formation of the β-type hydrogen-bonding bridge in dimethyl sulfoxide (DMSO)-d_6.[39]

The biphenyl-containing pseudo–amino acid **3** is used as a molecular scaffold in the backbone of the cyclic peptide **4** (Figure 4.5).[40] Although there are three interconverting (*R,R*)-, (*S,S*)-, and (*R,S*)-diastereomers in a solution based on the atropisomerism of the biphenyl units, the NMR studies and molecular dynamics calculation of **4** suggest the (*R,R*)-isomer as a major diastereomer to adopt the antiparallel β-sheet conformation in DMSO-d_6.

The epindolidione scaffold **5** is employed as a central strand of the β-sheet to orient the attached peptides appropriately. The NMR studies of **6** in DMSO-d_6 support the antiparallel β-sheet conformation of **6**, wherein the L-Pro-D-Ala dipeptide for the formation of a type II β-turn and a chain-reversing urea linking unit are incorporated (Figure 4.6).[41]

The distance between C4 and C6 (4.9 Å) of dibenzofuran is close to the distance between the strands of an antiparallel β-sheet. Not only the hydrogen bonding but also the hydrophobic interaction

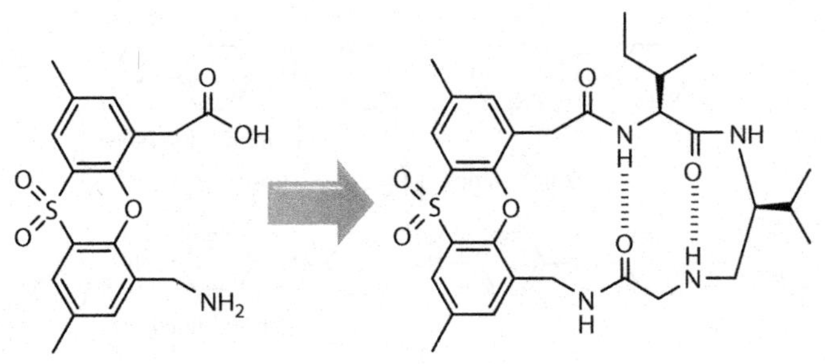

FIGURE 4.4 Chemical structures of the rigid aromatic spacer **1** and the cyclic peptide **2**.

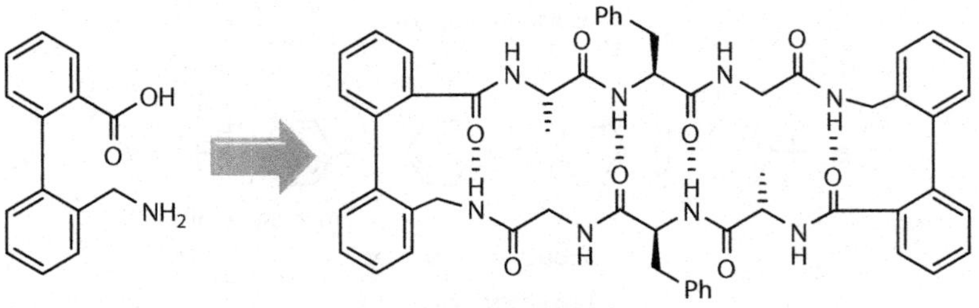

FIGURE 4.5 Chemical structures of the rigid aromatic spacer **3** and the cyclic peptide **4**.

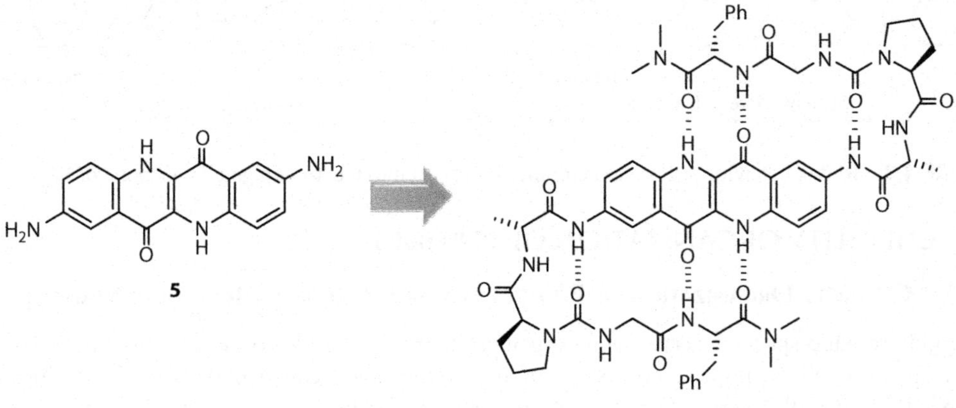

FIGURE 4.6 Chemical structures of the rigid aromatic spacer **5** and the cyclic peptide **6**.

plays an important role in protein folding. Heptapeptides **8** are able to form an antiparallel β-sheet structure only when the side chains of the flanking amino acid residues are hydrophobic, wherein the 4-(2-aminoethyl)-6-dibenzofuranpropionic acid scaffold **7** is employed to form a hydrophobic cluster composed of the dibenzofuran skeleton and the hydrophobic side chains of the flanking amino acids (Figure 4.7).[42]

The 2-amino-3-carboxy-*endo*-(2*S*,3*R*)-norborn-5-ene serves as a conformationally constrained β-amino acid scaffold. A urea linkage is introduced into pseudopeptide **9** to offset the two peptide chains from one another so as to induce interchain hydrogen bonding (Figure 4.8).[43] The NMR measurements of pseudopeptide **9** in CDCl$_3$ suggest the formation of a parallel β-sheet

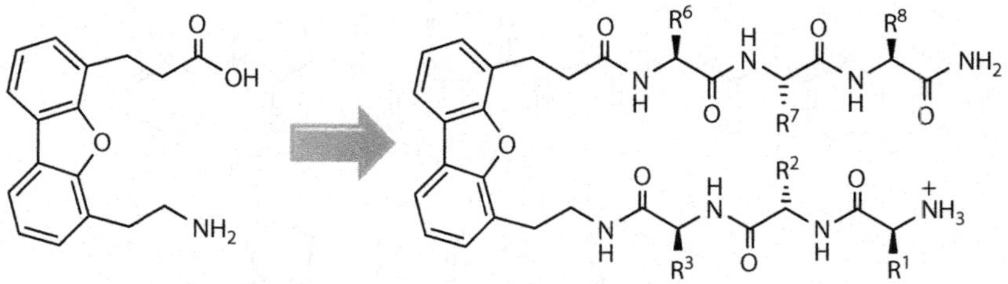

Chirality Organization of Peptides and π-Conjugated Polyanilines   137

FIGURE 4.7 Chemical structures of the rigid aromatic spacer **7** and the cyclic peptide **8**.

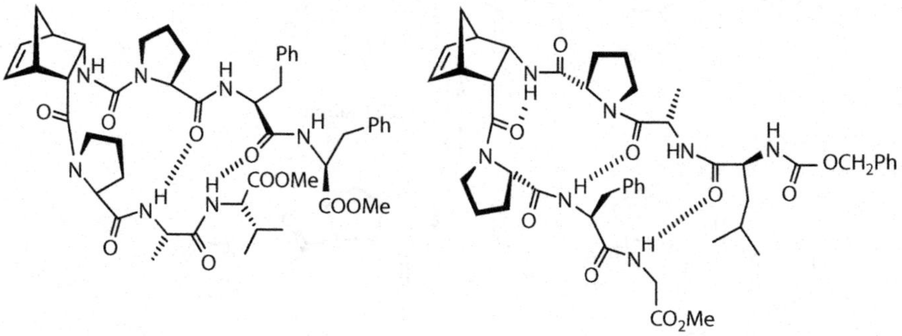

FIGURE 4.8 Chemical structures of the rigid aromatic spacer **9** and the cyclic peptide **10**.

conformation, which is stabilized by two intramolecular hydrogen bonds involving the NHs of Ala and Val units. On the other hand, the peptide **10** forms an antiparallel β-sheet conformation in CDCl$_3$ (Figure 4.8).[43]

An oligourea molecular scaffold **11** is utilized to hold multiple peptides in proximity. NMR studies of the peptide **12**, wherein a diurea molecular scaffold is introduced to juxtapose two peptide strands, indicate a parallel β-sheet conformation in CDCl$_3$ (Figure 4.9).[44]

A urea unit is introduced into peptides to allow the formation of the hydrogen-bonded duplex (Figure 4.10).[45] The single-crystal x-ray structure determination of the dipeptidyl urea **13** composed of two L-Ala-L-Pro dipeptide chains having the C-terminal pyridyl moieties reveals the formation of a hydrogen-bonded duplex by six intermolecular hydrogen bonds. A shuttle-like molecular

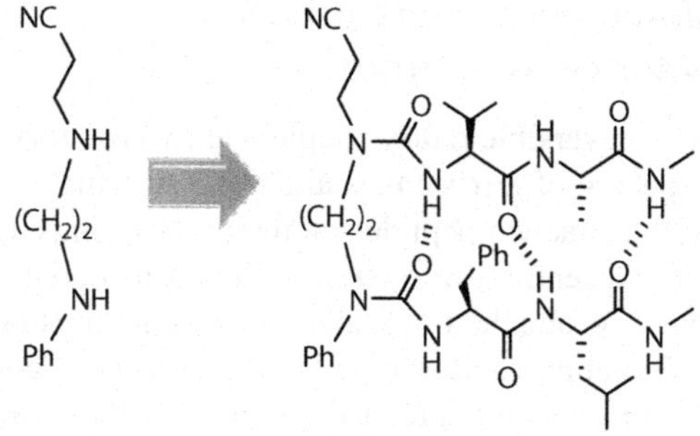

FIGURE 4.9 Chemical structures of the rigid aromatic spacer **11** and the cyclic peptide **12**.

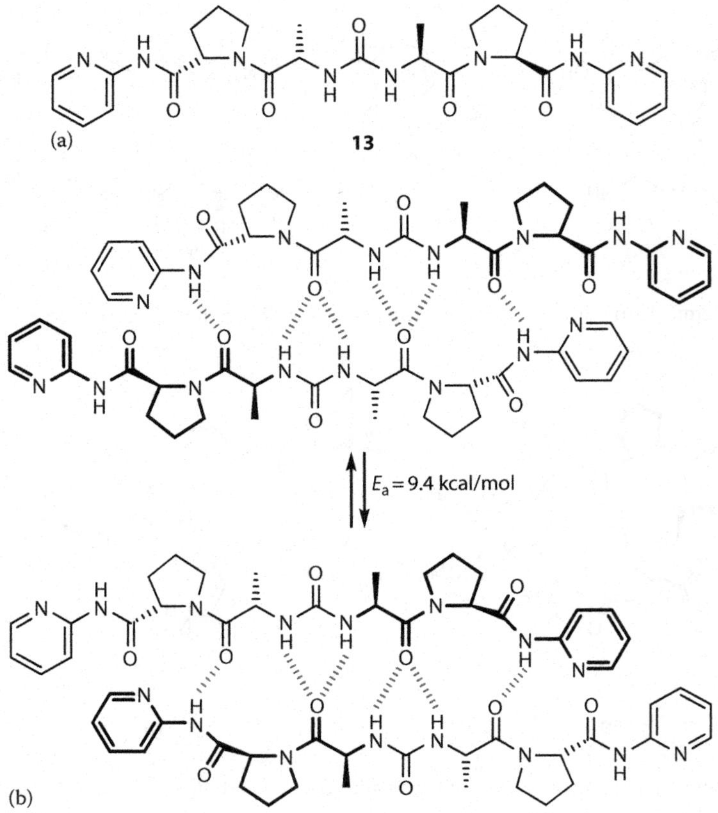

FIGURE 4.10 (a) Chemical structure of **13** and (b) a shuttle-like dynamic process of the dipeptidyl urea **13**.

dynamics based on the recombination of hydrogen bonds is possible in a solution state as shown in Figure 4.10b. The activation energy of this dynamic process is calculated as 9.4 kcal/mol from the Arrhenius equation.

The quadruply hydrogen-bonded duplex **14·15**, which has the complementary hydrogen-bonding sequences ADAA/DADD (A, hydrogen bond acceptor; D, hydrogen bond donor), is employed as a noncovalent scaffold for the nucleation of β-sheet structure (Figure 4.11).[46] The introduction of four tripeptide chains to the same end of a hydrogen-bonded duplex **14·15** induces the formation of four hybrid duplexes, **14a·15a**, **14a·15b**, **14b·15a**, and **14b·15b**, to form a β-sheet structure.

4.2.2 Chirality Organization of Peptides by Using Organometallic Molecular Scaffold

Organometallic ferrocene has a reversible redox couple and two rotatory coplanar cyclopentadienyl (Cp) rings.[47] The inter-ring spacing of ferrocene is about 3.3 Å, which is appropriate for hydrogen-bonding interaction between the attached peptides on the two Cp rings as observed in β-sheets. The utilization of a ferrocene unit as a central reverse-turn unit is considered to be one strategy to induce the β-sheet and turn structures of attached peptides. Ferrocene–peptide conjugates composed of Fc-dicarboxylic acid (Fc-cc), Fc-amino acid (Fc-ac), or Fc-diamine (Fc-aa) scaffold are designed to study the hydrogen-bonding ability of the attached peptides as shown in Figure 4.12, wherein the introduced peptide strands are regulated in the appropriate dimensions to induce a 10-membered hydrogen-bonded ring (Fc-cc scaffold), a 12-membered hydrogen-bonded ring (Fc-ac scaffold), and a 14-membered hydrogen-bonded ring (Fc-aa scaffold).

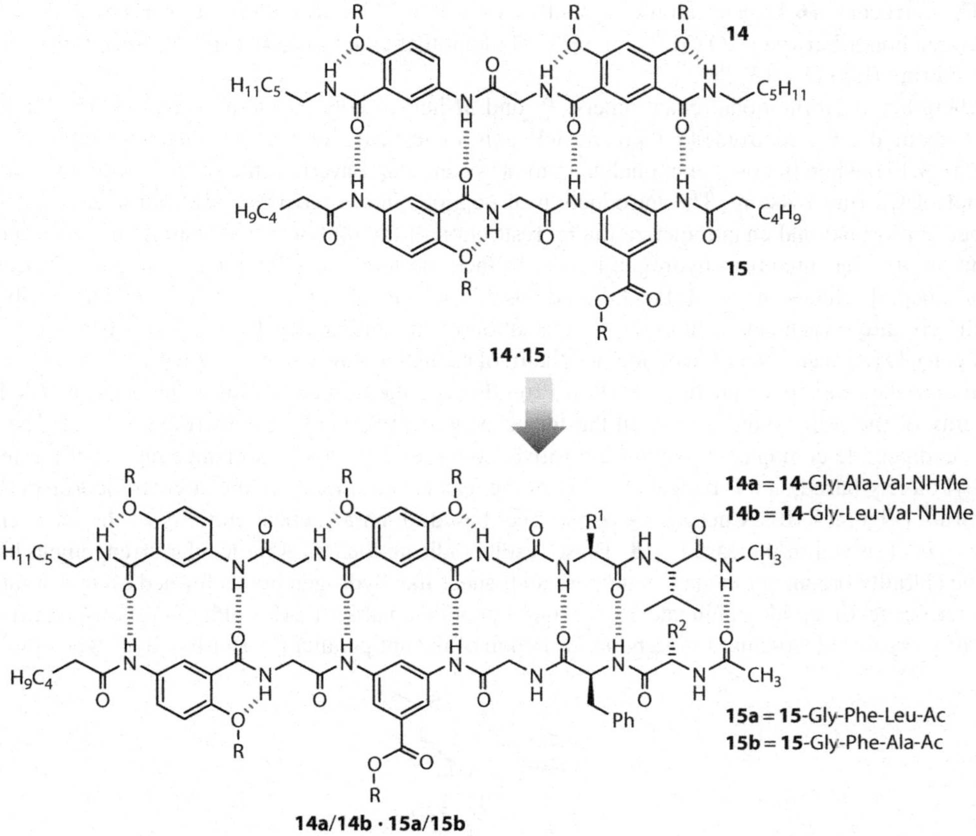

Chirality Organization of Peptides and π-Conjugated Polyanilines

FIGURE 4.11 Chemical structure of **14·15** and **14a/14b·15a/15b**.

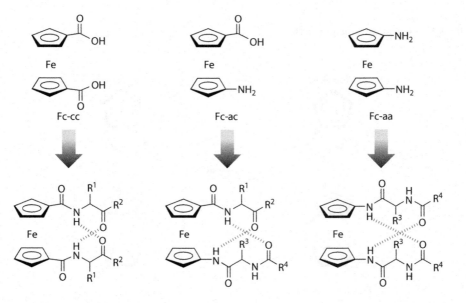

FIGURE 4.12 Ferrocene–peptide conjugates using Fc-dicarboxylic acid (Fc-cc), Fc-amino acid (Fc-ac), or Fc-diamine (Fc-aa) scaffold.

The ferrocene **16** bearing amino acid chains (-L-Val-OMe) forms the antiparallel β-sheet-like hydrogen bonds between CO (Val) and NH (Val of another chain) to give a 10-membered hydrogen-bonded ring (Figure 4.13).[48,49]

There are conformational enantiomers, *P*- and *M*-helical conformations, based on the torsional twist about the Cp(centroid)-Fe-Cp(centroid) axis in the case of the 1,1′-disubstituted ferrocene (Figure 4.14), wherein conformational enantiomers can interconvert with ease due to the low energy barrier of Cp ring twisting. The introduction of peptides into a ferrocene scaffold is envisioned to induce conformational enantiomerization by restriction of the torsional twist based on the formation of intramolecular interchain hydrogen bonds. In fact, the ferrocene–dipeptide conjugate **17** bearing the L-dipeptide chains (-L-Ala-L-Pro-OEt) adopts a *P*-helical chirality of the ferrocenoyl moiety based on the chirality organization through the formation of the antiparallel β-sheet-like hydrogen bonds between CO (Ala) and NH (Ala of another chain) of each dipeptide chain (Figure 4.15).[50–52] The introduction of the D-dipeptide chains (-D-Ala-D-Pro-OEt) into the ferrocene scaffold induces an *M*-helical chirality of the ferrocenoyl moiety in the ferrocene–dipeptide conjugate **18** (Figure 4.15). The ferrocene–dipeptide conjugate **17** shows a positive Cotton effect at the absorbance region of the ferrocenoyl moiety based on a *P*-helical chirality of the ferrocenoyl moiety in the circular dichroism (CD) spectrum (Figure 4.16). A negative Cotton effect based on an *M*-helical chirality of the ferrocenoyl moiety is observed in the case of **18**. These results indicate the chiral molecular arrangement based on the chirality organization via the antiparallel β-sheet-like hydrogen bonds formed even in solution. The ferrocene–dipeptide conjugate **19** bearing dipeptide chains (-L-Ala-L-Phe-OMe) also forms the chirality-organized structure through the formation of the antiparallel β-sheet-like hydrogen bonds.[53]

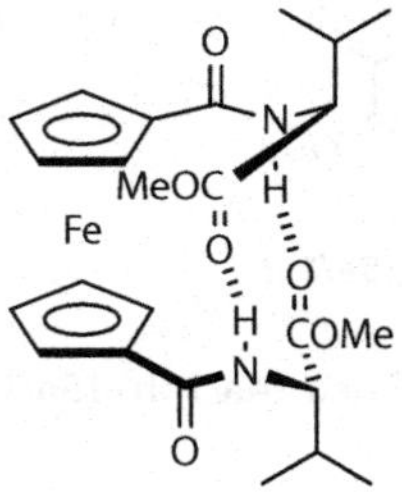

FIGURE 4.13 Chemical structure of **16**.

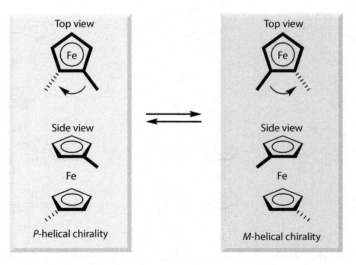

FIGURE 4.14 *P*- and *M*-helical conformations of 1,1′-disubstituted ferrocene.

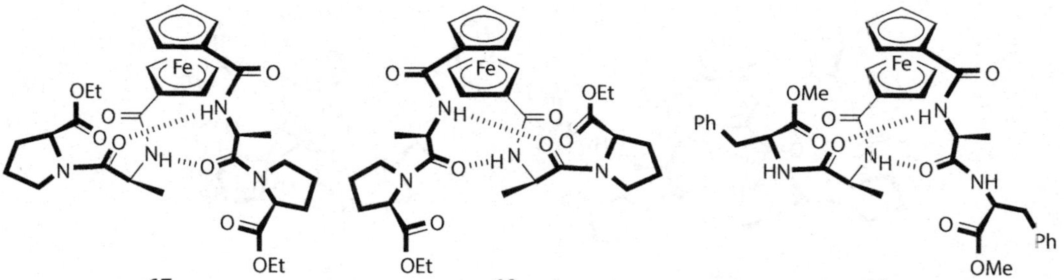

Chirality Organization of Peptides and π-Conjugated Polyanilines 141

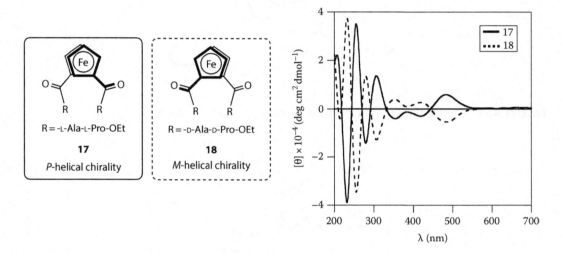

FIGURE 4.15 Chemical structures of the ferrocene–dipeptide conjugates **17–19**.

FIGURE 4.16 CD spectra of **17** and **18** in MeCN (1.0 × 10^{-4} M).

Configuration and sequence of amino acids play an important role in the construction of chirality-organized bioinspired systems. The single-crystal x-ray structure determination of the ferrocene–dipeptide conjugate **20** bearing dipeptide chains of the heterochiral sequence (-L-Ala-D-Pro-NH-2-Py) reveals the formation of the antiparallel β-sheet-like hydrogen bonds to create the chirality-organized structure as observed in **17**, resulting in a *P*-helical chirality of the ferrocenoyl moiety (Figure 4.17).[54] The *P*-helical chirality of the ferrocenoyl moiety is likely to be controlled by the configuration of the alanyl α-carbon atom,[55,56] because a similar type of helical chirality is observed with the ferrocene–dipeptide conjugate **17** bearing -L-Ala-L-Pro-OEt dipeptide chains. It should be noted that a type II β-turn-like structure is created by the intramolecular hydrogen bond between the NH adjacent to the pyridyl moiety and the CO adjacent to the ferrocene unit of the same dipeptide chain in each dipeptide chain. The combination of the ferrocene scaffold as a central reverse-turn unit with the L-Ala-D-Pro heterochiral dipeptide sequence permits the simultaneous formation of the antiparallel β-sheet-like and type II β-turn-like structures. Interestingly, the combination of the ferrocene scaffold with the L-Pro-L-Ala homochiral sequence induces the formation of the antiparallel β-sheet-like and inverse β-turn-like structures simultaneously in the ferrocene–dipeptide conjugate **21** (Figure 4.17).[57] The introduction of the diastereomeric dipeptide configurations shows different self-assembling properties. The NH adjacent to the pyridyl moiety of the ferrocene–dipeptide conjugate **22** bearing dipeptide chains (-L-Pro-D-Ala-NH-2-Py) participates in an intermolecular hydrogen bond with the CO (Pro) of the neighboring molecule to

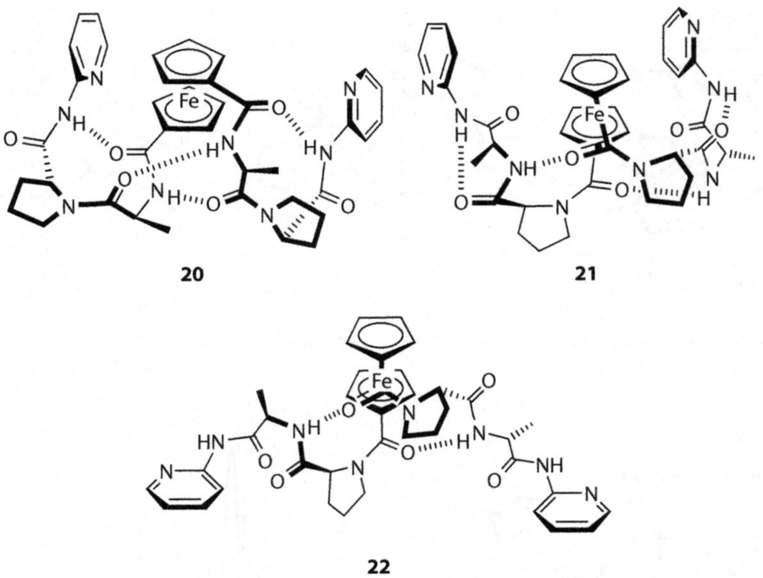

FIGURE 4.17 Chemical structures of the ferrocene–dipeptide conjugates **20–22**.

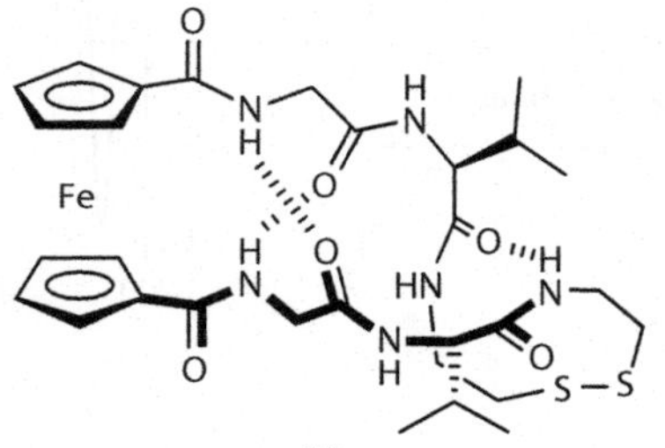

FIGURE 4.18 Chemical structure of the ferrocene–dipeptide conjugate **23**.

form a 14-membered intermolecular hydrogen-bonded ring, although the antiparallel β-sheet-like hydrogen bonds are formed between the NH of the Ala and the CO adjacent to the ferrocene unit of another chain.[57]

Cyclization of the ferrocene–dipeptide conjugate to arrange the two peptide chains in close proximity is a convenient approach to form a β-sheet structure. The ferrocene–peptide conjugate **23** bearing the cyclic peptide (-Gly-L-Val-CSA)$_2$ shows the antiparallel β-sheet-like hydrogen bonds and a *P*-helical chirality of the ferrocenoyl moiety (Figure 4.18).[58] Another interesting feature is that a β-barrel is formed by the interaction of four molecules of **23** through intermolecular hydrogen bonding.

The 1′-aminoferrocene-1-carboxylic acid (Fc-ac) also serves as a reliable organometallic molecular scaffold to allow the formation of a turn structure. The ferrocene–peptide conjugate **24** bearing the peptide chains (Boc-L-Ala-Fc-ac-L-Ala-L-Ala-OMe) shows intramolecular hydrogen bonds to form a 12-membered hydrogen-bonded ring in which a *P*-helical chirality of the Fc-ac moiety is induced (Figure 4.19).[59]

1,*n*′-Diaminoferrocene (Fc-aa) is also utilized as an organometallic molecular scaffold. The ferrocene conjugate **25** bearing amino acid chains (-L-Ala-Boc) composed of the Fc-aa scaffold displays an intramolecular 14-membered hydrogen-bonded ring as is observed in antiparallel β-sheet peptides to induce the chirality organization, wherein the Fc-diamine moiety adopts a *P*-helical structure (Figure 4.19).[60] The introduction of one amino acid substituent into the Fc-aa central unit

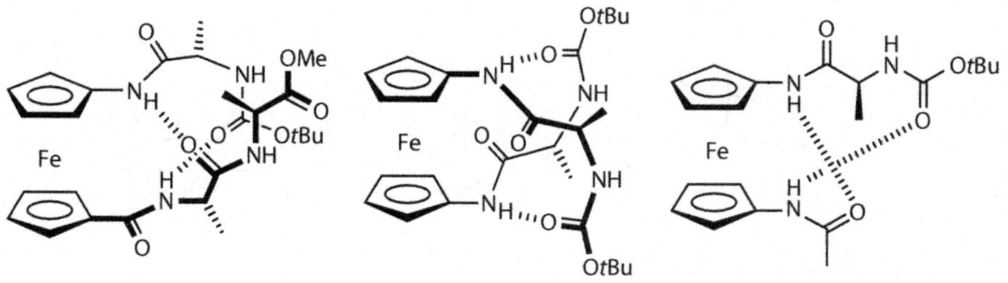

FIGURE 4.19 Chemical structures of the ferrocene–dipeptide conjugates **24–26**.

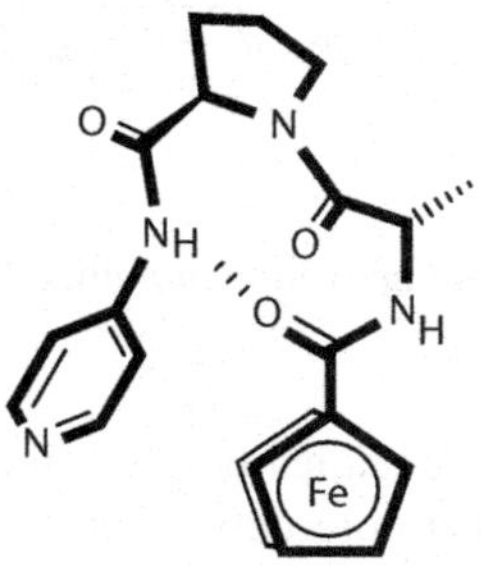

FIGURE 4.20 Chemical structure of the ferrocene–dipeptide conjugate **27**.

is demonstrated to form the intramolecular hydrogen bonds to induce a *P*-helical chirality of the Fc-diamine moiety in the ferrocene conjugate **26** (Figure 4.19).[61]

The formation of the β-turn mimic is performed by utilizing the minimum-sized peptide chain. The ferrocene–dipeptide conjugate **27** bearing one dipeptide chain of the heterochiral sequence (-L-Ala-D-Pro-NH-4-Py) forms the intramolecular hydrogen bonds between the NH adjacent to the pyridyl moiety and the CO adjacent to the ferrocene unit of the same dipeptide chain to induce a type II β-turn-like structure (Figure 4.20).[62]

4.3 SYNTHESIS OF OPTICALLY ACTIVE POLYANILINES

4.3.1 POLYMERIZATION OF ANILINES IN THE PRESENCE OF A CHIRAL ACID

PAn's can be prepared by electrochemical or chemical polymerization of anilines. The enantioselective electropolymerization of aniline in the presence of (1*S*)-(+)- or (1*R*)-(−)-10-camphorsulfonic acid (HCSA) is demonstrated to induce chirality into the π-conjugated backbone of PAn's.[63–65] The presence of a strong localized polaron absorption band at ca. 800 nm confirms the formation of the emeraldine salts as shown in Figure 4.21. Molecular models suggest that one helical screw sense of the polymer chain of the emeraldine salts is preferentially stabilized by enantiomeric CSA anions linking NH radical cation and NH amine centers along the polymer chain through electrostatic and hydrogen bonds. In PAn-(+)-HCSA or PAn-(−)-HCSA films, chiral holes are suggested to be formed in the polymer matrix during both redox and chemical dedoping/redoping cycles. Optically active PAn is also synthesized by the in situ deposition of films from aqueous solution of aniline in oxidative polymerization with ammonium persulfate in the presence of (+)- or (−)-HCSA.[66] Optically active PAn in the nondoped emeraldine oxidation state can be isolated by dedoping of the in situ deposited films.

Optical activity in the polymer chain is preserved when it is converted to the corresponding polymer doped with HCl.

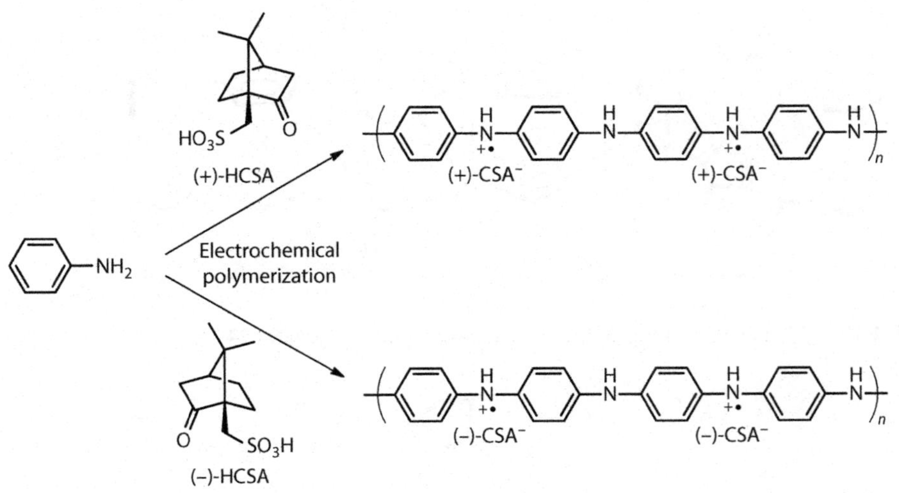

FIGURE 4.21 Enantioselective electropolymerization of aniline in the presence of (1S)-(+)- or (1R)-(−)-10-camphorsulfonic acid (HCSA).

The chiroptical properties of electrochemically deposited PAn-(+)-HCSA emeraldine salts are indicated to be dependent on the temperature employed during polymerization.[67] The CD spectra of films grown at elevated temperatures (35°C–65°C) are inverted compared to the CD spectra of analogous films grown at 0°C–25°C. A temperature-induced interconversion between the two diastereomeric emeraldine salts formed during the doping of the growing PAn chains with the chiral (+)-CSA anion is suggested (Figure 4.22).

The sulfonated PAn, poly(2-methoxyaniline-5-sulfonic acid), is prepared in optically active form via the electropolymerization of 2-methoxyaniline-5-sulfonic acid in the presence of (R)-(+)- or

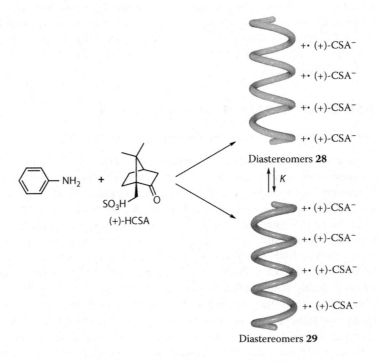

FIGURE 4.22 Interconversion of diastereomers **28** and **29** in equilibrium K.

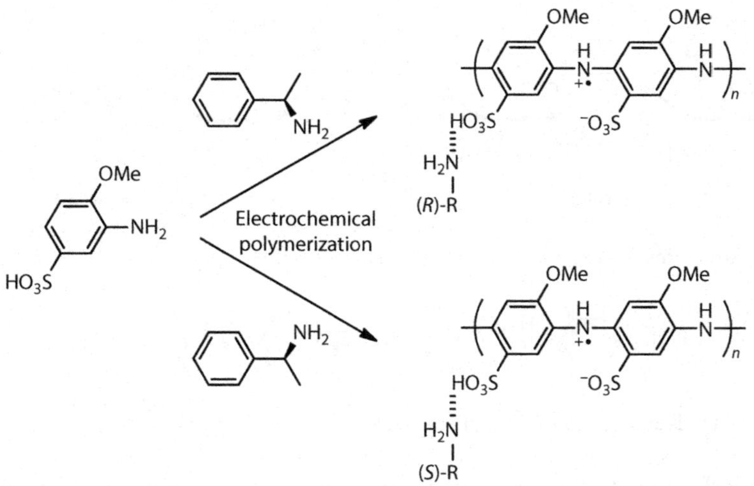

Chirality Organization of Peptides and π-Conjugated Polyanilines 145

FIGURE 4.23 Synthesis of optically active sulfonated PAn's in the presence of (R)-(+)- or (S)-(−)-1-phenylethylamine.

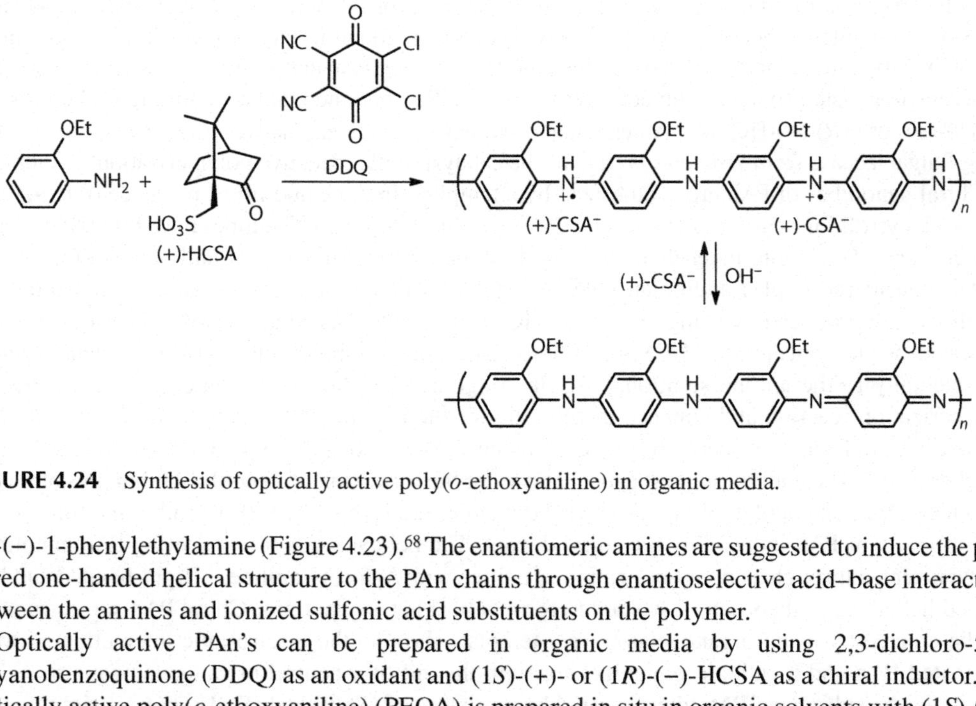

FIGURE 4.24 Synthesis of optically active poly(o-ethoxyaniline) in organic media.

(S)-(−)-1-phenylethylamine (Figure 4.23).[68] The enantiomeric amines are suggested to induce the preferred one-handed helical structure to the PAn chains through enantioselective acid–base interaction between the amines and ionized sulfonic acid substituents on the polymer.

Optically active PAn's can be prepared in organic media by using 2,3-dichloro-5,6-dicyanobenzoquinone (DDQ) as an oxidant and (1S)-(+)- or (1R)-(−)-HCSA as a chiral inductor.[69,70] Optically active poly(o-ethoxyaniline) (PEOA) is prepared in situ in organic solvents with (1S)-(+)- or (1R)-(−)-HCSA as shown in Figure 4.24. The optical activity can be memorized in the solid state during dedoping/redoping cycles. A solvent effect on the PEOA's chain conformation is observed. PEOA-(+)-HCSA loses its optical activity completely when dissolved in dimethyl sulfoxide despite maintenance of the emeraldine salt character. This result indicates that the electrostatic bonding of the CSA− sulfonate moiety to +·NH radical cation and hydrogen bonding of the CSA− carbonyl groups to NH amine centers are important to maintain a preferred one-screw sense of the polymer chain of the emeraldine salt.

The utilization of a molecular template is a convenient strategy to induce chirality-organized structures. Chirality induction into the π-conjugated backbone of the PAn main chains is induced by

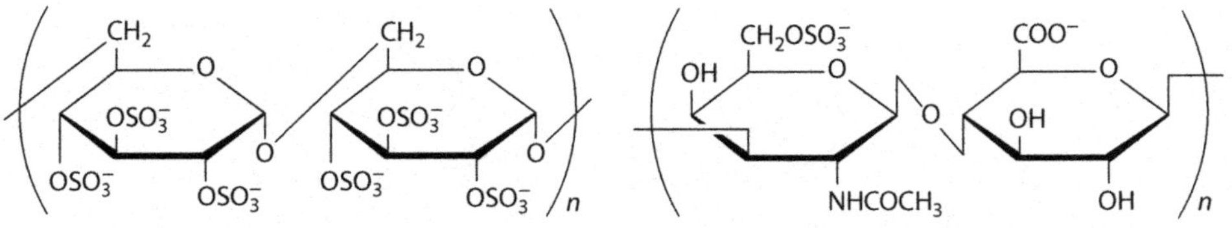

FIGURE 4.25 Structures of dextran sulfate and chondroitin sulfate.

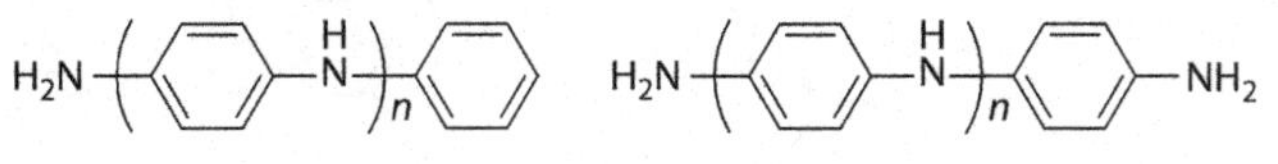

FIGURE 4.26 Phenyl–amine-capped oligomers and amine–amine-capped oligomers.

chemical polymerization of achiral aniline in the presence of a linear sulfated polysaccharide such as dextran sulfate[71,72] or chondroitin sulfate (Figure 4.25).[73]

The enzyme horseradish peroxidase (HRP) is used to catalyze a template-guided polymerization of aniline in the presence of a polyanionic template, sulfonated polystyrene.[74] The enzyme HRP catalyzes the polymerization of aniline in the presence of DNA to afford the electrically responsive conducting DNA–PAn complex, wherein wrapping of PAn on DNA is found to induce reversible changes in the secondary structure of DNA.[75] Water-soluble chiral conducting PAn nanocomposites are synthesized by the template-assisted (poly(acrylic acid)) enzymatic (HRP) polymerization of aniline in the presence of (1S)-(+)- or (1R)-(−)-HCSA.[76] An enzymatic synthesis of optically active PAn is also performed by using fungal laccase from *Trametes hirsuta* as the catalyst of the oxidative polymerization of aniline.[77]

Chiral nanotubes of PAn are synthesized by a template-free method in the presence of (S)-(−)- or (R)-(+)-2-pyrrolidone-5-carboxylic acid (PCA) as a chiral dopant.[78] The tubes have 80–220 nm outer diameter and 50–130 nm inner diameter. The formation of the micelles of aniline–(S)-PCA or (R)-PCA is considered to play a template role during the formation of the chiral nanotubes. Chiral PAn nanofibers are produced by using (+)- or (−)-HCSA in an aqueous solution, wherein polymerization is carried out in concentrated HCSA solutions, and ammonium persulfate as an oxidant is added incrementally to the aniline solution.[79] Aniline oligomers as shown in Figure 4.26 accelerate this polymerization reaction and enhance the chirality of the resulting PAn nanofibers. Right- and left-handed helical PAn nanofibers are prepared using (1S)-(+)- or (1R)-(−)-HCSA, respectively, as a chiral dopant.[80] The authors suggest a mechanism for the formation of nanofibrillar bundles of helical nanofibers. Helical nanofibers and nanofibrillar bundles are formed by self-assembling of one helical conformation of PAn induced by the chiral dopant. The helical sense of conducting PAn nanofibers can be inversed through a copolymerization of aniline with *N*-methylaniline.[81] Right-handed and left-handed helical nanofibers can be welded to form helical heterojunctions. Chiral PAn superstructures can be created by using organogelators as a template through the electrostatic interaction between sulfonated PAn and a cationic organogelator with the aid of a neutral organogelator matrix.[82] The helicity of the sulfonated PAn fibers created by a mixture of (S)-**30** and (S)-**31** shows a right-handed helical motif (Figure 4.27). The PAn films, which are obtained by oxidative electrochemical polymerization of aniline in the presence of (1S)-(+)- or (1R)-(−)-HCSA in water, display chiral electrochromism based on electrochemical reduction/oxidation in a 0.1 M sulfuric acid aqueous solution.[83]

An optically active copolyaniline, poly(aniline-*co*-*o*-toluidine), is prepared by chemical polymerization with (1S)-(+)-HCSA as the chiral inductor in organic media.[84] It is noteworthy that the helical conformation of PAn can be inverted by inserting *o*-toluidine units into the polymer main chain. Chiral helicity is induced in poly(2-methoxyaniline) by electrochemical polymerization of 2-methoxyaniline at pH = 2.5 in the presence of protonated β-cyclodextrin sulfate.[85] When NaCl is added to the reaction solution, the polymer backbone takes on an opposite excess of one-handed helicity.

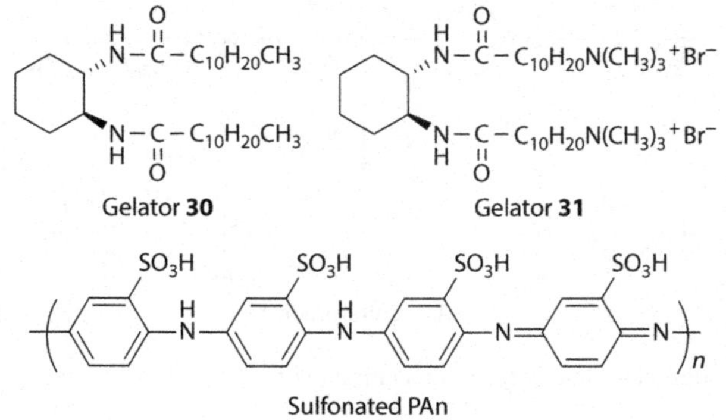

Chirality Organization of Peptides and π-Conjugated Polyanilines 147

FIGURE 4.27 Chemical structures of neutral organogelator **30**, cationic organogelator **31**, and sulfonated PAn.

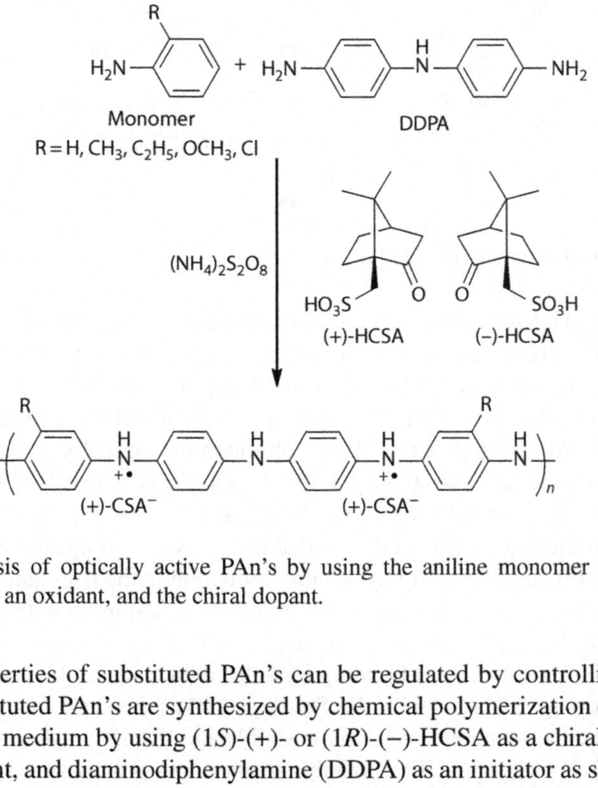

FIGURE 4.28 Synthesis of optically active PAn's by using the aniline monomer DDPA as an initiator, ammonium persulfate as an oxidant, and the chiral dopant.

Also, chiroptical properties of substituted PAn's can be regulated by controlling steric hindrance. Optically active substituted PAn's are synthesized by chemical polymerization of the corresponding monomers in aqueous medium by using (1S)-(+)- or (1R)-(−)-HCSA as a chiral dopant, ammonium persulfate as an oxidant, and diaminodiphenylamine (DDPA) as an initiator as shown in Figure 4.28. The substituent at the *ortho* position induces helical inversion of conformation in comparison with the parent PAn.[86] The effect of the *ortho* substituent on the chirality of the copolymers is increased with the increase of the steric hindrance of the substituent.

4.3.2 Doping of Emeraldine Bases with a Chiral Acid

Chirality induction into the π-conjugated backbone of the PAn main chains is demonstrated by doping with (1S)-(+)-HCSA.[87] Optically active PAn salts are also readily generated in solution via the enantioselective acid doping of the neutral emeraldine base with either (1S)-(+)- or (1R)-(−)-HCSA in a variety of organic solvents (*N*-methyl-2-pyrrolidone (NMP), dimethylformamide (DMF), DMSO, chloroform).[88] One helical screw of the polymer chain might be preferentially produced

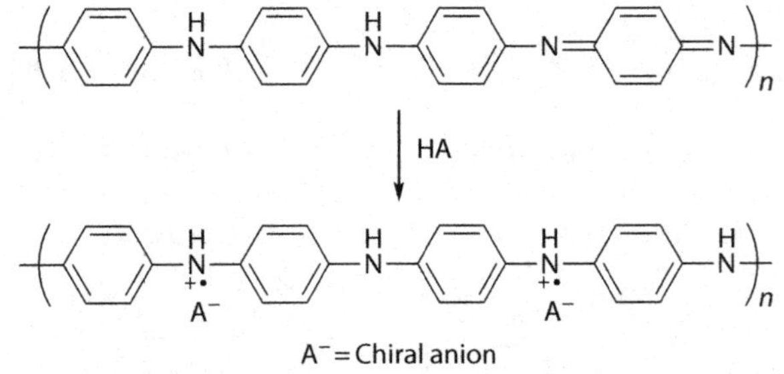

FIGURE 4.29 Doping of emeraldine bases with a chiral acid.

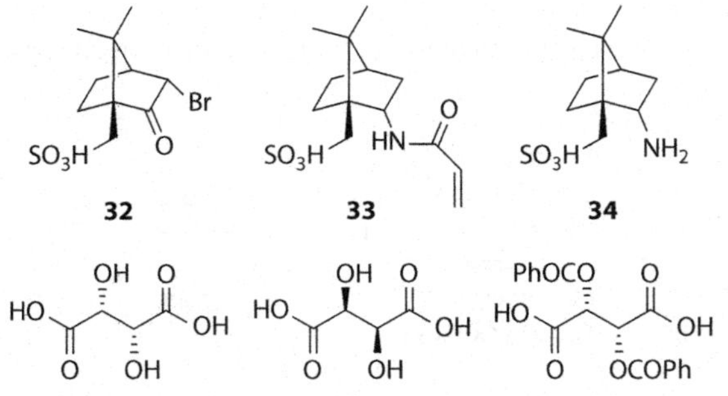

FIGURE 4.30 Chemical structures of chiral acids.

depending on the chiral CSA− anion (Figure 4.29). Optically active leucoemeraldine base, emeraldine base, pernigraniline base, and pernigraniline salt can be obtained by the chemical reduction or oxidation of optically active PAn-(+)-HCSA or by dedoping with NH$_4$OH.[89] The emeraldine base is doped with (1S)-(+)-3-bromocamphor-10-sulfonic acid[32] and chiral acrylamidesulfonic acid[33] in NMP and DMF to afford optically active PAn's, although the optically active aminosulfonic acid[34] does not dope the emeraldine base (Figure 4.30).[90,91] Optically active PAn's are also obtained by doping of the emeraldine base with the chiral dicarboxylic acids, (+)- or (−)-tartaric acid,[35,36] and O,O'-dibenzoyl-L-tartaric acid.[37] This indicates that the hydroxyl groups in tartaric acid are not critical participants in the processes leading to chirality induction in the resultant PAn's. The addition of water to a solution of the emeraldine base in NMP prior to acid doping with HCSA inverts the optical activity of the PAn-HCSA.[92,93] It has been suggested that the inversion of configuration for the PAn-HCSA may arise from hydrogen bonding of water molecules to the emeraldine base polymer chain prior to acid doping.[93] The metallic HCSA-doped PAn films cast from m-cresol show strong optical activity of the conducting electrons in the visible and IR regions. In contrast, the nonmetallic HCSA-doped PAn films cast from CHCl$_3$ exhibit weaker optical activity of the polymer backbone. When the films cast from CHCl$_3$ are exposed to secondary doping with m-cresol vapors, the PAn chains begin to straighten out and lose their limited optical activity.[94] The chiroptical properties of optically active PAn-(−)-HCSA in NMP are stabilized by the presence of potential hydrogen-bonding amino acids such as L-arginine, L-tyrosine, L-alanine, L-valine, and glycine in the precursor emeraldine base/NMP solution prior to acid doping with (1R)-(−)-HCSA.[95]

4.3.3 Introduction of Chiral Groups into Polyanilines

As mentioned in previous sections, optically active PAn's are readily prepared by the enantioselective acid doping of the neutral emeraldine base with either (1S)-(+)- or (1R)-(−)-HCSA.[63,64,87,88] However, the alkaline dedoping of the optically active emeraldine salt in organic solvent leads to racemization.[63,64]

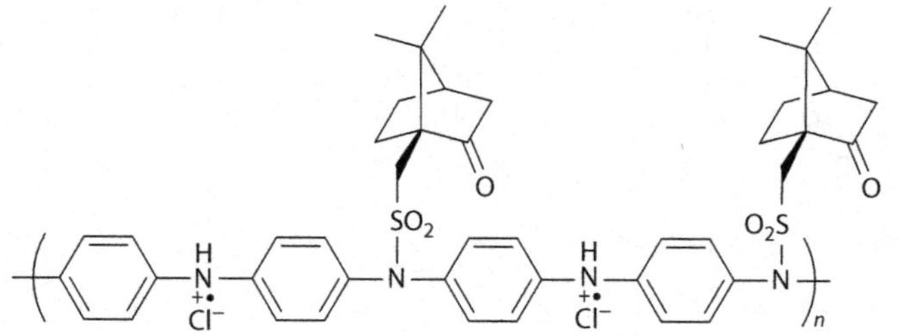

FIGURE 4.31 Optically active PAn salt bearing the covalent attachment of chiral groups.

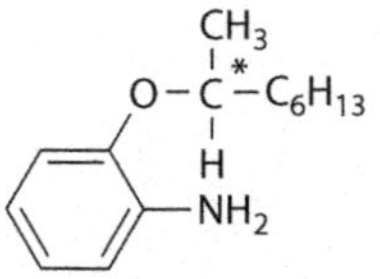

(R)-**38**
(S)-**38**

FIGURE 4.32 Chemical structures of aniline monomers with chiral ether substituents.

Optically active PAn salt can be obtained by the reaction of (1S)-(+)-10-camphorsulfonyl chloride with the emeraldine base through the covalent attachment of chiral groups to N centers along the polymer chain as shown in Figure 4.31.[96] Dedoping of the thus-obtained PAn salt with NaOH gives the optically active emeraldine base. An interfacial polymerization of aniline monomer (R)-**38** or (S)-**38** with chiral ether substituents at the *ortho* position with ammonium persulfate as an oxidant in a water/perchloric acid/chloroform system affords optically active PAn salts (Figure 4.32).[97]

Two redox-active π-conjugated units, a reduced form of phenylenediamine and an oxidized form of quinonediimine, are present in PAn's. The utilization of self-assembling properties of amino acids, which possess chiral centers and hydrogen bonding sites, is considered to be a convenient approach to a highly ordered system. The introduction of the amino acid pendant groups into phenylenediamines or quinonediimines is considered to be a convenient strategy to induce chirality into a π-conjugated backbone of PAn's, giving the hydrogen-bonded chiral polymers wherein redox species of PAn's are expected to be stabilized by hydrogen bonding. Chirality-organized aniline oligomers are prepared by the introduction of amino acid moieties, L- or D-Ala-OMe, into aniline oligomers (Figure 4.33), wherein the formation of intramolecular hydrogen bonds is demonstrated to play an important role to regulate the aniline oligomer moieties conformationally.[98] The aniline oligomer **39-L** exhibits an induced circular dichroism (ICD) at the absorbance region of the π-conjugated moiety as shown in Figure 4.34. This result indicates that the chirality induction of a π-conjugated backbone aniline oligomer is achieved by the chirality organization based on intramolecular hydrogen bonding. The mirror image of the CD signals observed with **39-L** is obtained in the CD spectrum of **39-D**, indicating a chiral molecular arrangement based on the regulated structures via intramolecular hydrogen bonding. The aniline oligomers **40** also show ICD based on the chirality-organized structures. The crystal structure of the aniline oligomer **41** bearing tetraethyl ester reveals the formation of the intramolecular hydrogen bonds between the amino NH and carbonyl oxygen, resulting in an *anti–anti–anti*-conformation of the π-conjugated moieties as depicted in Figure 4.35a. A *syn–anti–syn*-conformation of the π-conjugated moieties based on intramolecular hydrogen bonding is observed in the crystal structure of the diethyl ester **42** (Figure 4.35b). The terminal π-conjugated moieties of **42** are not regulated because of the absence of hydrogen bonds, indicating that the formation of the intramolecular hydrogen bonds plays an important role in the structural regulation of the π-conjugated moieties.

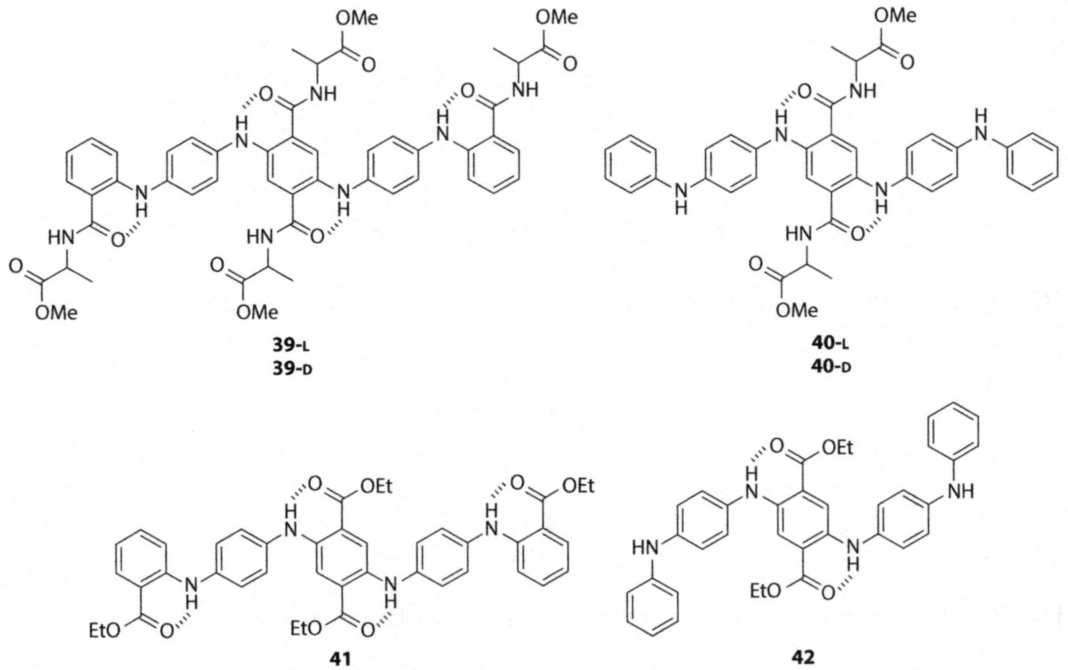

FIGURE 4.33 Chemical structures of the aniline oligomers.

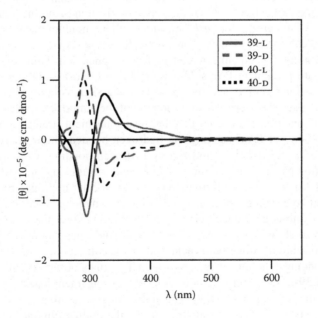

FIGURE 4.34 CD spectra of **39** and **40** in CH$_2$Cl$_2$ (5.0 × 10^{-5} M).

To get further insight into the effect of amino acid moieties on the chirality organization of the aniline oligomer, redox-active PAn unit molecules, a reduced form of phenylenediamine and an oxidized form of quinonediimine, are focused on. The introduction of the amino acid moieties into the phenylenediamine and quinonediimine derivatives is performed to induce chirality-organized structures and stabilize redox species by intramolecular hydrogen bonding (Figure 4.36).[99]

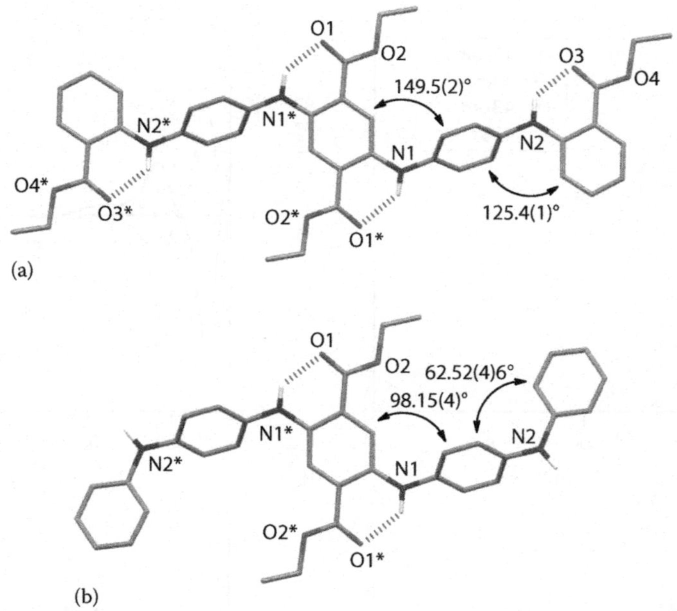

Chirality Organization of Peptides and π-Conjugated Polyanilines

FIGURE 4.35 Crystal structures of (a) **41** and (b) **42**.

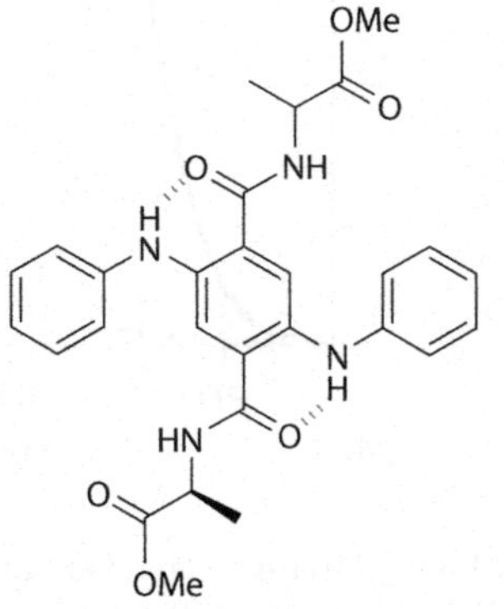

43red-L
43red-D

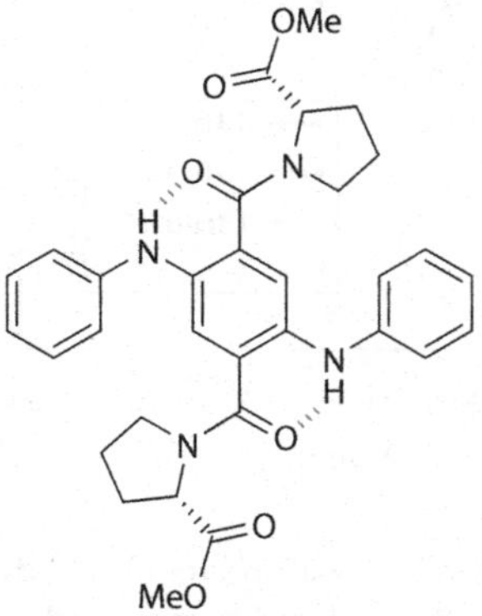

44red-L

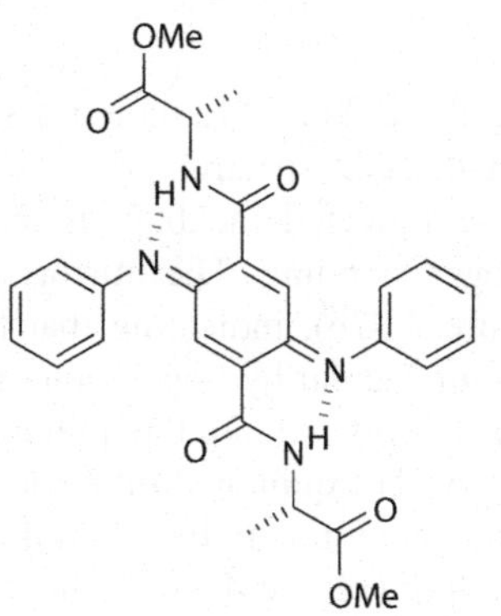

43ox-L

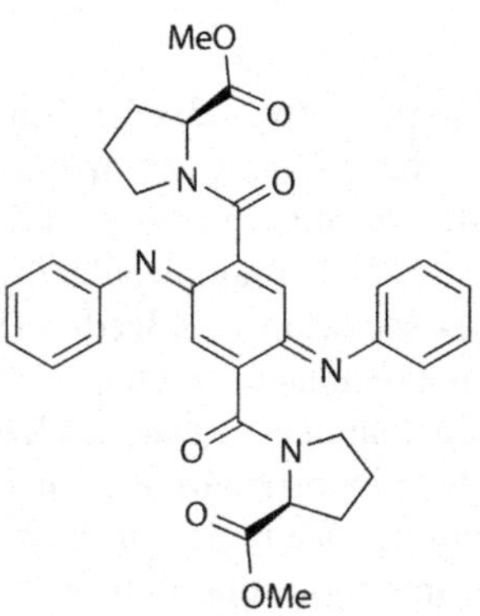

44ox-L

FIGURE 4.36 Chemical structures of the phenylenediamine and quinonediimine derivatives.

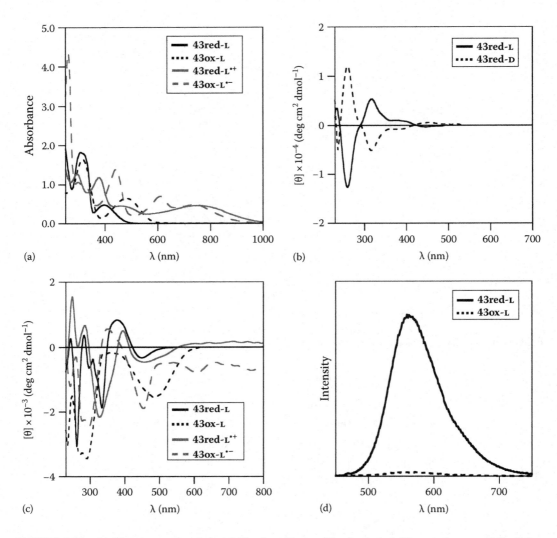

FIGURE 4.37 (a) The electronic spectra of **43red-L**, **43ox-L**, **43red-L**[·+], and **43ox-L**[·−] in acetonitrile (1.0 × 10[−4] M), (b) the CD spectra of **43red-L** and **43red-D** in dichloromethane (1.0 × 10[−4] M), (c) the CD spectra of **43red-L**, **43ox-L**, **43red-L**[·+], and **43ox-L**[·−] in acetonitrile (2.5 × 10[−4] M), and (d) the emission spectra of **43red-L** (λ_{ex} = 409 nm) and **43ox-L** (λ_{ex} = 409 nm) in dichloromethane (1.0 × 10[−4] M) at 298 K under nitrogen atmosphere.

The phenylenediamine derivative **43red-L** bearing the L-Ala-OMe moieties shows an ICD at the absorbance region of the π-conjugated moiety in the CD spectrum (Figure 4.37a and b). This result indicates that the chirality induction of the π-conjugated backbone is achieved by the chirality organization through the intramolecular hydrogen bonding. The mirror image of the signals is obtained in the CD spectrum of **43red-D** (Figure 4.37b), indicating that the chiral molecular arrangement based on a regulated structure via intramolecular hydrogen bonding is formed even in solution. The phenylenediamine derivative **43red-L** is oxidized into the quinonediimine derivatives **43ox-L** by treatment with iodosobenzene as an oxidant. The quinonediimine derivatives **43ox-L** also show an ICD at the absorbance region of the π-conjugated moiety based on the chirality-organized structure through the intramolecular hydrogen bonding (Figure 4.37c). The chirality organization of the phenylenediamine derivative **44red-L** bearing the L-Pro-OMe moieties through the intramolecular hydrogen bonding is supported by the appearance of ICD at the absorbance region of the π-conjugated moiety. The crystal structure of **43red-L** exhibits the formation of the intramolecular

Chirality Organization of Peptides and π-Conjugated Polyanilines

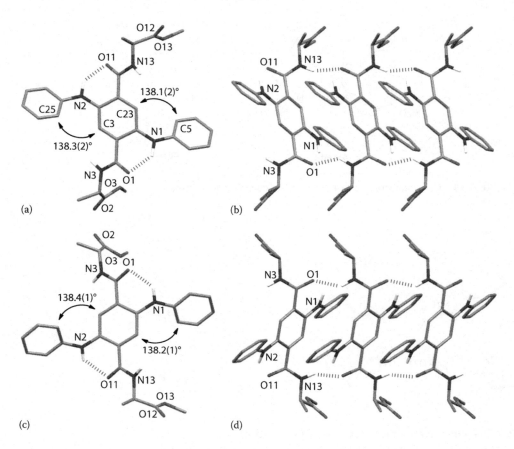

FIGURE 4.38 Crystal structure of (a) **43red-L** and (c) **43red-D**. A portion of a layer containing a sheet-like molecular arrangement through the intermolecular hydrogen bonding networks in the crystal packing of (b) **43red-L** and (d) **43red-D**.

hydrogen bonds between the amino NH of the phenylenediamine moiety and the carbonyl oxygen, resulting in an *anti*-conformation of the π-conjugated moiety (Figure 4.38a). The phenylenediamine derivative **43red-L** shows the sheet-like self-assembly through the intermolecular hydrogen bonding networks, wherein each molecule is bonded to two neighboring molecules by an 18-membered hydrogen-bonded ring as shown in Figure 4.38b. The crystal structure of **43red-D** composed of the corresponding D-amino acids (-D-Ala-OMe) is in a mirror image relationship with **43red-L**, indicating the conformational enantiomer (Figure 4.38c and d).

The phenylenediamine derivatives **43red-L** and **44red-L** show two one-electron redox waves based on the successive one-electron oxidation processes of the phenylenediamine moiety to give the corresponding oxidized species. The quinonediimine derivative **43ox-L** exhibits two one-electron redox waves assignable to the successive one-electron reduction processes of the quinonediimine moiety to give the corresponding reduced species. The reduction waves of the quinonediimine derivative **44ox-L** shift cathodically compared with **43ox-L**. The formation of intramolecular hydrogen bonding is considered to stabilize the generated reduced species. The chirality organization and the stabilization of the phenylenediamine radical cation **43red-L·+** and the semiquinonediimine radical anion **43ox-L·−** through the formation of intramolecular hydrogen bonds are confirmed by the appearance of an ICD at the absorbance region of the π-conjugated moieties in the CD spectra (Figure 4.37a and c) and the ESR spectra with hyperfine coupling.[99] The luminescent switching is achieved by changing the redox state of the π-conjugated moiety

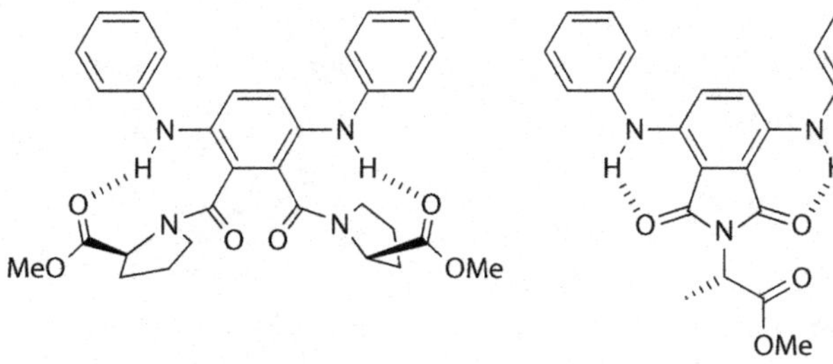

FIGURE 4.39 Chemical structures of the phenylenediamine derivatives.

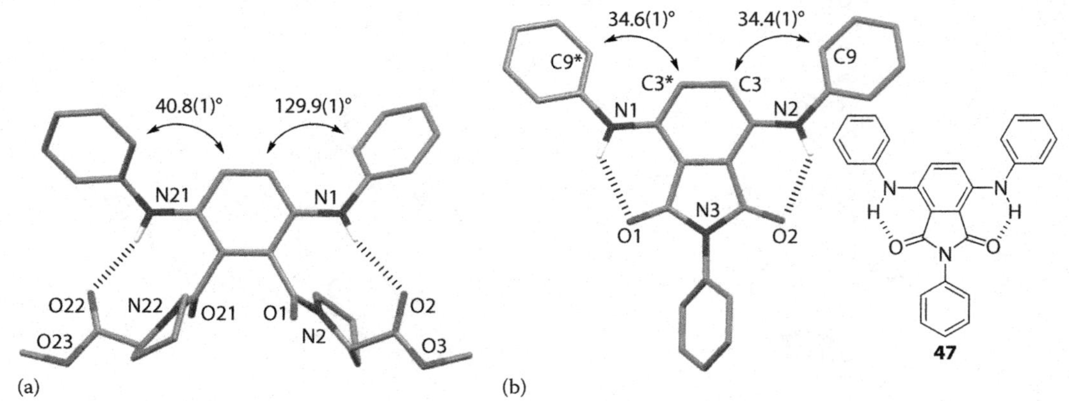

FIGURE 4.40 Crystal structure of (a) **45red-L** and (b) **47**.

of the phenylenediamine derivatives **43red-L** and **44red-L**.[99] The reduced form **43red-L** exhibits strong luminescence at 559 nm as shown in Figure 4.37d. On the contrary, weak luminescence is observed with the oxidized form **43ox-L**.

The introduction of the amino acid moieties into the phenylenediamine derivative at 2,3-positions results in a *syn*-conformation of the π-conjugated moiety (Figure 4.39).[100] An ICD at the absorbance region of the π-conjugated moiety based on the chirality-organized structure through the intramolecular hydrogen bonding is observed in the CD spectra of the phenylenediamine derivatives **45red-L** and **46red-L**. The crystal structure of **45red-L** confirms the chirality-organized structure based on the formation of the intramolecular hydrogen bonds between the amino NH of the phenylenediamine moiety and the carbonyl oxygen of Pro to form a nonpeptidic reverse-turn 9-membered hydrogen-bonded ring, resulting in a *syn*-conformation of the π-conjugated moiety as shown in Figure 4.40a. The formation of the intramolecular hydrogen bonds to induce a *syn*-conformation of the π-conjugated moiety is also observed in the crystal structure of the phenylenediamine derivative **47** (Figure 4.40b). Redox switching of the luminescent properties is also observed in the PAn unit molecules **45red-L** and **46red-L**.

4.4 CHIRAL COMPLEXATION OF EMERALDINE BASES WITH CHIRAL COMPLEXES

PAn's have the coordination properties of two nitrogen atoms of the quinonediimine moiety. The introduction of chiral complexes into the quinonediimine moiety of the neutral emeraldine base via coordination interaction is considered to be a convenient approach to induce chirality into a π-conjugated backbone of PAn's, giving chiral conjugated complexes. The reaction of the emeraldine

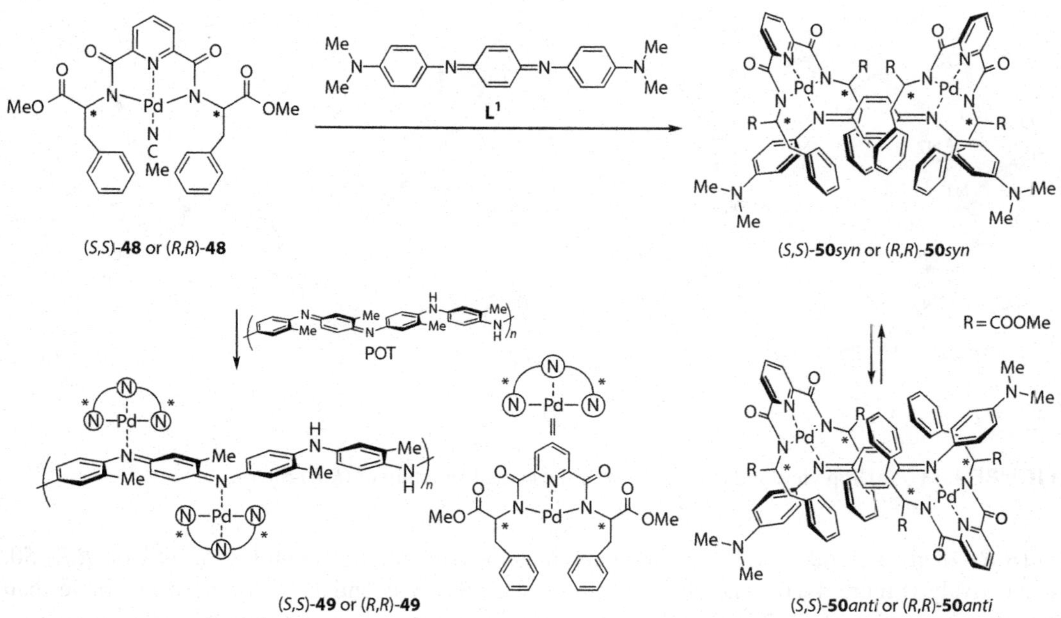

Chirality Organization of Peptides and π-Conjugated Polyanilines

FIGURE 4.41 Chiral complexation of POT or **L¹** with the chiral palladium(II) complex.

base form of poly(o-toluidine) (POT) with the chiral palladium(II) complex (S,S)-**48** having one interchangeable coordination site affords the chiral conjugated polymer complex (S,S)-**49** as shown in Figure 4.41.[101,102] The conjugated polymer complex (S,S)-**49** exhibits an ICD at around 500–800 nm based on the absorbance region of the π-conjugated moiety (Figure 4.42), suggesting the chirality induction of a π-conjugated backbone of POT through the chiral complexation. The mirror image of the CD signal is observed with (R,R)-**49**, which is obtained from the complexation of POT with (R,R)-**48**.

The chiral complexation of a model molecule of PAn, N,N′-bis(4′-dimethylaminophenyl)-1,4-benzoquinonediimine (**L¹**),[103] with two equimolar amounts of chiral palladium(II) complex (S,S)-**48**

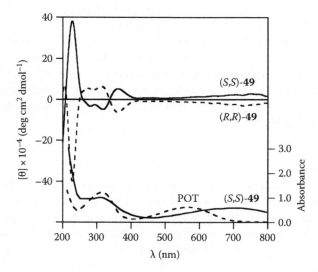

FIGURE 4.42 The electronic spectra of POT and (S,S)-**49** in THF (1.3×10^{-3} M of the monomer unit) and the CD spectra of (S,S)-**49** and (R,R)-**49** in THF (1.3×10^{-3} M of the monomer unit).

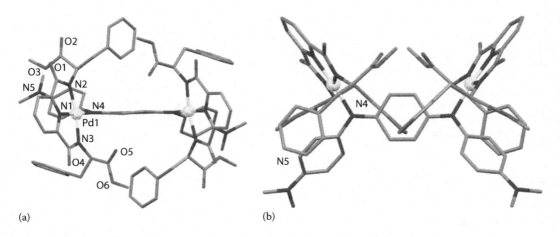

FIGURE 4.43 (a) Top view and (b) side view of the crystal structure of (*R,R*)-**50***syn*.

or (*R,R*)-**48** gives chiral 1:2 conjugated homobimetallic palladium(II) complex (*S,S*)-**50** or (*R,R*)-**50**, respectively (Figure 4.41). The chiral complex (*S,S*)-**50***syn* is enthalpically more favorable than the chiral complex (*S,S*)-**50***anti* in CD$_2$Cl$_2$ by 2.3 kcal mol^{-1}, but entropically less favorable by 11.0 cal mol^{-1} K^{-1}. The crystal structure of (*R,R*)-**50***syn* reveals that the two Pd units are bridged by the quinonediimine moiety of **L^1** to form the C_2-symmetrical 2:1 complex in a *syn*-conformation as depicted in Figure 4.43.

4.5 APPLICATION OF OPTICALLY ACTIVE POLYANILINES

Chiral dedoped PAn, which is prepared by removing (1*S*)-(+)- or (1*R*)-(−)-HCSA from the chiral doped PAn with (1*S*)-(+)- or (1*R*)-(−)-HCSA, can be used to separate racemic mixtures of DL-phenylalanine.[104–106] The dedoped PAn form of the chiral doped PAn with (1*R*)-(−)-HCSA preferentially complexes with L-phenylalanine. The enantiomer of phenylalanine can be released using isopropanol. Rapid and efficient enantiomeric detection of L- and D-phenylalanine is also observed with the chiral dedoped PAn using differential pulse voltammetry.[107] PAn films are prepared by electropolymerization of aniline under a magnetic field of 5 T parallel (+5 T) or antiparallel (−5 T) to the faradaic currents. These two electrodes have the opposite chirality of each other and possess the ability of chiral recognition for L-3-(3,4-dihydroxyphenyl)alanine, L-ascorbic acid, and D-isoascorbic acid.[108] The electrode column packed with L-aspartic acid (Asp) anion-templated PAn shows an excellent chiral selectivity toward isomers of Asp and its analog, such as glutamic acid.[109] The chiral recognition for the determination of the configuration of alanine via electrochemical methods is also performed by using a carbon paper electrode-loaded chiral PAn with (1*S*)-(+)- or (1*R*)-(−)-HCSA.[110]

4.6 CONCLUSION

The regulation of the attached peptide strands in the appropriate dimensions is achieved by using a variety of molecular scaffolds to form the β-sheet and turn structures of attached peptides, affording highly organized molecular structures. Designing protein secondary structure mimics will provide fundamental insight into the factors affecting the structure and stability of proteins. The utilization of organometallic ferrocene as a molecular scaffold with a central reverse-turn unit is a powerful strategy for the design of peptidomimetic chirality-organized systems. Chirality induction into the π-conjugated backbone of PAn's is demonstrated by the enantioselective polymerization of aniline in the presence of a chiral source or the enantioselective acid doping of the neutral emeraldine base. The introduction of chiral groups into PAn's is a convenient strategy to induce chirality into a π-conjugated backbone of PAn's through hydrogen bonding. The coordination interaction of the quinonediimine nitrogen of PAn with chiral complexes is performed to induce chirality into a π-conjugated backbone of PAn, giving chiral conjugated complexes. The control of structurally defined molecular arrangement in a solid state is a key factor in the design of functional materials. Structural control of polymers is of importance for the development of novel molecular devices. Optically active PAn's are envisioned to induce further functionalization based on redox properties of helical molecular coil.

REFERENCES

1. Braga D, Grepioni F, Desiraju, GR (1998) Crystal engineering and organometallic architecture. *Chem Rev* 98:1375–1406.
2. Balzani V, Credi A, Raymo FM, Stoddart JF (2000) Artificial molecular machines. *Angew Chem Int Ed* 39:3348–3391.
3. Swiegers GF, Malefetse TJ (2000) New self-assembled structural motifs in coordination chemistry. *Chem Rev* 100:3483–3538.
4. Sun, W-Y, Yoshizawa M, Kusukawa T, Fujita M (2002) Multicomponent metal–ligand self-assembly. *Curr Opin Chem Biol* 6:757–764.
5. Davis JT, Spada GP (2007) Supramolecular architectures generated by self-assembly of guanosine derivatives. *Chem Soc Rev* 36:296–313.
6. Tanaka K, Shionoya M (2007) Programmable metal assembly on bio-inspired templates. *Coord Chem Rev* 251:2732–2742.
7. Pijper D, Feringa BL (2008) Control of dynamic helicity at the macro- and supramolecular level. *Soft Matter* 4:1349–1372.
8. Horike S, Shimomura S, Kitagawa S (2009) Soft porous crystals. *Nat Chem* 1:695–704.
9. Forgan RS, Sauvage J-P, Stoddart JF (2011) Chemical topology: Complex molecular knots, links, and entanglements. *Chem Rev* 111:5434–5464.
10. Schmidbaur H, Schier A (2012) Aurophilic interactions as a subject of current research: An up-date. *Chem Soc Rev* 41:370–412.
11. Kyte J (1995) *Structure in Protein Chemistry*, Garland, New York.
12. Branden CI, Tooze J (1998) *Introduction to Protein Structure*, 2nd edn., Garland, New York.
13. Alcácer L (1987) *Conducting Polymers: Special Applications*, Reidel, Holland, the Netherlands.
14. Salaneck WR, Clark DT, Samuelsen EJ (1990) *Science and Application of Conductive Polymers*, Adam Hilger, New York.
15. Shirakawa H (2001) The discovery of polyacetylene film: The dawning of an era of conducting polymers (Nobel lecture). *Angew Chem Int Ed* 40:2574–2580.
16. MacDiarmid AG (2001) "Synthetic metals": A novel role for organic polymers (Nobel lecture). *Angew Chem Int Ed* 40:2581–2590.
17. Heeger AJ (2001) Semiconducting and metallic polymers: The fourth generation of polymeric materials (Nobel lecture). *Angew Chem Int Ed* 40:2591–2611.
18. Ritonga MTS, Sakurai H, Hirao T (2002) Synthesis and characterization of *p*-phenylenediamine derivatives bearing a thiadiazole unit. *Tetrahedron Lett* 43:9009–9013.
19. Sakurai H, Ritonga MTS, Shibatani H, Hirao T (2005) Synthesis and characterization of *p*-phenylenediamine derivatives bearing an electron-acceptor unit. *J Org Chem* 70:2754–2762.
20. Higuchi M, Imoda D, Hirao T (1996) Redox behavior of polyaniline—Transition metal complexes in solution. *Macromolecules* 29:8277–8279.
21. Hirao T, Yamaguchi S, Fukuhara S (1999) Controlled formation of synthetic metal—Transition metal conjugated complex systems. *Tetrahedron Lett* 40:3009–3012.
22. Hirao T, Yamaguchi S, Fukuhara S (1999) Construction of multinuclear heterobimetallic conjugated complex systems. *Synth Met* 106:67–70.
23. Hirao T, Higuchi M, Ikeda I, Ohshiro Y (1993) A novel synthetic metal catalytic system for dehydrogenative oxidation based on redox of polyaniline. *J Chem Soc Chem Commun* 194–195.
24. Hirao T, Higuchi M, Hatano B, Ikeda I (1995) A novel redox system for the palladium(II)-catalyzed oxidation based on redox of polyanilines. *Tetrahedron Lett* 36:5925–5928.
25. Higuchi M, Yamaguchi S, Hirao T (1996) Construction of palladium-polypyrrole catalytic system in the Wacker oxidation. *Synlett* 1996:1213–1214.

26. Higuchi M, Ikeda I, Hirao T (1997) A novel synthetic metal catalytic system. *J Org Chem* 62:1072–1078.
27. Amaya T, Saio D, Hirao T (2007) Template synthesis of polyaniline/Pd nanoparticle and its catalytic application. *Tetrahedron Lett* 48:2729–2732.
28. Amaya T, Nishina Y, Saio D, Hirao T (2008) Hybrid of polyaniline/iron oxide nanoparticles: Facile preparation and catalytic application. *Chem Lett* 37:68–69.
29. Saio D, Amaya T, Hirao T (2010) Redox-active catalyst based on poly(anilinesulfonic acid)-supported gold nanoparticles for aerobic alcohol oxidation in water. *Adv Synth Catal* 352:2177–2182.
30. Hirao T, Fukuhara S, Otomaru Y, Moriuchi T (2001) Conjugated complexes via oxidative complexation of polyaniline derivatives to vanadium(III). *Synth Met* 123:373–376.
31. Moriuchi T, Bandoh S, Miyaishi M, Hirao T (2001) A novel redox-active conjugated palladium homobimetallic complex. *Eur J Inorg Chem* 2001:651–657.
32. Moriuchi T, Miyaishi M, Hirao T (2001) Conjugated complexes composed of quinonediimine and palladium: Controlled formation of a conjugated trimetallic macrocycle. *Angew Chem Int Ed* 40:3042–3045.
33. Moriuchi T, Kamikawa M, Bandoh S, Hirao T (2002) Architectural formation of a conjugated bimetallic Pd(II) complex via oxidative complexation and a tetracyclic Pd(II) complex via self-assembling complexation. *Chem Commun* 2:1476–1477.
34. Hirao T (2002) Conjugated systems composed of transition metals and redox-active π-conjugated ligands. *Coord Chem Rev* 226:81–91.
35. Moriuchi T, Shen X, Saito K, Bandoh S, Hirao T (2003) Multimetallic complex composed of redox-active bridging quinonediimine ligand. *Bull Chem Soc Jpn* 76:595–599.
36. Shen X, Moriuchi T, Hirao T (2003) Redox-switchable π-conjugated systems bearing terminal ruthenium(II) complexes. *Tetrahedron Lett* 44:7711–7714.
37. Moriuchi T, Shiori J, Hirao T (2007) Redox-switchable conjugated bimetallic ruthenium(II) complexes. *Tetrahedron Lett* 48:5970–5972.
38. Ohmura SD, Moriuchi T, Hirao T (2013) Chiral homobimetallic palladium(II) complexes composed of chirality-organized quinonediimines bearing amino acid moieties. *J Inorg Organomet Polym* 23:251–255.
39. Feigel M (1986) 2,8-Dimethyl-4-(carboxymethyl)-6-(aminomethyl)phenoxathiin S-dioxide: An organic substitute for the β-turn in peptides? *J Am Chem Soc* 108:181–182.
40. Brandmeier V, Sauer WHB, Feigel M (1994) Antiparallel β-sheet conformation in cyclopeptides containing a pseudo-amino acid with a biphenyl moiety. *Helv Chim Acta* 77:70–85.
41. Kemp DS, Bowen BR (1988) Conformational analysis of peptide-functionalized diacylaminoepindolidiones ^{1}H NMR evidence for β-sheet formation. *Tetrahedron Lett* 29:5081–5082.
42. Díaz H, Tsang KY, Choo D, Espina JR, Kelly JW (1993) Design, synthesis, and partial characterization of water-soluble β-sheets stabilized by a dibenzofuran-based amino acid. *J Am Chem Soc* 115:3790–3791.
43. Jones IG, Jones W, North M (1998) Conformational analysis of peptides and pseudopeptides incorporating an *endo*-(2S,3R)-norborn-5-ene residue as a turn inducer. *J Org Chem* 63:1505–1513.
44. Nowick JS, Smith EM, Noronha G (1995) Molecular scaffolds. 3. An artificial parallel β-sheet. *J Org Chem* 60:7386–7387.
45. Moriuchi T, Tamura T, Hirao T (2002) Self-assembly of dipeptidyl ureas: A new class of hydrogen-bonded molecular duplexes. *J Am Chem Soc* 124:9356–9357.
46. Zeng H, Yang X, Flowers RA, II, Gong B (2002) A noncovalent approach to antiparallel β-sheet formation. *J Am Chem Soc* 124:2903–2910.
47. Togni A, Hayashi T (1995) *Ferrocenes*, Wiley-VCH, Weinheim, Germany.
48. Herrick RS, Jarret RM, Curran TP, Dragoli DR, Flaherty MB, Lindyberg SE, Slate RA, Thornton LC (1996) Ordered conformations in bis(amino acid) derivatives of 1,1′-ferrocenedicarboxylic acid. *Tetrahedron Lett* 37:5289–5292.
49. Oberhoff M, Duda L, Karl J, Mohr R, Erker G, Fröhlich R, Grehl M (1996) The isocyanate route to cyclopentadienyl-carboxamide- and cyclopentadienyl-amino ester-substituted metallocene complexes. *Organometallics* 15:4005–4011.
50. Nomoto A, Moriuchi T, Yamazaki S, Ogawa A, Hirao T (1998) A highly ordered ferrocene system regulated by podand peptide chains. *Chem Commun* 1963–1964.
51. Moriuchi T, Nomoto A, Yoshida K, Hirao T (1999) Characterization of ferrocene derivatives bearing podand dipeptide chains (-L-Ala-L-Pro-OR). *J Organomet Chem* 589:50–58.
52. Moriuchi T, Nomoto A, Yoshida K, Ogawa A, Hirao T (2001) Chirality organization of ferrocenes bearing podand dipeptide chains: Synthesis and structural characterization. *J Am Chem Soc* 123:68–75.
53. van Staveren DR, Weyhermüller T, Metzler-Nolte N (2003) Organometallic β-turn mimetics. A structural and spectroscopic study of inter-strand hydrogen bonding in ferrocene and cobaltocenium conjugates of amino acids and dipeptides. *Dalton Trans* 210–220.

54. Moriuchi T, Nagai T, Hirao T (2005) Chirality organization of ferrocenes bearing dipeptide chains of heterochiral sequence. *Org Lett* 7:5265–5268.
55. Kirin SI, Wissenbach D, Metzler-Nolte N (2005) Unsymmetrical 1,n'-disubstituted ferrocenoyl peptides: Convenient one pot synthesis and solution structures by CD and NMR spectroscopy. *New J Chem* 29:1168–1173.
56. Heinze K, Beckmann M (2005) Conformational analysis of chiral ferrocene-peptides. *Eur J Inorg Chem* 2005:3450–3457.
57. Moriuchi T, Nagai T, Hirao T (2006) Induction of γ-turn-like structure in ferrocene bearing dipeptide chains via conformational control. *Org Lett* 8:31–34.
58. Chowdhury S, Sanders DAR, Schatte G, Kraatz H-B (2006) Discovery of a pseudo β-barrel: Synthesis and formation by tiling of ferrocene cyclopeptides. *Angew Chem Int Ed* 45:751–754.
59. Barišić L, Dropučić M, Rapić V, Pritzkow H, Kirin SI, Metzler-Nolte N (2004) The first oligopeptide derivative of 1'-aminoferrocene-1-carboxylic acid shows helical chirality with antiparallel strands. *Chem Commun* 4:2004–2005.
60. Chowdhury S, Mahmoud KA, Schatte G, Kraatz H-B (2005) Amino acid conjugates of 1,1'-diaminoferrocene. Synthesis and chiral organization. *Org Biomol Chem* 3:3018–3023.
61. Djaković S, Siebler D, Semenčić MČ, Heinze K, Rapić V (2008) Spectroscopic and theoretical study of asymmetric 1,1'-diaminoferrocene conjugates of α-amino acids. *Organometallics* 27:1447–1453.
62. Moriuchi T, Fujiwara T, Hirao T (2009) β-Turn-structure-assembled palladium complexes by complexation-induced self-organization of ferrocene-dipeptide conjugates. *Dalton Trans* 4286–4288.
63. Majidi MR, Kane-Maguire LAP, Wallace GG (1994) Enantioselective electropolymerization of aniline in the presence of (+)- or (−)-camphorsulfonate ion: A facile route to conducting polymers with preferred one-screw-sense helicity. *Polymer* 35:3113–3115.
64. Majidi MR, Kane-Maguire LAP, Wallace GG (1998) Electrochemical synthesis of optically active polyanilines. *Aust J Chem* 51:23–30.
65. Innis PC, Norris ID, Kane-Maguire LAP, Wallace GG (1998) Electrochemical formation of chiral polyaniline colloids codoped with (+)- or (−)-10-camphorsulfonic acid and polystyrene sulfonate. *Macromolecules* 31:6521–6528.
66. Kane-Maguire LAP, MacDiarmid AG, Norris ID, Wallace GG, Zheng W (1999) Facile preparation of optically active polyanilines via the in situ chemical oxidative polymerisation of aniline. *Synth Met* 106:171–176.
67. Pornputtkul Y, Kane-Maguire LAP, Wallace GG (2006) Influence of electrochemical polymerization temperature on the chiroptical properties of (+)-camphorsulfonic acid-doped polyaniline. *Macromolecules* 39:5604–5610.
68. Strounina EV, Kane-Maguire LAP, Wallace GG (1999) Optically active sulfonated polyanilines. *Synth Met* 106:129–137.
69. Su S-J, Kuramoto N (2001) Optically active polyaniline derivatives prepared by electron acceptor in organic system: Chiroptical properties. *Macromolecules* 34:7249–7256.
70. Su S-J, Kuramoto N (2001) In situ synthesis of optically active poly(o-ethoxyaniline) in organic media and its chiroptical properties. *Chem Mater* 13:4787–4793.
71. Yuan G-L, Kuramoto N (2002) Chemical synthesis of optically active polyaniline in the presence of dextran sulfate as molecular template. *Chem Lett* 31:544–545.
72. Yuan G-L, Kuramoto N (2002) Water-processable chiral polyaniline derivatives doped and intertwined with dextran sulfate: Synthesis and chiroptical properties. *Macromolecules* 35:9773–9779.
73. Yuan G-L, Kuramoto N (2004) Synthesis of helical polyanilines using chondroitin sulfate as a molecular template. *Macromol Chem Phys* 205:1744–1751.
74. Liu W, Kumar J, Tripathy S, Senecal KJ, Samuelson L (1999) Enzymatically synthesized conducting polyaniline. *J Am Chem Soc* 121:71–78.
75. Nagarajan R, Liu W, Kumar J, Tripathy SK, Bruno FF, Samuelson LA (2001) Manipulating DNA conformation using intertwined conducting polymer chains. *Macromolecules* 34:3921–3927.
76. Thiyagarajan M, Samuelson LA, Kumar J, Cholli AL (2003) Helical conformational specificity of enzymatically synthesized water-soluble conducting polyaniline nanocomposites. *J Am Chem Soc* 125:11502–11503.
77. Vasil'eva IS, Morozova OV, Shumakovich GP, Shleev SV, Sakharov IY, Yaropolov AI (2007) Laccase-catalyzed synthesis of optically active polyaniline. *Synth Met* 157:684–689.
78. Yang Y, Wan M (2002) Chiral nanotubes of polyaniline synthesized by a template-free method. *J Mater Chem* 12:897–901.
79. Li W, Wang H-L (2004) Oligomer-assisted synthesis of chiral polyaniline nanofibers. *J Am Chem Soc* 126:2278–2279.

80. Yan Y, Yu Z, Huang Y, Yuan W, Wei Z (2007) Helical polyaniline nanofibers induced by chiral dopants by a polymerization process. *Adv Mater* 19:3353–3357.
81. Yan Y, Fang J, Liang J, Zhang Y, Wei Z (2012) Helical heterojunctions originating from helical inversion of conducting polymer nanofibers. *Chem Commun* 48:2843–2845.
82. Li C, Hatano T, Takeuchi M, Shinkai S (2004) Polyaniline superstructures created by a templating effect of organogels. *Chem Commun* 4:2350–2351.
83. Goto H (2007) Optically active electrochromism in polyanilines. *J Polym Sci, Part A: Polym Chem* 45:2085–2090.
84. Su S-J, Takeishi M, Kuramoto N (2002) Helix inversion of polyaniline by introducing *o*-toluidine units. *Macromolecules* 35:5752–5757.
85. Yuan G-L, Liu C, Kuramoto N, Yang Z-F (2010) Helical opposition in poly(2-methoxyaniline) by tuning the concentration of salts in reaction solution. *Polym Int* 59:1187–1190.
86. Anjum MN, Zhu L, Luo Z, Yan J, Tang H (2011) Tailoring of chiroptical properties of substituted polyanilines by controlling steric hindrance. *Polymer* 52:5795–5802.
87. Havinga EE, Bouman MM, Meijer EW, Pomp A, Simenon MMJ (1994) Large induced optical activity in the conduction band of polyaniline doped with (1*S*)-(+)-10-camphorsulfonic acid. *Synth Met* 66:93–97.
88. Majidi MR, Kane-Maguire LAP, Wallace GG (1995) Chemical generation of optically active polyaniline via the doping of emeraldine base with (+)- or (−)-camphorsulfonic acid. *Polymer* 36:3597–3599.
89. Majidi MR, Ashraf SA, Kane-Maguire LAP, Norris ID, Wallace GG (1997) Factors controlling the induction of optical activity in chiral polyanilines. *Synth Met* 84:115–116.
90. Ashraf SA, Kane-Maguire LAP, Majidi MR, Pyne SG, Wallace GG (1997) Influence of the chiral dopant anion on the generation of induced optical activity in polyanilines. *Polymer* 38:2627–2631.
91. Kane-Maguire LAP, Norris ID, Wallace GG (1999) Properties of chiral polyaniline in various oxidation states. *Synth Met* 101:817–818.
92. Egan V, Bernstein R, Hohmann L, Tran T, Kaner RB (2001) Influence of water on the chirality of camphorsulfonic acid-doped polyaniline. *Chem Commun* 801–802.
93. Boonchu C, Kane-Maguire LAP, Wallace GG (2003) The effect of added water on the conformation of optically active polysniline in organic solvents. *Synth Met* 135–136:241–242.
94. Tigelaar DM, Lee W, Bates KA, Saprigin A, Prigodin VN, Cao X, Nafie LA, Platz MS, Epstein AJ (2002) Role of solvent and secondary doping in polyaniline films doped with chiral camphorsulfonic acid: Preparation of a chiral metal. *Chem Mater* 14:1430–1438.
95. Mire CA, Kane-Maguire LAP, Wallace GG, in het Panhuis, MH (2009) Influence of added hydrogen bonding agents on the chiroptical properties of chiral polyaniline. *Synth Met* 159:715–717.
96. Reece DA, Kane-Maguire LAP, Wallace GG (2001) Polyanilines with a twist. *Synth Met* 119:101–102.
97. Goto H (2006) Synthesis of polyanilines bearing optically active substituents. *Macromol Chem Phys* 207:1087–1093.
98. Ohmura SD, Moriuchi T, Hirao T (2010) Chirality organization of aniline oligomers through hydrogen bonds of amino acid moieties. *J Org Chem* 75:7909–7912.
99. Moriuchi T, Ohmura SD, Morita K, Hirao T (2011) Hydrogen-bonding induced chirality organization and stabilization of redox species of polyaniline-unit molecules by introduction of amino acid pendant groups. *Chem Asian J* 6:3206–3213.
100. Moriuchi T, Ohmura SD, Morita K, Hirao T (2012) Chirality organization in phenylenediamines induced by a nonpeptidic reverse-turn. *Asian J Org Chem* 1:52–59.
101. Shen X, Moriuchi T, Hirao T (2004) Chirality induction of polyaniline derivatives through chiral complexation. *Tetrahedron Lett* 45:4733–4736.
102. Moriuchi T, Shen X, Hirao T (2006) Chirality induction of π-conjugated chains through chiral complexation. *Tetrahedron* 62:12237–12246.
103. Wei Y, Yang C, Ding T (1996) A one-step method to synthesize *N*,*N*′-bis(4′-aminophenyl)-1,4-quinonenediimine and its derivatives. *Tetrahedron Lett* 37:731–734.
104. Guo H, Knobler CM, Kaner RB (1999) A chiral recognition polymer based on polyaniline. *Synth Met* 101:44–47.
105. Kaner RB (2002) Gas, liquid and enantiomeric separations using polyaniline. *Synth Met* 125:65–71.
106. Huang J, Egan VM, Guo H, Yoon J-Y, Briseno AL, Rauda IE, Garrell RL, Knobler CM, Zhou F, Kaner RB (2003) Enantioselective discrimination of D- and L-phenylalanine by chiral polyaniline thin films. *Adv Mater* 15:1158–1161.
107. Sheridan EM, Breslin CB (2005) Enantioselective detection of D- and L-phenylalanine using optically active polyaniline. *Electroanalysis* 17:532–537.

108. Mogi I, Watanabe K (2006) Chiral recognition of magneto-electropolymerized polyaniline film electrodes. *Sci Technol Adv Mater* 7:342–345.
109. Kong Y, Ni J, Wang W, Chen Z (2011) Enantioselective recognition of amino acids based on molecularly imprinted polyaniline electrode column. *Electrochim Acta* 56:4070–4074.
110. Feng Z, Li M, Yan Y, Jihai T, Xiao L, Wei Q (2012) Several novel and effective methods for chiral polyaniline to recognize the configuration of alanine. *Tetrahedron: Asymmetry* 23:411–414.

5 Diastereoselective Syntheses of 3,4,5-Trihydroxypiperidine Iminosugars

Adam McCluskey and Michela I. Simone

CONTENTS

5.1 Introduction 163
5.2 Syntheses from the Chiral Pool 163
5.3 Bio-Catalyzed Syntheses 170
5.4 Asymmetric Syntheses 171
5.5 Conclusion and Future Directions 173
References 174

5.1 INTRODUCTION

Increasing the Fsp^3 index and decreasing the lipophilicity (log P) of drug leads can be linked to a reduction in toxicity (due to promiscuity), which is a preponderant factor in attrition during the later stages of drug discovery.[1–4] Recently, the presence of chiral centers has also been highlighted to impact success in the clinic favorably. The natural products 3,4,5-trihydroxypiperidines, scaffolds **1–4** (Figure 5.1), and their synthetic derivatives with their high Fsp^3 index, three chiral centers, and low log P represent a family of drug leads that display remarkable and often selective biological activities predominantly in the contexts of diabetes, lysosomal storage disorders, bacterial infections, immunosuppression, and inflammation.[5–44] Thus, efficient synthetic methodologies to 3,4,5-trihydroxypiperidines and their derivatizations are of importance to expansion of chemical space in this area and gathering of a more complete body of structure–activity relationship data.

These systems can be considered analogs of 1-deoxynojirimycin (1-DNJ), 1-deoxymannojirimycin (1-DMJ), 1-deoxygalactojirimycin (1-DGJ), lacking the C6-branching, and analogs of pentose monosaccharides (in their pyranose form). Hence the descriptors D/L-*xylo* and D/L-*ribo* for **1** and **2**, and L-*arabino*/L-*lyxo* and D-*arabino*/D-*lyxo* for **3** and **4** to designate the absolute stereochemical configurations of the three stereocenters (Figure 5.1). Furthermore, N-[5–10,12,13,15,18,20,22,27,28,35,41,42,45–63] and O-[11,14,17,18,64] derivatization have been demonstrated to produce improved selectivity in a number of biological contexts. Herein we report on the recent efforts to iminosugars **1–4**, which proceed either from the chiral pool or via asymmetric synthesis protocols, and by bio-catalyzed approaches.

5.2 SYNTHESES FROM THE CHIRAL POOL

Reports on the syntheses of **1–4** from the chiral pool are by far the most numerous due to the intrinsic presence of stereochemical information in the starting materials, usually monosaccharides.[6,7,9,22,23,25,26,29,41,42,54,56–61,63,65–76] Amine moiety introduction represents the main key step, followed by an N-alkylation step that allows the incorporation of the primary amine within the desired 6-membered ring structure. Both linear and divergent approaches have permitted access to all absolute stereochemical configurations of 3,4,5-trihydroxypiperidines **1–4**.[77] For example, iminosugar **1**

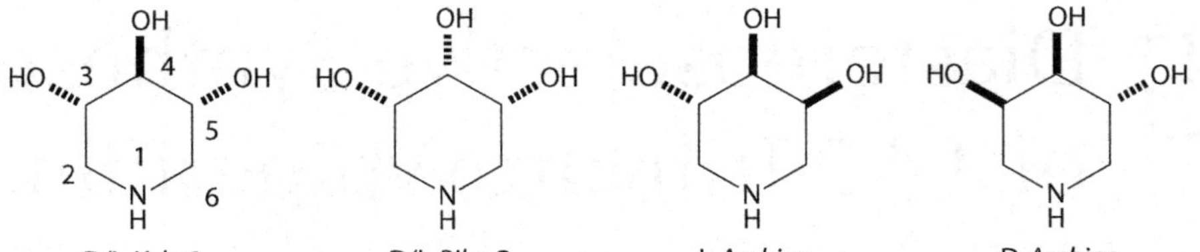

FIGURE 5.1 The four possible absolute stereochemical configurations for 3,4,5-trihydroxypiperidine iminosugars **1–4** with numbering system on structure **1**.

was successfully synthesized from D-xylose (linear[6,7,65,66]), **2** from D-ribose (linear,[28,41] divergent[25]), **3** from L-lyxose (divergent[25]), and **4** from D-arabinose (linear[25,26]).

Paulsen and coworkers reported the first synthetic approaches to 3,4,5-trihydroxypiperidine **1** from D-xylose derivatives, while studying the equilibria of sugar-derived hemiaminals.[65–67] Current approaches have built on Paulsen's pioneering efforts, using sodium azide for the introduction of a masked amine, thus allowing for simpler purification protocols.

Iminosugars **1** and **2** have been synthesized from D-glucose, which was firstly derivatized to 1,2-*O*-isopropylidene-3-*O*-benzyl-α-D-*xylo*-pentodialdose **5a** in 68% overall yield (Scheme 5.1).[23] Subsequent sodium borohydride reduction of the C-5 aldehyde, hydroxyl tosylation, and nucleophilic displacement of the tosyl group with sodium azide gave N-bearing intermediate **7a**. Catalytic reduction of the azide to the corresponding amine and Cbz protection gave final intermediate 5-(*N*-benzoxy-carbonylamino)-5-deoxy-1,2-*O*-isopropylidene-3-*O*-benzyl-α-D-xylofuranose **8a**, which upon deprotection with TFA/water and intramolecular reductive amination under palladium-catalyzed conditions afforded **1**. The same reaction sequence was effected using D-glucose-derived 1,2-*O*-isopropylidene-3-*O*-benzyl-α-D-*ribo*-pentodialdose **5b** as starting material, which afforded (3*S*,4*S*,5*R*)-piperidine triol **2** synthesized in a 62% overall yield.

D-*Xylo*-configured 3,4,5-trihydroxypiperidine **2** was obtained from D-xylose, via a cycle-opening protocol proceeding through open-chain intermediates **9–13** and subsequent amine introduction (Scheme 5.2).[25]

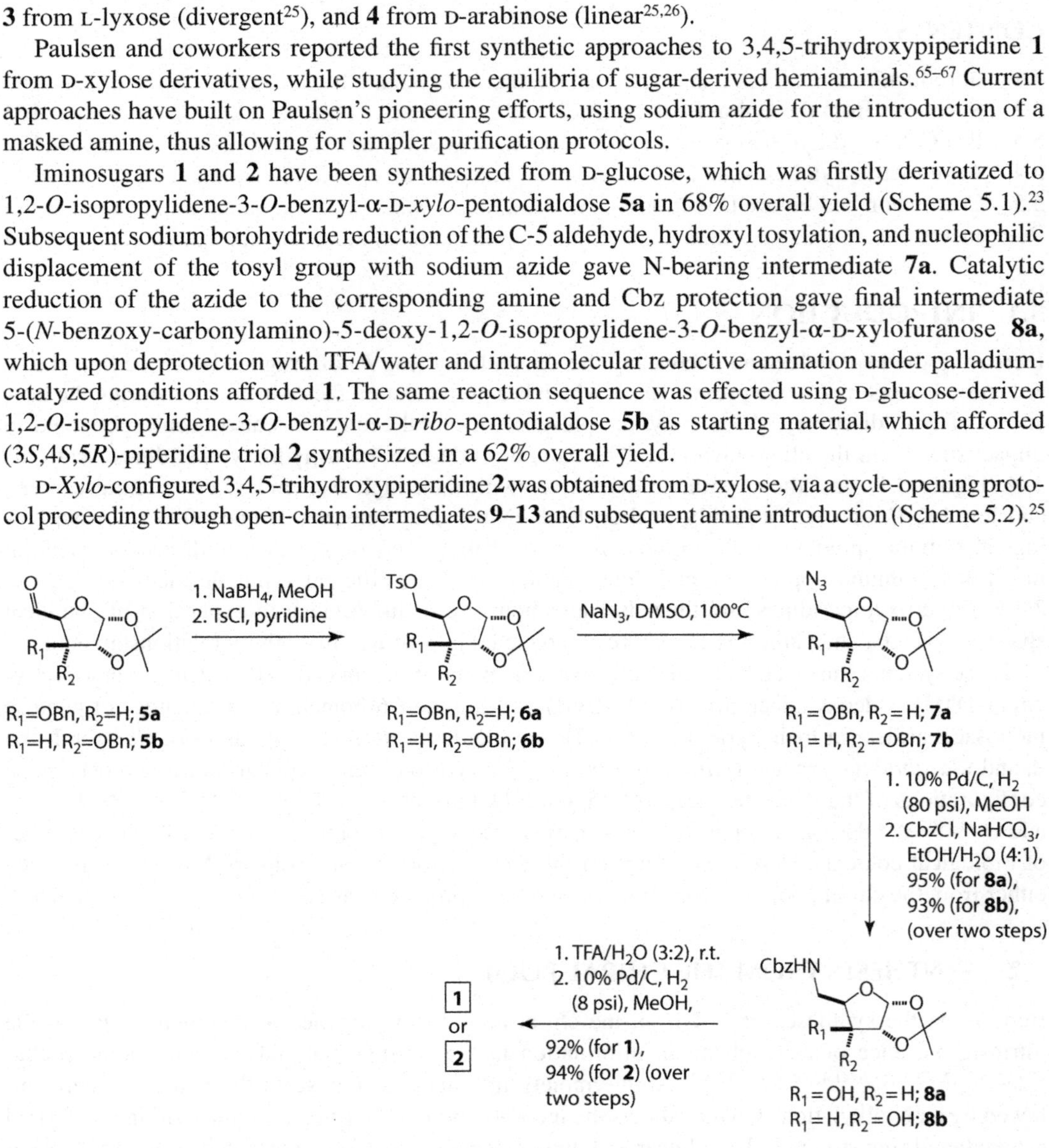

SCHEME 5.1 Synthesis of iminosugars **1** and **2** via azide introduction at C-5 of furanose monosaccharide derivatives.[23]

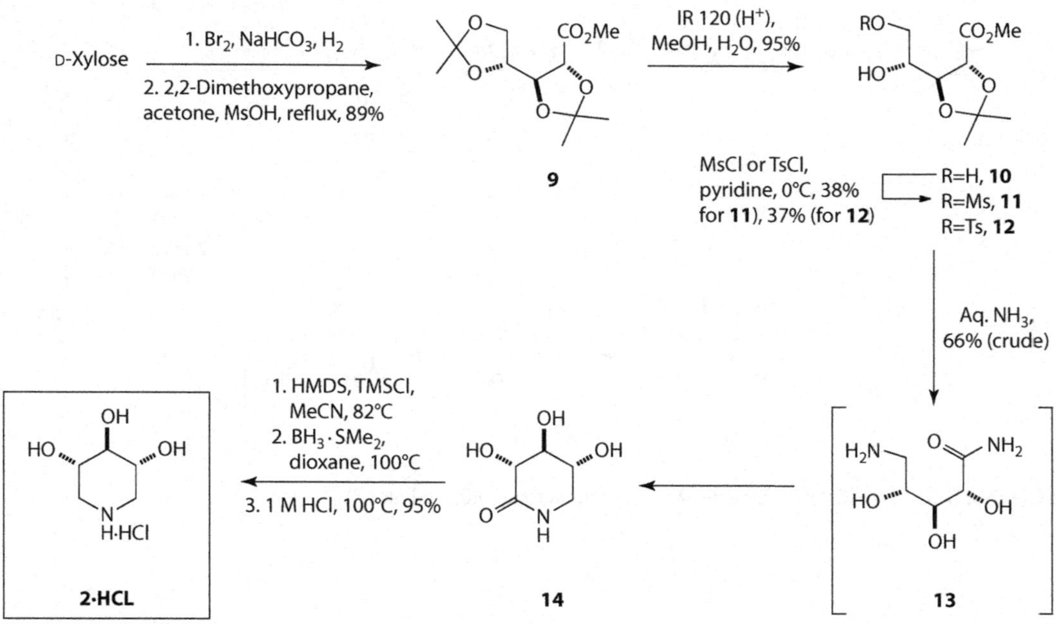

SCHEME 5.2 Trihydroxypiperidine **2** via the cycle-opening protocol of monosaccharide derivatives.[25]

Selective deprotection of the 4,5-acetal of **9** gave crystalline 2,3-*O*-isopropylidene-D-xylonic acid methyl ester **10** in 49% overall yield. Installation of a mesyl or tosyl leaving group afforded the corresponding 5-sulfonated esters **11** and **12**, respectively, in modest yields, followed by reaction with aqueous ammonia to install the amine. Deprotection (1% HCl in MeOH at 60°C) and lactamization gave 5-amino-5-deoxy-D-xylono-1,5-lactam **14**, via intermediate **13**. The lactam was silylated *in situ* and reduced using borane dimethylsulfide to give 1,5-dideoxy-1,5-imino-xylitol, isolated as the hydrochloride salt **2** in 95% yield over the final three steps.

The stereoisomeric 3,4,5-trihydroxypiperidines **2–4** were synthesized by Godskesen et al. from the corresponding aldonolactones or aldonic acid ester intermediates bearing a C-5 leaving group on treatment with aqueous ammonia.[25] From their work, synthesis of **4** started from D-arabinonolactone, which was transformed into 5-bromo-5-deoxy-D-arabinonolactone **15** on treatment with hydrogen bromide and acetic acid (Scheme 5.3).[69] Reaction of bromide **15** with aqueous ammonia produced **19**—precursor lactam to **4**—via reactive intermediate 5-bromo amide **16**. Monitoring by ^{13}C-NMR revealed the presence of epoxide **17** and 5-aminoamide **18** (in a ratio of 35:59:6). Subsequent acetonide protection of the *cis* hydroxyls and reduction of lactam **20** with borane dimethylsulfide gave the iminosugar **4**, whereas the use of a sodium borohydride/trifluoroacetic acid reagent combination afforded reduction of the lactam **20** to **4** and its isolation as the crystalline hydrochloride salt.

The key steps to target iminosugar **2**·HCl in another synthetic strategy involved a double displacement of two leaving groups on suitably protected open-chain sugar substrates to introduce the N atom to the monosaccharide backbone and close it into a cyclical structure, followed by a ring expansion rearrangement (Scheme 5.4).[41] Sodium borohydride reduction of 5-*O*-*tert*-butyldimethylsilyl-2,3-*O*-isopropylidene-D-ribose **21**, prepared from D-ribose in two steps, afforded the corresponding open-chain diol intermediate, which was doubly mesylated to afford **22**. Intermediate **22** underwent a double displacement with benzylamine at reflux in toluene to give *N*-benzyl-1,4-dideoxy-1,4-imino-2,3-*O*-isopropylidene-L-lyxitol. TBAF deprotection freed OH-6, which was mesylated to give pyrrolidine **23** in excellent yield. Reflux of intermediate **23** triggered intramolecular aziridine formation by displacement of the mesylate and production of salt **24**. Reactive intermediate **24** was opened with two types of nucleophiles: hydroxide and acetate. When hydroxide was used, piperidine

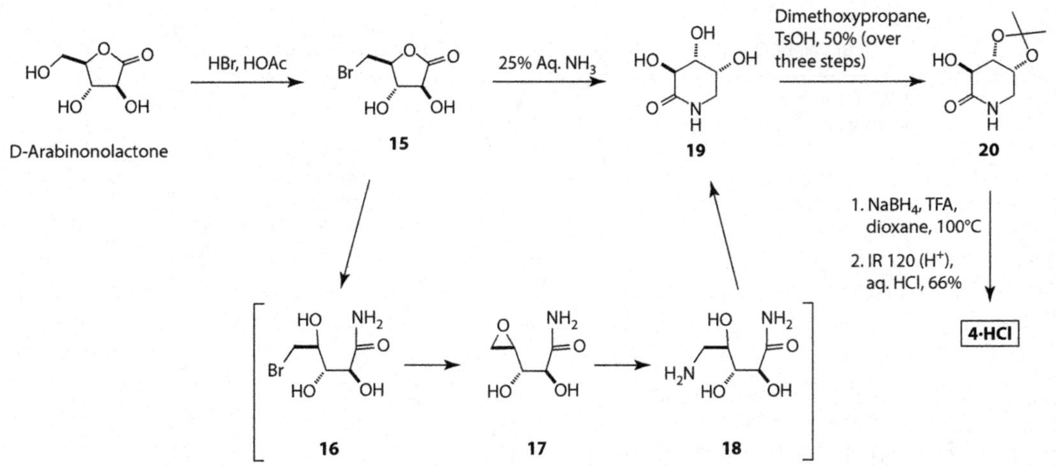

SCHEME 5.3 Trihydroxypiperidine **4** via aldonolactone intermediates.[25]

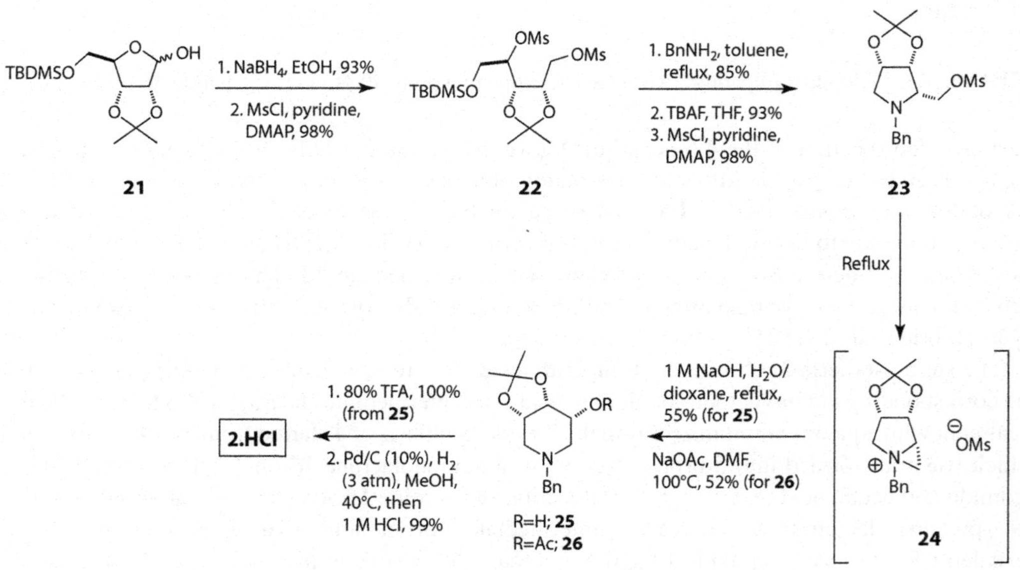

SCHEME 5.4 Iminosugar **2** via a double nucleophilic displacement of two leaving groups on suitably protected open-chain monosaccharides.[41]

derivative **25** and the pyrrolidine regioisomer were afforded in 55% and 40% yield, respectively. When acetate was used, the piperidine adduct **26** was isolated in 52% yield and its pyrrolidine regioisomer in 32% yield. Though the yields were modest for both products, these were formed as diastereoisomerically pure compounds. Acetonide (and acetate) hydrolysis with TFA, followed by reductive hydrogenation in the presence of Pd/C, afforded iminosugar **2** as the hydrochloride salt in excellent yield over the final two steps.

Another synthetic strategy employing the double displacement of two leaving groups situated on a monosaccharide backbone is outlined in Scheme 5.5a. Two leaving groups are installed on the primary hydroxyls of alditols; two intramolecular displacements by the adjacent secondary hydroxyls afford the corresponding *bis*-epoxide intermediate. The introduction of the N atom occurs via

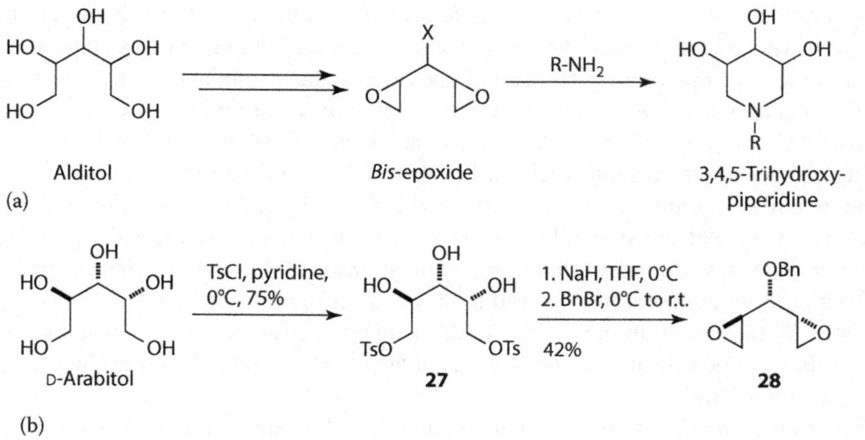

SCHEME 5.5 (a) Synthetic strategy to 3,4,5-trihydoxypiperidines employing *bis*-epoxide intermediates; (b) synthetic steps to *bis*-epoxide **28** from D-arabitol.[63]

double epoxide opening with suitable primary amines to produce focused libraries of N-alkylated 3,4,5-trihydroxypiperidines. The absolute stereochemical configuration of the three iminosugar stereocenters is retained from the chosen alditol starting material. Smith and Thomas synthesized all four pentitols via this *bis*-epoxide route.[57] Intermediate **28** was obtained in three steps via tosylation of the two primary hydroxyl groups, subsequent treatment with sodium hydride to the corresponding *bis*-epoxide, and benzylation of position C3–OH in a 31% overall yield (Scheme 5.5b).[63] The same synthetic protocol starting from L-arabitol (not shown) and xylitol gave the corresponding *bis*-epoxides in similar yields.

This *bis*-epoxide protocol did not work when ribitol was used as starting material, and in this case the benzylation step was effected before the *bis*-epoxide formation step (Scheme 5.6a). Hydroxyls on C1, C2, C4, and C5 were protected as the *bis*-acetonide and the hydroxyl of C3 were then benzylated. The acetonides were cleaved with acetic acid to tetrol **30**, which under Mitsunobu conditions gave *bis*-epoxide in 54% (32% overall yield over four steps).

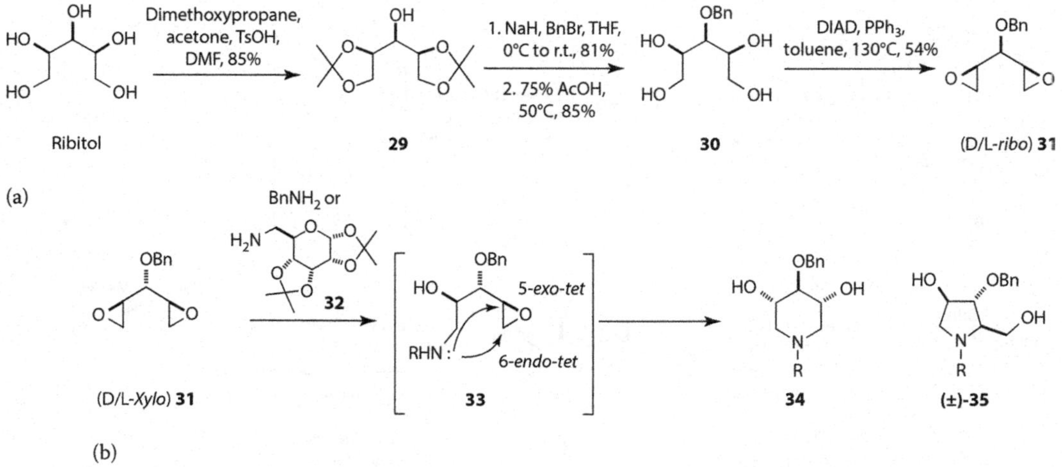

SCHEME 5.6 (a) Synthetic strategy to *bis*-epoxide (D/L-*ribo*) **31** from ribitol; (b) N-substituted trihydroxypiperidines and pyrrolidines from *bis*-epoxide (D/L-*xylo*) **31**.[57,78]

The *bis*-epoxides were opened with suitable primary amines (e.g., benzylamine or the bulkier D-allose derivative **32**) (Scheme 5.6b). The main synthetic challenge at this stage is the control of the intramolecular opening of the second epoxide by the secondary amine group in **33**. This could either occur via a 5-*exo*-tet mechanism to give an enantiomeric mixture of pyrrolidine iminosugars (±)-**35** or via the 6-*endo*-tet mechanism to afford the 3,4,5-trihydroxypiperidine **34**, a protected precursor to iminosugar **1**. Smith and Thomas[57] improved the selectivity of this second nucleophilic opening to maximize the yield of the piperidine *versus* the pyrrolidine products.[78] It was found that the slow addition of 6 equiv. of perchloric acid and 12 equiv. of small amines (such as benzylamine) to the *bis*-epoxide solution, followed by stirring for 2 h at 0°C, and for 16 h at room temperature, favored the 6-*endo*-tet in preference to a 5-*exo*-tet cyclization in a 3:1 ratio. With bulkier amines such as **32**, the best reaction conditions of the ones screened were to stir the *bis*-epoxide and amine in water at 50°C, which gave the piperidine to pyrrolidine adducts also in a 3:1 ratio.

Another strategy leading to the efficient syntheses of **1**, **3**, and **4** proceeds via bromopyranose sugar intermediates (e.g., **36** (D-*gluco*) to **1** (D/L-*xylo*)), initially derived from D-hexose monosaccharides (Scheme 5.7).[22] The key step involves a reductive ring fragmentation of bromopyranoses followed by *in situ* reductive amination of the resulting ω-alkenylaldehyde.[79,80] This gave access to alkene **37** in a one-pot reaction (70% yield) working directly on the unprotected starting material glycoside **36**. To prevent oxidation of the benzylic amine in the next step, ozonolysis of **37** was performed on the corresponding trifluoroacetate salt **37 · HCl**. When the first-formed ozonide was reductively cleaved using dimethylsulfide/sodium cyanoborohydride, N-derivatized trihydroxypiperidine **39** was obtained in 57% yield. Hydrogenolysis quantitatively gave the desired **1**. In a similar fashion, D-mannose and D-galactose were converted via intermediates **40** and **41** to enantiomerically pure **3** and **4**.

Allylic alcohol **42**, which supplies the five-carbon framework and two consecutive chiral centers of the final iminosugar **2**, was used as the starting material in the synthesis outlined in Scheme 5.8.[70] The key step was the introduction of the third chiral center (and hydroxyl group) via the diastereoselective epoxide formation that transforms **42** into **43**. This reaction was originally carried out under Sharpless conditions using (+)-DET; however, the optimized conditions use mCPBA in dichloromethane at −20°C to afford epoxides *anti* **43** and *syn* **43** with optimal balance of yield (83%) and selectivity (*anti*:*syn*, 5:1). Conversion to the MOM ethers (*anti* **44** and *syn* **44**) facilitated separation of the diastereoisomers at this stage. Opening of major epoxide *anti* **44** with sodium azide, followed by protecting group manipulations (a second MOM protection and terminal isopropylidene

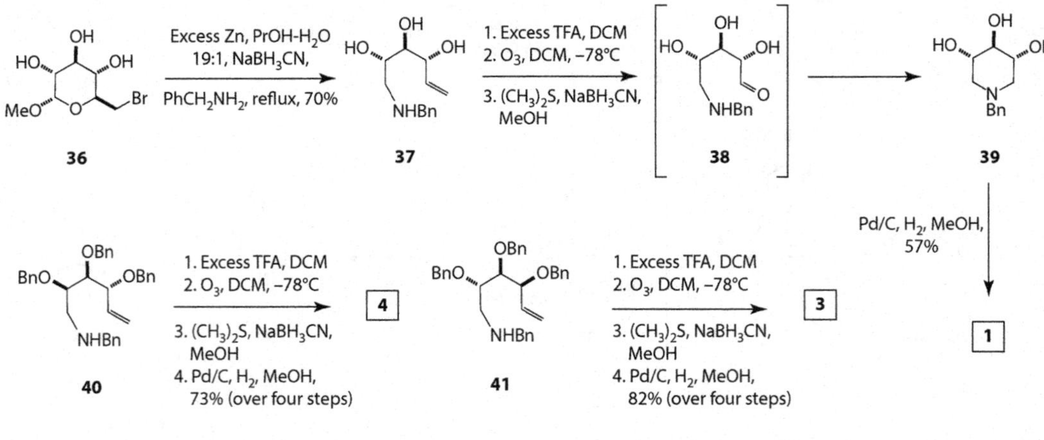

SCHEME 5.7 Iminosugars **1**, **3**, and **4** via a reductive ring fragmentation of bromopyranoses followed by *in situ* reductive amination of the resulting ω-alkenylaldehyde.[79,80]

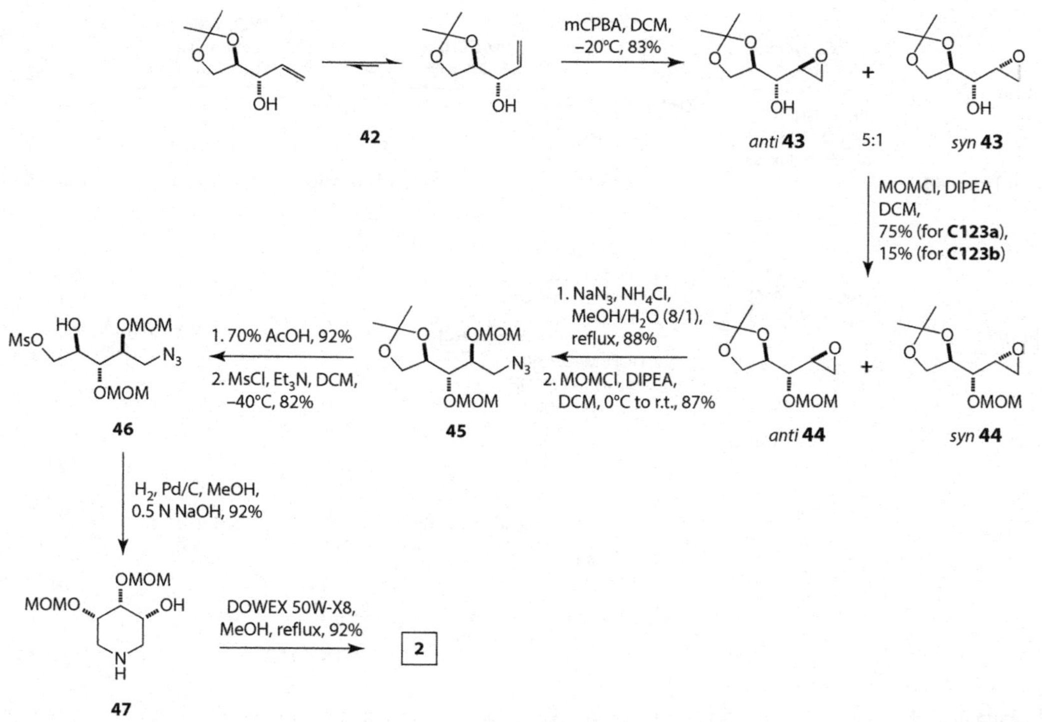

SCHEME 5.8 Iminosugar **2** via opening of a suitable epoxide intermediate with azide.[70]

hydrolysis), freed the primary alcohol, which upon mesylation gave **46**. This open-chain intermediate was subjected to reductive conditions using Pd/C and hydrogen to the corresponding amine, which cyclized intramolecularly to afford **47**. Stirring with DOWEX 50W-X8 resin afforded **2** as the free base.

A strategy starting from D-Threo **48** (a constitutional isomer of **42**) involved as key steps the introduction of N via nucleophilic displacement of a leaving group with PhCH$_2$NHCH$_2$TMS (D-Threo **48** → D-Threo **49**) and the chemical manipulation of C-5 via the use of the TMS in a photoinduced electron transfer (PET) cyclization reaction affording divergent intermediates D-Threo **49** → D-Threo **50** (Scheme 5.9) and L-Threo **49** → L-Threo **50** (not shown). These possess the correct stereochemistry at C-3 and C-4 to allow manipulation of the double bond at C-5, to ultimately control the stereochemical outcome at C-5 and allow access to a variety of iminosugars, including 3,4,5-trihydroxypiperidines **1**, **3**, and **4**. The PET cyclization protocol applied for these trihydroxypiperidine syntheses was developed earlier by the same authors,[58–61] who investigated this cyclization on α-trimethylsilylmethylamines tethered to π-functionality model systems.[61] The mechanism involves a three-centered amine radical cation species, delocalized between the nitrogen, and the silicon atom as the reactive intermediate, which results in the intramolecular addition of the π-electron of the tethered group with concomitant elimination of the TMS cation, and terminates with proton abstraction from 2-propanol and cyclization. In this report, a modified PET cyclization strategy afforded in moderate yield the D-Threo alkene **50** as a single diastereomer. Dihydroxylation with catalytic osmium tetroxide and NMO as co-oxidant proceeded in 95% yield to give D-Threo **51** also as a single diastereomer. Periodate oxidation of D-Threo **51** afforded the highly unstable corresponding ketone D-Threo **52** in 80% yield, which upon sodium borohydride reduction gave D-Threo alcohol **53** in a 9:1 diastereomeric ratio. Global deprotection in acidic reducing conditions gave 3,4,5-trihydroxypiperidine **4**.

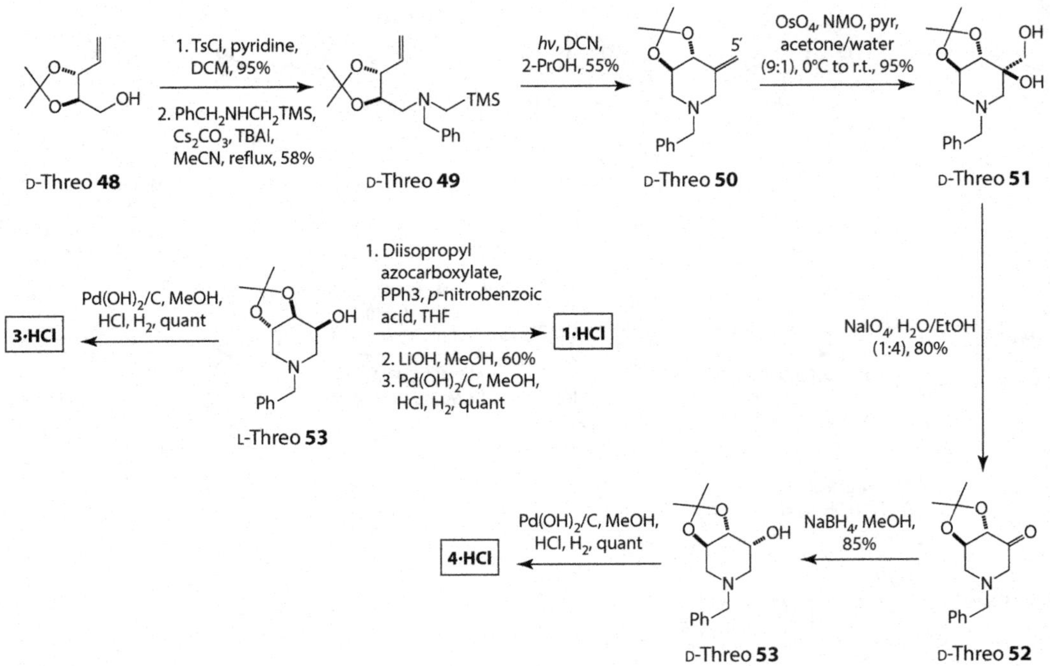

SCHEME 5.9 Trihydroxypiperidines **1**, **3**, and **4** via radical mechanism—the PET cyclization of suitable tertiary amines bearing a TMS group one carbon away.[58–61]

3,4,5-Trihydroxypiperidine **3** was synthesized according to the same synthetic route from L-Threo **53**. Under Mitsunobu conditions, 3,4,5-trihydroxypiperidine **1** was accessed from the same intermediate L-Threo **53**.

With the advent of an increase in automation in chemical synthesis,[81] protocols to automate natural product synthesis are coming to prominence in the research lab. Chang and coworkers elected to employ a combinatorial approach to a library of pyrrolidine and piperidine iminosugar analogs,[74] via key intermediates followed by their addition to nucleophiles in a diastereoselective fashion. These were prepared both on solid support and in solution phase. All reaction steps and work-ups were optimized to allow the use of automatic equipment, including a liquid handler and solution-phase synthesizer. Use of (4-methoxyphenyl)diisopropylsilylpropyl polystyrene resin was used as functionalized solid support to which four diverse saccharides were coupled and chemically modified.[75]

5.3 BIO-CATALYZED SYNTHESES

Bio-catalyzed steps allow for green protocols to be introduced in these syntheses with water as the reaction medium.[14,82,83] For example, a biocatalytic synthetic strategy to effect crossed aldol condensations allowed access to polyol structures, including 3,4,5-trihydroxypiperidines, and to avoid laborious isolation of unprotected hydroxyaldehydes usually obtained by chemical pathways.[83] Using 2-deoxyribose-5-phosphate aldolase (DERA), which catalyzes the stereoselective cross-aldol addition of acetaldehyde to other aldehydes, afforded the desired products, however in low conversion rates. DERA is unable to generate two consecutive hydroxylated centers with nonphosphorylated, unnatural substrates and is not efficient with nonphosphorylated, unnatural substrates. On the other hand, D-fructose-6-phosphate aldolase (FSA)[84,85] displays greater tolerance for donor substrate and gave access to 3,4,5-trihydroxypiperidines **1** and **4** from reaction between racemic **54** (also accepted by the enzyme) with glycolaldehyde **55** to produce iminosugar precursors **56** and **57**

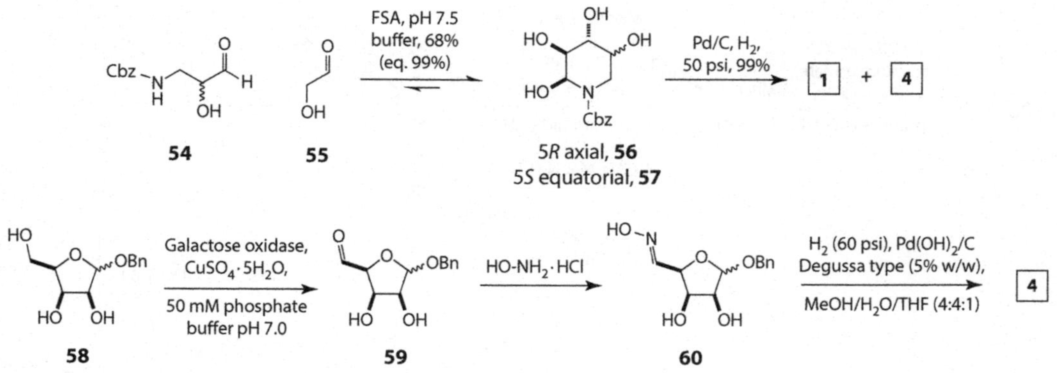

SCHEME 5.10 Iminosugars **1** and **4** via the use of bio-catalyzed step employing enzymes FSA[84,85] and GAO.[82]

in a combined yield of 69%. Removal of the Cbz-protecting group and reductive amination gave iminosugar **4** (Scheme 5.10).

Andreana and Wang[82] in their investigations into the enzymatic synthesis of polyhydroxyazepanes synthesized desired iminosugar **4** from substrate **58**. Pentose derivative **58** (D-*lyxo*) was fed to galactose oxidase (GAO). GAO resides extracellularly, where it utilizes molecular oxygen to convert the primary hydroxyl in position C-6 of nonreducing terminal galactose to the corresponding aldehyde. Due to its broad substrate specificity, GAO also transformed the primary hydroxyl on C-5 of **58** into aldehyde **59**. The pH was then adjusted to 5.2 and further reacted with hydroxylamine hydrochloride to give the corresponding oxime, which, after hydrogenolysis with a Degussa-type Pd catalyst at a pressure of 60 psi, gave final compound **4**.

5.4 ASYMMETRIC SYNTHESES

Syntheses from asymmetric starting materials generally employ as key steps ring-closing metatheses to afford the heterocycle and one or two dihydroxylation steps to introduce the hydroxyl groups on consecutive carbons of the cycle.[12,13,26,86–96] In this manner, Han devised a general methodology for the asymmetric synthesis of 3,4,5-trihydroxypiridines **2** and **3** from the readily available achiral olefin **61** (Scheme 5.11).[91]

The achiral olefin **61** was chosen as starting material for the regioselective aminohydroxylation reaction based on the postulate that aryl–aryl stacking interaction between the *p*-methoxyphenyl group (PMP) of **61** and the catalyst could increase the enantioselectivity of this reaction,[97] and the observation that the aminohydroxylation reaction of terminal olefins has been seen to occur so that a nitrogen atom predominantly adds to the terminal carbon of the olefins. Thus, the osmium-catalyzed aminohydroxylation reaction of **61** gave the aminoalcohol **62** in 9:1 regioselectivity, which was subsequently isolated via flash column chromatography followed by recrystallization. MOM protection followed and the nitrogen deprotected and reprotected as Boc derivative, which gave a better yield in the dihydroxylation step (**65** → **66**). N-Allylation, deprotection of the PMP group by ceric ammonium nitrate, and Swern oxidation of the resulting alcohol transformed **62** into **63**. Olefination of the aldehyde **63** was optimized to provide intermediate **64** in 89% yield under modified Horner–Wadsworth–Emmons conditions. Ring-closing metathesis (RCM) was effected using second-generation Grubbs' catalyst, followed by dihydroxylation of the cyclic olefin **65** under the Upjohn conditions to **66** with high diastereoselectivity (exclusive *anti*-addition), which upon hydrolysis provided target iminosugar **3** as its hydrochloride salt. Iminosugar **2** was obtained via cyclic sulfate chemistry on diol **66**, to give **67**, via treatment with thionyl chloride followed by oxidation of the resulting cyclic sulfite with RuCl₃ and NaIO₄. Ring opening of

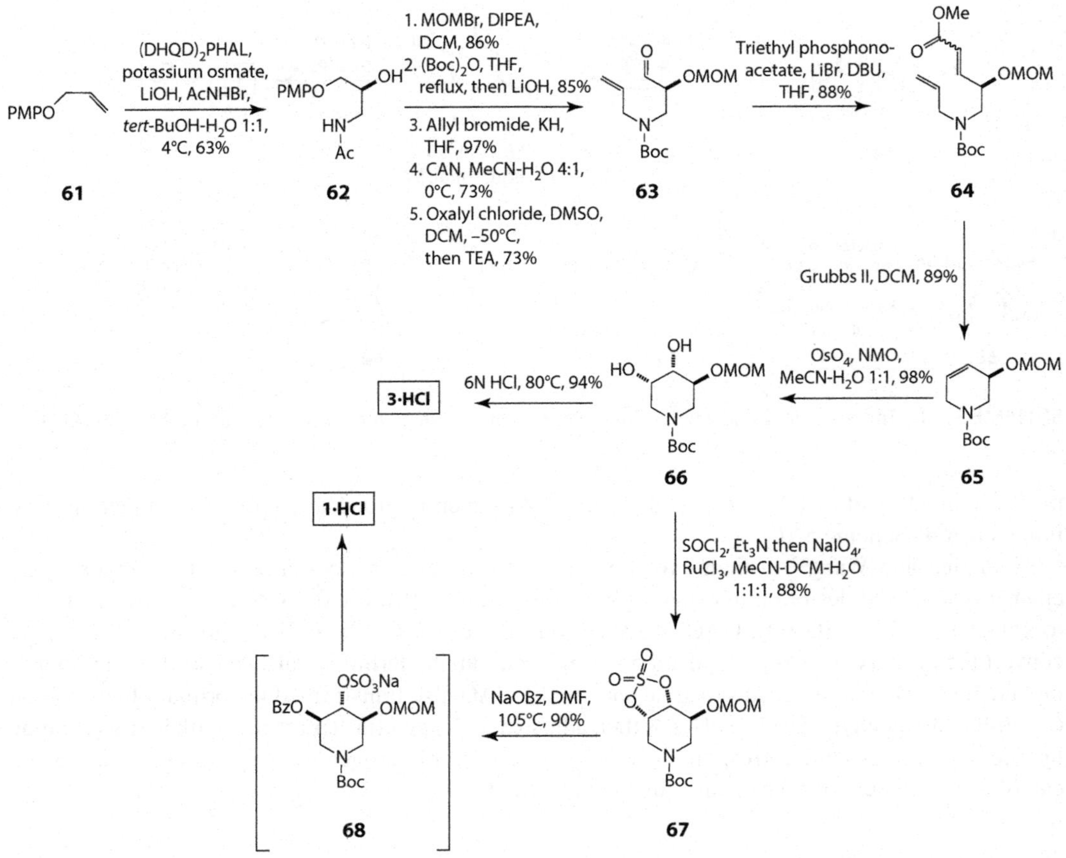

SCHEME 5.11 Iminosugars **1** and **3** via aminohydroxylation reaction of terminal olefins, followed by RCM and dihydroxylation.[97]

67 preferentially via C-3 (*vs.* at C-4 in a 9:1 ratio) with sodium benzoate and acidic hydrolysis allowed isolation of **1** as the corresponding hydrochloride salt.

Another asymmetric synthesis of substituted piperidines **1** and **4** starting from readily available achiral 4-methylphenacyl bromide **69** was achieved via three key steps: β-cyclodextrin- or oxazaborolidine-catalyzed asymmetric reduction of α-azidoaryl ketones to the corresponding alcohols, followed by a ring-closing metathesis, and a selective dihydroxylation (Scheme 5.12).[92]

Treatment of 4-methylphenacyl bromide **69** with NaN$_3$ in the presence of β-cyclodextrin to give azide **70** is the first-reported β-cyclodextrin-mediated azidation of a bromo compound. Asymmetric reduction of azido aryl ketone **70**/β-cyclodextrin complex with sodium borohydride in water produced the corresponding alcohol in 95% yield and 80% *ee*. However, reaction optimization showed that this reduction catalyzed in the presence of oxazaborolidine gave the alcohol in 94% yield and 100% *ee*.[93] MOM ether protection, followed by oxidative cleavage with RuCl$_3$ and NaIO$_4$, gave methyl ester **72**. Reduction of the azide group with β-cyclodextrin and triphenyl phosphine (TPP) gave the corresponding amine, which was Boc-protected *in situ*. NaBH$_4$- and LiCl-mediated reduction of the methyl ester group of **73** yielded alcohol **74**, which upon oxidation to the corresponding aldehyde and Wittig olefination produced alkene **75**. N-Allylation of **75** with allyl bromide in the presence of NaH and TBAI afforded **76**, the substrate for the second key reaction, the RCM, which was effected with Grubbs I catalyst. Dihydroxylation of **77** with OsO$_4$ and NMO, followed by Boc hydrolysis in 6 M HCl in MeOH, resulted in the hydrochloride salt of **4**.

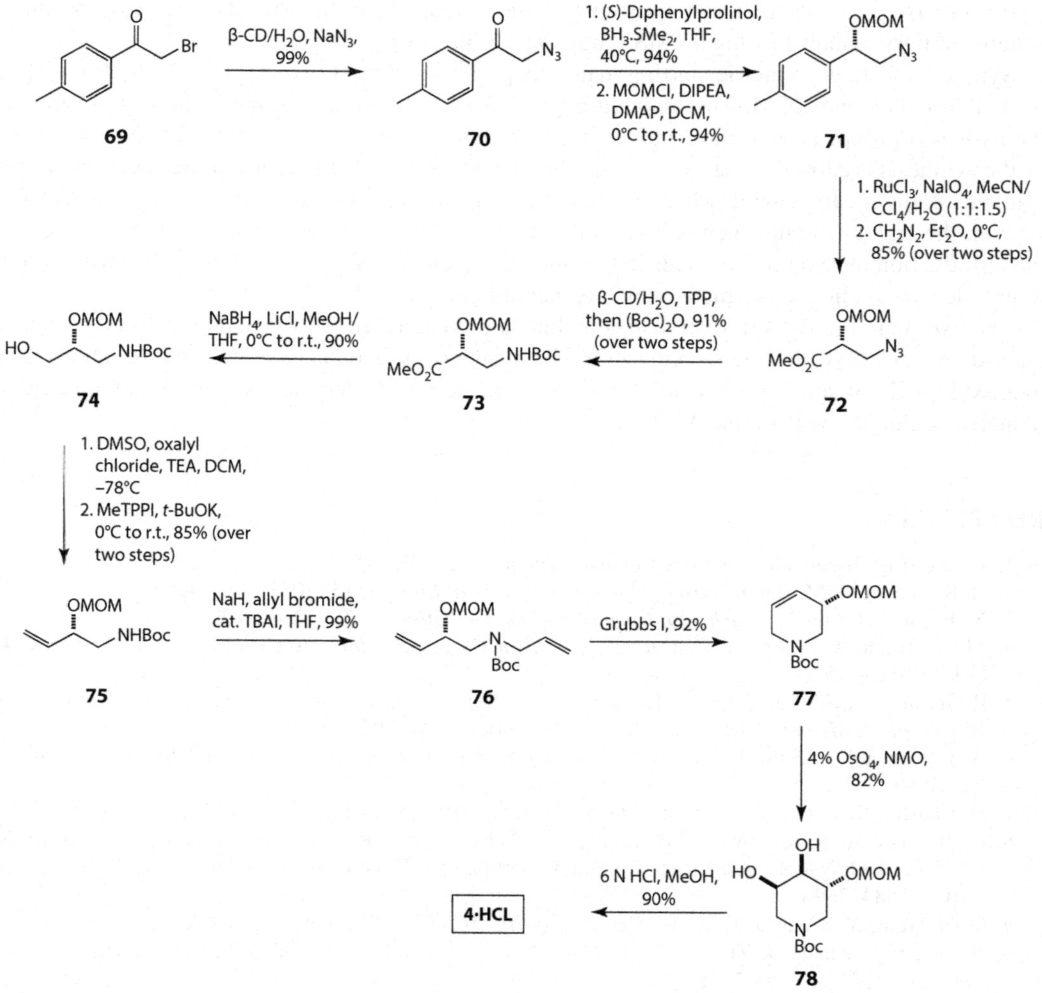

SCHEME 5.12 Iminosugar 4 via β-cyclodextrin- or oxazaborolidine-catalyzed asymmetric reduction of α-azidoaryl ketones, followed by RCM and a selective dihydroxylation.[92]

5.5 CONCLUSION AND FUTURE DIRECTIONS

3,4,5-Trihydropiperidines (and their N- and O-derivatives) constitute a family of natural product analogs with wide-ranging and often selective biological activities. Library building in this region of chemical space (high Fsp³ index, presence of three consecutive chiral centers, and low LogP values) is advantageous to tuning selectivities and potencies through gathering of a more complete body of structure–activity relationship data. Efficient syntheses to these target molecules have been achieved via strategies starting from the chiral pool, asymmetric starting materials, and via biocatalyzed protocols. One of the main synthetic challenges is the introduction of the N atom. This has been successfully conducted in various ways, which have been described in this chapter. One effective strategy is to introduce the amine masked as an azide at C-5 of pentofuranose derivatives via nucleophilic displacement, or via opening of γ-lactones, or via a double displacement of two leaving groups on suitably protected open-chain sugar substrates by ammonia or a primary amine, or by imine formation at C-1 on solid support using flow chemistry protocols. Alternatively, the endocyclic nitrogen is introduced at the end of the synthetic pathway by opening of suitable *mono-* or *bis-*epoxides. Another approach involves a radical mechanism, the PET cyclization of suitable tertiary amines bearing a TMS group one carbon away.

Syntheses from asymmetric starting materials primarily exploit two reactions: dihydroxylation of double bonds and RCM. The main stumbling block in these syntheses is the introduction of the hydroxyl groups around the heterocycle, since the introduction of the N is either not needed (if the synthesis is carried out from pyridine derivatives) or easily introduced at the beginning of the synthetic pathway. In general, when pyridine-derived starting materials are subjected to dihydroxylation, the endocyclic nitrogen is either protected or carries a chiral inducer. In other strategies, the introduction of two double bonds in the ring allows the RCM key step to be performed in order to produce the cyclic intermediate, which is then dihydroxylated.

The two main bio-catalyzed strategies include a bio-catalyzed aldol reaction to form the cyclic piperidine scaffold (which then is reduced by chemical means) and the *bio*-catalyzed oxidation of the hydroxyl on C5 of suitably protected furanose monosaccharide derivatives (which then undergoes reductive amination with amines).

REFERENCES

1. F. Lovering, *Medicinal Chemistry Communications*, 4 (2013): 515–519.
2. A. R. Leach, M. M. Hann, *Current Opinion in Chemical Biology*, 15 (2011): 489–496.
3. X. Jalencas, J. Mestres, *Medicinal Chemistry Communications*, 4 (2013): 80–87.
4. M. M. Hann, A. R. Leach, G. Harper, *Journal of Chemical Information and Computer Sciences*, 41 (2001): 856–864.
5. P. Greimel, H. Häusler, I. Lundt, K. Rupitz, A. E. Stütz, C. A. Tarling, S. G. Withers, T. M. Wrodnigg, *Bioorganic & Medicinal Chemistry Letters*, 16 (2006): 2067–2070.
6. A. J. Steiner, A. E. Stütz, C. A. Tarling, S. G. Withers, T. M. Wrodnigg, *Australian Journal of Chemistry*, 62 (2009): 553–557.
7. H. Häusler, K. Rupitz, A. E. Stütz, S. G. Withers, *Monatshefte für Chemie*, 133 (2002): 555–560.
8. D. Rejman, A. Rabatinová, A. R. Pombinho, S. Kovačková, R. Pohl, E. Zborníková, M. Kolař, K. Bogdanová, O. Nyč, H. Šanderová, T. Látal, P. Bartůněk, L. Krásný, *Journal of Medicinal Chemistry*, 54 (2011): 7884–7898.
9. G.-N. Wang, Y. Xiong, J. Ye, L.-H. Zhang, X.-S. Ye, *ACS Medicinal Chemistry*, 2 (2011): 682–686.
10. X. Wu, F.-Y. Zhang, J. Zhu, C. Song, D.-C. Xiong, Y. Zhou, Y. Cui, X.-S. Ye, *Chemistry: An Asian Journal*, 9 (2014): 2260–2271.
11. S. Front, N. Court, M.-L. Bourigault, S. Phanie Rose, B. Ryffel, F. Erard, V. F. J. Quesniaux, O. R. Martin, *ChemMedChem*, 6 (2011): 2081–2093.
12. F. Backenstrass, J. Streith, T. Tschamber, *Tetrahedron Letters*, 31 (1990): 2139–2142.
13. T. Tschamber, F. Backenstrass, M. Neuburger, M. Zehnder, J. Streith, *Tetrahedron*, 50 (1994): 1135–1152.
14. R. Uchida, A. Nasu, S. Tokutake, K. Kasai, K. Tobe, N. Yamaji, *Chemical and Pharmaceutical Bulletin*, 47 (1999): 187–193.
15. M. G. Szczepina, B. D. Johnston, Y. Yuan, B. Svensson, B. M. Pinto, *Journal of the American Chemical Society*, 126 (2004): 12458–12469.
16. P. Compain, O. R. Martin, C. Boucheron, G. Godin, L. Yu, K. Ikeda, N. Asano, *ChemBioChem*, 7 (2006): 1356–1359.
17. F. Oulaïdi, S. Front-Deschamps, E. Gallienne, E. Lesellier, K. Ikeda, N. Asano, P. Compain, O. R. Martin, *ChemMedChem*, 6 (2011): 353–361.
18. C. Matassini, S. Mirabella, X. Ferhati, C. Faggi, I. Robina, A. Goti, E. Moreno-Clavijo, A. J. Moreno-Vargas, F. Cardona, *European Journal of Organic Chemistry*, 2014 (2014): 5419–5432.
19. C. Parmeggiani, S. Catarzi, C. Matassini, G. D'Adamio, A. Morrone, A. Goti, P. Paoli, F. Cardona, *ChemBioChem*, 16 (2015): 2054–2064.
20. C. Matassini, S. Mirabella, A. Goti, I. Robina, A. J. Moreno-Vargas, F. Cardona, *Beilstein Journal of Organic Chemistry*, 11 (2015): 2631–2640.
21. T. Sekioka, M. Shibano, G. Kusano, *Natural Medicines*, 49 (1995): 332–335.
22. R. C. Bernotas, G. Papandreou, J. Urbach, B. Ganem, *Tetrahedron Letters*, 31 (1990): 3393–3396.
23. N. T. Patil, S. John, S. G. Sabharwal, D. D. Dhavalea, *Bioorganic & Medicinal Chemistry*, 10 (2002): 2155–2160.

24. B. Winchester, C. Barker, S. Baines, G. S. Jacob, S. K. Namgoong, G. Fleet, *Biochemical Journal*, 265 (1990): 277–282.
25. M. Godskesen, I. Lundt, R. Madsen, B. Winchester, *Bioorganic & Medicinal Chemistry*, 4 (1996): 1857–1865.
26. G. Legler, A. E. Stiitz, H. Immich, *Carbohydrate Research*, 272 (1995): 17–30.
27. J. Lehmann, B. Rob, *Carbohydrate Research*, 272 (1995): C11–C13.
28. Y. Igarashi, M. Ichikawa, Y. Ichikawa, *Bioorganic & Medicinal Chemistry Letters*, 6 (1996): 553–558.
29. Y. Ichikawa, Y. Igarashi, M. Ichikawa, Y. Suhara, *Journal of the American Chemical Society*, 120 (1998): 3007–3018.
30. S. Subramaniyan, P. Prema, *Critical Reviews in Biotechnology*, 22 (2002): 33–64.
31. M. E. Flores, R. Perez, C. Huitrón, *Letters in Applied Microbiology*, 24 (1997): 410–416.
32. D. B. Jordan, K. C. Wagschal, *Applied Microbiology and Biotechnology*, 86 (2010): 1647–1648.
33. J. E. Wraith, *Journal of Inherited Metabolic Disease*, 29 (2006): 442–447.
34. T. D. Butters, R. A. Dwek, F. M. Platt, *Glycobiology*, 15 (2005): 43R–52R.
35. A. J. Rawlings, H. Lomas, A. W. Pilling, M. J.-R. Lee, D. S. Alonzi, J. S. S. Rountree, S. F. Jenkinson, G. W. J. Fleet, R. A. Dwek, J. H. Jones, T. D. Butters, *ChemBioChem*, 10 (2009): 1101–1105.
36. K. J. Valenzano, R. Khanna, A. C. Powe, R. Boyd, G. Lee, J. J. Flanagan, E. R. Benjamin, *Assay and Drug Development Technologies*, 9 (2011): 213–235.
37. R. A. Steet, S. Chung, B. Wustman, A. Powe, H. Do, S. A. Kornfeld, *Proceedings of the National Academy Sciences USA*, 103 (2006): 13813–13818.
38. G.-N. Wang, G. Twigg, T. D. Butters, S. Zhang, L. Zhang, L.-H. Zhang, X.-S. Ye, *Organic & Biomolecular Chemistry*, 10 (2012): 2923–2927.
39. Y. Suzuki, *Proceedings of the Japanese Academy, Series B*, 90 (2014): 145–162.
40. G. A. Grabowski, *Lancet*, 372 (2008): 1263–1271.
41. D.-K. Kim, G. Kim, Y.-W. Kim, *Journal of the Chemical Society, Perkin Transactions 1*, 1996 (1996): 803–808.
42. G. Pandey, K. C. Bharadwaj, M. I. Khan, K. S. Shashidhara, V. G. Puranik, *Organic & Biomolecular Chemistry*, 6 (2008): 2587–2595.
43. C. Kuriyama, O. Kamiyama, K. Ikeda, F. Sanae, A. Kato, I. Adachi, T. Imahori, H. Takahata, T. Okamoto, N. Asano, *Bioorganic & Medicinal Chemistry*, 16 (2008): 7330–7336.
44. A. L. Lovering, S. S. Lee, Y.-W. Kim, S. G. Withers, N. C. J. Strynadka, *The Journal of Biological Chemistry*, 280 (2005): 2106–2116.
45. E. D. Faber, L. A. van den Broek, E. E. Oosterhuis, B. P. Stok, D. K. Meijer, *Drug Delivery*, 5 (1998): 3–12.
46. H. J. Ahr, M. Boberg, E. Brendel, H. P. Krause, W. Steinke, *Arzneimittel-Forschung*, 47 (1997): 734–745.
47. C. Ficicioglu, *Therapeutics and Clinical Risk Management*, 4 (2008): 425–431.
48. N. Asano, M. Nishida, A. Kato, H. Kizu, K. Matsui, Y. Shimada, T. Itoh, M. Baba, A. A. Watson, R. J. Nash, P. M. d. Q. Lilley, D. J. Watkin, G. W. J. Fleet, *Journal of Medicinal Chemistry*, 41 (1998): 2565–2571.
49. N. Asano, H. Kizu, K. Oseki, E. Tomioka, K. Matsui, M. Okamoto, M. Baba, *Journal of Medicinal Chemistry*, 38 (1995): 2349–2356.
50. P. Elías-Rodríguez, E. Moreno-Clavijo, S. Carrión-Jiménez, A. T. Carmona, A. J. Moreno-Vargas, I. Caffa, F. Montecucco, M. Cea, A. Nencioni, I. Robina, *ARKIVOC*, 2014 (2014): 197–214.
51. G. Godin, G. Compain, O. R. Martin, K. Ikeda, L. Yu, N. Asano, *Bioorganic & Medicinal Chemistry Letters*, 14 (2004): 5991–5995.
52. R. G. Spiro, *Part II: Biology of Saccharides*, Wiley-VCH, Weinheim, Germany, 2000.
53. A. D. Elbein, R. J. Molyneux, *Iminosugars as Glycosidase Inhibitors*, Wiley-VCH, Weinheim, Germany, 1999.
54. A. J. Steiner, A. E. Stütz, C. A. Tarling, S. G. Withers, T. M. Wrodnigg, *Carbohydrate Research*, 342 (2007): 1850–1858.
55. A. R. Katritzky, S. Rachwal, G. J. Hitchings, *Tetrahedron*, 47 (1991): 2683–2732.
56. C. Matassini, S. Mirabella, A. Goti, F. Cardona, *European Journal of Organic Chemistry*, 2012 (2012): 3920–3924.
57. R. D. Smith, N. R. Thomas, *Synlett*, 2 (2000): 193–196.
58. G. Pandey, G. Kumaraswamy, *Tetrahedron*, 50 (1994): 8185–8194.
59. G. Pandey, G. Kumaraswamy, U. T. Bhalerao, *Tetrahedron Letters*, 30 (1989): 6059–6062.
60. G. Pandey, M. Kapur, *Synthesis*, 8 (2001): 1263–1267.
61. G. Pandey, M. Kapur, *Tetrahedron Letters*, 41 (2000): 8821–8824.
62. B. B. Shankar, M. P. Kirkup, S. W. McCombie, A. K. Ganguly, *Tetrahedron Letters*, 34 (1993): 7171–7174.

63. B. Chenera, J. C. Boehm, G. B. Dreyer, *Bioorganic & Medicinal Chemistry Letters*, 1 (1991): 219–222.
64. Y.-W. Kim, H. Chen, S. G. Withers, *Carbohydrate Research*, 340 (2005): 2735–2741.
65. H. Paulsen, F. Leupold, K. Todt, *Liebigs Annalen der Chemie*, 692 (1966): 200–214.
66. H. Paulsen, G. Steinert, *Chemische Berichte*, 100 (1967): 2467–2473.
67. H. Paulsen, *Liebigs Annalen der Chemie*, 683 (1965): 187–198.
68. A. de Raadt, M. Ebner, C. W. Ekhart, M. Fechter, A. Lechner, M. Strobl, A. E. Stütz, *Catalysis Today*, 22 (1994): 549–561.
69. K. Bock, I. Lundt, C. Pedersen, *Carbohydrate Research*, 104 (1982): 79–85.
70. K. C. Jang, I.-Y. Jeong, M. S. Yang, S. U. Cjoi, K. H. Park, *Heterocycles*, 53 (2000): 887–896.
71. J. Mulzer, R. Becker, E. Brunner, *Journal of the American Chemical Society*, 111 (1989): 7500–7504.
72. K. E. Harding, S. R. Burks, *Journal of Organic Chemistry*, 49 (1984): 40–44.
73. S. S. Chirke, A. Rajender, B. V. Rao, *Tetrahedron*, 70 (2014): 103–109.
74. Y. -F. Chang, C. -W. Guo, T. -H. Chan, Y. -W. Pan, E. -L. Tsou, W. -C. Cheng, *Molecular Diversity*, 15 (2011): 203–214.
75. J. A. Tallarico, K. M. Depew, H. E. Pelish, N. J. Westwood, C. W. Lindsley, M. D. Shair, S. L. Schreiber, M. A. Foley, *Journal of Combinatorial Chemistry*, 3 (2001): 312–318.
76. Y. Suman Reddy, P. K. Kancharla, R. Roy, Y. D. Vankar, *Organic and Biomolecular Chemistry*, 10 (2012): 2760–2773.
77. G. Pandey, M. Kapur, *Organic Letters*, 4 (2002): 3883–3886.
78. L. Poitout, Y. Le Merrer, J.-C. Depezay, *Tetrahedron Letters*, 35 (1994): 3293–3296.
79. R. C. Bernotas, B. Ganem, *Tetrahedron Letters*, 26 (1985): 1123–1126.
80. B. Bernet, A. Vasella, *Helvetica Chimica Acta*, 62 (1979): 1990–2016.
81. S. V. Ley, D. E. Fitzpatrick, R. J. Ingham, R. M. Myers, *Angewandte Chemie International Edition*, 54 (2015): 3449–3464.
82. P. R. Andreana, T. Sanders, A. Janczuk, J. I. Warrick, P. G. Wang, *Tetrahedron Letters*, 43 (2002): 6525–6528.
83. X. Garrabou, J. A. Castillo, C. Gurard-Hlaine, T. Parella, J. Joglar, M. Lemaire, P. Claps, *Angewandte Chemie International Edition*, 48 (2009): 5521–5525.
84. J. A. Castillo, J. Calveras, J. Casas, M. Mitjans, M. P. Vinardell, T. Parella, T. Inoue, G. A. Sprenger, J. Joglar, P. Clapés, *Organic Letters*, 21 (2006): 6067–6070.
85. A. L. Concia, C. Lozano, J. A. Castillo, T. Parella, J. Joglar, P. Clapés, *Chemistry—A European Journal*, 15 (2009): 3808–3816.
86. M. Amat, M. Llor, M. Huguet, E. Molins, E. Espinosa, J. Bosch, *Organic Letters*, 3 (2001): 3257–3260.
87. M. Amat, J. Bosch, J. Hidalgo, M. Cantó, M. Pérez, N. Llor, E. Molins, C. Miravitlles, M. Orozco, J. Luche, *Journal of Organic Chemistry*, 65 (2000): 3074–3084.
88. M. Amat, M. Huguet, N. Llor, O. Bassas, A. M. Gomez, J. Bosch, J. Badia, L. Baldoma, J. Aguilar, *Tetrahedron Letters*, 45 (2004): 5355–5358.
89. M. D. Groaning, A. I. Meyers, *Tetrahedron*, 56 (2000): 9843–9873.
90. H. Ouchi, Y. Mihara, H. Takahata, *Journal of Organic Chemistry*, 70 (2005): 5207–5214.
91. H. Han, *Tetrahedron Letters*, 44 (2003): 1567–1569.
92. M. Somi Reddy, M. Narender, K. Rama Rao, *Tetrahedron*, 63 (2007): 331–336.
93. J. S. Yadav, P. Thirupathi Reddy, S. R. Hashim, *Synlett*, 2000 (2000): 1049–1051.
94. I. G. Rosset, A. C. B. Burtoloso, *Journal of Organic Chemistry*, 78 (2013): 9464–9470.
95. B. Bernardim, V. D. Pinho, A. C. B. Burtoloso, *The Journal of Organic Chemistry*, 77 (2012): 9926–9931.
96. V. D. Pinho, A. C. B. Burtuloso, *Tetrahedron Letters*, 53 (2012): 876–878.
97. E. J. Corey, A. Guzman-Perez, M. C. Noe, *Journal of the American Chemical Society*, 117 (1995): 10805–10816.

6 Use of Specific New Artificial or Semisynthetic Biocatalysts for Synthesis of Regio- and Enantioselective Compounds

Marco Filice, Oscar Romero, and Jose M. Palomo

CONTENTS

6.1 Introduction 177
6.2 Regioselective Preparation of Monodeprotected Esters 178
    6.2.1 Regioselective Monodeprotection of Peracetylated Carbohydrates 178
    6.2.2 Regioselective Preparation of Monodeprotected Nucleosides 180
6.3 Regio- and Enantioselective Preparation of Chiral Alcohols and Amines 182
6.4 Regioselective Artificial Hydrolase-Catalyzed C–C Bond Formation 185
References 187

6.1 INTRODUCTION

Enzymes are versatile biocatalysts and find increasing application in many areas, including organic synthesis. The major advantages of using enzymes in biotransformations are their chemo-, regio-, and stereoselectivity as well as the very mild reaction conditions that can be used, where the generation of the side products is minimized.[1–3] Their stereoselectivity (the ability to selectively act on a single enantiomer) their regioselectivity (the possibility to recognize one position in a molecule), and, finally, their selectivity toward defined functional group among others quite similar in reactivity allowed distinguishing them precisely. Each type of selectivity shows advantages that can accrue to chemical processes because of the special properties of the enzymes.

Indeed, regioselective and enantiomerically pure molecules are key compounds for many industrially relevant processes. The application of biocatalysts ranges from the synthesis of regio- and chiral intermediates (usually building blocks in drug chemistry), *via* neuromodulators: antibacterial, antifungal, and antiviral agents: enzymes inhibitors; chiral-labeled compounds UP to natural surfactant analogs.

In particular, lipases (acyl hydrolyses) are one of the most used enzymes for this purpose.[4–6] Although nature created them to hydrolyze fats, their peculiar catalytic mechanism based on the movement of an oligopeptide chain, which generates a closed or an open lipase conformation depending on the absence or the presence of a hydrophobic interface, respectively,[7] gives these enzymes a high versatility. They have been successfully used in aqueous media and also in organic solvents for hydrolysis of esters and amides besides esterification and amidation processes.[5,6] Some of the most interesting processes to easily produce chiral compounds with these enzymes are based on the use of resolution of racemic mixtures,[6] dynamic kinetic resolutions (DKR),[8] or desymmetrization reactions.[9] The lipase-catalyzed regioselective product formation is commonly based on the deprotection or protection strategies.[6,10]

However, there are still many cases where isolated enzymes lack enough stability—out of the natural environment—and activity or selectivity—toward non-natural substrates—limiting their applicability. Therefore, in the last two decades, researchers have tried to solve this difficulty by developing strategies improving stability, activity, and selectivity or even developing catalysts with unexpected synthetic catalytic activities.[11]

This chapter is focused on the application of these new artificial biocatalysts in the synthesis of regioselective monoacetylated carbohydrates and nucleosides, the DKR of alcohols and amines, and the generation by peculiar lipase-catalyzed C–C bond formation reactions.

6.2 REGIOSELECTIVE PREPARATION OF MONODEPROTECTED ESTERS

6.2.1 Regioselective Monodeprotection of Peracetylated Carbohydrates

Carbohydrates represent complex and structurally diverse molecules containing multiple chiral centers, and they can exist naturally as oligo- and polysaccharides and coupled to, for example, peptides and lipids being key molecules in many biological processes.[12,13] Generally, the complex carbohydrates cannot be easily extracted from natural sources in acceptable purity and quantity. On the other hand, their synthesis intrinsically expects two major challenges, namely, the regioselective protection and deprotection of polyhydroxy groups as well as the stereoselective assembly of glycosidic linkages. Thus, to overcome these drawbacks, significant efforts have been devoted to the development of chemical, enzymatic, or chemo-enzymatic methods for the synthesis of structurally well-defined carbohydrates and glycoconjugates.[13] Nevertheless, many challenges still follow unsolved. For example, chemical approaches for oligosaccharide or glycoderivative synthesis involve several protection and glycosylation steps finally impacting on overall yields. As an alternative, the preparation of selectively protected monosaccharide units bearing a single strategically positioned free hydroxyl group (a nucleophilic acceptor) together with a stereoselective glycosylation could symbolize a breakthrough in carbohydrate synthesis.[14] Due to the high regio- and enantioselectivity of the enzymes, mainly lipases represent a promising alternative to chemical methods to overcome all these drawbacks.[15] Lipases are quite particular enzymes with a complex catalytic mechanism, which manages their catalysis on the basis of the conformational changes between a "closed" and an "open" form.[7] Thus, by applying different immobilization protocols (involving different areas, rigidity, microenvironments, etc.) for each specific lipase, it is possible to obtain different catalysts of the same enzyme finally altering its activity and regioselectivity.[16]

In the last years, a chemo-enzymatic lipase-based strategy for highly efficient regioselective deprotection of single hydroxyls in peracetylated monosaccharides and disaccharides has been described.[10,17] Starting from a fully protected carbohydrate, using only acetyl ester as protecting group (monoprotective approach),[18] it is possible to deprotect different unique positions of peracetylated substrates applying a high regioselective biocatalytic process combined (when required) with a subsequent chemical reaction (Scheme 6.1). These building blocks can be successfully used in stereoselective chemical glycosylations to obtain target compounds protected with one chemical group (acetyl ester) that can be easily removed in only one step at the end of the synthesis.[19]

For example, in Table 6.1, the immobilization effect on the hydrolysis of peracetylated β-galactose (**1**) catalyzed by three different *Thermomyces lanuginose* lipase (TLL) derivatives is reported.[20] When TLL is immobilized on hydrophobic support (octyl-Sepharose-TLL), the biocatalyst was very specific and regioselective toward hydrolysis in C-6 (**2**) with a 99% yield. On the other hand, the CNBr-activated TLL immobilized biocatalyst (CNBr-TLL, covalent attachment by N-terminus) showed high regioselectivity toward the C-1 position (**3**). Moreover, when adsorbed on a cationic exchanger (PEI-TLL), the biocatalyst was poorly active and not selective at all against one of these positions. These findings have been successfully extended to many other lipases demonstrating the potential as general methodology.[17,21] Using this strategy, several carbohydrates were transformed in the corresponding 6-OH products with good or excellent overall yields (from 60% to >95%).[17]

Use of Specific New Artificial or Semisynthetic Biocatalysts

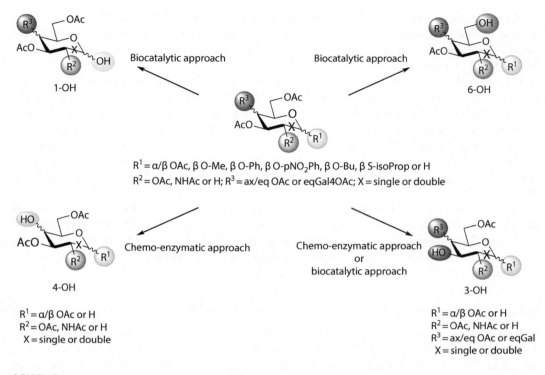

SCHEME 6.1 General scheme of the regioselective monodeprotection of per-*O*-acetylated carbohydrates.

TABLE 6.1
Specificity and Regioselectivity of Different Immobilized Preparation of Lipases in the Hydrolysis of 1

Biocatalysts	Yield (%) 2	Yield (%) 3
Octyl-TLL	99	0
CNBr-TLL	0	95
PEI-TLL	7	14

A complementary interesting strategy is represented by the use of chemical acyl migration to produce other building blocks bearing free hydroxyl groups in positions different than that previously achieved by the first enzymatic step.[22] Generally, this process has been described as a secondary undesired reaction where an acyl group could migrate to an adjacent free hydroxyl group in aqueous neutral or basic medium.[22] By a proper fine-tuning of critical parameters such as pH, time, and temperature and depending on the steric configuration (glucosidic or galactosidic) of substrate, the shift of the acyl group from C-3 or the C-4 position to the 6-OH free group can be easily controlled (Scheme 6.1).[22]

Therefore, the combination of the biocatalytic strategy together with the controlled medium engineering allowed to efficiently prepare—in "one-pot" step and very mild conditions—a small combinatorial library of different monohydroxy acetylated monosaccharides in good overall yields (from 30% to 95%) (Scheme 6.1).[7]

Very recently, an elegant strategy combining directed mutagenesis and site-directed chemical modification permitted a great improvement in the regioselective hydrolysis of peracetylated glucal.[23] This strategy is based on the creation of semisynthetic lipase by incorporation of tailor-made peptides on the lipase-lid site by means of a fast thiol–disulfide exchange ligation. Hence, by this protocol, a cysteine variant of the lipase of *Bacillus thermocatenolatus* (BTL-A193C) was site selective modified with a hydrophobic peptide (Ac-Phe-Cys-Phe-Gly-Phe-CONH$_2$: **p1**). This semisynthetic lipase yielded C-3 deprotected glucal in 94% compared to the 70% yield achieved using the native enzyme.[23]

These regioselective monodeprotected monosaccharides were successfully used as scaffolds in the synthesis of several biologically active molecules.[18–24] In particular, in the synthesis of peracetylated naphtyl-lactosamine derivative (antitumoral prodrug), Scheme 6.2 shows the advantage of the chemo-enzymatic route (20% overall yield of product in 4 steps)[24] against the best standard chemical route (5% overall yield and 11 synthetic steps).[25]

6.2.2 REGIOSELECTIVE PREPARATION OF MONODEPROTECTED NUCLEOSIDES

Natural nucleosides and their analogs have attracted intense interest to the chemical industry due to their potential as fungicidal, antitumor, and antiviral agents.[26] In the past 25 years, several nucleoside analog reverse-transcriptase inhibitors have been licensed and used in the treatment of HIV/AIDS, for example, Zidovudine.[27] For nucleoside analogs, several chemical synthesis approaches have been reported employing natural nucleosides as starting materials or involving condensation of the carbohydrate precursors and the heterocyclic modified bases. In both cases, to ensure chemo-, regio-, and stereoselectivity of the final products, tedious protecting and deprotecting steps are necessary, finally hampering the overall yield.[28] Hence, suitable introduction and removal of protecting groups is often a critical step. In nucleoside chemistry, particularly in the sugar moieties, this problem is accentuated by the presence of multiple hydroxyl functions with similar reactivity. As previously mentioned, lipases are excellent catalysts for this task. In fact, several lipase-catalyzed strategies to produce mono-protected nucleosides on the sugar moieties have been reported: acylation of free nucleosides[29] and deacylation by hydrolysis[30] or alcoholysis[31] of peracylated nucleosides.

Nowadays, several examples of lipase-catalyzed deacylation of protected nucleosides have been reported.[23,29,32] The immobilized lipases on hydrophobic matrixes as biocatalysts have been successfully used for the regioselective monoprotection of different fully acetylated nucleosides in aqueous media.[26] Some of the results of the deprotection of 3,5-*O*-diacetylated thymidine (**4**) by immobilized lipases are shown in Table 6.2. The best result was obtained using immobilized *C. rugosa* lipase, with an excellent 91% yield of C-5 monoprotected acetylated thymidine (**5**).[30]

Recently, the previously mentioned semisynthetic lipase–peptide conjugates were also used as catalysts in this reaction. Depending on the enzyme variant and the peptide used, the results strongly varied (Table 6.2).[23] Particularly, the cysteine incorporation in position 196 caused an inverted regioselectivity, hydrolyzing selectively the acetylated C-3 position in the ribose of thymidine. The incorporation of the peptide Ac-Cys-(Asp)$_4$-Asp-COOH (**p2**) on BTL-S196C variant enhanced the regioselectivity of the enzyme, producing the C-3 derivative (**6**) in 97% yield.

Both **5** and **6** are excellent building blocks for the preparation of 3′- or 5′-functionalized nucleosides and mononucleotides. The potential of this approach was demonstrated by the chemo-enzymatic synthesis of doxifluridine (a cancer chemotherapy prodrug[33] with an overall yield of 12% starting from 2,3,5-tri-*O*-acetyluridine) (Scheme 6.3).[30]

Use of Specific New Artificial or Semisynthetic Biocatalysts

181

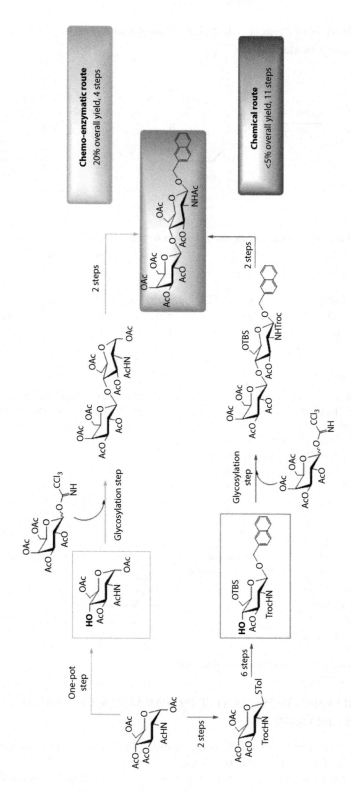

SCHEME 6.2 Comparison between the chemo-enzymatic and chemical pathways for the preparation of β-O-napthylmethyl-lactosamine peracetate.

TABLE 6.2
Regioselective Hydrolysis of Peracetylated Thymidine 4 Catalyzed by Different Heterogeneous Lipase Biocatalysts

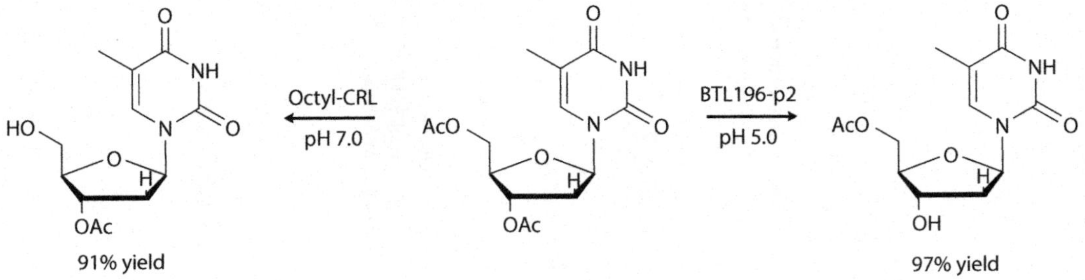

Biocatalyst	pH	Initial Rate[a]	Reaction Time (h)	Conversion (%)	5[b] (%)	6[b] (%)	T[b] (%)	Reference
CRL	7.0	0.10	24	98	91	4	3	[30]
PFL	7.0	1.07	2	96	46	4	46	[30]
PCL	7.0	0.28	24	96	73	11	12	[30]
PPL	7.0	0.02	48	34	5	26	3	[30]
BTL	7.0	0.13	70	100	52	5	43	[23]
BTL193	7.0	0.13	93	100	79	8	13	[23]
BTL193-p1	7.0	0.17	48	100	88	4	8	[23]
BTL196	5.0	0.20	50	100	0	86	14	[23]
BTL196-p2	5.0	0.02	100	100	0	97	3	[23]

T, thymidine; CRL, *Candida rugose* lipase; PFL, *Pseudomonas fluorescens* lipase; PCL, *Pseudomonas cepacia* lipase; PCL, porcine pancreas lipase; BTL, *Bacillus thermocatenolatus* lipase; **p1**, Ac-Cys-Phe-Gly-Phe-Gly-Phe-CONH$_2$; **p2**, Ac-Cys-(Asp)$_4$-Asp-COOH.

[a] The initial rate in μmol × mgprot^{-1} × h^{-1}.
[b] Yield of the corresponding product at 100% conversion.

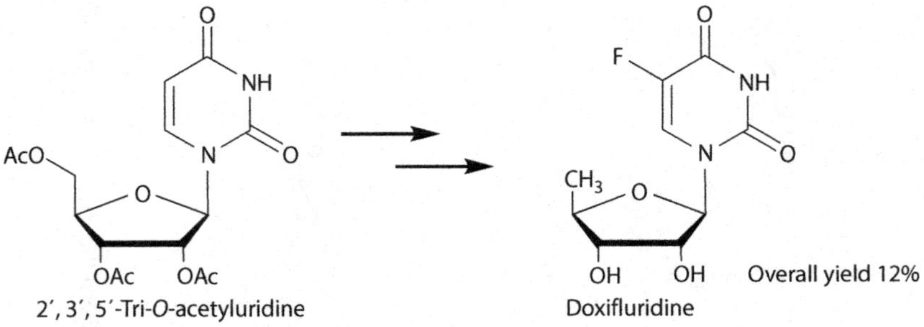

SCHEME 6.3 Chemo-enzymatic synthesis of doxifluridine.

6.3 REGIO- AND ENANTIOSELECTIVE PREPARATION OF CHIRAL ALCOHOLS AND AMINES

The synthesis of optically pure compounds is increasingly in demand in the pharmaceutical, fine chemicals, cosmetic, and agroalimentary industries.

Undoubtedly, the complete transformation of a racemic mixture of amine or alcohols into a single enantiomer is one of the challenging problems in chiral synthesis. For these reasons, many different approaches targeting this problem have been developed. For example, various methods as diastereomeric crystallization and asymmetric hydrogenation of imines, enamines, and oximes[6,34] have been described for chiral amine synthesis.

- Together with such classic synthetic processes, due to the high chemo-, regio-, and enantioselectivities commonly displayed by enzymes, the biotransformations field has acquired more interest for the production of chiral building blocks, as well as biologically active compounds,[35,36] offering the development of more environmentally and economically attractive processes.[37,38]
- Among all the biocatalytic strategies reported,[39] the kinetic resolution (KR) of racemates has become a widespread methodology for the synthesis of optically active molecules, which could be useful for industrial purposes. Commonly, KR is defined as a process in which one of the enantiomers of a racemic mixture is recognized by the biocatalyst faster than the other one (Scheme 6.4a) leading to its complete conversion into the product while the other remains unchanged. Therefore, by KR, it is possible to reach a maximum 50% conversion and an enantiomeric excess value higher than 99%. In particular, the use of serine hydrolases as catalysts for the enantioselective hydrolysis, alcoholysis, or aminolysis of esters and amides[40,41] together with stereoselective transesterification processes[42] is a common strategy to afford chiral compounds through KR processes. Nonetheless, these strategies present an important intrinsic limitation, due to the fact that only 50% of the maximum yield of the desired enantiomer can be achieved. To overcome this drawback, different methodologies have been developed such as a further racemization[43] (or the *in situ* Mitsunobu stereo-inversion of the unreacted enantiomer[44]). Hence, a combination of continuous racemization of the remaining enantiomer with kinetic resolution, called DKR, can provide the desired enantiomer in yields theoretically close to 100% (Scheme 6.4b). The coupled racemization process of the remnant substrate can be performed either spontaneously or by the employment of a chemo- or biocatalyst, which must be compatible with the reaction conditions used for the KR reaction. In fact, to perform an optimal DKR process, the two catalysts should be able to work together in the same reaction conditions and the reaction rates of the two processes should present a similar order of magnitude.[45] In this way, different KR racemization processes have been combined to synthesize many chiral secondary alcohols and amines in the last decade.[46–50] Thanks to all these features, DKR has received great attention during the last years with many research groups devoted to its optimization. For example, the combination of lipase-catalyzed enantioselective transesterification of secondary alcohols and primary amines together with the racemization reaction mediated by a transition metal catalyst is probably the most employed strategy to develop a DKR. In this process, the synthesis of many chiral building blocks as useful intermediates in the synthesis of different drugs has been achieved.[51–53]

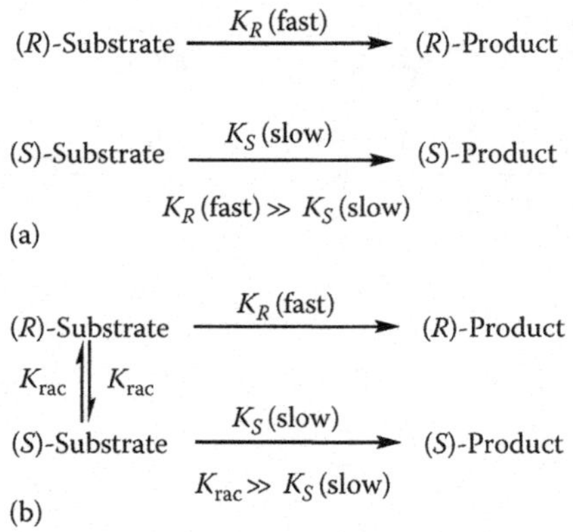

(a)
(R)-Substrate $\xrightarrow{K_R \text{ (fast)}}$ (R)-Product

(S)-Substrate $\xrightarrow{K_S \text{ (slow)}}$ (S)-Product

$K_R \text{ (fast)} \gg K_S \text{ (slow)}$

(b)
(R)-Substrate $\xrightarrow{K_R \text{ (fast)}}$ (R)-Product

$K_{rac} \updownarrow K_{rac}$

(S)-Substrate $\xrightarrow{K_S \text{ (slow)}}$ (S)-Product

$K_{rac} \gg K_S \text{ (slow)}$

SCHEME 6.4 General concept of KR (a) and DKR (b) processes.

For example, Backvall and coworkers optimized an innovative and scalable route to prepare chiral Me-imidacloprid (a chloronicotinyl insecticide) employing ruthenium complex (Shvo's catalyst) in combination with *Candida antarctica* lipase B (CAL-B) as catalysts for the DKR of key intermediate 1-(6-chloropyridin-3-yl)ethanol (Scheme 6.5).[54] DKR parameters were optimized, reaching maximum conversion (100% conversion and 91% yield) and enantiomeric excess (99% *ee*) at 50°C in toluene after 36 h. Although the ruthenium catalyst works at room temperature, increasing the temperature enhanced the reaction rate and the enantioselectivity.

Another meaningful example of the application of DKR to produce pure enantiomer of amines as one of the stereo-defining steps in a practical synthetic process is represented by the new synthesis of norsertraline (an antidepressant of the selective serotonin reuptake inhibitor [SSRI] class) developed by Thalen and coworkers (Scheme 6.6).[55] Firstly, they optimized the coupled racemization–kinetic resolution process catalyzed by different Ru catalysts and *Candida antarctica* lipase B, respectively, allowing a variety of functionalized primary amines to be transformed into one enantiomer in high yield and with high enantioselectivity. The optimized process (Shvo's catalyst, CAL-B, and isopropyl acetate as acyl donor) was then applied on the readily available *rac*-1,2,3,4-tetrahydro-1-naphthylamine to obtain the key chiral building block subsequently used in the new synthesis of target drug (Scheme 6.6).

Recently, a one-pot tandem catalytic process was performed by using heterogeneous Pd nanoparticles (PdNPs)-protein biohybrid.[56] This catalyst successfully showed both activities, enzymatic and

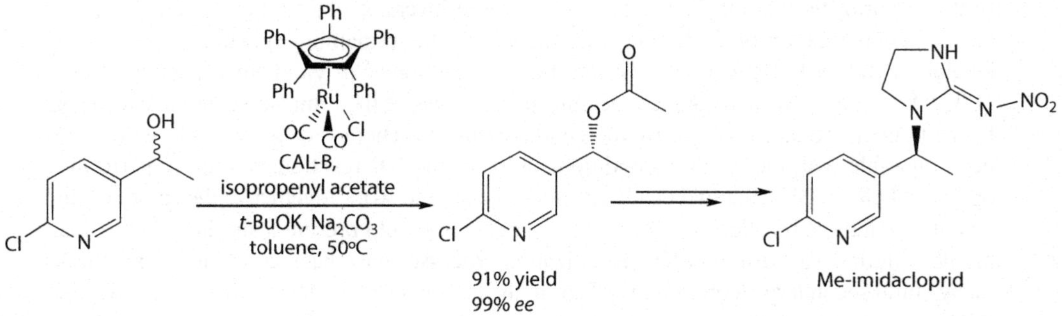

SCHEME 6.5 Enzyme–ruthenium organometallic combo DKR of 1-(6-chloropyridin-3-yl)ethanol as for the synthesis of chiral Me-imidacloprid.

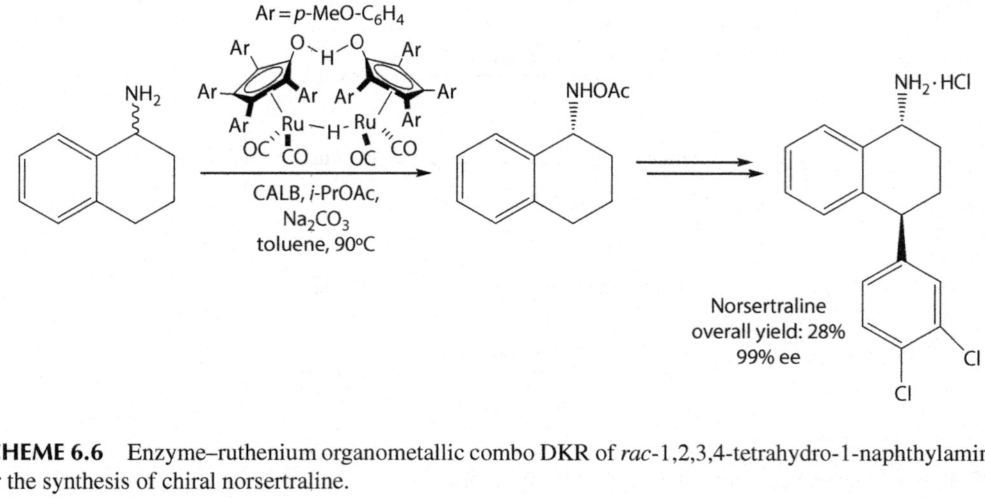

SCHEME 6.6 Enzyme–ruthenium organometallic combo DKR of *rac*-1,2,3,4-tetrahydro-1-naphthylamine as for the synthesis of chiral norsertraline.

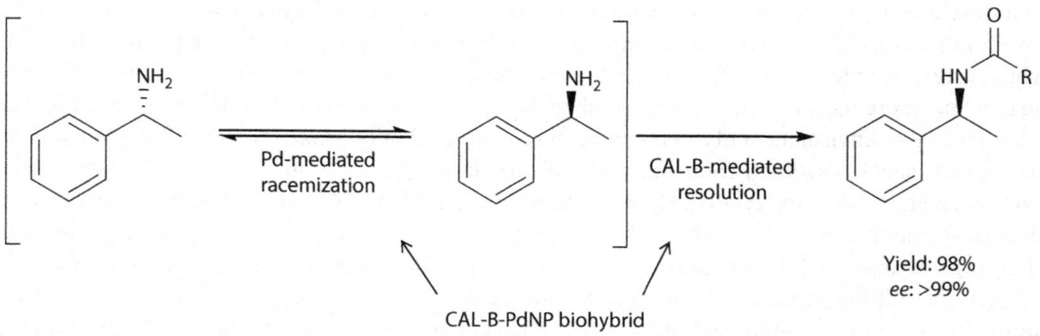

SCHEME 6.7 Dynamic kinetic resolution of racemic aryl amines catalyzed by Pd-enzyme nanohybrid.

metallic, to perform the DKR of *rac*-phenylethylamine in organic medium. The lipase catalyzed the enantioselective transesterification of *S*-arylamine, whereas the PdNPs catalyzed the racemization of unreacted *R*-enantiomer (Scheme 6.7). After a fine optimization of many reaction parameters, the quantitative formation of enantiopure (*R*)-benzylamide with *ee* >99% was achieved. This artificial catalyst also demonstrated high operational stability and recyclability.[56]

6.4 REGIOSELECTIVE ARTIFICIAL HYDROLASE-CATALYZED C–C BOND FORMATION

As previously stated, during the last three decades, the use of hydrolases for the catalysis of environmentally friendly organic processes under mild reaction conditions has been well documented. Due to their high stability, catalytic efficiency, commercial availability, and broad substrate specificity in a wide spectrum of biocatalyzed processes, hydrolases have shown themselves to be ideal tools for the acceleration of synthetic transformations. Besides the hydrolytic and transacylating abilities, lipases have shown to be able to catalyze various other reactions such as aldol additions,[57] conjugate additions,[58,59] direct epoxidation reactions,[60] Mannich reactions,[61] or Markovnikov additions[62] in a stereoselective manner. This nonconventional property, defined as catalytic promiscuity, proceeds through transition state geometries and reaction mechanisms different from that of the conventional ones. The formed or broken bond should differ from that of the native transformation and so should lead to an alternative transition state structure. Generally, this phenomenon can be promoted when offering non-natural substrates to the wild-type enzyme (accidental catalytic promiscuity) or in a mutated enzyme variant (induced catalytic promiscuity).[63] The latter promiscuity can be achieved either through binding of a non-natural cofactor or as a result of protein engineering (rational design, semirational design, or directed evolution).

Despite the fact that enantioselectivities displayed by hydrolases are not always very high and numerous factors must be taken into account in order to improve current results, this recent research line—besides its primary synthetic purpose—could also help scientists to shine a light on the rational understanding of the mechanistic aspects of enzymatic catalysis.[64,65]

Undoubtedly, if compared to the induced one, the non-natural substrate-promoted catalytic promiscuity has been largely investigated and its synthetic application has been extensively reviewed.[66,67] As one of the representative examples reported in the literature, the enzyme-catalyzed asymmetric aldol addition between acetone and different aromatic aldehydes with strong electron-withdrawing groups employing pig pancreas lipase (PPL) as catalyst in wet reaction media will be described.[57] Li and coworkers firstly extensively studied the influence of the concentration of water on the reaction rate and on the enantioselectivity of the process, observing that the optimal rate was achieved when the reaction was performed with 20% water content (96% yield), although the best values of enantioselectivity (up to 44% *ee*) were obtained in near-anhydrous media. These results suggest a strong influence of the water concentration on the spatial conformation of the enzyme active site, which is responsible not only for the reaction kinetics but also for the enantioselectivity. Following on this line, more recently, the same research group reported the role of *Candida antarctica* lipase B (CALB) in two environmentally benign decarboxylative addition protocols (decarboxylative aldol and Knoevenagel reactions) (Scheme 6.8).[68] In the first case, the authors reported an optimized CAL-B-catalyzed decarboxylative aldol addition between β-ketoesters and aromatic aldehydes with electron-withdrawing groups (Scheme 6.8).[68] It is noteworthy how the addition of a catalytic amount of an organic base (1,4,7,10-tetraazacyclododecane or cyclen) reduced the reaction time (from 72 to 20 h), permitting to isolate the decarboxylative aldol adducts in excellent yields (Scheme 6.8). Surprisingly, during the optimization studies of such reaction, the authors found that when a primary amine was added to the reaction medium, α,β-unsaturated ketones were isolated due to the occurrence of a decarboxylative Knoevenagel reaction (Scheme 6.8).[68] Other reaction parameters were studied (i.e., presence of different concentrations of water) to finally obtain excellent yields with a wide panel of aromatic aldehydes.

As an example of induced catalytic promiscuity, very recently, Berglund and coworkers reported that the CAL-B Ser105Ala mutant was able to catalyze the Michael addition between methyl acrylate and acetyl acetone in aqueous media instead of the native hydrolytic process (Scheme 6.9).[59] In fact, the absence of the nucleophilic Ser105 residue in its active site prevented the formation of the acyl enzyme complex consequently suppressing the native hydrolytic activity and favoring the promiscuous Michael addition.

Furthermore, the creation of artificial metalloenzymes opens the door to the application of enzymes and organometallic compounds in C–C bond formation. Indeed, recently we described the application of the nanohybrids of Pd nanoparticles created using CAL-B in the Suzuki reaction.[56] The use of a hydrolase for the preparation of heterogeneous Pd catalyst makes possible a successful synthesis of biphenyl in aqueous media in excellent yields (>99%) (Scheme 6.10).

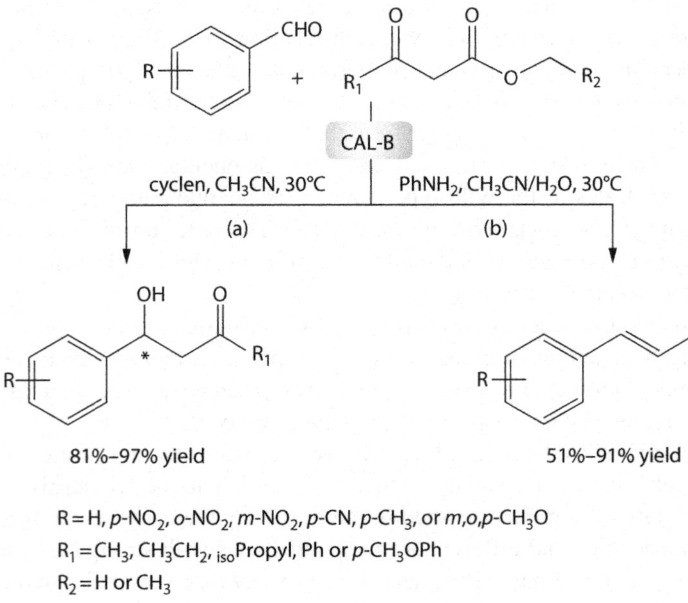
CHO
R
+
R1
O
O
O
R2
CAL-B
cyclen, CH3CN, 30°C
PhNH2, CH3CN/H2O, 30°C
(a)
(b)
OH
O
*
R1
R
O
R1
R
81%–97% yield
51%–91% yield
R = H, p-NO2, o-NO2, m-NO2, p-CN, p-CH3, or m,o,p-CH3O
R1 = CH3, CH3CH2, isoPropyl, Ph or p-CH3OPh
R2 = H or CH3

SCHEME 6.8 (a) CAL-B-catalyzed decarboxylative aldol reaction and (b) CAL-B-catalyzed decarboxylative Knoevenagel reaction.

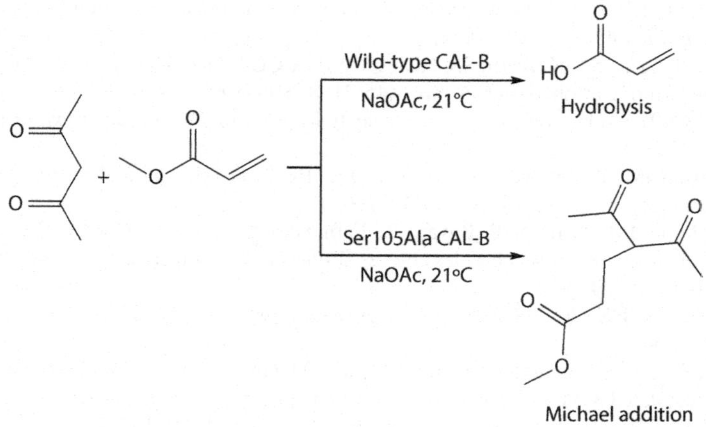

SCHEME 6.9 Michael addition reaction catalyzed by an artificial CAL-B lipase.

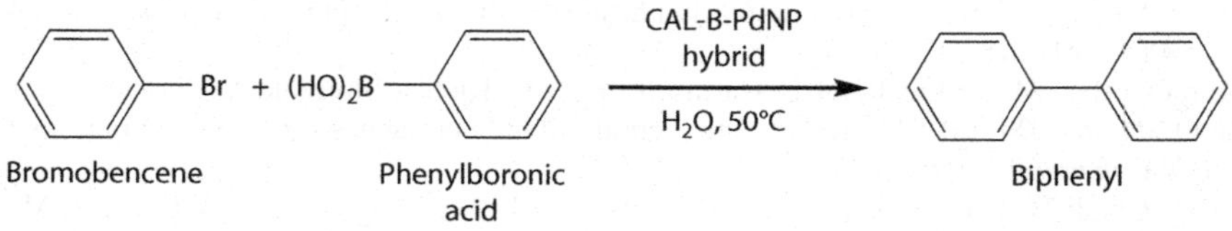

SCHEME 6.10 Suzuki–Miyaura reaction in aqueous media catalyzed by lipase-PdNP hybrid catalyst.

REFERENCES

1. Busto E, Gotor-Fernandez V, Gotor V (2011) Hydrolases in the stereoselective synthesis of N-heterocyclic amines and amino acid derivatives. *Chemical Reviews* 111(7):3998–4035.
2. Domínguez de María P, Maugeri Z (2011) Ionic liquids in biotransformations: From proof-of-concept to emerging deep-eutectic-solvents. *Current Opinion in Chemical Biology* 15(2):220–225.
3. Zymańczyk-Duda E, Klimek-Ochab M (2012) Stereoselective biotransformations as an effective tool for the synthesis of chiral compounds with P-C bond—Scope and limitations of the methods. *Current Organic Chemistry* 16(11):1408–1422.
4. Palomo JM (2008) Lipases enantioselectivity alteration by immobilization techniques. *Current Bioactive Compounds* 4(2):126–138.
5. Reetz MT (2002) Lipases as practical biocatalysts. *Current Opinion in Chemical Biology* 6(2):145–150.
6. Faber K (2011) *Biotransformation in Organic Chemistry*. Springer-Verlag, Berlin, Germany.
7. Verger R (1997) 'Interfacial activation' of lipases: Facts and artifacts. *Trends in Biotechnology* 15(1):32–38.
8. Pàmies O, Bäckvall JE (2004) Chemoenzymatic dynamic kinetic resolution. *Trends in Biotechnology* 22(3):130–135.
9. García-Urdiales E, Alfonso I, Gotor V (2005) Enantioselective enzymatic desymmetrizations in organic synthesis. *Chemical Reviews* 105(1):313–354.
10. Filice M, Palomo JM (2012) Monosaccharide derivatives as central scaffolds in the synthesis of glycosylated drugs. *RSC Advances* 2(5):1729–1742.
11. Martínez-Montero S, Fernández S, Sanghvi YS, Gotor V, Ferrero M (2011) An expedient biocatalytic procedure for abasic site precursors useful in oligonucleotide synthesis. *Organic and Biomolecular Chemistry* 9(17):5960–5966.
12. Weymouth-Wilson AC (1997) The role of carbohydrates in biologically active natural products. *Natural Product Reports* 14(2):99–110.
13. Varki A, Cummings DC, Esko JD, Freeze HH, Stanley P, Bertozzi CR, Hart GW, Etzler ME (2009) *Essentials of Glycobiology*, 2nd edn. Cold Spring Harbor Press, New York.

14. Zhu X, Schmidt RR (2009) New principles for glycoside-bond formation. *Angewandte Chemie—International Edition* 48(11):1900–1934.
15. Wang CC, Lee JC, Luo SY, Kulkarni SS, Huang YW, Lee CC, Chang KL, Hung SC (2007) Regioselective one-pot protection of carbohydrates. *Nature* 446(7138):896–899.
16. Palomo JM (2009) Modulation of enzymes selectivity via immobilization. *Current Organic Synthesis* 6(1):1–14.
17. Filice M, Guisan JM, Terreni M, Palomo JM (2012) Regioselective monodeprotection of peracetylated carbohydrates. *Nature Protocols* 7(10):1783–1796.
18. Filice M, Palomo JM, Bonomi P, Bavaro T, Fernandez-Lafuente R, Guisan JM, Terreni M (2008) Preparation of linear oligosaccharides by a simple monoprotective chemo-enzymatic approach. *Tetrahedron* 64(39):9286–9292.
19. Wuts PGM, Greene TW (2007) *Greene's Protective Groups in Organic Synthesis*, 4th edn. John Wiley & Sons, Hoboken, NJ.
20. Palomo JM, Filice M, Fernandez-Lafuente R, Terreni M, Guisan JM (2007) Regioselective hydrolysis of different peracetylated β-monosaccharides by immobilized lipases from different sources. Key role of the immobilization. *Advanced Synthesis and Catalysis* 349(11–12):1969–1976.
21. Filice M, Fernandez-Lafuente R, Terreni M, Guisan JM, Palomo JM (2007) Screening of lipases for regioselective hydrolysis of peracetylated β-monosaccharides. *Journal of Molecular Catalysis B: Enzymatic* 49(1–4):12–17.
22. Filice M, Bavaro T, Fernandez-Lafuente R, Pregnolato M, Guisan JM, Palomo JM, Terreni M (2009) Chemo-biocatalytic regioselective one-pot synthesis of different deprotected monosaccharides. *Catalysis Today* 140(1–2):11–18.
23. Romero O, Filice M, De Las Rivas B, Carrasco-Lopez C, Klett J, Morreale A, Hermoso JA, Guisan JM, Abian O, Palomo JM (2012) Semisynthetic peptide-lipase conjugates for improved biotransformations. *Chemical Communications* 48(72):9053–9055.
24. Filice M, Ubiali D, Fernandez-Lafuente R, Fernandez-Lorente G, Guisan JM, Palomo JM, Terreni M (2008) A chemo-biocatalytic approach in the synthesis of β-O-naphtylmethyl-N-peracetylated lactosamine. *Journal of Molecular Catalysis B: Enzymatic* 52–53(1–4):106–112.
25. Mong TKK, Lee LV, Brown JR, Esko JD, Wong CH (2003) Synthesis of N-acetyllactosamine derivatives with variation in the aglycon moiety for the study of inhibition of sialyl Lewis × expression. *ChemBioChem* 4(9):835–840.
26. Herdewijn P (2008) Biologically active nucleosides. *Current Protocols in Nucleic Acid Chemistry* (S34):14.10.11–14.10.16.
27. Esté JA, Cihlar T (2010) Current status and challenges of antiretroviral research and therapy. *Antiviral Research* 85(1):25–33.
28. Vorbrüggen H, Ruh-Pohlenz C (2001) *Handbook of Nucleoside Synthesis*. John Wiley & Sons, New York.
29. Ferrero M, Gotor V (2000) Chemoenzymatic transformations in nucleoside chemistry. *Monatshefte fur Chemie* 131(6):585–616.
30. Bavaro T, Rocchietti S, Ubiali D, Filice M, Terreni M, Pregnolato M (2009) A versatile synthesis of 5′-fenctionalized nucleosides through regioselective enzymatic hydrolysis of their peracetylated precursors. *European Journal of Organic Chemistry* 2009(12):1967–1975.
31. Zinni MA, Iglesias LE, Iribarren AM (2007) Preparation of potential 3-deazauridine and 6-azauridine prodrugs through an enzymatic alcoholysis. *Journal of Molecular Catalysis B: Enzymatic* 47(1–2):86–90.
32. Iglesias LE, Zinni MA, Gallo M, Iribarren AM (2000) Complete and regioselective deacetylation of peracetylated uridines using a lipase. *Biotechnology Letters* 22(5):361–365.
33. Jyonouchi H, Zhang-Shanbhag L, Tomita Y, Yokoyama H (1994) Nucleotide-free diet impairs T-helper cell functions in antibody production in response to T-dependent antigens in normal C57Bl/6 mice. *Journal of Nutrition* 124(4):475–484.
34. Breuer M, Ditrich K, Habicher T, Hauer B, Keßeler M, Stürmer R, Zelinski T (2004) Industrial methods for the production of optically active intermediates. *Angewandte Chemie—International Edition* 43(7):788–824.
35. De Carvalho CCCR (2011) Enzymatic and whole cell catalysis: Finding new strategies for old processes. *Biotechnology Advances* 29(1):75–83.
36. Straathof AJJ, Panke S, Schmid A (2002) The production of fine chemicals by biotransformations. *Current Opinion in Biotechnology* 13(6):548–556.
37. Wohlgemuth R (2010) Biocatalysis-key to sustainable industrial chemistry. *Current Opinion in Biotechnology* 21(6):713–724.

38. Tao J, Xu JH (2009) Biocatalysis in development of green pharmaceutical processes. *Current Opinion in Chemical Biology* 13(1):43–50.
39. Ahn Y, Ko SB, Kim MJ, Park J (2008) Racemization catalysts for the dynamic kinetic resolution of alcohols and amines. *Coordination Chemistry Reviews* 252(5–7):647–658.
40. Liese A, Seelbach K, Wandrey C (2006) *Industrial Biotransformations*, 2nd edn. Wiley-VCH, Weinheim, Germany.
41. Buchholz K, Kasche V, Bornscheuer UT (2012) *Biocatalysts and Enzyme Technology*, 2nd edn. Wiley-VCH, Weinheim, Germany.
42. Gotor-Fernandez V, Brieva R, Gotor V (2006) Lipases: Useful biocatalysts for the preparation of pharmaceuticals. *Journal of Molecular Catalysis B: Enzymatic* 40(3–4):111–120.
43. Vänttinen E, Kanerva LT (1994) Lipase-catalysed transesterification in the preparation of optically active solketal. *Journal of the Chemical Society, Perkin Transactions* 1(23):3459–3463.
44. Wallner A, Mang H, Glueck SM, Steinreiber A, Mayer SF, Faber K (2003) Chemo-enzymatic enantio-convergent asymmetric total synthesis of (S)-(+)-dictyoprolene using a kinetic resolution—Stereoinversion protocol. *Tetrahedron: Asymmetry* 14(16):2427–2432.
45. Kitamura M, Tokunaga M, Noyori R (1993) Mathematical treatment of kinetic resolution of chirally labile substrates. *Tetrahedron* 49(9):1853–1860.
46. Pellissier H (2011) Recent developments in dynamic kinetic resolution. *Tetrahedron* 67(21):3769–3802.
47. Kim Y, Park J, Kim MJ (2011) Dynamic kinetic resolution of amines and amino acids by enzyme-metal cocatalysis. *ChemCatChem* 3(2):271–277.
48. Lee JH, Han K, Kim MJ, Park J (2010) Chemoenzymatic dynamic kinetic resolution of alcohols and amines. *European Journal of Organic Chemistry* 6:999–1015.
49. Turner NJ (2010) Deracemisation methods. *Current Opinion in Chemical Biology* 14(2):115–121.
50. Pàmies O, Bäckvall JE (2003) Combination of enzymes and metal catalysts. A powerful approach in asymmetric catalysis. *Chemical Reviews* 103(8):3247–3261.
51. Berkessel A, Jurkiewicz I, Mohan R (2011) Enzymatic dynamic kinetic resolution of oxazinones: A new approach to enantiopure β²-amino acids. *ChemCatChem* 3(2):319–330.
52. Krumlinde P, Bogár K, Bäckvall JE (2010) Asymmetric synthesis of bicyclic diol derivatives through metal and enzyme catalysis: Application to the formal synthesis of sertraline. *Chemistry—A European Journal* 16(13):4031–4036.
53. Fischer T, Pietruszka J (2010) Key building blocks via enzyme-mediated synthesis. *Topics in Current Chemistry* 297:1–43.
54. Krumlinde P, Bogár K, Bäckvall JE (2009) Synthesis of a neonicotinoide pesticide derivative via chemoenzymatic dynamic kinetic resolution. *Journal of Organic Chemistry* 74(19):7407–7410.
55. Thalén LK, Zhao D, Sortais JB, Paetzold J, Hoben C, Bäckvall JE (2009) A chemoenzymatic approach to enantiomerically pure amines using dynamic kinetic resolution: Application to the synthesis of norsertraline. *Chemistry—A European Journal* 15(14):3403–3410.
56. Filice M, Marciello M, Morales MP, Palomo JM (2013) Synthesis of heterogeneous enzyme–metal nanoparticle biohybrids in aqueous media and their applications in C–C bond formation and tandem catalysis. *Chemical Communications* 49, 6876–6878.
57. Li C, Feng XW, Wang N, Zhou YJ, Yu XQ (2008) Biocatalytic promiscuity: The first lipase-catalysed asymmetric aldol reaction. *Green Chemistry* 10(6):616–618.
58. Torre O, Alfonso I, Gotor V (2004) Lipase catalysed Michael addition of secondary amines to acrylonitrile. *Chemical Communications* 10(15):1724–1725.
59. Svedendahl M, Jovanović B, Fransson L, Berglund P (2009) Suppressed native hydrolytic activity of a lipase to reveal promiscuous Michael addition activity in water. *ChemCatChem* 1(2):252–258.
60. Svedendahl M, Carlqvist P, Branneby C, Allnér O, Frise A, Hult K, Berglund P, Brinck T (2008) Direct epoxidation in *Candida antarctica* lipase B studied by experiment and theory. *ChemBioChem* 9(15):2443–2451.
61. Li K, He T, Li C, Feng XW, Wang N, Yu XQ (2009) Lipase-catalysed direct Mannich reaction in water: Utilization of biocatalytic promiscuity for C–C bond formation in a "one-pot" synthesis. *Green Chemistry* 11(6):777–779.
62. Lou FW, Liu BK, Wu Q, Lv DS, Lin XF (2008) *Candida antarctica* lipase B (CAL-B)-catalyzed carbon-sulfur bond addition and controllable selectivity in organic media. *Advanced Synthesis and Catalysis* 350(13):1959–1962.
63. Hult K, Berglund P (2007) Enzyme promiscuity: Mechanism and applications. *Trends in Biotechnology* 25(5):231–238.

64. Khersonsky O, Roodveldt C, Tawfik DS (2006) Enzyme promiscuity: Evolutionary and mechanistic aspects. *Current Opinion in Chemical Biology* 10(5):498–508.
65. Tokuriki N, Tawfik DS (2009) Protein dynamism and evolvability. *Science* 324(5924):203–207.
66. Busto E, Gotor-Fernandez V, Gotor V (2010) Hydrolases: Catalytically promiscuous enzymes for non-conventional reactions in organic synthesis. *Chemical Society Reviews* 39(11):4504–4523.
67. Humble MS, Berglund P (2011) Biocatalytic promiscuity. *European Journal of Organic Chemistry* 2011(19):3391–3401.
68. Feng XW, Li C, Wang N, Li K, Zhang WW, Wang Z, Yu XQ (2009) Lipase-catalysed decarboxylative aldol reaction and decarboxylative Knoevenagel reaction. *Green Chemistry* 11(12):1933–1936.

7 Bioactive Natural Products and Their Structure–Activity Relationships Studies

Athar Ata and Hadeel Alhazmi

CONTENTS

7.1 Introduction 191
7.2 Antimicrobial Natural Products 191
7.3 Glutathione S-Transferase Inhibitors 194
7.4 Acetylcholinesterase Inhibitors 196
7.5 α-Glucosidase Inhibitors 198
7.6 Antirenin Natural Products 201
References 202

7.1 INTRODUCTION

The field of natural products has generated a significant and growing number of compounds with promising biomedical activities.[1] Recent estimates indicate that ca. 60% of antitumor and anti-infective agents that are commercially available on the market are of natural product origin and 25% of them are of plant origin.[2–5] Plants are a rich source of lead bioactive compounds, and less than 5% of medicinally important plants have been chemically investigated.[6–7] Ethnomedicinally important plants have shown a 74% success rate in discovering plant-based pharmaceuticals compared to other approaches of plant selections including previously reported bioactivities, bioinformatic, and phylogenetic approaches.[6–7] Enzymes are necessary for human life in mediating/regulating biochemical processes including metabolism, catabolism, cellular signal transduction, and cell cycling and development. Malfunctioning in these biochemical systems leads to diseases that are caused by the dysfunction of enzymes as well as overexpression or hyperactivation of enzymes involved.[8] An understanding of diseases at the molecular level has led to the discovery of effective enzyme inhibitors that are used in clinical practice. For instance, huperzine A, one of the potent acetylcholinesterase inhibitors, is a prescribed drug for the treatment of Alzheimer's disease.[9] Our recent phytochemical studies on medicinally important plants have resulted in the isolation of novel natural products exhibiting various potent bioactivities that are summarized as follows.

7.2 ANTIMICROBIAL NATURAL PRODUCTS

Medicinal plants have been ignored for discovering new antimicrobial agents though they produce antimicrobial natural products for their survival. Plants are being used to treat wounds in folk medicines. These medicinal plants may be helpful for discovering new antimicrobial compounds to overcome the microbial drug resistance problems, a real and significant threat to human life. A very little work is being carried out in discovering new lead antimicrobial agents, as during the last 45 years only two new classes of antibiotics, oxazolidinone (linezolid) and lipopeptide (daptomycin), were introduced in clinics. One of them, linezolid, developed resistance 1 year after its approval

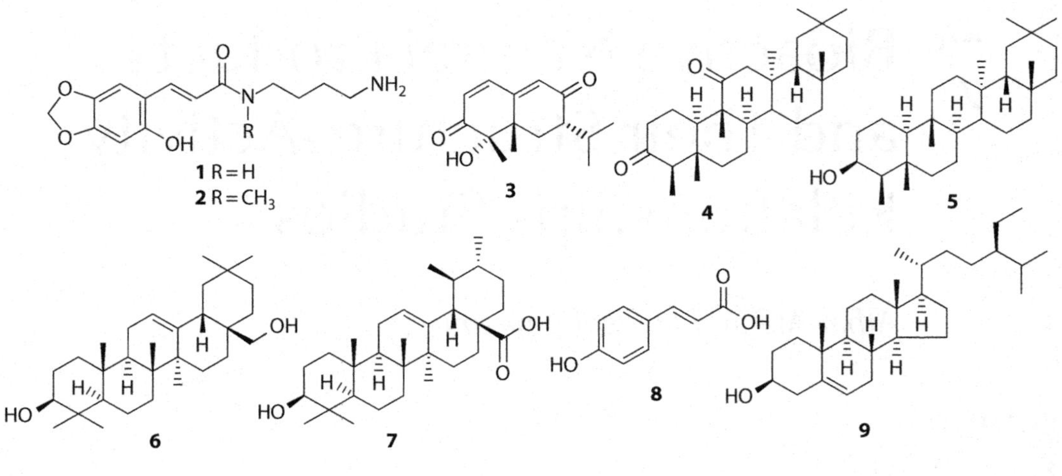

FIGURE 7.1 Structures of compounds 1–9.

by the FDA.[10,11] There is thus an urgent need to discover new antimicrobial agents having diverse structures and novel modes of action. It would be worthwhile to investigate medicinal plants used to treat wounds in folk medicines for the identification of new lead antimicrobial compounds. Toward this end, we have carried out phytochemical studies on stems and bark of *Drypetes staudtii*, collected from Nigeria, based on its reported wound-healing applications in folk medicines. Its crude methanolic extract exhibited antimicrobial activity against *Escherichia coli*, *Staphylococcus aureus*, *Streptococcus agalactiae*, and *Pseudomonas aeruginosa*. The crude extract was also active against *Candida albicans*. Our antimicrobial-directed phytochemical investigation of the methanolic extracts of *D. staudtii* afforded nine antimicrobial compounds: 4,5-(methylenedioxy)-*o*-coumaroylputrescine (**1**), 4,5-(methylenedioxy)-*o*-coumaroyl-4′-*N*-methylputrescine (**2**), 4α-hydroxyeremophila-1,9-diene-3,8-dione (**3**), drypemolundein B (**4**), friedelan-3β-ol (**5**), erythrodiol (**6**), ursolic acid (**7**), *p*-coumaric acid (**8**), and β-sitosterol (**9**).[12] Structures of compounds **1–9** are shown in Figure 7.1. All of the isolates exhibited antibacterial activity against Gram-positive and Gram-negative bacteria with minimum inhibitory concentration (MIC) in the range of 8–128 μg/mL (Table 7.1). Compounds **1–2** also exhibited moderate activity against *C. albicans* (MIC value of 32 μg/mL).

Sphaeranthus indicus, a medicinally important plant of Sri Lankan origin, was active in our antibacterial assay against Gram-positive and Gram-negative bacteria. Our antibacterial activity-guided fractionations of the crude extract resulted in the isolation of a sesquiterpenoid, 7α-hydroxyfrullanolide,[10] and this compound was significantly active in antibacterial assay against Gram-positive bacteria. For structure–activity relationships studies, we carried out a combination of chemical and microbial reactions. For microbial reactions, we used whole-cell cultures of fungi. Microorganisms perform several chemical reactions including oxidation, aldol condensation, Michael addition, and umpolung-type reactions.[13] These reactions are also helpful in predicting the fate of new pharmaceuticals as the results obtained from whole-cell microbial culture reactions are quite often similar to those obtained from mammal biotransformations.[14] This relationship between mammal and microbial reaction is due to the presence of a common enzyme, cytochrome P450 monooxygenase.[15] These reactions also help to generate lead bioactive compounds using natural products as main substrates. The metabolic products often have improved bioactivity with minimal toxicity compared to the parent natural products. Microbial reactions on **10** using the whole-cell cultures of *Cunninghamella echinulata* and *Curvularia lunata* yielded three compounds: 1β,7α-dihydroxyfrullanolide (**11**), 1-oxo-7α-hydroxyfrullanolide (**12**), and 7α-hydroxy-4,5-dihydrofrullanolide (**13**). Incubation of compound **10** with the liquid cultures of *Aspergillus niger* and *Rhizopus circinans* yielded three metabolites: 17α-hydroxy-11,13-dihydrofrullanolide (**14**), 13-acetyl-7α-hydroxyfrullanolide (**15**), and

TABLE 7.1
Minimum Inhibitory Concentrations (MIC) in μg/mL of Compounds 1–18 in Antibacterial Assay

Compounds	Staph. aureus	S. agalactiae	E. coli	P. aeruginosa
1	8	8	16	16
2	8	8	16	16
3	64	64	64	64
4	32	32	64	64
5	16	16	32	32
6	32	32	64	64
7	32	32	64	64
8	128	128	128	128
9	128	128	128	128
10	32	32	128	128
11	8	8	64	64
12	8	8	64	64
13	128	128	128	128
14	128	128	128	128
15	128	128	128	128
16	128	128	128	128
17	128	128	128	128
18	128	128	128	128
Thymol	8	8	16	16
Penicillin G	1	1	1	8

Thymol and penicillin G are positive controls.

2α,7α-dihydroxy-sphaerantholide (**16**).[16] We prepared 4α,5α-epoxy-7α-hydroxyfrullanolide (**17**) and 4α,5α-epoxy-7α-hydroxyfrullanolide (**18**) by carrying out an epoxidation reaction on **10** using *meta* chloroperbenzoic acid.[16] Structures of compounds **10–18** are shown in Figure 7.2. Compounds **10–18** exhibited antibacterial activity (Table 7.1). Structure–activity relationships studies on compounds **10–18** indicated that the double bonds $\Delta^{4,5}$, Δ^{11-13} and a γ-lactone moiety are the required pharmacophores for the expression of the antibacterial activity of compound **10**.

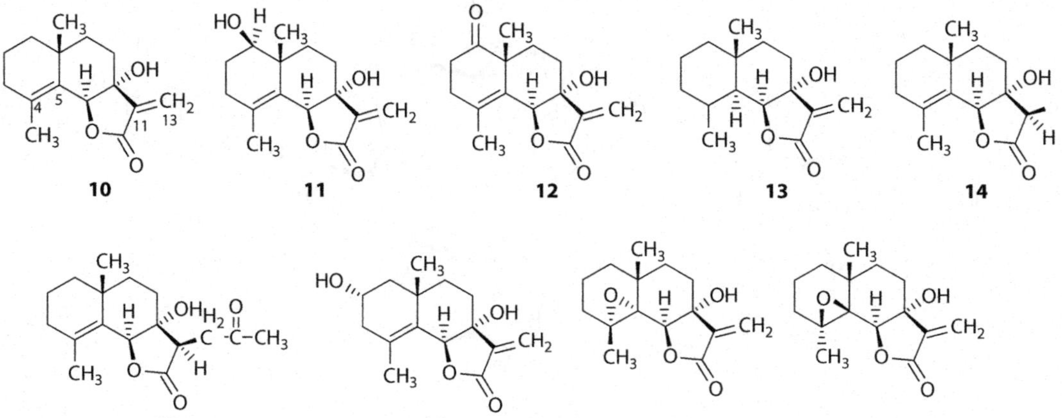

FIGURE 7.2 Structures of compounds **10–18**.

7.3 GLUTATHIONE S-TRANSFERASE INHIBITORS

Currently, anticancer and antiparasitic pharmaceuticals are developing resistance and an active research is going on to understand the mechanism of acquired drug resistance during anticancer and antiparasitic chemotherapy. Isozymes, glutathione S-transferase (GST), have been reported to play an active role in acquired drug resistance during these chemotherapies. GSTs are phase II detoxification isozymes that catalyze reactions of exogenous or endogenous electrophilic substrates with glutathione to make water-soluble adduct. This adduct can easily be excreted from the body.[17] Anticancer drugs containing electrophilic centers make glutathione adduct and are excreted from the body, thus lowering the effectiveness of anticancer pharmaceuticals. There are various dimerized isoenzyme classes of human GSTs: α (A), μ (M), ω, π (P), θ (T), ζ (Z), and σ classes. These different forms provide broad substrate specificities to improve detoxification of many toxic substances. It has been reported in the literature that GSTs are overexpressed in various cancer cells compared to normal tissues. A two fold increase in GST activity is reported in the literature in lymphocytes obtained from chronic lymphocytic leukemia (CLL) patients, resistant to chlorambucil when compared with untreated CLL patients.[18] The use of GST inhibitors as an adjuvant during cancer chemotherapy might help to overcome the drug resistance problem. Ethacrynic acid (**19**), first generation of GST inhibitor, exhibits significant in-vitro potentiating activity but has diuretic side effects and lack of isoenzyme specificity. These limitations restricted its use as an adjuvant during cancer chemotherapy.[19] Toward this end, we screened several medicinally important plants for their GST inhibitory activity and discovered that the crude methanolic extracts of *Barleria prionitis*, *Nauclea latifolia*, *Artocarpus nobilis*, and *Caesalpinia bonduc* were active in this assay with IC$_{50}$ (concentration required to inhibit 50% activity of enzyme) values of 160.0, 10.5, 125, and 83.0 μg/mL, respectively. Due to the colorimetric nature of enzyme inhibition assays, polyphenols, present in crude plant extracts, often give false-positive results. In order to rule out the possibility of false-positive results, it is worthwhile to remove polyphenols by filtering plant extracts through MN-polyamide SC6 resin before screening for enzyme inhibition activities.[20] Our recent phytochemical studies on the crude methanolic extract of *B. prionitis*, collected from Sri Lanka, yielded barlerinoside (**20**) that showed GST inhibitory activity with an IC$_{50}$ value of 12.4 μM.[21] The potency of compound **20** to inhibit the GST activity was same as that of ethacrynic acid (**19**), a substrate GST inhibitor (IC$_{50}$ = 16.5 μM). Structures of compounds **19–20** are shown in Figure 7.3.

Chemical analysis of the ethanolic extract of *N. latifolia*, collected from Nigeria, resulted in the isolation of five known indole alkaloids exhibiting GST inhibitory activity: strictosamide **21** (IC$_{50}$ = 20.3 μM), naucleamides A **22** (IC$_{50}$ = 37.2 μM), naucleamide F **23** (IC$_{50}$ = 23.6 μM), quinovic

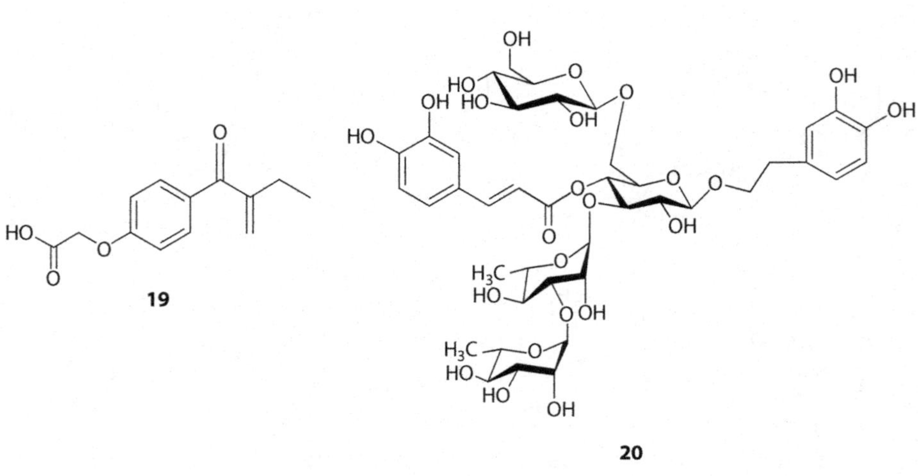

FIGURE 7.3 Structures of compounds **19** and **20**.

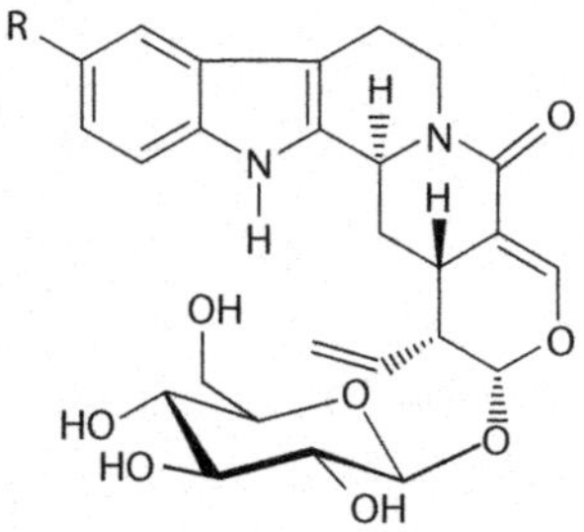

21 R=H
26 R=OH
27 R=β-D-glucose

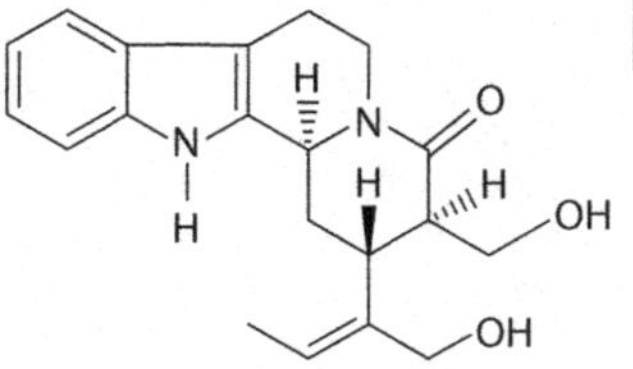

22

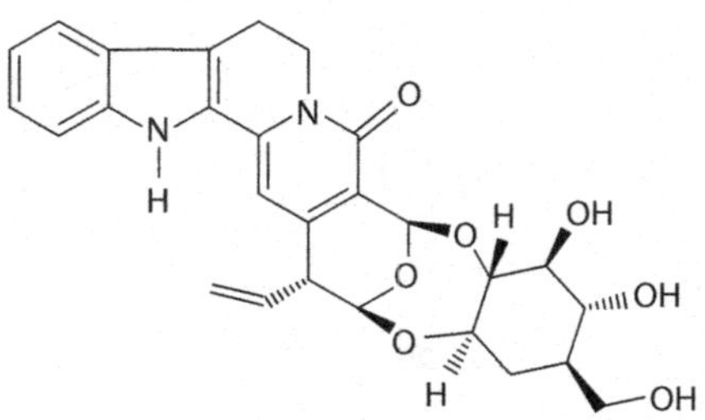

23

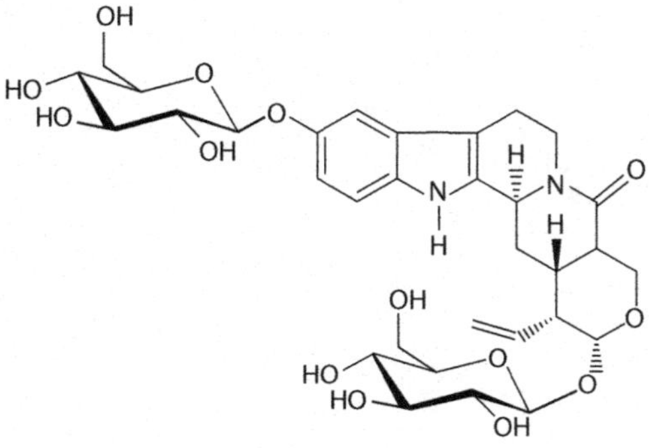

28

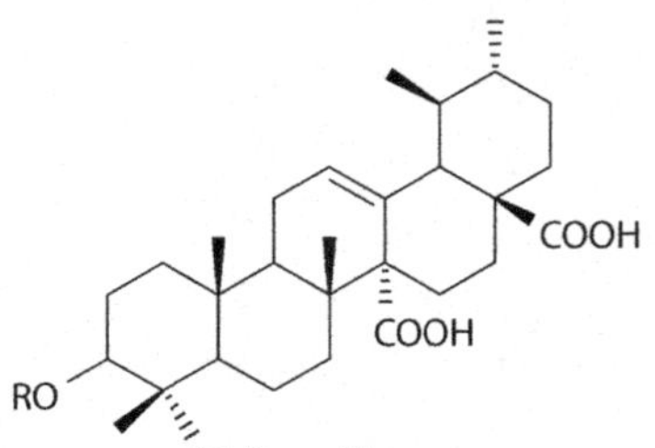

24 R=α-Rhamnose

25 R=β-Fucose

FIGURE 7.4 Structures of compounds **21–28**.

acid-3-*O*-β-rhamnosylpyranoside **24** (IC_{50} = 143.8 μM), and quinovic acid 3-*O*-β-fucosylpyranoside **25** (IC_{50} = 53.5 μM).[22] Compound **21** showed significant anti-GST property and was isolated in a large quantity. It was, therefore, decided to prepare analogs by using microbial reactions and to evaluate them for GST inhibitory activity in order to study their structure–activity relationships. Microbial reactions of compound **21** using the fungi *Cunninghamella blakesleeana* and *R. circinans* afforded three metabolites: 10-hydroxystrictosamide (**26**), 10-β-glucosyloxyvincoside lactam (**27**), and 16,17-dihydro-10-β glucosyloxyvincoside lactam (**28**). This is the first time that we have discovered that *R. circinans* is capable of performing microbial hydroxylation reaction on the aromatic compound. Structures of compounds **21–28** are shown in Figure 7.4. The sequence for the formation of metabolites **26–28** was determined by carrying out time-dependent biotransformation experiments by incubating compounds **21** with the liquid culture of *R. circinans*. These studies showed that first of all compound **26** was produced, which further underwent glycosylation, followed by reduction of the $\Delta^{16,17}$ double bond to give compounds **27** and **28**, respectively. Compounds **26–28** were also active in GST inhibition assay with IC_{50} values of 18.6, 12.3, and 46.6 μM, respectively. The bioactivity of **27** was approximately twofold higher compared to the parent compound (**21**), suggesting that the higher potency might be due to the presence of a sugar moiety at C-10. This moiety might have increased its solubility in water for its better interactions with GST enzyme.

A. nobilis, of Sri Lankan origin, was also active in GST inhibition assay, and our GST-inhibition-directed studies on the crude extract of this plant yielded five known anti-GST flavonoids: artonins E (**29**), artobiloxanthone (**30**), artoindonesianin U (**31**), cyclocommunol (**32**), and multiflorins A (**33**).[23] Compounds **29–33** were active in this bioassay with IC_{50} values of 2.0, 1.0, 6.0, 3.0, and 14.0 μM, respectively. The in-vitro anti-GST activity data suggest that all of these flavonoids are more potent GST inhibitors compared to ethacrynic acid (**19**). Flavonoids are reported to exhibit GST inhibitory activity with the IC_{50} values in the range of 15–30 μM.[24]

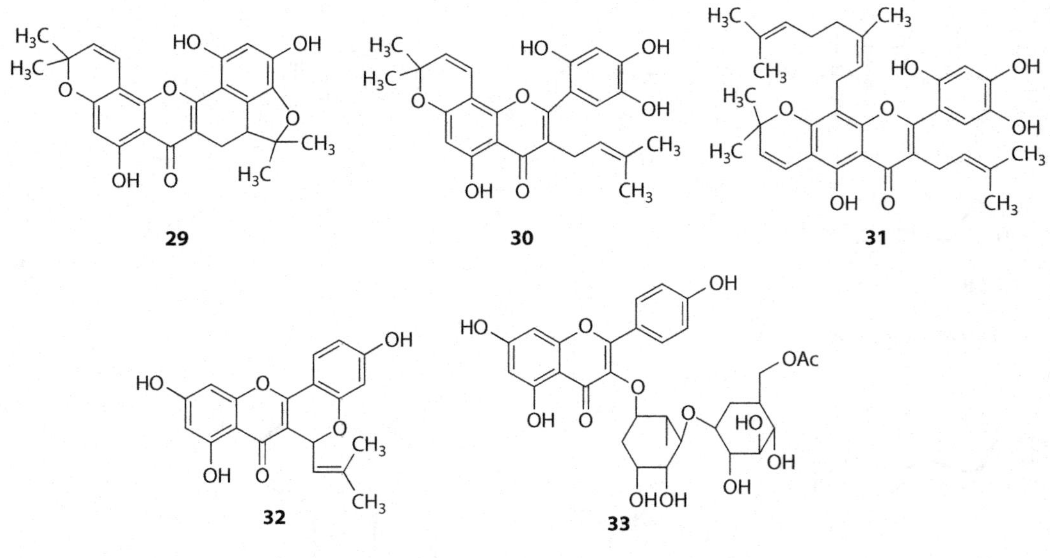

FIGURE 7.5 Structures of compounds **29–33**.

34 R = H
35 R = CH₃

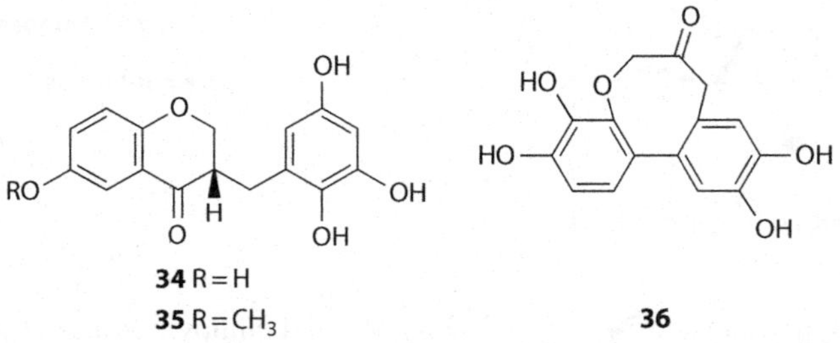

FIGURE 7.6 Structures of compounds **34–36**.

Compounds **29–33** are members of flavonoid class of natural products, and their higher potency might be due to the presence of prenyl group incorporated in their structures.[23] Figure 7.5 lists the structures of compounds **29–33**.

Phytochemical studies on the bioactive fraction of *Caesalpinia bonduc*, of Sri Lankan origin, afforded three GST-inhibiting natural products: caesalpinianone (**34**), 6-*O*-methylcaesalpinianone (**35**), and hematoxylol (**36**).[25] Compounds **34–36** exhibited anti-GST activity with IC₅₀ values of 16.5, 17.1, and 23.6 μM, respectively. Structures of compounds **34–36** are shown in Figure 7.6.

The comparison of the structures of all GST-inhibiting natural products reveals the presence of α,β-unsaturated carbonyl group, which is a required pharmacophore for the expression of this bioactivity. The α,β-unsaturated carbonyl group may form a glutathione adduct of these compounds through Michael addition to inhibit the activity of GST.[26]

7.4 ACETYLCHOLINESTERASE INHIBITORS

Alzheimer's disease (AD) is a neurodegenerative disorder, which causes severe health problems in elderly people.[27] Acetylcholinesterase (AChE) is reported to be involved in the pathogenesis of AD, which breaks down the acetylcholine into acetic acid and choline causing the deficiency of acetylcholine in the brain, resulting in memory loss in AD patients.[28,29] Enhancement of acetylcholine levels in the brain is an effective approach to treat the early symptoms of AD.[30,31] AChE inhibitors not only help to prevent the degradation of acetylcholine but also play a role in preventing proaggregating activity

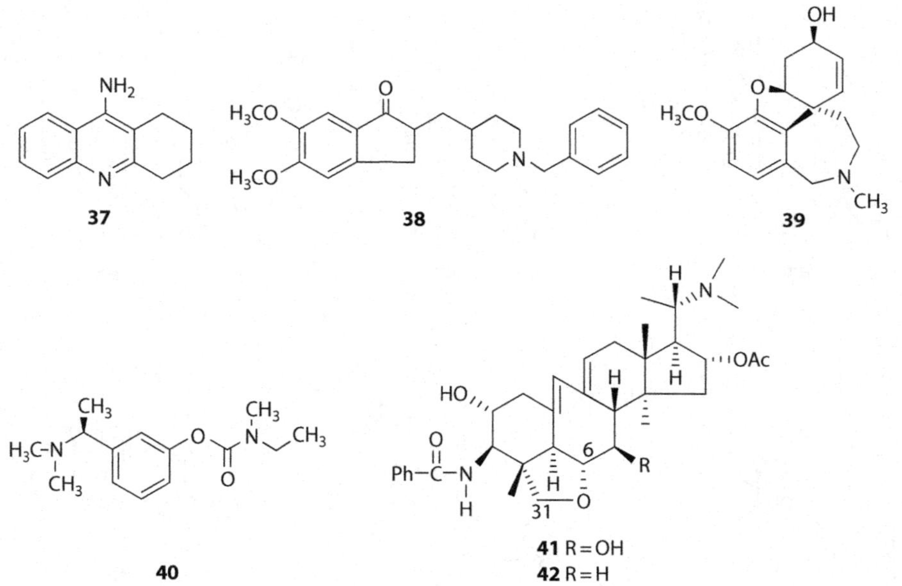

FIGURE 7.7 Structures of compounds 37–42.

of AChE leading to the deposition of β-amyloid, another cause of AD.[32] Compounds with AChE-inhibiting activity are also used to treat senile dementia, ataxia, myasthenia gravis, and Parkinson's disease.[33] Currently, four AChE inhibitors, tacrine (**37**), donepezil (**38**), galanthamine (**39**), and rivastigmine (**40**), are approved by the FDA to be used in clinics to cure AD.[34] These pharmaceuticals have limited effectiveness and a number of side effects.[35] For example, tacrine shows hepatotoxic liability and rivastigmine has a short half-life. In this area, we have explored plants of the genus *Buxus* for the identification of new AChE inhibitors. Steroidal alkaloids isolated from *B. hyrcana* have shown their potential as AChE inhibitors. These alkaloids are O^6-buxafurandiene (**41**) and 7-deoxy-O^6-buxafurandiene (**42**), exhibiting this bioactivity with IC_{50} values of 17.0 and 13.0 μM, respectively.[36] Structures of compounds **37–42** are shown in Figure 7.7.

Based on these results, we collected *B. natalensis* and *B. macowanii* from South Africa based on their ethnomedicinal use to enhance memory in elderly people by the local traditional healers.[37] The crude methanolic extracts of these plants were active in AChE inhibition assay: *B. natalensis* (IC_{50} = 28 μg/mL) and *B. macowanii* (IC_{50} = 30 μg/mL). Our AChE-inhibition-guided chemical investigation of *B. natalensis* resulted in the isolation of four bioactive natural products: O^2-natafuranamine (**43**), O^{10}-natafuranamine (**44**), buxafuranamide (**45**), and buxalongifolamidine (**46**). All of these compounds exhibited anti-AChE activity with IC_{50} values of 3.0, 8.5, 14, and 30.2 μM, respectively.[38] Phytochemical studies on *B. macowanii* resulted in the isolation of seven bioactive compounds: 31-hydroxybuxatrienone (**47**), macowanioxazine (**48**), 16α-hydroxymacowanitriene (**49**), macowanitriene (**50**), macowamine (**51**), N_b-demethylpapillotrienine (**52**), and moenjodaramine (**53**). Compounds **47–53** exhibited anti-AChE activity with IC_{50} values of 17, 32.5, 11.4, 10.8, 45, 19, and 27 μM, respectively. Compounds **47** and **53** were also moderately active in BACE1 inhibitory assay.[39] Structures of all of these compounds (**43–52**) are shown in Figure 7.8.

The bioactivity of compound **43** was almost identical to huperzine A. Compounds **44**, **45**, **49**, and **50** were equally potent in AChE inhibition assay, suggesting that the bioactivity of these compounds might be due to the presence of tetrahydrofuran or tetrahydrooxazine ring incorporated in their structures. The structural analysis of these compounds further indicated that the location of an ether linkage in these compounds does not play any role in enzyme inhibition activity, as **44** contains ether linkage between C-31 and C-10, and **45** has an ether linkage between C-31 and C-6. Compounds **49** and **50**

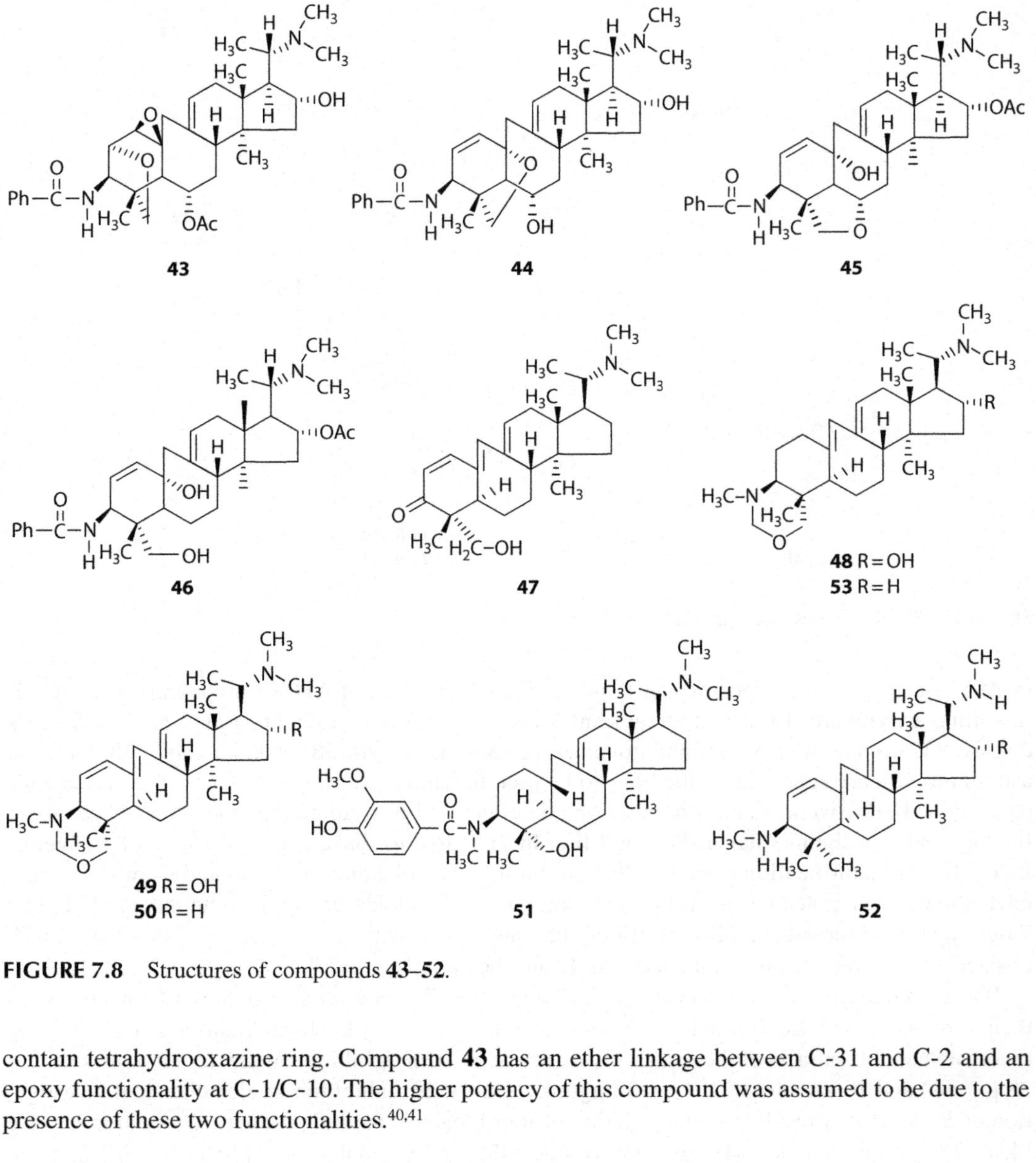

FIGURE 7.8 Structures of compounds **43–52**.

contain tetrahydrooxazine ring. Compound **43** has an ether linkage between C-31 and C-2 and an epoxy functionality at C-1/C-10. The higher potency of this compound was assumed to be due to the presence of these two functionalities.[40,41]

7.5 α-GLUCOSIDASE INHIBITORS

α-Glucosidase, an intestinal membrane-bound enzyme, catalyzes the carbohydrate digestion by hydrolyzing the glycosidic bonds in carbohydrates to liberate free glucose causing postprandial hyperglycemia. This results in type 2 diabetes mellitus that affects approximately over two billion people worldwide.[42] This health-related problem can be overcome by suppressing hyperglycemia that includes reduction of glucose absorption in the gut. The potent α-glucosidase inhibitors are reported to accomplish this task.[43] For instance, acarbose is one of the currently used drugs and it works by inhibiting the activity of α-glucosidase. These inhibitors are also used to treat sugar-mediated diseases including obesity, cancer, HIV, etc.[44] Natural products are reported to exhibit a novel mode of action in inhibiting the activity of α-glucosidase. For instance, aegeline (**54**), a hydroxyl amide alkaloid, is reported to suppress both blood glucose and plasma triglyceride levels.[45] Natural products have shown great potential in discovering novel α-glucosidase inhibitors.

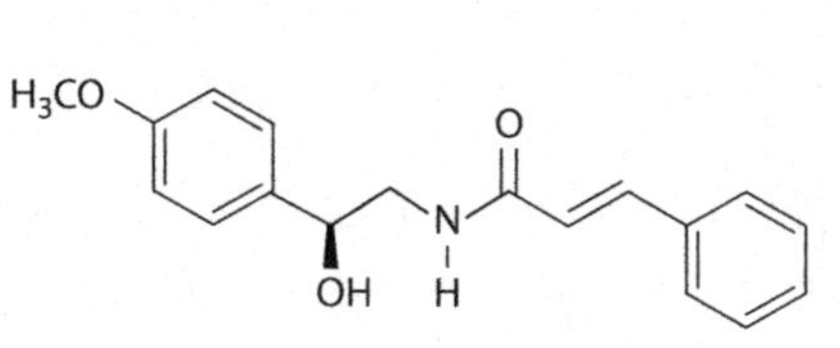

FIGURE 7.9 Structures of compounds **54–57**.

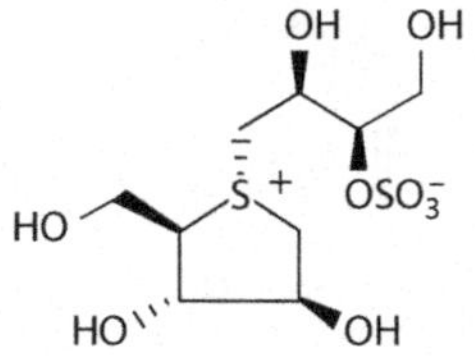

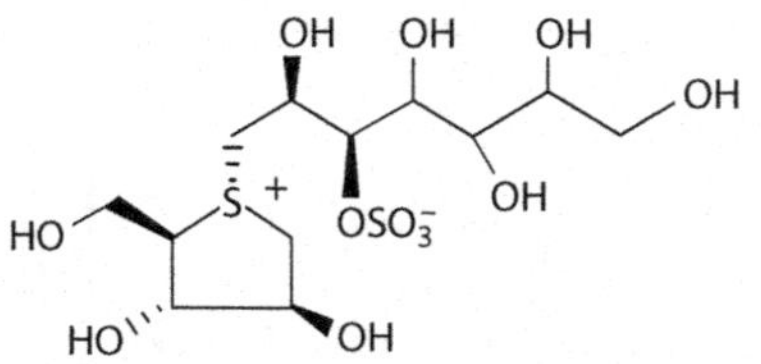

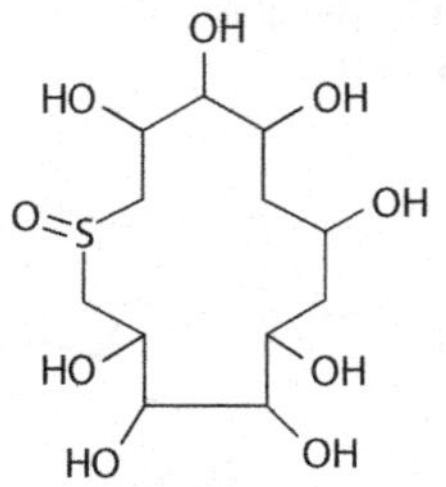

Salacia reticulata (Kothala himbutu) is an ethnomedicinal plant that has applications in treating diabetes mellitus in Sri Lanka.[46] α-Glucosidase-inhibition-guided chemical studies on the crude extract of this plant resulted in the isolation of three bioactive natural products: salacinal (**55**), kotalanol (**56**), and a polyhydroxylated cyclic 13-membered sulfoxide (**57**).[47,48] These compounds exhibited anti-α-glucosidase activity: **55** (IC$_{50}$ maltase 9.58 µM, sucrase 2.51 µM, isomaltase 1.77 µM), **56** (IC$_{50}$ maltase 6.60 µM, sucrase 1.37 µM, isomaltase 4.48 µM), and **57** (IC$_{50}$ maltase 0.227 µM, sucrose 0.186 µM, isomaltase 0.099 µM), respectively. The high potency of compounds **57** is hypothesized due to the orientation of hydroxyl, ring structure, and sulfoxide groups.[47,48] Structures of compounds **54–57** are shown in Figure 7.9.

In this area, we have explored a few medicinal plants for discovering naturally occurring α-glucosidase inhibitors. For instance, our chemical investigation of the methanolic extract of *Drypetes gossweileri*, of South African origin, resulted in the isolation of bioactive natural products: *N*-β-D-glucopyranosyl-*p* hydroxyphenylacetamide (**58**), *p*-hydroxyphenyl-acetic acid (**59**), *p*-hydroxyphenylacetonitrile (**60**), *p*-hydroxy-acetophenone (**61**), 3,4,5-trimethoxyphenol (**62**), dolichandroside A (**63**), and β-amyrone (**64**). Compounds **58–64** (all of these structures are shown in Figure 7.10) exhibited anti-α-glucosidase activity with IC$_{50}$ values of 12, 50, 48, 50, 56, 20, and 25 µM, respectively.[49]

Compound **58** exhibited potent bioactivity compared to the rest of the phytochemicals. This compound represents the first example of the plant natural product containing *N*-glucose moiety incorporated in its structure, and its bioactivity was comparable with that of acarbose (IC$_{50}$ = 15.2 µM). In order to carry out structure–activity relationships studies on compound **58**, we carried out its acidic hydrolysis to afford compound (**65**; Figure 7.10) that exhibited weak α-glucosidase inhibition activity (IC$_{50}$ = 60.0 µM), suggesting that the higher potency of compound (**58**) might be due to the presence of *N*-glucose moiety. Due to the poor yield of this natural product, we were unable to carry out more reactions on it. At this stage, we are in the process of synthesizing this compound in our lab to allow us to carry out its detailed structure–activity relationships studies.

Our phytochemical investigation of the chloroform fraction of *Swertia corymbosa*, of Indian origin, resulted in the isolation of five bioactive xanthones, 3-allyl-2,8-dihydroxy-1,6-dimethoxy xanthen-9-one (**66**), xanthones gentiacaulein (**67**), norswertianin (**68**), 1,3,6,8-tetrahydroxy xanthone (**69**), and 1,3-dihydroxy xanthone (**70**). Compounds **66–70** (Figure 7.11) were found to exhibit anti-α-glucosidase activities with IC$_{50}$ values of 26.3, 44.5, 23.2, 39.0, and 35.2 µM, respectively.[50]

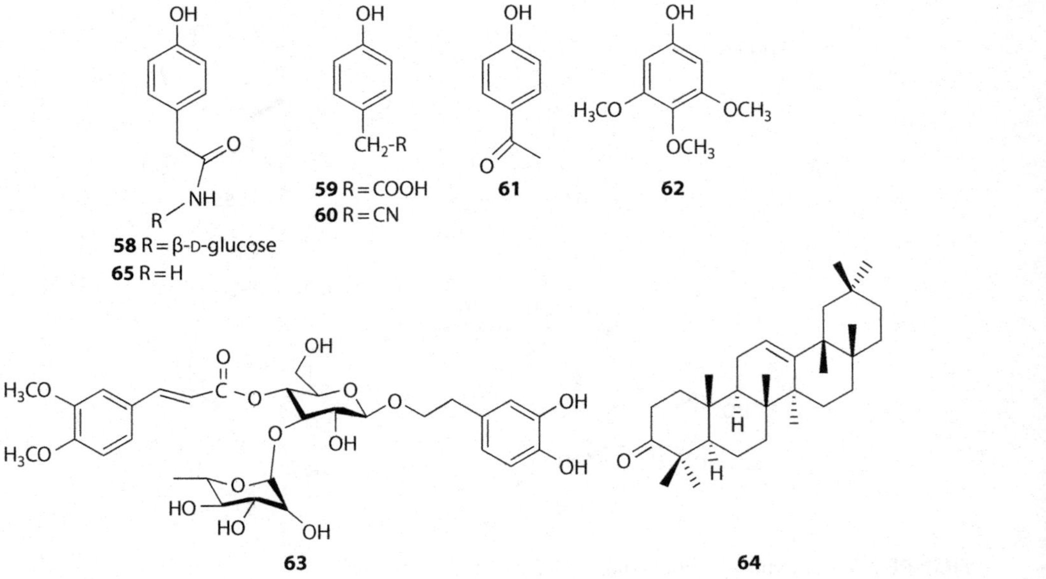

FIGURE 7.10 Structures of compounds **58–65**.

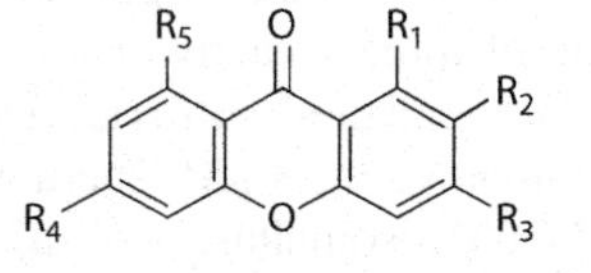

66 R₁, R₄ = OMe, R₂, R₅ = OH, R₃ = Allyl,
67 R₁, R₄ = OMe, R₂, R₅ = OH, R₃ = H,
68 R₁, R₂, R₄, R₅ = OH, R₃ = H
69 R₁, R₃, R₄, R₅ = OH, R₂ = H
70 R₁, R₃ = OH, R₂, R₄, R₅ = H

FIGURE 7.11 Structures of compounds **66–70**.

Phytochemical investigation of the crude extract of *Epilobium angustifolium* of Manitoban origin resulted in the isolation of two new potent α-glucosidase inhibitors: **71** (IC$_{50}$ = 120 nM) and **72** (IC$_{50}$ = 122 nM). Compounds **71** and **72** (Figure 7.12) were identified as minor secondary metabolites of *E. angustifolium*. For the detailed in-vitro and in-vivo bioactivity studies, we have carried out chemo-enzymatic synthesis of compound **71** from commercially available betulin (**73**). Compound **73** on reaction with TMSCl and DMAP yielded compound **74**. Incubation of compound **74** with the whole-cell liquid culture of *C. blakesleeana* afforded 21-hydroxybetulin (**75**), which on reaction with Hg(CF$_3$COO)$_2$ and NaBH$_4$ followed by treatment with HF yielded compound **71**. Figure 7.13 shows the chemo-enzymatic synthesis of **71**. This chemo-enzymatic synthesis of natural products using a combination of chemical and microbial reactions will be an important tool in natural product chemistry-based drug discovery program to scale up minor lead bioactive compounds from commercially available compounds or major secondary metabolites, isolated during bioassay-directed fractionation method. This green chemistry approach can help to synthesis bioactive compounds that are difficult to synthesize using traditional synthetic chemistry to supply minor lead bioactive natural products for structure–activity relationships studies and detailed in-vivo and clinical testing.

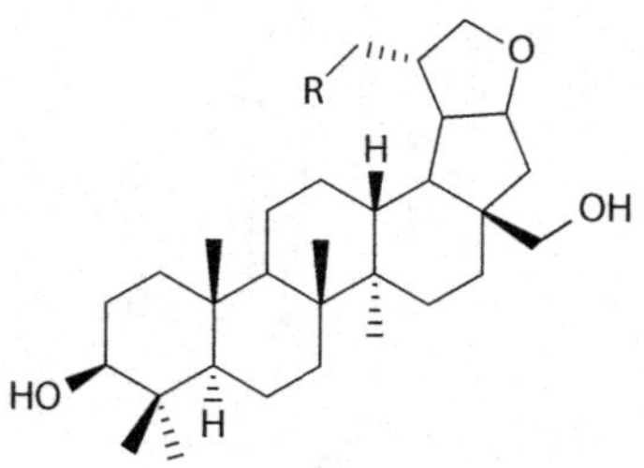

FIGURE 7.12 Structures of compounds **71** and **72**.

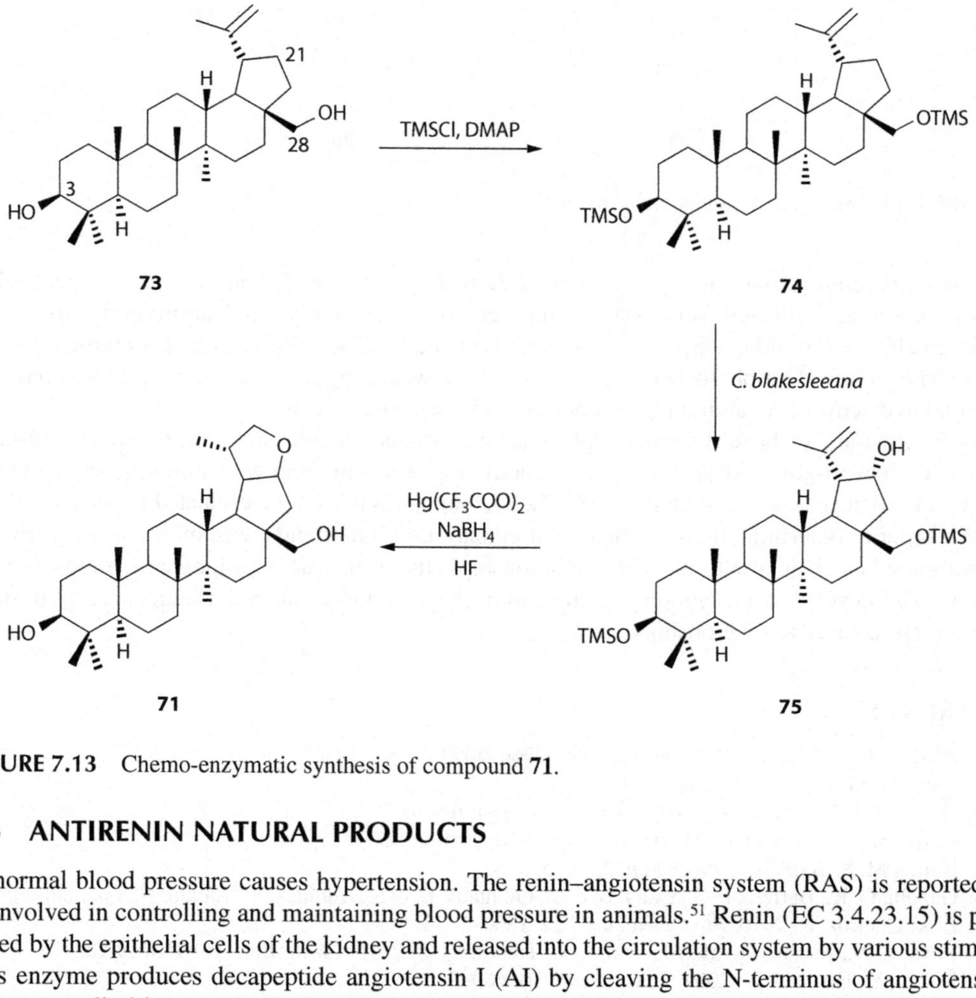

FIGURE 7.13 Chemo-enzymatic synthesis of compound **71**.

7.6 ANTIRENIN NATURAL PRODUCTS

Abnormal blood pressure causes hypertension. The renin–angiotensin system (RAS) is reported to be involved in controlling and maintaining blood pressure in animals.[51] Renin (EC 3.4.23.15) is produced by the epithelial cells of the kidney and released into the circulation system by various stimuli. This enzyme produces decapeptide angiotensin I (AI) by cleaving the N-terminus of angiotensinogen, a rate-limiting step in RAS. AI is an inactive peptide that is transformed by the angiotensin-converting enzyme (ACE) to angiotensin II (AII), which acts on the arterial smooth muscle cells to maintain blood pressure and stimulates the synthesis and release of aldosterone from the adrenal cortex. Compounds with RAS inhibition activity can help to treat hypertension. *N. latifolia* is reported to exhibit antihypertensive activity,[52] and our detailed phytochemical study on the methanolic extract of this plant, collected from Nigeria, resulted in the isolation of five bioactive indole alkaloids, latifoliamides A–E (**76–80**). Structures of compounds **76–80** are shown in Figure 7.14. Compounds **76–80**

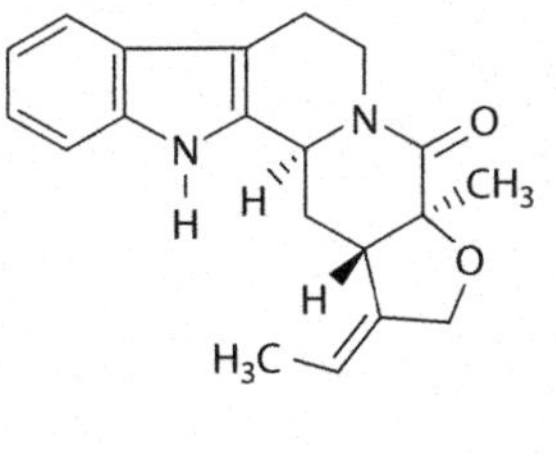

FIGURE 7.14 Structure of compounds **76–80**.

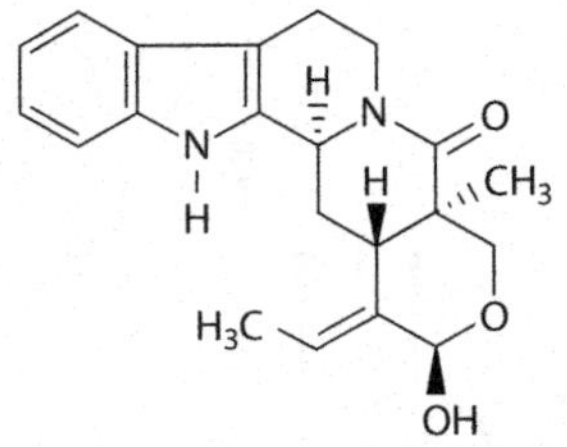

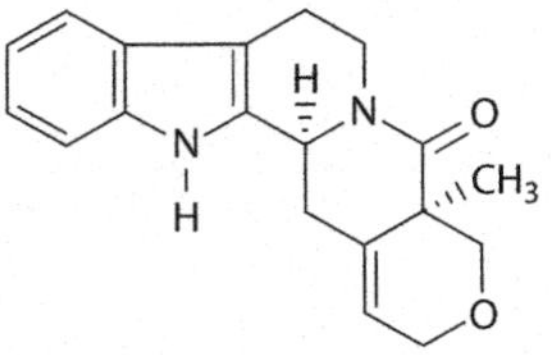

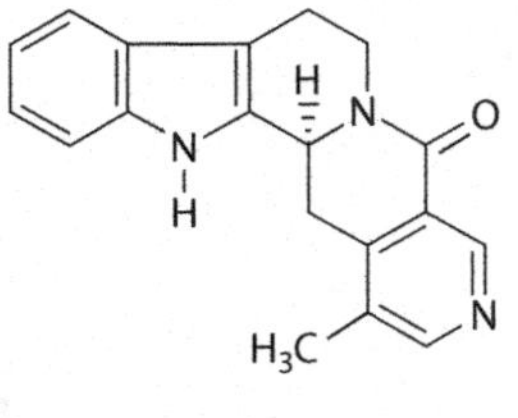

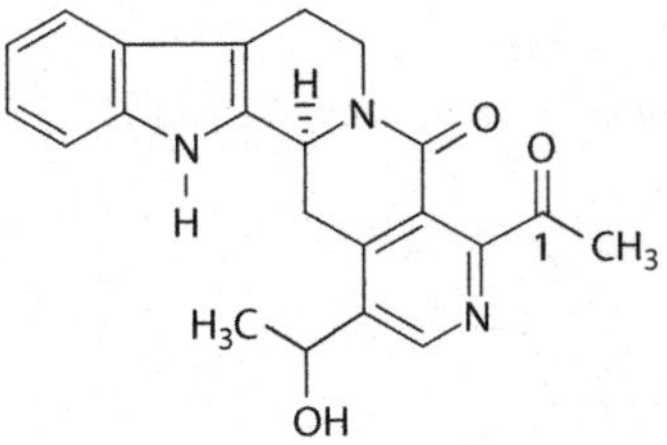

exhibited antirenin activity with IC$_{50}$ values of 32.6, 11.3, 95.0, 94.5, and 16.3 µM, respectively.[53] This antirenin activity was very weak compared to the currently used antihypertensive drug, aliskiren (IC$_{50}$ = 0.6 nM), which works by inhibiting the RAS activity. Aliskiren is reported to have several side effects. Our phytochemical results, though weak, suggest that natural products have the potential to discover lead antirenin compounds with fewer side effects.

In conclusion, we have discovered plant natural products exhibiting antimicrobial, anti-GST, anti-AChE, anti-α-glucosidase, and antirenin activities. A few of them were potently active in GST, AChE, and α-glucosidase inhibition assays. These compounds need to be evaluated for in vivo bioassays in order to determine their biomedical applications. Compounds with moderate bioactivities are warranted for their structure–activity relationships studies in order to improve their bioactivities. We have also developed a chemo-enzymatic approach to synthesize natural product analogs to study their structure–activity relationships.

REFERENCES

1. Phillipson, D. J. 2007. *Phytochemistry* 68: 2960–2972.
2. Rates, S. M. K. 2001. *Toxicon* 39: 603–613.
3. Hasskarl, J. 2010. *Recent Results Cancer Res.* 184: 61–70.
4. Newman, D. J., Cragg, G. M. 2007. *J. Nat. Prod.* 70: 461–477.
5. Butler, M. S. 2008. *Nat. Prod. Rep.* 25: 475–516.
6. Gandhi, G. R., Barreto, P. G., Lima, B. d. S., Quintans, B. d. S., Antunes, A., Narain, N., Quintans-Junior, L. J., Queiroz, R. 2016. *Phytomedicine* 123: 1830–1842.
7. Silva, L. N., Zimmer, K. R., Macedo, A. J., Trentin, D. S. 2016. *Chem. Rev.* 116: 9162–9236.
8. Nakao, Y., Fusetani, N. 2007. *J. Nat. Prod.* 70: 689–710.
9. Hamilton, L. R., Schachter, S. C., Myers, T. M. 2016. *Neurochemical Research* 42: 1962–1971.
10. Taubes, G. 2008. *Science* 321: 356–361.
11. Shi, R. Itagaki, N. Sugawara, I. 2007. *Mini Rev. Med. Chem.* 7: 1177–1185.
12. Grace, D., Khan, M. S., Friesen, K., Ata, A. 2016. *Chem. Biodivers.* 13: 913–917.
13. Ata, A., Conci, L. J., Betteridge, J., Orhan, I., Sener, B. 2007. *Chem. Pharm. Bull.* 55: 118.
14. Ata, A., Nachtigall, J. A. 2004. *Z. Naturforsch.* 59c: 209.
15. Orabi, K. Y. 2000. *J. Nat. Prod.* 63: 1709; Rosazza, J. P. 1978. *J. Nat. Prod.* 41: 297.

16. Ata, A., Betteridge, J., Schaub, E., Kozera, D. J., Holloway, P., Samerasekera, R. 2009. *Chem. Biodivers.* 6: 1453–1462.
17. Ata, A., Diduck, C., Udenigwe, C. C., Zahid, S., Decken, A. 2007. *ARKIVOC* 13: 195–203.
18. Schisselbauer, J. C., Silber, R., Papadopoulos, E., Abrams, K., LaCreta, F. P., Tew, K. D. 1990. *Cancer Res.* 50: 3562–3568.
19. Al-Qattan, M. N., Mordi, M. N., Mansor, S. M. 2016. *Comput. Biol. Chem.* 64: 237–249.
20. Slanc, P., Doljak, B., Kreft, S., Lunder, M., Janeš, D., Štrukelj, B. 2009. *Phytother. Res.* 23: 874–877.
21. Ata, A., Kalhari, K. S., Samarasekera, R. 2009. *Phytochem. Lett.* 2: 37–40.
22. Ata, A., Udenigwe, C. C., Matochko, W., Holloway, P., Eze, M. O., Uzoegwu, P. N. 2009. *Nat. Prod. Commun.* 4: 1185–1188.
23. Iverson, C. D., Zahid, S., Li, Y., Shoqafi, A. H., Ata, A., Samarasekera, R. 2010. *Phytochem. Lett.* 3: 207–211.
24. Ata, A., Udenigwe, C. C. 2008. *Curr. Bioact. Compd.* 4: 41–50.
25. Ata, A., Gale, E. M., Samarasekera, R. 2009. *Phytochem. Lett.* 2: 106–109.
26. Udenigwe, C. C., Ata, A., Samarasekera, R. 2007. *Chem. Pharm. Bull.* 55: 442–445.
27. Ata, A., Conci, L. J., Orhan, I. 2006. *Heterocycles* 68: 2097–2106.
28. Rosenberry, T. L. 1975. *Adv. Enzymol. Relat. Areas Mol. Biol.* 43: 103–218.
29. Enz, A., Amstutz, R., Boddeke, H., Gmelin, G., Malonowski, J. 1993. *Prog. Brain Res.* 98: 431–445.
30. Fang, L., Chen, Y., Zhang, Y. 2009. *Yaoxue Jinzhan* 33: 289–296.
31. Tumiatti, V., Minarini, A., Bolognesi, M. L., Milelli, A., Rosini, M., Melchiorre, C. 2010. *Curr. Med. Chem.* 17: 1825–1838.
32. Ribeiz, S. R. I., Bassitt, D. P., Arrais, J. A., Avila, R., Steffens, D. C., Bottino, C. M. C. 2010. *CNS Drugs* 24: 303–317.
33. Heinrich, M. 2010. Galanthamine from Galanthus and other Amaryllidaceae—Chemistry and biology based on traditional use. In *The Alkaloids*, Cordell, G. A., Ed., Elsevier Sciences, San Diego, CA, Vol. 68, pp. 157–165.
34. Ata, A., Naz, S., Elias, E. M. 2011. *Pure Appl. Chem.* 83: 1741–1749.
35. Kaur, J., Zhang, M.-Q. 2000. *Curr. Med. Chem.* 7: 273–294.
36. Babar, Z. U., Ata, A., Meshkatalsadat, M. H. 2006. *Steroids* 71: 1045–1051.
37. Kaikabo, A. 1990. MSc thesis, University of Pretoria, Pretoria, South Africa.
38. Matochko, W. L., James, A., Lam, C. W., Kozera, D. J., Ata, A., Gengan, R. M. 2010. *J. Nat. Prod.* 73: 1858–1862.
39. Lam, C. W., Wakeman, A., James, A., Ata, A., Gengan, R. M., Ross, S. A. 2015. *Steroids* 95: 73–79.
40. Ata, A. 2012. Novel plant-derived biomedical agents and their biosynthetic origin. In *Studies in Natural Products Chemistry*, Atta-ur-Rahman, Ed., Elsevier, Amsterdam, the Netherlands, Vol. 38, pp. 225–245.
41. Mollataghi, A., Coudiere, E., Hadi, A. H. A., Mukhtar, M. R., Awang, K., Litaudon, M., Ata, A. 2012. *Fitoterapia* 83: 298–302.
42. Atta-ur-Rahman, Zareen, S., Choudhary, M. I., Akhtar, M. N., Khan, S. N. 2008. *J. Nat. Prod.* 71: 910–913.
43. Kim, G.-N., Kwon, Y.-I., Jang, H.-D. 2011. *J. Med. Food* 14: 712–717.
44. Atta-ur-Rahman, Choudhary, M. I., Basha, F. Z., Abbad, G., Khan, S. N., Shah, S. A. A. 2007. *Pure Appl. Chem.* 70: 2263–2268.
45. Phuwapraisirisan, P., Puksasook, T., Jong-aramruang, J., Kokpol, U. 2008. *Bioorg. Med. Chem. Lett.* 18: 4956–4958.
46. Zhang, L., Wang, Y., Wang, Z. 2008. *Huagong Shikan* 22: 42–44.
47. Choubdar, N., Bhat, R. G., Stubbs, K. A., Yuzwa, S., Pinto, B. M. 2008. *Carbohydr. Res.* 343: 1766–1777.
48. Ozaki, S., Oe, H., Kitamura, S. 2008. *J. Nat. Prod.* 71: 981–984.
49. Ata, A., Tan, D. S., Matochko, W. L., Adesanwo, J. K. 2011. *Phytochem. Lett.* 4: 34–37.
50. Uvarani, C., Arumugasamy, K., Chandraprakash, K., Sankaran, M., Ata, A., Mohan, P. S. 2011. *Chem. Biodivers.* 12: 358–370.
51. Takahashi, S., Tokiwano, T., Hata, K., Kodama, I., Hokari, M., Suzuki, N., Yoshizawa, Y., Gotoh, T. 2010. *Biosci. Biotechnol. Biochem.* 74: 1713.
52. Nworgu, Z. A. M., Eferakeya, A. E., Onwukaeme, D. N., Afolayan, A. J., Ameachina, F. C., Ayinde, B. A. 2009. *J. Appl. Sci. Res.* 5: 2208.
53. Agomuoh, A. A., Ata, A., Udenigwe, C. C., Aluko, R. E., Irenus, I. 2013. *Chem. Biodivers.* 10: 401–410.

8 Asymmetric Biocatalysis in Organic Synthesis of Natural Products

Renata Kołodziejska, Aleksandra Karczmarska-Wódzka, and Agnieszka Tafelska-Kaczmarek

CONTENTS

8.1 Introduction 206
8.2 Prochirality *Meso* Compounds 206
8.3 Enzymatic Desymmetrization 209
8.4 Factors Affecting an Enzymatic Reaction Enantioselectivity 212
    8.4.1 Solvent 212
        8.4.1.1 Stability and Reactivity 212
        8.4.1.2 Selectivity 215
    8.4.2 Effect of pH 217
    8.4.3 Effect of Temperature 217
    8.4.4 Additives 218
8.5 Hydrolases 218
8.6 Application of Esterases and Proteases in Enzymatic Desymmetrization of Symmetrical Compounds 219
8.7 Application of Lipases in Enzymatic Desymmetrization of Symmetrical Compounds 222
8.8 Oxidoreductases 235
8.9 Application of Dehydrogenases in Enzymatic Reduction of Symmetrical Compounds 237
8.10 Application of Ene-Reductases in Enzymatic Reduction of Symmetrical Compounds 251
8.11 Application of Dehydrogenases in Enzymatic Oxidation Reactions of Symmetrical Compounds 254
8.12 Application of Monooxygenases in Enzymatic Oxidation Reactions of Symmetrical Compounds 255
8.13 Application of Dioxygenases in Enzymatic Oxidation Reactions of Symmetrical Compounds 259
8.14 Application of Oxidases in Enzymatic Oxidation Reactions of Symmetrical Compounds 261
8.15 Application of Peroxidases in Enzymatic Oxidation Reactions of Symmetrical Compounds 261
References 264

8.1 INTRODUCTION

Stereoselective biotransformation can be performed by asymmetric synthesis as well as kinetic resolution (KR) of racemic mixtures. The two concepts differ significantly from each other. According to the definition, an asymmetric synthesis is a reaction in which an achiral molecule is converted to a chiral molecule in such a manner that the formation of the stereoisomeric products occurs in unequal amounts. However, in KR, only one of the enantiomers of a racemic mixture undergoes transformation and one stereoisomer is produced.[1–4]

The kinetic resolution method is based on differences in reaction rates at which a biocatalyst reacts with enantiomers. One enantiomer undergoes a more rapid reaction than the one of the opposite configurations. However, the disadvantage of this method is that it allows obtaining a maximum yield of 50%, provided the enzyme exhibits high selectivity. To improve the performance of enantioselective biotransformation, the dynamic kinetic resolution (DKR) method was developed. It combines an enzymatic reaction and a simultaneous *in situ* racemization. Continuous racemization enables the maintenance of the equilibrium between two enantiomers during the reaction, which generates the selective conversion of only one of them, namely, the more responsive one. This prevents a decrease in the optical purity, while the degree of conversion exceeds 50% (Scheme 8.1).[1–8]

With regard to the subject of this study, enzymatic asymmetric synthesis will be discussed in detail. In the enzymatic asymmetric synthesis, the enzyme allows the desymmetrization of achiral compounds resulting in chiral compounds of high optical purity. Therefore, this type of biotransformation is known as enantioselective enzymatic desymmetrization (EED).[6] EED, which is an interesting alternative to the previously mentioned methods of stereoselective synthesis, allows achieving high enantioselectivity with quantitative yield.[5–11] This method is related to the generation of an asymmetry (loss of symmetry elements) in prochiral molecules (most often an sp^3- or sp^2-hybridized carbon atom), in *meso* synthons, and centrosymmetric compounds.

8.2 PROCHIRALITY *MESO* COMPOUNDS

A prochiral carbon atom can be either a tetragonal carbon atom with two identical substituents or a trigonal carbon atom, that is, a carbonyl group. An achiral center of the tetrahedral system is defined as a prochiral one if it becomes chiral as a result of one of the two substituents' replacement, which, when separated from the particles, are indistinguishable. An sp^2-hybridized carbon atom is prochiral if the addition to a double bond can occur with the differentiation of faces, which allows the formation of a chiral product. The nomenclature of prostereoisomerism depends on the prochiral carbon atom hybridization.

In the prochiral molecule of $R_1R_2CX_2$ type, identical X substituents become enantiotopic or diastereotopic under the influence of a chiral agent (the diastereotopic one, if apart from the prochiral group a chirality center in a molecule appears). They become distinguishable and one substituent can be marked as X_a and the other one X_b. The molecule should be oriented in such a way that one of the substituents X, for example, X_a, is directed away from a viewer and for the other substituents a priority should be assigned according to the Cahn–Ingold–Prelog (CIP) system. If they appear in a clockwise manner (e.g., $R_1 > R_2 > X_b$), the prochiral substituent X_a is designated as the *pro-R* one. The opposite arrangement is described as *pro-S* (Scheme 8.2).[1–4,9]

This can be illustrated by the enzymatic stereoselective transesterification of a prochiral pyrimidine acyclonucleoside and the hydrolysis of its corresponding diacetylated derivatives. The Amano PS lipase from *Burkholderia cepacia* (BCL) introduces an acyl group onto one of the primary hydroxyl groups of an acyclonucleoside in the transesterification reaction. If the acylation occurs on the *pro-R* group, a monoester with an absolute *R* configuration is formed, while the acylation of the *pro-S* hydroxyl group delivers a product of the reverse configuration (Scheme 8.3). In this case, the lipase BCL allows for the enantiotopic differentiation of prochiral hydroxyl groups of pyrimidine acyclonucleoside and selectively acetylates the *pro-R* group.[12]

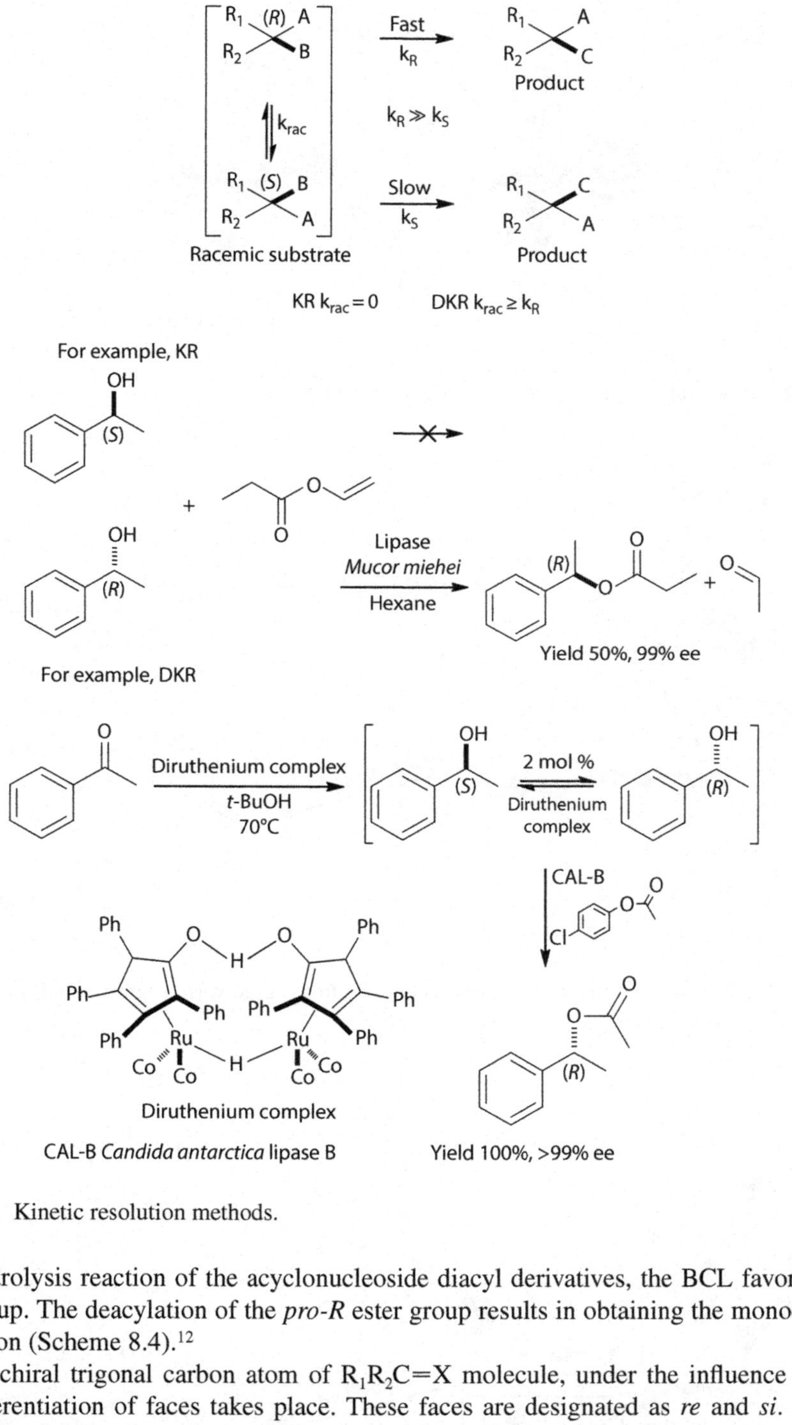

Asymmetric Biocatalysis in Organic Synthesis of Natural Products 207

SCHEME 8.1 Kinetic resolution methods.

In the hydrolysis reaction of the acyclonucleoside diacyl derivatives, the BCL favors the same prochiral group. The deacylation of the *pro-R* ester group results in obtaining the monoester of the S-configuration (Scheme 8.4).[12]

In the prochiral trigonal carbon atom of $R_1R_2C=X$ molecule, under the influence of a chiral agent, a differentiation of faces takes place. These faces are designated as *re* and *si*. In order to indicate the face of the trigonal atom, the priority of ligands R_1, R_2, and X should be determined according to the CIP rules. If the priority of the ligands is anticlockwise, this face is designed as *re* (alternatively to the other one—*si*) (Scheme 8.5).[1-4]

The reduction of ketones with the use of microorganisms to secondary alcohols is an example of an enzymatic desymmetrization of the compound containing the sp²-hybridized carbon atom. The enzymes produced in the microorganism *Geotrichum candidum* (IFO 5767) enable the performance

208 Asymmetric Synthesis of Drugs and Natural Products

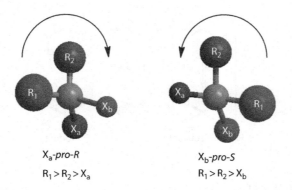

X_a-*pro-R*
$R_1 > R_2 > X_a$

X_b-*pro-S*
$R_1 > R_2 > X_b$

SCHEME 8.2 Prochiral tetragonal carbon atom of $R_1R_2CX_2$ molecule.

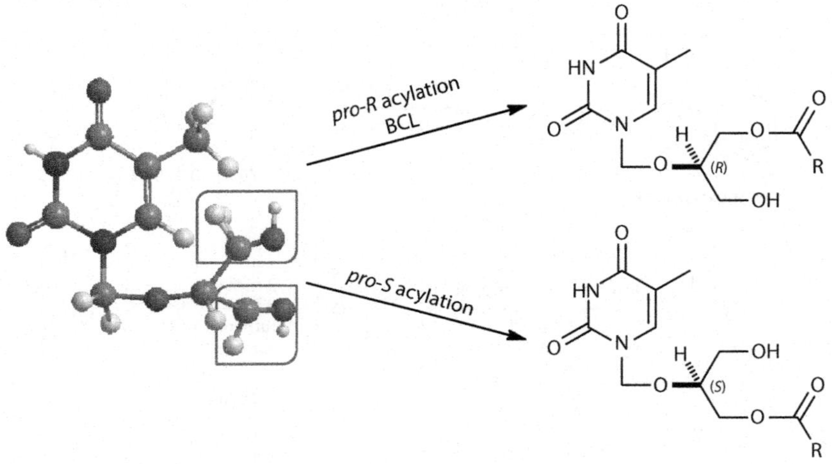

SCHEME 8.3 Enantiotopic differentiation of prochiral hydroxyl groups *via* enzymatic reaction.

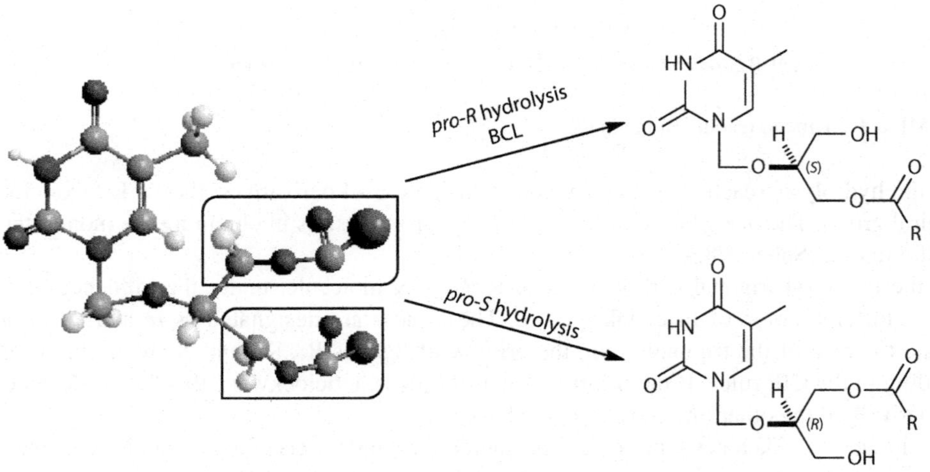

SCHEME 8.4 Enantiotopic differentiation of prochiral ester groups *via* enzymatic reaction.

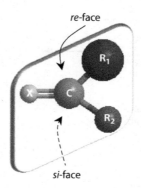

SCHEME 8.5 Prochiral trigonal carbon atom of R₁R₂C=X molecule.

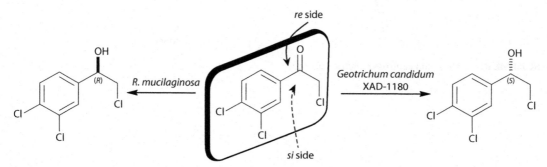

SCHEME 8.6 Differentiation of carbonyl group sides.

of the reduction with a differentiation of carbonyl group sides in the prochiral substrate. The reaction involves an internal asymmetric induction takes place, and the (S)-alcohol is obtained as a result of a selective attack on the *re*-face. However, the reduction with the application of the microorganism *Rhodotorula mucilaginosa* (CBS 2378) provides an opposite enantiomer. In this case, the attack on the *si* side is preferable (Scheme 8.6).[13]

Similar to the prochiral compounds, possessing tetrahedral carbon atoms, breaking the symmetry in *meso* and centrosymmetric substrates leads to the optically active products (Scheme 8.7).[14,15]

8.3 ENZYMATIC DESYMMETRIZATION

The point of enzymatic desymmetrization is the production of chiral compounds of the highest optical purity. Asymmetric synthesis is enantioselective when one of the enantiotopic (diastereotopic) groups or faces of an optically inactive compound is biotransformed faster than the other. To achieve the high level of enantioselectivity of an enzymatic reaction, two different diastereoisomorphic transition states have to be formed in this reaction. The enantioselectivity of the reaction is determined by the free energy difference ($\Delta\Delta G^{\ddagger}$) of these transition states (Equation 8.1):

$$\Delta\Delta G^{\ddagger} = -RT \ln E \qquad (8.1)$$

The higher the $\Delta\Delta G^{\ddagger}$ value, the higher the enantioselectivity of the enzymatic reaction. Due to the logarithmic relationship between E (enantiomeric ratio) and $\Delta\Delta G^{\ddagger}$, a small increase in $\Delta\Delta G^{\ddagger}$ causes a dramatic change in E, for example, when $\Delta\Delta G^{\ddagger}$ increases by about 1 kcal/mol, the enantiomeric

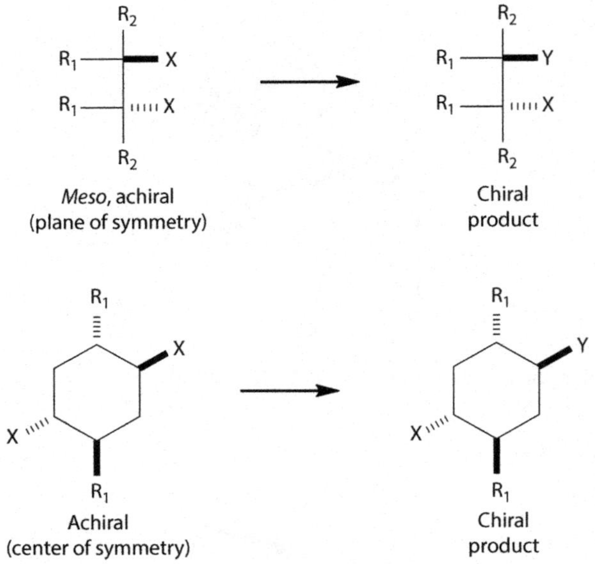

SCHEME 8.7 *Meso* and centrosymmetric substrates.

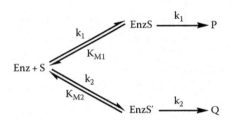

SCHEME 8.8 Enzymatic desymmetrization.

excess of the product rises from 80% to 95%.[1,14] In the enzymatic reaction, the E parameter corresponds to the ratio of the reaction rates of two enantiomeric products formation (Scheme 8.8).

The P and Q products are formed from diastereomeric Michaelis complexes (Enz-S and Enz-S′) as a result of irreversible reactions. The complexes are obtained by the noncovalent binding of a substrate to the active site of an enzyme. The rate of P and Q product formation at infinitely low concentrations is defined as the ratio of specific constants (Equation 8.2):

$$E = \frac{P}{Q} = \frac{V_1}{V_2} = \frac{(k/K_M)_1}{(k/K_M)_2} \qquad (8.2)$$

The E value is the parameter describing the selectivity of an enzymatic reaction. That value depends on the ratio of two constants (k/K_M) and is independent of concentration. In the case of enzymatic kinetic resolution, in order to evaluate the enantioselectivity of a reaction, it is not sufficient to provide enantiomeric excess values of a substrate and product (ee_S and ee_P) because those values are the functions of concentration. In contrast to the KR, the ee_P value remains constant throughout the EED.

Meso diacetate catalyzed by esterase is a representative example of the enzymatic desymmetrization (Scheme 8.9).[16]

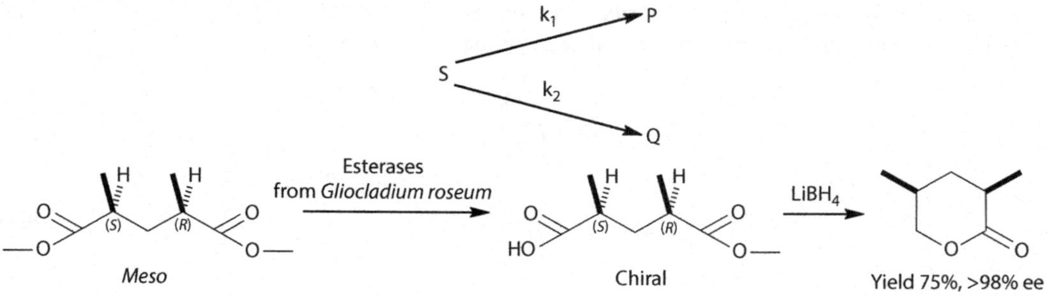

Asymmetric Biocatalysis in Organic Synthesis of Natural Products

SCHEME 8.9 Enzymatic desymmetrization—single-step kinetics.

The reaction is virtually irreversible and terminates after the initial stage of the carboxylate monoester formation. In the EED reaction, the E value is defined as the ratio of the apparent first-order rate constants of the enantiomeric product (k_1/k_2) formation. The E value can be determined from Equation 8.3, in which E is the function of the ee_P, the magnitude easily measured experimentally:

$$E = \frac{1+ee_P}{1-ee_P} \quad (8.3)$$

The ee_P parameter cannot be used to quantify the EED when the product of the reaction is not stable and it is further transformed (i.e., the ee_P is the function of concentration). In practice, many desymmetrization reactions undergo further cleavage in the second step to yield an achiral diol. This happens because the monoester formed in the first step becomes the substrate for hydroxylases (Scheme 8.10).[10,11] Despite the fact that the second step is usually slower, it cannot be neglected as it influences the yield and enantioselectivity of the reaction.

The ratio of the first and second step [$(k_1 + k_2)/(k_3 + k_4)$] reaction rates has an impact on the chemical yield of the products (P + Q). In order to obtain high yield, the first step should be faster than the second one [$(k_1 + k_2) \gg (k_3 + k_4)$]. Similarly, the optical purity of the chiral products also depends on four rate constants. The reaction is enantioselective when $k_1 \gg k_2$ and $k_4 \gg k_3$. In other words, let us assume that the P product is a fast-forming enantiomer whereas the Q one is a slowly forming enantiomer. Then, if the S substrate is converted faster into P rather than into Q, the enzyme exhibits constant enantioselectivity during the reaction in which the R diol is preferentially formed from Q. The combination of enantioselective desymmetrization of the prochiral substrate followed by the kinetic resolution of an enantiomeric mixture provides optical purity enrichment with more than 50% yield.[10,11]

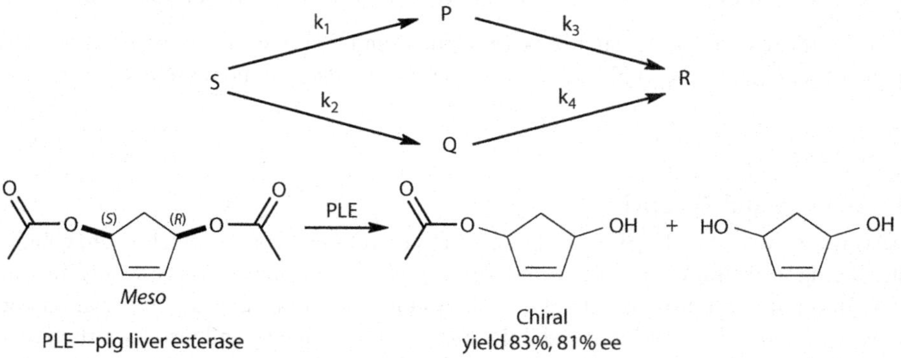

SCHEME 8.10 Enzymatic desymmetrization—double-step kinetics.

To optimize the yield and optical purity of a monoester fraction in a two-step desymmetrization, two kinetic parameters are additionally indicated (Equations 8.4 and 8.5):

$$E_1 = \frac{k_3}{(k_1 + k_2)} \quad (8.4)$$

$$E_1 = \frac{k_4}{(k_1 + k_2)} \quad (8.5)$$

In many biochemical catalyses applied in organic synthesis, apart from commercially available enzymes, microorganisms are employed. These can possess either a single enzyme of a determined enantioselectivity or more, but belonging to the same class and showing diverse preferences. For instance, baker's yeast (*Saccharomyces cerevisiae*) containing two oxidoreductases generates chiral products of opposite configurations.

Enantioselectivity is independent of a substrate concentration when microorganisms contain a single enzyme. Otherwise, competing enzymatic reactions take place.

Therefore, the ee_P is the function of conversion. At low substrate concentration, the relative rates of the competing reactions depend on V/K, and the ee_P is determined by Equation 8.6, whereas at high substrate concentration by Equation 8.7:

$$ee = \frac{V_R/K_R - V_S/K_S}{V_R/K_R + V_S/K_S} \quad (8.6)$$

$$ee = \frac{V_R - V_S}{V_R + V_S} \quad (8.7)$$

where
V_R, V_S are maximal reaction rates
K_R, K_S are Michaelis constants

It is often possible to minimize the competing reactions of enzymes contained in the preparation by carrying out a reaction at low substrate concentration, so that the V/K value is not high. Moreover, an enhancement of the reaction enantioselectivity can be reached by the addition of a selective inhibitor or a substrate modification by changing the size of substituents.[10,11,17]

8.4 FACTORS AFFECTING AN ENZYMATIC REACTION ENANTIOSELECTIVITY

An effective enzymatic catalysis should be performed under conditions optimal for a biocatalyst's performance. Hence, it is essential to select an appropriate reaction medium, pH, and temperature.

8.4.1 SOLVENT

8.4.1.1 Stability and Reactivity

In the past, water was considered to be the optimal environment for an effective enzymatic catalysis. Water, being a natural evolutionary and physiological environment of enzymes, maintains the proper conformation of a protein. Interactions between water molecules and polar amino acid residues of an enzyme are involved in the stabilization of its secondary and tertiary structures. Due to a limited water solubility of most organic compounds, reactions were carried out in water with the use of water-miscible organic cosolvents. It soon became evident that an organic cosolvent not only increases the solubility of organic substrates but also improves the stereoselectivity of a biocatalyst. In the 1970s and 1980s, Bryan Jones pioneered in this area of research. He was the first scientist who conducted a selective hydrolysis catalyzed by pig liver esterase (PLE) in water with the addition of methanol. The deacylation reaction of prochiral 3-methyl glutarate diester in 20% aqueous methanol provides the R-monoester with the highest enantiomeric excess.[14,18,19]

Generally, the conversion of an enzymatic reaction medium from water systems to organic solvents is associated with a number of new opportunities, among which the following are most relevant:

- Increasing the solubility of organic substrates
- Ability to carry out reactions impossible in an aqueous environment because of thermodynamic and kinetic limitations
- Increasing the stability of an enzyme
- Ease of product recovery from organic solvents as compared to water
- Ease of enzyme recovery and the elimination of immobilization necessity because of a reduced solubility of a protein in an organic solvent

Klibanov and coworkers were the first researchers to move an enzymatic reaction from an aqueous to a hydrophobic environment. Solid enzyme preparations (immobilized or lyophilized on a support) are suspended in an organic solvent in the presence of a small amount of buffer to ensure catalytic activity. In enzymatic reactions carried out in organic solvents, it is essential to maintain a certain amount of water required to keep proper protein conformation. Water molecules form a monolayer on the enzyme surface, which stabilizes its structure and blocks the interactions between polar residues and solvent molecules. Zaks and Klibanov, in their studies on oxidoreductase activity in various organic solvents, found that the maximal enzymatic activity is attained at about 1000 molecules of water per enzyme molecule.[20–22]

In a nonaqueous environment (considering the minimum necessary amount of water), the charged sides of the surface residues are found to fold back onto the surface of a protein, thus reducing the surface area.[23,24] The consequence of this phenomenon is an increase in enzyme stability in an organic solvent. Lack of protein–solvent interactions causes the impossibility of the protein unfolding. At the same time, conditions favorable in the formation of a larger number of bonds within a biocatalyst molecule (hydrogen bonds and salt bridges) are generated, which results in a decrease in protein size and increase in its stability.[23] The hydrophobicity of organic solvents is the feature that affects the stability of enzymes. The logarithm of the partition coefficient log P is a measure of substance lipophilicity. It is calculated as the ratio of the substance concentrations in two immiscible solvents that are in equilibrium. Generally, it is determined for water and n-octanol and then represented by (Equation 8.8)

$$\log P_{oct/wat} = \log \left(\frac{[\text{solute}]_{octanol}}{[\text{solute}]_{water}} \right) \tag{8.8}$$

The growth in the environment hydrophobicity is associated with an increase in protein rigidity. Spanning water network surrounding the protein is another important element that affects the stability of the biocatalyst.[23] The element is dependent on hydrophobicity. Water molecules, stabilizing the conformation of a protein in a nonaqueous medium, show lower mobility in a hydrophobic environment. In more polar solvents, an interaction between the molecules of these solvents and water molecules occurs. This leads to an increase in the monolayer mobility and formation of amino acid residues–solvent interactions and, thus, disorders in protein conformation. Unfavorable conformational changes of a protein in hydrophilic solvents decrease the catalytic enzyme activity, which is correlated to the mole fraction of water (χ_w) and water activity coefficient (γ_w). It was demonstrated that together with increasing hydrophobicity, the catalytic activation of an enzyme increases due to a decrease in the activity coefficient. Therefore, to attain the same level of enzyme hydration, less water is necessary in hydrophobic solvents than in hydrophilic ones.[25–28]

Typical organic solvents for enzymatic syntheses include mainly alkanes (hexane, cyclohexane, log P > 3), aromatics (toluene, benzene, log P 2.5 and 2, respectively), haloalkanes (chloroform, dichloromethane, log P 2 and 0.85, respectively), and ethers (diisopropyl ether, *t*-butylmethyl ether, diethyl ether, log P 1.9, 1.3, 0.85, respectively).[14] In most cases, an enzymatic catalysis becomes less effective with an increase in solvent polarity. Solvents such as dimethylformamide or dimethyl sulfoxide are able to dissolve and denature the protein. DMSO dissolves the protein by forming a hydrogen bond with amide hydrogen. This, together with the disintegration of water monolayer, favors protein denaturation. The choice of a solvent for an enzymatic synthesis depends mainly on the protein type. For example, *Candida antarctica* lipase B (CAL-B) not only tolerates the polar environment (tetrahydrofuran, acetonitrile, log P 0.49 and 0.33, respectively) but exhibits an exceptional stability under these conditions.[14]

Ionic liquids and supercritical fluids are particularly interesting solvents used in the biotransformation. Ionic liquids are organic salts composed of large asymmetric cations and inorganic or organic anions (Scheme 8.11).[29,30]

Their unique physical properties such as high boiling point, low melting point, low vapor pressure, nonflammability, and thermal and chemical stability have made them an attractive alternative to conventional organic solvents. Ionic liquids are "designed solvents" due to the modification possibility within either cations or anions. This allows the preparation of solvents exhibiting specific physical properties for the purpose of particular reactions, for example, enzymatic reactions.[29,30] Ionic liquids may be an ideal microenvironment for an enzyme, leading to more compact conformation showing high activity and stability. Furthermore, in an enzymatic reaction, an ionic liquid by immobilizing the enzyme by means of hydrogen bonds, van der Waals interactions, and ionic interactions stabilizes the active conformation of a catalytic protein even under unfavorable deactivating circumstances.[31–36] The first successful enzymatic synthesis in an ionic liquid in the presence of CAL-B was reported in 2000.[37] Sheldon and coworkers carried out transesterification, ammonolysis, and oxidation. The results of these reactions were comparable with, or better than, those obtained in conventional organic solvents. For example, the reaction of octanoic acid with ammonia in [Bmim]

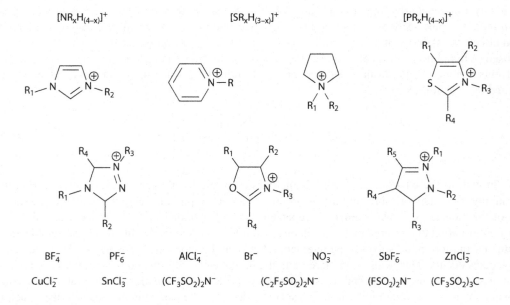

SCHEME 8.11 Structure of ionic liquids.

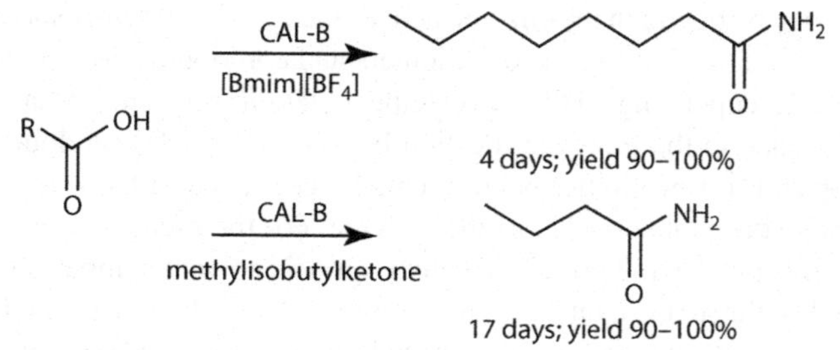

SCHEME 8.12 Enzymatic transformation in [Bmim][BF$_4$].

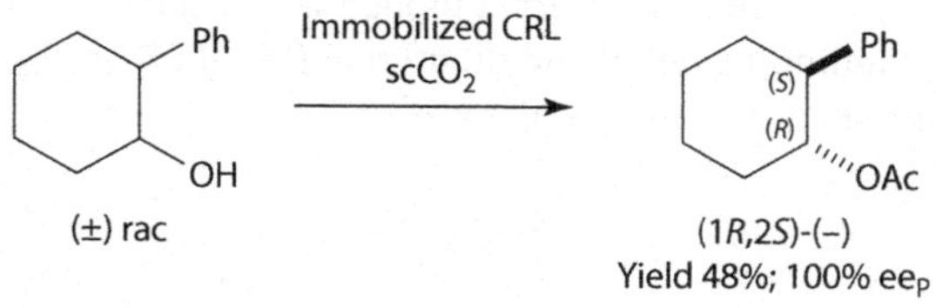

SCHEME 8.13 Kinetic resolution of (±)-*trans*-2-phenyl-1-cyclohexanol in scCO$_2$.

[BF$_4$] at 40°C proceeded quantitatively to the corresponding amide in 4 days, whereas the complete conversion in methylisobutylketone was achieved in 17 days (Scheme 8.12).[37,38]

Supercritical fluids (a transitional state of matter between the gaseous and liquid state) may also act as the reaction medium. Supercritical state is characterized by a viscosity similar to that typical of a gas, density similar to that of a liquid, and high diffusivity. Carbon dioxide (scCO$_2$), hydrophobicity of which is comparable to that of hexane, is the supercritical fluid most frequently used in biotransformation. Supercritical CO$_2$ is environmentally friendly, nonflammable, and essentially nontoxic. Moreover, it is easy to recover and recycle in the process.[39] The kinetic resolution of (±)-*trans*-2-phenyl-1-cyclohexanol in transesterification by vinyl acetate as acyl donor in the presence of *Candida rugosa* (CRL) is an example of an enzymatic reaction in scCO$_2$. The reaction in scCO$_2$ was compared to the reaction in conventional organic solvents. However, the best result was obtained in scCO$_2$; an optically pure ester was formed with 48% yield (Scheme 8.13).[40]

Enzymatic reactions in a biphasic system, that is, ionic liquid/scCO$_2$, were also examined. The addition of an ionic liquid to scCO$_2$ for both butyl butyrate synthesis and the kinetic resolution of 1-phenylethanol processes by transesterification increase the activity and selectivity of CAL-B. Most probably, the ionic liquid protects the lipase against thermal and scCO$_2$ deactivation.[32]

8.4.1.2 Selectivity

Optimization of the reaction conditions in terms of an appropriate solvent selection is effective and most frequently the simplest way to modify the enzyme selectivity. One of the most important criteria for solvent selection is its nature (polarity).[41] The enzyme selectivity is conditioned by its conformational rigidity, which increases in more hydrophobic medium (typical hydrophobic solvents, scCO$_2$). A hydrophobic solvent decreases biocatalyst lability, which does not allow the connection between the structurally mismatched substrate and the active side of an enzyme. Ionic liquids are a separate group of solvents that, despite their high hydrophobicity (log P ≪ 0) and polarity, can constitute an ideal medium for the biotransformation reactions.

In addition, in hydrophobic solvents, a significant change in protein conformation can be achieved by the use of a reversible inhibitor. If an inhibitor is involved in the enzyme active side, it induces the conformational changes that affect enzyme selectivity. A removal of an inhibitor does not result in the amino acid residues returning to their previous position. This would involve exposing these groups to the hydrophobic solvent.[23]

Solvent molecules can be bound in the enzyme active side to form enzyme–solvent complexes. They interfere in the transformation of one of the enantiomers and contribute to different enzyme enantioselectivity.[42–45] For example, depending on the solvent, the *Aspergillus oryzae* protease (protease A) shows a different stereopreference in the transesterification between D or L *N*-acetylphenylalanine 2-chloroethyl ester and 1-propanol. L-Enantiomer is preferentially converted in the active center if the benzyl group is hidden in the hydrophobic pocket of the enzyme. For the D-enantiomer, catalysis is effective when the benzyl group is outside the pocket. The hydrophobic solvent competes with the substrate for a space in the pocket. For this reason, only D-enantiomer is selectively transformed in hydrophobic solvents. Consequently, the L-enantiomer is transesterified in the hydrophilic solvents (Scheme 8.14).[46,47]

The substrate properties and interactions between the enzyme and solvent should also be taken into account when the solvent effect on the selectivity is considered.

Klibanov and coworkers proposed two distinct modes, a nonstereoselective and a stereoselective one, for the 2-(1-naphthoylamino)trimethylene dibutyrate (Figure 8.1) bonding to the active center of the *Pseudomonas* sp. lipase in the hydrolysis reaction.[48]

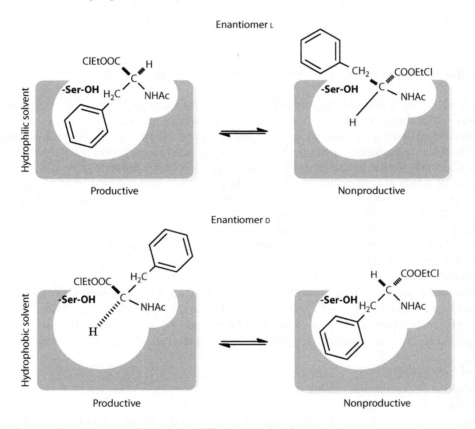

SCHEME 8.14 Enzymatic transformation in different organic solvents.

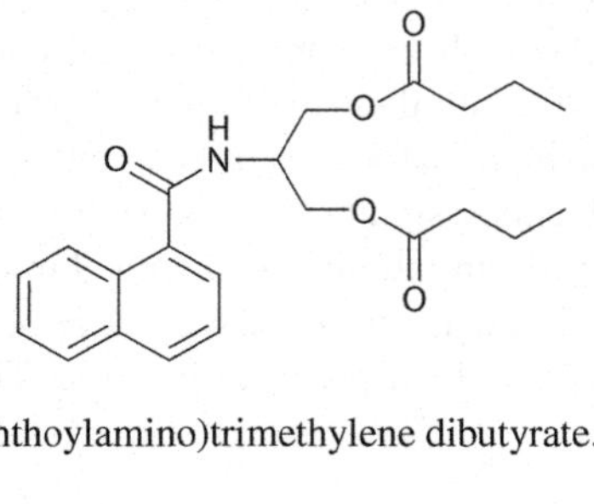

FIGURE 8.1 Structure of 2-(1-naphthoylamino)trimethylene dibutyrate.

Asymmetric Biocatalysis in Organic Synthesis of Natural Products

In the first mode, both monoester enantiomers can be equally obtained. In the stereoselective mode, the naphthyl moiety of a diester is directed to the hydrophobic enzyme pocket. In this case, only a pro-*S* substrate can match the active center allowing the catalysis. The authors noticed that the selectivity of this hydrolysis increased with a decrease in the hydrophobic properties of a solvent (stereoselective mode). The explanation of this relationship can be found in the substrate structure. In the hydrophobic environment, it is more advantageous for the naphthyl moiety to partition into the hydrophobic enzyme pocket, which favors the selective binding of substrate. The hydrophobic solvent favors the nonselective binding mode, where the naphthyl group is exposed to the solvent. The correlation between the reaction selectivity and solvent hydrophilicity was obvious only for the substrate possessing the naphthyl group. When this substituent was replaced with the propyl group, this correlation was not so clear. It indicates that, apart from the environmental influence on the enzyme, also the substrate–environment interactions are a significant factor affecting the enzyme selectivity.[48]

8.4.2 Effect of pH

Enzymatic catalysis depends on the acid–base properties of environment in the case of a reaction carried out in an aqueous solution. Experimental data show that an optimal pH value for most enzymes is in the range of 5.0–9.0 (apart from a few exceptions, e.g., pepsin, the proteolytic enzyme found in a human stomach). A decrease in an enzyme activity in extreme pH values is caused by enzyme denaturation and its conformational changes. The change of the amino acid group charges, required to maintain the active secondary and tertiary structure of the enzyme, leads to the lability change. Enzymatic catalysis in the nonaqueous environment is related to the concept of the enzyme "pH memory." The enzyme, which was precipitated from the aqueous solution at a specific pH value, and placed in an organic medium, exhibits the same ionization state. This feature allows controlling the enzyme activity by the selection of the optimum pH value of an aqueous solution.[14,20]

8.4.3 Effect of Temperature

Generally, the reaction rate increases together with increasing temperature, as a result of a molecule kinetic energy growth. This dependence also refers to an enzymatic catalysis, but only to a certain extent (characteristic temperature for the enzyme). When the enzymatic kinetic energy exceeds a certain energy threshold, hydrogen bonds and hydrophobic interactions stabilizing the protein structure can break. It leads to enzyme denaturation and loss of its catalytic properties.

Enzyme activity increases with temperature. However, an inverse relationship applies to selectivity. Enzymatic catalysis occurs slower at a lower temperature value. At the slow motion of the substrate molecules, the possibility of a substrate binding by the enzyme is determined by the structural cross-matching of the enzyme and substrate. While optimizing the enzymatic catalysis conditions, it is essential to set temperature at such a level that will enable obtaining both the highest yield and selectivity.[49] In order to adjust the optimum temperature, the so-called racemic temperature can be determined. Racemic temperature (T_{rac}) is the temperature value at which an enzymatic reaction proceeds without stereochemical discrimination since there is no difference in free energy of both diastereomeric intermediates. Hence, the racemate is formed:

$$\Delta\Delta G^{\ddagger} = \Delta\Delta H^{\ddagger} - T\Delta\Delta S^{\ddagger} \tag{8.9}$$

If $\Delta\Delta G^{\ddagger} = 0$, then

$$T = T_{rac} = \frac{\Delta\Delta H^{\ddagger}}{\Delta\Delta S^{\ddagger}} \tag{8.10}$$

where T_{rac} is the racemic temperature.

If the temperature is lower than T_{rac}, the stereochemical outcome of the enzymatic reaction is influenced by the activation enthalpy difference. In this case, the optical purity of a product decreases together with an increase in temperature. On the other hand, at a temperature higher than T_{rac}, the reaction is controlled by the activation entropy difference. Therefore, the optical purity of the product increases with an increase in temperature. In the enzymatic synthesis carried out at the temperature $T > T_{rac}$, an opposite enantiomer is obtained in contrast to the reaction at $T < T_{rac}$.[50,51]

8.4.4 ADDITIVES

An enzyme enantioselectivity can be enhanced by the addition of other substances. The additives are placed in the reaction medium together with the organic solvent, the enzyme, and the reagents, co-lyophilized with the enzyme, or complexed with the substrate before their addition to the organic medium.[14] For example, the increase in the enzyme selectivity was observed in the KR of racemic propionic acid derivatives, with the addition of a small amount of an aqueous solution containing the metal ions LiCl and $MgCl_2$. The enantioselectivity of CRL lipase remarkably enhanced up to 200 times. Presumably, the presence of metal ions increased the reaction rate of one of the enantiomers (R-enantiomer).[52]

Enhanced enantioselectivity and catalytic activity were observed for *Pseudomonas cepacia* lipase (PCL), which was co-lyophilized with crown ethers and cyclodextrins before the enzymatic reaction. The cyclodextrin effect is ascribed to the preservation of the enzyme active site.[53] Co-lyophilizing horseradish peroxidase with amino acids affects the selectivity improvement in the sulfoxydation of methyl phenyl sulfide. The greatest effect was obtained with D-proline.[54] Higher selectivity of catalytic proteins is observed also by forming chiral complexes between a substrate and an additive. Racemic salts are prepared by the addition of the Brønsted–Lowry acids or bases to the equimolar mixture of chiral substrates. Due to the steric hindrance, only one of these salts is selectively transformed. Ke and Klibanov reported on the enzymatic transesterification of racemic salts to be more selective than free substrates. The growth of enantioselectivity takes place only in organic solvents; in water, the substrate–additive complexes dissociate.[55]

8.5 HYDROLASES

Hydrolases (3.1.1.3–3.9.1.1.) represent a significant class of enzymes extensively used in organic synthesis. Their availability, the lack of sensitive cofactors that would have to be recycled, and a broad substrate specificity are particularly important. Hydrolysis of ester and amide bonds is a natural function of hydrolases. However, they also found an application in the catalysis of Diels–Alder, Baeyer–Villiger, and Michael addition reactions.[9,15]

The mechanism of enzymatic hydrolysis catalyzed by hydrolases is similar to that observed in the chemical hydrolysis with the use of base. A nucleophilic group from the enzyme active side attacks the carbon atom of the substrate carbonyl group forming an acyl-enzyme intermediate. This nucleophilic group initiating the biotransformation can be either a serine hydroxyl group (e.g., pig liver esterase, subtilisin, and the majority of lipases), an aspartic acid carboxyl group (e.g., pepsin), or a cysteine thiol functionality (e.g., papain).[15,56–59]

Herein, the "bi-bi ping-pong" mechanism of a serine hydrolase–catalyzed ester hydrolysis is discussed. The serine hydrolase at the actual active side contains the catalytic triad consisting of the serine, histidine, and aspartic or glutamic acid residues (Ser-His-Asp/Glu). In the first stage, the serine attacks the carbonyl group and the acyl-enzyme complex is created. Next, one of the three potential nucleophilic molecules, that is, water, alcohol, or ester, attacks this complex. Finally, the product is formed by tetrahedral intermediate deacylation and the enzyme is regenerated. The resulting product depends on the nucleophile used; it can be a carboxylic acid or an ester (Scheme 8.15).[6]

Among all hydrolases, esterases (3.1.X.X.), proteases (3.4.X.X.), lipases (3.1.X.X), and nitrilases (3.5.X.X.) are most frequently used catalysts in enantioselective organic synthesis.

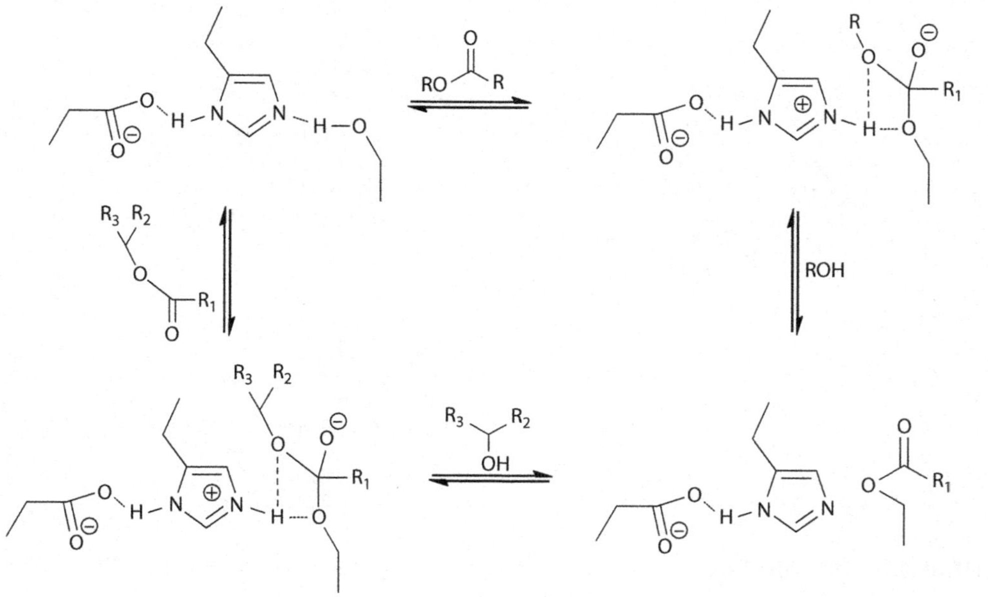

SCHEME 8.15 The "bi-bi ping-pong" mechanism of a serine hydrolase.

8.6 APPLICATION OF ESTERASES AND PROTEASES IN ENZYMATIC DESYMMETRIZATION OF SYMMETRICAL COMPOUNDS

Esterases and proteases are most frequently used as catalysts in the hydrolysis of carboxylic esters. However, the number of esterases that effectively catalyzes these enzymatic reactions is limited. The structural features of the substrates should be similar to those of natural reagents. Among all the esterases, pig liver esterase (PLE; 3.1.1.1) is a versatile enzyme, which exhibits an exceptionally wide substrate tolerance and high stereoselectivity. In the case of proteases, the most commonly used members of this group are α-chymotrypsin (3.4.21.1), subtilisin (3.4.21.62), trypsin (3.4.21.4), pepsin (3.4.23.1), and papain (3.4.22.2).[15]

As depicted in Scheme 8.16, PLE (or α-chymotrypsin) selectively transformed the prochiral α,α-disubstituted malonic diesters to the corresponding chiral monoesters. These transformations may serve as an example of an "alternative fit" of substrates with different steric requirements. When the substrate possessed a position substituent (R) of a small size (ethyl, n-propyl, n-butyl, phenyl) at C-2, PLE hydrolyzed the pro-S ester group. On the other hand, an increase in the steric bulkiness of R (n-hexyl, n-heptyl, p-substituted benzyl) forced the substrate to enter the enzyme active side in the opposite orientation. Thus, the pro-R ester group was preferentially cleaved and the monoesters of (R)-configuration were formed (Scheme 8.16).[60–63]

The PLE-catalyzed enzymatic desymmetrization of an α-alkyl α-aminomalonate derivative led to the corresponding monoacetate, which is an intermediate in the synthesis of α-substituted serine amino acids and glycopeptides. Enzymatic hydrolysis of one of the prochiral groups (pro-S) gave the product of (S)-configuration in 80% ee (Scheme 8.17).[64]

Glutaric acid derivatives are examples of prochiral compounds having identical substituents in β-positions in relation to the prochiral carbon atom. One of the monoethylglutarates, obtained in the α-chymotrypsin-catalyzed reaction, was used as an intermediate for the synthesis of atorvastatin, a drug reducing the level of lipids, mainly cholesterol. This enzyme provided a high selectivity of the reaction resulting in the product with the (S)-configuration.[65] Paroxetine, an antidepressant drug, was also synthesized in a chemoenzymatic reaction, in which chirality is introduced by means of the enzymatic modification of dimethyl 3-(4′-fluorophenyl)glutarate catalyzed by PLE (Scheme 8.18).[66]

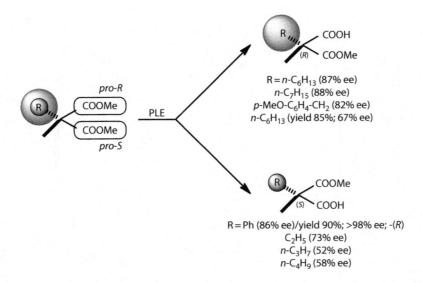

SCHEME 8.16 Selective hydrolysis of α,α-disubstituted malonic diesters in the presence of PLE.

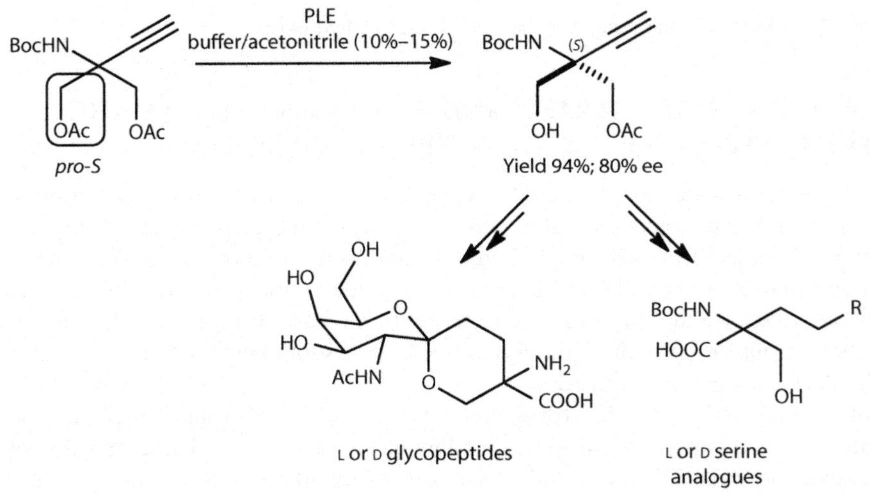

SCHEME 8.17 Desymmetrization of α-alkyl α-aminomalonate derivative in the presence of PLE.

Catalytic properties of esterase and protease also found an application in the hydrolysis reaction of *meso* compounds, such as succinic and glutaric acid esters,[67,68] cyclic,[69–73] and bicyclic[74,75] diesters. In the case of cyclic *meso* diesters, in dependence on the ring size, PLE hydrolyzed either the *R*-center or *S*-center ester group. When the rings were small (n = 1, 2), (*S*)-ester was selectively modified, whereas the (*R*)-counterpart preferentially reacted when the rings were larger (n = 4).[69,70] Analogous to the enzymatic desymmetrization of prochiral compounds, the reversal of esterase stereospecificity was the result of different substrate hydrophobic group orientation in the enzyme active side (Scheme 8.19).

One of the intermediates in the synthesis of oseltamivir phosphate (Tamiflu), a prodrug used in the prevention and treatment of influenza infections, was obtained via an enzymatic desymmetrization of a *meso* 1,3-cyclohexanedicarboxylic acid diester. Enantioselective hydrolysis with pig liver esterase enabled obtaining the enantiomerically pure (*S*)-monoacid (Scheme 8.20).[74]

Cyclopentene *meso* diesters were hydrolyzed in a similar manner. The obtained monoesters constitute the most important chiral synthons for prostaglandins and their derivatives. Two esterases,

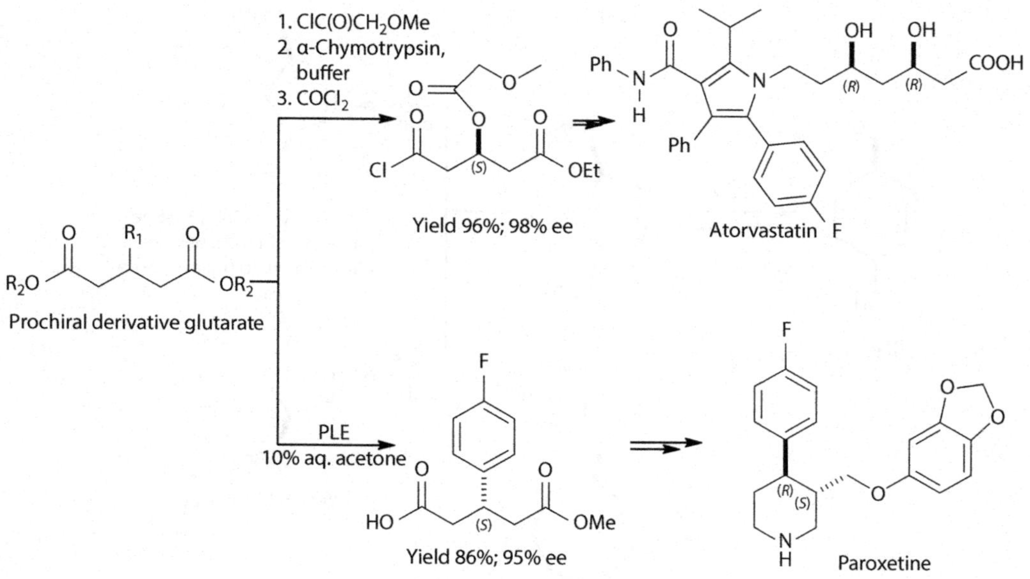

SCHEME 8.18 Enzymatic transformation of glutaric acid derivatives.

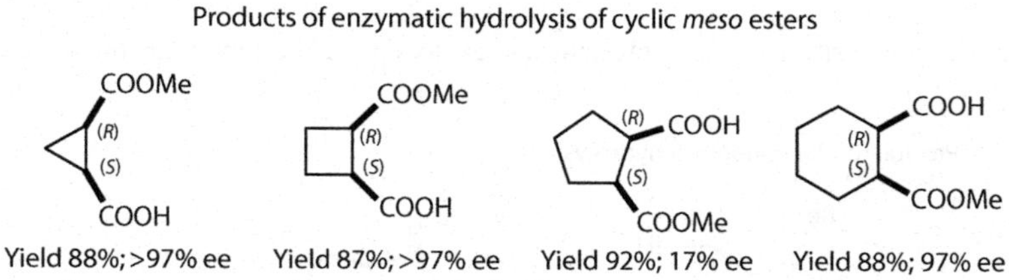

SCHEME 8.19 Enzymatic desymmetrization of cyclic *meso* diesters.

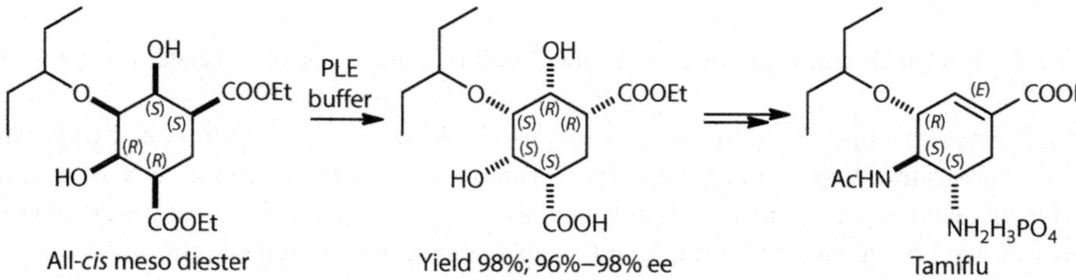

SCHEME 8.20 Enzymatic desymmetrization of a *meso* 1,3-cyclohexanedicarboxylic acid diester catalyzed by PLE.

PLE and ACE (acetylcholinesterase), were used as the catalysts. ACE hydrolyzed the cyclopentene diesters of stereoselectivity and the opposite stereopreference higher than that of PLE. The desymmetrization of cyclohexene and cycloheptene diesters in the presence of ACE was also examined. Only the seven-membered compound gave the optically pure product, while the six-membered analogue led to the racemate (Scheme 8.21).[15,71,72]

Meso 1,3-imidazolin-2-one derivative diacetates proved to be excellent substrates for PLE. The product of enzymatic hydrolysis served as a starting material for the synthesis of the vitamin (+)-biotin (Scheme 8.22).[73]

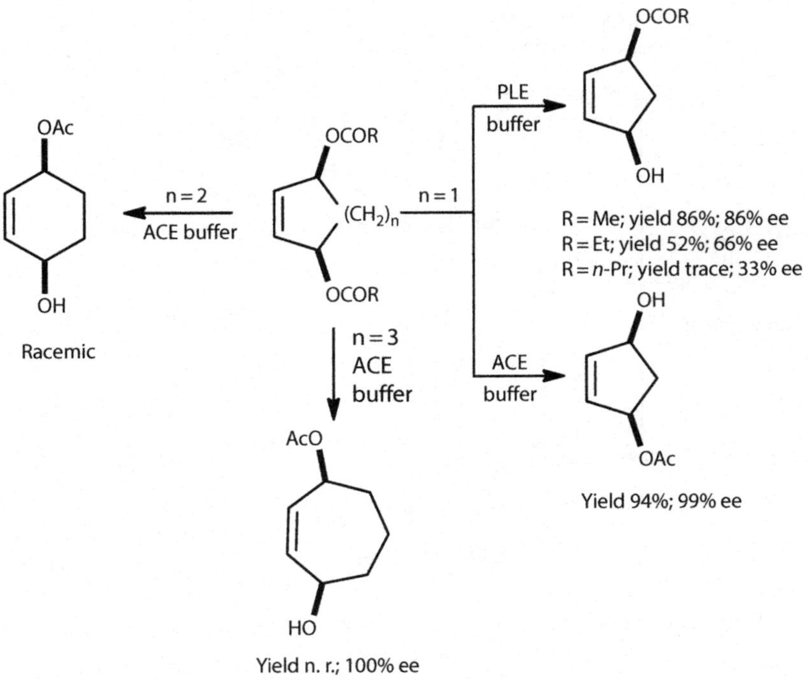

SCHEME 8.21 Desymmetrization of cyclohexene and cycloheptene diesters in the presence of ACE.

Product of PLE-catalyzed hydrolysis

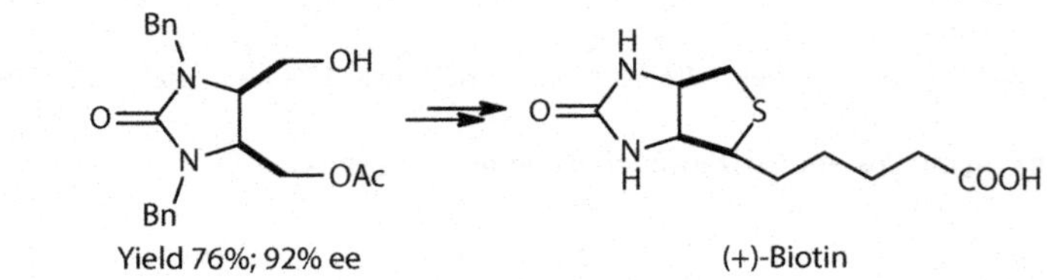

SCHEME 8.22 Enzymatic hydrolysis of *meso* 1,3-imidazolin-2-one derivative diacetates catalyzed by PLE.

When it comes to bulky bicyclic *meso* diesters, PLE selectively hydrolyzed only the *exo*-configurated substrates. *Endo*-isomer was transformed with lower enantioselectivity at a significantly reduced reaction rate. The β-ketodiester was subjected to the PLE-catalyzed asymmetric hydrolysis followed by decarboxylation, resulting in a monoester in high enantiomeric excess. This monoester was an intermediate in the tropane-type alkaloid ferruginine synthesis (Scheme 8.23).[76]

The hydrolysis of enantiotopic ester groups of *meso* 1,4-diacetoxy-2,5-dimethylcyclohexane is one of a few examples of centrosymmetric molecule desymmetrization. The corresponding monoester was isolated with an excellent optical and chemical yield using PLE as the catalyst (Scheme 8.24).[77]

8.7 APPLICATION OF LIPASES IN ENZYMATIC DESYMMETRIZATION OF SYMMETRICAL COMPOUNDS

Lipases (E.C. 3.1.1.3), according to the International Union of Biochemistry and Molecular Biology, are hydrolases of glycerol esters. They catalyze the acyl group transfer process from the corresponding donor to the acceptor, allowing the performance of esterification, transesterification, amides and peptides synthesis, as well as esters, amides, and peptides hydrolysis. Lipases are widespread in

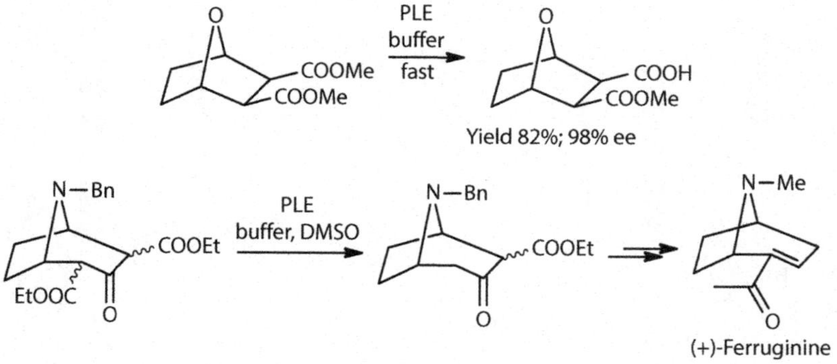

SCHEME 8.23 Enzymatic transformation of bicyclic *meso* diesters in the presence of PLE.

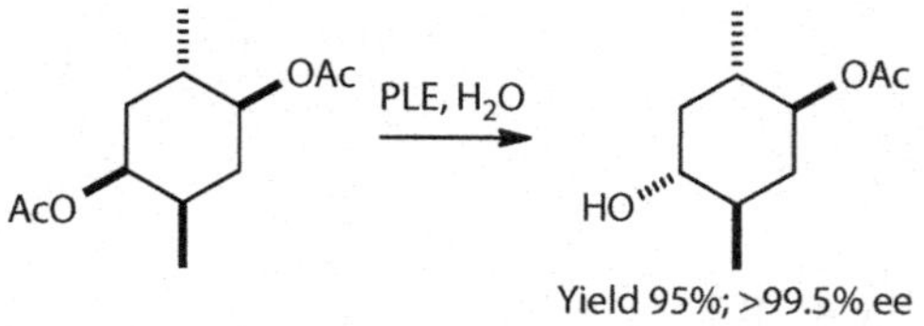

SCHEME 8.24 Desymmetrization of *meso* 1,4-diacetoxy-2,5-dimethylcyclohexane using PLE as the catalyst.

nature; they occur in seeds, many vegetative plant organs, in certain microorganisms, and in human and animal organs (pancreas, liver, intestines, stomach wall). They have found an application in cosmetics, detergents, pharmaceutical, oleochemical, and dairy industry. Lipases are enzymes of highest importance in stereoselective organic synthesis, mainly due to their exceptionally broad substrate tolerance, stability, activity in unphysiological systems, and relatively low price.

In contrast to esterases, the lipase active site is hidden under the hydrophobic helical lid. This explains the need for the biocatalyst reorientation at the interface (lipase interface activation).[77–80] However, there are lipases, such as a lipase from *Fusarium solani*, a lipase from *Pseudomonas aeruginosa*, and CAL-B, which do not display interfacial activation.[80–82] A specific structure of the lipase active site allows carrying out the catalysis in both aqueous environment and organic solvents, which extends the range of their applicability for organic synthesis.

The selectivity of enzymatic catalysis depends on the substrate orientation in the enzyme active site. The active site, for example, in CAL-B, can be considered as being built of two pockets (labeled as the acyl and alcohol pockets), separated by several hydrophobic amino acid residues. In efficient catalysis, the acyl part of the tetrahedral intermediate must be introduced into pocket II (acyl pocket), and the alcohol must be placed in pocket I (alcohol pocket). An insertion of the alcohol part of the tetrahedral intermediate in the enzyme active site is not random either. There are two modes of the chiral alcohol moiety orientation. In the first mode, the large substituent (the acyl residue of alcohol moiety of the tetrahedral intermediate) is placed outside the catalytic center. The medium substituent (the chiral part of alcohol moiety of the tetrahedral intermediate) is located in the restricted space in the active site called the "stereospecificity pocket." This orientation is favorable for the steric reasons. In the second mode, the large substituent is in the "stereospecificity pocket"; the unfavorable steric interactions take place (Scheme 8.25).[83–88]

The hydrolytic enzymes, including lipases, catalyze the reactions, mainly in the aqueous environment, under physiological conditions. Water molecules are the substrate in the hydrolysis reaction. However, proteins are capable of carrying out the reverse reactions. The elimination of water from the reaction system shifts the equilibrium toward esterification. In the lipase-catalyzed enzymatic hydrolysis, in order to shift the equilibrium toward the product formation, an excess of a solvent

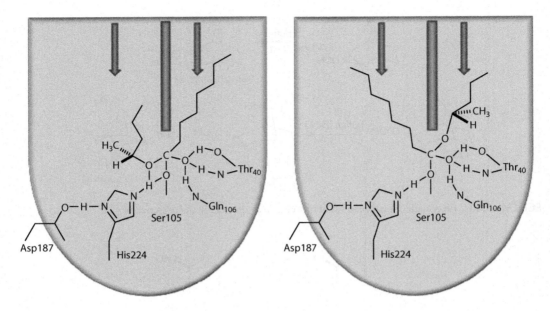

SCHEME 8.25 Active side of CAL-B.

is used. Simultaneously, the solvent is a nucleophilic agent. The active esters are used in the acylation reactions. The most popular acyl donors are enol esters, such as vinyl acetate (VA), isopropenyl acetate (IPA), and 1-etoxyvinyl esters. The leaving group is an enol that immediately tautomerizes to the stable carbonyl compound (acetaldehyde, acetone, ethyl acetate). Acetaldehyde, formed as a by-product in the transesterification with vinyl acetate, may have unfavorable effects on the enzyme catalytic activity. A reaction between this aldehyde and an amino group of a lysine side chain produces an imine that causes a gradual catalyst deactivation. However, acetone and ethyl acetate do not have such properties. Lipases exhibit a varied tolerance with respect to the acyl moiety of the acyl donor. CAL-B proved to be one of the few lipases reacting with most acylating agents. Apart from esters, acyclic, cyclic, and carboxyl-carbonate anhydrides were used in the lipase-catalyzed reactions.[14,89]

Lipases were successfully used for the desymmetrization of different *meso* and prochiral diesters and alcohols. The acylation of free alcohols by means of a transesterification reaction or hydrolysis of appropriate acyl derivatives of alcohols leads to the chiral hemiesters through the differentiation of the same, in chemical terms, groups. Most lipases preferentially convert the same enantiomers or prochiral groups in the types of reaction discussed. This allows the preparation of both enantiomers of the product in high chemical and optical yield.[9,14] There is an example of a reaction in which the enzyme shows a different enantiopreference; it is the desymmetrization of 2-ethylpropane-1,3-diol and its di-*O*-acetate. *Pseudomonas fluorescens* lipase (PFL) selectively acylates the pro-*S* hydroxyl group in the transesterification reaction. During deacylation, it hydrolyzes the pro-*R* ester group. Both processes lead to the monoacetate of (*R*)-configuration (Scheme 8.26).[90]

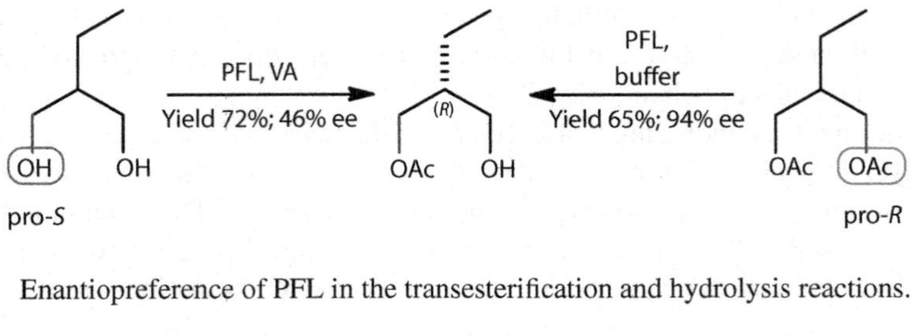

SCHEME 8.26 Enantiopreference of PFL in the transesterification and hydrolysis reactions.

Optically pure propane-1,3-diol derivatives are an interesting family of substrates since, in most cases, they are precursors in the synthesis of biologically active compounds. For instance, derivatives **1**, **2a**, and **2b** prepared by enzymatic catalysis were regarded as chiral building blocks in antitumor antibiotic syntheses. Optical purity was provided by desymmetrization in the presence of *Pseudomonas* sp. lipase (PSL). This lipase selectively acylates the pro-*R* group forming the (*S*)-monoesters in high enantiomeric excess (Scheme 8.27a, compounds **1**, **2a**, **2b**).[91,92] 2-Substituted-propane-1,3-diol is the key intermediate in the synthesis of SCH51048, an antifungal agent. The desymmetrization of the corresponding diol and diester, catalyzed by Novo SP435 and Amano lipase CE, respectively, furnished the (*S*)-monoester, a precursor in a new efficient synthesis of SCH51048. Novo SP435 shows a significant pro-*S* selectivity for the acylation of the hydroxyl group, when Amano CE pro-*R* exhibits selectivity in the deacylation of ester (Scheme 8.27a, comp. **3**).[93]

The desymmetrization of other propane-1,3-diol derivatives catalyzed by *Pseudomonas cepacia* lipase (PSA), Amano PS lipase, and porcine pancreatic lipase (PPL) leads to the formation of (*R*)-hemiesters (Scheme 8.27a and b, comp. **4–6**) through the selective acylation of one of the prochiral hydroxyl groups.[94–96] Compound **4** is an intermediate in the synthesis of the major component isolated from the hair pencils of male *Danaus chrysippus* (African monarch), compound **5** is a precursor of new β-amino acids, while chiral allendiols derivatives **6** are employed as versatile synthetic building blocks.

Optically pure N-protected serinol derivatives (Scheme 8.27b, comp. **7a**, **7b**, **8**), prepared by enzymatic desymmetrization with PPL and *Pseudomonas* sp., were used for the synthesis of a chiral Evans auxiliary, as building blocks for various amino acid derivatives (e.g., 2-hydroxymethylaziridines and N-Boc-N,O-isopropylidene-α-methylserinal), and as an intermediate in the synthesis of a novel immunosuppressant 4-methyl-4-[(2-(thiophen-2-yl)ethyl]oxazolidin-2-one. In all these cases, the monoesters of (*R*)-configuration were formed at the desymmetrization stage.[97–99]

The structure of a substrate has the major impact on the stereochemical preference of an enzymatic catalyst. To exemplify, Amano PS exhibits different enantiopreference for the prochiral hydroxyl groups of 2-carbamoylomethyl-1,3-propanediols (Scheme 8.27b, comp. **9a** and **9b**). For the *N*-monoalkylcarbamoyl group **9a**, the possibility of hydrogen binding results in the stabilization of the transition product in the active site, and the lipase selectively acylates the pro-*S* group. The presence of an additional alkyl substituent in the amide **9b** forces a different orientation of the substrate. Therefore, in this case, (*R*)-enantiomer was formed by the transesterification of a pro-*R* group.[100]

The chiral fluorinated analogues of 1-aminocyclopropane-1-carboxylic acid (ACC) were also prepared in chemoenzymatic synthesis. ACC is a biologically active compound, used as an intermediate in the biosynthesis of the ripening hormone ethylene, a component of bacterial phytotoxins, and an intermediate in the biosynthesis of azetidine-2-carboxylic acid. The desymmetrization of prochiral diol and its diacetate in the presence of Amano PS is the stage at which two enantiomerically pure monoesters are obtained (Scheme 8.27b, comp. **10**).[101]

Apart from the prochiral diols, lipases are able to biotransform prochiral diamines. For example, 2-substituted propan-1,3-diamines were desymmetrized by using from *Pseudomonas cepacia* (PCL) lipase as a catalyst. Enzymatic alkoxycarbonylation allows the preparation of optically active nitrogenated derivatives of (*R*)-configuration. The best substrates for this reaction were 2-phenylpropan-1,3-diamin and its *ortho*- and *para*-substituted analogues (Scheme 8.27b, comp. **11**).[102]

2-Substituted glycerine derivatives are the next set of compounds biotransformed by means of lipases. In enzymatic transesterification, 2-*O*-acylglycerols are quite a troublesome group of substrates because during the reaction the acyl group tends to migrate contributing to the decrease in enantioselectivity. The introduction of bulky substituents into the C-2 position reduces the lability of substrates. The application of 2-benzoyloxy-1,3-propanediol for the enzymatic desymmetrization with PPL leads to the optically pure monoester **1** (Scheme 8.28, comp. **1**).[103] Similarly, 2,2-disubstituted propane-1,3-diols, used as a starting material in enzymatic synthesis, give the products having a stereogenic quaternary carbon center with high

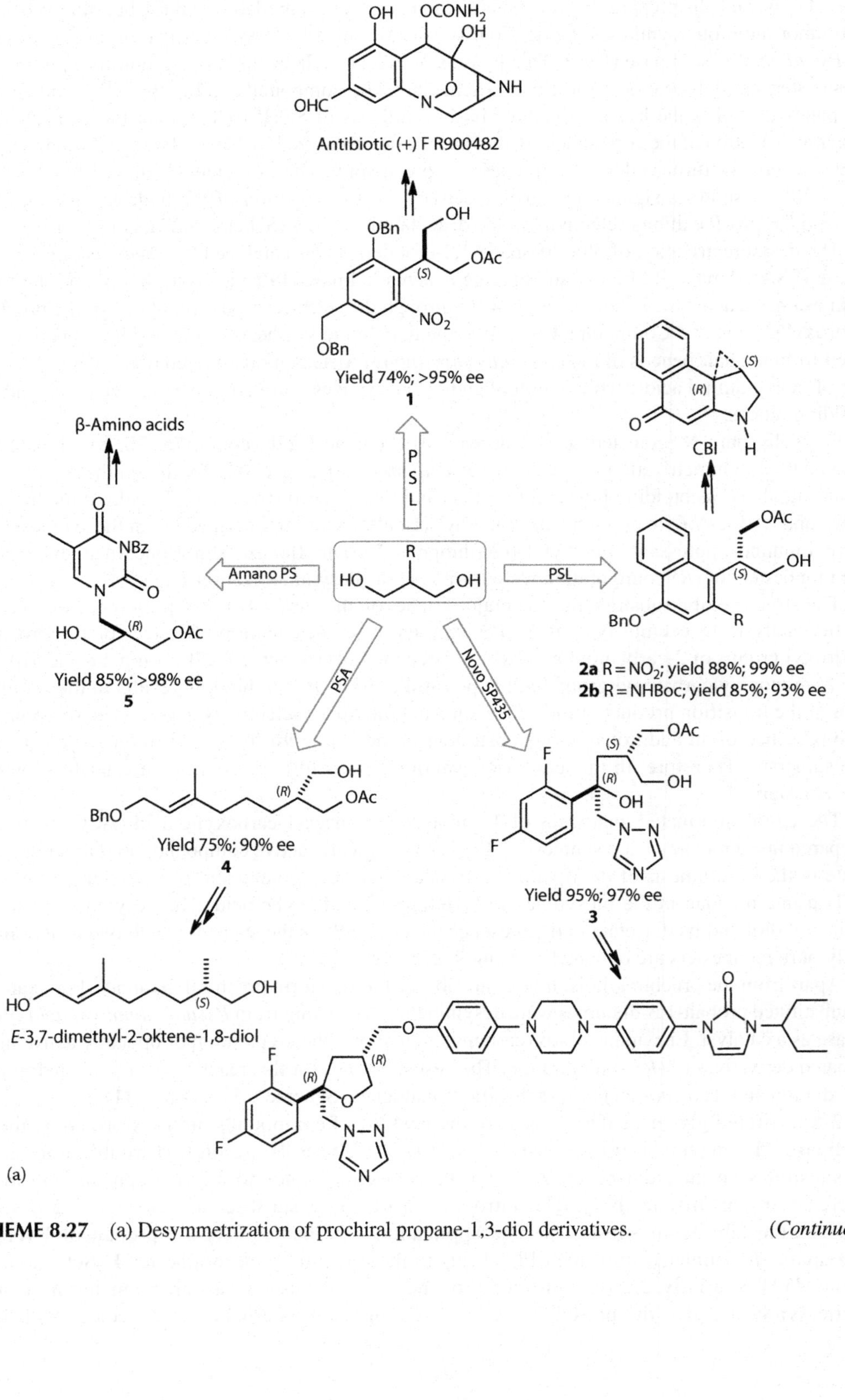

SCHEME 8.27 (a) Desymmetrization of prochiral propane-1,3-diol derivatives. *(Continued)*

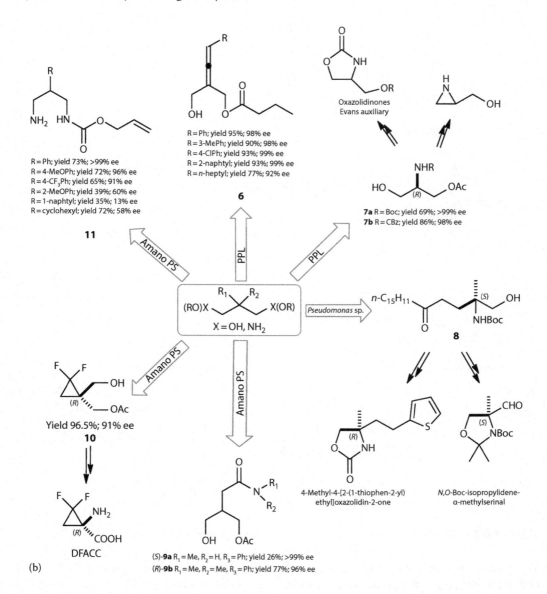

SCHEME 8.27 (Continued) (b) Desymmetrization of prochiral propane-1,3-diol derivatives.

chemical and optical yield. For instance, the stereoselective acylation of the chiral chromane derivative in the presence of CAL-B provided the (S)-monoester 2 with high enantiomeric excess. In the multistep synthesis, both enantiomers of α-tocotrienol (components of vitamin E) were prepared following two different paths (Scheme 8.28, comp. 2).[104]

Leustroducsin B (isolated from *Streptomyces platensis* SANK 60191) is another example of a natural compound obtained through a chemoenzymatic reaction from a 2,2-disubstituted glycerine derivative. The optical purity of the final product depends on the enzymatic desymmetrization stage in which, in the presence of AK lipase, chiral (S)-monoester 3 was isolated in high ee (Scheme 8.28, comp. 3).[105]

Starting from 1,3-dichloroacetone in an eight-step synthesis, (S)-phosphonotrixin was obtained with 93% ee and 11% overall yield. This compound is a natural product isolated from the microorganism *Saccharothrix* sp. ST-888, used as a herbicidal antibiotic (Scheme 8.28, comp. 4).

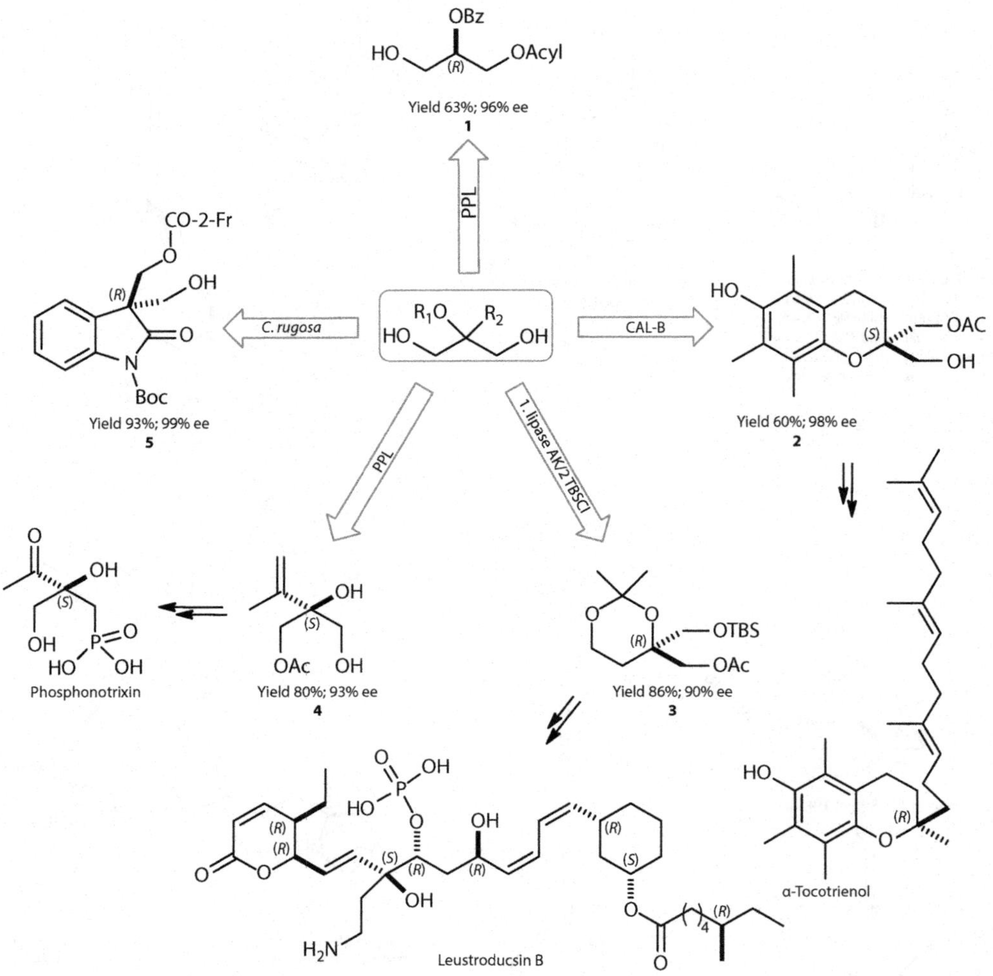

SCHEME 8.28 Desymmetrization of prochiral 2-substituted glycerine derivatives.

The desymmetrization of oxindoles having a stereogenic quaternary carbon center at the C-3 position allows the preparation of natural compounds used as chiral building blocks in indole synthesis. For example, enzymatic acylation of the corresponding diol catalyzed by CRL provided an optically pure N-Boc-protected derivative (Scheme 8.28, comp. **5**).[106]

Prochiral pentane-1,5-diol derivatives are useful precursors of important biologically active compounds and chiral auxiliaries. For example, (*R*)-3-*t*-butyldimethylsilyloxy-5-acetoxy-1-phenyl (Scheme 8.29), used as a chiral building block for the synthesis of a cyanobacterial heterocyst glycolipid, was prepared by the enzymatic hydrolysis of diacetyl ester of 1,3,5-pentanetriol in the presence of *Pseudomonas fluorescens* lipase (PFL).[107]

Apart from alcohol esters, lipases can also desymmetrize prochiral carboxylic acid derivatives. The enzymatic synthesis of 3-substituted glutaric acid, which is an excellent building block in the synthesis of many important biologically active organic compound derivatives, is an example of this reaction. The enzymatic desymmetrization of the prochiral diethyl 3-(3′,4′-dichlorophenyl)glutarate, used as an intermediate in the synthesis of a series of neurokinin receptor antagonists, in the presence of most hydrolases resulted in a product with the (*R*)-configuration. CAL-B is one of a few lipases demonstrating enantiopreference for pro-*S* ester group and its deacylation leads to the (*S*)-monoester (Scheme 8.30, comp. **1**).[108]

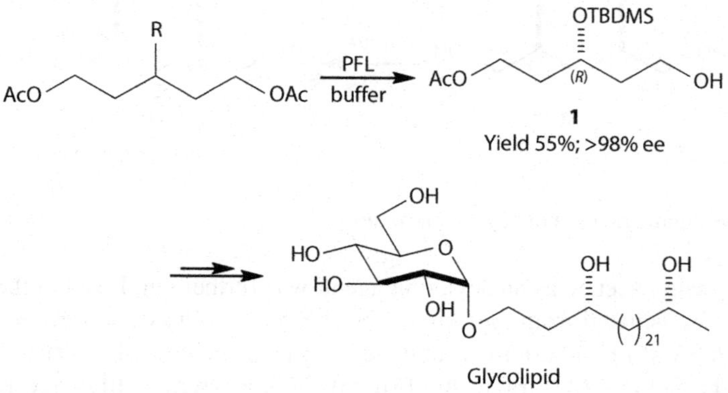

SCHEME 8.29 Desymmetrization of prochiral pentane-1,5-diol derivatives.

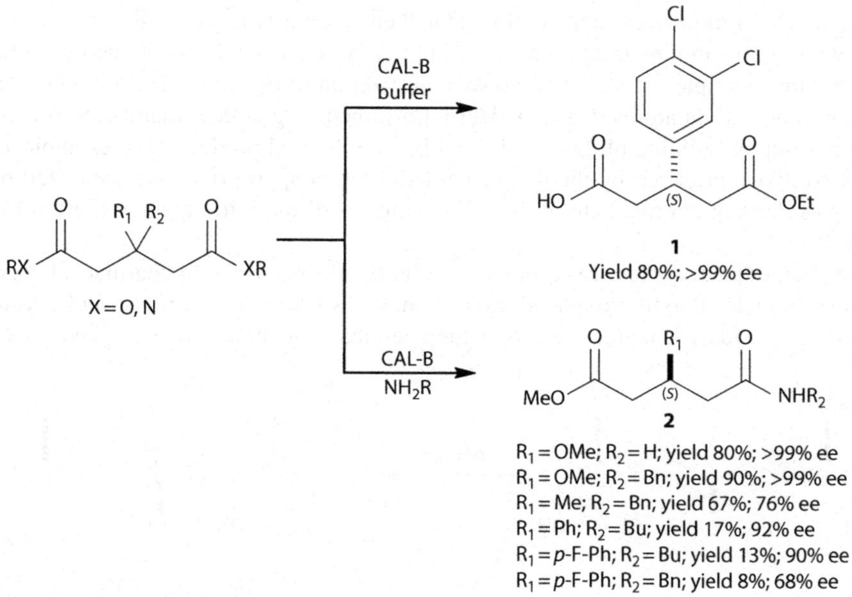

SCHEME 8.30 Desymmetrization of prochiral 3-substituted glutarates.

Enzymatic ammonolysis and aminolysis reactions of 3-substituted glutarates have been widely used for the preparation of monoamides—chiral synthons used for the synthesis of biologically active amino acids and β-lactams. CAL-B lipase has presented its ability to catalyze the amide bond formation as a result of the prochiral ester group differentiation. The highest chemical and optical yields were obtained for the derivatives with a heteroatom at C-3 position, while aliphatic or aromatic substituents extended the reaction time and lowered the yields and ee's. Probably, the presence of heteroatom in the C-3 position allows for hydrogen bonding stabilization of the substrate in the enzyme active site. Although the degree of enantioselection changes for different substrates, noteworthy is the fact that monoamides are always obtained of the same absolute (*S*)-configuration (Scheme 8.30, comp. **2**).[109]

Meso compounds (bearing a plane of symmetry) are another group of compounds used in EEDs. Similar to prochiral compounds, selective acylation or hydrolysis of *meso* substrates leads to optically active products. One of the acyclic *meso* diols, after enzymatic modification catalyzed by CRL

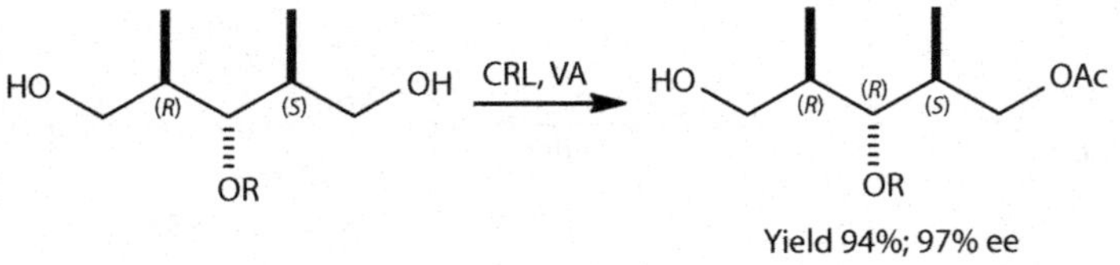

SCHEME 8.31 Desymmetrization of acyclic *meso* diols.

and application of vinyl acetate as an acylating agent, was further employed in the synthesis of the secondary metabolite isolated from the skin of the aspidean mollusk *Dolabrifera dolabrifera*. The corresponding (2R,3R,5S)-monoester was obtained in an excellent yield and ee (Scheme 8.31).[110]

Additionally, the C(19)–C(27) fragment of rifamycin S, a natural antibiotic originating from the microorganism *Amycolatopsis mediterranei*, was successfully synthesized by the stereoselective acylation of the *meso* polyol by vinyl acetate (used as a solvent and acyl donor) in the presence of PPL. This enzyme was highly regioselective for one of the primary hydroxyl end groups giving (2R,3R,4S,6R,7S,8S)-monoacetate in good yield and enantiopurity (Scheme 8.32).[111]

Cyclic alcohols are another important class of optically active alcohols obtained from EEDs. The following are the examples of enzymatic desymmetrization of cyclic *meso* compounds containing from three to nine carbon atoms in a ring. *Meso* diols, including a three-membered ring, constitute a significant group of building blocks for the synthesis of natural products. For example, one of the aziridine derivatives, prepared by the monoacetylation of an appropriate diol catalyzed by Amano PS lipase, was the key intermediate in the total synthesis of the mitomycin antibiotic FR-900482 (Scheme 8.33).[112]

Five-membered heterocycles are common structural units of synthetic pharmaceutical and agrochemical compounds. Polyhydroxylated pyrrolidines, also known as imino- and azasugars, are inhibitors of glycosidase enzymes showing therapeutical potential in the treatment of diabetes,

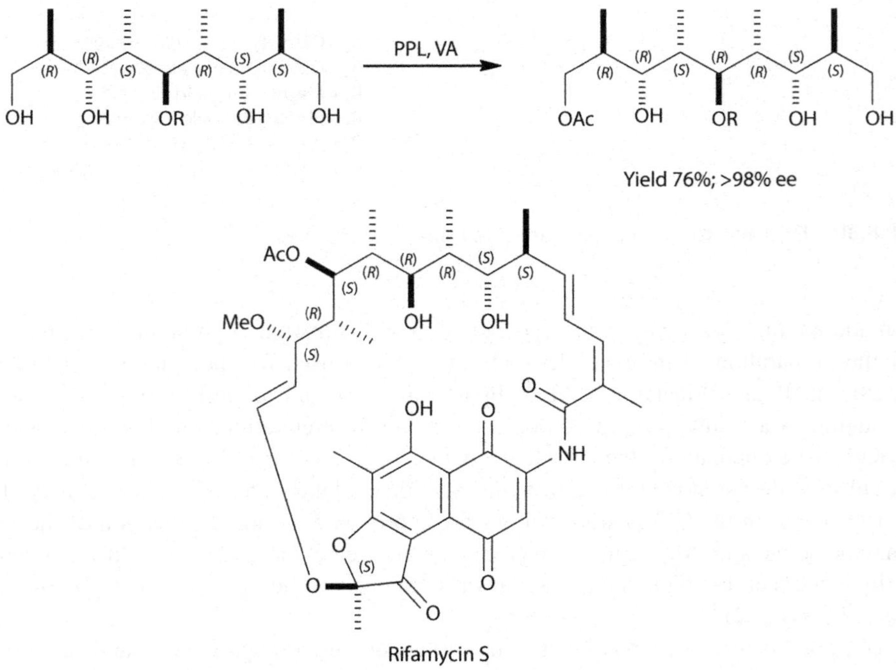

SCHEME 8.32 Desymmetrization of the *meso* polyol.

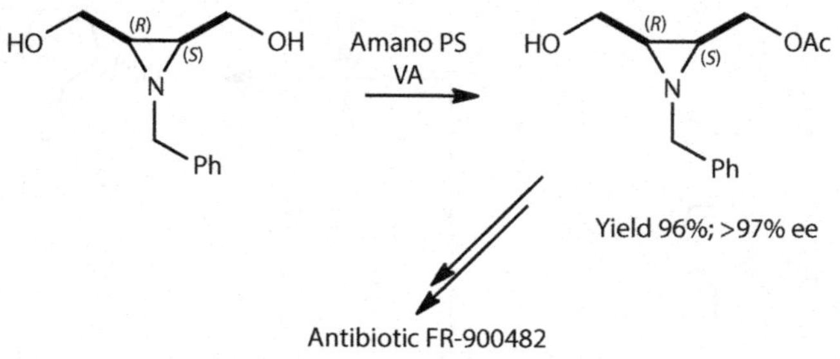

SCHEME 8.33 Desymmetrization of cyclic *meso* diols.

AIDS, and cancer. Moreover, the chiral pyrrolidines are used as organocatalysts and chiral auxiliaries. Various *N*-Boc-2,5-*cis*-substituted pyrrolidines and pyrrolines were prepared by enzymatic desymmetrization. Independent of a catalyst used, the hydrolysis of diesters, like the transesterification of diols, produced the monoesters in high ee (Scheme 8.34).[113]

The desymmetrization of *meso* 5-*t*-butyldimethylsilyloxy-2-cyclopentene-1,4-diol and its diacetate catalyzed by Amano AK lipase leads to optically pure monoesters. Both enantiomers are applied in the synthesis of macromolecular chromoprotein antitumor antibiotics such as neocarzinostatin chromophore (Scheme 8.35).[114]

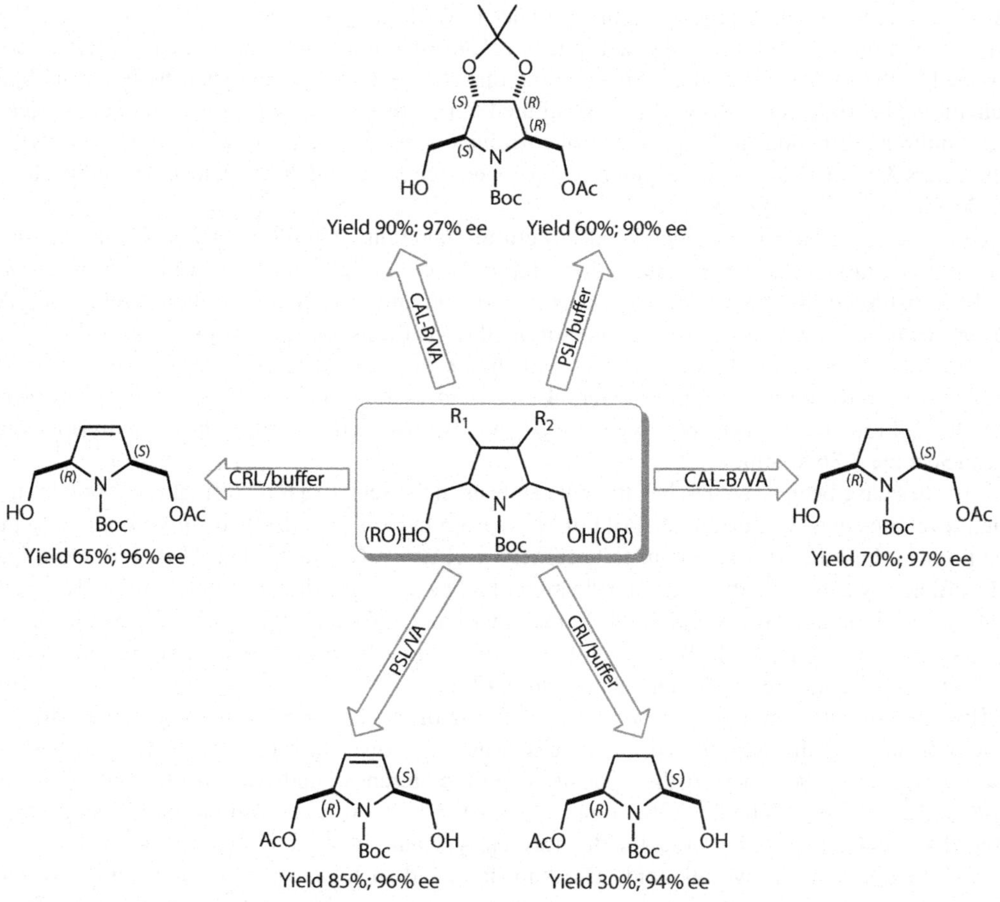

SCHEME 8.34 Desymmetrization of *meso* five-membered heterocycles.

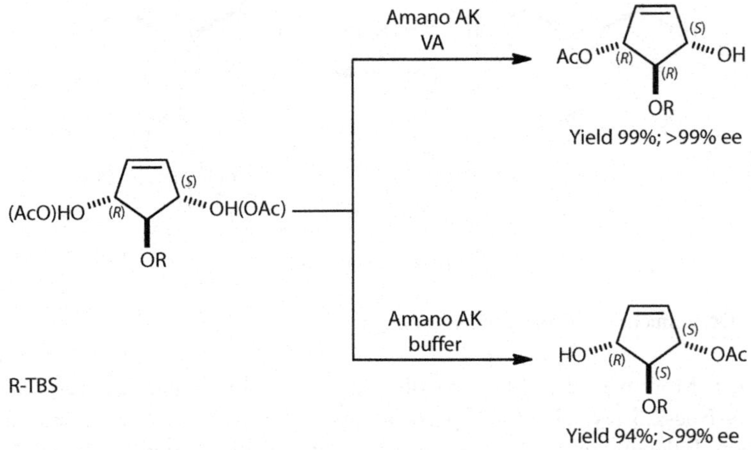

SCHEME 8.35 Desymmetrization of *meso* 5-*t*-butyldimethylsilyloxy-2-cyclopentene-1,4-diol.

Meso tetrahydropyrans and *meso* piperidines, representatives of six-membered ring systems, were also successfully desymmetrized. In many bioactive natural products, such as phorboxazoles, bryostatins, kendomycin, and ratjadone A, a tetrahydropyran moiety is found. The stereoselective acylation of *meso* tetrahydropyrans in the presence of CAL gave the (2R,3S,6S)-monoesters of high enantiomeric purity. The opposite enantiomers, (2S,4R,6R)-monoesters with equally high ee, were provided in the hydrolysis of the corresponding diacetate derivatives catalyzed by the same lipase (Scheme 8.36, comp. **1**).[115] *Meso* 3,4,5-substituted tetrahydropyranyl 2,6-diols were also subjected to the biotransformation in the presence of *Rhizomucor miehei* lipase (RML). Tetrahydropyranyl monoesters **2**, with up to five stereogenic centers, were prepared with high yield and enantioselectivity (Scheme 8.36, comp. **2**).[116]

The piperidine ring is a widespread structural fragment of biologically active compounds. In this sense, monoester **3** was obtained in high ee through the stereoselective acylation of *meso cis*-2,6-substituted piperidines by vinyl acetate (solvent and acyl donor) in the presence of CAL. (2S,6R)-Hemiester was used as a precursor in the synthesis of (5S,9S)-(+)-indolizidine 209D (Scheme 8.36, comp. **3**).[117] The hydrolysis of the appropriate diesters catalyzed by *Aspergillus niger* lipase produced (2R,6S)-monoesters **4** that were used as starting materials in the synthesis of both enantiomers of *cis*-6-(hydroxymethyl)- and *cis,cis*-4-hydroxy-6-(hydroxymethyl)pipecolic acids (Scheme 8.36, comp. **4**).[118]

Enzymes have been proven to be efficient catalysts in the selective transformation of polyhydroxylated cyclohexanes. For example, EED of *meso* 1,3,5-cyclohexane triols results in optically pure enantiomers, building blocks useful to access 19-*nor*, *des*-C,D vitamin D_3. *Alcaligenes* sp. lipase (QL) efficiently catalyzed the transesterification of *meso* diols, yielding the enantiopure (1R,3S,5S)-monoesters (Scheme 8.36, comp. **5**).[119] On the other hand, OF lipase (commercially available from CRL) catalyzed well the hydrolysis of only the *trans* isomer from a *cis/trans* mixture of triacetates providing (1R,3R)-diester **6** (Scheme 8.36, comp. **6**).[120]

The desymmetrization of *meso* all-*cis*-3,5-dihydroxy-1-(methylcarbonyl)cyclohexane and 4-methyl- and 4-ethyl-substituted analogues was also investigated. The PPL and SAM II (*Pseudomonas*) lipases showed the opposite enantiopreferences and two enantiomers, (−)-**7** and (+)-**8**, were obtained (Scheme 8.36, comp. **7** and **8**). The hydrolyses of dibutyrates using PSL or SAM II as catalysts gave the same: the dextrorotatory enantiomers of monoacetates.[121]

(−)-Lobeline, a bioactive alkaloid, also known as Indian tobacco, is another compound containing a six-membered ring. It was synthesized *via* the stereoselective desymmetrization of *meso*

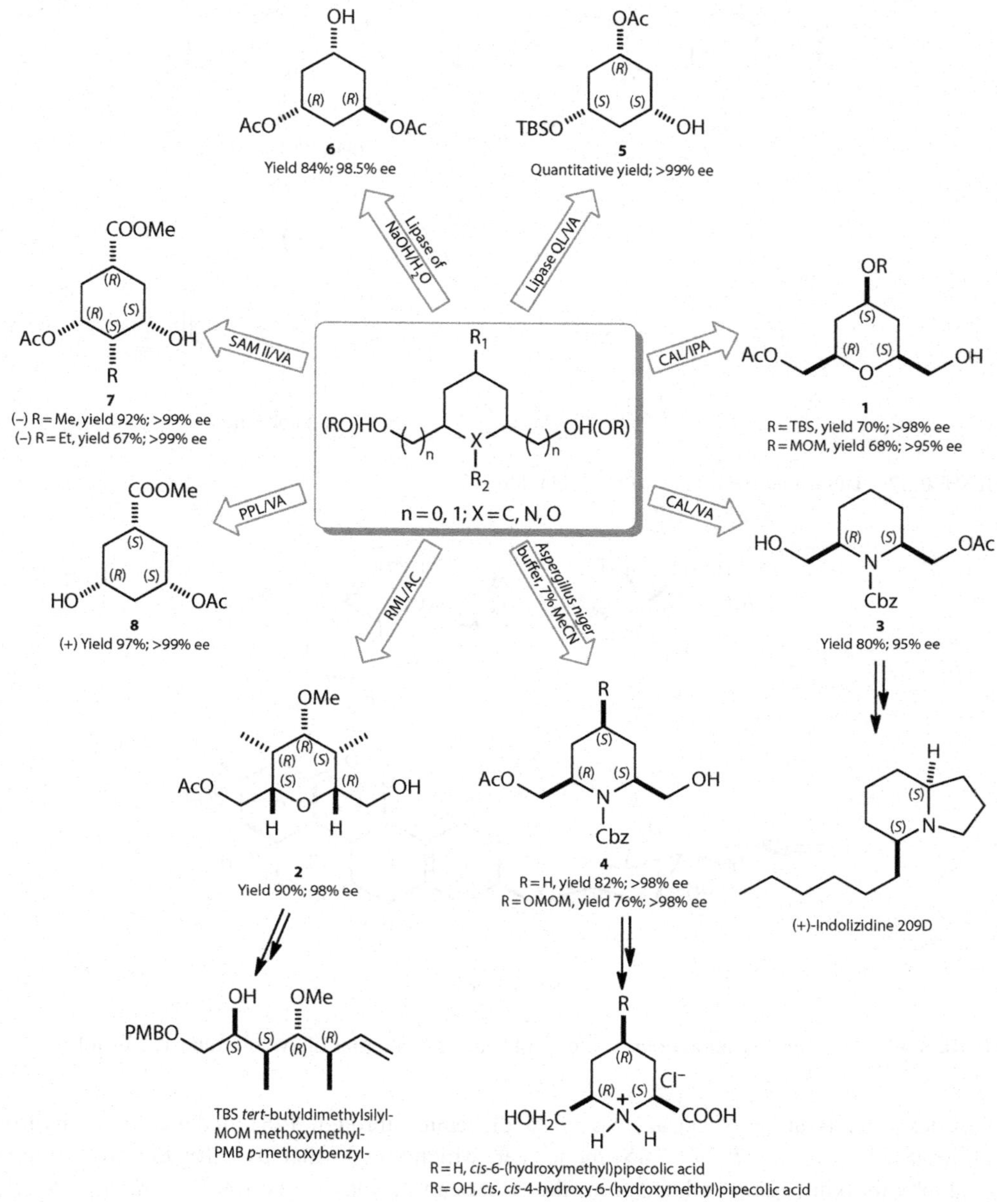

SCHEME 8.36 Desymmetrization of *meso* six-membered ring systems.

lobelanidine, catalyzed by CAL-B. This enzyme allowed the preparation of (*S*)-monoester of high optical purity (Scheme 8.37).[122]

The 2-substituted monoester of cyclohexane-1,2,3-triol, used as the starting material in the total synthesis of antibiotics called aquayamicin, was also prepared by the enzymatic means (Scheme 8.38).[123]

Primary *meso* diols attached to systems larger than six-membered rings were also successfully desymmetrized. For instance, the enantioselective transesterification of cyclic polyols by vinyl acetate gave the corresponding monoacetates in high yields and ee's. These compounds were used for the synthesis of the D(E) and F rings of ciguatoxin, a polyether marine toxin (Scheme 8.39).[124]

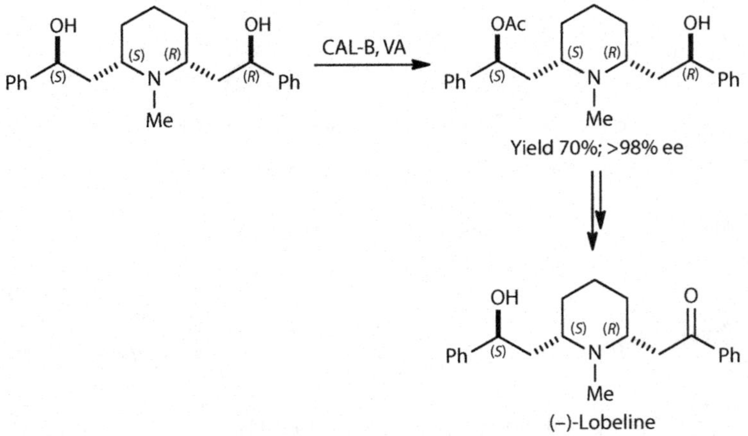

SCHEME 8.37 Desymmetrization of *meso* lobelanidine.

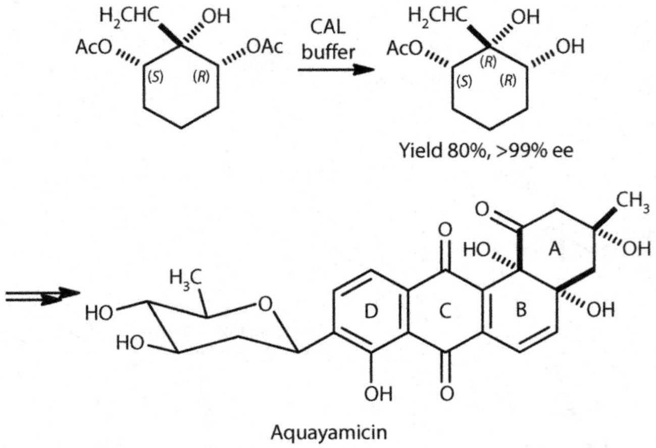

SCHEME 8.38 Desymmetrization of *meso* the 2-substituted monoester of cyclohexane-1,2,3-triol.

The desymmetrization of tetracyclic *meso* diacetates through PPL-mediated ester hydrolysis allowed achieving a (1R,2R,3S,4S)-monoester, which was further used for the preparation of (−)-podophyllotoxin (lignin isolated from the American mayapple tree [*Podophyllum peltatum*]), a potent antimitotic and its C$_2$-epimer (Scheme 8.40).[125]

Chemoenzymatic processes have also been applied in the chiral bicyclo[2.2.1]heptane derivative syntheses. This structural motif is found embedded in a range of compounds employed in medicinal chemistry, organic synthesis, and peptide chemistry. The enantioselective acylation of *meso* 7-azabicyclo[2.2.1]heptanediol in the presence of CAL-B gave the corresponding (1R,2R,3S,4S)-monoacetate in high ee (Scheme 8.41a).[126] On the other hand, CCL lipase selectively transformed the *meso* diol and diester, 1,4,5,6,7,7-hexachlorobicyclo[2.2.1]hepta-2,5-diene derivatives (Scheme 8.41b).[127]

Lipases, as similar to esterases, selectively hydrolyzed the *meso* carboxylic acid derivatives. For instance, *meso cis*-cyclohexane-1,3-dicarboxylic acid diesters were selectively hydrolyzed with PS-30 from *Pseudomonas cepacia* and AY-30 from CRL. Both enantiomers of the monoesters were formed with high ee due to the opposite enantiopreferences of these enzymes (Scheme 8.42a).[128] In turn, cyclohexene diester was deacetylated in the presence of CAL-B (Novozym 435), resulting in

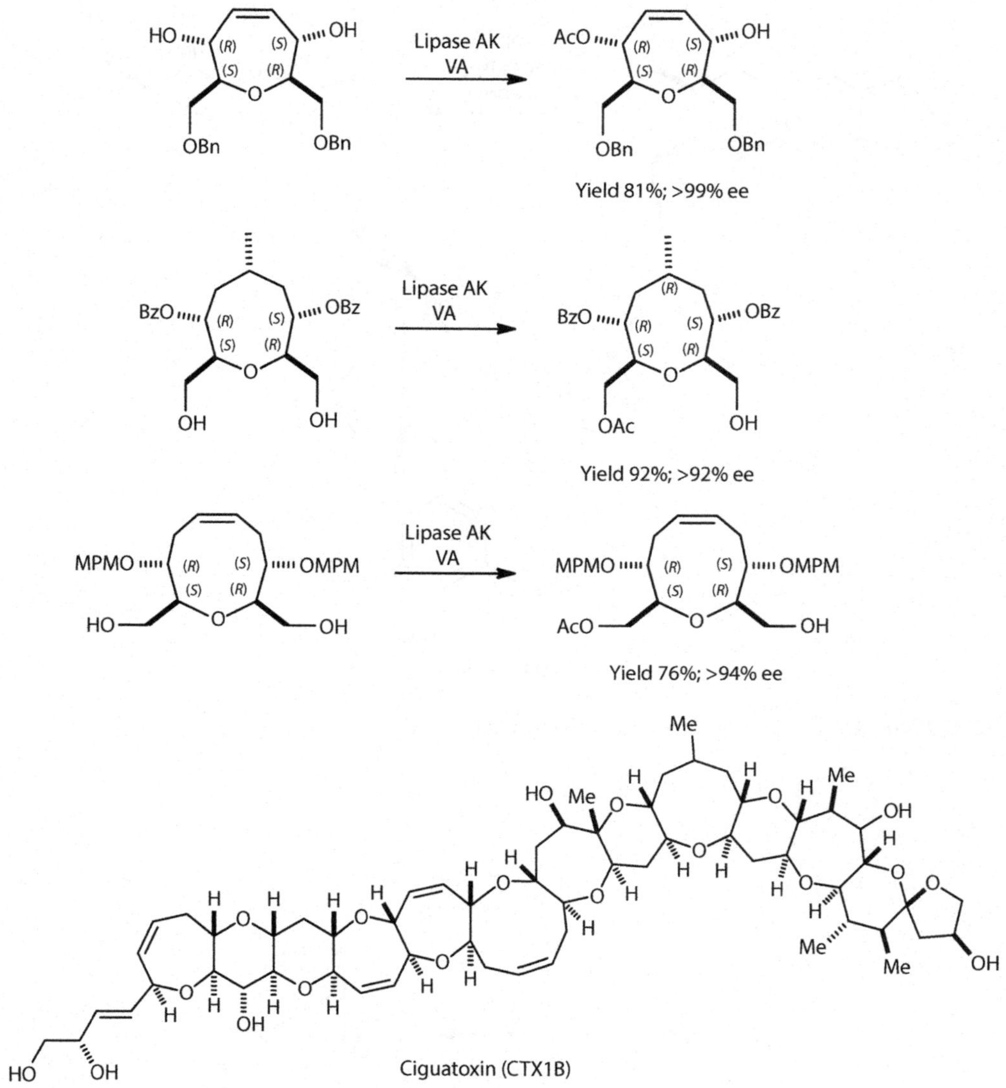

SCHEME 8.39 Desymmetrization of primary *meso* diols.

(1*S*,2*R*)-2-(methoxycarbonyl)cyclohex-4-ene-carboxylic acid, the key intermediate for the synthesis of a drug for the modulation of chemokine receptor activity (Scheme 8.42b).[129]

8.8 OXIDOREDUCTASES

Oxidoreductases (EC 1.1.1.1–1.18.6.1) are enzymes employed in redox reactions. They are classified into three categories. Dehydrogenases constitute one of those categories. Among them, alcohol dehydrogenases (ADH), also named carbonyl reductases (KRED), can be distinguished. ADHs are widely used for the primary and secondary alcohol oxidation and the carbonyl compound reduction. They are classified as enzymes of 1.1.X.X class (EC 1.1.X.X).

Oxygenases, named so because they use molecular oxygen as cosubstrate, are another category of oxidoreductases. Due to their catalytic properties in the oxidation of C–H and C=C bonds, they proved to be particularly useful for hydroxylation and epoxidation reactions.

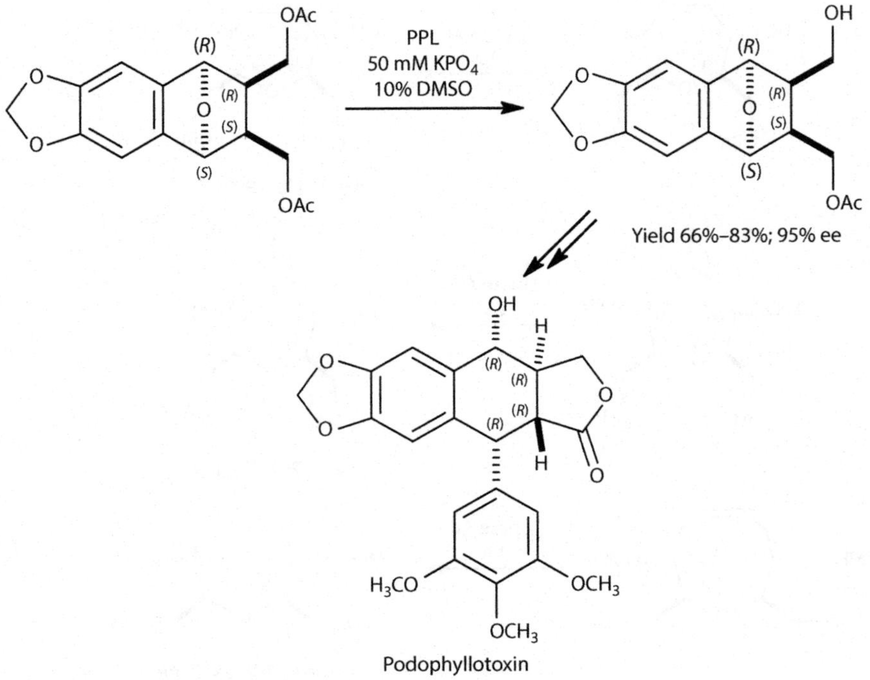

SCHEME 8.40 Desymmetrization of tetracyclic *meso* diacetates.

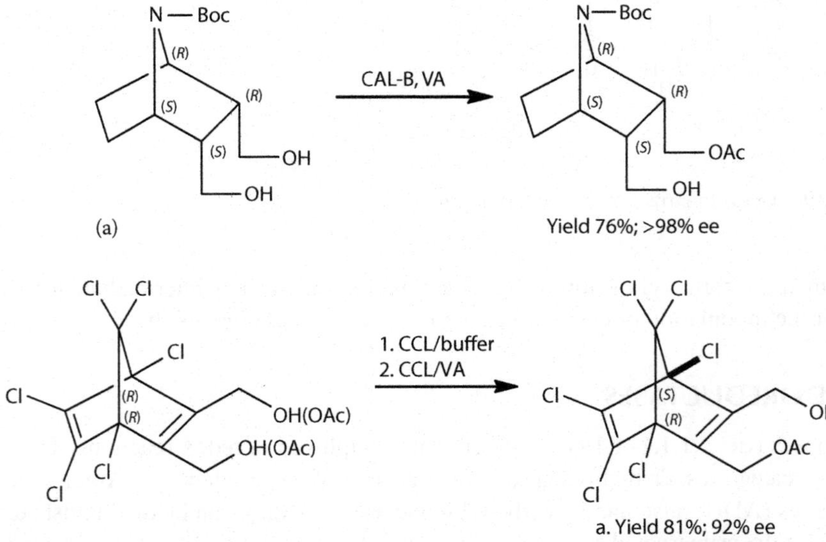

SCHEME 8.41 (a) Desymmetrization of *meso* 7-azabicyclo[2.2.1]heptanediol. (b) Desymmetrization of *meso* diol and diester of 1,4,5,6,7,7-hexachlorobicyclo[2.2.1]hepta-2,5-diene derivatives.

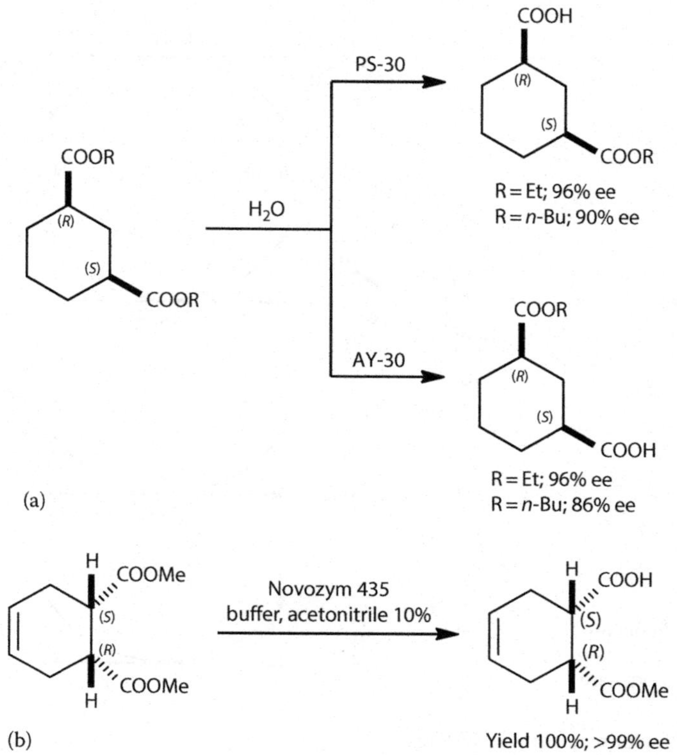

SCHEME 8.42 (a) Desymmetrization of *meso* cis-cyclohexane-1,3-dicarboxylic acid diesters. (b) Desymmetrization of *meso* cyclohexene-1,3-dicarboxylic acid diesters.

Oxidases (EC 1.1.3.X), the last group of oxidoreductases, are responsible for the transfer of electrons. They play a minor role in unnatural organic compound biotransformation, but their use has been growing recently.

8.9 APPLICATION OF DEHYDROGENASES IN ENZYMATIC REDUCTION OF SYMMETRICAL COMPOUNDS

Catalytic potential of oxidoreductases is most commonly used in reduction reactions. Dehydrogenases and reductases catalyze the reversible desymmetrization reactions of *meso* and prochiral carbonyl compounds and alkenes. The oxidoreductase-catalyzed reactions require cofactors to initiate catalysis. In most cases, it is nicotinamide adenine dinucleotide (NADH) or its phosphorylated derivative (NADPH), which acts as a hydride donor. The necessity of employing expensive cofactors was, for a long time, one of the main limitations to the use of dehydrogenases. This problem was solved by developing a regeneration system of a cofactor in the reaction environment. Various systems are used for cofactor recycling (Scheme 8.43). In the case of a carbonyl compound reduction, an irreversible oxidation of formic acid to carbon dioxide is most frequently used.

The reduction of a carbonyl compound by means of an alcohol dehydrogenase takes place when the transfer of one of the diastereotopic hydrogen atoms from the dihydropyridine ring of the coenzyme to a particular face of the carbonyl group occurs. In the cofactor molecule, there are two equally acceptable hydrogen atoms. One of them, a hydride, bonds with the carbon atom, whereas the second one "attacks" the oxygen atom of C=O. The pro-(*S*)- or pro-(*R*)-hydride (H$_S$ and H$_R$, respectively) of the cofactor (NAD(P)H) attacks the *si*- or *re*-face of an sp^2-hybridized carbon atom

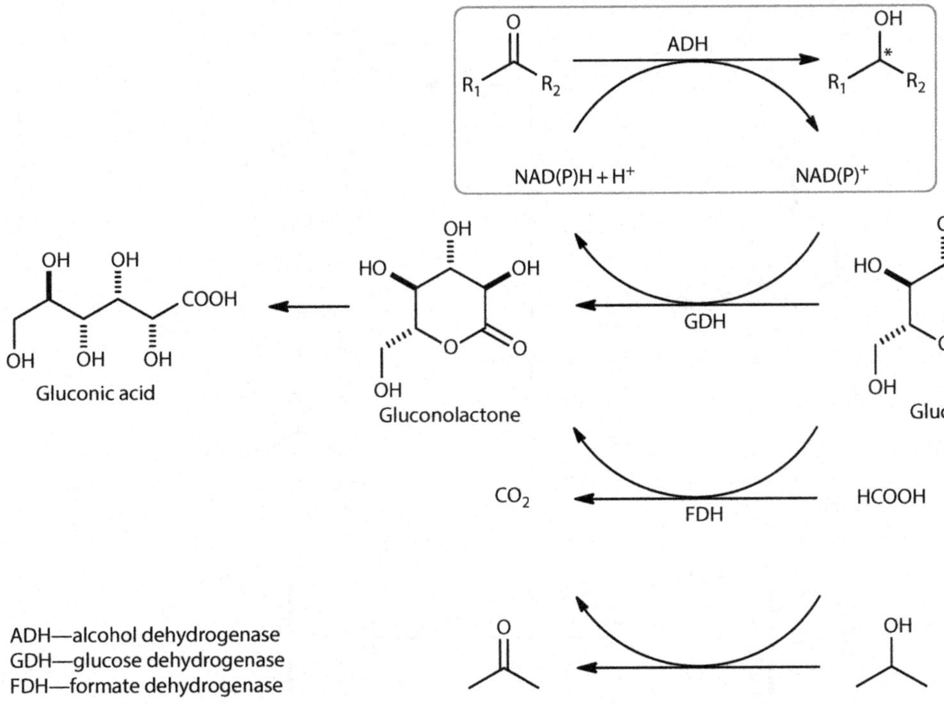

SCHEME 8.43 Cofactor regeneration systems.

of the carbonyl group. In E1 and E2 stages, the hydride attacks the *si*-face; in E3 and E4, it attacks the *re*-face, which results in the formation of (*R*)- and (*S*)-alcohols, respectively.[130] Dehydrogenases can selectively transfer the pro-(*R*)- or pro-(*S*)-hydride of the coenzyme.

For example, alcohol dehydrogenases from *Pseudomonas* sp. and *Lactobacillus kefiri* transfer the pro-(*R*)-hydride to the *si*-face of the carbonyl group (E1), while dehydrogenases from baker's yeast, horse liver, and *Moraxella* sp. pass the pro-(*R*)-hydride to the *re*-face (E3). On the other hand, the glycerol dehydrogenase from *Geotrichum candidum* and dehydroxyacetone reductase from *Mucor javanicus* selectively transfer the pro-(*S*)-hydride to the *si*-face (E2).[130]

Cofactors can be used in catalytic or larger amounts and then recycled *in situ*. However, in organic synthesis, mostly whole cells of microorganisms, plants, or animals are used as catalysts.

Baker's yeast is the most popular whole-cell biocatalyst for the asymmetric reduction of prochiral ketones due to their unlimited availability, ease of growing, and low costs. A broad substrate tolerance of baker's yeast results from the abundance of dehydrogenases present. Some of them have various stereopreferences, which often lower the optical purity of alcohol products. Nevertheless, there are various means of avoiding enzymatic incompatibility, for example, selective disabling of the disturbing biocatalyst, variation in a substrate structure and concentration, use of an organic solvent, lyophilization, and immobilization of the cells, and genetic manipulation.

In the presence of baker's yeast, the reaction can be carried out both in an aqueous medium and in organic solvents. The living cells synthesize and regenerate enzymes and cofactors under fermentation conditions. However, this method can be applied only to a small group of substrates soluble in water.

Apart from baker's yeast, enzymes contained in other various microorganisms are used in asymmetric synthesis, for example, in other yeast (*Rhodotorula rubra*, *Geotrichum candidum*, *Saccharomyces montanus*, *Rhodotorula glutinis*, *Zygosaccharomyces bailii*), in bacteria (*Lactobacillus kefir*, *Lactobacillus fermentum*, *Corynebacterium*), and in whole plant cells (*Daucus carota*, *Datura stramonium*). The best bioreduction results, especially the ee of products, are obtained for isolated

Asymmetric Biocatalysis in Organic Synthesis of Natural Products   239

enzymes (dehydrogenases, reductases). Prevalently, due to economic reasons, syntheses carried out in the presence of isolated enzymes are small-scale processes performed in scientific laboratories. In biotechnological processes, only cheap and easy synthesis methods are applied.

In this section, selected examples of whole-cell and isolated enzyme applications in the carbonyl compound reduction are discussed. The application of baker's yeast in the EED of prochiral ketones leads to a broad spectrum of chiral alcohols used as intermediates in the syntheses of many pharmaceuticals and compounds presenting a potential biological activity.

In Scheme 8.44, baker's yeast–catalyzed asymmetric dialkyl, alkyl aryl, and cyclic ketones are presented.[131–137]

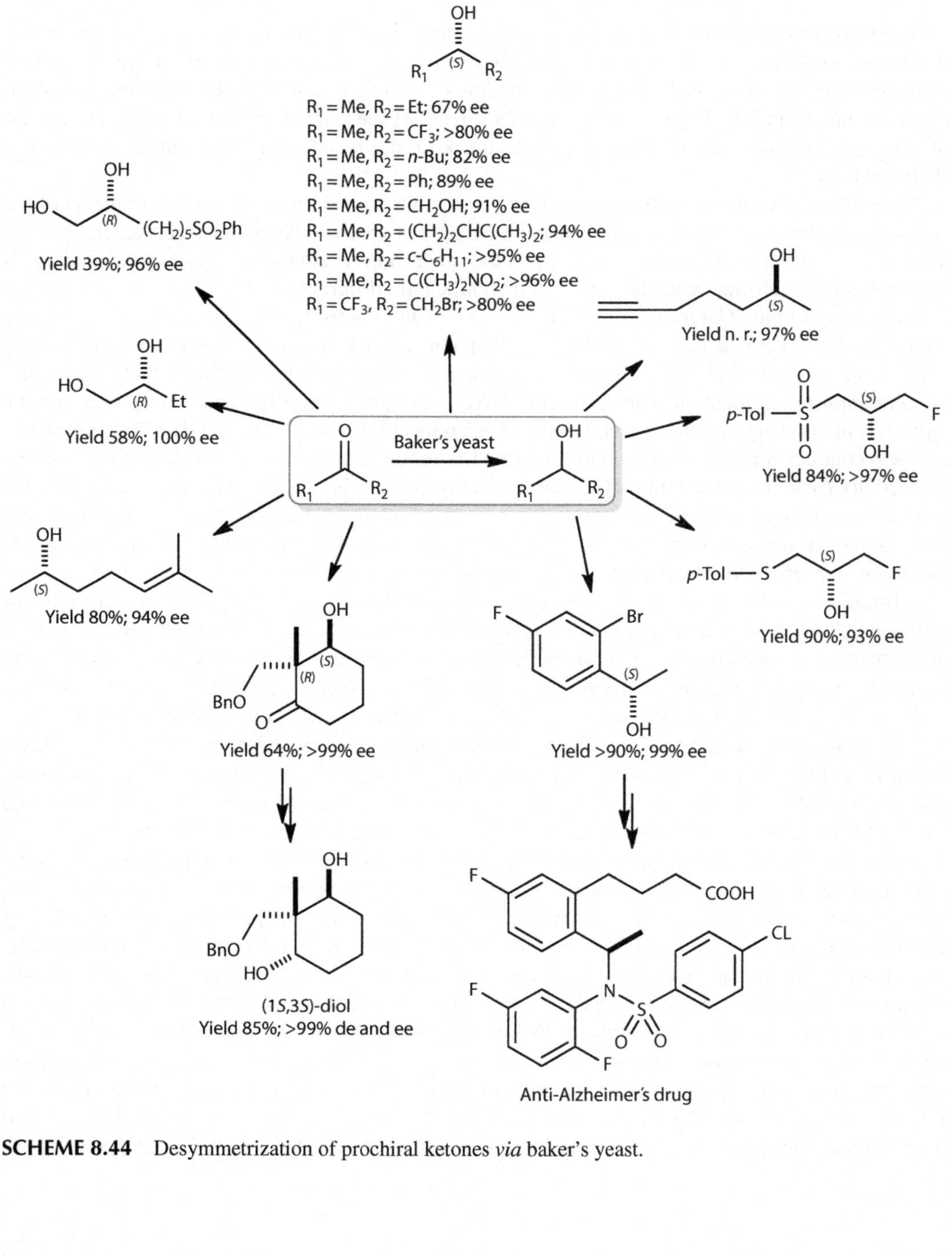

R_1 = Me, R_2 = Et; 67% ee
R_1 = Me, R_2 = CF$_3$; >80% ee
R_1 = Me, R_2 = n-Bu; 82% ee
R_1 = Me, R_2 = Ph; 89% ee
R_1 = Me, R_2 = CH$_2$OH; 91% ee
R_1 = Me, R_2 = (CH$_2$)$_2$CHC(CH$_3$)$_2$; 94% ee
R_1 = Me, R_2 = c-C$_6$H$_{11}$; >95% ee
R_1 = Me, R_2 = C(CH$_3$)$_2$NO$_2$; >96% ee
R_1 = CF$_3$, R_2 = CH$_2$Br; >80% ee

SCHEME 8.44 Desymmetrization of prochiral ketones *via* baker's yeast.

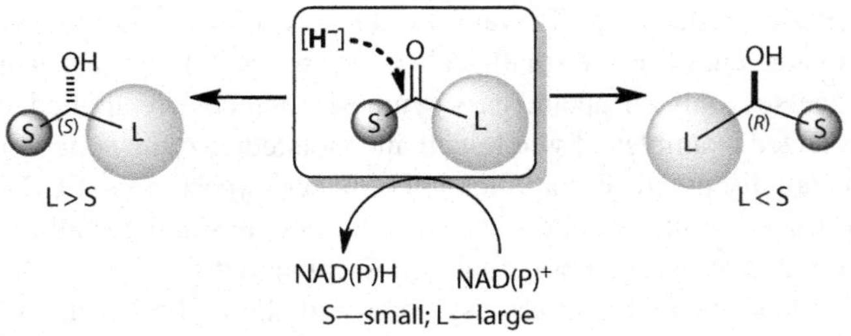

SCHEME 8.45 Stereochemistry of the product in the enzymatic reduction.

The stereochemical outcome of the simple ketone reduction follows Prelog's rule. Under fermentation conditions, in the presence of baker's yeast, the hydride is delivered from the *re*-face of the carbonyl group delivering the corresponding (*S*)-configured alcohol, if the small substituent (S) is minor and the large (L) is a major (S < L).[138] If the priority is changed (e.g., in the case of α-hydroxyketones) as a result of a hydride attack on the *re* side, the (*R*)-enantiomer is formed (Scheme 8.45).[139,140]

The efficiency of the microbial catalysis is mainly determined by the steric requirements of the substrate. For instance, baker's yeast does not tolerate long-chain dialkyl ketones; however, one long alkyl chain is accepted if the methyl group is the other moiety. Highly stereoselective catalysis is achieved for the substrate with substituents of significantly different sizes.[141]

Dimorphic fungus *Geotrichum candidum* is one of the microorganisms that effectively reduce simple aliphatic ketones (e.g., 2-pentanone, 2-butanone, 3-hexanone). Similar to the reduction by baker's yeast, (*S*)-alcohols were obtained at excellent enantioselectivities (Scheme 8.46, comp. **1a**). *G. candidum* in conjunction with Amberlite XAD-7 polymer was also successfully used for the reduction of alkyl aryl ketones providing the (*S*)-alcohols (Scheme 8.46, comp. **1a**). When the reaction was conducted under aerobic conditions, the (*R*)-products were obtained at similar high enantioselectivity (Scheme 8.46, comp. **1b**). Most probably, the aerobic conditions activate the oxidation from the (*S*)-alcohol to the ketone (reversible process), and anaerobic conditions inhibit the oxidation. Moreover, the reduction from the ketone to the (*R*)-alcohol is irreversible, and thus, in aerobic conditions the (*R*)-product is formed preferentially.[142]

β-Bromoalcohols of the (*R*)-configuration at high ee and yield were obtained by the reduction of corresponding ketones with the yeast *Rhodotorula rubra* (Scheme 8.46, comp. **2a** and **2b**). The analysis of different reaction conditions revealed that the best results were obtained in the presence of sodium lauryl sulfate, an anion surfactant, in argon atmosphere. **2a** and **2b** compounds were employed as intermediates in the syntheses of β-adrenoreceptor agonists, (*R*)-denopamine, and (*R*)-salmeterol, respectively.[143,144]

The employment of carrot root (*Daucus carota*) whole cells for the desymmetrization of prochiral ketones is not only an economical but also an effective method of chiral alcohol synthesis. These cells selectively catalyze the bioreduction of aliphatic, aromatic, cyclic, and other various functionalized ketones yielding the corresponding products in high ee (Scheme 8.46, comp. **3a, 3b, 4**).[145]

Trifluoromethyl-substituted alkyl aryl ketones were successfully reduced in the presence of various microorganisms obtaining secondary alcohols of (*R*)- and (*S*)-configuration, depending on the enzyme used. Fluorinated alcohols are valuable building blocks in the preparation of several drug candidates. For example, **5a** and **5b** compounds were used as chiral intermediates in the synthesis of antiviral CCR5 antagonists; **5c** was applied in the synthesis of an antidepressant NK1 receptor antagonist and antimuscarinic M2 receptor antagonist (Scheme 8.46, comp. **5a, 5b, 5c**).[146]

A series of *o*-, *m*-, and *p*-substituted α-hydroxyphenylethanones were reduced to the corresponding 1,2-diols using *Geotrichum* sp. cells as catalyst. The optical purity and the absolute configuration of the products were dependent on the position of the substituent in a phenyl ring. *Ortho*-substituted ketones generally gave the (*R*)-1,2-diols in moderate ee's, whereas unsubstituted or *m*- and *p*-substituted ones provided (*S*)-1,2-diols in good to excellent ee's (Scheme 8.46, comp. **6**).[147]

Asymmetric Biocatalysis in Organic Synthesis of Natural Products 241

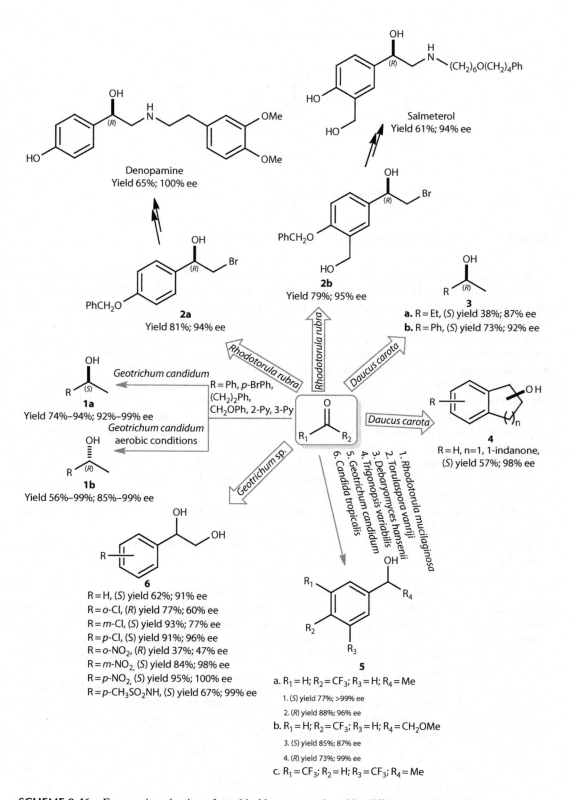

SCHEME 8.46 Enzymatic reduction of prochiral ketones catalyzed by different microorganisms.

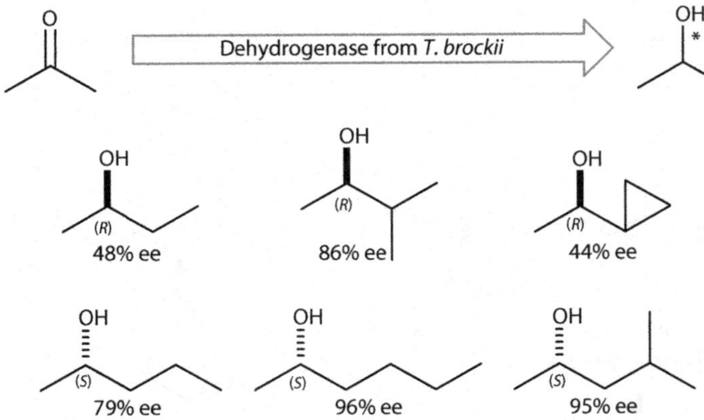

SCHEME 8.47 Desymmetrization of prochiral ketones *via* ADH-TB.

Isolated enzymes, as well as whole cells, can stereoselectively reduce different ketones to furnish secondary chiral alcohols. Unlike the microorganisms or whole plant cells, the isolated dehydrogenases exhibit a specific stereopreference allowing for a completely selective synthesis of one enantiomer of the product. Among the broad range of dehydrogenases, the (R)- and (S)-stereospecific enzymes can be selected for the preparation of Prelog and anti-Prelog. The majority of commercially available dehydrogenases used for the reduction of ketones are (R)-stereospecific, for example, dehydrogenase from *Pseudomonas* sp., *Lactobacillus kefir*, yeast alcohol dehydrogenase, horse liver alcohol dehydrogenase, *Moraxella* sp., *Thermoanaerobium brockii* alcohol dehydrogenase. On the other hand, microbial dehydrogenases from, for example, *Geotrichum candidum*, *Mucor javanicus*, and *Candida parapsilosis* show (S)-stereopreference. Nevertheless, the biocatalyst enantiopreference largely depends on the substrate structure and the reaction conditions. For instance, an interesting substrate size-inducted reversal of enantioselectivity was observed for alcohol dehydrogenase from *Thermoanaerobium brockii* (ADH-TB). The reduction of smaller substrates, such as methyl ethyl, methyl isopropyl, or cyclopropyl ketones provided the (R)-alcohols, whereas higher ketones produced the (S)-enantiomers (Scheme 8.47).[15]

In Scheme 8.48, some examples of the use of selected dehydrogenases for bioreduction of simple ketones are presented.[148–156]

Chiral alcohols containing nitrogen in their structure are found to be potential synthetic precursors of pharmaceutically applicable molecules. For example, the reductive EED of the 2-azido-1-aryl ketones catalyzed by the enzymes from the *Daucus carota* root in the aqueous medium provided the corresponding (R)-chiral azido alcohols, the key intermediates in the syntheses of (R)-(−)-tembamide and (R)-(−)-aegeline. These compounds are used in traditional Indian medicine and show a good hypoglycemic activity (Scheme 8.49, comp. **1**).[157] Yadav and coworkers reported on the stereoselective reduction of 2-azido-1-aryl ketones using baker's yeast in the presence of allyl alcohol to inhibit the (S)-oxidoreductase enzyme. Thus, they exclusively obtained optically pure (R)-isomers of 2-azido-1-aryl ethanols.

Bulky ketones were also successfully desymmetrized by means of microbe-catalyzed reductions. 1-Phenyl-1-(2-phenylthiazol-5-yl)methanone was enantioselectively reduced using *Saccharomyces montanus* CBS 6772 and *Rhodotorula glutinis* var. *dairenensis* MUCL 30607; (S)- and (R)-alcohols were achieved, respectively, in high yields and ee's (Scheme 8.49, comp. **2**).[158]

Similarly, the chiral β-chlorohydrin derivatives were obtained by microbial reduction. From 2-chloro-1-[6-(2,5-dimethylpyrrol-1-yl)pyridine-3-yl]ethanone, in the presence of *Zygosaccharomyces bailii* ATCC No. 38924, the (R)-enantiomer of the chlorohydrin was formed. It can be used in the synthesis of β-adrenergic receptor agonists (Scheme 8.49, comp. **3**).[159]

Asymmetric Biocatalysis in Organic Synthesis of Natural Products 243

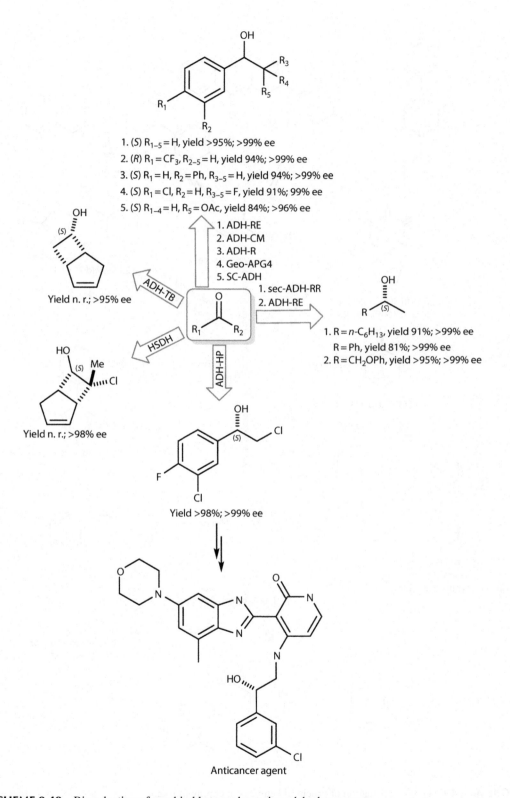

SCHEME 8.48 Bioreduction of prochiral ketones by various dehydrogenases.

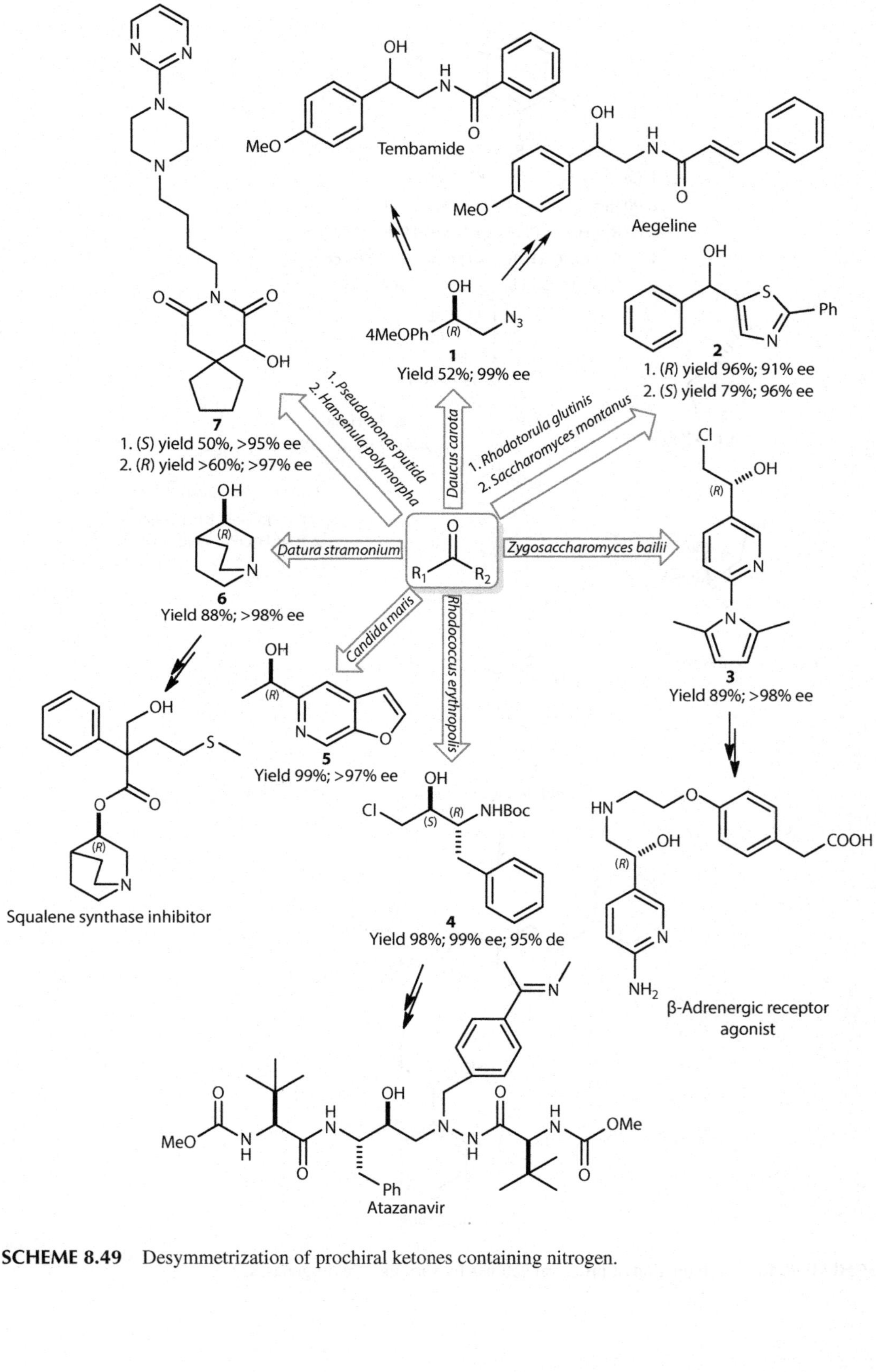

SCHEME 8.49 Desymmetrization of prochiral ketones containing nitrogen.

Another chlorohydrin, (1S,2R)-[3-chloro-2-hydroxy-1-(phenylmethyl)propyl]carbamic 1,1-dimethylethyl ester, was prepared in the diastereoselective reduction of an appropriate α-chloroketone using microbial cultures, among which three strains of *Rhodococcus* sp. and *Streptomyces nodosus* SC 13149 gave the highest yields and diastereoselectivity. The compound **4** is the key chiral intermediate required for the total synthesis of the HIV protease inhibitor, atazanavir (Scheme 8.49, comp. **4**).[160,161] (R)-5-(1-Hydroxyethyl)furo[2,3-c]pyridine, which was obtained by enzymatic reduction of the ketone in the presence of *Candida maris* IFO 1003 (Scheme 8.49, comp. **5**), is the next important chiral compound applied in the synthesis of the HIV reverse-transcriptase inhibitor and often used as a chiral building block in asymmetric synthesis.[162,163]

Employing the tropinone reductase-I derived from the plant *Datura stramonium*, (R)-3-quinuclidinol was prepared in high chemical and optical yield (Scheme 8.49, comp. **6**).[164] This compound is used as an intermediate to produce a variety of physiologically or pharmacologically active agents, for example, squalene synthase inhibitor. Patel and coworkers screened about 150 microorganism for the enantioselective reduction of 6-oxobuspirone. The use of *Rhizopus stolonifer* SC 13898, *R. stolonifer* SC 16199, *Neurospora crassa* SC 13816, *Mucor racemosus* SC 16198, and *Pseudomonas putida* SC 13817 led to the (S)-hydroxybuspirone in >95% ee, while the yeast strains *Hansenula polymorpha* SC 13845 and *Candida maltose* SC 16112 gave the (R)-enantiomer in >60% reaction yield and >97% ee (Scheme 8.49, comp. **7**).[165]

Ketoesters represent an extensive group of selectively bioreduced compounds. Chiral hydroxyesters and, subsequently, hydroxy acids are valuable intermediates in the syntheses of many biologically active compounds. Acyclic α- and β-ketoesters are transformed to the corresponding (R)- and (S)-hydroxyesters by using a specific dehydrogenases. The whole-cell enzymes, for example, baker's yeast, may exhibit a different catalytic activity depending on the substrate structure. In most cases, the (R)-specific enzyme of baker's yeast shows a higher activity toward β-ketoesters bearing the short-chain acyl residues, such as methyl esters. In contrast, the (S)-specific dehydrogenase is more active on long-chain counterparts, for example, octyl esters. For example, baker's yeast–catalyzed reduction of ethyl 4-chloro-3-oxobutanoate led to the (S)-hydroxyester, whereas its long-chain analogue, *n*-hexyl ester, gave the corresponding (R)-enantiomer. However, in both reactions, the products were obtained in moderate ee (Scheme 8.50, comp. **1**). Inhibition of one of the dehydrogenases, contained in baker's yeast, causes an increase in the formation of an appropriate product. For instance, the addition of allyl alcohol or methyl vinyl ketone inhibits the (S)-enzyme, which results in an increase in the (R)-β-hydroxyester optical purity. On the other hand, various haloacetates, thioethers, and allyl bromide proved to be selective inhibitors for the (R)-enzyme. Various microorganisms were used in the selective reduction of 4-chloro-3-oxobutanoic acid ethyl ester. Among them, *Candida magnoliae*,[166] *Geotrichum candidum* SC 4569,[167] *Lactobacillus kefir*,[168] *Kluyveromyces lactis*,[169] and GDH from *Bacillus megaterium* allowed obtaining the (S)-enantiomer of the corresponding alcohol, while *D. carota*,[170] *Sporobolomyces salmonicolor*,[171] *Lactobacillus fermentum*,[172] *Candida parapsilosis*,[173] and carbonyl reductase from *Rhodococcus erythropolis*[174] and phenylacetaldehyde reductase *Corynebacterium* strain ST-10[175] resulted in the product showing opposite configuration. Chiral 4-chloro-3-hydroxybutanoic acid and its analogues found broad applications as the promising building blocks in the syntheses of pharmaceuticals, such as L-carnitine hydrochloride and a 3-hydroxy-3-methylglu-taryl-CoA reductase inhibitor.[176,177]

Enantiomerically pure (R)- and (S)-hydroxyesters of butanoic acid were prepared by reduction with the enzymes obtained from *Lactobacillus brevis* and yeast YDR389w, YPL275w, and YPR070c (Scheme 8.50, comp. **2**). The bioreduction of the N-acetylcysteamine β-ketothioesters based on ketide synthase–deficient strains of *Streptomyces coelicolor* A3 (2) (CH999), soil bacteria, resulted in the corresponding β-hydroxy acids of (S)-configuration. In order to facilitate the purification of the reaction mixture, these hydroxy acids were conventionally esterified to the methyl esters (Scheme 8.50, comp. **3**), which were isolated in moderate yields and ee's. Analogously, the (S)-lactone (Scheme 8.50, comp. **4**) was obtained as a single enantiomer by the cyclization of the initially formed corresponding (S)-β-hydroxy acid.[178]

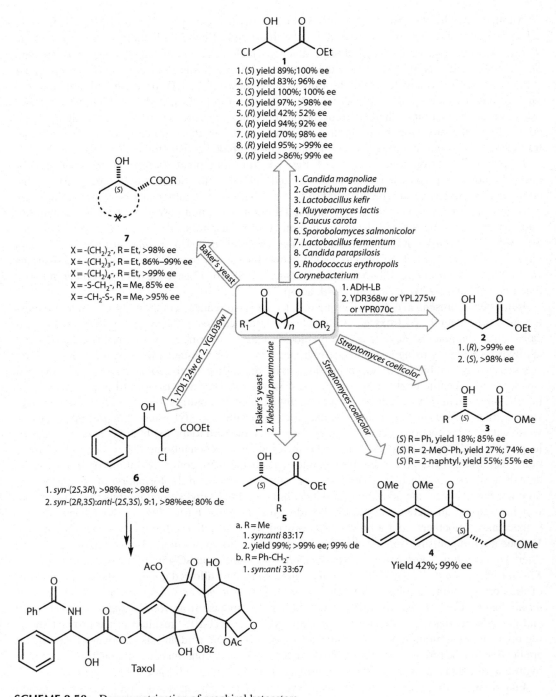

SCHEME 8.50 Desymmetrization of prochiral ketoesters.

β-Ketoesters bearing an additional stereogenic center in the α-position are stereochemically labile due to possible enolization leading to *in situ* racemization of the substrate enantiomers. The asymmetric microbial reduction of α-monosubstituted β-ketoesters allows the formation of diastereomeric *syn*- and *anti*-β-hydroxyesters. The formation of one of them in superiority is primarily determined by the substituent size in the α-position. With small α-substituents, the *syn*-diastereomers are predominant. With the substituent size increasing, the stereoselectivity changes and the *anti*-product is the major one. For example, the yeast-catalyzed reduction of ethyl 2-methyl-3-oxobutanoate led to the mixture of *syn/anti* β-hydroxyesters in 83:17 ratio. However, the reduction of ethyl 2-benzyl-3-oxobutanoate under similar conditions gave mostly the *anti*-diastereomer product (*syn:anti* = 33:67) (Scheme 8.50, comp. **5a** and **5b**).[179,180]

Klebsiella pneumoniae IFO 3319 also selectively reduced the ethyl ester of 2-methylo-3-oxobutanoic acid, resulting in (2R,3S)-hydroxyester, presenting excellent enantio- and diastereoselectivity (Scheme 8.50, comp. **5a**). The absolute configuration on the carbon atom in 3-position was determined by the stereoselectivity of the biocatalyst used; for baker's yeast as well as *Klebsiella pneumoniae* IFO 3319, the same isomer was formed.

If an α-substituent of β-ketoesters is an electron-withdrawing group, *in situ* racemization does not take place due to the impossibility of enolate formation and, consequently, the product of high diastereomeric excess is obtained. The selective reduction of α-chloro-β-ketoester, depending on baker's yeast enzymes used, led to (R)- or (S)-enantiomers at high ee and de. In the presence of YDL124w reductase, *syn*-(2S,3R)-hydroxyester is prepared, whereas a short-chain YGL039w dehydrogenase gave a mixture of two diastereomers, *syn*-(2R,3S)- and *anti*-(2S,3S)-alcohol, in 9:1 ratio (Scheme 8.50, comp. **6**). Both products are the precursors of the *N*-benzoyl phenylisoserine Taxol side chain.[181]

Baker's yeast enzymes selectively reduce the cyclic β-ketoesters, providing mainly *anti* diastereomers due to the lack of rotation around the single α,β carbon–carbon bond. The enzymatic reduction of the esters, cyclopentanone, and cyclohexanone derivatives gave the optically active *anti*-alcohol enantiomers (Scheme 8.50, comp. **7**).[15] However, the YDR368w and YGL039w enzymes accepted ethyl 2-ketocycloheptanecarboxylate yield mainly in the *cis*-(1R,2S)-alcohol.[182]

The reductive EED of prochiral α-ketoesters as well as β-ketoesters is an interesting transformation in organic chemistry due to the importance of the resulting chiral α-hydroxy acids and their derivatives functioning as building blocks. Baker's yeast–catalyzed reduction of alkyl esters derived from pyruvate and benzoylformate allows the preparation of the (R)-alcohols (Scheme 8.51, comp. **1**). As in the case of β-ketoester reduction, the employment of well-known yeast dehydrogenase inhibitors, for example, methyl vinyl ketone, increases the enantioselectivity and the yield of the reduction.[183]

In the presence of *Aureobasidium pullulans* SC 13849, the reduction of another α-ketoester was carried out, affording the chiral (R)-hydroxyester (Scheme 8.51, comp. **2**).[184] This compound and its related optically active α-hydroxy acid are used in the synthesis of a retinoic acid receptor gamma-specific agonist. Optionally, this compound can also be directly obtained from the corresponding α-ketoacid by reduction with *Candida maltose* SC 16112 and two strains of *Candida utilis* (SC 13983, SC 13984).

The synthesis of (S)-E-2-{3-[3-[2-(7-chloro-2-quilolinyl)ethenyl]phenyl]-3-hydroxypropyl}-benzoic acid methyl ester is an interesting example of the microbial reduction application in pharmacology (Scheme 8.51, comp. **3**).[184] This compound is the key intermediate in the synthesis of an anti-asthma drug (montelukast) and was prepared by the reduction of the corresponding ketoester with *Microbacterium campoquemadoensis* (MB5614) or *Mucor hiemalis*.

Polyketones can also be subjected to the reductive EED to give different compounds bearing the quaternary stereogenic centers, which are broadly applied in asymmetric synthesis. α- and β-diketones are successfully reduced in the presence of baker's yeast. In the case of β-diketones, microbial reduction leads mainly to β-hydroxyketones, whereas α-diketones, depending on the reaction condition, give monohydroxy or dihydroxy products. When reducing cyclic β-diketones with baker's yeast, the replacement of the highly acidic protons on the α-carbon atom by substituents is significant in order to avoid product racemization. For example, 2,2-disubstituted cycloalkanediones such as 2-benzyloxymethyl-2-methylcyclohexane-1,3-dione, 2,2-dimethylcyclohexa-1,3-dione, and 2-methyl-2-propylcyclopenta-1,3-dione were selectively reduced in the presence of baker's yeast to the corresponding enantiopure (3S)-hydroxyketones (Scheme 8.52, comp. **1a–c**).[133] The highly functionalized chiral intermediates obtained by this type of process are the interesting building blocks in the syntheses of some terpenoids. The optically active δ-hydroxyketoesters **2a–b** were obtained in the rec-ADH-LB-catalyzed reduction of the appropriate diketoesters (Scheme 8.52, comp. **2**).[185] These products easily cyclize to δ-lactones. In particular, the (R)-lactone of the compound **2** was described

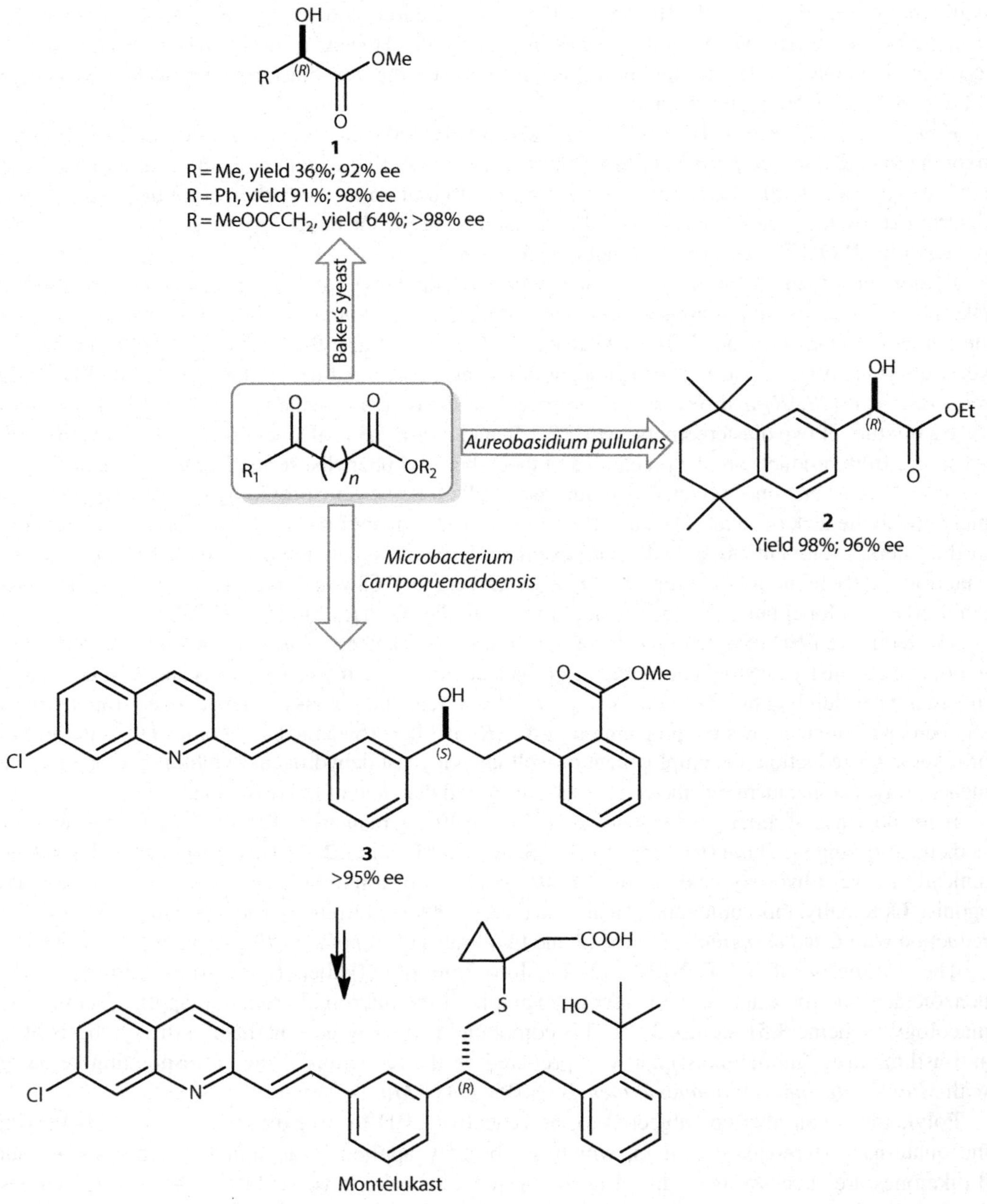

SCHEME 8.51 Desymmetrization of prochiral ketoesters.

as a natural fragrance,[186] while the lactone of (S)-configuration was employed in the syntheses of (−)-callystatin A, (−)-20-*epi*-callystatin A,[187,188] (S)-argentilactone, and (S)-goniothalamin.[189] It is worth mentioning that (R)-**2b** with an equally high ee can also be prepared through bioreduction with the use of baker's yeast.

The transformation of 3,5-diketo-6-(benzyloxy)hexanoic acid ethyl ester with the use of ketoreductases from *Acinetobacter* sp. SC13874 led to the *syn*-diol **3**. In dependence on the microorganism strain, the (R)- or (S)-enantiomers were formed (Scheme 8.52, comp. **3**).[137] Compound **3** allowed synthesizing the Kaneka alcohol, the precursor of the HMG-CoA reductase inhibitors, in a few stages.

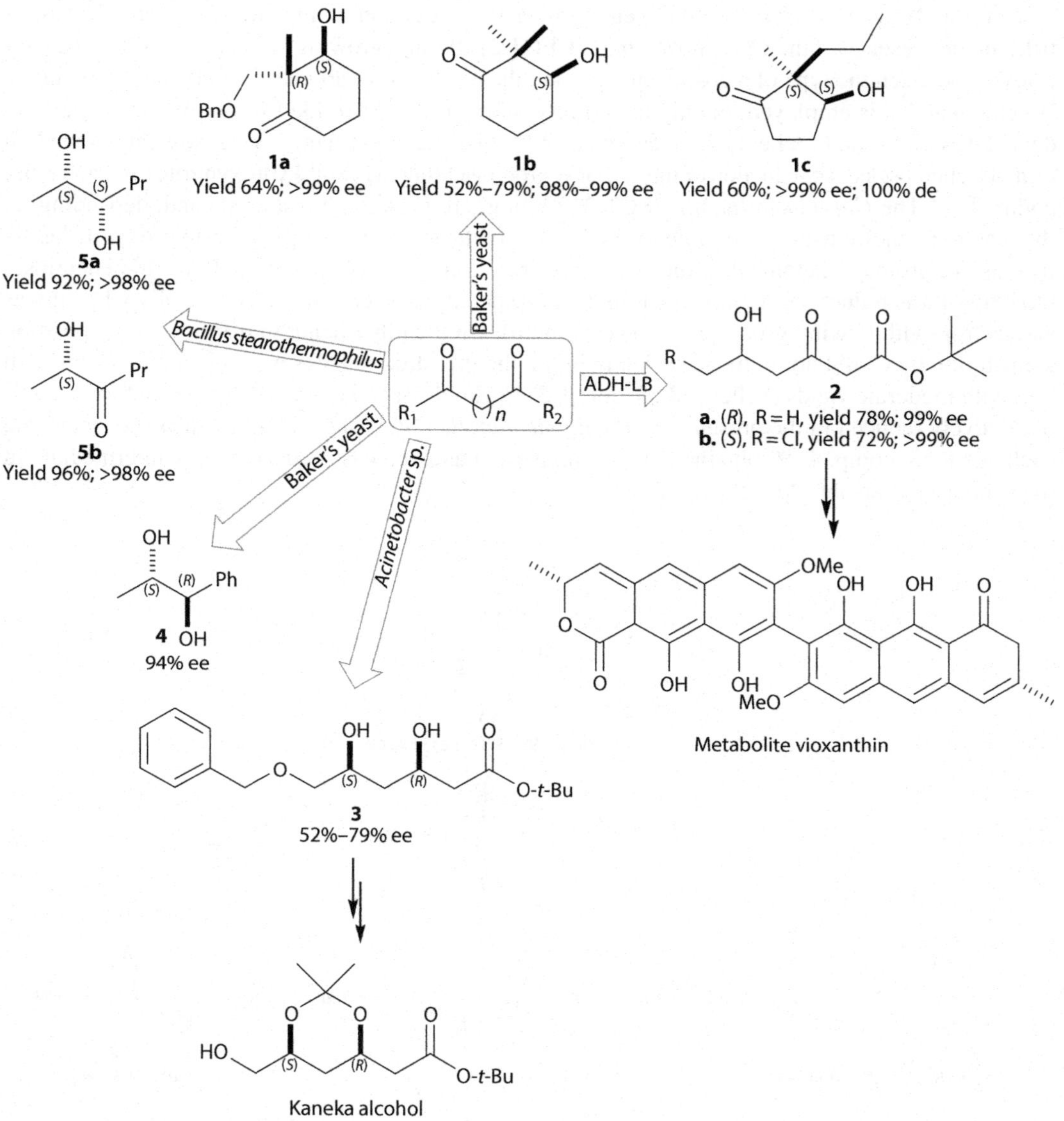

Asymmetric Biocatalysis in Organic Synthesis of Natural Products   249

SCHEME 8.52 Enzymatic reduction of prochiral polyketones.

The situation is different in the case of α-diketone reduction. Enzymatic reduction, in which enzymes from microorganisms are particularly commonly used as catalysts, leads to the formation of diols. In the first step, the less hindered carbonyl group is reduced. Further reduction of the (usually more sterically hindered) remaining carbonyl group yields the corresponding diol predominantly in the *anti*-configuration. For instance, the reduction of 1-phenylpropane-1,2-dione in the presence of baker's yeast gave the *anti*-1,2-diol at 94% ee as the final product (Scheme 8.52, comp. **4**).[190,191] However, the reduction of the same α-diketone to the corresponding diol and hydroxyketone in the presence of reductase from *Bacillus stearothermophilus* can be controlled. Both carbonyl groups were reduced when *endo*-bicyclo[3.2.0]hept-2-en-6-ol was used as a hydrogen source for coenzyme recycling. On the other hand, α-hydroxyketone was achieved using glucose-6-phosphate/glucose-6-phosphate dehydrogenase for coenzyme recycling (Scheme 8.52, comp. **5a–b**).[192]

Gotor and coworkers researched the enantioselective reduction of aliphatic and aromatic ketonitriles in the presence of the *Curvularia lunata* CECT 2130 fungi growing cells. They observed a relationship between the type of a cosolvent used and the product (nonalkylated or alkylated) obtained. When ethanol was employed, mainly the corresponding (2R,1'R)-α-alkyl-β-hydroxy nitrile at high de and ee was formed (Scheme 8.53, comp. **1**).[193,194] The use of methanol as a cosolvent allowed, in turn, the chemoselective reduction of the carbonyl group yielding the (S)-β-hydroxynitrile (Scheme 8.53, comp. **2**).[195] The *Curvularia lunata* CECT 2130 fungi are versatile biocatalysts and, depending on the reaction conditions, it is possible to obtain the alkylated and nonalkylated β-hydroxynitriles. In the case of aromatic ketonitriles, the substituent position in the phenyl ring affects the enantioselectivity of this reduction. A substituent in the *para* position lowered the ee, especially when it was an electron-withdrawing group (e.g., chlorine atom). On the other hand, in *ortho*- or *meta*-position substitution by Cl did not influence selectivity; all the β-hydroxynitriles were obtained at very high ee's with moderate yields (Scheme 8.53, comp. **3**).[195] Ethyl-5-ketohexanenitrile was, in turn, enantioselectively reduced in the presence of *Pichia methanolica* SC 16116. (S)-5-Hydroxyhexanenitrile (Scheme 8.53, comp. **4**),[196] obtained at very high yield and ee, was used as the key intermediate in the synthesis of an anti-Alzheimer's drug.

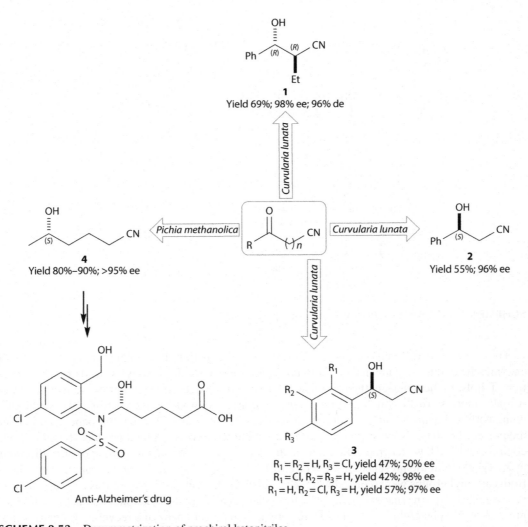

SCHEME 8.53 Desymmetrization of prochiral ketonitriles.

8.10 APPLICATION OF ENE-REDUCTASES IN ENZYMATIC REDUCTION OF SYMMETRICAL COMPOUNDS

In asymmetric synthesis, similar to carbon–oxygen double bonds, carbon–carbon double bonds of prochiral alkanes can be reduced to obtain the optically active saturated compounds. The reduction of alkenes is catalyzed by both the whole cells (microorganisms, plant cells) as well as isolated enzymes belonging to the oxidoreductases, the so-called ene-reductases (EC 1.3.1.31). The whole-cell catalysts are suited, mostly, for the preparative scale syntheses, but they are less chemoselective in comparison to the isolated reductases. In the case of polyfunctionalized alkenes, microorganisms can cause the additional side reaction, reducing the desired product yield.[9]

For a long time, the old yellow enzyme family of flavoproteins (OYE, EC 1.6.99.1) extracted from the yeast were the only kind of isolated enzymes used in bioreduction. Their natural function was the reduction of a double bond in carboxylic compounds. In the early 1980s, Simon et al. isolated enoate reductases (EC 1.3.1.31) from anaerobes such as *Clostridium* spp. (*C. kluyveri*, *C. tyrobutyricum*, and *C. sporogenes*). OYE and enoate reductase belong to different subclasses of oxidoreductases; nevertheless, they both catalyze the selective reduction of activated carbon–carbon double bonds. Enoate reductases are air sensitive; therefore, they are less commonly applied in bioreduction reactions.[197]

The overall addition of a hydrogen molecule to the carbon–carbon double bond catalyzed by enoate reductases proceeds in a *trans*-fashion with absolute stereospecificity. A hydride (derived from a reduced flavin cofactor) is stereoselectively transferred onto the β carbon, while a proton (derived from the solvent) onto the α carbon from the opposite side of the double bond. Cofactor regeneration is carried out with the participation of NAD(P)H, which becomes a hydride donor. This stereoselective addition of hydrogen in the presence of ene-reductases allows the preparation of reduction products at high enantio- or diastereoselectivity. The reductive EED of alkenes is successful if they possess the electron-withdrawing substituent (EWG) in their structures. EWG increases the acidity of the hydrogen atom at the α-position to this substituent. Microorganisms, the whole cells of plants, and isolated OYE from the flavoprotein family allow the selective reduction of unsaturated compounds bearing the additional groups such as carbonyl, carboxyl, ester, imide, nitrile, and nitro ones. The bioreduction of alkenes conjugated with a carbonyl group in the presence of OYE is most often studied. Unsaturated aldehydes in the presence of microorganisms can be reduced to the optically active saturated primary alcohols. For example, in the biotransformation of a citral derivative catalyzed by baker's yeast, both the double bond in the α-position to the carbonyl group and the aldehyde group were reduced giving the primary alcohol (Scheme 8.54, comp. **1a**). In contrast, the bioreduction of citral with the use of the OPR3 ene-reductase (12-ketophytodienoic acid reductase isolated from *Lycopersicon esculentum*) proceeded in a chemoselective manner yielding the fragrant compound, (*R*)-citronellal, at high chemical and optical yields (Scheme 8.54, comp. **1b**).[198,199]

Similarly, the nonracemic α-methyl dihydrocinnamaldehyde derivatives were prepared by the reduction with ene-reductases. The highest enantioselectivities were obtained in chemoenzymatic reactions where reductases from *Zymomonas mobilis* (NCR) and isoenzymes 1–3 from yeast OYE (OYEs 1–3) were applied. The resulting product compounds were used as olfactory principles in perfumes (Lilial, Helional).[200] Also, the aromatic unsaturated propenals can be converted into the saturated (*S*)-alcohols by using fermenting cells of baker's yeast as the catalyst (Scheme 8.54, comp. **3**). The enantiopure products are chiral building blocks useful in the synthesis of bisabolane sesquiterpenes, such as (*S*)-(+)-curcuphenol, (*S*)-(+)-xanthorrhizol, (*S*)-(−)curcuquinone, and (*S*)-(+)-curcuhydroquinone.[201] The baker's yeast–mediated reduction of sulfur-functionalized methacroleins led to the unsaturated and saturated alcohols that are suitable precursors for the preparation of bifunctional chiral synthons (Scheme 8.54, comp. **4a–b**). The stereochemistry of this biohydrogenation depends on the double bond position and also on the oxidation state of sulfur. Thus, the reduction of 2-methyl-3-(phenylthio)acrylaldehyde and 2-methyl-3-(phenylsulfonyl)acrylaldehyde gave the saturated alcohols (*R*)-**4a** and (*S*)-**4a**, respectively, in good ee's but low yields. This fact was due to the incomplete reduction of the starting material, the allylic alcohols, thus being obtained at

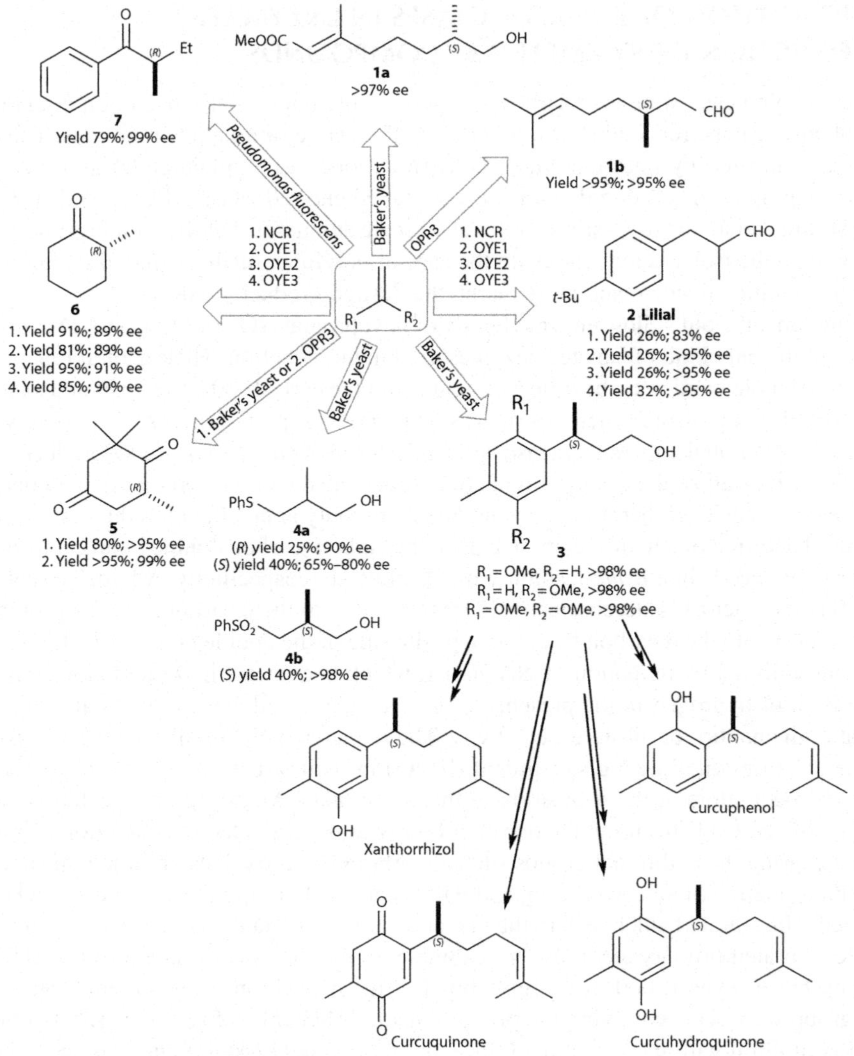

SCHEME 8.54 Asymmetric bioreduction of prochiral unsaturated compounds using ene-reductases.

significant yields. On the other hand, the reduction of 2-[(phenylthio)methyl]acrylaldehyde gave the corresponding saturated alcohol (*S*)-**4a** at moderate ee.[202]

In contrast to the unsaturated aldehydes, the reduction of α,β-unsaturated ketones in the presence of baker's yeast is more selective. For instance, from ketoisophorone nonracemic levodione was obtained at 80% yield and >95% ee (with negligible amount of two other products arising from the over-reduction of the carbonyl moieties) (Scheme 8.54, comp. **5**). This compound is a precursor for the synthesis of carotenoids (astaxanthin, zeaxanthin).[203] On the other hand, no trace of the carbonyl reduction was observed using the ene-reductase OPR3.[204] Most of the ene-reductases (NCR, OYE 1–3, YqiM from *Bacillus subtilis*, OPR1) allowed reducing 2-methylcyclohexenone to the (*R*)-configured products at very good yields and ee's (Scheme 8.54, comp. **6**).[205,206] Another example of an unsaturated ketone bioreduction is the biotransformation of methyleneketone in the presence of *Rhizopus arrhizus*, *P. fluorescens*, and immobilized *S. cerevisiae*. The best result was obtained for *P. fluorescens* used as a biocatalyst; the optically pure ketone **7** was formed at good yield (Scheme 8.54, comp. **7**).[207]

Asymmetric Biocatalysis in Organic Synthesis of Natural Products   253

The next important series of selectively bioreduced compounds are α,β-unsaturated nitriles. As a result of these reactions, saturated nitriles that can easily be converted to the carboxylic acids, amides, and aldehydes are obtained. In turn, saturated nitro products are used in the syntheses of amines, carboxylic acids, and hydrocarbons. Various Z-configured *para*-substituted phenyl butenenitriles were reduced in the presence of the ene-reductase ERED to the corresponding (*R*)-products at ee up to 99% (Scheme 8.55, comp. **1**). The same enzyme was also active in the reduction of 6-chloro-5-methylspiro[1*H*-indene-1,4′-piperidine]-3-carbonitrile, a chiral pharmaceutical building block (Scheme 8.55, comp. **2**).[205,206]

Bioreduction by ene-reductases of simple monocarboxylic esters is impossible; the presence of an additional activating group, for example, a halogen atom, a nitro group, or a second ester, is required. For example, β-nitroacrylates were successfully reduced in the presence of OYE1 (Scheme 8.56, comp. **1**). In some cases, the *R*/*S*-configuration of the products is dependent on the substrate *E*/*Z*-configuration. The size of the substituent in the β-position also plays a significant role. A bulky group hinders the hydride attack and, consequently, may lower this reduction selectivity. The reduction of methyl 2-chloro-2-alkenoates with the use of baker's yeast, in dependence on the substrate configuration, allows the preparation of both the product enantiomers. The Z-configured substrate was selectively reduced to the optically pure (*S*)-product. Despite the fact that the *E*-alkenoate gave the corresponding (*R*)-product, the selectivity of this reduction was definitely reduced (Scheme 8.56, comp. **2**).

α-Substituted butenedioic esters were enzymatically reduced to both product enantiomers. The stereochemical outcome of this reduction could be controlled by the choice of the ene-reductase or the use of a *E*- or *Z*-configurated substrate. (*R*)-2-Methylsuccinate was obtained from the *E*-substrate in the reaction catalyzed by reductases OPR1 or SYE-4 (OYE homologue), while the oppositely configurated product was formed (from the same substrate) in the presence of YqjM (Scheme 8.56, comp. **3a–b**). Both the enantiomers can also be prepared from the reduction of the *Z*/*E* substrate mixture catalyzed by YqjM due to the selective transformation of the *Z*-isomer to the (*R*)-enantiomer.[205–210]

(*R*)-3-Hydroxy-2-methylpropanoate, known as the Roche ester, was synthesized by the ene-reductase-catalyzed reduction as the key step. The protection of a hydroxyl group allowed the preparation of a product at high optical purity. The Roche ester is a popular chiral building block for the synthesis of

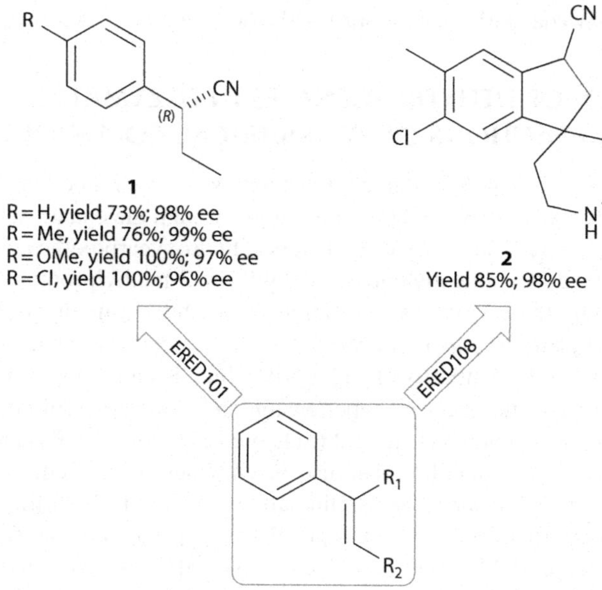

SCHEME 8.55 Bioreduction of prochiral α,β-unsaturated nitriles.

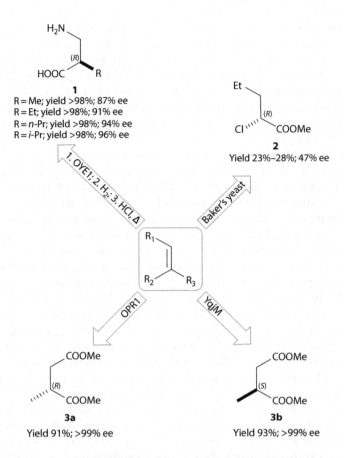

SCHEME 8.56 Desymmetrization of prochiral unsaturated carboxylic acid derivatives.

vitamins (e.g., α-tocopherol), fragrance components (e.g., muscone), antibiotics (e.g., calcimycin, palinurin, rapamycin, 13-deoxytedanolide, dictyostatin), and other natural products (e.g., spiculoic acid A).[211]

8.11 APPLICATION OF DEHYDROGENASES IN ENZYMATIC OXIDATION REACTIONS OF SYMMETRICAL COMPOUNDS

Dehydrogenases are not only able to catalyze the enantioselective reduction of prochiral ketones, but they can also desymmetrize *meso* and prochiral diols through the enantioselective oxidation. As a result of this process, optically active hydroxyketones, hydroxycarboxylic acids, and their derivatives are obtained. For example, optically pure (*R*)-3-hydroxy-2-methylpropionic acid (Scheme 8.57, comp. **1**) was obtained by the asymmetric oxidation of prochiral 2-methyl-1,3-propanediol in the presence of *Acetobacter pasteurianus* DSM 8937, one of the acetic acid bacterial strains. The acid **1** is the key intermediate in the synthesis of Captopril, which is an angiotensin-converting enzyme (ACE) inhibitor used for the treatment of hypertension and some types of congestive heart disorders.[212] The optically active α-hydroxyketone **2** (Scheme 8.57, comp. **2**)[213] can be obtained by the biotransformation of *mezo*-2,3-butandiol. The use of enzymes from acetic acid bacteria belonging to the *Acetobacter* and *Gluconobacter* strains allows obtaining both the enantiomers. In the reaction involving *G. cerinus* DSM 9534, the optically pure product of the (*R*)-configuration was observed; the use of *G. asaii* MIM 1000/9 afforded the (*S*)-hydroxyketone. However, *meso* and prochiral primary diols were transformed into hydroxyaldehydes by means of the horse liver alcohol dehydrogenase (HLADH)-catalyzed oxidation. Accordingly, stable five-, six-, and seven-membered

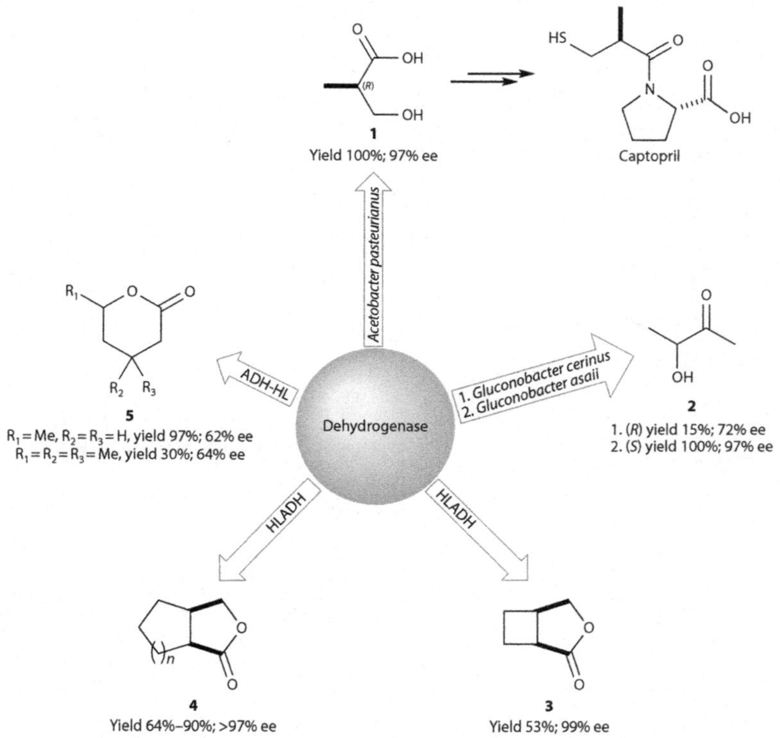

Asymmetric Biocatalysis in Organic Synthesis of Natural Products   255

SCHEME 8.57 Oxidation reaction catalyzed by dehydrogenase isolated from horse liver.

hemiacetals were formed. Further oxidation of these hemiacetals by HLADH resulted in the corresponding lactones. HLADH preferentially oxidizes only one of the primary hydroxyl groups, pro-*S* in the prochiral compounds and (*S*)-OH group in the *meso* diols. For instance, optically pure γ-lactones 3 and 4 can be obtained *via* the selective biotransformation of the hydroxyl group of the (*S*)-configuration in *meso*-1,4-diols (Scheme 8.57, comp. **3** and **4**).[214,215] The regioselective oxidation of acyclic primary and secondary 1,5- and 1,6-diols by HLADH afforded δ-i ε-lactones, respectively (Scheme 8.57, comp. **5**).[216] The highest ee's were observed for the oxidation reaction of 1,6-diols.

8.12 APPLICATION OF MONOOXYGENASES IN ENZYMATIC OXIDATION REACTIONS OF SYMMETRICAL COMPOUNDS

Cytochrome P450 monooxygenases (CYPs) constitute a family of heme-containing enzymes that exhibits a variety of catalytic activities. They catalyze different reactions, such as hydroxylation, epoxidation, oxidative deamination, or N- and (S)-oxidation. In the oxidation reaction with monooxygenases, the whole cells are commonly used as catalysts. The use of monooxygenases in the oxidation reaction of prochiral alkanes provides the optically active alcohols. It is very significant that these transformations are still difficult to carry out by chemical methods. The relative reactivity of carbon atoms in biohydroxylation reactions decreases in the following order: secondary > tertiary > primary carbon atom; therefore, it is one of the alternative methods for the preparation of mainly secondary alcohols. For example, the hydroxylation of nonactivated C–H bonds in *N*-benzyl-pyrrolidin-2-one by *Sphingomonas* sp. cells gave a 75% conversion of enantiopure (*S*)-*N*-benzyl-4-hydroxypyrrolidin-2-one (Scheme 8.58, comp. **1**).[217] The reaction was carried out in the presence of 2% glucose solution, which contributed to the cofactor regeneration. The product is a useful intermediate in the synthesis of antibiotics.

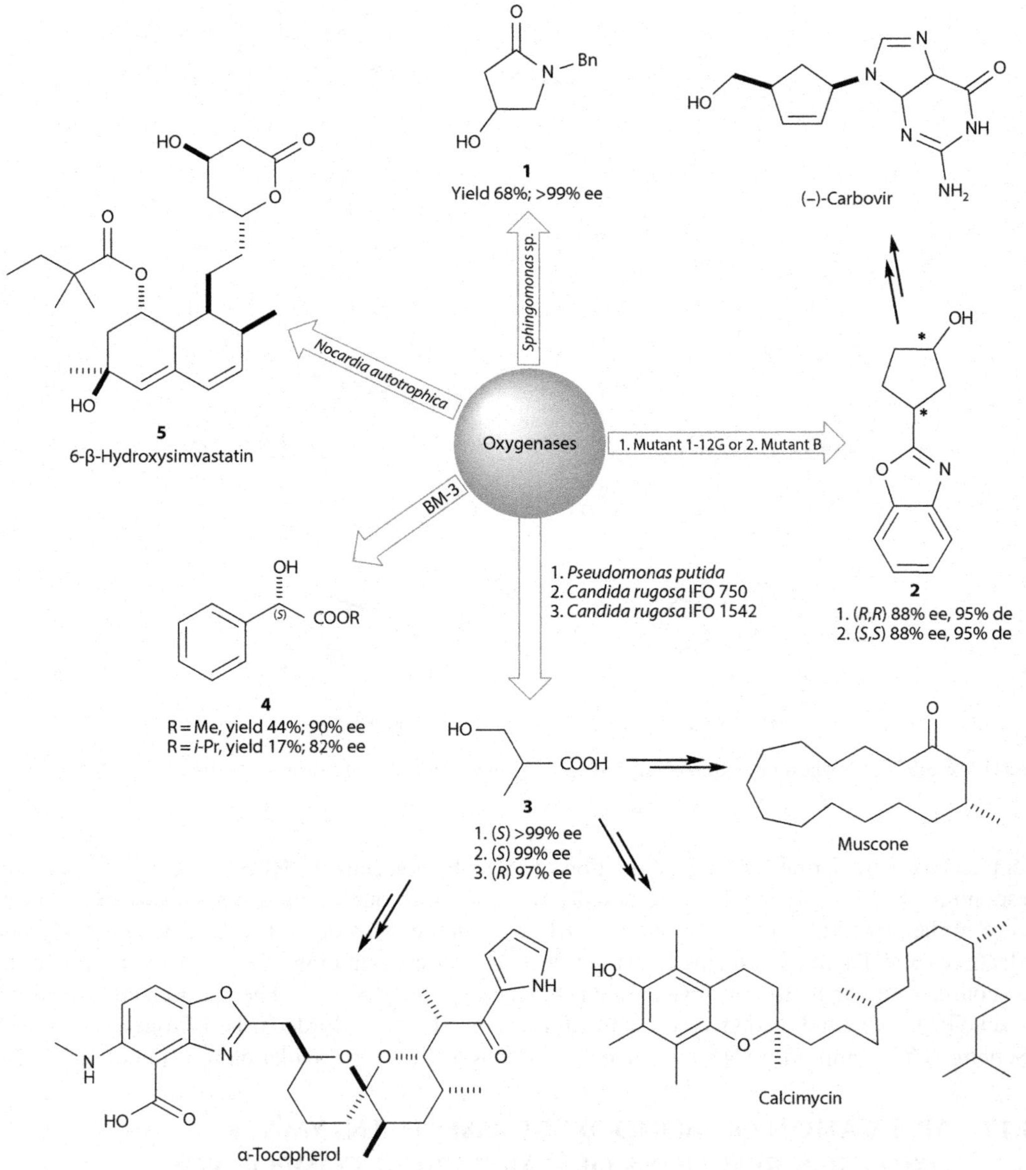

SCHEME 8.58 Examples of application of oxygenase in the oxidation reaction.

The mutant enzymes BM-3 from *Bacillus megaterium* were used in the hydroxylation of 2-cyclopentylbenzoxazole, which is used as a starting material in the synthesis of Carbovir, a carbocyclic nucleoside presenting a potential activity in fighting HIV (Scheme 8.58, comp. **2**).[218] In the presence of the 1-G mutant, the product of the (R,R)-configuration was observed, whereas the B mutant allowed obtaining the (S,S)-enantiomer. In all the cases, the reaction proceeded showing high enantio- and diastereoselectivity.

The optically active 3-hydroxyisobutyric acid (Scheme 8.58, comp. **3**) was obtained in the asymmetric hydroxylation reaction of isobutyric acid. Depending on the microorganisms used, both enantiomers were formed. *Pseudomonas putida* ATCC 21244 and *Candida rugosa* IFO 750 led to the (S)-enantiomer of optical purity, while *Candida rugosa* IFO 1542 gave the (R)-product of high

Asymmetric Biocatalysis in Organic Synthesis of Natural Products

enantiomeric excess. The resulting chiral 3-hydroxyisobutyric acid was used as a starting material for the synthesis of vitamins (e.g., α-tocopherol),[219] fragrances (e.g., muscone),[220] and antibiotics (e.g., calcimycin).[221] The selective hydroxylation of various 2-aryl acetic acid esters yielded the corresponding (S)-mandelic acid esters at high chemical and optical yields (Scheme 8.58, comp. 4). For instance, the methyl and propyl esters of (S)-mandelic acid were obtained in 90% and 82% ee, respectively. The mandelic acid derivatives were found to act as thrombin inhibitors and anticoagulants.[222]

Simvastatin, a potent cholesterol-lowering drug, was transformed into the hydroxyl counterpart by regioselective oxidation in the presence of *Nocardia autotrophica*. Preferential allyl group hydroxylation allowed obtaining 6-β-hydroxysimvastatin (Scheme 8.58, comp. 5).[223]

The enzymatic epoxidation of olefins can also be performed by monooxygenases. In the epoxidation reactions of styrenes, the use of recombinant *E. coli* growing cells containing overexpressed styrene monooxygenase yielded in the corresponding optically active oxiranes in good to excellent yields (Scheme 8.59, comp. **1a–h**). In the case of **1a–c** and **1e** compounds, (S)-enantiomers of products were obtained; the **1d, f–h** compounds gave diastereomers of (1S,2R)-configuration. **1c, 1e**, and **1h** products of excellent yields and ee's were also obtained in the epoxidation reaction of the corresponding styrenes catalyzed by styrene monooxygenase from *Pseudomonas* sp.[224,225] The asymmetric hydrolysis of *meso*-epoxides catalyzed by *Sphingomonas* sp. delivered the *trans*-diols at high ee (Scheme 8.59, comp. **2** and **3**). Various natural epoxide hydrolases were capable

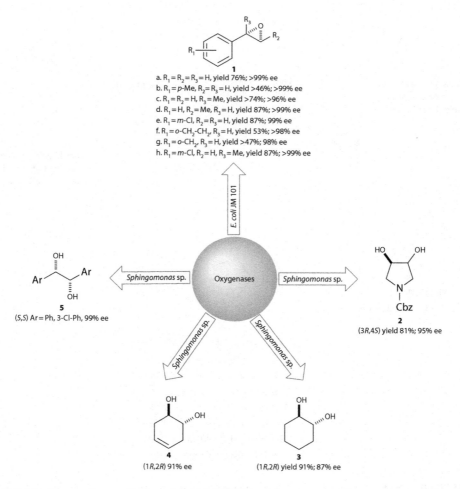

SCHEME 8.59 Enzymatic epoxidation of prochiral olefins.

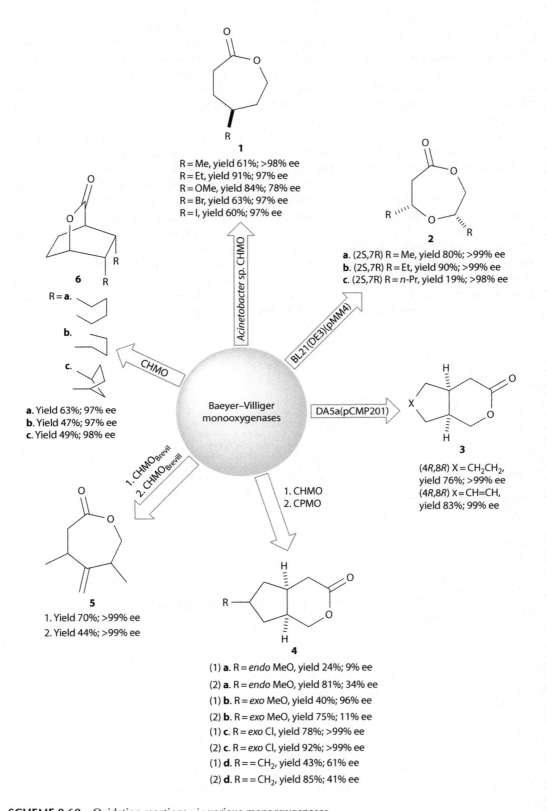

SCHEME 8.60 Oxidation reactions *via* various monooxygenases.

of selective hydrolyzing a wide range of cyclic and aryl *meso*-epoxides. From these processes, the corresponding chiral (*R,R*)- or (*S,S*)-diols were obtained at high ee's (Scheme 8.59, comp. **4** and **5**).[226–228]

Baeyer–Villiger monooxygenases (BVMO, EC 1.14.13.X) effectively catalyze the nucleophilic and electrophilic oxidation reactions of various functional groups. BVMO are highly regio- and stereoselective enzymes, and their catalytic potential is used in the synthesis of optically pure lactones and esters. From among many bacterial species having the monooxygenases capable of performing Baeyer–Villiger oxidation, the cyclohexanone monooxygenase (CHMO) from *Acinetobacter calcoaceticus* NCIMB 9871 is the most frequently investigated.

Whole cells of an *E. coli* strain that overexpresses *Acinetobacter* sp. NCIB 9871 CHMO were employed for the Baeyer–Villiger oxidation of different 4-mono- and 4,4-disubstituted cyclohexanones giving (*S*)-lactones (Scheme 8.60, comp. **1**).[229]

The recombinant cells of *E. coli* BL21 (DE3)/pMM4 CHMO isolated from *A. calcoaceticus* NCIMB 9871 were applied in the oxidation reaction of *cis*-2,6-dialkylperhydropyrans to the corresponding lactones (Scheme 8.60, comp. **2a–c**).[230] The size of the R group had a substantial impact on the substrate conversion. Ketones containing a short alkyl chain, such as methyl (**2a**) and ethyl (**2b**), were oxidized with excellent yields, while an extension of this substituent resulted in a significant decrease in conversions or even no reaction was observed. Biotransformation of the prochiral bicyclo[4.3.0]ketones in the presence of recombinant *E. coli* overexpressing *Comamonas* sp. NCIMB 9872 CPMO provided the bicyclic lactones at excellent yields and enantioselectivities (Scheme 8.60, comp. **3**).[231] Other bicyclic ketones, such as pentalenones, were subjected to the Baeyer–Villiger oxidation with both types of enzymes: recombinant *E. coli* overexpressing *Comamonas* sp. NCIMB 9872 CPMO and *Acinetobacter* sp. NCIMB 9871 CHMO. In the case of CHMO biotransformations, the *exo* geometry in the substrates determined the high optical purity of the lactone products. Furthermore, the presence of a less-polar functional group (Cl < MeO) was favorable to obtain high yield and ee.[232] Similarly, the oxidation of *meso* and prochiral bicyclic ketones with recombinant *E. coli* cells expressing two new monooxygenases from *Brevibacterium* (CHMO$_{BreviI}$ and CHMO$_{BreviII}$) led to enantiomerically pure lactones in synthetically useful yields. In the case of unsaturated ketones, oxidation occurred chemoselectively without the concomitant epoxidation of the double bond (Scheme 8.60, comp. **5**).[233] Recombinant whole cells expressing the Baeyer–Villiger monooxygenase from *Xanthobacter* sp. ZL5 were used in the biotransformation of various prochiral ketones. Not only cyclic and bicyclic but also polycyclic substrates were selectively oxidized by this enzymatic catalyst. Cyclohexanone monooxygenase (CHMO) was able to oxidize the substrates containing a large group of excellent enantioselectivity (Scheme 8.60, comp. **6**).[234]

8.13 APPLICATION OF DIOXYGENASES IN ENZYMATIC OXIDATION REACTIONS OF SYMMETRICAL COMPOUNDS

Aromatic dioxygenases such as toluene dioxygenase (TDO), naphthalene dioxygenase (NDO), and biphenyl dioxygenase (BPDO) found in prokaryotic microorganisms are enzymes belonging to the dioxygenase class. They are most commonly used in organic synthesis.

Washed-cell preparations of recombinant *E. coli* JM109 (pDTG141), engineered to express the naphthalene dioxygenase (NDO) gene from *Pseudomonas* sp. NCIB 9816-4, were used for the biooxidation of a series of diakyl and aryl alkyl sulfides (Scheme 8.61).[235]

The alkyl aryl sulfides possessing either *n*- or *iso*-alkyl groups with ≥3 carbon atoms (**1b–d**) and dialkyl sulfides with the alkyl substituent containing ≥7 carbon atoms (**1e–g**) were oxidized to the sulfoxides of (*R*)-configuration. The selectivity of alkyl aryl sulfide sulfoxidation depended on the substituent position in the aromatic ring. The closer the substituent approached the sulfur atom of the substrate, the lower the ee of the resultant sulfoxide was observed.

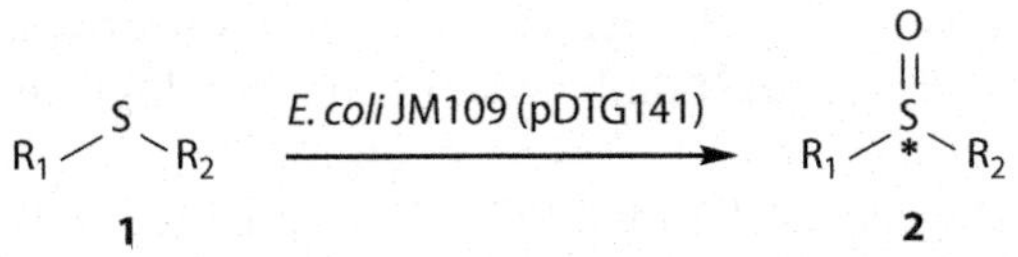

SCHEME 8.61 Oxidation of aryl alkyl sulfides by *E. coli* JM109 (pDTG141). (From Garcia-Urdiales, E. et al., *Chem. Rev.*, 105, 313, 2005.)

Sulfide	R₁	R₂	Yield (%)	ee (%)	Config.
a	Ph	Me	98	98	S
b	Ph	Pr	58	76	R
c	Ph	iPr	69	74	R
d	Ph	Bu	25	97	R
e	Me	Hep	<5	3	R
f	Me	Oct	<5	4	R
g	Me	Nom	<5	5	R
h	4-Cl-Ph	Me	10	90	S
i	3-Cl-Ph	Me	21	75	S
j	2-Cl-Ph	Me	12	31	Unknown

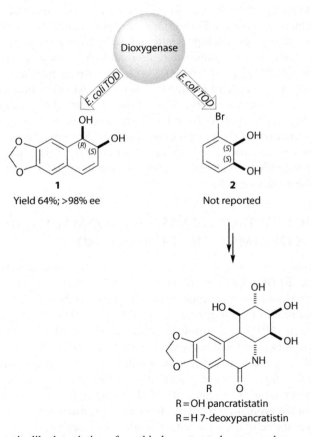

SCHEME 8.62 Enzymatic dihydroxylation of prochiral unsaturated compounds.

The enzymatic dihydroxylation of dioxole catalyzed by recombinant *E. coli* cells expressing a naphthalene dioxygenase from *Pseudomonas putida* G7 provided the corresponding polycyclic (5R,6S)-diol at good yield and ee (Scheme 8.62, comp. **1**). This compound can be applied in the synthesis of a pancratistatin analogue.

Another example of the carbon–carbon double bond *cis*-dihydroxylation is the conversion of dibromobenzenes (1,2-, 1,3-, and 1,4-isomers) to the corresponding *cis*-cyclohexadienediols by whole-cell fermentation with *E. coli* cells overexpressing TOD. One of the products, (1S,2S)-3-bromocyclohexa-3,5-diene-1,2-diol (Scheme 8.62, comp. **2**), was also used for the total synthesis of (+)-pancratistatin and (+)-7-deoxypancratistatin, promising antitumor agents.[236]

8.14 APPLICATION OF OXIDASES IN ENZYMATIC OXIDATION REACTIONS OF SYMMETRICAL COMPOUNDS

The α-oxidation of various fatty acids in the presence of an α-oxidase from germinating peas is one of a few examples of oxidase application in asymmetric organic synthesis. The intermediary α-hydroxyperoxyacids can undergo two competing reactions: the decarboxylation of the corresponding aldehydes or reduction to the (R)-2-hydroxy acids. In order to eliminate the competitive decarboxylation reaction, tin(II) chloride was used as an *in situ* reducing agent (Scheme 8.63).[237]

8.15 APPLICATION OF PEROXIDASES IN ENZYMATIC OXIDATION REACTIONS OF SYMMETRICAL COMPOUNDS

Peroxidases are the redox enzymes found in various sources such as animals, plants, and microorganisms. Due to the fact that, in contrast to monooxygenases, no additional cofactors are required, peroxidases are highly attractive for preparative biotransformation. Oxidation reactions catalyzed by (halo)peroxidases are often used in organic synthesis. N-Oxidation in amines, for instance, leads to the formation of the corresponding aliphatic N-oxides, or aromatic nitro- or nitroso-compounds. From a preparative synthesis standpoint, however, sulfoxidation of thioether is important since it was proven to proceed in a highly stereo- and enantioselective manner. Furthermore, depending on the source of the haloperoxidase, chiral sulfoxides of opposite configurations can be obtained. For example, the chloroperoxidase from *Caldariomyces fumago* is a selective catalyst for the oxidation reaction of methylthioethers to give (R)-sulfoxides (Scheme 8.64, comp. **1**). Haloperoxidase from the marine *Corallina officinalis* algae had the opposite enantiopreference forming the (S)-sulfoxides. In order to achieve a high optical purity of sulfoxides, the concentration of H_2O_2 was kept at a constant low level.[238–241] Enantioselective sulfoxidation of an organic sulfide series catalyzed by *Pseudomonas frederiksbergensis* sp. led to the corresponding (S)-sulfoxides at good yield and good to excellent enantioselectivity (Scheme 8.64, comp. **2**).[242] It was observed that the reaction

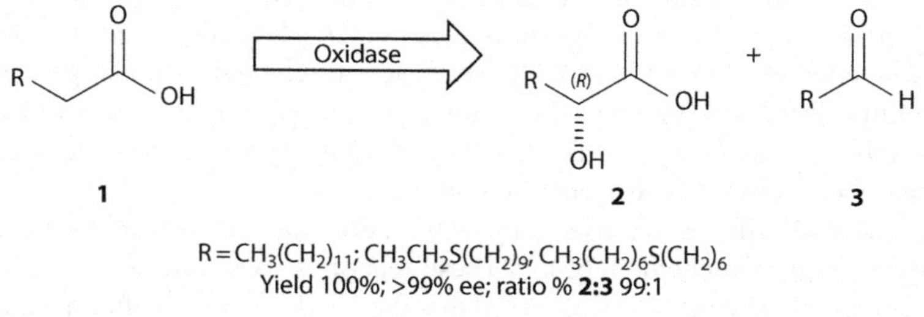

R = $CH_3(CH_2)_{11}$; $CH_3CH_2S(CH_2)_9$; $CH_3(CH_2)_6S(CH_2)_6$
Yield 100%; >99% ee; ratio % **2:3** 99:1

SCHEME 8.63 α-Oxidation of various fatty acids in the presence of an α-oxidase. (From Adam, W. et al., *J. Am. Chem. Soc.*, 120, 11044, 1998.)

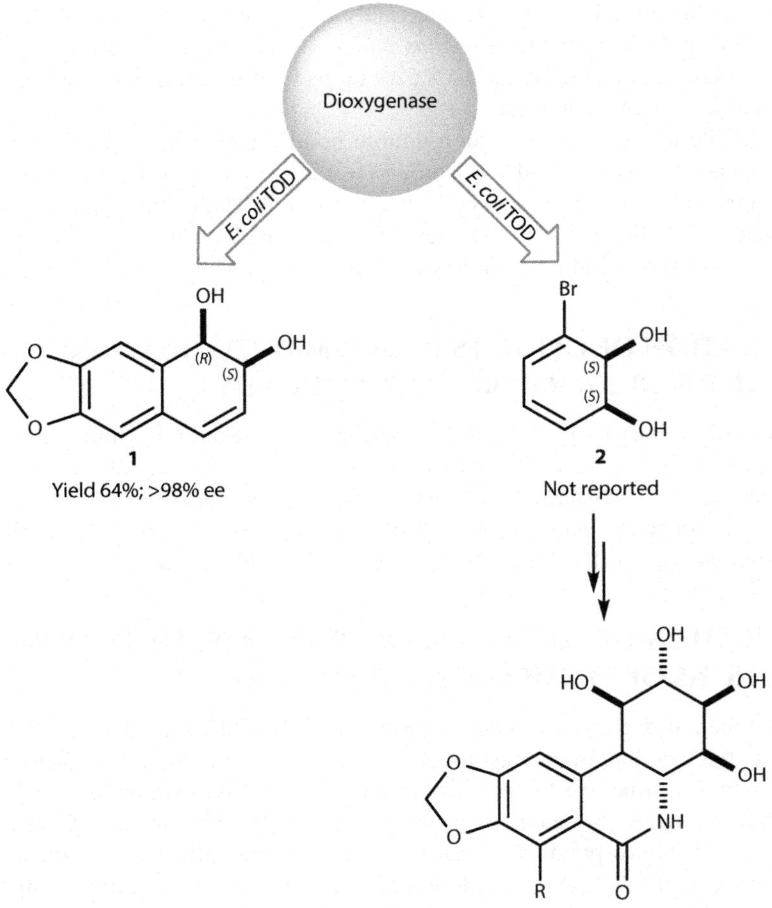

SCHEME 8.64 Enzymatic reaction of prochiral compounds with peroxidases.

time, pH, and substrate structure influenced the yield and the enantioselectivity of these reactions. The best results were obtained for alkyl aryl sulfides. It was noted that the substitution of the phenyl ring in 1,4-position affected the increase in enantioselectivity. However, a decrease in both the reaction yield and stereoselectivity took place due to an increase in the alkyl substituent size. The desymmetrization of other aliphatic aromatic sulfides in the presence of *Aspergillus terreus* CCT 3320 cells was carried out, yielding the (*S*)-sulfoxides (Scheme 8.64, comp. **3**).[243] In most cases, this biotransformation led to ee's higher than 95%. The reaction selectivity, to a large extent, depended on the reaction time. The ee product was the result of a two-stage process: enantiotopic desymmetrization followed by kinetic resolution. The first step was gaining moderate selectivity, but the latter allowed obtaining the enantiomerically enriched products.

The hydroxylation of alkynes catalyzed by chloroperoxidase (CPO) allowed the preparation of (*R*)-propargylic alcohols (Scheme 8.65).[244] These reactions were carried out in the presence of hydrogen peroxide as an oxidant. It was observed that the yields increased after the addition of more enzyme and oxidant. Generally, most of the products obtained were characterized by a good or very

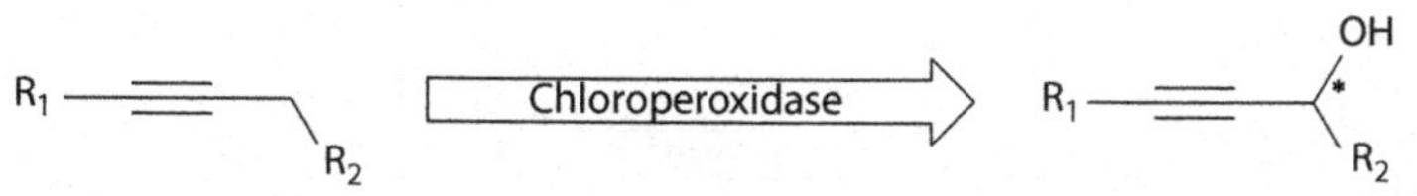

Asymmetric Biocatalysis in Organic Synthesis of Natural Products

R₁—≡—R₂ →[Chloroperoxidase]→ R₁—≡—C*(OH)(R₂)

R₁	R₂	Yield (%)	ee (%)
Me	Me	7	57
Et	Me	26	91
Pr	Me	30	87
Bu	Me	8	78
Ph	Me	15	86
CH₂OAc	Me	52	95
CH₂Br	Me	65	94
(CH₂)₂OAc	Me	26	83
(CH₂)₂Br	Me	25	94
CH₂OAc	Et	8	87

SCHEME 8.65 Hydroxylation of alkynes using chloroperoxidase. (From Garcia-Urdiales, E. et al., *Chem. Rev.*, 105, 313, 2005.)

good enantiomeric purity but at different chemical yields. The worst results were obtained for the smallest-size substrates (e.g., pent-2-yl).

Chloroperoxidase also allows carrying out the epoxidation of unsaturated compounds. The highest ee's were obtained for unfunctionalized *cis*-alkenes[245] and 1,1-disubstituted olefins (Scheme 8.66).[246,247]

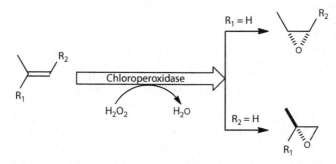

R₁	R₂	ee (%)
H	n-C₄H₉	96
H	(CH₃)₂CH–CH₂	94
H	Ph	96
Ph	H	89
CH₂–COOEt	H	93–94
(CH₂)₂–Br	H	85
n-C₅H₁₁	H	95

SCHEME 8.66 Asymmetric epoxidation of alkenes using chloroperoxidase.

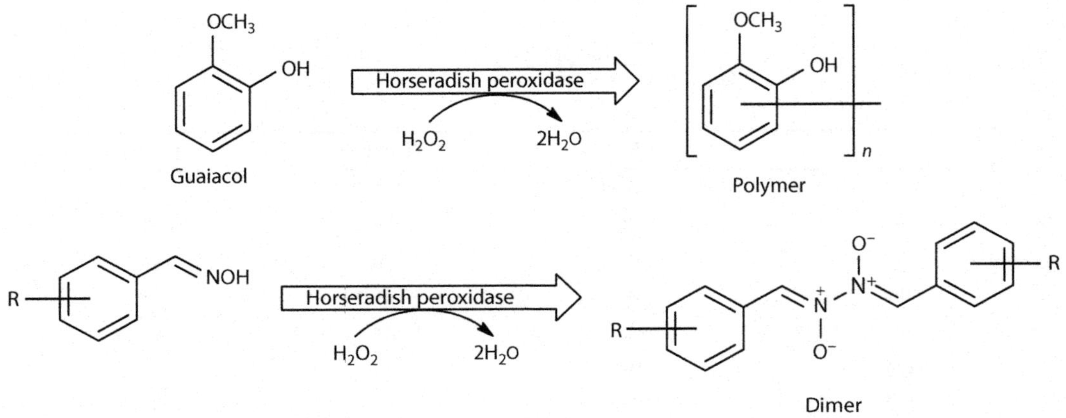

SCHEME 8.67 Horseradish peroxidase–catalyzed oxidation reaction of aromatic compounds.

The final example of peroxidase application is the oxidation of aromatic compounds. This type of reactions is commonly denoted as the "classical peroxidase reaction" since it was the first peroxidase reaction discovered. The oxidation of phenols (e.g., guaiacol, resorcin) and anilines (e.g., aniline, 1,2-dianisidine) in the presence of the horseradish peroxidase led to the formation of oligomers and polymers under mild conditions.[248–250] In some cases, the dimers[251] (e.g., aldoximes) and biaryls[252] were obtained (Scheme 8.67).

REFERENCES

1. Gawroński, J.; Gawrońska, K. *Stereochemia w syntezie organicznej*, PWN, Warszawa, Poland, 1988.
2. Siemion, I. Z. *Biostereochemia*, PWN, Warszawa, Poland, 1985.
3. Carey, F. A.; Sundberg, R. J. *Advanced Organic Chemistry*, Springer, New York, 2007.
4. Chen, C.-S.; Fujimoto, Y.; Girdaukas, G.; Sih, C. J. *J. Am. Chem. Soc.* 1982, *104*, 7294.
5. Gotor-Fernández, V.; Busto, E.; Gotor, V. *Adv. Synth. Catal.* 2006, *384*, 797.
6. Frings, K.; Koch, M.; Hartmeier, W. *Enzyme Microb. Technol.* 1999, *25*, 303.
7. Kim, M.-J.; Ann, Y.; Park, J. *Bull. Korean Chem. Soc.* 2005, *26*, 515.
8. Persson, B. A.; Larsson, A. L. E.; Ray, M. L.; Bäckvall, J.-E. *J. Am. Chem. Soc.* 1999, *121*, 1645.
9. Garcia-Urdiales, E.; Alfonso, I.; Gotor, V. *Chem. Rev.* 2005, *105*, 313.
10. Wang, Y.-F.; Chen, C.-S.; Girdaukas, G.; Sih, C. J. *J. Am. Chem. Soc.* 1984, *106*, 3695.
11. Sih, C. J.; Shieh, W.-R.; Chen, C.-S.; Wu, S.-H.; Girdaukas, G. *Int. Symp. Bioorg. Chem.* 1986, *471*, 239.
12. Kołodziejska, R.; Górecki, M.; Frelek, J.; Dramiński, M. *Tetrahedron: Asymmetry* 2012, *23*, 623.
13. Barbieri, C.; Caruso, E.; D'Arrigo, P.; Fantoni, G. P.; Servi, S. *Tetrahedron: Asymmetry* 2003, *14*, 43.
14. Gotor, V.; Alfonso, I.; Garcia-Urdiales, E. *Asymmetric Organic Synthesis with Enzymes*, Wiley-VCH Verlag GmbH & Co. KGaA, Weinheim, Germany, 2008.
15. Faber, K. *Biotransformations in Organic Chemistry*, Springer, Berlin, Germany, 1992.
16. Chen, C.-S.; Fujimoto, Y.; Sih, C. J. *J. Am. Chem. Soc.* 1981, *103*, 3580.
17. Shieh, W.-R.; Gopalan, A. S.; Sih, C. J. *J. Am. Chem. Soc.* 1985, *107*, 2993.
18. Jones, J. B.; Mehes, M. M. *Can. J. Chem.* 1979, *57*, 2245.
19. Lam, L. K. P.; Hui, R. A.; Jones, J. B. *J. Org. Chem.* 1986, *51*, 2047.
20. Zaks, A.; Klibanov, A. M. *Proc. Natl. Acad. Sci. USA* 1985, *82*, 3192.
21. Klibanov, A. *Acc. Chem. Res.* 1990, *23*, 114.
22. Zaks, A.; Klibanov, A. M. *J. Biol. Chem.* 1988, *263*, 8017.
23. Colombo, G.; Ottolino, G.; Carrea, G. *Monatsh. Chem.* 2000, *131*, 527.
24. Trodler, P.; Pleiss, J. *BMC Struct. Biol.* 2008, *8*(9), 1.
25. Halling, P. J. *Enzyme Microb. Technol.* 1994, *16*, 178.
26. Halling, P. J. *Trends Biotechnol.* 1989, *7*, 50.
27. Halling, P. J. *Enzyme Microb. Technol.* 1984, *6*, 513.

28. Goderis, H. L.; Ampe, G.; Feyten, M. P.; Fouwe, B. L.; Guffenes, W. M.; Van Cauwenbergh, S. M.; Tobback, P. P., *Biotechnol. Bioeng.* 1987, *30*, 258.
29. Jain, N.; Kumar, A.; Chauhan, S.; Chauhan, S. M. S. *Tetrahedron* 2005, *61*, 1015.
30. Yang, Z.; Pan, W. *Enzyme Microb. Technol.* 2005, *37*, 19.
31. Lozano, P.; De Diego, T.; Carrié, D.; Vaultier, M.; Iborra, J. L. *J. Mol. Catal. B: Enzym.* 2003, *21*, 9.
32. Lozano, P.; De Diego, T.; Carrié, D.; Vaultier, M.; Iborra, J. L. *Chem. Commun.* 2002, 692.
33. Reetz, M. T.; Wiesenhöfer, W.; Franciò, G.; Leitner, W. *Chem. Commun.* 2002, 992.
34. Lozano, P.; De Diego, T.; Gmouh, S.; Vaultier, M.; Iborra, J. L. *Biotechnol. Prog.* 2004, *20*, 661.
35. Keskin, S.; Kayrak-Talay, D.; Akman, U.; Hortaçsu, Ö. *J. Supercrit. Fluids* 2007, *43*, 150.
36. Miyawaki, O.; Tatsuno, M. *J. Biosci. Bioeng.* 2008, *105*, 61.
37. Lau, R. M.; Van Rantwijk, F.; Seddon, K. R.; Sheldon, R. A. *Org. Lett.* 2000, *2*, 4189.
38. Sheldon, R. A.; Lau, R. M.; Sorgedrager, M.; Van Rantwijk, F.; Seddon, K. R. *Green Chem.* 2002, *4*, 147.
39. Mastuda, T.; Harada, T.; Nakamura, K.; Ikariya, T. *Tetrahedron: Asymmetry* 2005, *16*, 909.
40. Celia, E. C.; Cerina, E.; D'Acquarica, I.; Palocci, C.; Soro, S. *J. Mol. Catal. B: Enzym.* 1999, *6*, 495.
41. Fitzpatrick, P. A.; Klibanov, A. M. *J. Am. Chem. Soc.* 1991, *113*, 3166.
42. Secundo, F.; Riva, S.; Carrea, G. *Tetrahedron: Asymmetry* 1992, *3*, 267.
43. Hirose, Y.; Kariya, K.; Sasaki, J.; Kurono, Y.; Ebike, H.; Achiwa, K. *Tetrahedron Lett.* 1992, *33*, 7157.
44. Nakamura, K.; Takebe, J.; Kitayama, T.; Ohno, A. *Tetrahedron Lett.* 1991, *32*, 4941.
45. Nakamura, K.; Kinoshita, M.; Ohno, A. *J. Am. Chem. Soc.* 1992, *114*, 8799.
46. Tawaki, S.; Klibanov, A. M. *J. Am. Chem. Soc.* 1992, *1143*, 1882.
47. Ottosson, J.; Fransson, L.; King, J. W.; Hult, K. *Biochim. Biophys. Acta* 2002, *1574*, 325.
48. Terradas, F.; Teston-Henry, M.; Fitzpatrick, P. A.; Klibanov, A. M. *J. Am. Chem. Soc.* 1993, *115*, 390.
49. Köhler, J.; Wünsch, B. *Tetrahedron: Asymmetry* 2006, *17*, 3091.
50. Phillips, R. S. *Enzyme Microb. Technol.* 1992, *14*, 417.
51. Phillips, R. S. *Trends Biotechnol.* 1996, *14*, 13.
52. Okamoto, T.; Yasuhito, E.; Ueji, S.-I. *Org. Biomol. Chem.* 2006, *4*, 1147.
53. Mine, Y.; Fukunaga, K.; Itoh, K.; Yoshimoto, M.; Nakao, K.; Sugimura, Y. *J. Biosci. Bioeng.* 2003, *95*, 441.
54. Yu, J.-H.; Klibanov, A. M. *Biotechnol. Lett.* 2006, *28*, 555.
55. Ke, T.; Klibanov, A. M. *J. Am. Chem. Soc.* 1999, *121*, 3334.
56. Lee, H. C.; Ko, Y. H.; Baek, S. B.; Kim, D. H. *Bioorg. Med. Chem. Lett.* 1998, *8*, 3379.
57. Otto, H.-H.; Schirmeister, T. *Chem. Rev.* 1997, *97*, 133.
58. Harrison, M. J.; Burton, N. A.; Hillier, I. H.; Gould, I. R. *Chem. Commun.* 1996, *24*, 2769.
59. Fersht, A. *Enzyme Structure and Mechanism*, 2nd edn., Freeman, New York, 1985.
60. Hæffner, F.; Norin, T. *Chem. Pharm. Bull.* 1999, *47*, 591.
61. Schneider, M.; Engel, N.; Boensmann, H. *Angew. Chem. Int. Ed.* 1984, *23*, 66.
62. Luyten, M.; Müller, S.; Herzog, B.; Keese, R. *Helv. Chim. Acta* 1987, *70*, 1250.
63. Iosub, V.; Haberl, A. R.; Leung, J.; Tang, M.; Vembaiyan, K.; Parvez, M.; Back, T. G. *J. Org. Chem.* 2010, *75*, 1612.
64. Honda, T.; Koizumi, T.; Komatsuzaki, Y.; Yamashita, R.; Kanai, K.; Nagase, H. *Tetrahedron: Asymmetry* 1999, *10*, 2703.
65. Lane, J. W.; Halcomb, R. L. *Org. Lett.* 2003, *5*, 4017.
66. Öhrlein, R.; Baisch, G. *Adv. Synth. Catal.* 2003, *345*, 713.
67. Yu, M. S.; Lantos, I.; Peng, Z.-Q.; Yu, J.; Cacchio, T. *Tetrahedron Lett.* 2000, *41*, 5647.
68. Mohr, P.; Waespe-Sarcevic, N.; Tamm, C.; Gawrońska, K.; Gawroński, J. K. *Helv. Chim. Acta* 1983, *66*, 2501.
69. Schregenberger, C.; Seebach, D. *Liebigs Ann. Chem.* 1986, 2081.
70. Sabbioni, G.; Jones, J. B. *J. Org. Chem.* 1987, *52*, 4565.
71. Laumen, K.; Schneider, M. *Tetrahedron Lett.* 1984, *25*, 5875.
72. Deardorff, D. R.; Mathews, A. J.; McMeekin, D. S.; Craney, C. L. *Tetrahedron Lett.* 1986, *27*, 1255.
73. Wang, Y. F.; Sih, C. J. *Tetrahedron Lett.* 1984, *25*, 4999.
74. Zutter, U.; Iding, H.; Spurr, P.; Wirz, B. *J. Org. Chem.* 2008, *73*, 4895.
75. Bloch, R.; Guibe-Jampel, E.; Girard, G. *Tetrahedron Lett.* 1985, *26*, 4087.
76. Katoh, T.; Kakiya, K.; Nakai, T.; Nakamura, S.; Nishide, K.; Node, M. *Tetrahedron: Asymmetry* 2002, *13*, 2351.
77. Böhm, C.; Austin, W. F.; Trauner, D. *Tetrahedron: Asymmetry* 2003, *14*, 71.
78. Uppenberg, J.; Öhrner, N.; Norin, M.; Hult, K.; Kleywegt, G. J.; Spener, F.; Schimd, R. D.; Schoburg, D. *J. Mol. Biol.* 1996, *259*, 704.

79. Brzozowski, A. M.; Derewenda, U.; Derewenda, Z. S.; Dodson, G. G.; Lawson, D. M.; Turkenburg, J. P.; Björkling, F.; Huge-Jensen, B.; Patkar, S.; Thim, L. *Nature* 1991, *351*, 491.
80. Derewenda, U.; Brzozowski, A. M.; Lawson, D. M.; Derewenda, Z. S. *Biochemistry* 1992, *31*, 1532.
81. Martinez, C.; Nicolas, A.; Van Tilbeurgh, H.; Egloff, M. P.; Cudrey, C.; Verger, R.; Cambillau, C. *Biochemistry* 1994, *15*, 29.
82. Jaeger, K. E.; Ransac, S.; Dijkstra, B. W.; Colson, C.; Van Heuvel, M.; Misset, O. *FEMS Microbiol. Rev.* 1994, *15*, 29.
83. Martinelle, M.; Holmquist, M.; Hult, K. *Biochim. Biophys. Acta* 1995, *1258*, 272.
84. Orrenius, J.; Hæffner, F.; Rotticci, D.; Öhrner, N.; Norin, T.; Hult, K. *Biocatal. Biotransformation* 1998, *16*, 1.
85. Uppenberg, J.; Öhrner, N.; Norin, M.; Hult, K.; Kleywegt, G. J.; Patkar, S.; Waagen, V.; Anthonsen, T.; Jones, T. A. *Biochemistry* 1995, *34*, 16838.
86. Ottosson, J., Rotticci-Mulder, J. C., Rotticci, D., Hult, K. *Protein Sci.* 2001, *10*, 1769.
87. Faber, K.; Ottolina, G.; Riva, S. *Biocatal. Biotransformation* 1993, *8*, 91.
88. Holmquist, M.; Hæffner, F.; Norin, T.; Hult, K. *Protein Sci.* 1996, *5*, 83.
89. Chênevert, R.; Pelchat, N.; Morin, P. *Tetrahedron: Asymmetry* 2009, *20*, 1191.
90. Izquierdo, I.; Plaza, M. T.; Rodríguez, M.; Tamayo, J. *Tetrahedron: Asymmetry* 1999, *10*, 449.
91. Fellows, I. M.; Kaelin, D. E.; Martin, S. F. *J. Am. Chem. Soc.* 2000, *122*, 10781.
92. Kastrinsky, D. B.; Boger, D. L. *J. Org. Chem.* 2004, *69*, 2284.
93. Morgan, B.; Dodds, D. R.; Zaks, A.; Andrews, D. R.; Klesse, R. *J. Org. Chem.* 1997, *62*, 7736.
94. Takabe, K.; Mase, N.; Hashimoto, H.; Tsuchiya, A.; Ohbayasi, T.; Yoda, H. *Bioorg. Med. Chem. Lett.* 2003, *13*, 1967.
95. Yokomatsu, T.; Takada, K.; Yasumoto, A.; Yuasa, Y.; Shibuya, S. *Heterocycles* 2002, *56*, 545.
96. Sapu, C. M.; Bäckvall, J.-E.; Deska, J. *Angew. Chem. Int. Ed.* 2011, *50*, 9731.
97. Neri, C.; Williams, J. M. *J. Adv. Synth. Catal.* 2003, *345*, 835.
98. Choi, J. Y.; Borch, R. F. *Org. Lett.* 2007, *18*, 215.
99. Tsuji, T.; Iio, Y.; Takemoto, T.; Nishi, T. *Tetrahedron: Asymmetry* 2005, *16*, 3139.
100. Takabe, K.; Iida, Y.; Hiyoshi, H.; Ono, M.; Hirose, Y.; Fukui, Y.; Yoda, H.; Mase, N. *Tetrahedron: Asymmetry* 2000, *11*, 4825.
101. Kirihara, M.; Kawasaki, M.; Takuwa, T.; Kakuda, H.; Wakikawa, T.; Takeuchi, Y.; Kirk, K. L. *Tetrahedron: Asymmetry* 2003, *14*, 1753.
102. Ríos-Lombardía, N.; Busto, E.; García-Urdiales, E.; Gotor-Fernández, V.; Gotor, V. *J. Org. Chem.* 2009, *74*, 2571.
103. Bódai, V.; Novák, L.; Poppe, L. *Synlett* 1999, 759.
104. Chênevert, R.; Courchesne, G.; Pelchat, N. *Bioorg. Med. Chem.* 2006, *14*, 5389.
105. Shimada, K.; Kauragi, Y.; Fukuyama, T. *J. Am. Chem. Soc.* 2003, *125*, 4048.
106. Chênevert, R.; Simard, M.; Bergeron, J.; Dasser, M. *Tetrahedron: Asymmetry* 2004, *15*, 1889.
107. Akai, S.; Tsujino, T.; Akiyama, E.; Tanimoto, K.; Naka, T.; Kita, Y. *J. Org. Chem.* 2004, *69*, 2478.
108. Homann, M. J.; Vail, R.; Morgan, B.; Sabesan, V.; Levy, C.; Dodds, D. R.; Zaks, A. *Adv. Synth. Catal.* 2001, *343*, 744.
109. López-García, M.; Alfonso, I.; Gotor, V. *Tetrahedron: Asymmetry* 2003, *14*, 603.
110. Chênevert, R.; Courchesne, G.; Caron, D. *Tetrahedron: Asymmetry* 2003, *14*, 2567.
111. Chênevert, R.; Rose, Y. S. *J. Org. Chem.* 2000, *65*, 1707.
112. Davoli, P.; Caselli, E.; Bucciarelli, M.; Forni, A.; Torre, G.; Prati, F. *J. Chem. Soc., Perkin Trans. 1* 2002, 1948.
113. Chênevert, R.; Jacques, F.; Giguère, P.; Dasser, M. *Tetrahedron: Asymmetry* 2008, *19*, 1333.
114. Toyama, K.; Iguchi, S.; Sakazaki, H.; Oishi, T.; Hirama, M. *Bull. Chem. Soc. Jpn.* 2001, *74*, 997.
115. Chênevert, R.; Goupil, D.; Rose, Y. S.; Bèdard, E. *Tetrahedron: Asymmetry* 1998, *9*, 4285.
116. Candy, M.; Aundran, G.; Bienaymé, B. C.; Pons, J.-M. *Org. Lett.* 2009, *11*, 4950.
117. Chênevert, R.; Ziarani, G. M.; Morin, M. P.; Dasser, M. *Tetrahedron: Asymmetry* 1999, *10*, 3117.
118. Chênevert, R.; Morin, M.-P. *J. Org. Chem.* 1999, *64*, 3178.
119. Hilpert, H.; Wirz, B. *Tetrahedron* 2001, *57*, 681.
120. Wirz, B.; Iding, H.; Hilpert, H. *Tetrahedron: Asymmetry* 2000, *11*, 4171.
121. Zhao, Y.; Wu, Y.; De Clerq, P.; Vandewalle, M.; Maillos, P.; Pascal, J.-C. *Tetrahedron: Asymmetry* 2000, *11*, 3887.
122. Chênevert, R.; Morin, P. *Bioorg. Med. Chem.* 2009, *17*, 1837–1839.
123. Matsumoto, T.; Konegawa, T.; Yamaguchi, H.; Nakamura, T.; Sugai, T.; Suzuki, K. *Synlett* 2001, 1650.
124. Oishi, T.; Maruyama, M.; Shoji, M.; Maeda, K.; Kumahara, N.; Tanaka, S.-I.; Hirama, M. *Tetrahedron* 1999, *55*, 7471.

125. Berkowitz, D. B.; Choi, S.; Maeng, J. H. *J. Org. Chem.* 2000, *65*, 847.
126. Michaud, A.; Lévesque, C.; Fila, M.; Morin, P.; Pelchat, N.; Chênevert, R. *Tetrahedron: Asymmetry* 2011, *22*, 919.
127. Tanyeli, C.; Karadağ, T.; Mecitoğlu, A. *Tetrahedron: Asymmetry* 2004, *15*, 3071.
128. Goswani, A.; Kissick, T. P. *Org. Process Res. Dev.* 2009, *13*, 483.
129. Kashima, Y.; Liu, J.; Takenami, S.; Niwayama, S. *Tetrahedron: Asymmetry* 2002, *13*, 953.
130. De Wildeman, S. M. A.; Sonke, T.; Schoemaker, T. S.; May, O. *Acc. Chem. Res.* 2007, 1260.
131. Sato, T.; Okumura, Y.; Itai, J.; Ujisawa, T. *Chem. Lett.* 1988, 1537.
132. MacLeod, R.; Prosser, H.; Fikentscher, L.; Lanyi, J.; Mosher, H. S. *Biochemistry* 1964, 3838.
133. Bernardi, R.; Bravo, P.; Cardillo, R.; Ghiringhelli, D.; Resnati, G., *J. Chem. Soc. Perkin Trans. 1* 1988, 283.
134. Iwamoto, M.; Kawada, H.; Tanaka, T.; Nakada, M. *Tetrahedron Lett.* 2003, *44*, 7239.
135. Patel, R. N.; Goswami, A.; Chu, L.; Donovan, M. J.; Nanduri, V.; Goldberg, S.; Johnston, R.; Siva, P. J.; Nielsen, B.; Fan, J.; He, W. X.; Shi, Z.; Wang, K. Y.; Eiring, R.; Cazzulino, D.; Singh, A.; Mueller, R. *Tetrahedron: Asymmetry* 2004, *15*, 1247.
136. Prasad, C. V. C.; Vig, S.; Smith, D. W.; Gao, Q.; Polson, C. T.; Corsa, J. A.; Guss, V. L.; Loo, A.; Barten, D. M.; Zheng, M.; Felsenstein, K. M.; Roberts, S. B. *Bioorg. Med. Chem. Lett.* 2004, *14*, 3535.
137. Schenk, D.; Games, D.; Seubert, P. *J. Mol. Neurosci.* 2001, *17*, 259.
138. Guo, Z.; Chen, Y.; Goswami, A.; Hanson, R. L.; Patel, R. N. *Tetrahedron: Asymmetry* 2006, *17*, 1589.
139. Kawano, S.; Horikawa, M.; Yasohara, Y.; Hasegawa, J. *Biosci. Biotechnol. Biochem.* 2003, *67*, 809.
140. Patel, R. N.; Chu, L.; Mueller, R. *Tetrahedron: Asymmetry* 2003, *14*, 3105.
141. Patel, R. N. *Adv. Synth. Catal.* 2001, *343*, 527.
142. Nakamura, K.; Takenaka, K.; Fujii, M.; Ida, Y. *Tetrahedron Lett.* 2002, *43*, 3629.
143. Goswami, A.; Bezbaruah, R. L.; Goswami, J.; Borthakur, N.; Dey, D.; Hazarika, A. K. *Tetrahedron: Asymmetry* 2000, *11*, 3701.
144. Goswami, J.; Bezbaruah, R. L.; Goswami, A.; Borthakur, N. *Tetrahedron: Asymmetry* 2001, *12*, 3343.
145. Yadav, J. S.; Nanda, S.; Reddy, P. T.; Rao, A. B. *J. Org. Chem.* 2002, *67*, 3900.
146. Homann, M. J.; Vail, R. B.; Previte, E., Tamarez, M., Morgan, B., Dodds, D. R., Zaks, A. *Tetrahedron* 2004, *60*, 789.
147. Wei, Z.-L.; Lin, G.-Q.; Li, Z.-Y. *Bioorg. Med. Chem.* 2000, *8*, 1129.
148. Stampfer, W.; Kosjek, B.; Moitzi, C.; Kroutil, W.; Faber, K. *Angew. Chem. Int. Ed.* 2002, *41*, 1014.
149. Wittman, M.; Carboni, J.; Attar, R.; Balasubramanian, B.; Balimane, P.; Brassil, P.; Beaulieu, F.; Chang, C.; Clarke, W.; Dell, J.; Eummer, J.; Frennesson, D.; Gottardis, M.; Greer, A.; Hansel, S.; Hurlburt, W.; Jacobson, B.; Krishnananthan, S.; Lee, F. Y.; Li, A.; Lin, T.-A.; Liu, P.; Ouellet, C.; Sang, X.; Saulnier, M. G.; Stoffan, K.; Sun, Y.; Velaparthi, U.; Wong, H.; Yang, Z.; Zimmermann, K.; Zoeckler, M.; Vyas, D. *J. Med. Chem.* 2005, *48*, 5639.
150. Hanson, R. L.; Goldberg, S.; Goswami, A.; Tully, T. P.; Patel, R. N. *Adv. Synth. Catal.* 2005, *347*, 1073.
151. Butt, S.; Davies, H. G.; Dawson, M. J.; Lawrence, G. C.; Leaver, J.; Roberts, S. M.; Turner, M. K.; Wakefield, B. J.; Wall, W. F.; Winders, J. A. *J. Chem. Soc. Perkin Trans. 1* 1987, 903.
152. Hummel, W.; Abokitse, K.; Drauz, K.; Rollmann, C.; Gröger, H. *Adv. Synth. Catal.* 2003, *345*, 153.
153. Zhu, D.; Yang, Y.; Hua, L. *J. Org. Chem.* 2006, *71*, 4202.
154. Burda, E.; Hummel, W.; Gröger, H. *Angew. Chem. Int. Ed.* 2008, *47*, 9551.
155. Matsuda, T.; Harada, T. *J. Org. Chem.* 2000, *65*, 157.
156. Ema, T.; Moriya, H.; Kofukuda, T.; Ishida, T.; Maehara, K.; Utaka, M.; Sakai, T. *J. Org. Chem.* 2001, *66*, 8682.
157. Yadav, J. S.; Reddy, P. T.; Nanda, S.; Rao, A. B. *Tetrahedron: Asymmetry* 2001, *12*, 3381.
158. Roy, S.; Alexandre, V.; Neuwels, M.; Le Texier, L. *Adv. Synth. Catal.* 2001, *343*, 738.
159. Burns, M. P.; Wong, J. W. *Chem. Abstr.* 2002, *137*, 2466.
160. Bold, G.; Fässler, A.; Capraro, H.-G.; Cozens, R.; Klimkait, T.; Lazdins, J.; Mestan, J.; Poncioni, B.; Rösel, J.; Stover, D.; Tintelnot-Blomley, M.; Acemoglu, F.; Beck, W.; Boss, E.; Eschbach, M.; Hürlimann, T.; Masso, E.; Roussel, S.; Ucci-Stoll, K.; Wyss, D.; Lang, M. *J. Med. Chem.* 1998, *41*, 3387.
161. Robinson, B. S.; Riccardi, K. A.; Gong, Y.-F.; Guo, Q.; Stock, D. A.; Blair, W. S.; Terry, B. J.; Deminie, C. A.; Djang, F.; Colonno, R. J.; Lin, P.-F. *Antimicrob. Agents Chemother.* 2000, *2093*, 44.
162. Kadnikova, E. N.; Kostić, N. M. *J. Non-Cryst. Solids* 2001, *283*, 63.
163. Kadnikova, E. N.; Kostić, N. M. *J. Org. Chem.* 2003, *68*, 2600.
164. Yamamoto, H.; Ueda, M.; Ritsuzui, P.; Hamatani, T. E. *Chem. Abstr.* 2003, *139*, 35200.
165. Patel, R. N.; Chu, L.; Nanduri, V. N.; Jianqing, L.; Kotnis, A.; Parker, W. L.; Liu, M.; Mueller, R. *Tetrahedron: Asymmetry* 2005, *16*, 2778.

166. Kizaki, N.; Yasohara, Y.; Hasegawa, J.; Wada, M.; Kataoka, M.; Shimizu, S. *Appl. Microbiol. Biotechnol.* 2001, *55*, 590.
167. Patel, R. N.; McNamee, C. G.; Banerjee, A.; Howell, J. M.; Robison, R. S.; Szarka, L. *J. Enzyme Microb. Technol.* 1992, *14*, 731.
168. Aragozzini, F.; Valenti, M.; Santaniello, E.; Ferraboschi, P.; Grisenti, P. *Biocatalysis* 1992, *5*, 325.
169. Yamamoto, H.; Kimoto, N.; Matsuyama, A.; Kabayashi, Y. *Biosci. Biotechnol. Biochem.* 2002, *66*, 1775.
170. Akakabe, Y.; Takahashi, M.; Kamezawa, M.; Kikuchi, K.; Tachibana, H.; Ohtani, T.; Naoshima, Y. *J. Chem. Soc., Perkin Trans. 1* 1995, 1295.
171. Kataoka, M.; Yamamoto, K.; Kawabata, H.; Wada, M.; Kita, K.; Yanase, H.; Shimizu, S. *Appl. Microbiol. Biotechnol.* 1999, *51*, 486.
172. Yamamoto, H.; Matsuyama, A.; Kabayashi, Y. *Biosci. Biotechnol. Biochem.* 2002, *66*, 481.
173. Zelinski, T.; Kula, M.-R. *Bioorg. Med. Chem. Lett.* 1994, *2*, 421.
174. Itoh, N.; Matsuda, M.; Mabuchi, M.; Dairi, T.; Wang, J. *Eur. J. Biochem.* 2002, *269*, 2394.
175. Yasohara, Y.; Kizaki, N.; Hasegawa, J.; Takahashi, S.; Wada, M.; Kataoka, M.; Shimizu, S. *Appl. Microbiol. Biotechnol.* 1999, *51*, 847.
176. Wada, M.; Kataoka, M.; Kawabata, H.; Yasohara, Y.; Kizaki, N.; Hasegawa, J.; Shimizu, S. *Biosci. Biotechnol. Biochem.* 1998, *62*, 280.
177. Wada, M.; Kawabata, H.; Yoshizumi, A.; Kataoka, M.; Nakamori, S.; Yasohara, Y.; Kizaki, N.; Hasegawa, J.; Shimizu, S. *J. Biosci. Bioeng.* 1999, *87*, 144.
178. Anson, C. E.; Bibb, M. J.; Booker-Milburn, K. I.; Clissold, C.; Haley, P. J.; Hopwood, D. A.; Ichinose, K.; Revill, W. P.; Stephenson, G. R.; Surti, C. M. *Angew. Chem. Int. Ed.* 2000, *39*, 224.
179. Fujisawa, T.; Itoh, T.; Sato, T. *Tetrahedron Lett.* 1984, *25*, 5083.
180. Buisson, D.; Henrot, S.; Larcheveque, M.; Azerad, R. *Tetrahedron Lett.* 1987, *28*, 5033.
181. Feske, B. D.; Kaluzna, I. A.; Stewart, J. D. *J. Org. Chem.* 2005, *70*, 9654.
182. Padhi, S. K.; Kaluzna, I. A.; Buisson, D.; Azerad, R.; Stewart, J. D. *Tetrahedron: Asymmetry* 2007, *18*, 2133.
183. Kayser, M. M.; Mihovilovic, M. D.; Kearns, J.; Feicht, A.; Stewart, J. D. *J. Org. Chem.* 1999, *64*, 6603.
184. Patel, R. N.; Chu, L.; Chidambaram, R.; Zhu, J.; Kant, J. *Tetrahedron: Asymmetry* 2002, *13*, 349.
185. Wolberg, M.; Hummel, W.; Müller, M. *Chem. Eur. J.* 2001, *7*, 4562.
186. Wolberg, M.; Ji, A.; Hummel, W.; Müller, M. *Synthesis* 2001, 937.
187. Enders, D.; Vicario, J. L.; Job, A.; Wolberg, M.; Müller, M. *Chem. Eur. J.* 2002, *8*, 4272.
188. Vicario, J. L.; Job, A.; Wolberg, M.; Müller, M.; Enders, D. *Org. Lett.* 2002, *4*, 1023.
189. Job, A.; Wolberg, M.; Müller, M.; Enders, D. *Synthesis* 2001, 1796.
190. Fujisawa, T.; Kojima, E.; Sato, T. *Chem. Lett.* 1987, 2227.
191. Takeshita, M.; Sato, T. *Chem. Pharm. Bull.* 1989, *37*, 1085.
192. Bortolini, O.; Fantin, G.; Fogagnolo, M.; Giovannini, P. P.; Guerrini, A.; Medici, A. *J. Org. Chem.* 1997, *62*, 1854.
193. Gotor, V.; Dehli, J. R.; Rebolledo, F. *J. Chem. Soc., Perkin Trans. 1* 2000, 307.
194. Dehli, J. R.; Gotor, V. *Tetrahedron: Asymmetry* 2001, *12*, 1485.
195. Dehli, J. R.; Gotor, V. *Tetrahedron: Asymmetry* 2000, *11*, 3693.
196. Prasad, C. V. C.; Wallace, O. B.; Noonan, J. W.; Sloan, C. P.; Lau, W.; Vig, S.; Parker, M. F.; Smith, D. W.; Hansel, S. B.; Polson, C. T.; Barten, D. M.; Felsenstein, K. M.; Roberts, S. B. *Bioorg. Med. Chem. Lett.* 2004, *14*, 3361.
197. Hall, M.; Bommarius, A. S. *Chem. Rev.* 2011, *111*, 4088.
198. Gramatica, P.; Manitto, P.; Monti, D.; Speranza, G. *Tetrahedron* 1987, *43*, 4481.
199. Gramatica, P.; Manitto, P.; Monti, D.; Speranza, G. *Tetrahedron* 1988, *44*, 1299.
200. Stueckler, C.; Mueller, N. J.; Winkler, C. K.; Glueck, S. M.; Gruber, K.; Steinkellner, G.; Faber, K. *Dalton Trans.* 2010, *39*, 8472.
201. Fuganti, C.; Serra, S. *J. Chem. Soc. Perkin Trans. 1* 2000, 3758.
202. Serra, S.; Fuganti, C. *Tetrahedron: Asymmetry* 2001, *12*, 2191.
203. Leuenberger, H. G.; Boguth, W.; Widmer, E.; Zell, R. *Helv. Chim. Acta* 1976, *59*, 1832.
204. Stueckler, C.; Reiter, T. C.; Baudendistel, N.; Faber, K. *Tetrahedron* 2010, *66*, 663.
205. Hall, M.; Stueckler, C.; Kroutil, W.; Macheroux, P.; Faber, K. *Angew. Chem. Int. Ed.* 2007, *46*, 3934.
206. Hall, M.; Stueckler, C.; Hauer, B.; Stuermer, R.; Friedrich, T.; Breuer, M.; Kroutil, W.; Faber, K. *Eur. J. Org. Chem.* 2008, *9*, 1511.
207. De Mancilha, M.; De Conti, R.; Moran, P. J. S.; Rodrigues, J. A. R. *Arkivoc* 2001, 85.
208. Utaka, M.; Konishi, S.; Mizuoka, A.; Ohkubo, T.; Sakai, T.; Tsuboi, S.; Takeda, A. *J. Org. Chem.* 1989, *54*, 4989.

209. Muller, A.; Hauer, B.; Rosche, B. *Biotechnol. Bioeng.* 2007, *98*, 22.
210. Stueckler, C.; Hall, M.; Ehammer, H.; Pointner, E.; Kroutil, W.; Macheroux, P.; Faber, K. *Org. Lett.* 2007, *9*, 5409.
211. Stueckler, C.; Winkler, C. K.; Bonnekessel, M.; Faber, K. *Adv. Synth. Catal.* 2010, *352*, 2663.
212. Molinari, F.; Gandolfi, R.; Villa, R.; Urban, E.; Kiener, A. *Tetrahedron: Asymmetry* 2003, *14*, 2041.
213. Romano, A.; Gandolfi, R.; Nitti, P.; Rollini, M.; Mollinari, F. *J. Mol. Catal. B: Enzym.* 2002, *17*, 235.
214. Hertweck, C.; Boland, W. *J. Prakt. Chem.-Chem.-Ztg.* 1997, *339*, 754.
215. Schoffers, E.; Golebiowski, A.; Johnson, C. R. *Tetrahedron* 1996, *52*, 3769.
216. Boratynski, F.; Kielbowicz, G.; Wawrzenczyk, C. *J. Mol. Catal. B: Enzym.* 2010, *65*, 30.
217. Chang, D.; Feiten, H.-J.; Witholt, B.; Li, Z. *Tetrahedron: Asymmetry* 2002, *13*, 2141.
218. Munzer, D. F.; Meinhold, P.; Peters, M. W.; Feichtenhofer, S.; Griengl, H.; Arnold, F. H.; Glieder, A.; De Raadt, A. *Chem. Commun.* 2005, 2597.
219. Cohen, N.; Eichel, W. F.; Lopersti, R. J.; Neukom, C.; Saucy, G. *J. Org. Chem.* 1976, *41*, 3505.
220. Branca, Q.; Fischli, A. *Helv. Chim. Acta* 1977, *60*, 925.
221. Evans, D. A.; Sacks, C. E.; Kleschick, W. A.; Taber, T. R. *J. Am. Chem. Soc.* 1979, *101*, 6789.
222. Landwehr, M.; Hochrein, L.; Otey, C. R.; Kasrayan, A.; Bäckvall, J.-E.; Arnold, F. H. *J. Am. Chem. Soc.* 2006, *128*, 6058.
223. Gbewonyo, K.; Buckland, B. C.; Lilly, M. D. *Biotechnol. Bioeng.* 1991, *37*, 1101.
224. Schmid, A.; Hofstetter, K.; Feiten, H.-J.; Hollmann, F.; Witholt, B. *Adv. Synth. Catal.* 2001, *343*, 73.
225. Hofstetter, K.; Lutz, J.; Lang, I.; Witholt, B.; Schmid, A. *Angew. Chem. Int. Ed.* 2004, *43*, 2163.
226. Chang, D.; Wang, Z.; Heringa, M. F.; Wirthner, R.; Witholt, B.; Li, Z. *Chem. Commun.* 2003, 960.
227. Chang, D.; Heringa, M. F.; Witholt, B.; Li, Z. *J. Org. Chem.* 2003, *68*, 8599.
228. Zhao, L.; Han, B.; Huang, Z.; Miller, M.; Huang, H.; Malashock, D. S.; Zhu, Z.; Milan, A.; Robertson, D. E.; Weiner, D. P.; Burk, M. J. *J. Am. Chem. Soc.* 2004, *126*, 11156.
229. Mihovilovic, M. D.; Chen, G.; Wang, S.; Kyte, B.; Rochon, F.; Kayser, M. M.; Stewart, J. D. *J. Org. Chem.* 2001, *66*, 733.
230. Mihovilovic, M. D.; Rudroff, F.; Kandioller, W.; Grötzl, B.; Stanetty, P.; Spreitzer, H. *Synlett* 2003, 1973.
231. Mihovilovic, M. D.; Müller, B.; Kayser, M. M.; Stanetty, P. *Synlett* 2002, 700.
232. Mihovilovic, M. D.; Müller, B.; Schulze, A.; Stanetty, P.; Kayser, M. M. *Eur. J. Org. Chem.* 2003, *12*, 2243.
233. Mihovilovic, M. D.; Rudroff, F.; Müller, B.; Stanetty, P. *Bioorg. Med. Chem. Lett.* 2003, *13*, 1479.
234. Rial, D. V.; Bianchi, D. A.; Kapitanova, P.; Lengar, A.; Van Beilen, J. B.; Mihovilovic, M. D. *Eur. J. Org. Chem.* 2008, *7*, 1203.
235. Kerridge, A.; Willetts, A.; Holland, H. *J. Mol. Catal. B: Enzym.* 1999, *6*, 59.
236. Phung, A. N.; Zannetti, M. T.; Whited, G.; Fessner, W.-D. *Angew. Chem. Int. Ed.* 2003, *42*, 4821.
237. Adam, W.; Boland, W.; Hartmann-Schreier, J.; Humpf, H.-U.; Lazarus, M.; Saffert, A.; Saha-Möller, C. R.; Schreier, P. *J. Am. Chem. Soc.* 1998, *120*, 11044.
238. Kobayashi, S.; Nakano, M.; Kimura, T.; Schaap, A. P. *Biochemistry* 1987, *26*, 5019.
239. Andersson, M.; Willetts, A.; Allenmark, S. *J. Org. Chem.* 1997, *62*, 8455.
240. Allenmark, S. G.; Andersson, M. A. *Tetrahedron: Asymmetry* 1996, *7*, 1089.
241. Andersson, M. A., Allenmark, S. G. *Tetrahedron* 1998, *54*, 15293.
242. Adam, W.; Heckel, F.; Saha-Möller, C. R.; Taupp, M.; Screier, P. *Tetrahedron: Asymmetry* 2004, *15*, 983.
243. Porto, A. L. M.; Cassiola, F.; Dias, S. L. P.; Joekes, I.; Gushikem, Y.; Rodrigues, J. A. R.; Moran, P. J. S.; Manfio, G. P.; Marsaioli, A. J. *J. Mol. Catal. B: Enzym.* 2002, *327*, 19.
244. Hu, S.; Hager, L. P. *J. Am. Chem. Soc.* 1999, *121*, 872.
245. Allain, E. J.; Hager, L. P.; Deng, L.; Jacobsen, E. N. *J. Am. Chem. Soc.* 1993, *115*, 4415.
246. Dexter, A. F.; Lakner, F. J.; Campbell, R. A.; Hager, L. P. *J. Am. Chem. Soc.* 1995, *117*, 6412.
247. Lakner, F. J.; Hager, L. P. *J. Org. Chem.* 1996, *61*, 3923.
248. Dordick, J. S. *Trends Biotechnol.* 1992, *10*, 287.
249. Uyama, H.; Kurioka, H.; Sugihara, J.; Kobayashi, S. *Bull. Chem. Soc. Jpn.* 1996, *69*, 189.
250. Kobayashi, S.; Shoda, S.; Uyama, H. *Adv. Polym. Sci.* 1995, *121*, 1.
251. Fukunishi, K.; Kitada, K.; Naito, I. *Synthesis* 1991, 237.
252. Schmitt, M. M.; Schüler, E.; Braun, M.; Häring, D.; Schreier, P. *Tetrahedron Lett.* 1998, *39*, 2945.

9 Asymmetric Synthesis of Biaryls and Axially Chiral Natural Products

Renata Kołodziejska and Agnieszka Tafelska-Kaczmarek

CONTENTS

9.1 Introduction 271
9.2 Chirality in Biaryl Compounds 273
9.3 Bridged and Nonbridged Biaryl Compounds 273
9.4 Selective Construction of Biaryl Axes 274
    9.4.1 Atroposelective Synthesis of Axially Chiral Biaryls 274
    9.4.2 Intramolecular Coupling with Chiral Tethers 274
    9.4.3 Intermolecular Coupling with Chiral *Ortho* Substituents 277
    9.4.4 Intermolecular Coupling with Chiral Leaving Groups 279
    9.4.5 Intermolecular Coupling with the Element of Planar Chirality 279
9.5 Oxidative Homocoupling in the Presence of Chiral Additives 281
9.6 Redox-Neutral Cross-Coupling Catalyzed by Chiral Metal Complexes 282
9.7 Transformations of Stable but Achiral Biaryls and Conformationally Unstable but Chiral Biaryls 284
9.8 Desymmetrization of Prostereogenic Biaryl Compounds 284
9.9 Conversion of Axially Chiral but Conformationally Unstable Biaryl Compounds—Introduction of Additional Substituents in the *Ortho* Position 285
9.10 Conversion of Axially Chiral but Conformationally Unstable Biaryl Compounds—Bridge Formation 285
9.11 Conversion of Axially Chiral but Conformationally Unstable Biaryl Compounds—Cleavage of a Bridge 286
9.12 Asymmetric Axially Chiral Biaryl Synthesis by the Construction of an Aromatic Ring 287
References 290

9.1 INTRODUCTION

In the early twentieth century, it was already known that chemical compounds might be chiral without containing the chiral atoms. The presence of the stereogenic center is a sufficient but not necessary condition that the molecule appears in two forms, which are mirror images. In certain cases, the limit of free rotation in the molecule may result in asymmetry, for example, inhibition of rotation around single bond leads to axial isomers. This is the kind of conformational isomerism, which according to the nomenclature is called atropisomerism.[1,2] For the first time, the axially chiral compounds were described in 1922, while the term "atropisomerism" was introduced 11 years later by Richard Kuhn.

The most often optically active molecules without stereogenic atoms possessing an axial chirality are biaryls, which are commonly found in nature. In most cases, pharmacological activity of biaryls is associated with the presence of axial chirality. The famous antibiotic heptapeptide vancomycin is one of the representatives of this class of compounds (Figure 9.1).

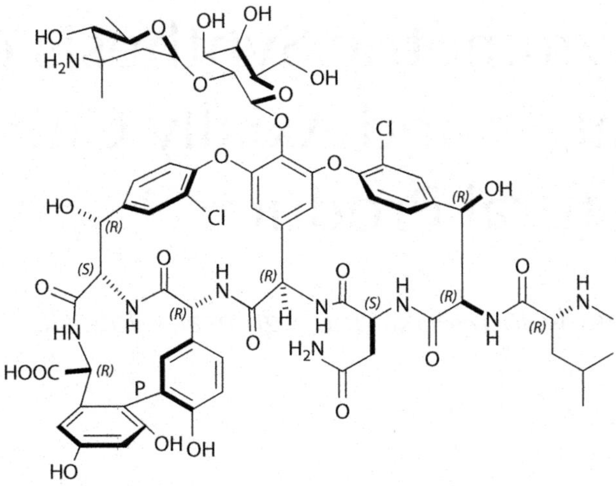

FIGURE 9.1 Structure of vancomycin.

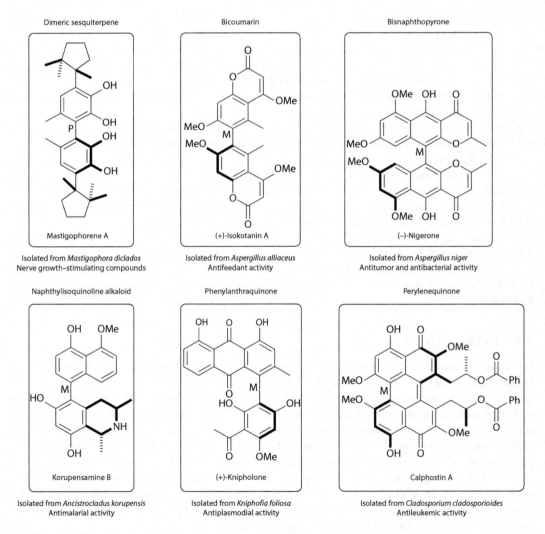

FIGURE 9.2 Selected examples of naturally occurring axially chiral biaryl compounds.

Asymmetric Synthesis of Biaryls and Axially Chiral Natural Products

To the group of natural axially chiral biarylic compounds belong also dimeric sesquiterpenes like mastigophorene B, bicoumarins like (+)-isokotanin A, bisnaphtylopyrone-nigeron, perylenequinone-calphostin A-D, and also unsymmetric biaryls, for example, phenylanthraquinone–knipholone, naphthylisoquinoline alkaloids like dioncopeltine A, ancistroealaine A, korupensamine B (Figure 9.2). The biscarbazole murrastifoline-F and the naphthylisoquinoline ancisheynine present very rare natural N,C-bonded biaryls. Furthermore, such compounds are used as reagents and catalysts in asymmetric synthesis.[3]

9.2 CHIRALITY IN BIARYL COMPOUNDS

Axially chiral biaryls have different substituents on both sides of the axis, for example, A ≠ B and A' ≠ B'. In order to determine the configuration, the priority of substituents (A(A') or B(B')) should be assigned according to the Cahna-Ingolda-Preloga rules. Assuming that the A(A') is a substituent of higher priority than B(B'), and following the path from the substituent of higher priority at the closest to the observer ring (proximal ring) to the higher-ranking one at the distal ring (ring that is farther from observer), that is, A → A', the turn is clockwise, the absolute configuration of biaryl compound is P (plus). In contrast, the 90° turn in the opposite direction (counterclockwise direction) gives isomer of the absolute configuration of M (minus) (Scheme 9.1). For chiral biaryl compounds in which A = A' and B = B', the absolute configuration also can be determined. In this case, for two pairs of the same substituents, an additional rule was introduced: the groups located in the proximal ring have priority over the substituents at the distal ring.[1,2,4]

9.3 BRIDGED AND NONBRIDGED BIARYL COMPOUNDS

Generally, chiral biaryls are divided into bridged biaryls (the presence of the additional ring bridging), and biaryls, which do not contain the additional ring. The thermal stability of both enantiomeric/diastereomeric forms is an essential precondition for atropisomerism. For a given temperature, the stable conformational isomers may coexist when their half-life is at least 1000 s, which gives the minimum energy barrier of 93 kJ mol^{-1} (22 kcal mol^{-1}) at 300 K (27°C). The required minimum energy barrier is temperature dependent. Even conformationally labile chiral biaryls with a low degree of steric hindrance can be separated into two atropisomers at low temperature. However, at higher temperature, atropisomerization (racemization) frequently occurs, despite the isomers' stability at room temperature. In the case of nonbridged biaryl compounds, atropisomerism depends on the presence of bulky substituents next to the axis, whereas bridged biaryls should have the rigid and long chains of rings. In some cases, stable chiral biaryls can undergo atropisomerization under the influence of chemical and photochemical factors. Therefore, particular attention should be paid to the proper storage of the samples in order to avoid the racemization.

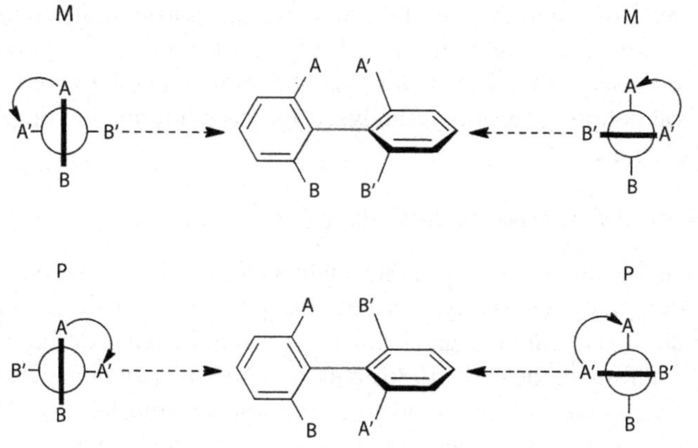

SCHEME 9.1 Assignment of absolute configuration in chiral biaryl compounds. (From Bringmann, G. et al., *Angew. Chem. Int. Ed.*, 44, 5384, 2005.)

Nonbridged biaryls (i.e., open chain), both mono- and disubstituted, are unstable at room temperature. Tri-*ortho*-substituted biaryl compounds can form stable atropisomers, for example, naphthylisoquinoline alkaloid dioncophylline E, which is stable for a few hours due to the presence of bulky substituents. However, tetra-*ortho*-substituted biaryls are particularly conformationally stable, even if the substituents are small. Stability of *ortho*-atropisomers is the result of steric effects that increase with the increase of van der Waals radii of the substituents (I > Br > Me > Cl > NO$_2$ > CO$_2$H > OMe). Also, substituents in the *meta* and *para* positions influence the increase of rotational barrier; both the size and chemical character of these substituents play a role (electron-withdrawing groups decrease the electron density at the carbon atom C-1).[1,5,6]

Conformational stability of bridged biaryl systems depends on the bridge size. The presence of a single bridging atom (a five-membered ring is formed) does not lead to the stable atropisomers at room temperature. A six-membered ring insignificantly hinders axial rotation; in this case, the isomers' stability depends on the size of the substituents at the *ortho* position. Among the bridged biaryls, the most stable are compounds containing seven-membered rings, despite low steric hindrance in the proximity of the axis. Their stability is comparable to those of nonbridged tetra-*ortho*-substituted analogues. Most frequently, this type of compounds is additionally stabilized by hydrogen bonding.[1,5,6]

9.4 SELECTIVE CONSTRUCTION OF BIARYL AXES

Axially chiral biaryls can be obtained by

1. The classic atroposelective aryl–aryl coupling reaction between the appropriately modified two aryl units
2. Atroposelective conversion of achiral and conformationally unstable biaryls
3. The specific methods in which nonaromatic parts of one aryl ring are converted to a second aryl ring[1]

9.4.1 Atroposelective Synthesis of Axially Chiral Biaryls

A direct aryl–aryl coupling reaction is the most common method of synthesis of axially chiral biaryls. Diastereoselective or enantioselective coupling can be accomplished by three main strategies (Scheme 9.2)[1,7]:

1. Intramolecular coupling reaction between two aryl substrates by the use of the chiral tether as a source of asymmetric information.
2. Intermolecular reaction of the modified aryl compounds containing a chiral auxiliary. A source of chiral information can be a planar-chiral element (transition metal complex), the chiral leaving group, and the chiral *ortho* substituent close to the coupling site.
3. Biaryl coupling in the presence of chiral additives, for example, stoichiometric or catalytic oxidation in the presence of the transition metal complexes containing chiral ligands, and the redox-neutral coupling reactions catalyzed by transition metal complexes with chiral bidentate N,P-ligands.

9.4.2 Intramolecular Coupling with Chiral Tethers

Intramolecular coupling reaction of two aryl units with a chiral tether leads to both homo- and cross-coupled products with high optical purity. In the first step the target bond is formed; next the tether (which served as an auxiliary to induce an asymmetry) is eliminated. For the first time this method was used for biaryl synthesis in the early 1980s employing (M)-binol connected via diester bridges to the aryl halides.[8–10] Intramolecular aromatic nucleophilic substitution (S$_N$2Ar) of aryl halides in the presence of copper (i.e., Ullmann coupling) allows to create the new C–C bonds with the simultaneous asymmetric induction. Hydrolysis of the ester bonds results in the optically pure biaryl diacids (with the newly formed C–C bond) with recovery of the auxiliary biaryl-(M)-binol (Scheme 9.3).

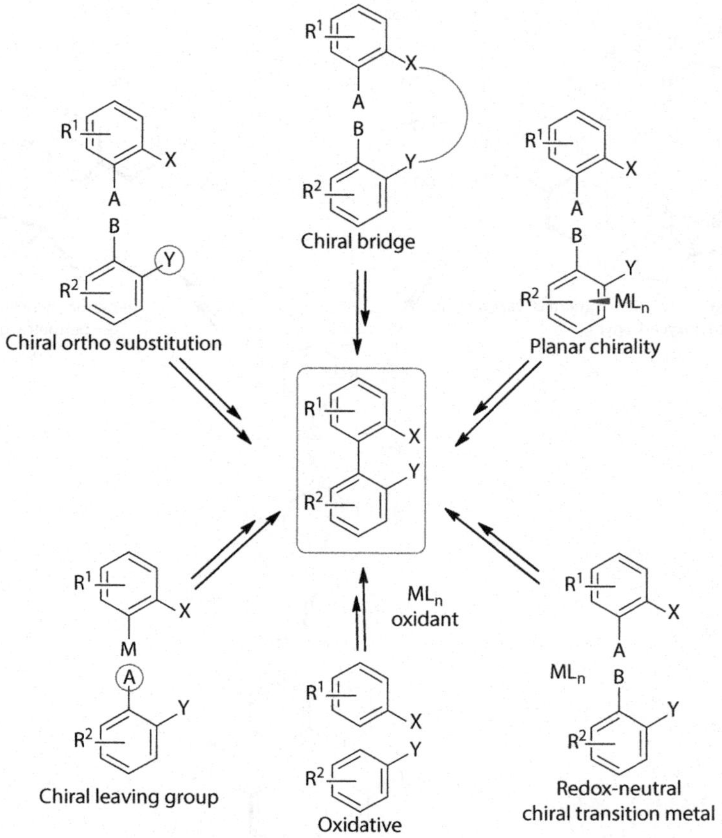

SCHEME 9.2 Atroposelective synthesis of biaryl axes. (From Bringmann, G. et al., *Angew. Chem. Int. Ed.*, 44, 5384, 2005.)

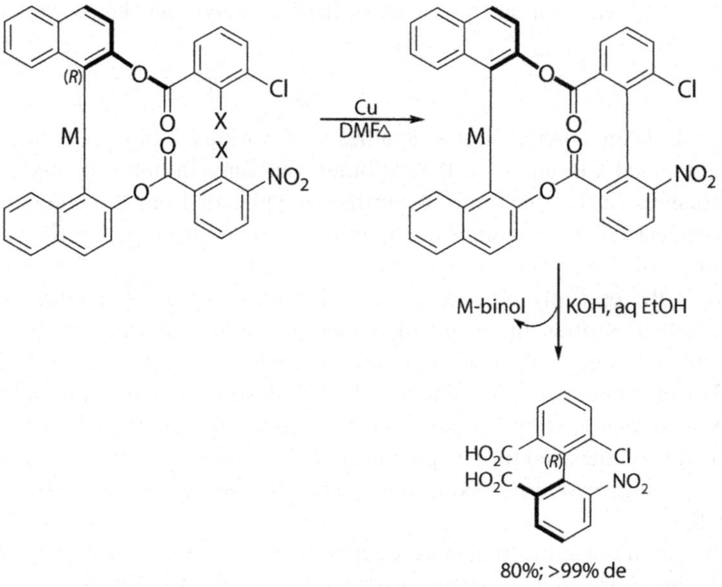

SCHEME 9.3 Atroposelective preparation of the biaryl diacids by Ullmann coupling of the (M)-binol-bridged diesters.

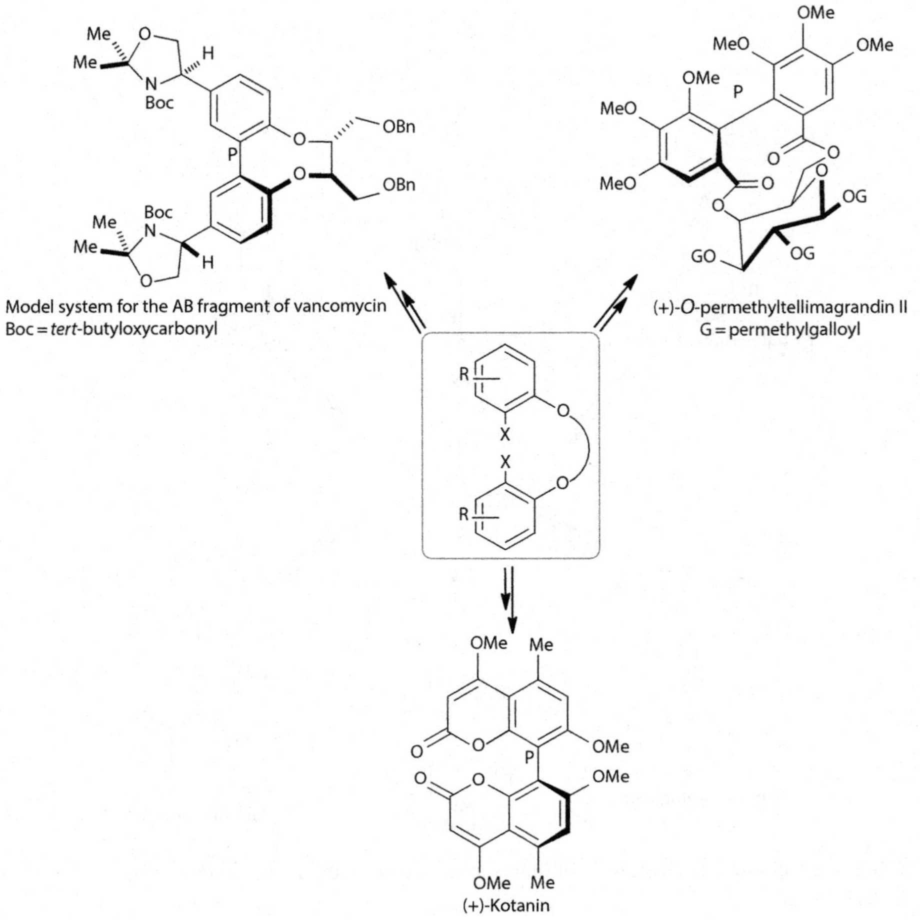

SCHEME 9.4 Use of diethers in the atroposelective synthesis of biaryl natural products.

This method has been applied in the synthesis of various biologically active compounds. Lipshutzet et al. obtained a fragment of the antibiotic vancomycin in a P-configurated form,[11] and O-permethyltellimagrandin II,[12] naturally occurring polyphenol from the ellagitannin group. Lin and coworkers synthesized (P)-kotanin,[13] the natural metabolite from *Aspergillus glaucus* possessing insecticidal properties (Scheme 9.4).

The synthesis of the optically pure axially chiral biaryls by the intramolecular coupling with the use of chiral tether is often the result of oxidative coupling reaction. This type of transformation is very attractive due to the use of stable and environmentally friendly molecular oxygen as the oxidant. For example, perylenequinone—the calphostin[14] derivative, an inhibitor of protein kinase C (PKC)—was obtained by oxidative coupling reaction. For this purpose, (R,R)-bis(*ortho*-naphthalenequinone) (synthesized from naphthalenes bridged with phthaloyl chloride) was oxidized and (M)-perylenequinone, naturally produced by pathogenic fungi, was obtained as a single diastereomer (Scheme 9.5).

Atroposelective intramolecular oxidative coupling reaction also allows to synthesize natural compounds belonging to the lignans group. Waldvogel and coworkers[15] received one of the lignans of steganacin derivative, a cytostatic drug, as a single M-configurated diastereomer (Scheme 9.6). Molybdenum pentachloride was used instead of oxygen as an oxidant.

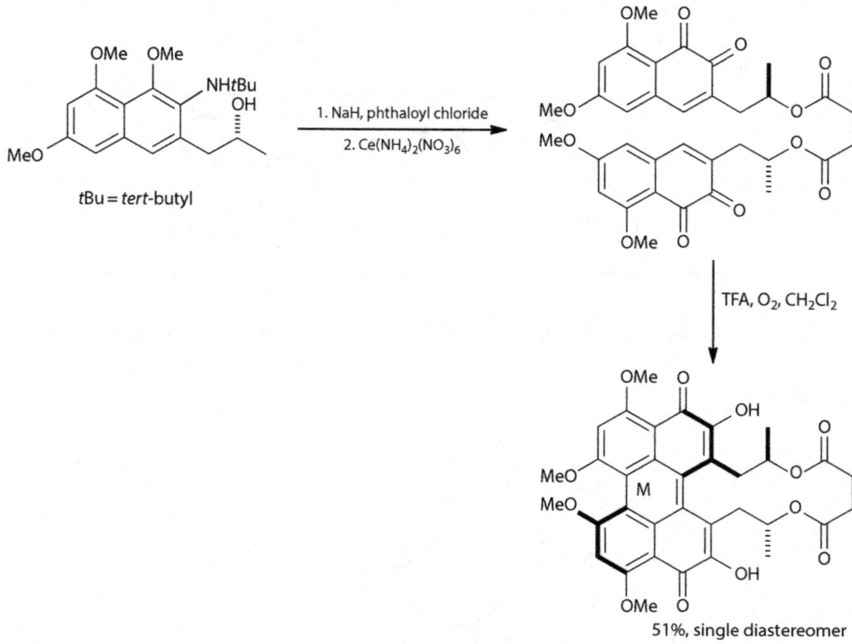

SCHEME 9.5 Atropodiastereoselective synthesis of calphostin. (From Merlic, C.A. et al., *J. Org. Chem.*, 66, 1297, 2001.)

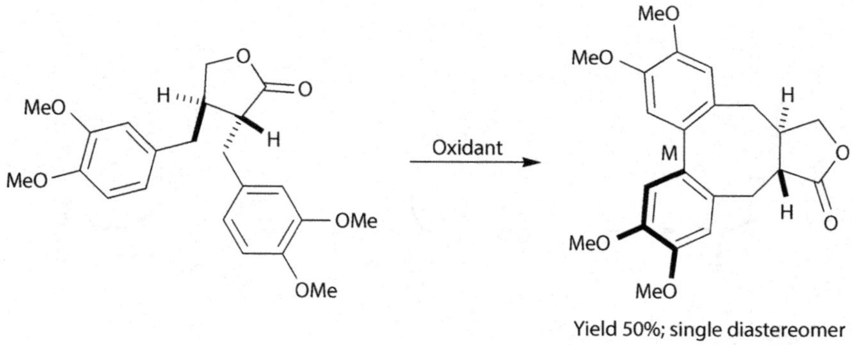

SCHEME 9.6 Atropodiastereoselective synthesis of lignin. (From Kramer, B. et al., *Angew. Chem.*, 114, 3103, 2002.)

9.4.3 Intermolecular Coupling with Chiral Ortho Substituents

Intermolecular coupling reaction of aryl substrates with the chiral *ortho* substituents in comparison with intramolecular coupling reaction does not require any chemical modification of the two aryl compounds. Moreover, the chiral information is located next to the axis, in the *ortho* position of the coupling site. The reaction proceeds according to the S$_N$2Ar mechanism. For example, coupling of the chiral *ortho*-methoxyarene with the aryl Grignard reagent provides the biaryl product with high atroposelectivity, with a predominance of one of the possible diastereoisomers (Scheme 9.7). The chiral oxazoline moiety, located in *ortho* position to the methoxy group (coupling position), induces the chirality in the biaryl product and facilitates the nucleophilic attack by stabilizing the developing negative charge.[16–18]

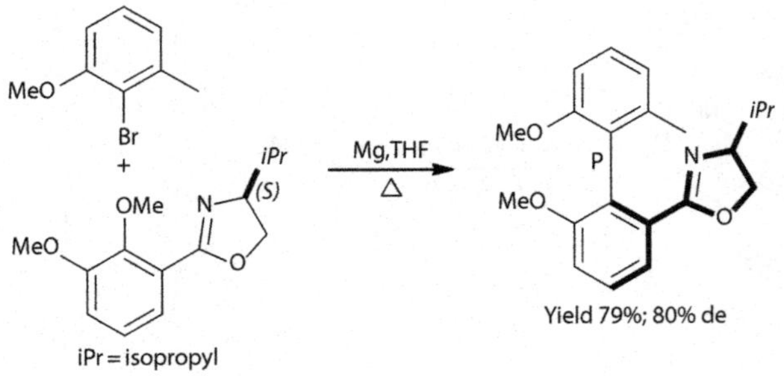

SCHEME 9.7 Synthesis of axially chiral biphenyl by S_N2Ar reaction of oxazolinylbenzene derivative.

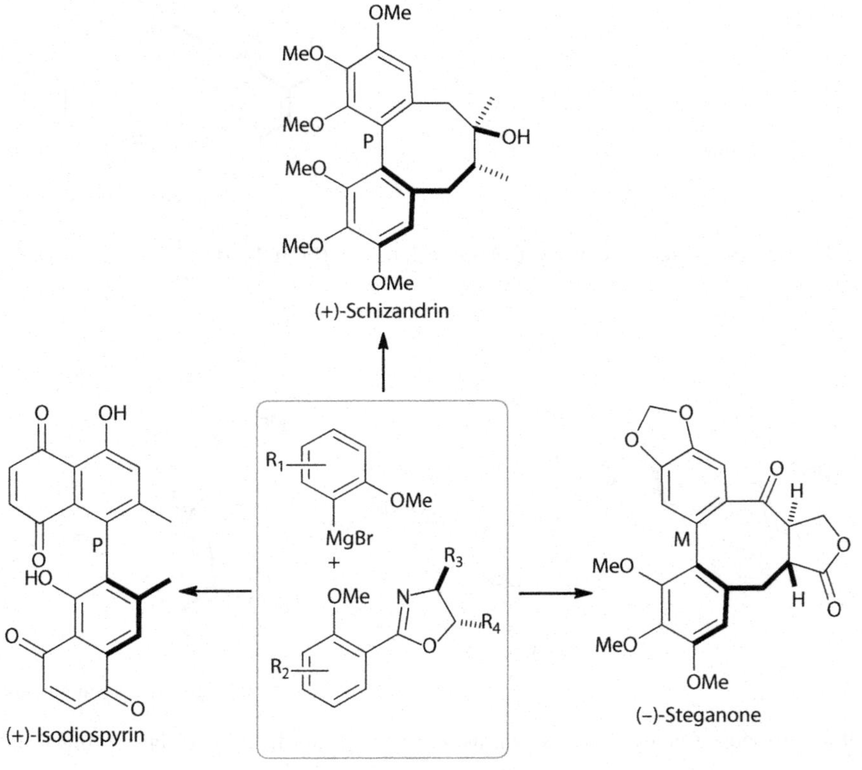

SCHEME 9.8 Biaryl natural products prepared by nucleophilic aromatic substitution of aryl Grignard reagents.

The method has found application for the synthesis of several natural products and their derivatives, some of which are shown in Scheme 9.8. For example, schizandrin[19] is a natural substance found in the fruits *Schisandra chinensis*, used for a long time in Chinese medicine as a revitalizing tonic for the kidneys and brain, while isodiospyrin[20] is a natural compound derived from plants *Diospyros morrisiana*, exhibiting cytotoxic activity against tumor cell lines. The coupling of the corresponding aryl Grignard reagents and aryl halides allows to obtain atropisomers P-configurated, which do not occur naturally. Following the same technique, (M)-steganone[21] was synthesized, a compound that was isolated from the *Steganotaenia*

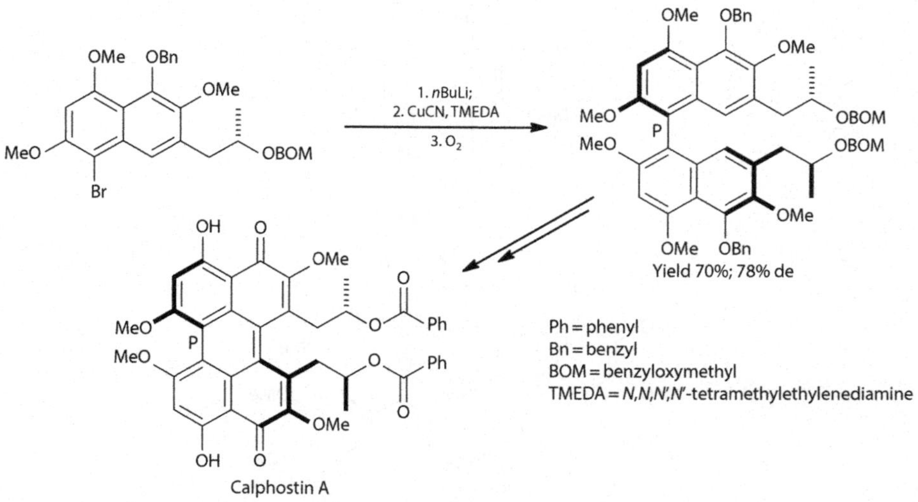

SCHEME 9.9 Atroposelective synthesis of calphostin.

araliacea having *in vivo* activity against P-388 leukemia in mice, and *in vitro* against tumor cells derived from human nasopharynx.

Coleman et al. in the oxidative dimerization reaction of aryl with chiral *ortho* substituents (in the presence of *n*BuLi and CuCN) obtained the biaryl product in good yield, which was used as a precursor in the synthesis of calphostin A (Scheme 9.9).[22] Similar to the schizandrin and isodiospyrin syntheses, also in this case, the enantiomer with opposite configuration than that naturally occurring was formed predominately.

Meyers and coworkers[23,24] applied the Ullmann homocoupling for the synthesis of the natural compound gossypol—a pigment found in plants of the genus *Gossypium*. *tert*-Butyloxazoline was used as the chiral *ortho* substituent close to the coupling site; the biaryl product was obtained in 80% yield and 89% de, which was next transformed into the target (P)-gossypol, a compound possessing antiviral, antitumor, and antifungal activity (Scheme 9.10).

9.4.4 INTERMOLECULAR COUPLING WITH CHIRAL LEAVING GROUPS

Synthesis of biaryls by nucleophilic aromatic substitution of aryl units utilizes the chiral leaving groups as the source of the asymmetric information. Generally, the coupling of aryl Grignard compounds with chirally modified alkoxyarenes or aryl sulfoxides leads to axially chiral biaryls, wherein as the chiral leaving group (*R*)-menthol is often used. For example, the reaction of alkyl 1-(*R*)-menthoxynaphthalene-2-carboxylate with the naphthyl Grignard reagent delivered the bisnaphthyl product in high enantiomeric excess (ee) (Scheme 9.11).[25]

9.4.5 INTERMOLECULAR COUPLING WITH THE ELEMENT OF PLANAR CHIRALITY

Another strategy of the C–C bond formation in the direct atroposelective synthesis is intermolecular coupling reaction with the use of planar-chiral transition metal complexes. Uemura and coworkers applied a chiral [(arene)Cr(CO)$_3$] complex to the coupling with the phenylboronic acid (Suzuki coupling) catalyzed by [Pd(PPh)$_4$], receiving the appropriate biaryls *syn* and *anti* with excellent atropodiastereoselectivities (Scheme 9.12).[26–28]

Planar-chiral [(arene)Cr(CO)$_3$] complex was used in the atropoenantioselective synthesis of the AB unit of vancomycin, of the naturally occurring (M)-steganone, and (P)-korupensamines.[29–32]

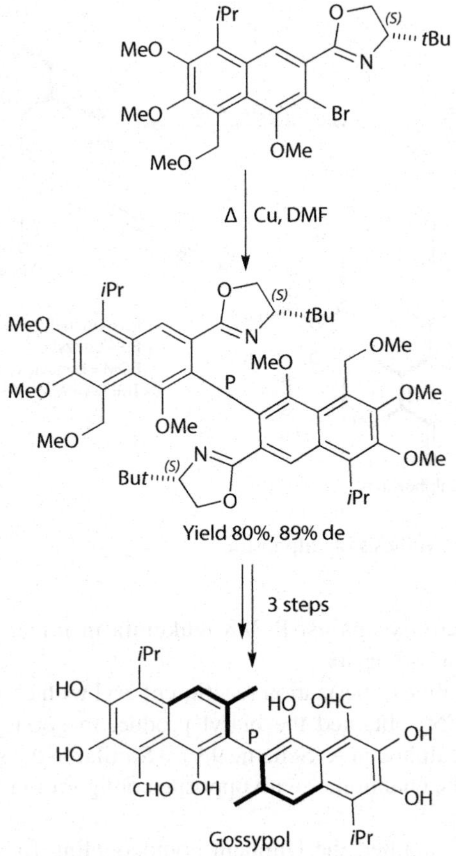

SCHEME 9.10 Atroposelective synthesis of gossypol.

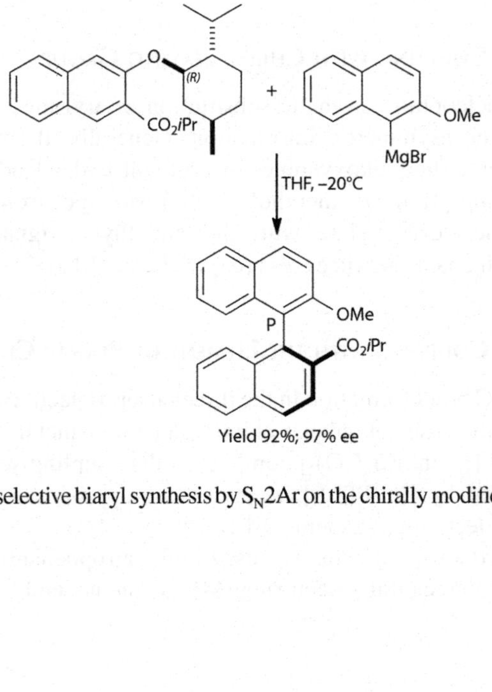

SCHEME 9.11 Atropostereoselective biaryl synthesis by S_N2Ar on the chirally modified 1-(R)-menthoxynaphthalene.

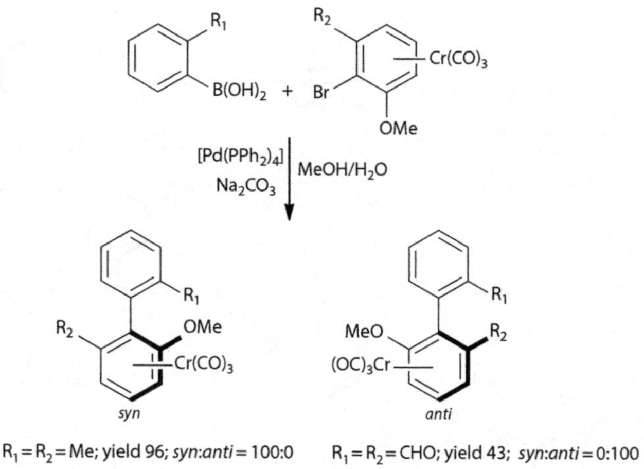

SCHEME 9.12 Atropodiastereoselective Suzuki cross-couplings of the phenylboronic acids with planar-chiral [(arene)Cr(CO)₃] complexes.

9.5 OXIDATIVE HOMOCOUPLING IN THE PRESENCE OF CHIRAL ADDITIVES

The oxidative coupling of phenols is one of the most common ways for the formation of homogeneous biaryls. The reaction proceeds through the step of forming the phenoxy radical. The asymmetric information may be contained in the chiral *ortho* substituents, the chiral tether, or in the chiral catalysts added to the reaction system. The synthesis of biaryls, published by Kozlowski and coworkers,[33,34] is an example of a coupling reaction, wherein the chiral catalyst was introduced into the reaction mixture. The reaction was carried out in the presence of (S,S)-5-diazadecalin and CuI, and oxygen as the oxidant. Under these conditions, the homocoupling of the 2-naphthol derivative resulted in the M-configured product in good yields and high ee (Scheme 9.13).

The copper catalyst used by Kozlowski proved to be very effective in the atroposelective oxidative coupling. Nevertheless, another variant of the copper catalyst is the dicopper-salen complex, which was used in the coupling of simple 2-naphthols, for example, Gao et al. obtained binol in high yield (95%; 88% ee).[35] More recently, besides the catalysts based on copper, in the oxidative coupling of naphthols, vanadium and iron catalysts have been used. In general, the copper catalysts can be used for the electron-deficient substrates, with different substituents in the aryl ring, while coupling of 2-naphthols in the presence of iron or vanadium catalysts requires the electron-rich substrates.[2] The oxidative coupling of phenols allows to obtain intermediate compounds in the synthesis of natural alkaloids belonging to the *Erythrina* family, the morphine derivatives,[36,37] and the natural precursor of bisnaphthopyrone, nigerone (Scheme 9.14).[38] The homocoupling of appropriate naphthol derivatives is a key step, due to the stereochemistry, in perylenequinone synthesis.[39]

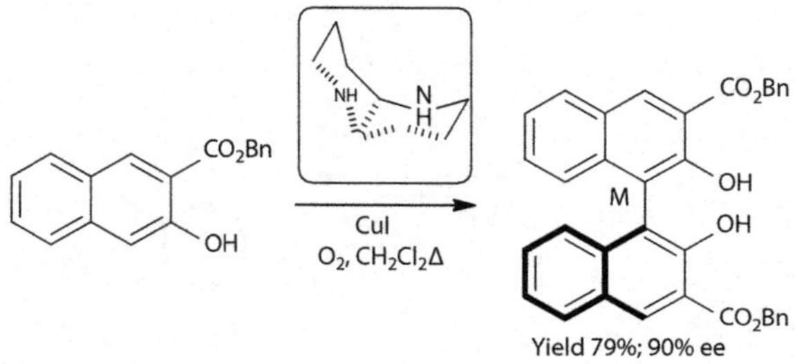

SCHEME 9.13 Copper-catalyzed oxidative homocoupling of 2-naphthols in the presence of chiral diamine and O₂.

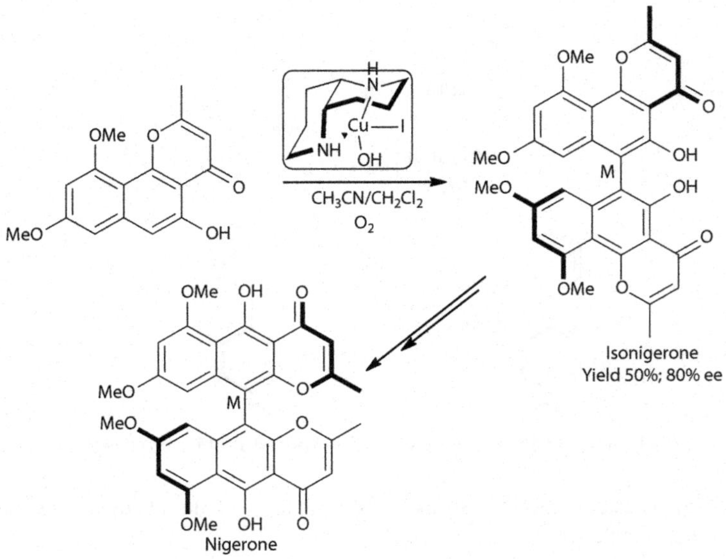

282 Asymmetric Synthesis of Drugs and Natural Products

SCHEME 9.14 Atroposelective synthesis of nigerone.

9.6 REDOX-NEUTRAL CROSS-COUPLING CATALYZED BY CHIRAL METAL COMPLEXES

Redox-neutral biaryl coupling is commonplace in the nonselective synthesis of biaryls, primarily because the reaction proceeds under mild conditions, and allows regional selective cross-coupling of two different aromatic units. The addition of the chiral information carrier in the form of chiral P,N-ligands induces asymmetry and thus leads to the atroposelective coupling reaction. The most common reactions take place between aryl halides and aryl Grignard reagents in the presence of nickel catalysts (Kumada cross-coupling), and boroorganic aryl compounds and aryl halides catalyzed by palladium complexes (Suzuki cross-coupling). For example, in Scheme 9.15 is depicted the highly enantioselective synthesis of binaphthyl compound in the Kumada coupling; the reaction proceeded in the presence of ≤5 mol% NiBr$_2$ and chiral modifier ($_p$S,S)-ferrocenylphosphine (Scheme 9.15).[40,41]

The bis-*ortho*-substituted binaphthyl derivative with good (85%) atroposelectivity was obtained by Suzuki coupling of the cyclic arylboronic ester and *ortho*-substituted aryl iodide, catalyzed by PdCl$_2$ with the same chiral bidentate iron complex (Scheme 9.16).[42,43]

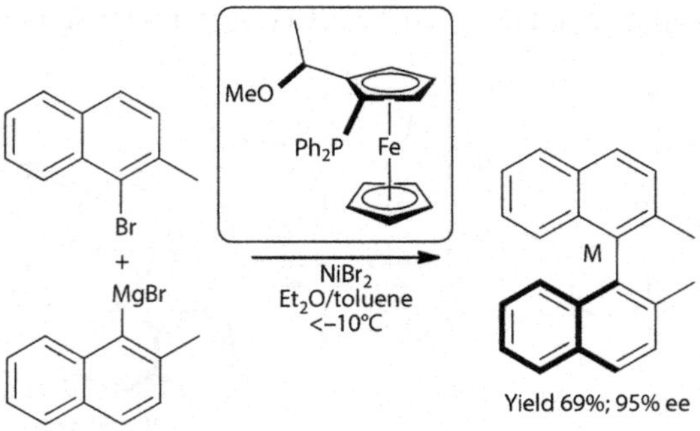

SCHEME 9.15 Nickel-catalyzed Kumada cross-coupling of 1-bromonaphthalene with the 2-methyl-1-naphthyl Grignard reagent in the presence of chiral ferrocenylphosphine.

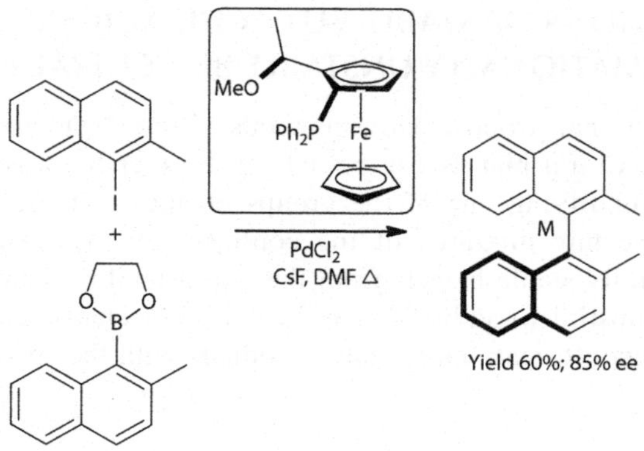

SCHEME 9.16 Palladium-catalyzed Suzuki cross-coupling of 1-iodonaphthalene with cyclic phenylboronic acids in the presence of biaryl aminophosphine.

One of the notable applications of the asymmetric Suzuki cross-coupling is the synthesis of the biaryl vancomycin fragment. The reaction was carried out between the appropriate iodophenyl and the cyclic boronic acid in the presence of Pd(OAc)$_2$ with a chiral ligand, (M)- or (P)-binap.[44] Depending on the chiral modifier configuration, one of the atropisomers was obtained predominately (Scheme 9.17).

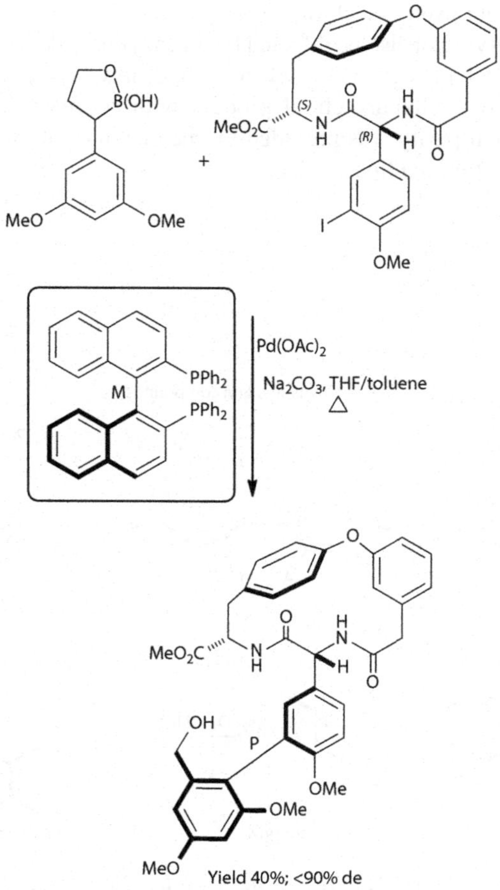

SCHEME 9.17 Atroposelective synthesis of the biaryl fragment of vancomycin.

9.7 TRANSFORMATIONS OF STABLE BUT ACHIRAL BIARYLS AND CONFORMATIONALLY UNSTABLE BUT CHIRAL BIARYLS

In addition to the conventional atroposelective methods of biaryl synthesis, alternative strategies have also been applied. Chiral biaryls can be achieved either by desymmetrization of stable but achiral biaryls by modifying one of the groups on the aromatic moiety or by dynamic kinetic resolution of racemic mixtures of the conformationally unstable chiral substrates. The creation of the chirally stable biaryls from the chiral labile substrates is most frequently the result of the extra substituent addition, and formation or cleavage of a bridge. Regardless of the used method, the conformationally stable products with the specified configuration[1] are formed.

9.8 DESYMMETRIZATION OF PROSTEREOGENIC BIARYL COMPOUNDS

Desymmetrization of the prostereogenic aryl compounds proceeds in two steps. At first, the achiral biaryls are prepared in nonselective coupling reactions. Subsequently, the symmetric biaryl compounds are converted to the axially chiral products by modification of atropoenantiotopos substituents in the arene moiety. For example, in an achiral tetra-*ortho*-hydroxybiphenyl formation of an eight-membered diether bridge led to the optically pure M-configurated biaryl (Scheme 9.18).[1,45] Etherification was carried out using the chiral dimesylate.

Asymmetrization of biaryls can also be effected by an enzymatic desymmetrization. Symmetrical 2,6-diacetoxy-2′-alkyl biaryl in the enantioselective hydrolysis of one of the prochiral groups (*pro-P*) was converted into the axially chiral product with configuration M in high ee (≥96%) (Scheme 9.19).[46] The optical purity was provided by the enzyme catalysts: *Pseudomonas cepacia* lipase (PCL) or *Candida antarctica* lipase (CAL).

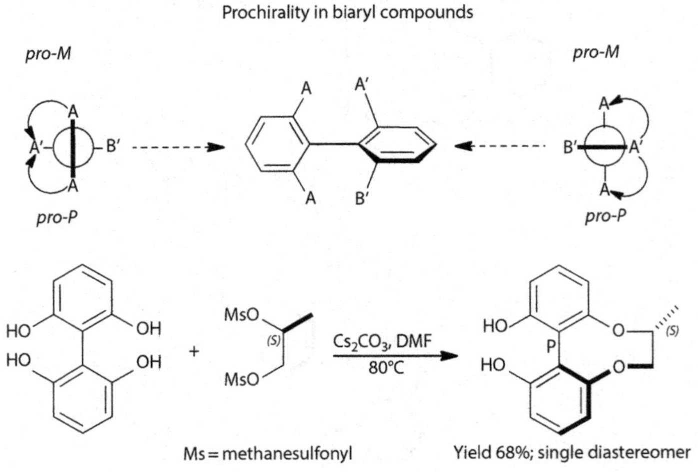

SCHEME 9.18 Desymmetrization of configurationally stable but achiral biaryl compound by atropodiastereoselective formation of a diether bridge.

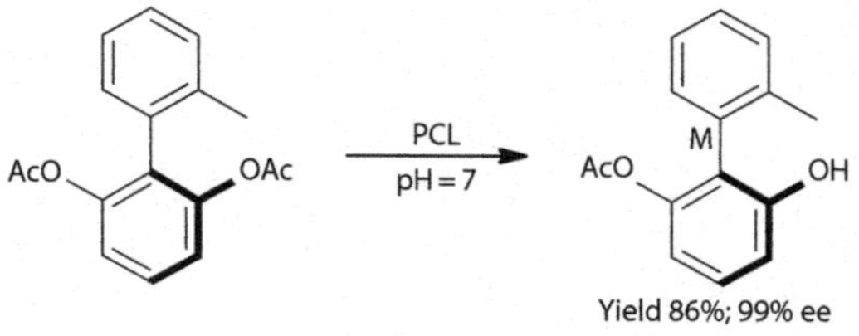

Asymmetric Synthesis of Biaryls and Axially Chiral Natural Products   285

Yield 86%; 99% ee

SCHEME 9.19 Atropoenantiotopos-differentiating enzymatic *O*-deacetylation using PCL (*Pseudomonas cepacia* lipase).

9.9 CONVERSION OF AXIALLY CHIRAL BUT CONFORMATIONALLY UNSTABLE BIARYL COMPOUNDS—INTRODUCTION OF ADDITIONAL SUBSTITUENTS IN THE *ORTHO* POSITION

As mentioned earlier, atroposelective conversion of axially chiral but conformationally unstable biaryl compounds may occur through the introduction of an additional substituent in the *ortho* position. Murai et al. reported the alkylation of 1-(naphthalen-1-yl)isoquinoline in the presence of the chiral rhodium catalyst, wherein the product was obtained with moderate enantioselectivity (22% ee) (Scheme 9.20).[47]

9.10 CONVERSION OF AXIALLY CHIRAL BUT CONFORMATIONALLY UNSTABLE BIARYL COMPOUNDS—BRIDGE FORMATION

One of the methods of generating the axially stable biaryls from labile substrates is the introduction of the additional ring, which, in some cases, decreases the rotation possibility in comparison to the open-chain system. For example, treatment of racemic, unstable 2,2′-dibeznoyl dichloride with the glycopyranose derivative gave the sugar-bridge biaryl product with a P configuration as a single atropisomer (Scheme 9.21).[48]

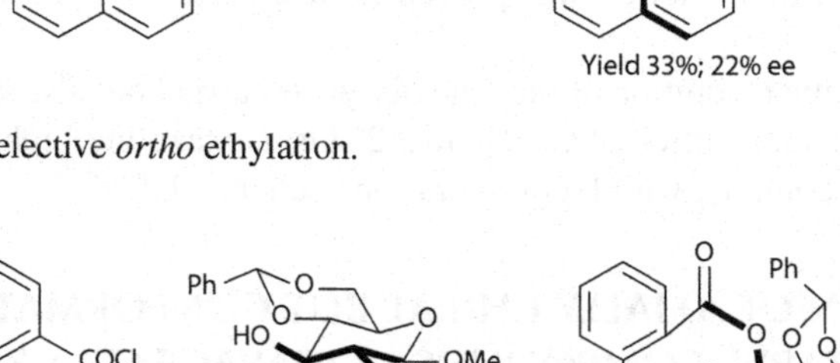

Yield 33%; 22% ee

SCHEME 9.20 Atroposelective *ortho* ethylation.

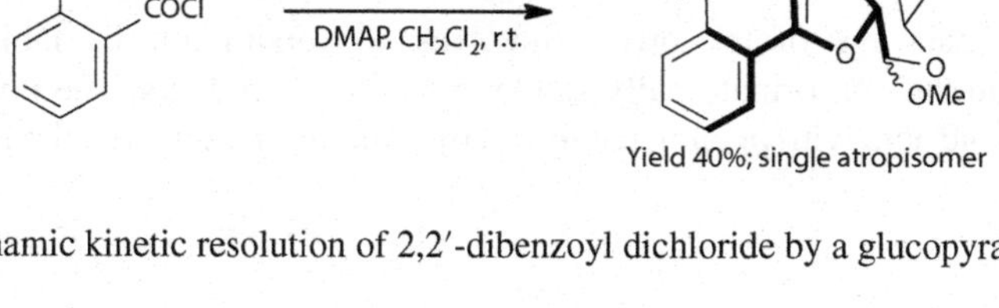

Yield 40%; single atropisomer

SCHEME 9.21 Dynamic kinetic resolution of 2,2′-dibenzoyl dichloride by a glucopyranose-based scaffold.

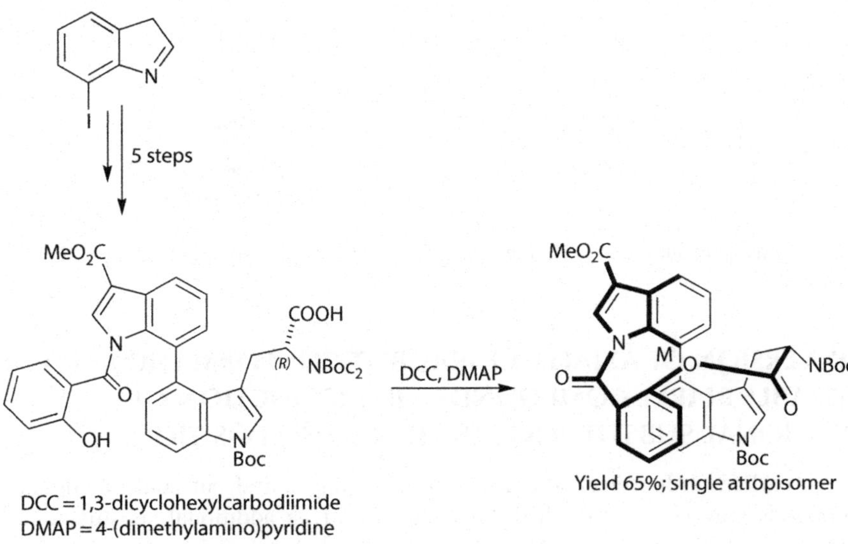

SCHEME 9.22 Atropodiastereoselective synthesis of the diazonamide A by macrolactonization.

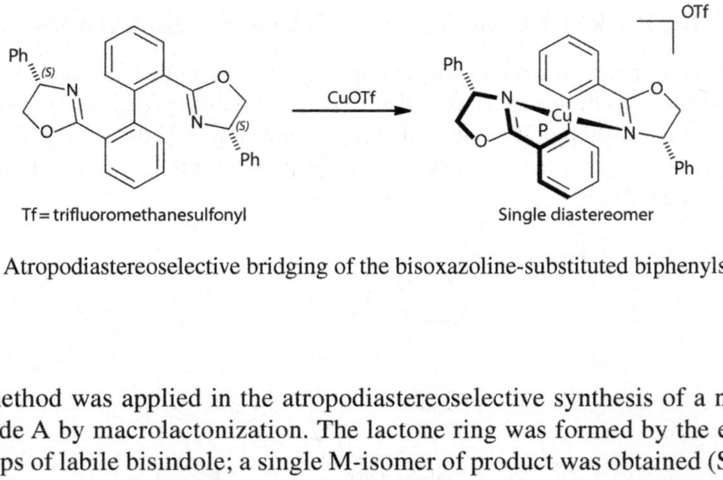

SCHEME 9.23 Atropodiastereoselective bridging of the bisoxazoline-substituted biphenyls by CuOTf.

The same method was applied in the atropodiastereoselective synthesis of a naturally occurring diazonamide A by macrolactonization. The lactone ring was formed by the esterification of functional groups of labile bisindole; a single M-isomer of product was obtained (Scheme 9.22).[49] The unique selectivity of this reaction was proved to be an effect of the bulky NBoc$_2$ substituent presence.

Alternatively, the kinetic resolution of racemic biaryls is carried out also by chelation with transition metals. For example, an addition of CuOTf to 2,2′-bis(2-oxazoline)biphenyl resulted in a single diastereomer as a chiral complex with P configuration (Scheme 9.23).[50,51]

9.11 CONVERSION OF AXIALLY CHIRAL BUT CONFORMATIONALLY UNSTABLE BIARYL COMPOUNDS—CLEAVAGE OF A BRIDGE

The stereoselective cleavage of the bridge involves conversion of the labile bridged biaryls, primarily lactones, in the conformationally stable open-chain products. The nonselective coupling of *ortho*-bromobenzoic acid with phenol, catalyzed by palladium, delivered the labile biaryl lactone,

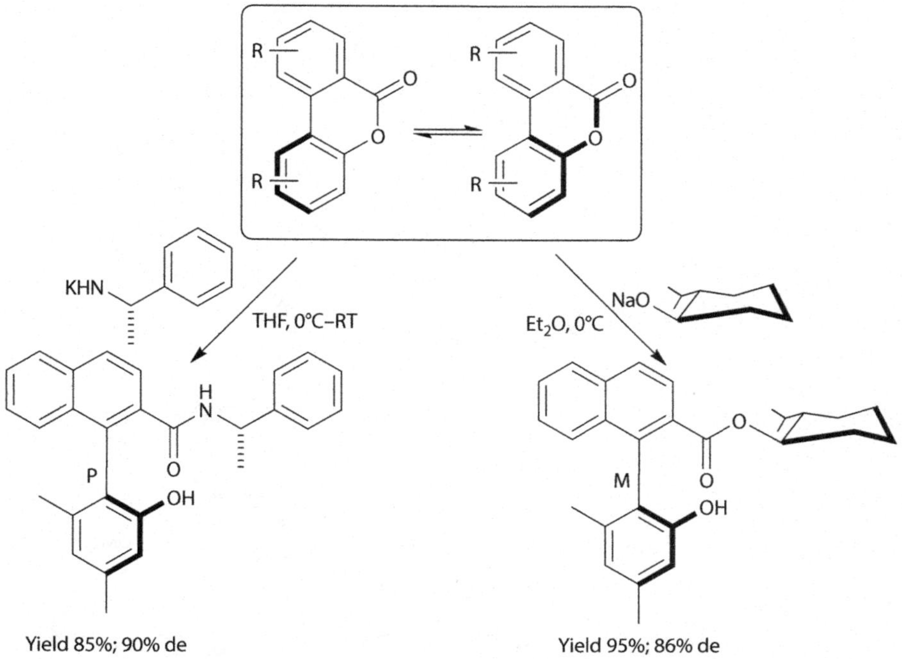

SCHEME 9.24 Atroposelective ring opening of the lactones.

which was next converted to the nonbridged analogue via the atroposelective ring opening reaction (Scheme 9.24). Chiral nucleophiles, such as potassium (S)-1-phenylethylamide and sodium (R)-menthoxide, were used for the ring opening delivering, respectively, the biaryl amide and ester, in high yields and diastereoselectivities.[52–54]

The strategy of stereoselective cleavage of the bridge has found an application in the synthesis of many biaryl natural products. For example, the stereoselective reduction of the seven-membered lactones with borane in the presence of (S)-CBS gave the M-configured biaryl diol with 75% ee, a key intermediate in the synthesis of (+)-isokotanin A-bicoumarin from the sclerotia of *Aspergillus alliaceus*. While the unreacted P-lactone with high enantiomeric purity was used as a precursor of P-4,4'-bisorcinol (Scheme 9.25).[55]

Application of the lactone method allows to obtain a number of biologically active natural compounds, such as stegonone,[56] isoschizandrin,[57] knipholone,[58,59] mastigophorene A,[60,61] korupensamine A,[62,63] dioncopeltine C,[64] and a precursor in the vankomycin[65] synthesis (Scheme 9.26).

9.12 ASYMMETRIC AXIALLY CHIRAL BIARYL SYNTHESIS BY THE CONSTRUCTION OF AN AROMATIC RING

The latest method of the asymmetric biaryl synthesis is the atroposelective transformation of the alkyl substituent of the arene ring into a second aromatic ring in the presence of an organometallic catalyst. The Gutnov and Heller group has developed the synthesis of axially chiral 2-aryl pyridines by asymmetric [2+2+2] cycloaddition. The reaction of 1-naphthyl diyne with alkyl or aryl nitriles, catalyzed by chiral cobalt complex, led to the (M)-pyridines in good yields and ee's (Scheme 9.27).[66]

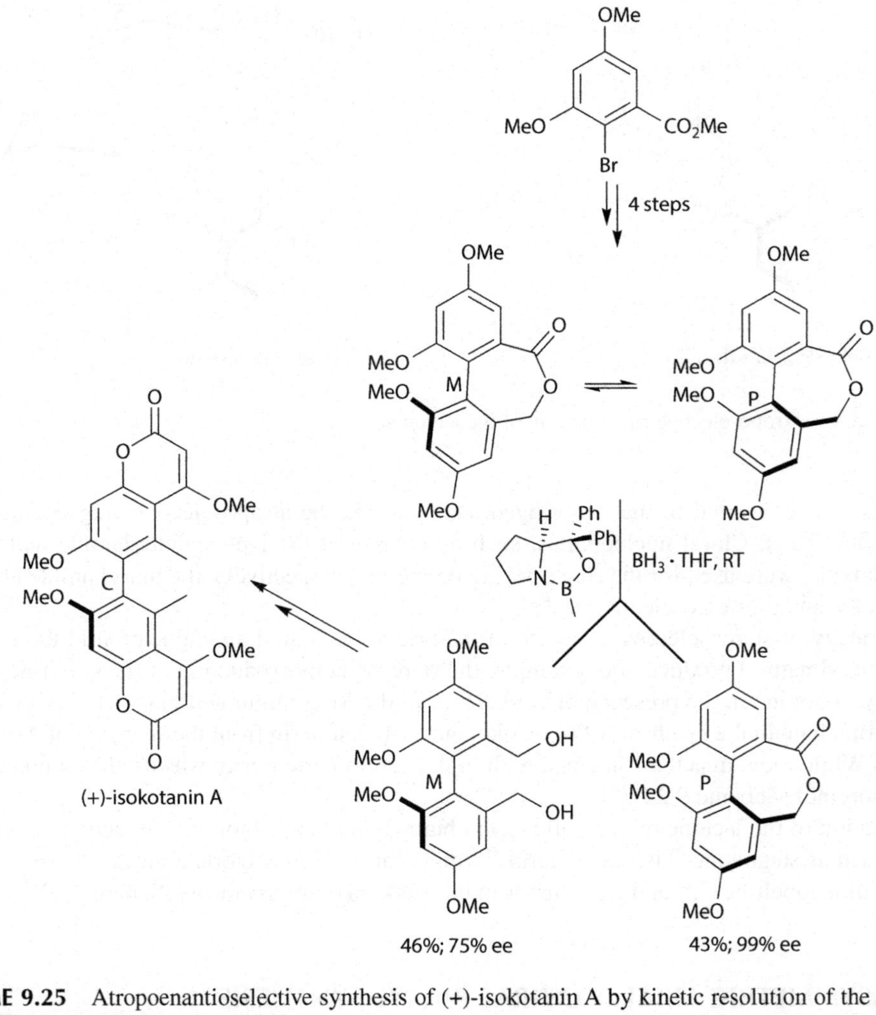

SCHEME 9.25 Atropoenantioselective synthesis of (+)-isokotanin A by kinetic resolution of the configurationally stable biaryl lactone.

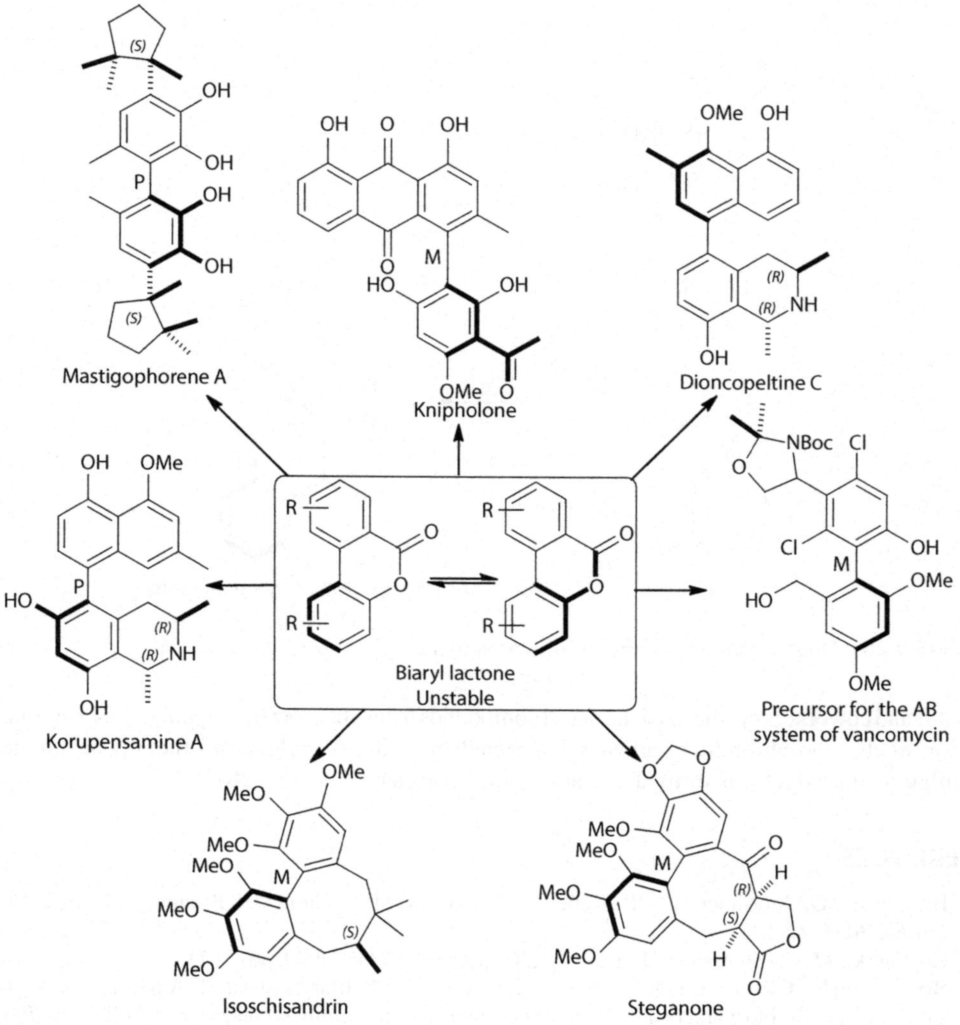

Asymmetric Synthesis of Biaryls and Axially Chiral Natural Products 289

SCHEME 9.26 Synthesis of axially chiral biaryl natural products by the lactone method.

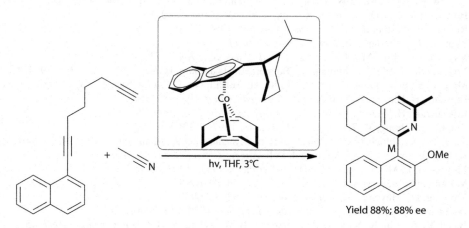

SCHEME 9.27 Atropoenantioselective synthesis of the 1-aryl-5,6,7,8-tetrahydroisoquinolines by asymmetric [2+2+2] cycloaddition.

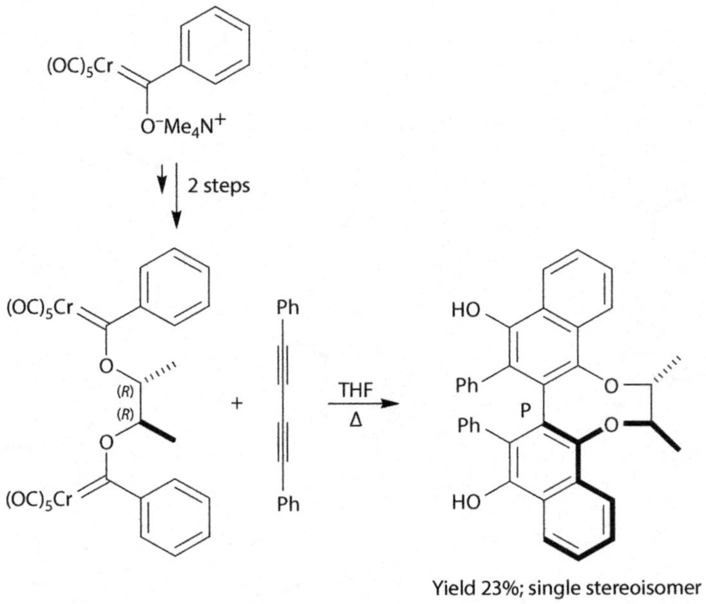

SCHEME 9.28 Double benzannulation of a bis(chromium carbene) species.

Wulff and coworkers synthesized the biaryl compounds using the Dötz benzannulation. In the one-step reaction of the two chromium carbenes connected by a chiral bridge with the 1,3-butadiyne, the P-configurated product was formed as a single stereoisomer (Scheme 9.28).[67]

REFERENCES

1. Bringmann, G.; Mortimer, A. J. P.; Keller, P. A.; Gresser, M. J.; Garner, J.; Breuning, M. *Angew. Chem. Int. Ed.* 2005, *44*, 5384.
2. Kozlowski, M. C.; Morgan, B. J.; Linton, E. C. *Chem. Soc. Rev.* 2009, *38*, 3193.
3. Bringmann, G.; Günther, C.; Ochse, M.; Schupp, O.; Tasler, S. Biaryls in nature: A multi-facetted class of stereochemically, biosynthetically, and pharmacologically intriguing secondary metabolites. In: *Progress in the Chemistry of Organic Natural Products* (Herz, W.; Falk, H.; Kirby, G. W.; Moore, R. E., eds.), Vol. 82, Springer-Verlag, Wein, Austria, 2001.
4. Mislow, K.; Glass, M. A. W.; Hopps, H. B.; Simon, E.; Wahl, G. H. Jr. *J. Am. Chem. Soc.* 1964, *86*, 1710.
5. Bott, G.; Field, L. D.; Sternhell, S. *J. Am. Chem. Soc.* 1980, *102*, 5618.
6. Rieger, M.; Westheimer, F. H. *J. Am. Chem. Soc.* 1950, *72*, 19.
7. Zhou, F.; Cai, Q. *Beilstein J. Org. Chem.* 2015, *11*, 2600.
8. Miyano, S.; Tobita, M.; Hashimoto, H. *Bull. Chem. Soc. Jpn.* 1981, *54*, 3522.
9. Miyano, S.; Handa, S.; Shimizu, K.; Tagami, K.; Hashimoto, H. *Bull. Chem. Soc. Jpn.* 1984, *57*, 1943.
10. Miyano, S.; Fukushima, H.; Handa, S.; Ito, H.; Hashimoto, H. *Bull. Chem. Soc. Jpn.* 1988, *61*, 3249.
11. Lipshutz, B. H.; Liu, Z.-P.; Kayser, F. *Tetrahedron Lett.* 1994, *35*, 5567.
12. Lipshutz, B. H.; Muller, P.; Leinweber, D. *Tetrahedron Lett.* 1999, *40*, 3677.
13. Lin, G.-Q.; Zhong, M. *Tetrahedron: Asymmetry* 1997, *8*, 1369.
14. Merlic, C. A.; Aldrich, C. C.; Albaneze-Walker, J.; Saghatelian, A.; Mammen, J. *J. Org. Chem.* 2001, *66*, 1297.
15. Kramer, B.; Averhoff, A.; Waldvogel, S. R. *Angew. Chem.* 2002, *114*, 3103.
16. Meyers, A. I.; Meier, A.; Rawson, D. J. *Tetrahedron Lett.* 1992, *33*, 853.
17. Moorlag, H.; Meyers, A. I. *Tetrahedron Lett.* 1993, *34*, 6989.
18. Moorlag, H.; Meyers, A. I. *Tetrahedron Lett.* 1993, *34*, 6993.
19. Warshawsky, A. M.; Meyers, A. I. *J. Am. Chem. Soc.* 1990, *112*, 8090.
20. Baker, R. W.; Liu, S.; Sargen, M. V.; Skelton, B. W.; White, A. H. *Chem. Commun.* 1997, 451.
21. Meyers, A. I.; Flisak, J. R.; Aitken, R. A. *J. Am. Chem. Soc.* 1987, *109*, 5446.
22. Coleman, R. S.; Grant, E. B. *J. Am. Chem. Soc.* 1995, *117*, 10889.

23. Meyers, A. I.; Willemsen, J. J. *Tetrahedron* 1998, *54*, 10493.
24. Meyers, A. I.; Willemsen, J. J. *Tetrahedron Lett.* 1996, *37*, 791.
25. Suzuki, T.; Hotta, H.; Hattor, T.; Miyano, S. *Chem. Lett.* 1990, 807.
26. Kamikawa, K.; Watanabe, T.; Uemura, M. *J. Org. Chem.* 1996, *61*, 1375.
27. Kamikawa, K.; Watanabe, T.; Uemura, M. *Synlett* 1995, 1040.
28. Tanaka, Y.; Sakamoto, T.; Kamikawa, K.; Uemura, M. *Synlett* 2003, 519.
29. Watanabe, T.; Tanaka, Y.; Shoda, R.; Sakamoto, R.; Kamikawa, K.; Uemura, M. *J. Org. Chem.* 2004, *69*, 4152.
30. Watababe, T.; Shakadou, M.; Uemura, M. *Synlett* 2000, 1141.
31. Watanabe, T.; Uemura, M. *Chem. Commun.* 1998, 871.
32. Kamikawa, K.; Watanabe, T.; Daimon, A.; Uemura, M. *Tetrahedron* 2000, *56*, 2325.
33. Li, X.; Yang, J.; Kozlowski, M. C. *Org. Lett.* 2001, *3*, 1137.
34. Li, X.; Hewgley, J. B.; Mulrooney, C. A.; Yang, J.; Kozlowski, M. C. *J. Org. Chem.* 2003, *68*, 5500.
35. Gao, J.; Reibenspies, J. H.; Martell, A. E. *Angew. Chem. Int. Ed.* 2003, *42*, 6008.
36. Pal, T.; Pal, A. *Curr. Sci.* 1996, *71*, 106.
37. Keseru, G. M.; Nogradi, M. In: *Studies in Natural Products Chemistry* (Atta-ur-Rahman, ed.), Elsevier Science, Vol. 20, New York, 1998.
38. DiVirgilio, E. S.; Dugan, E. C.; Mulrooney, C. A.; Kozlowski, M. C. *Org. Lett.* 2007, *9*, 385.
39. Mulrooney, C. A.; Li, X.; DiVirgilio, E. S.; Kozlowski, M. C. *J. Am. Chem. Soc.* 2003, *125*, 6856.
40. Colletti, S. L.; Halterman, R. L. *Tetrahedron Lett.* 1989, *30*, 3513.
41. Harris, J. M.; McDonald, R.; Vederas, J. C. *J. Chem. Soc. Perkin Trans. 1* 1996, 2669.
42. Cammidge, A. N.; Crepy, K. V. L. *Chem. Commun.* 2000, 1723.
43. Cammidge, A. N.; Crepy, K. V. L. *Tetrahedron* 2004, *60*, 4377.
44. Nicolaou, K. C.; Li, H.; Boddy, C. N. C.; Ramanjulu, J. M.; Yue, T.-Y.; Natarajan, S.; Chu, X.-J.; Bräse, S.; Rübsam, F. *Chem. Eur. J.* 1999, *5*, 2584.
45. Harada, T.; Ueda, S.; Yoshida, T.; Inoue, A.; Takeuchi, M.; Ogawa, N.; Oku, A.; Shiro, M. *J. Org. Chem.* 1994, *59*, 7575.
46. Matsumoto, T.; Konegawa, T.; Nakamura, T.; Suzuki, K. *Synlett* 2002, 122.
47. Kakiuchi, F.; Le Gendre, P.; Yamada, A.; Ohtaki, H.; Murai, S. *Tetrahedron: Asymmetry* 2000, *11*, 2647.
48. Capozzi, G.; Ciampi, C.; Delogu, G.; Menichetti, S.; Nativi, C. *J. Org. Chem.* 2001, *66*, 8787.
49. Feldman, K. S.; Eastman, K. J.; Lessene, G. *Org. Lett.* 2002, *4*, 3525.
50. Imai, Y.; Zhang, W.; Kida, T.; Nakatsuji, Y.; Ikeda, I. *Tetrahedron Lett.* 1997, *38*, 2681.
51. Imai, Y.; Zhang, W.; Kida, T.; Nakatsuji, Y.; Ikeda, I. *J. Org. Chem.* 2000, *65*, 3326.
52. Bringmann, G.; Hartung, T.; Göbel, L.; Schupp, O.; Ewers, C. L. J.; Schöner, B.; Zagst, R.; Peters, K.; Von Schnering, H. G.; Burschka, C. *Liebigs Ann. Chem.* 1992, 225.
53. Kamikawa, K.; Norimura, K.; Furusyo, M.; Uno, T.; Sato, Y.; Konoo, A.; Bringmann, G.; Uemura, M. *Organometallics* 2003, *22*, 1038.
54. Kamikawa, K.; Furusyo, M.; Uno, T.; Sato, Y.; Konoo, A.; Bringmann, G.; Uemura, M. *Org. Lett.* 2001, *3*, 3667.
55. Bringmann, G.; Hinrichs, J.; Henschel, P.; Kraus, J.; Peters, K.; Peters, E.-M. *Eur. J. Org. Chem.* 2002, 1096.
56. Abe, H.; Takeda, S.; Fujita, T.; Nishioka, K.; Takeuchi, Y.; Harayama, T. *Tetrahedron Lett.* 2004, *45*, 2327.
57. Molander, G. A.; George, K. M.; Monovich, L. G. *J. Org. Chem.* 2003, *68*, 9533.
58. Bringmann, G.; Menche, D. *Angew. Chem.* 2001, *113*, 1733.
59. Bringmann, G.; Menche, D.; Kraus, J.; Mühlbacher, J.; Peters, K.; Peters, E.-M.; Brun, R.; Bezabih, M.; Abegaz, B. M. *J. Org. Chem.* 2002, *67*, 5595.
60. Bringmann, G.; Hinrichs, J.; Pabst, T.; Henschel, P.; Peters, K.; Peters E.-M. *Synthesis* 2001, 155.
61. Bringmann, G.; Pabst, T.; Henschel, P.; Kraus, J.; Peters, K.; Peters, E.-M.; Rycroft, D. S.; Connolly, J. *J. Am. Chem. Soc.* 2000, *122*, 9127.
62. Bringmann, G.; Ochse, M. *Synlett* 1998, 1294.
63. Bringmann, G.; Ochse, M.; Götz, R. *J. Org. Chem.* 2000, *65*, 2069.
64. Bringmann, G.; Holenz, J.; Weirich, R.; Rübenacker, M.; Funke, C.; Boyd, M. R.; Gulakowski, R. J.; François, G. *Tetrahedron* 1998, *54*, 497.
65. Bringmann, G.; Menche, D.; Mühlbacher, J.; Reichert, M.; Saito, N.; Pfeiffer, S. S.; Lipshutz, B. H. *Org. Lett.* 2002, *4*, 2833.
66. Gutnov, A.; Heller, B.; Fischer, C.; Drexler, H.-J.; Spannenberg, A.; Sundermann, B.; Sundermann, C. *Angew. Chem.* 2004, *116*, 3883.
67. Bao, J.; Wulff, W. D.;Fumo, M. J.; Grant, E. B.; Heller, D. P.; Whitcomb, M. C.; Yeung, S.-M. *J. Am. Chem. Soc.* 1996, *118*, 2166.

10 Palladium-Catalyzed Asymmetric Transformations of Natural Products and Drug Molecules

Mariette M. Pereira, Carolina S. Vinagreiro, Fábio M.S. Rodrigues, and Rui M.B. Carrilho

CONTENTS

10.1 Introduction 293
10.2 Palladium-Catalyzed Asymmetric Allylic Alkylations 297
10.3 Asymmetric Intramolecular Cyclization Reactions 304
10.4 Palladium-Catalyzed Carbonylation Reactions 311
10.5 Conclusion and Future Perspectives 319
References 319

10.1 INTRODUCTION

The efficient synthesis of chiral natural products, synthons, and drugs using palladium metal complexes as catalysts is currently a great challenge for modern synthetic chemistry, due to its well-known economic and ecological advantages. These approaches contribute to a significant lowering of energy consumption, solvents, and reagents when compared with classic organic syntheses.

The application of palladium salts as key reagents in organic synthesis started in the 1960s, when Smidt,[1] at *Wacker Chemie AG*, discovered that, in the presence of $PdCl_2$ and $CuCl_2$, ethylene could be oxidized to acetaldehyde, under air atmosphere. Since then, palladium catalysts have been used in a vast number of synthetic processes, being currently considered one of the most powerful tools for the formation of new C–C, C–H, and C–heteroatom bonds.[2–4] The importance of such reactions has been demonstrated by their multiple industrial applications and was recognized with the attribution of Nobel Prize to Richard Heck,[5] Ei-ichi Negishi,[6] and Akira Suzuki[7] (Figure 10.1), in 2010, for their research contributions in the field of palladium-catalyzed cross-coupling reactions for organic synthesis.[8,9]

The relevance of this topic is clearly evidenced by the large number of publications related to "palladium complexes in organic synthesis," as well as the increasing number of citations over the last 20 years, in WEB OF SCIENCE™ platform (Figure 10.2).

Palladium-catalyzed coupling reactions usually involve a Pd(0) precursor, which is able to activate a R–X substrate toward nucleophiles, alkenes, or organometallic reagents. In general, R is usually a sp^2-hybridized carbon atom, while X can be a halide or a pseudohalide (Scheme 10.1).

Such coupling reactions may have different designations, depending on the metal (or metalloid) used in the transmetallation step, for instance, Stille coupling[11,12] (with Sn), Negishi coupling[6,13] (with Zn), Kumada coupling[14] (with Mg), Hiyama coupling[15] (with Si), Takagi cyanation[16]

294 Asymmetric Synthesis of Drugs and Natural Products

FIGURE 10.1 Photographs of 2010 Chemistry Nobel Prize laureates: (a) Richard Heck from University of Delaware, USA (Born: August 15, 1931, Springfield, MA, USA–Died: October 9, 2015, Manila, Philippines); (b) Ei-ichi Negishi from Purdue University, West Lafayette, USA (Born: July 14, 1935, Changchun, China); (c) Akira Suzuki from Hokkaido University, Sapporo, Japan (Born: September 12, 1930, Mukawa, Japan).[10]

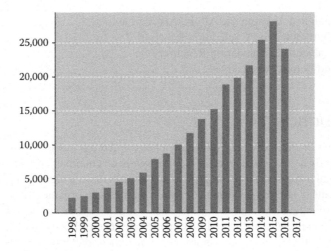

FIGURE 10.2 Citations related to "palladium in organic synthesis," from WEB OF SCIENCE™.

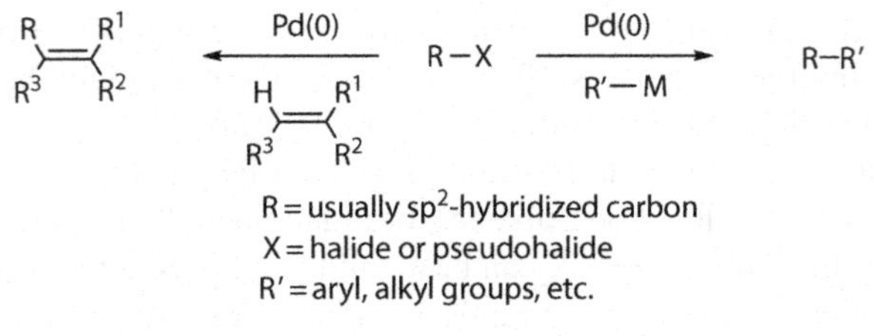

SCHEME 10.1 General palladium-catalyzed coupling reactions.

(with Zn), and Suzuki–Miyaura coupling[17] (with boron). Moreover, new palladium-catalyzed carbon–carbon bond formation can be achieved *via* functionalization of alkynes (Sonogashira couplings),[18] amines (Buchwald–Hartwig reactions),[19,20] and alkenes (Heck reactions)[21] (Scheme 10.2, A). The scope of palladium-catalyzed reactions has been expanded to a diversity of other asymmetric reactions, such as allylic alkylations, carbonylations, and intramolecular cyclizations, whose applications in the functionalization of natural products and in the asymmetric synthesis of enantiomerically pure synthons/drugs[22] will be highlighted in this chapter (Scheme 10.2, B).

The intensive academic research on this area has led to the establishment of a well-accepted mechanism for palladium-catalyzed C–C coupling processes, which is presented, in a simplified way, in Scheme 10.3. The general catalytic cycle is initiated with the oxidative addition of an activated substrate R-X to a Pd(0) complex (**A**). The resulting intermediate can react with an organometallic reagent (R′-M), inducing the occurrence of transmetallation step (**B**). Then, reductive elimination reaction (**D**) gives rise to the formation of new C–C bond (Scheme 10.3, pathway 1). Otherwise, the intermediate of the oxidative addition (**A**) can coordinate with an olefin, followed by 1,2-insertion (**E**), and, finally, β-elimination reaction (**F**) giving rise to a new C–C bond, with concomitant regeneration of the Pd(0) catalyst through reductive elimination (**G**) (Scheme 10.3, pathway 2).

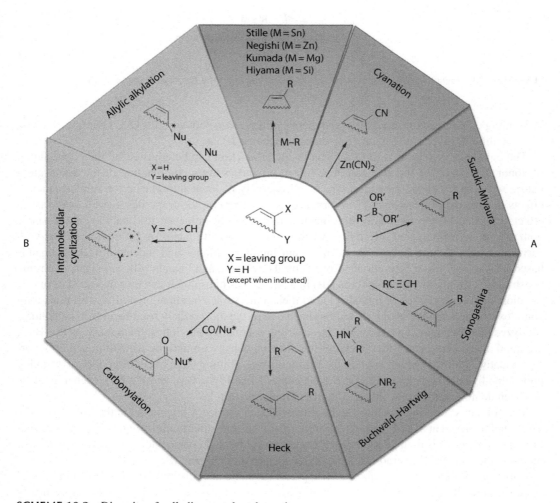

SCHEME 10.2 Diversity of palladium-catalyzed reactions.

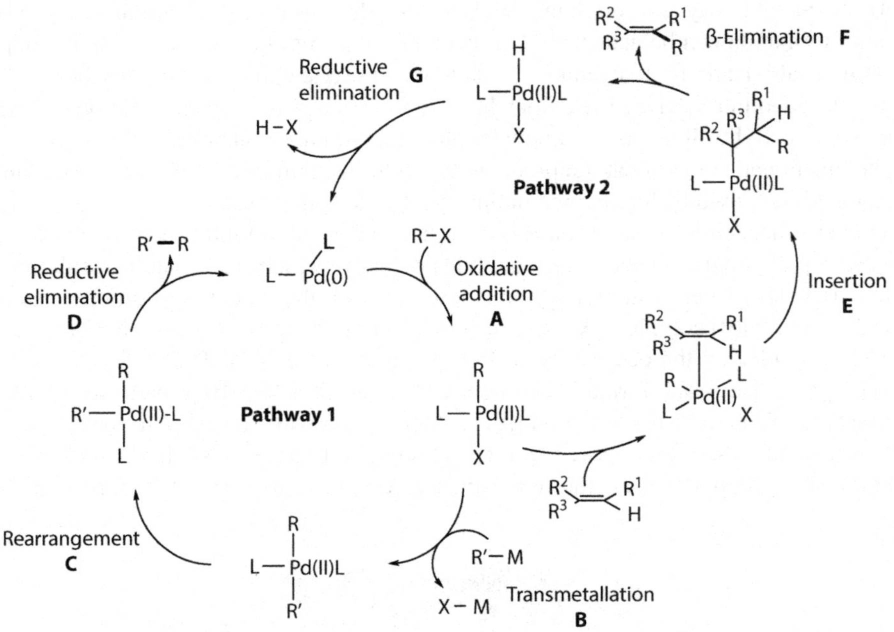

SCHEME 10.3 General mechanisms of palladium-catalyzed coupling reactions (adapted from[23]).

Depending on the reaction conditions and metals of these two pathways for C–C coupling, we may have the different types of reactions presented in Scheme 10.2a.

The palladium reactivity toward a given catalytic reaction can be adjusted by the coordination of donor ligands to the metal center. The ligand properties are a combination of the donor atom nature with the backbone structure, which is frequently of great importance for stereo-inductive effects in asymmetric reactions. Therefore, σ-donor and π-acceptor properties as well as the steric effects imposed by the ligands on the metal center strongly influence its catalytic performance, both in terms of activity and selectivity. An enormous variety of ligands of different types and with variable structures have been reported in recent literature: mono- and bidentate ligands, ligands based on single donor or heterodonor atoms, chiral or achiral, and ligands with steric and electronic constraints.[2] Among them, phosphorus (III) ligands such as phosphines can be considered as one of the most important classes of ligands, due to their remarkable coordination ability with palladium complexes. Triphenylphosphine (PPh$_3$) has been one of the most widely used ligands in palladium-catalyzed coupling reactions. However, basic strong σ-donating alkylphosphines, such as P(n-Bu)$_3$, are good alternatives, since they increase the electron density around the palladium center, favoring oxidative addition, which is often the reaction's rate-limiting step. In addition, the use of bulky phosphine ligands, with large Tolman's cone angles[24] like PPh(t-Bu)$_2$ and P(t-Bu)$_3$ (Scheme 10.4), has been demonstrated to favor the reductive elimination step.[25,26]

The Tolman's cone angle values can be obtained using a space-filling model of the M(PR$_3$) group. The metal is placed at 2.28 Å from the phosphorous atom and, after optimization of the ligand structure, the cone angle is defined as the apex angle of a cylindrical cone centered in the metal and just touching the van der Waals radii of the outermost atom of the ligand (Scheme 10.5).[25]

Therefore, for each reaction and for each particular substrate, it is important to perform a careful selection of the ligand or, in some cases, it is necessary to synthesize appropriate ligands specifically designed for the desired reaction, particularly when enantioselective catalysis is the goal.[28]

Despite the huge number of books, chapters, and review papers in the recent literature describing the application of palladium catalysts in organic synthesis,[29–37] the application of palladium-catalyzed

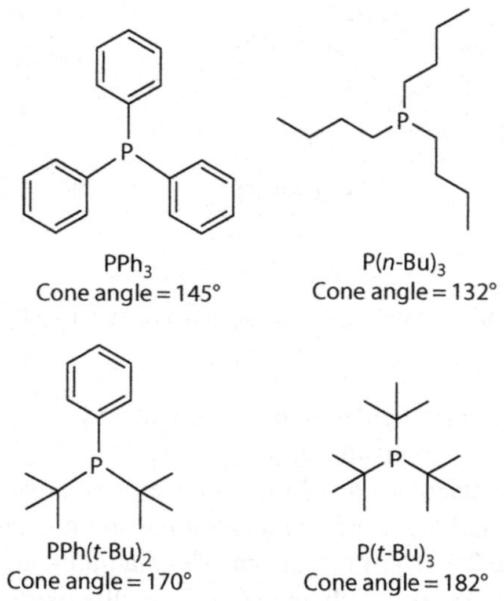

PPh₃
Cone angle = 145°

P(n-Bu)₃
Cone angle = 132°

PPh(t-Bu)₂
Cone angle = 170°

P(t-Bu)₃
Cone angle = 182°

SCHEME 10.4 Examples of commercially available phosphine-based ligands used in palladium catalysis.[27]

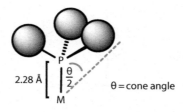

SCHEME 10.5 Definition of Tolman's cone angle.

strategies in asymmetric synthesis of drug compounds and in the functionalization of natural products with pharmaceutical interest is still an emerging area.[38–47]

In this chapter, selected examples from the last 3 years are presented, related to the application of palladium metal complexes in the synthesis and functionalization of enantiomerically pure synthons, drugs, and natural products, particularly focused on asymmetric allylic alkylations, intramolecular cyclizations, and carbonylation reactions. A brief description of each palladium-catalyzed strategy is given accompanied by the general reaction mechanism and illustrated with an example of its application to the synthesis of chiral synthons and, in some cases, by the retrosynthetic approach for the target natural product/drug molecule.

10.2 PALLADIUM-CATALYZED ASYMMETRIC ALLYLIC ALKYLATIONS

Palladium-catalyzed asymmetric allylic alkylations represent a considerable challenge in synthetic organic chemistry since it is a powerful C–C bond–forming process that allows the construction of new stereogenic centers.[48,49] The general catalytic cycle of palladium-catalyzed asymmetric allylic alkylation involves the oxidative generation of a π-allylpalladium intermediate from an allyl electrophile, followed by nucleophilic attack on the allyl terminus and subsequent reduction of the metal center to Pd(0) (Scheme 10.6).[49]

Particularly, palladium-catalyzed decarboxylative asymmetric allylic alkylation is a reliable approach to generate quaternary stereocenters.[50] This class of reactions was developed in the 1980s by Tsuji, employing allyl enol carbonates[51] or β-keto esters as substrates.[52] Since then, several

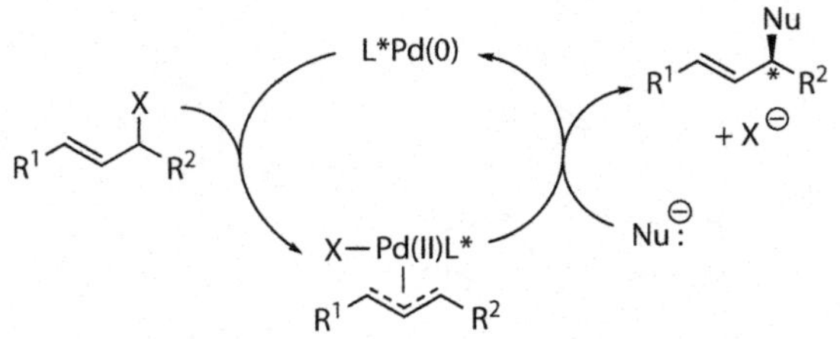

SCHEME 10.6 General mechanism of palladium-catalyzed asymmetric allylic alkylation.

applications of this technology in the field of natural product and drug synthesis have been described, highlighting the power and broad applicability of this reaction.[53–56]

Stoltz[57] reported a highly efficient protocol for the decarboxylative enantioselective allylic alkylation of β-keto esters using palladium acetate and chiral phosphinooxazoline ligands (S)-t-BuPHOX **1**[58] and (S)-(CF$_3$)$_3$-t-BuPHOX **2**.[59] The mechanism of palladium-catalyzed decarboxylative asymmetric allylic alkylation is presented in Scheme 10.7. First, the oxidative addition of Pd(0) species **A** to the allyl–O bond of β-keto ester **B** generates complex **C**, which undergoes decarboxylation to

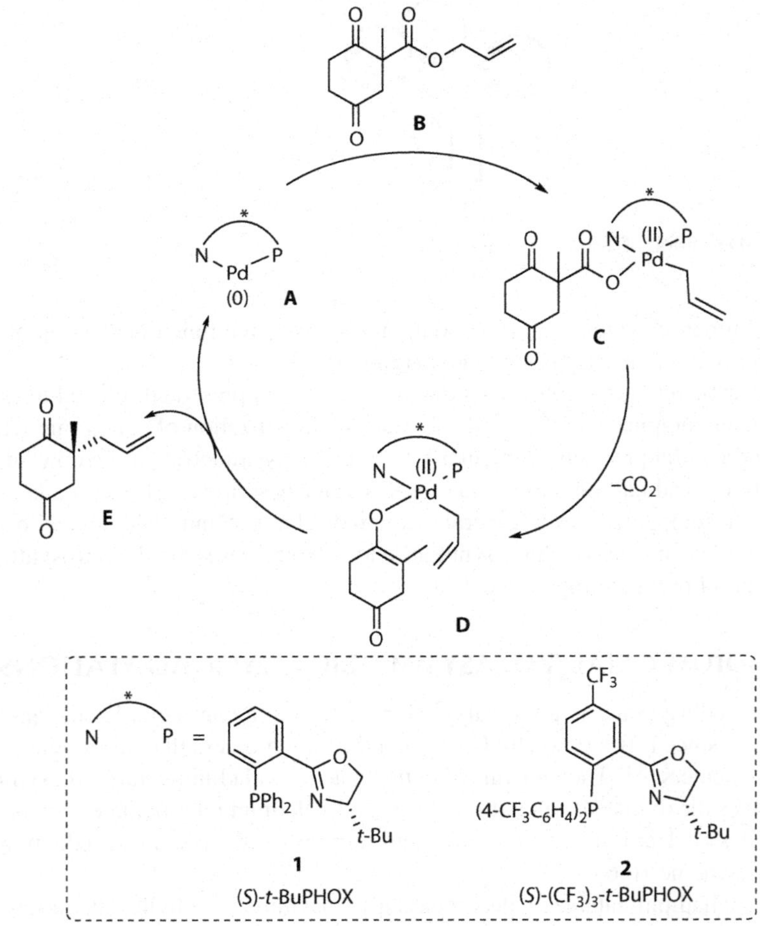

SCHEME 10.7 Catalytic cycle of palladium-catalyzed decarboxylative allylic alkylation.

form allyl palladium enolate **D**. Finally, reductive elimination furnishes quaternary ketone product **E** and regenerates the active Pd(0) catalyst **A**.

After optimization of several critical reaction parameters, such as temperature, concentration, ligand, and solvent, a broad range of β-keto esters were alkylated with the use of low metal loadings (0.075 mol%), generating the desired chiral ketone products in high yields (TON of up to 1320) and excellent enantioselectivities (up to 99% ee).

This palladium-catalyzed strategy was successfully applied to prepare (2*R*,5*R*)-2,5-diallyl-2,5-dimethylcyclohexane-1,4-dione **4**, a critical intermediate in the synthesis of (−)-cyanthiwigin F, a drug belonging to the class of cyathins, with known cytotoxic activity against human primary tumor cells (IC$_{50}$ = 3.1 μg/mL).[60] The retrosynthetic approach to achieve (−)-cyanthiwigin F from the initial substrate **3** is presented in Scheme 10.8.[61] The five-membered ring construction occurs at the latest stage, leading back to bicyclic ketone **7**, which may be prepared through intramolecular metathesis of ketone **6**. In turn, this ketone could be afforded via Negishi cross-coupling of vinyl triflate **5**. The third intermediate could be obtained from monoanionic desymmetrization of cyclohexadione **4**. Finally, cyclohexadione **4** may be obtained from bis(β-keto ester) **3** *via* palladium-catalyzed enantioselective double allylic alkylation.

This palladium-catalyzed step was elegantly applied by Stoltz[57] for the preparation of (2*R*,*R*)-2,5-diallyl-2,5-dimethylcyclohexane-1,4-dione **4**, in high yield (97%) and excellent enantioselectivity (99% ee), using only 0.25 mol% of palladium acetate (Scheme 10.9). This first critical step allows the early establishment of both quaternary stereocenters present at the ring junctions of the natural product (−)-cyanthiwigin F.

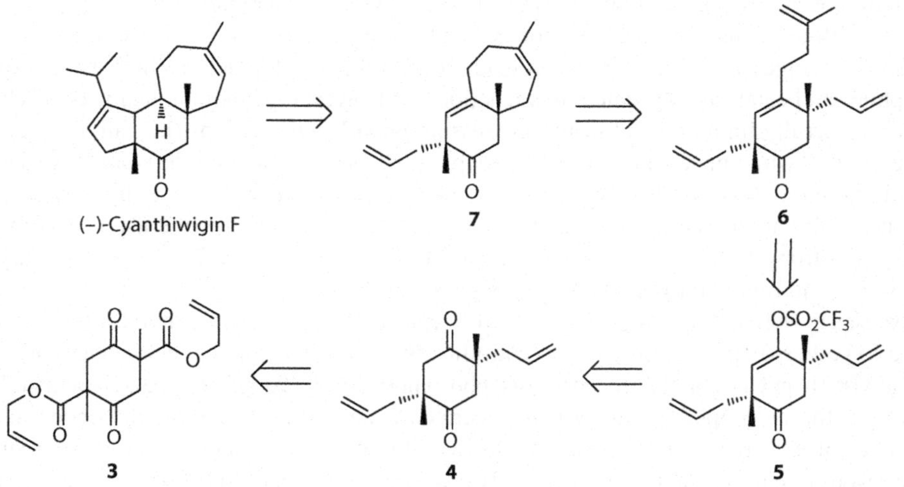

SCHEME 10.8 Retrosynthetic approach to (−)-cyanthiwigin F.

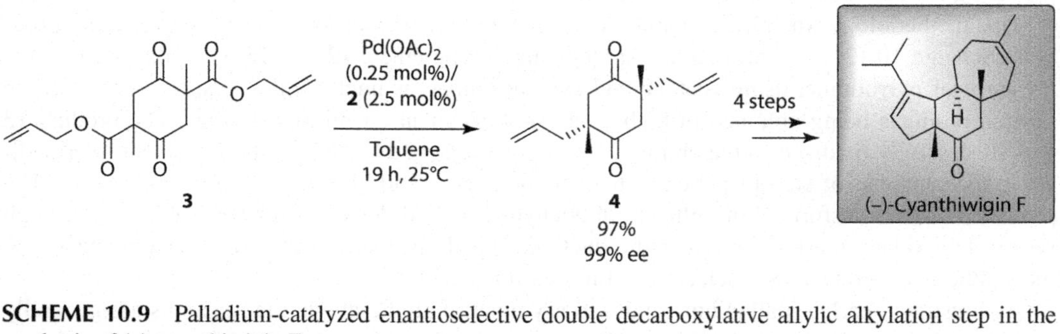

SCHEME 10.9 Palladium-catalyzed enantioselective double decarboxylative allylic alkylation step in the synthesis of (−)-cyanthiwigin F.

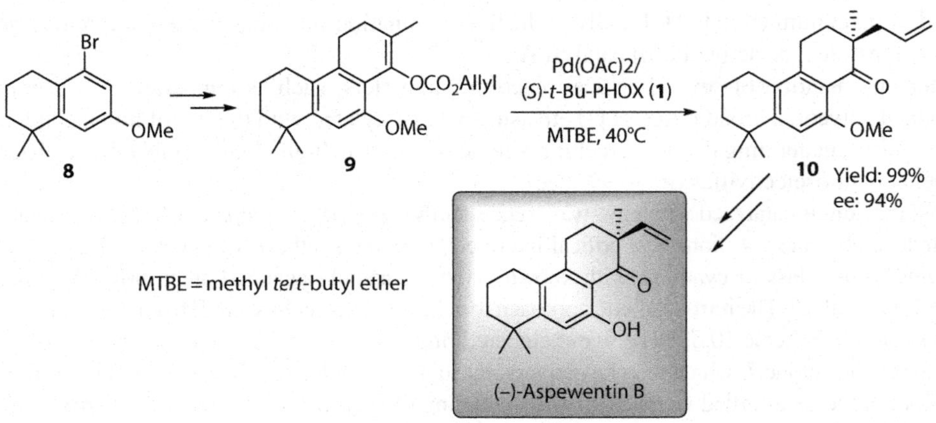

SCHEME 10.10 Synthesis of (−)-aspewentin B *via* palladium-catalyzed enantioselective allylic alkylation.

A similar palladium-catalyzed decarboxylative allylic alkylation strategy was used for the preparation of intermediate **10**, in the first enantioselective total synthesis of (−)-aspewentin B (Scheme 10.10),[62] whose biological activity was evaluated in growth inhibition tests against a marine zooplankton (*Artemia salina*), demonstrating its high cytotoxicity (LC_{50} = 6.36 µM).[63]

As mentioned earlier, palladium-catalyzed allylic alkylation is a reliable and widely used method for the formation of multiple types of bonds[64] and has been extensively used in a variety of total syntheses.[65,66] Huge progresses have been made toward the development of increasingly elaborate nucleophiles and catalysts to promote this reaction.[67,68] However, palladium-catalyzed allylic alkylations are generally limited to substrates possessing good leaving groups (i.e., allylic acetates and carbonates),[64–67] while unactivated leaving groups, such as ethers and amines, are less used due to difficulties in promoting the cleavage of C–O bonds.[69–71] Due to their high stability, the use of allylic ethers in palladium-catalyzed allylic alkylation reactions often requires the use of stoichiometric quantities of strong Lewis acids[70] or specific substrates,[72] such as vinyl epoxides and allylic aryl ethers, which can limit their potential use in organic synthesis.

These issues were addressed by Zhang,[73] who reported a convenient method for the direct use of allylic alkyl ethers in palladium-catalyzed allylic alkylations *via* hydrogen bond activation, using [Pd(η³-allyl)Cl]₂ and the chiral ferrocene-based phosphinooxazoline dppf **11** as ligand (Scheme 10.11a). In this case, the nucleophile is an *in situ*–generated enamine, formed through condensation of a ketone reagent with pyrrolidine (Scheme 10.11b). Studies on the effects of the solvent suggested that the reaction requires the use of protic solvents, such as MeOH, EtOH, and *n*-PrOH, to activate the less reactive moiety allyl ether (Scheme 10.11a). The procedure was applied to a variety of functionalized allylic ether substrates, under mild conditions, and proceeded with high regioselectivity, providing the desired allyl-ketone products in excellent yields (up 99%, 29 examples).

This methodology was further applied in the asymmetric synthesis of enantiopure natural products (Scheme 10.12).[74,75] Through this strategy, the allylic ethers **12** and **15** were alkylated, in the presence of pyrrolidine, using acetone and cyclohexanone as nucleophiles, respectively, with the desired products being obtained in high yields and excellent enantioselectivities. The product **13** was successfully oxidized to the chiral γ-oxocarboxylic acid **14**, which is an important intermediate in the synthesis of various peptides, therapeutic agents, and bioactive natural products.[76,77] The product **16** was transformed into the chiral phenylacetic acid derivative **17** (with 97% ee) through sequential oxidation, esterification, and reduction. This derivative was then used as a precursor for the synthesis of product **18**, a selective antimuscarinic agent.[78]

As illustrated earlier, palladium-catalyzed asymmetric allylic alkylation reactions are generally performed with several stabilized or "soft" carbon and heteroatom nucleophiles (from conjugate

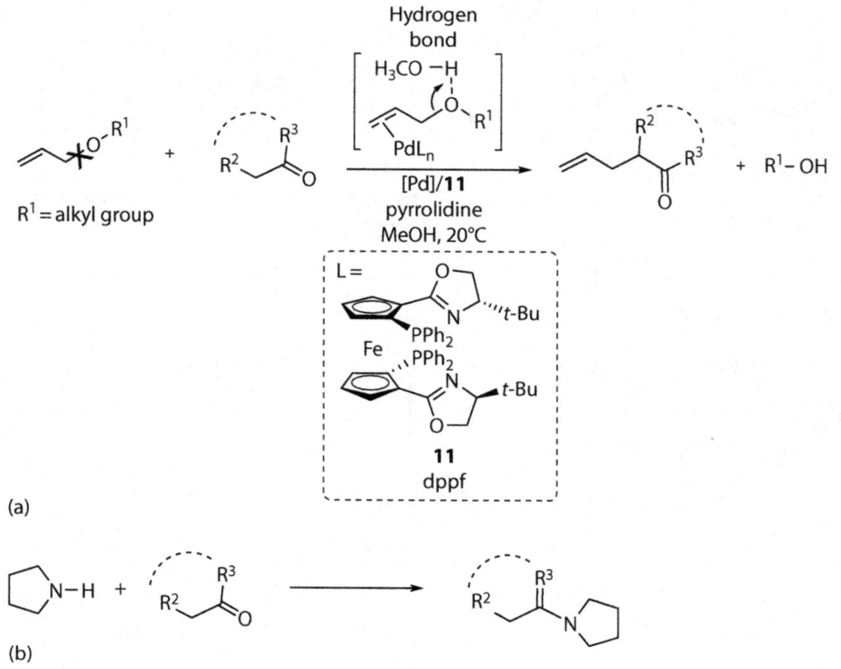

SCHEME 10.11 (a) Palladium-catalyzed alkylation of allylic ethers *via* hydrogen bond activation; (b) *in situ* enamine generation.

acids with a pK_a lower than 25).[79] However, palladium-catalyzed allylic substitution reactions using "hard" nucleophiles (from conjugate acids with a pK_a higher than 25) are often more demanding.[80,81] Some strategies to turn "hard" nucleophiles into "softer" were tested successfully, such as the use of BF_3.[82] However, this strategy appeared as inadequate for pronucleophiles bearing less acidic C–H units, namely, toluene derivatives (pK_a ≈ 14). To overcome this problem, Walsh[83] reported the first highly enantioselective palladium-catalyzed allylic alkylations using (η^6-$C_6H_5CH_3$)Cr(CO)$_3$ as activated benzylic pronucleophile (Scheme 10.13a). The methodology was applied to a variety of cyclic and acyclic allylic electrophiles, furnishing the products in excellent yields and enantioselectivities of up to 92%. This type of reaction enables the enantioselective synthesis of α-2-propenyl benzyl motifs, which are known to be important scaffolds in natural products and pharmaceuticals, as those represented in Scheme 10.13b.

The relevance of this palladium-catalyzed strategy was illustrated by its application to the synthesis of a nonsteroidal anti-inflammatory drug (NSAID) analogue.[84] Beginning with the pyridyl-containing toluene complex **19**, the asymmetric allylic alkylation, followed by demetallation, provided the allylated product **22** (>99% ee), which was subsequently converted into the enantioenriched α-arylalkanoic acid **23** in 61% yield, with 99% ee (Scheme 10.14a).

Palladium-catalyzed allylic alkylation has also been shown to be a distinctive method for the oxidation of allylic esters, yielding cycloalkenones when *meso*-1,4-allylic dibenzoates are used as substrates.[85] Such chiral cycloalkenones are important building blocks in both natural product and pharmaceutical synthesis. In this context, Trost[86] reported a strategy for the synthesis of a set of chiral cycloalkenone derivatives *via* asymmetric palladium catalysis (Scheme 10.15). Using optimized procedures, the oxidative desymmetrization of different *meso*-dibenzoates afforded the corresponding cycloalkenones in good yields and excellent enantioselectivities. For example, when *meso*-diester **24** was subjected to this reaction, in the presence of chiral ligand **25**, an oxidative desymmetrization proceeded, providing the respective cycloalkenone **30** in good yield and 99% ee. In this process,

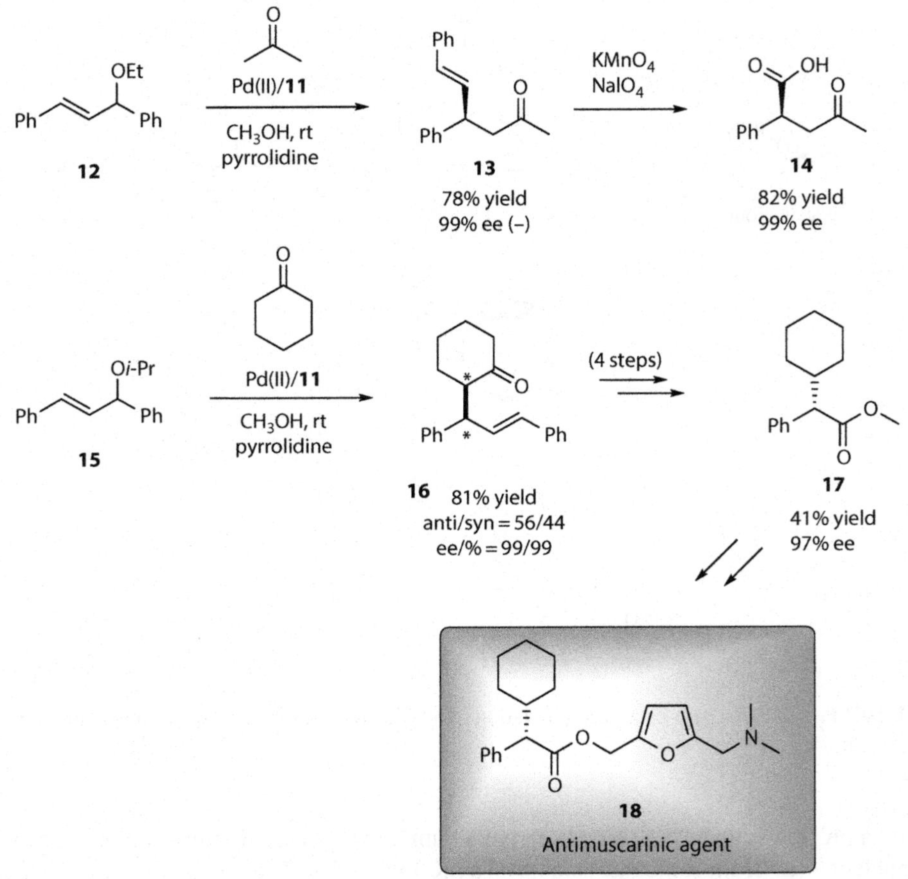

SCHEME 10.12 Synthesis of antimuscarinic agent **18** by palladium-catalyzed asymmetric alkylation of allylic ethers *via* hydrogen bond activation.

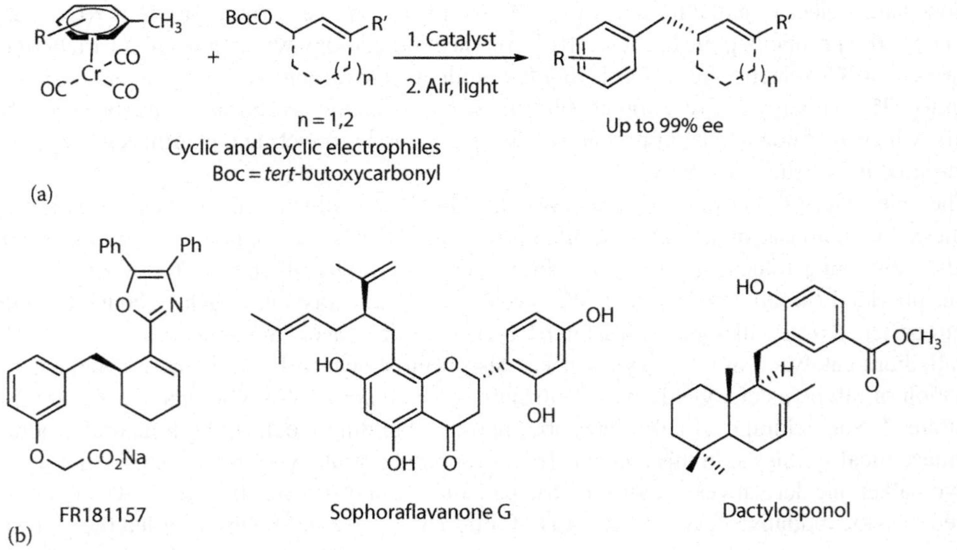

SCHEME 10.13 (a) Enantioselective palladium-catalyzed allylic alkylation with "softened" hard pronucleophiles; (b) examples of natural products containing α-2-propenyl benzyl motifs.

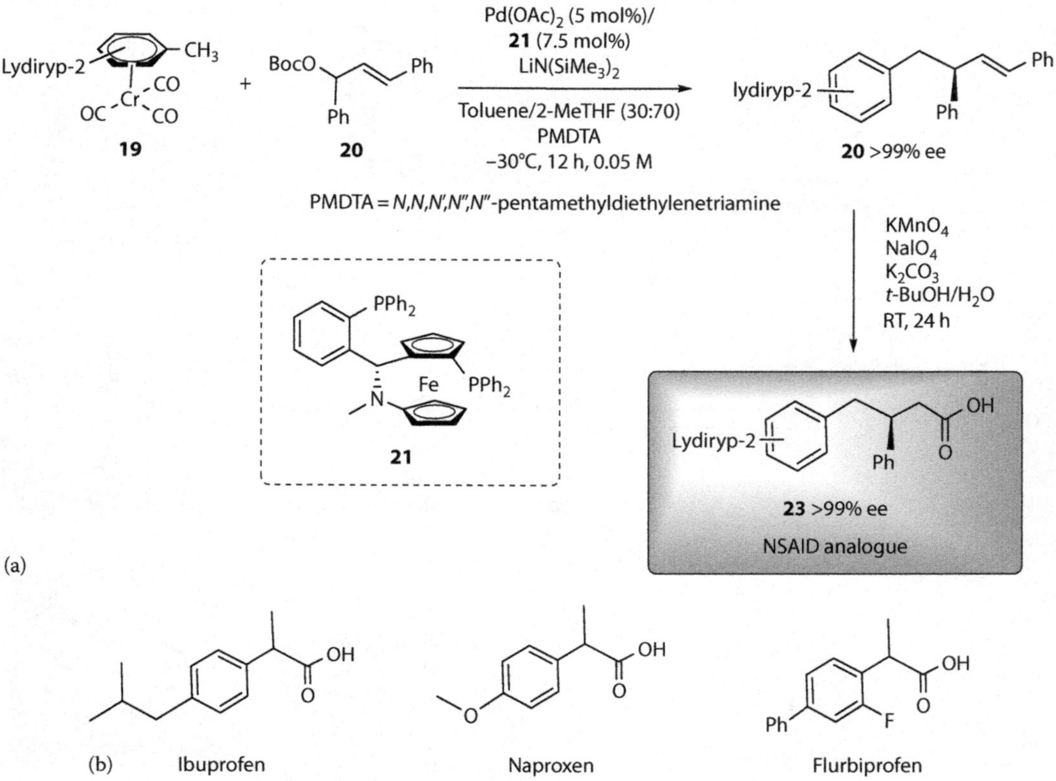

SCHEME 10.14 (a) Synthesis of α-arylalkanoic acid 23 *via* enantioselective palladium-catalyzed allylic alkylation with benzylic nucleophiles. (b) Examples of aryl propionic acid NSAIDs.

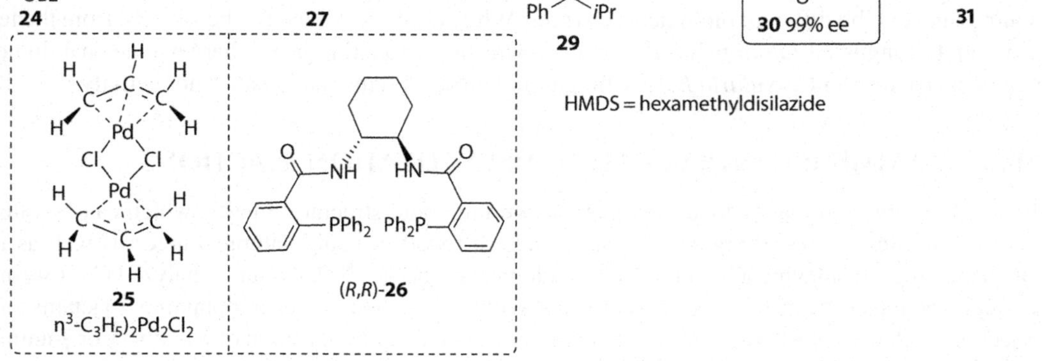

SCHEME 10.15 Asymmetric synthesis of cycloalkenone derivatives *via* palladium-catalyzed allylic alkylation.

304   Asymmetric Synthesis of Drugs and Natural Products

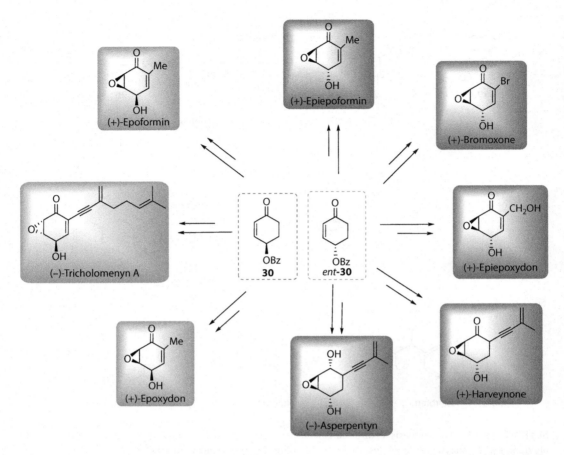

SCHEME 10.16 Examples of cycloalkanone-based natural products obtained by palladium-catalyzed allylic alkylation.

a π-allylpalladium intermediate **27** is generated, which undergoes selective O-alkylation with nitronate **28** to yield **29**. Subsequent fragmentation provides α,β-unsaturated product **30** and the oxime by-product **31**, which can be recycled to **28** (Scheme 10.15).

In addition, starting from the building block **30** mentioned earlier, more elaborate cyclohexanone-derived natural products could be prepared, such as the epoxyquinoid natural products presented in Scheme 10.16, with excellent levels of enantio- and diastereoselectivity. Among the many examples, harveynone is a paradigmatic one, since both enantiomers can be obtained from different natural sources and exhibit different biological activities: While (+)-harveynone can be isolated from the tea gray blight fungus *Pestalotiopsis theae* and is a known phytotoxin, the (−)-harveynone enantiomer can be isolated from *Curvularia harvey* fungus and presents remarkable antitumor activity.[87]

10.3 ASYMMETRIC INTRAMOLECULAR CYCLIZATION REACTIONS

It is well established that lactones and lactams are important structural motifs, which can be often used as building blocks in organic synthesis and are found in many natural sources as well as in synthetic drugs, displaying a broad range of biological properties.[88] Palladium-catalyzed C3–C4 ring closure reactions have been reported as efficient synthetic methodologies of obtaining γ-lactams and γ-lactones, alternatively to the conventional intramolecular O- or N-acylation of γ-hydroxy or γ-amino acid derivatives.[89] These intramolecular palladium-catalyzed allylic alkylations have been widely explored by Poli.[90] For instance, it was successfully applied to the synthesis of *trans* pyrrolidone **33**,

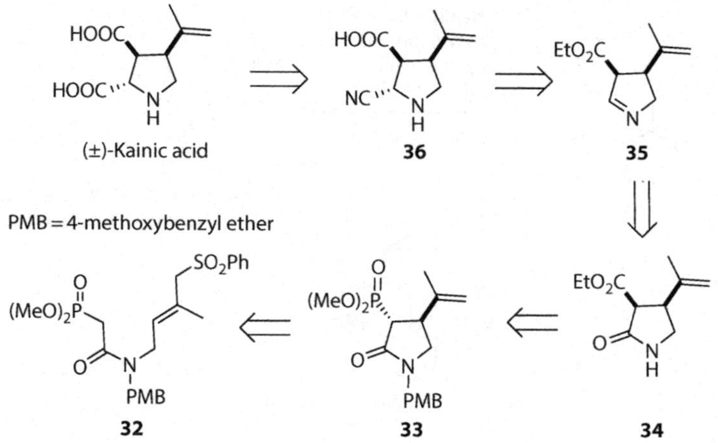

SCHEME 10.17 Retrosynthetic approach to (±)-kainic acid.

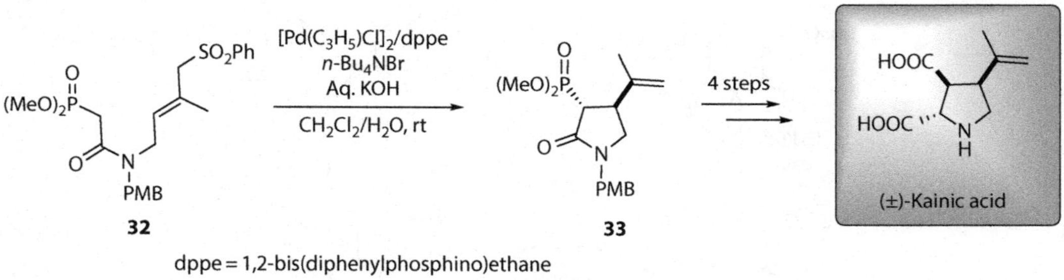

SCHEME 10.18 Synthesis of (±)-kainic acid *via* intramolecular palladium-catalyzed allylic alkylation.

a critical intermediate in the synthesis of (±)-α-kainic acid (Scheme 10.18), a potent central nervous system excitant that is used as a model in epilepsy and Alzheimer's disease research.[91] The retrosynthetic approach to achieve (±)-kainic acid from the initial substrate **32** is presented in Scheme 10.17. The five-membered ring construction occurs on the first step, affording *trans* pyrrolidone **33**. Further functionalization including a Horner–Wadsworth–Emmons olefination[92] followed by a diastereoselective conjugated hydride addition of the resulting electron-poor alkene yielded the advanced precursor **34** of kainic acid, previously reported by Xia and Ganem.[93]

This first critical step that allows the early establishment of the desired stereocenters of the five-membered heterocycle present in kainic acid was reported by Poli[90] *via* intramolecular allylic alkylation of the unsaturated phosphono-allylacetamide **32**, yielding quantitatively the desired *trans* pyrrolidine **33** (Scheme 10.18).

The same authors transposed this cyclization strategy to other allylamides to obtain enantiomerically pure lactams, using η³-allylpalladium complexes in the presence of the chiral diphosphine **38**. Treatment of (*E*)-**37** and (*Z*)-**37** under phase-transfer catalytic conditions provided the desired *trans* pyrrolidone **39** in up to 84% ee (Scheme 10.19).[94]

The general accepted mechanism for intramolecular allylic alkylation, based on DFT studies, is presented in Scheme 10.20. The η³-allyl palladium intermediates can exist as *syn* or *anti* isomers, which can originate the *syn*-5-*exo* or *anti*-5-*exo* isomers and cyclize to form the desired γ-lactam or lactone.

A different approach for the synthesis of biologically relevant chiral cyclopropane-fused chiral lactams was reported by Cramer,[95] through enantioselective palladium-catalyzed C–H functionalizations

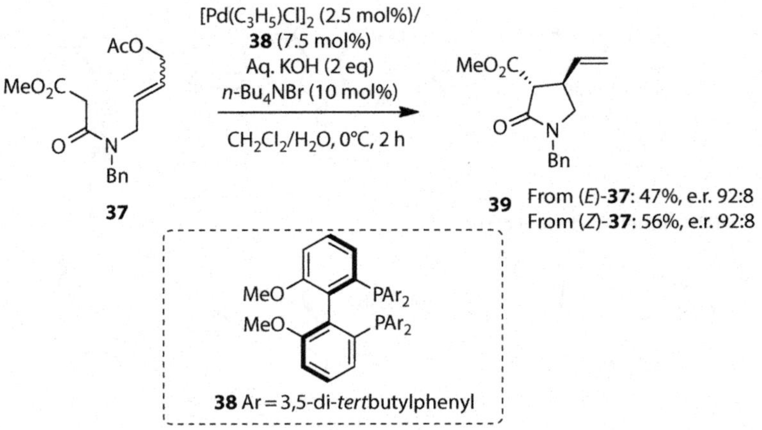

SCHEME 10.19 Palladium-catalyzed asymmetric intramolecular allylic alkylation of allylamides.

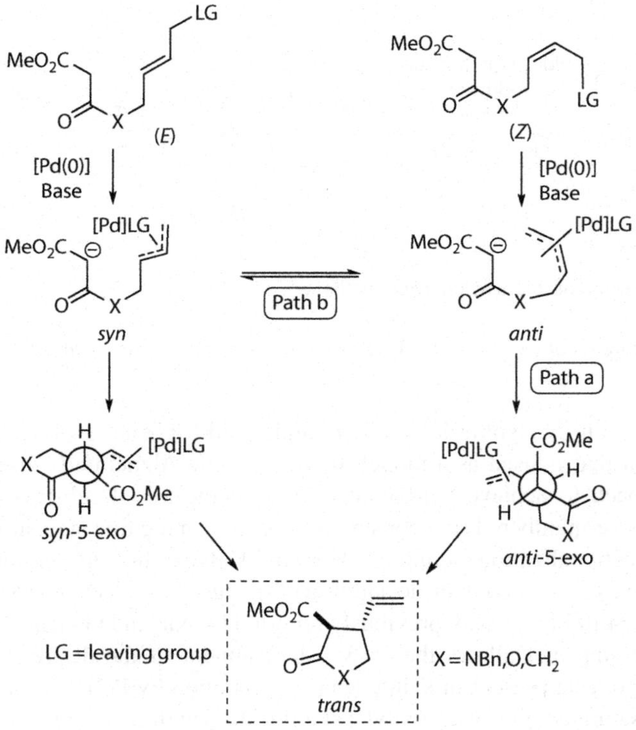

SCHEME 10.20 Cyclization pathways of palladium-catalyzed C3–C4 ring closure reactions.

of chloroacetamide substrates, using a bulky taddol-based phosphonite ligand **41** in combination with adamantane-1-carboxylic acid (Ad-CO$_2$H) as cocatalyst (Scheme 10.21).

This synthetic strategy was efficiently applied to the preparation of a variety of valuable 2-azabicyclo[3.1.0]hexane derivatives in excellent yields (up to 99%) and enantiomeric excesses of up to 96%. Such cyclopropane-fused γ-lactam products are analogues of saxagliptin, an oral hypoglycemic (antidiabetic drug) from the dipeptidyl peptidase-4 (DPP-4) inhibitor class (Scheme 10.21).[96] According to the accepted reaction's mechanism (Scheme 10.22),[97,98] the chloroacetamide group of **40** is the electrophile, which after selective C–H activation of one of the two enantiotopic C–H

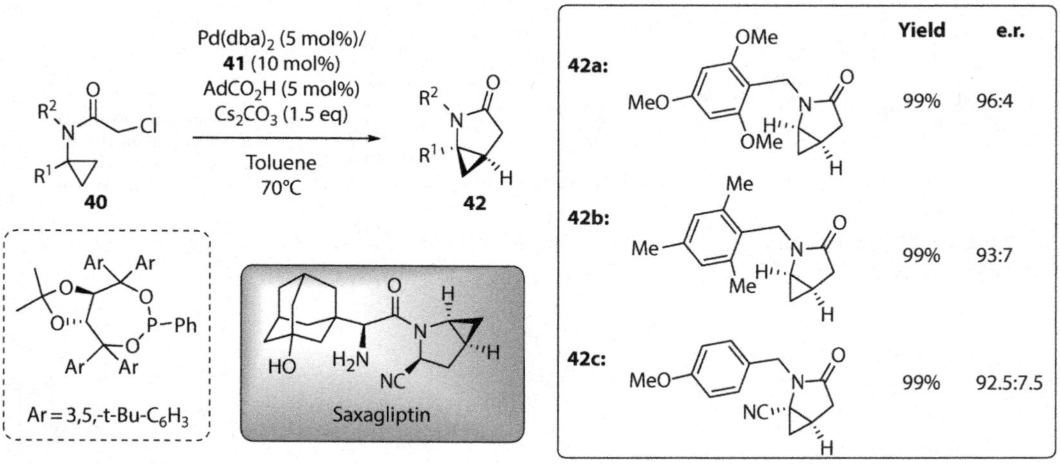

SCHEME 10.21 Synthesis of cyclopropane-fused chiral lactams *via* palladium-catalyzed C–H functionalizations.

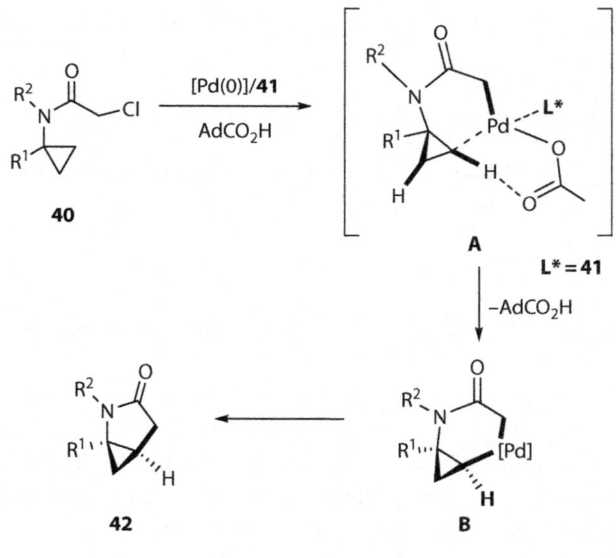

SCHEME 10.22 Mechanism of palladium-catalyzed C–H functionalization to cyclopropane-fused chiral γ-lactams.

bonds of the cyclopropane ring leads to the formation of intermediate **A**. This is transformed during the concerted metalation–deprotonation step, to give the six-membered palladacycle **B**. Subsequent reductive elimination yields the target cyclopropane-fused γ-lactam **42**.

In parallel, Baudoin[99] also reported the synthesis of mono- and bicyclic α-alkylidene-γ-lactams, through an intramolecular alkenylation reaction of unactivated primary C(sp^3)–H bonds using acyclic bromoalkenes (Scheme 10.23).

This approach was particularly promising to obtain a set of diverse five-membered nitrogen heterocycles, which are common motifs in numerous bioactive natural products such as pyrrolidine, pyrrolizidine, indolizidine, and *Stemona* alkaloids. For instance, the monocyclic α-alkylidene-γ-lactam **45** was obtained from substrate **43** in the presence of allylpalladium(II) chloride dimer precursor and chiral phosphine ligand **44**, with a moderate enantioselectivity (58% ee) (Scheme 10.24). Although modest, this result constitutes an important proof of concept for the enantioselective

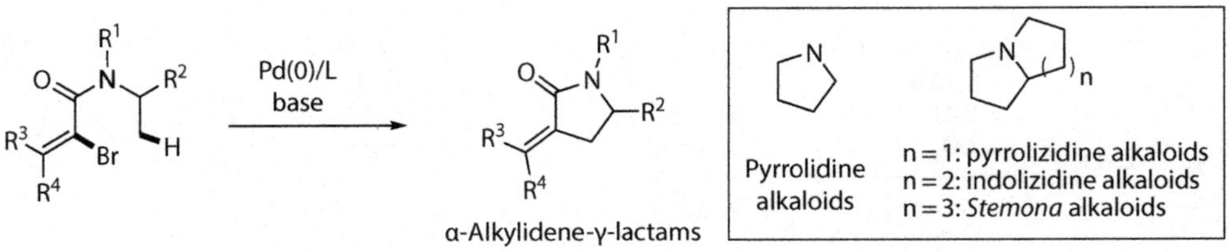

SCHEME 10.23 Palladium-catalyzed intramolecular alkenylation of unactivated primary C(sp³)–H bonds of acyclic bromoalkenes.

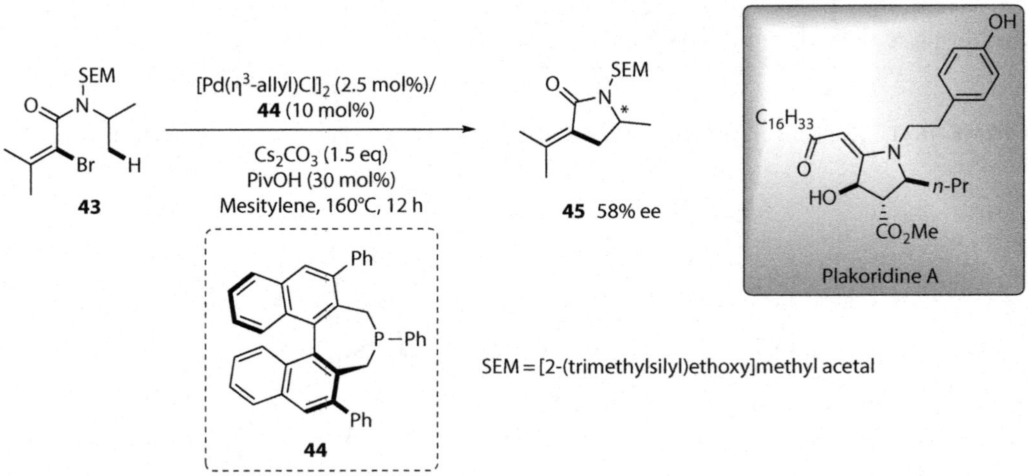

SCHEME 10.24 Asymmetric palladium-catalyzed intramolecular alkenylation, leading to a pyrrolizidine alkaloid analogue.

synthesis of γ-lactam through this C(sp³)–H alkenylation approach. This class of compounds are analogues of plakoridine A,[100] a tyramine-containing pyrrolidine alkaloid, extracted from the marine sponge *Plakortis* sp., which was demonstrated to be an effective cytotoxic agent against murine lymphoma L1210 cell lines.[101]

Additionally, a palladium-catalyzed intramolecular reductive Heck cyclization reaction was reported by Xie,[102] as one of the main steps in the total synthesis of two naturally occurring polyketide–terpenoids, such as PI-220, an inhibitor of rabbit platelet aggregation, and (+)-3-*epi*-furaquinocin C, an important cytotoxic agent against melanoma cells (Scheme 10.25).

The ester (*R*)-**46** was cyclized to (2*R*,3*S*)-**47** in 92% yield, which shows the ability of this palladium-catalyzed intramolecular reductive Heck reaction for the selective synthesis of 2,3-*cis*-dimethyldihydrobenzofurans. The reaction's stereoselectivity was explained by the authors through DFT calculations on the basis of transition state models, in which the energy differences between the two transition states, *cis*-**TS** and *trans*-**TS**, explain the favorable formation of the isomer with *cis*-oriented methyl groups (Scheme 10.26).

Another relevant synthetic methodology involving a palladium-catalyzed cyclization reaction was applied by Tietze to perform the first enantioselective total synthesis of (−)-blennolide A[103] and secalonic acid E.[104] Secalonic acids display a wide range of biological activities that include anticancer, antimicrobial, antitumor, and anti-HIV properties, as well as DNA topoisomerase I and protein

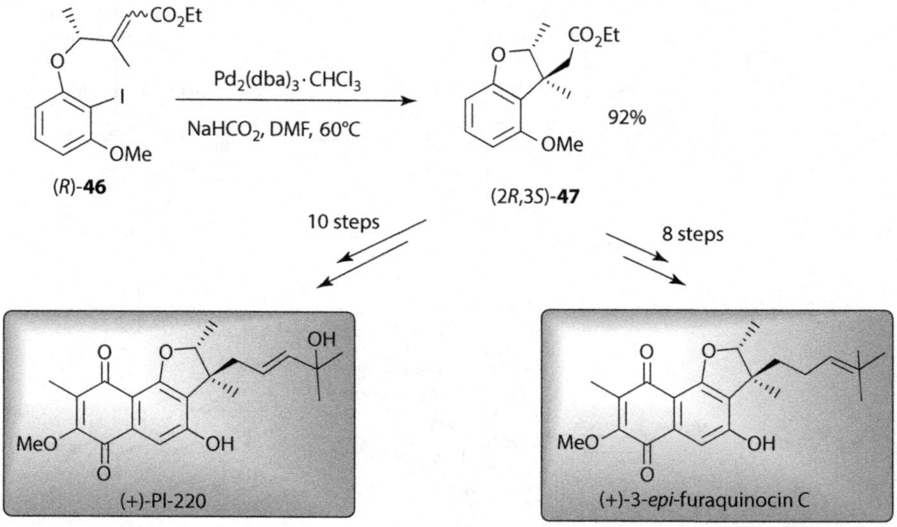

SCHEME 10.25 Synthesis of chiral 2,3-*cis*-dimethyldihydrobenzofuran (2*R*,3*S*)-**47** *via* palladium-catalyzed intramolecular reductive Heck reaction.

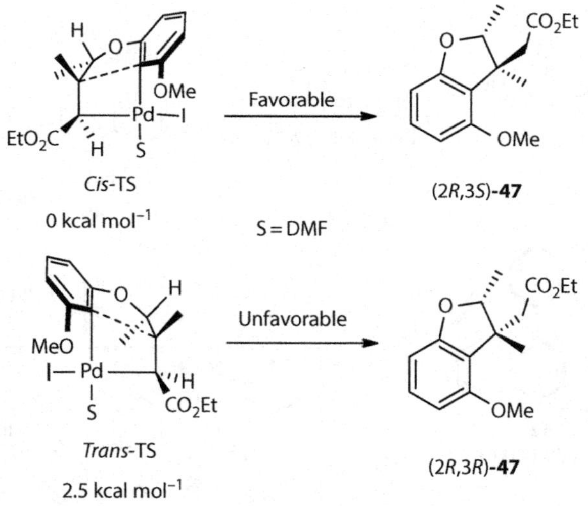

SCHEME 10.26 Stereoselectivity of 2,3-*cis*-dimethyldihydrobenzofurans, obtained *via* palladium-catalyzed intramolecular reductive Heck cyclization.

kinase C inhibition.[104] The retrosynthetic approach for secalonic acid E starts with its desymmetrization by cleavage of the 2,2′-biphenol linkage, which leads to the iodinated monomeric tetrahydroxanthenone synthon **51**. This may be obtained from **50** by an intramolecular Dieckmann cyclization, which can be further disassembled to give the chiral chroman **49** with a quaternary stereogenic center. Finally, the chroman **49** can be obtained from **48** by an enantioselective palladium-catalyzed Wacker-type cyclization (Scheme 10.27).

The intramolecular asymmetric Wacker-type cyclization of methoxyphenolic compound **48** was performed using palladium(II)-trifluoroacetate (TFA)/(*S*,*S*)-*i*Pr-BOXAX (2,2′-bis(oxazolyl)-1,1′-binaphthyl **52**) complex as catalyst and *p*-benzoquinone to promote the oxidation of Pd(0) to Pd(II).

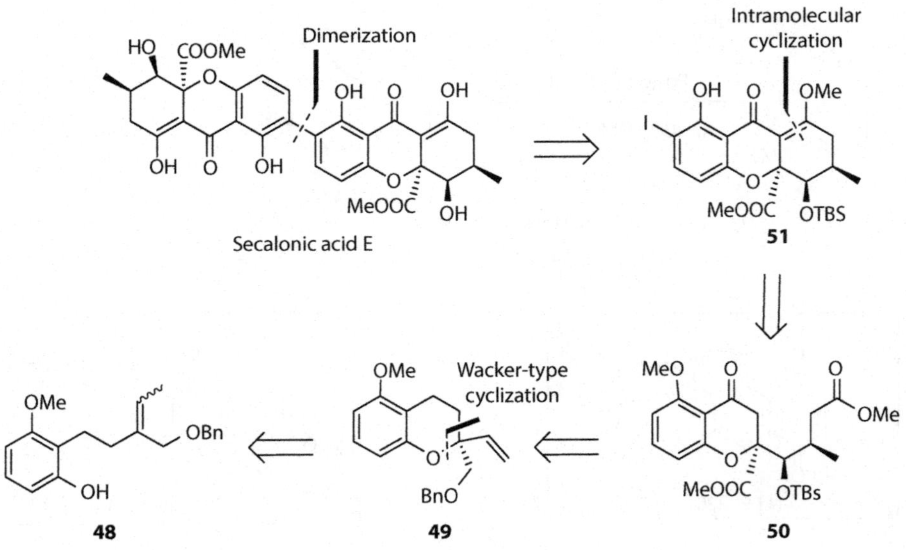

SCHEME 10.27 Retrosynthetic approach to secalonic acid E.

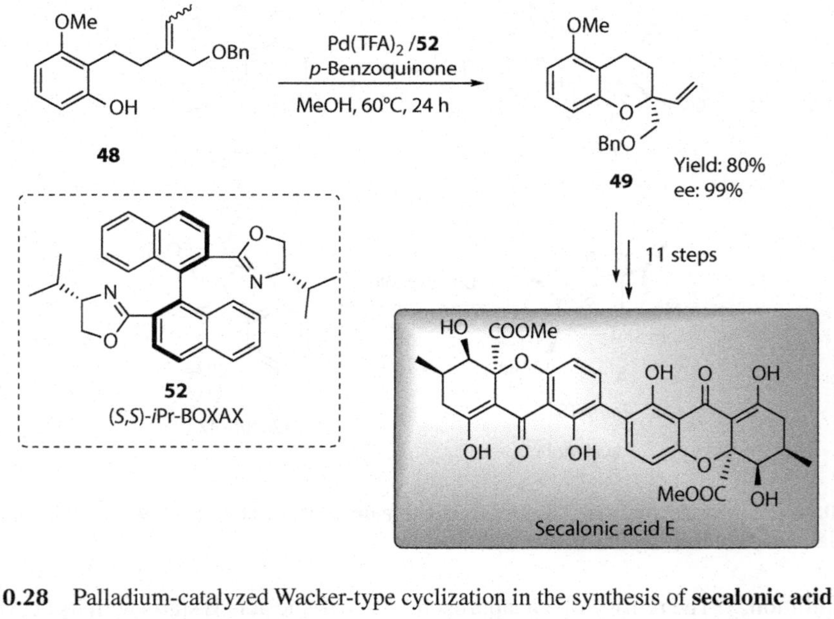

SCHEME 10.28 Palladium-catalyzed Wacker-type cyclization in the synthesis of **secalonic acid E**.

The chiral chroman **49**, containing a quaternary carbon stereogenic center, was obtained with 80% yield and an enantiomeric excess higher than 99% (Scheme 10.28).

The proposed mechanism for this Wacker-type cyclization is presented in Scheme 10.29. The initial alkene coordinates to the electron-deficient $L_nPd(II)X_2$ complex and activates the olefin toward intramolecular nucleophilic attack by the hydroxyl group. A subsequent β-hydride reductive elimination produces acid (HX) and $L_nPd(0)$, which, after oxidation promoted by *p*-benzoquinone, regenerates the $L_nPd(II)X_2$ species that enters again in the catalytic cycle.

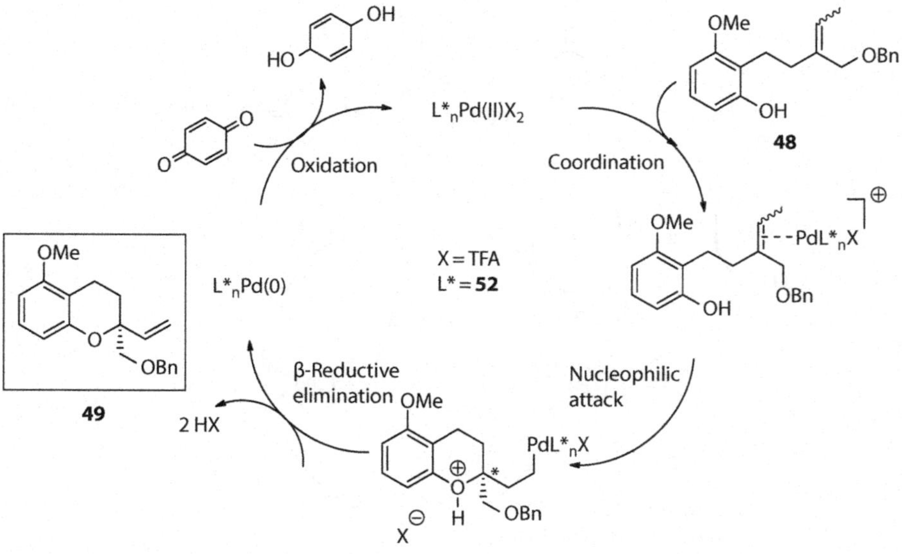

SCHEME 10.29 Mechanism of palladium-catalyzed Wacker-type cyclization.

10.4 PALLADIUM-CATALYZED CARBONYLATION REACTIONS

Palladium-catalyzed carbonylation of organic halides was first reported by Heck[105,106] in the 1970s. Since then, this reaction has become a valuable tool for organic synthesis, being extensively applied to the preparation of carbonyl compounds with high synthetic relevance, such as biologically active natural products and derivatives.[107–110] The process follows a general pathway in which the organic halide undergoes oxidative addition with palladium(0), resulting in a palladium(II) complex. The coordination of carbon monoxide and subsequent migratory insertion form an acylpalladium species, which is susceptible to attack by nucleophiles (alcohols, water, or amines). In the presence of base, the reaction yields the coupled products, forming esters from alcohols (*alkoxycarbonylation*), carboxylic acids from water (*hydroxycarbonylation*), or amides in the presence of amine nucleophiles (*aminocarbonylation*), while the reductive elimination of HX regenerates the palladium(0) catalyst (Scheme 10.30, cycle A). Typically, the reactions require a stoichiometric amount of base to regenerate the catalyst. In line with the C–X bond energy, the rate of the oxidative addition of the organic halide to an electronically unsaturated Pd complex decreases in the following order: C–I > C–OTf ≥ C–Br >> C–Cl >> C–F.[111] In the case of double carbonylation,[112] which is usually a consequence of higher CO pressures, the mechanism proceeds *via* CO insertion into the Pd–R bond before reductive elimination occurs (Scheme 10.30, cycle B).

Palladium-catalyzed aminocarbonylation reactions, in which the nucleophile is an amine, have become an easy and practical method for the synthesis of amides.[113] Amides constitute one of the most important functional groups in contemporary chemistry, being essential for sustaining life, for instance, linking the amino acids in proteins. They are found in numerous natural products and are some of the most prolific moieties in modern pharmaceutical molecules.[114,115] Particularly, steroidal dicarboxamides were found to be of great importance as molecular umbrellas in drug delivery, as antifungal, and as cell antiproliferative agents.[116,117]

In this context, Pereira and Kollár[118] reported the synthesis of a set of bioactive steroid dimers, linked by structurally different dicarboxamide spacers at C-17, through a palladium-catalyzed carbonylation strategy. The classic catalytic system Pd(OAc)$_2$/PPh$_3$ was used, under moderate reaction conditions (30 bar CO, 100°C), to efficiently perform the diaminocarbonylation of 17-iodo-5α-androst-16-ene **53**, leading to the target chiral androstene-dicarboxamides **54**, in good to excellent yields (Scheme 10.31).

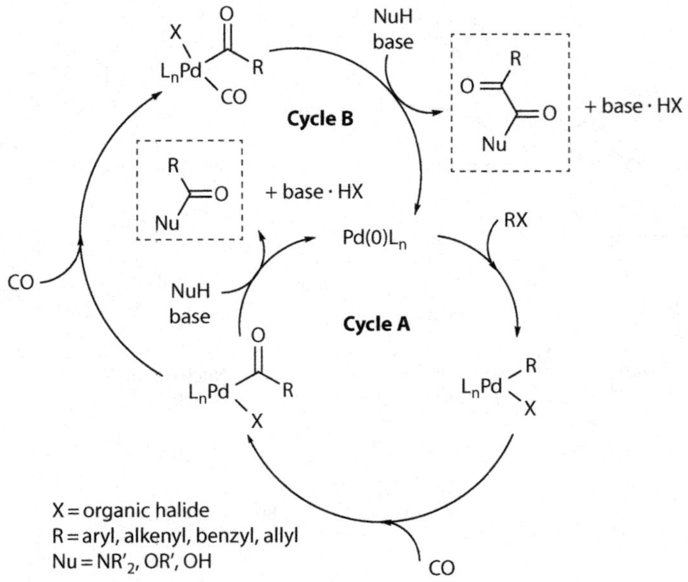

SCHEME 10.30 General mechanism of palladium-catalyzed carbonylation of organic halides.

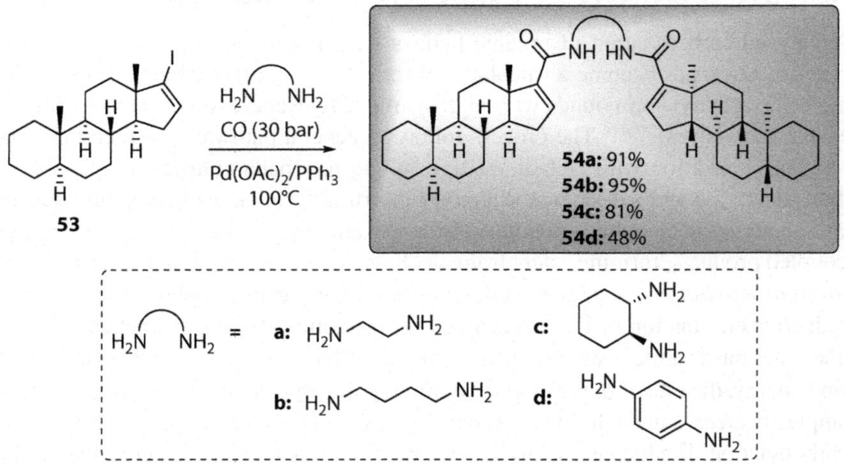

SCHEME 10.31 Synthesis of androst-16-ene-17,17′-dicarboxamides *via* palladium-catalyzed diaminocarbonylation.

The reaction's mechanism is presented in Scheme 10.32. The oxidative addition of the alkenyl-iodide substrate **53** to *in situ*–formed palladium(0) species results in alkenyl-palladium(II) intermediate **A**, which is able to coordinate with carbon monoxide, **B**. The acyl complex **C** is then formed by CO migratory insertion and undergoes a nucleophilic attack by the amino-monocarboxamide **D**, formed previously in the catalytic cycle, resulting in the formation of dicarboxamides **54** and regenerating the Pd(0) catalytic resting state, by reductive elimination.[119,120]

Preliminary investigations on the biological activity of these steroid dicarboxamides were performed, through *in vitro* cytotoxicity tests against human lung carcinoma A549 cells,[121] in which the dicarboxamides **54b** and **54d** were shown to be the most efficient cytotoxic agents, with similar IC50 values, of approximately 40 μM. In sum, the use of diamines as nucleophiles, under carbonylation

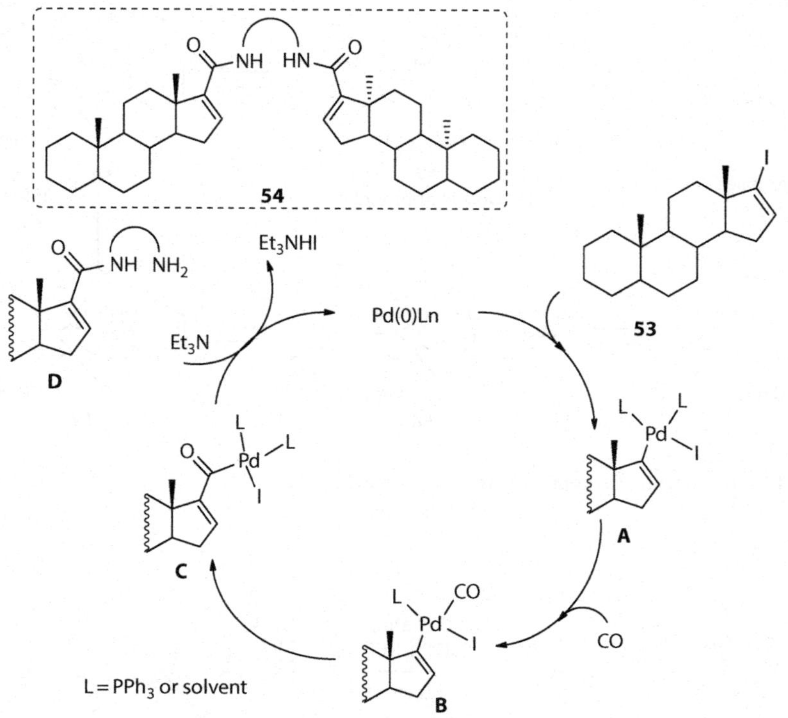

SCHEME 10.32 Mechanism of palladium-catalyzed diaminocarbonylation of 17-iodosteroids, leading to androst-16-ene-17,17'-dicarboxamides.

conditions, provided an advantageous straightforward route for the synthesis of steroid-based dicarboxamides, providing a promising advance in the chemical synthesis of bioactive molecules.

On the other hand, palladium-catalyzed alkoxycarbonylation reactions, in which the nucleophile is an alcohol, have also been applied as main synthetic steps in the total synthesis of biologically relevant natural products. For instance, Makabe[122] described a methodology for the total synthesis of (+)-boronolide and (+)-deacetylboronolide, in which one of the main steps is a palladium-catalyzed intramolecular alkoxycarbonylation. Such boronolides consist of natural α,β-unsaturated δ-lactones, which can be extracted from plants[123,124] and whose biological activity as antimalarial agents has been reported.[124,125]

The retrosynthetic approach to (+)-boronolide is outlined in Scheme 10.33. First, the deacetylation of (+)-boronolide would afford **59**, which can be obtained through deprotection of both acetonide and methoxymethyl (MOM) groups of the δ-lactone **58**. This should be accessible from CO insertion and cyclization of **57**.[126,127] The cyclization precursor **57** can be obtained from terminal alkyne **56**, which can be synthesized from **55**.

The intramolecular alkoxycarbonylation reaction of **57** was performed with $Cl_2Pd(PPh_3)_2$ catalyst, in the presence of 1 atm CO, using K_2CO_3 as base and hydrazine hydrate as an additive, in THF, leading to the desired lactonization product **58** in 71% yield (Scheme 10.34).[122]

In this case, the oxidative addition of the alkenyl-iodide substrate **57** to the *in situ*–formed palladium(0) species results in alkenyl-palladium(II) intermediate **A**, which upon coordination with carbon monoxide leads to species **B**. The subsequent CO migratory insertion results in complex **C**, which undergoes an intramolecular nucleophilic attack by the hydroxyl group, resulting in the intramolecular cyclization with formation of δ-lactone **58** and regenerating the Pd(0) catalytic resting state, through reductive elimination (Scheme 10.35).

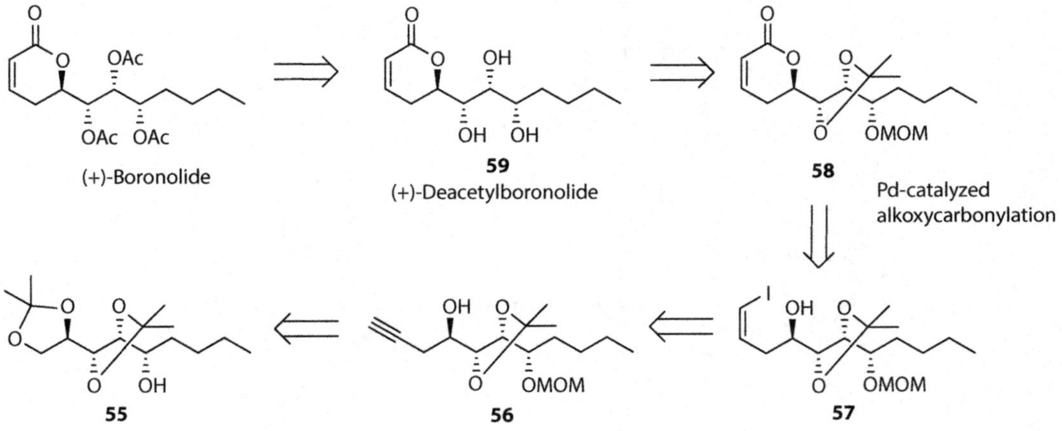

SCHEME 10.33 Retrosynthetic approach to (+)-boronolide.

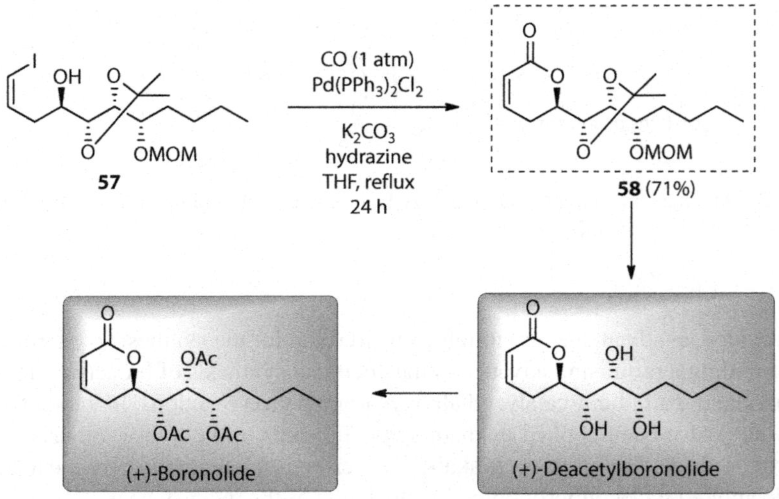

SCHEME 10.34 Palladium-catalyzed intramolecular alkoxycarbonylation as a key step in the synthesis of (+)-boronolide and (+)-deacetylboronolide.

A similar palladium-catalyzed intramolecular alkoxycarbonylation reaction has been reported as one of the key steps in the total synthesis of (+)-monocerin.[128] Monocerins can be extracted from several fungal sources, and they are known to have remarkable antifungal, insecticidal, antimalarial, and phytotoxic properties.[129] Thus, they are regarded as highly attractive targets in synthetic organic chemistry.

The retrosynthetic analysis of (+)-monocerin is presented in Scheme 10.36. The lactone skeleton of the monocerin could be readily obtained from the aryl iodide containing a hydroxyl group **63** *via* palladium-catalyzed intramolecular alkoxycarbonylation. The *cis*-fused tetrahydrofuran **63** can be prepared by a tandem dihydroxylation-S$_N$2 cyclization sequence from allyl alcohol **62**. Finally, the allyl alcohol **62** can be obtained by cross metathesis of allyl alcohol **61** and vinylbenzene **60**, using a Grubbs catalyst.

The synthesis of the intermediate **65** was accomplished through alkoxycarbonylation reaction using Pd(OAc)$_2$ as a catalyst, Cs$_2$CO$_3$ as a base, and dimethylacetamide (DMA) as a solvent, at 130°C

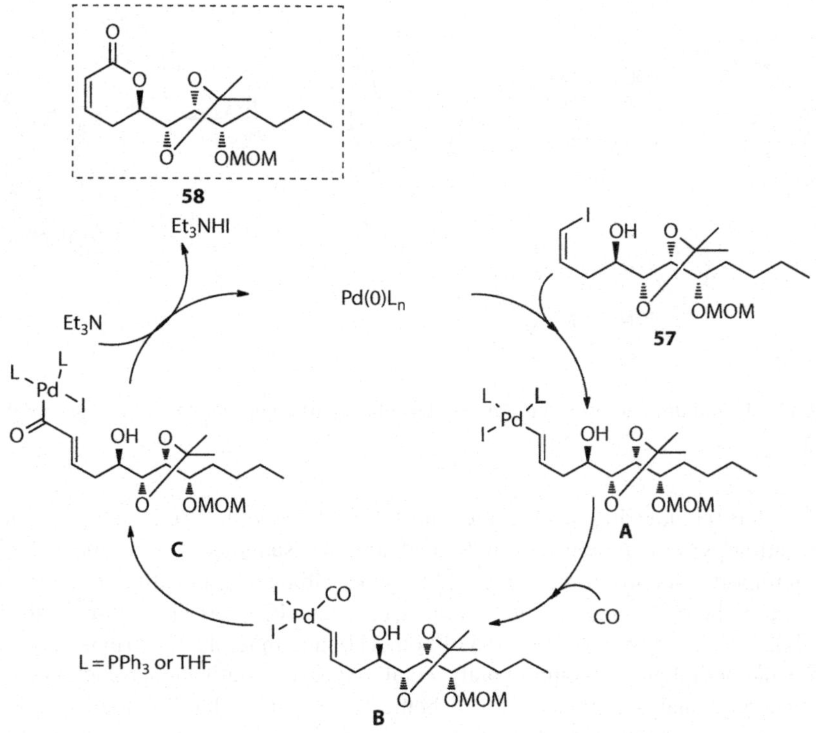

SCHEME 10.35 Mechanism of palladium-catalyzed intramolecular alkoxycarbonylation, leading to δ-lactone **58**, an intermediate of (+)-boronolide synthesis.

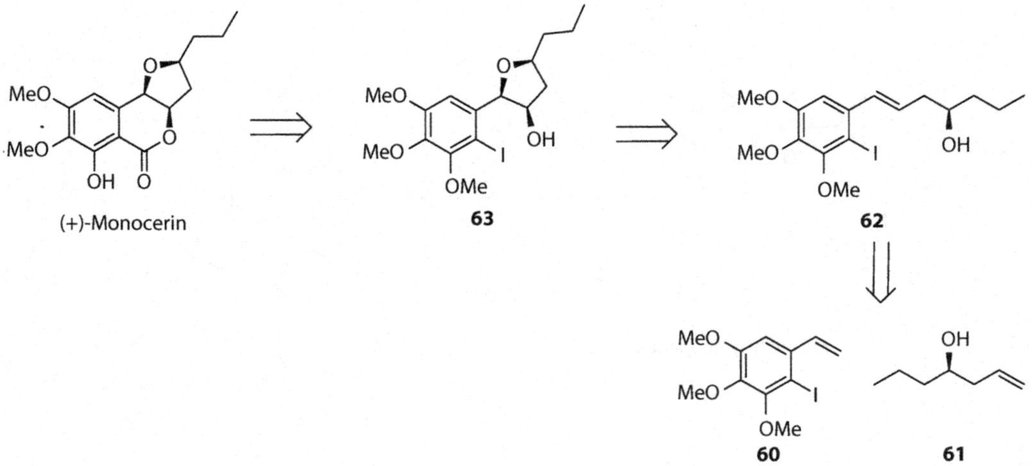

SCHEME 10.36 Retrosynthetic approach to (+)-monocerin.

and 1 atm CO pressure. Remarkably, the authors have applied several nitrogen ligands instead of the commonly used phosphines, and the best yield (72%) was obtained with use of ligand bathocuproine **64**. Finally, (+)-monocerin was obtained from **65** by partial demethylation with boron tribromide in 55% yield and 20% of the starting material recovered (Scheme 10.37).

Macrocyclic structural motifs, such as tetrahydropyran (THP)-containing macrolides, are often present in several bioactive compounds.[130,131] Such motifs can be prepared by a variety of synthetic

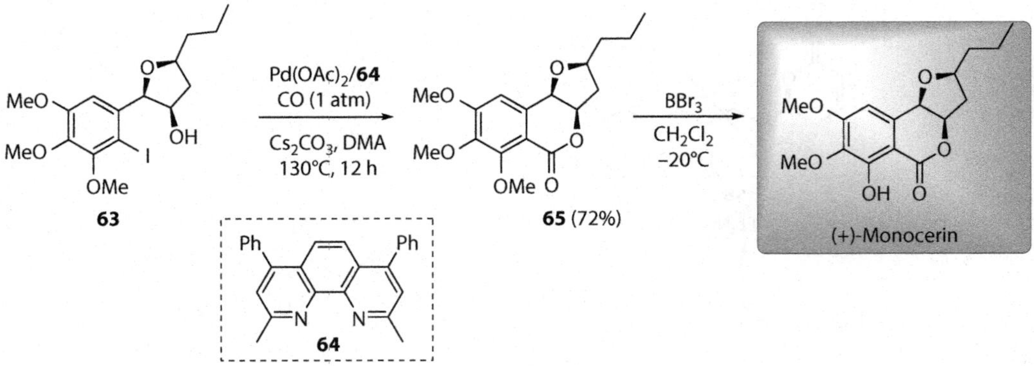

SCHEME 10.37 Palladium-catalyzed intramolecular alkoxycarbonylation as a key step in the synthesis of (+)-monocerin.

methodologies that include ring-closing metathesis,[132] macrocyclic Prins-type cyclization,[133] dual macrolactonization/pyran-hemiketal formation,[134] and transannular oxa-Michael cyclizations.[135] Dai[136] gave an important contribution in this field, by describing an innovative strategy for the preparation of 9-demethylneopeltolide, a potent anticancer agent (IC$_{50}$ = 0.813 nM against P388 murine leukemia cells),[137] in which the construction of both THP and macrolactone rings occurs in one step from a simple alkenediol *via* palladium-catalyzed alkoxycarbonylative macrolactonization.

The retrosynthetic analysis, presented in Scheme 10.38, shows that 9-demethylneopeltolide can be accessed from (Z)-5-(2-(Z)-(3-((methoxycarbonyl)amino)prop-1-en-1-yl)oxazol-4-yl)pent-2-enoic acid and the intermediate **68** by Mitsunobu reaction. In turn, this intermediate can be prepared from **67** by ketal removal and reduction. Finally, the second intermediate may be obtained from **66** *via* palladium-catalyzed alkoxycarbonylative macrolactonization.

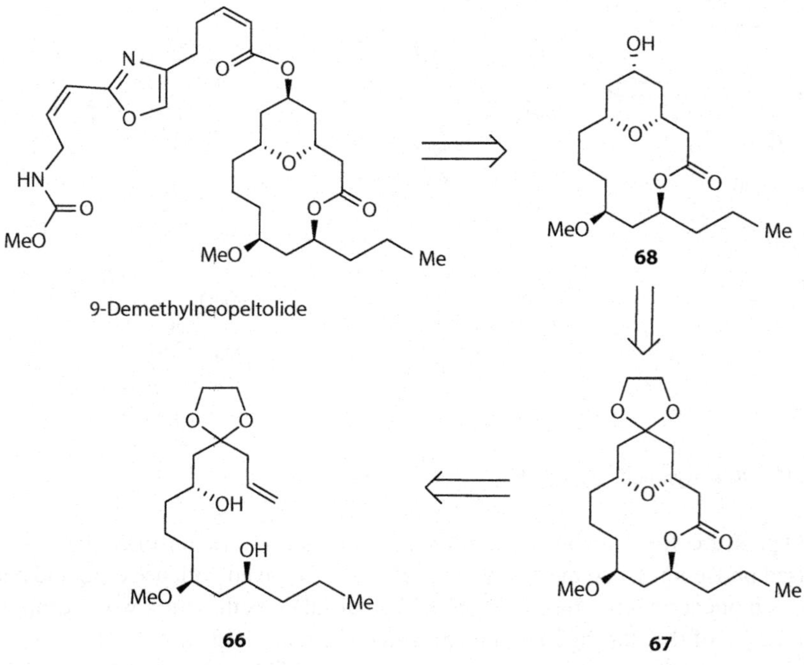

SCHEME 10.38 Retrosynthetic approach to 9-demethylneopeltolide.

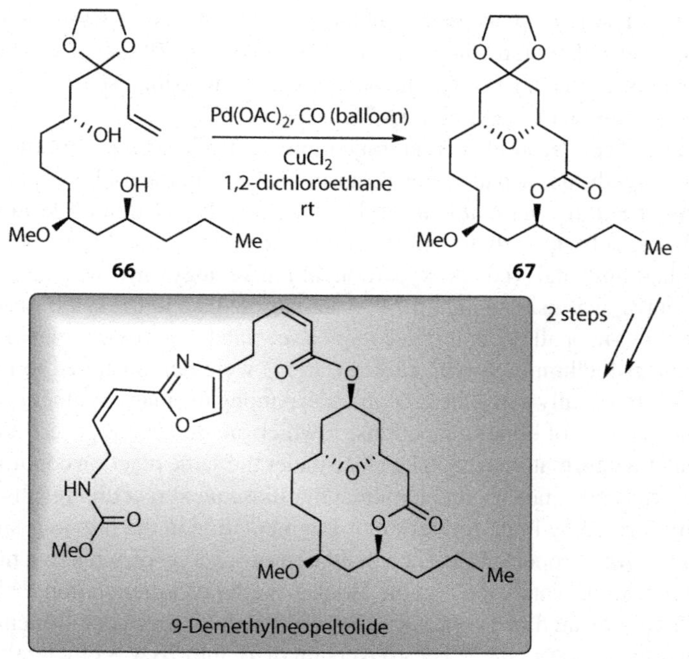

SCHEME 10.39 Synthesis of 9-demethylneopeltolide *via* alkoxycarbonylative macrolactonization.

This synthetic methodology reported by Dai[136] was successfully applied to the palladium-catalyzed alkoxycarbonylative macrolactonization of epimer **66**, which yielded the macrolactone **67** in a reasonable yield (58%) and full *cis*-selectivity (Scheme 10.39). This first critical step allows the early establishment of the desired stereochemistry at both rings of this natural product.

Mechanistic studies on the oxypalladation step, under carbonylative macrolactonization conditions, were performed using **69** as a model substrate with a *cis*-double bond (Scheme 10.40).

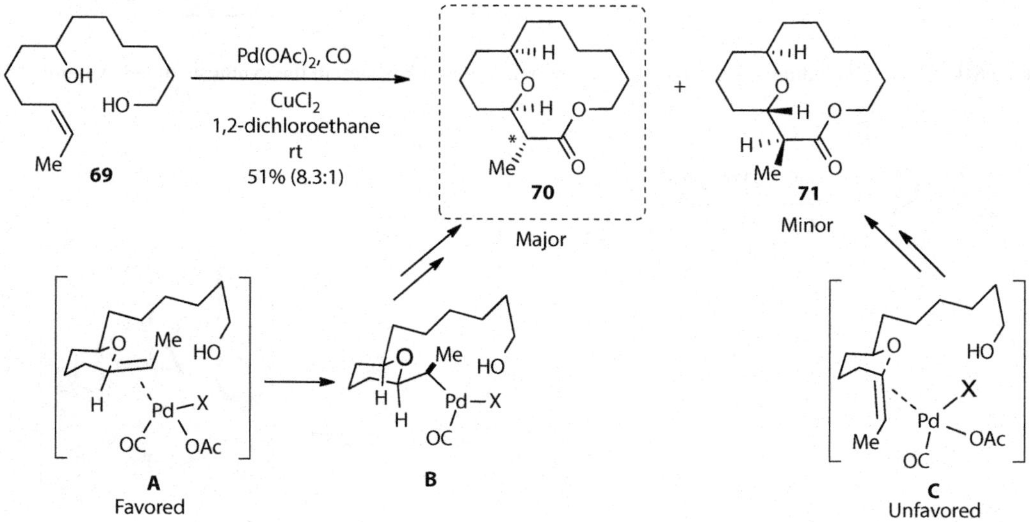

SCHEME 10.40 Mechanism of the oxypalladation step in palladium-catalyzed alkoxycarbonylative macrolactonization.

The products **70** and **71** were obtained in a combined yield of 51%, in 8.3:1 diastereomeric ratio. The stereochemical outcome with preferential formation of stereoisomer **70** supports a *trans*-oxypalladation process *via* a chair-like intermediate **B**. On the other hand, the transition state **C**, which is less favored than **A**, leads to the formation of minor product **71**.

A similar palladium-catalyzed alkoxycarbonylation reaction was reported by Carreira as one of the key steps in the synthesis of pallambin B, an important diterpenoid compound, isolated from the Chinese liverwort *Pallavicina ambigua*.[138] The alkoxycarbonylation of **72** afforded the desired product **73**, which was subsequently transformed into the target pallambin B (Scheme 10.41).

Moreover, a palladium-catalyzed alkoxycarbonylation strategy, involving an iron carbonyl complex as carbonyl source, was used to obtain **74**,[139] an intermediate in the synthesis of **75**, which is an important subunit of (−)-neopallavicinin,[140] a complex secolabdane-type diterpenoid (Scheme 10.42).

As shown earlier, palladium-catalyzed carbonylations were demonstrated to be an efficient tool for the synthesis of biologically active heterocycles.[141] Among the many synthetic strategies reported so far, some of them consist of domino reactions,[142] which are defined as processes involving two or more bond-forming transformations that take place under the same reaction conditions, without adding additional reagents and catalysts, and in which the subsequent reaction results as a consequence of the functionality formed by bond formation or fragmentation in the previous step.[143]

In this context, Tietze[144] reported the total synthesis of a series of bioactive molecules, through the application of palladium-catalyzed domino Wacker/methoxycarbonylation,[145–147] whose efficient applicability was demonstrated in the preparation of the natural product (−)-blennolide C (an antifungal and antibacterial compound).[148] The conversion of **76** into **78** was efficiently performed using palladium(II)-trifluoroacetate/BOXAX **77** as catalyst and *p*-benzoquinone as the oxidant, under

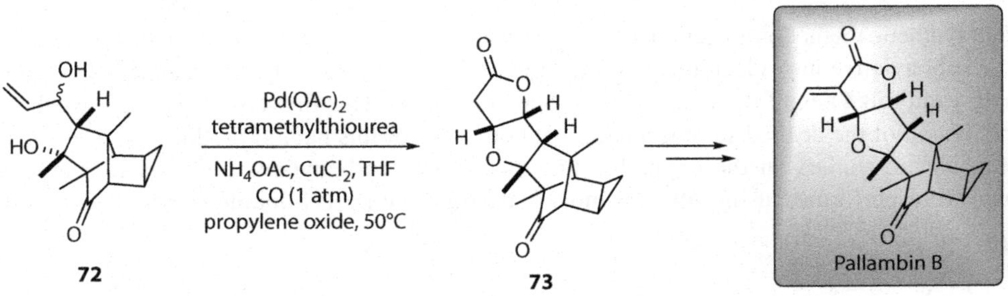

SCHEME 10.41 Palladium-catalyzed alkoxycarbonylation as a key step in the synthesis of (+)-pallambin B.

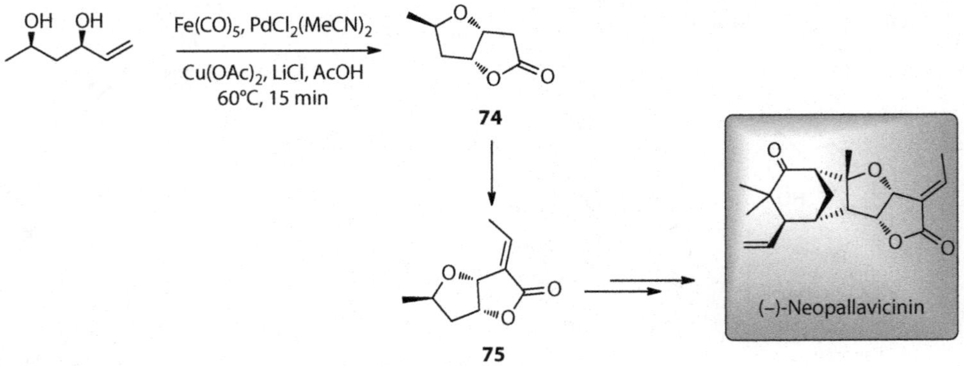

SCHEME 10.42 Palladium-catalyzed alkoxycarbonylation as a key step in the synthesis of **75**, a subunit of (−)-neopallavicinin.

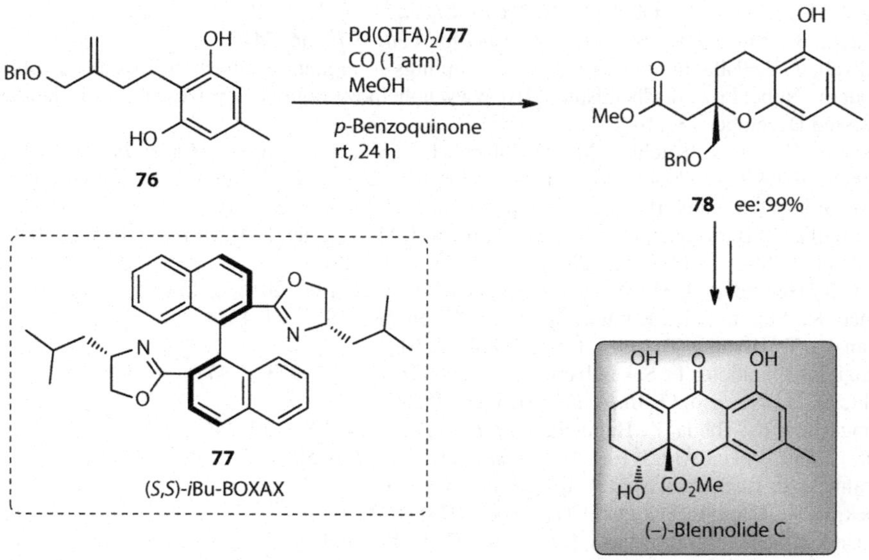

SCHEME 10.43 Palladium-catalyzed enantioselective domino Wacker/carbonylation/methoxylation, leading to chroman **78**.

1 atm CO pressure in methanol, which allowed the formation of the chroman **78**, with a new stereogenic center, in 68% yield and 99% ee (Scheme 10.43). In sum, the enantioselective domino Wacker/methoxycarbonylation reaction was a key tool for the successful establishment of the stereogenic centers in (−)-blennolide C. The reaction follows a similar mechanism to that presented in Scheme 10.29.

10.5 CONCLUSION AND FUTURE PERSPECTIVES

The examples presented herein on the application of palladium complexes in the synthesis of drug intermediates and/or in the functionalization of natural products give a clear evidence of the importance of palladium-catalyzed reactions for the development of new drugs, regarding the replacement of the conventional classic synthetic organic strategies by more efficient, atom-economic, and environmentally sustainable processes. Nowadays, palladium-catalyzed reactions are a topic with utmost relevance, and still in huge expansion for synthetic organic chemistry, mainly due to the development of chiral ligands that make palladium chemistry an excellent tool for the preparation of enantiopure chemical entities. Although the substitution of palladium by cheaper and nontoxic alternative metals is one of the greatest challenges for future research, palladium complexes are still the catalysts of choice, with successful applications in a wide range of organic reactions. Furthermore, the immobilization of homogeneous palladium catalysts in solid supports, in order to allow their recovery and reuse, can also be considered one of the most challenging areas for both academia and pharmaceutical industry.

REFERENCES

1. Smidt, J.; Hafner, W.; Jira, R.; Sieber R.; Sedlmeier, J.; Sabel, A. *Angew. Chem. Int. Ed.* 1962, *1*, 80.
2. Molnár, A. (Ed.) *Palladium-Catalyzed Coupling Reactions: Practical Aspects and Future Developments*, Wiley-VCH, Weinheim, Germany, 2013.
3. de Vries, J. G. *Top. Organomet. Chem.* 2012, *42*, 1.
4. Kambe, N.; Iwasakia, T.; Terao. *J. Chem. Soc. Rev.* 2011, *40*, 4937.
5. Heck, R. F.; Nolley, J. P. *J. Org. Chem.* 1972, *37*, 2320.

6. Negishi, E.; King, A. O.; Okukado, N. *J. Org. Chem.* 1977, *42*, 1821.
7. Miyaura, N.; Yamada, K.; Suzuki, A. *Tetrahedron Lett.* 1979, *36*, 3437.
8. Bäckvall, J.-E. Palladium-catalyzed cross couplings in organic synthesis in Advanced Information of About the Nobel Prize in Chemistry, 2010, www.nobelprize.org/nobel_prizes/chemistry/laureates/2010/. Accessed December 13, 2016.
9. Seechurn, C. C. C. J.; Kitching, M. O.; Colacot, T. J.; Snieckus V. *Angew. Chem. Int. Ed.* 2012, *51*, 5062.
10. Nobel Media AB. The Nobel Prize in chemistry 2010, 2014. Nobelprize.org. Accessed on December 13, 2016, http://www.nobelprize.org/nobel_prizes/chemistry/laureates/2010/.
11. Cordovilla, C.; Bartolomé, C.; Martínez-Ilarduya, J. M.; Espinet, P. *ACS Catal.* 2015, *5*, 3040.
12. Milstein, D.; Stille, J. K. *J. Am. Chem. Soc.* 1978, *100*, 3636.
13. Haas, D.; Hammann, J. M.; Greiner, R.; Knochel, P. *ACS Catal.* 2016, *6*, 1540.
14. Tamao, K.; Sumitani, K.; Kumada, M. *J. Am. Chem. Soc.* 1972, *94*, 4374.
15. Hatanaka, Y.; Hiyama, T. *J. Org. Chem.* 1988, *53*, 918.
16. Takagi, K.; Okamoto, T.; Sakakibara, Y.; Oka, S. *Chem. Lett.* 1973, *2*, 471.
17. Maluenda, I.; Navarro, O. *Molecules* 2015, *20*, 7528.
18. Sonogashira, K.; Tohda, Y.; Hagihara, N. *Tetrahedron Lett.* 1975, *50*, 4467.
19. Paul, F.; Patt, J.; Hartwig, J. F. *J. Am. Chem. Soc.* 1994, *116*, 5969.
20. Guram, A. S.; Buchwald, S. L. *J. Am. Chem. Soc.* 1994, *116*, 7901.
21. Dieck, H. A.; Heck, R. F. *J. Am. Chem. Soc.* 1974, *96*, 1133.
22. Fernández-Ibañez, M. A.; Maciá, B.; Alonso, D. A.; Pastor, I. M. *Molecules* 2013, *18*, 10108.
23. Pereira, M. M.; Figueiredo, J. L.; Faria, J. (Eds.) *Catalysis from Theory to Application*, Imprensa da Universidade de Coimbra, Coimbra, Portugal, 2008.
24. Tolman, C. A. *Chem. Rev.* 1977, *77*, 313.
25. Galardon, E.; Ramdeehul, S.; Brown, J. M.; Cowley, A.; Hii, K. K.; Jutand, A. *Angew. Chem. Int. Ed.* 2002, *41*, 1760.
26. Fleckenstein, C. A.; Plenio, H. *Chem. Soc. Rev.* 2010, *39*, 694.
27. Bilbrey, J. A.; Kazez, A. H.; Locklin, J.; Allen, W. D. *J. Comput. Chem.* 2013, *34*, 1189.
28. Demchuk, O. M.; Kapłon, K.; Kącka, A.; Pietrusiewicz, K. M. *Phosphorus, Sulfur Silicon Relat. Elem.* 2016, *191*, 180.
29. Negishi, E.-I. *Handbook of Organopalladium Chemistry for Organic Synthesis*, vols. 1 and 2, John Wiley & Sons, Inc., New York, 2002.
30. Tietze, L. F.; Ila, H.; Bell, H. P. *Chem. Rev.* 2004, *104*, 3453.
31. Wang, Z. (Ed.) *Comprehensive Organic Name Reactions and Reagents*, John Wiley & Sons, Inc., New York, 2010, p. 1350.
32. McDonald, R. I.; Liu, G.; Stahl, S. S. *Chem. Rev.* 2011, *111*, 2981.
33. Tsuji, J. *Palladium Reagents and Catalysts: New Perspectives for the 21st Century*, John Wiley & Sons, Ltd, Chichester, U.K., 2004.
34. Sore, H. F.; Galloway, W. R. J. D.; Spring, D. R. *Chem. Soc. Rev.* 2012, *41*, 1845.
35. Nicolaou, K. C.; Bulger, P. G.; Sarlah, D. *Angew. Chem. Int. Ed.* 2005, *44*, 4442.
36. Campagne, J.-M.; Prim, D.; Genêt, J.-P. *Les complexes de palladium en synthèse organique*, CNRS Éditions, Paris, France, 2001.
37. Kollár, L. (Ed.) *Modern Carbonylation Methods*, Wiley-VCH, Weinheim, Germany, 2008.
38. Mohr, J. T.; Krout, M. R.; Stoltz, B. M. *Nature* 2008, *455*, 323.
39. Cannon, J. S.; Overman, L. E. *Acc. Chem. Res.* 2016, *49*, 2220.
40. Zeng, X.-P.; Cao, Z.-Y.; Wang, Y.-H.; Zhou, F.; Zhou, J. *Chem. Rev.* 2016, *116*, 7330.
41. Sivanandan, S. T.; Shaji, A.; Ibnusaud, I.; Seechurn, C. C. C. J.; Colacot, T. J. *Eur. J. Org. Chem.* 2015, *2015*, 38.
42. Shockley, S. E.; Holder, J. C.; Stoltz, B. M. *Org. Process Res. Dev.* 2015, *19*, 974.
43. Affron, D. P.; Davis, O. A.; Bull, J. A. *Org. Lett.* 2014, *16*, 4956.
44. Hellmuth, T.; Frey, W.; Peters, R. *Angew. Chem. Int. Ed.* 2015, *54*, 2788.
45. Friis, S. D.; Pirnot, M. T.; Buchwald, S. L. *J. Am. Chem. Soc.* 2016, *138*, 8372.
46. Bao, H.; Bayeh, L.; Tambar, U. K. *Synlett* 2013, *24*, 2459.
47. Aikawa, K.; Yoshida, S.; Kondo, D.; Asai, Y.; Mikami, K. *Org. Lett.* 2015, *17*, 5108.
48. Mohr, J. T.; Stoltz, B. M. *Chem. Asian J.* 2007, *2*, 1476.
49. Liu, Y.; Han, S.-J.; Liu, W.-B.; Stoltz, B. M. *Acc. Chem. Res.* 2015, *48*, 740.
50. Trost, B. M.; Schäffner, B.; Osipov, M.; Wilton, D. A. A. *Angew. Chem. Int. Ed.* 2011, *50*, 3548.
51. Tsuji, J.; Minami, I.; Shimizu, I. *Tetrahedron Lett.* 1983, *24*, 1793.

52. Shimizu, I.; Yamada, T.; Tsuji, J. *Tetrahedron Lett.* 1980, *21*, 3199.
53. Liu, Y.; Liniger, M.; McFadden, R. M.; Roizen, J. L.; Malette, J.; Reeves, C. M.; Behenna, D. C.; Seto, M.; Kim, J.; Mohr, J. T.; Virgil, S. C.; Stoltz, B. M. Beilstein, J. *Org. Chem.* 2014, *10*, 2501.
54. Hong, A. Y.; Stoltz, B. M. *Angew. Chem. Int. Ed.* 2012, *51*, 9674.
55. Xu, Z.; Wang, Q.; Zhu, J. *J. Am. Chem. Soc.* 2013, *135*, 19127.
56. Wei, Y.; Zhao, D.; Ma, D. *Angew. Chem. Int. Ed.* 2013, *52*, 12988.
57. Marziale, A. N.; Duquette, D. C.; Craig II, R. A.; Kim, K. E.; Liniger, M.; Numajiri, Y.; Stoltz, B. M. *Adv. Synth. Catal.* 2015, *357*, 2238.
58. Krout, M. R.; Mohr, J. T.; Stoltz, B. M. *Organic Synth.* 2009, *86*, 181.
59. McDougal, N. T.; Streuff, J.; Mukherjee, H.; Virgil, S. C.; Stoltz, B. M. *Tetrahedron Lett.* 2010, *51*, 5550.
60. Peng, J.; Walsh, K.; Weedman, V.; Bergthold, J. D.; Lynch, J.; Lieu, K. L.; Braude, I. A.; Kelly, M.; Hamann, M. T. *Tetrahedron*, 2002, *58*, 7809.
61. Enquist Jr, J. A.; Stoltz, B. M. *Nature* 2008, *453*, 1228.
62. Liu, Y.; Virgil, S. C.; Grubbs, R. H.; Stoltz, B. M. *Angew. Chem. Int. Ed.* 2015, *54*, 11800.
63. Miao, F.-P.; Liang, X.-R.; Liu, X.-H.; Ji, N.-Y. *J. Nat. Prod.* 2014, *77*, 429.
64. Milhau, L.; Guiry, P. J. *Top. Organomet. Chem.* 2012, *38*, 95.
65. Weaver, J. D.; Recio, A.; Grenning, A. J.; Tunge, J. A. *Chem. Rev.* 2011, *111*, 1846.
66. Huters, A. D.; Styduhar, E. D.; Garg, N. K. *Angew. Chem. Int. Ed.* 2012, *51*, 3758.
67. Ma, G.; Afewerki, S.; Deiana, L.; Palo-Nieto, C.; Liu, L.; Sun, J.; Ibrahem, I.; Córdova, A. *Angew. Chem. Int. Ed.* 2013, *52*, 6050.
68. Liu, W.-B.; Reeves, C. M.; Virgil, S. C.; Stoltz, B. M. *J. Am. Chem. Soc.* 2013, *135*, 10626.
69. Zhao, X.; Liu, D.; Guo, H.; Liu, Y.; Zhang, W. *J. Am. Chem. Soc.* 2011, *133*, 19354.
70. Mukai, R.; Horino, Y.; Tanaka, S.; Tamaru, Y.; Kimura, M. *J. Am. Chem. Soc.* 2004, *126*, 11138.
71. Trost, B. M.; Jiang, C. *J. Am. Chem. Soc.* 2001, *123*, 12907.
72. Trost, B. M.; Bunt, R. C.; Lemoine, R. C.; Calkins, T. L. *J. Am. Chem. Soc.* 2000, *122*, 5968.
73. Huo, X.; Quan, M.; Yang, G.; Zhao, X.; Liu, D.; Liu, Y.; Zhang, W. *Org. Lett.* 2014, *16*, 1570.
74. Zhang, W.; Adachi, Y.; Hirao, T.; Ikeda, I. *Tetrahedron: Asymmetry* 1996, *7*, 451.
75. Zhang, W.; Hirao, T.; Ikeda, I. *Tetrahedron Lett.* 1996, *37*, 4545.
76. Williams, R. M. *Synthesis of Optically Active a-Amino Acids*, Pergamon, Oxford, U.K., 1989.
77. Cregge, R. J.; Durham, S. L.; Farr, R. A.; Gallion, S. L.; Hare, C. M.; Hoffman, R. V.; Janusz, M. J.; Kim, H.-O.; Koehl, J. R.; Mehdi, S.; Metz, W. A.; Peet, N. P.; Pelton, J. T.; Schreuder, H. A.; Sunder, S.; Tradif, C. *J. Med. Chem.* 1998, *41*, 2461.
78. Feriani, A.; Gaviraghi, G.; Toson, G.; Mor, M.; Barbieri, A.; Grana, E.; Boselli, C.; Guarneri, M.; Simoni, D.; Manfredini, S. *J. Med. Chem.* 1994, *37*, 4278.
79. Trost, B. M.; Crawley, M. L. *Chem. Rev.* 2003, *103*, 2921.
80. Ardolino, M. J.; Morken, J. P. *J. Am. Chem. Soc.* 2014, *136*, 7092.
81. Misale, A.; Niyomchon, S.; Luparia, M.; Maulide, N. *Angew. Chem. Int. Ed.* 2014, *53*, 7068.
82. Trost, B. M.; Thaisrivongs, D. A. *J. Am. Chem. Soc.* 2008, *130*, 14092.
83. Mao, J.; Zhang, J.; Jiang, H.; Bellomo, A.; Zhang, M.; Gao, Z.; Dreher, S. D.; Walsh, P. J. *Angew. Chem. Int. Ed.* 2016, *55*, 2526.
84. Landoni, M. F.; Soraci, A. *Curr. Drug Metab.* 2001, *2*, 37.
85. Trost, B. M.; Richardson, J.; Yong, K. *J. Am. Chem. Soc.* 2006, *128*, 2540.
86. Trost, B. M.; Masters, J. M.; Lumb, J.-P.; Fateen, D. *Chem. Sci.* 2014, *5*, 1354.
87. Marco-Contelles, J.; Molina, M. T.; Anjum, S. *Chem. Rev.* 2004, *104*, 2857.
88. Ye, L.-W.; Shu, C.; Gagosz, F. *Org. Biomol. Chem.* 2014, *12*, 1833.
89. Bellus, D.; Jacobsen, E. N.; Ley, S. U.; Noyari, R.; Regitz, M.; Reider, P. J.; Schaumann, E.; Shinkai, I.; Thomas, E. J.; Trost, B. M. *Science of Synthesis: Houben-Weyl Methods of Molecular Transformations*, Vol. 21, Georg Thieme Verlag, Leipzig, Germany, 2005, p. 647.
90. Kammerer, C.; Prestat, G.; Madec, D.; Poli, G. *Acc. Chem. Res.* 2014, *47*, 3439.
91. Thuong, M. B. T.; Sottocornola, S.; Prestat, G.; Broggini, G.; Madec, D.; Poli, G. *Synlett* 2007, *10*, 1521.
92. Bisceglia, J. A.; Orelli, L. R. *Curr. Org. Chem.* 2015, *19*, 744.
93. Xia, Q.; Ganem, B. *Org. Lett.* 2001, *3*, 485.
94. Bantreil, X.; Prestat, G.; Moreno, A.; Madec, D.; Fristrup, P.; Norrby, P.-O.; Pregosin, P. S.; Poli, G. *Chem. Eur. J.* 2011, *17*, 2885.

95. Pedroni, J.; Cramer, N. *Angew. Chem. Int. Ed.* 2015, *54*, 11826.
96. Augeri, D. J.; Robl, J. A.; Betebenner, D. A.; Magnin, D. R.; Khanna, A.; Robertson, J. G.; Wang, A.; Simpkins, L. M.; Taunk, P.; Huang, Q.; Han, S.-P.; Abboa-Offei, B.; Cap, M.; Xin, L.; Tao, L.; Tozzo, E.; Welzel, G. E.; Egan, D. M.; Marcinkeviciene, J.; Chang, S. Y.; Biller, S. A.; Kirby, M. S.; Parker, R. A.; Hamann, L. G. *J. Med. Chem.* 2005, *48*, 5025.
97. Pedroni, J.; Boghi, M.; Saget, T.; Cramer, N. *Angew. Chem. Int. Ed.* 2014, *53*, 9064.
98. Ackermann, L. *Chem. Rev.* 2011, *111*, 1315.
99. Holstein, P. M.; Dailler, D.; Vantourout, J.; Shaya, J.; Millet, A.; Baudoin, O. *Angew. Chem. Int. Ed.* 2016, *55*, 2805.
100. Takeuchi, S.; Ishibashi, M.; Kobayashi, J. *J. Org. Chem.* 1994, *59*, 3712.
101. Ma, D.; Sun, H. *Tetrahedron Lett.* 2000, *41*, 1947.
102. Pu, L.-Y.; Chen, J.-Q.; Li, M.-L.; Li, Y.; Xie, J.-H.; Zhou, Q.-L. *Adv. Synth. Catal.* 2016, *358*, 1229.
103. Tietze, L. F.; Ma, L.; Reiner, J. R.; Jackenkroll, S.; Heidemann, S. *Chem. Eur. J.* 2013, *19*, 8610.
104. Ganapathy, D.; Reiner, J. R.; Löffler, L. E.; Ma, L.; Gnanaprakasam, B.; Niepötter, B.; Koehne, I.; Tietze, L. F. *Chem. Eur. J.* 2015, *21*, 16807.
105. Schoenberg, A.; Heck, R. F. *J. Org. Chem.* 1974, *39*, 3327.
106. Schoenberg, A.; Heck, R. F. *J. Am. Chem. Soc.* 1974, *96*, 7761.
107. Grigg, R.; Mutton, S. P. *Tetrahedron* 2010, *66*, 5515.
108. Brennführer, A.; Neumann, H.; Beller, M. *Angew. Chem. Int. Ed.* 2009, *48*, 4114.
109. Barnard, C. F. J. *Organometallics* 2008, *27*, 5402.
110. Wu, X.-F.; Neumann, H.; Beller, M. *Chem. Soc. Rev.* 2011, *40*, 4986.
111. Cornils, B.; Herrmann, W. A. (Eds.) *Applied Homogeneous Catalysis with Organometallic Compounds*, Vol. 1, Wiley-VCH, Weinheim, Germany, 1996.
112. de la Fuente, V.; Godard, C.; Zangrando, E.; Claver, C.; Castillón, S. *Chem. Commun.* 2012, *48*, 1695.
113. Xu, T.; Alper, H. *Tetrahedron Lett.* 2013, *54*, 5496.
114. Malawska, B. *Curr. Top. Med. Chem.* 2005, *5*, 69.
115. Luszczki, J. J.; Swiader, M. J.; Swiader, K.; Paruszewski, R.; Turski, W. A.; Czuczwar, S. *J. Fund. Clin. Pharmacol.* 2008, *22*, 69.
116. Janout, V.; Jing, B.; Regen, S. L. *Bioconjug. Chem.* 2002, *13*, 351.
117. Salunke, D. B.; Hazra, B. G.; Pore, V. S.; Bhat, M. K.; Nahar, P. B.; Despande, M. V. *J. Med. Chem.* 2004, *47*, 1591.
118. Carrilho, R. M. B.; Pereira, M. M.; Moreno, M. J. S. M.; Takács, A.; Kollár, L. *Tetrahedron Lett.* 2013, *54*, 2763.
119. Amatore, C.; Carré, E.; Jutand, A.; M'Barki, M. A.; Meyer, G. *Organometallics* 1995, *14*, 5605.
120. Csákai, Z.; Skoda-Földes, R.; Kollár, L. *Inorg. Chim. Acta* 1999, *286*, 93.
121. Carrilho, R. M. B.; Almeida, A. R.; Kiss, M.; Kollár, L.; Skoda-Földes, R.; Dabrowski, J. M.; Moreno, M. J. S. M.; Pereira, M. M. *Eur. J. Org. Chem.* 2015, *2015*, 1840.
122. Kurogome, Y.; Hattori, Y.; Makabe, H. *Tetrahedron Lett.* 2014, *55*, 2822.
123. Davies-Coleman, M. T.; Rivett, D. E. A. *Phytochemistry* 1987, *26*, 3047.
124. Van Puyvelde, L.; De Kimpe, N.; Dube, S.; Chagnon-Dube, M.; Boily, Y.; Borremans, F.; Schamp, N.; Anteunis, M. J. O. *Phytochemistry* 1981, *20*, 2753.
125. Watt, J. M.; Brandwijk, M. G. B. *The Medicinal and Poisonous Plants of Southern and Eastern Africa*, E. & S. Livingston Ltd., Edinburgh, Scotland, 1962.
126. Cowell, A.; Stille, J. K. *J. Am. Chem. Soc.* 1980, *102*, 4193.
127. Makabe, H.; Okajima, M.; Konno, H.; Kamo, T.; Hirota, M. *Biosci. Biotechnol. Biochem.* 2003, *67*, 2658.
128. Punganuru, S. R.; Aviraboina, S.; Srivenugopal, K. S. *J. Chem. Res.* 2016, *40*, 375.
129. Scott, F. E.; Simpson, T. J.; Trimble, L. A.; Vederas, J. C. *J. Chem. Soc. Chem. Commun.* 1984, 756.
130. Kopp, F.; Stratton, C. F.; Akella, L. B.; Tan, D. S. *Nat. Chem. Biol.* 2012, *8*, 358.
131. Parenty, A.; Moreau, X.; Niel, G.; Campagne, J.-M. *Chem. Rev.* 2013, *113*, PR1.
132. Cossy, J.; Arseniyadis, S.; Meyer, C. (Eds.) *Metathesis in Natural Product Synthesis*, Wiley-VCH, Weinheim, Germany, 2010.
133. Custar, D. W.; Zabawa, T. P.; Scheidt, K. A. *J. Am. Chem. Soc.* 2008, *130*, 804.
134. Hoye, T. R.; Danielson, M. E.; May, A. E.; Zhao, H. *J. Org. Chem.* 2010, *75*, 7052.
135. Kanematsu, M.; Yoshida, M.; Shishido, K. *Angew. Chem. Int. Ed.* 2011, *50*, 2618.
136. Bai, Y.; Davis, D. C.; Dai, M. *Angew. Chem. Int. Ed.* 2014, *53*, 6519.
137. Fuwa, H.; Saito, A.; Naito, S.; Konoki, K.; Yotsu-Yamashita, M.; Sasaki, M. *Chem. Eur. J.* 2009, *15*, 12807.

138. Ebner, C.; Carreira, E. M. *Angew. Chem. Int. Ed.* 2015, *54*, 11227.
139. Markovič, M.; Ďuranova, M.; Koóš, P.; Szolcsányi, P.; Gracza, T. *Tetrahedron* 2013, *69*, 4185.
140. Li, Z.-J.; Lou, H.-X.; Yu, W.-T.; Fan, P.-H.; Ren, D.-M.; Ma, B.; Ji, M. *Helv. Chim. Acta* 2005, *88*, 2637.
141. Wu, X.-F.; Neumann, H.; Beller, M. *Chem. Rev.* 2013, *113*, 1.
142. Pellissier, H. *Chem. Rev.* 2013, *113*, 442.
143. Tietze, L. F.; Beifuss, U. *Angew. Chem. Int. Ed.* 1993, *32*, 131.
144. Tietze, L. F.; Jackenkroll, S.; Hierold, J.; Ma, L.; Waldecker, B. *Chem. Eur. J.* 2014, *20*, 8628.
145. Tietze, L. F.; Spiegl, D. A.; Stecker, F.; Major, J.; Raith, C.; Große, C. *Chem. Eur. J.* 2008, *14*, 8956.
146. Tietze, L. F.; Jackenkroll, S.; Raith, C.; Spiegl, D. A.; Reiner, J. R.; Campos, M. C. O. *Chem. Eur. J.* 2013, *19*, 4876.
147. Tietze, L. F.; Ma, L.; Jackenkroll, S.; Reiner, J. R.; Hierold, J.; Gnanaprakasam, B.; Heidemann, S. *Heterocycles* 2014, *88*, 1101.
148. Zhang, W.; Krohn, K.; Zia-Ullah; Flörke, U.; Pescitelli, G.; Di Bari, L.; Antus, S.; Kurtán, T.; Rheinheimer, J.; Draeger, S.; Schulz, B. *Chem. Eur. J.* 2008, *14*, 4913.

11 Enantioselective Organocatalysis from Concepts to Applications in the Synthesis of Natural Products and Pharmaceuticals

Marta Meazza and Ramon Rios

CONTENTS

11.1 Introduction 325
11.2 Enamine Chemistry 326
    11.2.1 Modes of Activation: Enamine Catalysis 326
    11.2.2 Applications in Total Synthesis: Synthesis of Hirsutene 330
11.3 Modes of Activation: Iminium Catalysis 332
    11.3.1 Applications in Total Synthesis: Synthesis of Paroxetine 335
11.4 Cascade Reactions Based on Iminium/Enamine Chemistry 338
    11.4.1 Iminium/Enamine Organocascades 338
11.5 Application in Total Synthesis of Organocascade Reactions: Synthesis of Strychnine 338
11.6 Conclusions 342
References 342

11.1 INTRODUCTION

The synthesis of natural products and pharmaceuticals has been one of the driving forces to develop new methodologies in organic chemistry.[1] As the major part of natural products is chiral, the development of asymmetric methodologies has become of huge importance for the organic chemistry community. During the last century, the plethora of enantioselective methodologies that have been reported was mainly based on organometallic complexes. Metal ion complexes with chiral ligands resulted in almost total control of the enantioselectivity of the reactions. However, as we were more aware of sustainable issues associated with metals (especially transition metals), it is clear that there is a need to develop more benign and green procedures that will fulfill the requirements of sustainability.

In nature, enzymes are the most efficient catalysts and almost half of them are metal free. However, in organic chemistry, only very few examples of enantioselective metal-free catalysis can be found in the last century. Metal-free catalysis can be traced back to the earlier works of Emil Knoevenagel.[2] Knoevenagel studied the use of primary and secondary amines, as well as their salts, as catalysts for the aldol condensation of β-ketoesters or malonates with aldehydes or ketones.

An important contribution to the field of organocatalysis was made by Gilbert Stork with his work on enamine chemistry. Most of the subsequent work done in catalytic enamine chemistry was first conducted by Stork's research group with preformed enamines (Scheme 11.1).[3]

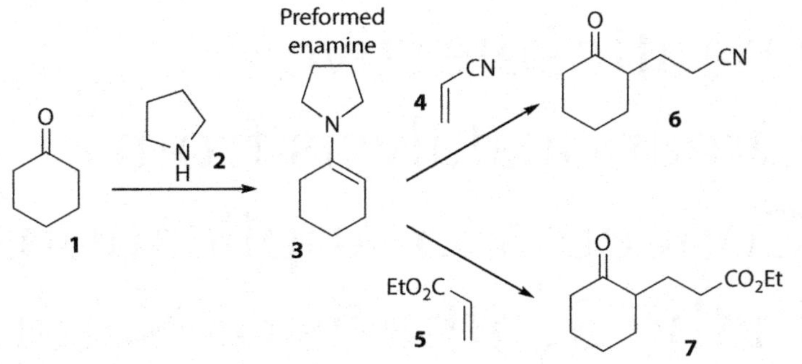

SCHEME 11.1 Reactions developed by Stork with preformed enamines.

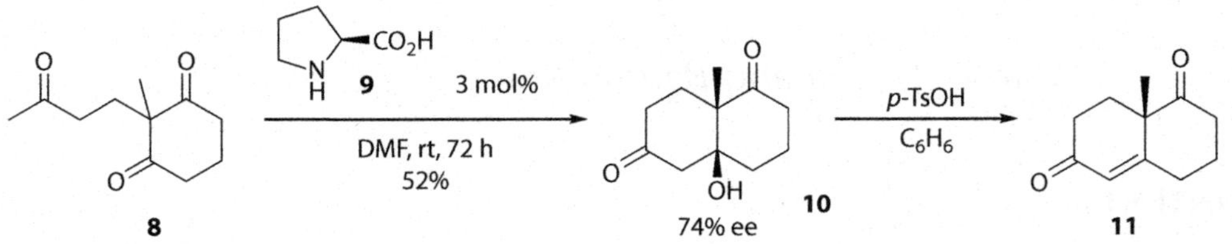

SCHEME 11.2 Aldol reaction performed by Hajos–Parrish in 1974.

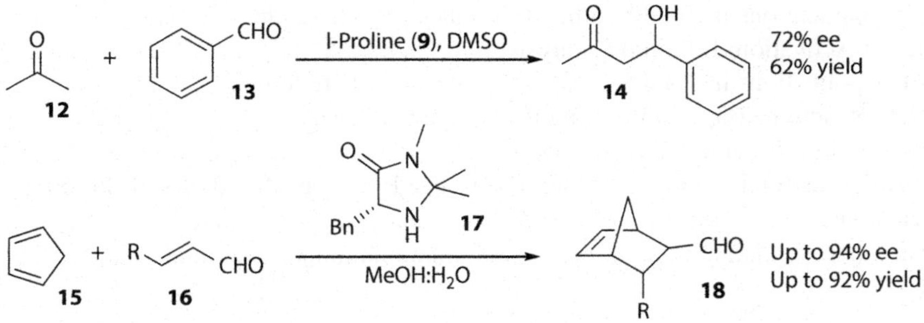

SCHEME 11.3 First examples of catalytic enamine and iminium chemistry.

These studies and findings arguably led to one of the most famous highlights in organocatalysis, the Hajos–Parrish–Eder–Sauer–Wiechert reaction,[4] which used proline as the catalyst to perform a desymmetrization aldol reaction to form the bicyclic ketone **11** (Scheme 11.2).

In 2000, the term "organocatalysis" was coined by David W. C. MacMillan, and it was the starting line for a breathtaking progress in this area over the last decade. During the last years, this area has grown into one of the three pillars of asymmetric catalysis, complementing and, sometimes, improving bio- and metal catalysis. This "renaissance" of organocatalysis started with the works of List, Barbas, and Lerner[5] in enamine chemistry and the works of D. W. C. MacMillan[6] in iminium chemistry in 2000 (Scheme 11.3).

11.2 ENAMINE CHEMISTRY

11.2.1 MODES OF ACTIVATION: ENAMINE CATALYSIS

Enamine catalysis has become one of the most used organocatalytic modes of activation, allowing the enantioselective α-functionalization of enolizable aldehydes and ketones with a variety of electrophiles.[7]

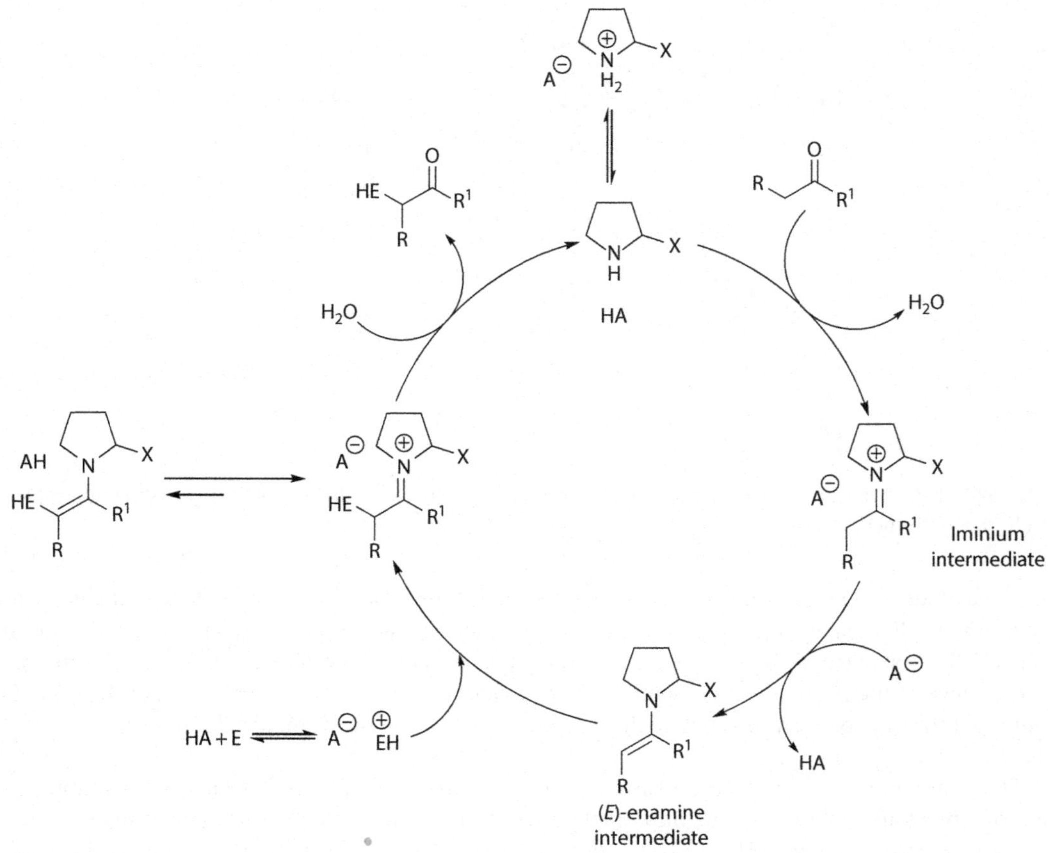

FIGURE 11.1 Generalized mechanism for the amine-catalyzed α-functionalization of carbonyls.

The general mechanism is presented in Figure 11.1. A chiral 2–substituted pyrrolidine is the most representative catalyst, acting with a Brønsted acid cocatalyst AH. The acid can be a protic solvent (water, alcohols), an added external acid, or a functional group present in the amine catalyst. The first step is the acid-promoted condensation of the carbonyl compound with the secondary amine forming an iminium ion. One of the α-acidic protons of the iminium ion is removed by the conjugate base of the Brønsted acid, forming the nucleophilic enamine intermediate. The reaction with an electrophile generates an iminium ion that, after hydrolysis, liberates the final product, the acid, and the amine catalyst. The efficiency of this catalytic cycle is based on the fast and quantitative generation of the iminium ion, its interconversion in the (E)-enamine intermediate, and the high stereochemical control over the electrophilic attack. Finally, it is important to note that the possible reaction between the amine and the electrophile is slow or reversible.

The stereochemical outcome of the reaction can be predicted based on the substituent present in the position 2 of the pyrrolidine. If the chiral amine bears a hydrogen bond–directing group (carboxylic acid, amide), the attack of the electrophile takes place *via* a cyclic transition state (List–Houk model; Figure 11.2a). Seebach and Eschenmoser[8] proposed an alternative transition state based on the protonation of the electrophile, followed by an electrophilic attack directed by an intramolecular reaction of the conjugate base of the substituent on the amine (Figure 11.2b).[9] If the amine substituent is bulky and without acidic protons, the attack of the electrophile is directed by steric effect, leading to the opposite facial stereoselectivity (Figure 11.2c).[10]

Proline is arguably the most important asymmetric organic catalyst used in enamine catalysis. Particularly important is the mechanism of proline-catalyzed aldol reactions. This mechanism has

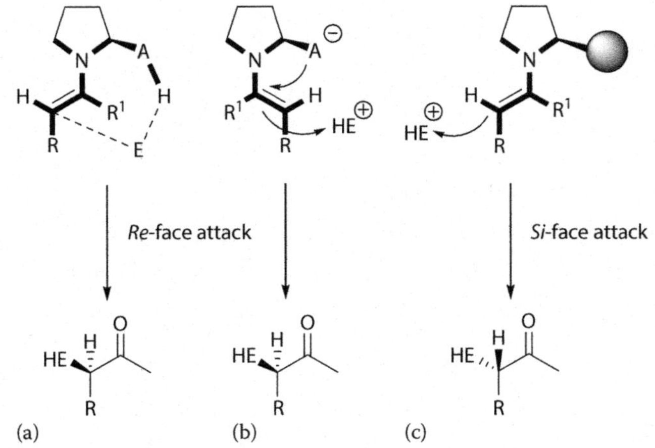

FIGURE 11.2 Transition state models for the electrophilic attack to the enamine. (a) List–Houk model, (b) Seebach–Eschenmoser model, and (c) steric model.

been the object of numerous experimental and theoretical investigations. The first mechanistic studies on the proline-catalyzed (intramolecular) aldol reaction were reported by Hajos and Parrish in 1974. In the first place, these authors demonstrated that both the secondary amine and the carboxylic acid moiety of the proline were essential for the reaction (Scheme 11.4). The secondary amine is important for its capability to generate the enamine and, in the case of the carboxylic acid, to induce enantioselectivity.

The commonly accepted mechanism for the proline-catalyzed intermolecular aldol reaction was reported by Houk and List.[9] Some important observations from Houk and List concerning the aldol reaction regarded the perfect linearity existing between the enantiomeric excess of the proline catalyst and one of the ketol products. Based on this evidence, they predicted the presence of only one molecule of proline in the transition state and they proposed a Zimmerman–Traxler cyclic transition state that led to the anti-aldol product observed in these reactions (Scheme 11.5).

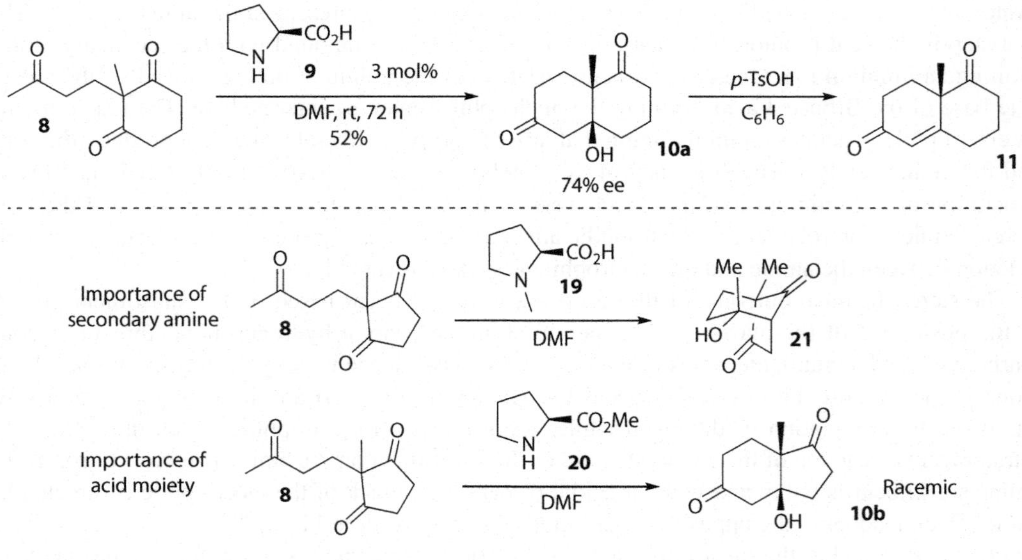

SCHEME 11.4 Hajos–Parrish aldol reaction and the importance of the secondary amine and acid moieties.

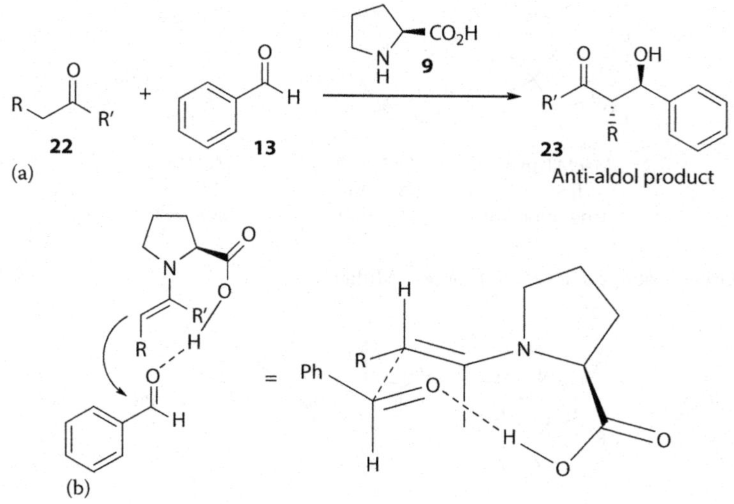

SCHEME 11.5 (a) Intermolecular aldol reaction and (b) Houk–List transition state.

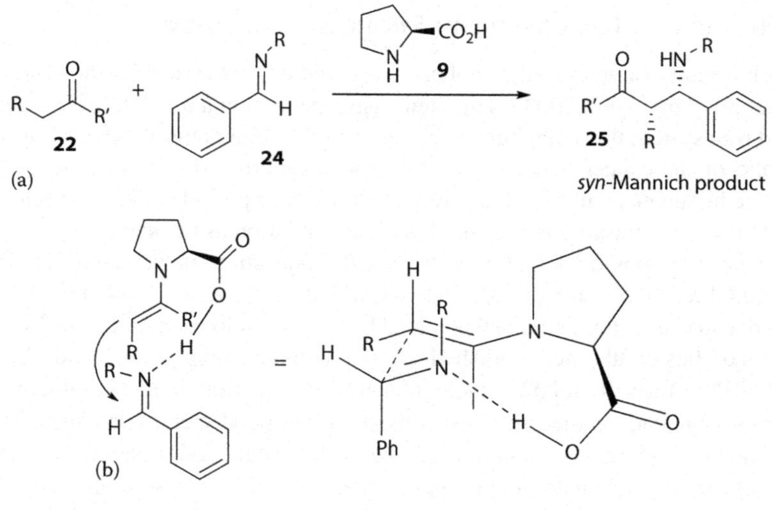

SCHEME 11.6 (a) Intermolecular Mannich reaction and (b) transition state.

This model can be easily applied to the intermolecular Mannich reaction that led to the *syn*-Mannich product. The only difference this time is the orientation of the imine due to its *E*-configuration (Scheme 11.6).

Another important type of catalyst in enamine chemistry is the so-called MacMillan catalyst. In 2000, MacMillan developed a secondary amine derived from amino acids that showed an extraordinary enantioselective induction.[6] Years later, the structure was improved leading to the development of MacMillan second- and third-generation catalysts (Figure 11.3).

The most important feature of these catalysts is the excellent facial differentiation of the enamine due to the lock of the conformation. As can be seen in Figure 11.4, the enamine formed has always an *E*-configuration to avoid steric interactions. The bulky group of the catalyst is positioned away of the enamine and this group and/or R[1] are shielding efficiently the bottom face of the enamine (Figure 11.4).

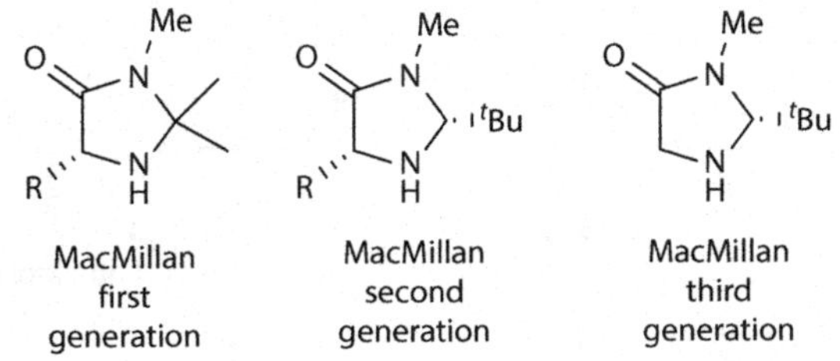

FIGURE 11.3 Organocatalysts developed for MacMillan.

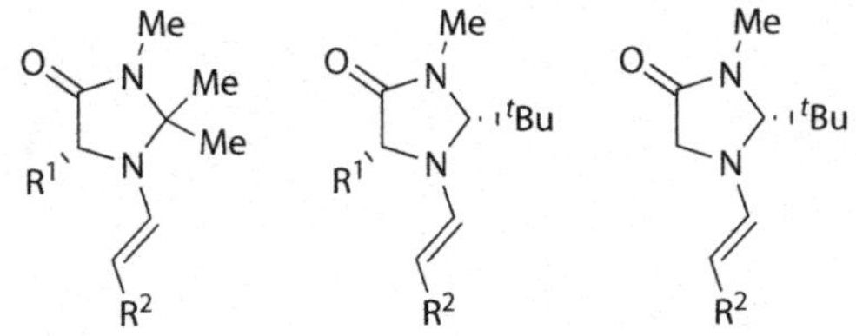

FIGURE 11.4 Enamines derived from MacMillan catalysts.

11.2.2 Applications in Total Synthesis: Synthesis of Hirsutene

One of the most elegant examples of the application of enamine chemistry for the synthesis of natural products was reported by List in 2008.[11] Hirsutene, isolated by Shibata in 1976, is a linearly fused tricyclopentanoid ring system that constitutes the skeleton of several naturally occurring sesquiterpenes having antibiotic or anticancer activity. Several groups reported organometallic or cycloaddition approaches to the hirsutene synthesis. List overcame the major part of the issues related to hirsutene synthesis by the use of an intramolecular aldol desymmetrization as the key step.

The retrosynthesis proposed by List consists in a disconnection of the fused 5.5.0 bicycle by an intramolecular aldol reaction leading to diketone **28**, which can be easily formed by the cyclization of diacid **29**. This diacid can, in turn, be easily prepared by radical cyclization of diester **30** (Scheme 11.7).

The synthesis of diester **30** is accomplished starting with the commercial diacid **31** that is reduced with borane in THF to furnish diol **32**, which, after Swern oxidation, renders dialdehyde **33**. Next, a double Wittig reaction leads to the synthesis of diester **30** in good yields (Scheme 11.8).

The next step is the radical cyclization to get the cyclic product **34**. List and coworkers used Mg in MeOH to generate the radical that undergoes cyclization with the unsaturated ester to furnish the dimethyl ester **34** in excellent yield, 88%. It should be noticed that the transesterification took place using MeOH as solvent (Scheme 11.9).

Next, hydrolysis to the diacid, followed by acyl chloride formation and the addition of TMSCH$_2$N$_2$, generates the bisdiazo compound **38** that, under ruthenium catalysis, produces the mesobicylic compound after reduction with Pd/C **28** in 91% yield (Scheme 11.10).

The mechanism of the last reaction is shown in Figure 11.5. First, the insertion of ruthenium into one of the diazo groups generates the carbine **41** that, after intramolecular nucleophilic attack, generates ruthenium complex **42**. Next, elimination of N$_2$, and subsequent discoordination, leads to the bicyclic compound **44**.

With the diketone in hand, List's group did the intramolecular aldol reaction as shown in Scheme 11.11. The proline derivative catalyst forms the enamine in such a way that the carboxylic acid moiety coordinates with the other carbonyl group, leading to the formation of the desired configuration in the tricyclic compound in good yields and excellent diastereo- and enantioselectivities.

Once the tricyclic ketone is obtained, E1cb elimination followed by Birch-type methylation and Wittig reaction furnishes the (+)-hirsutene in good overall yield (Scheme 11.12).

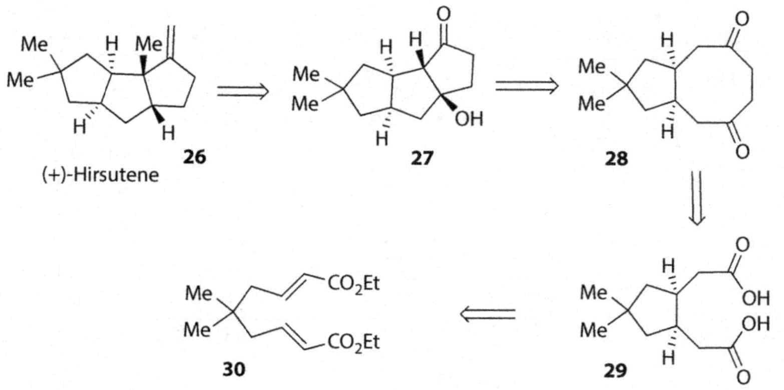

SCHEME 11.7 Hirsutene retrosynthesis reported by List.

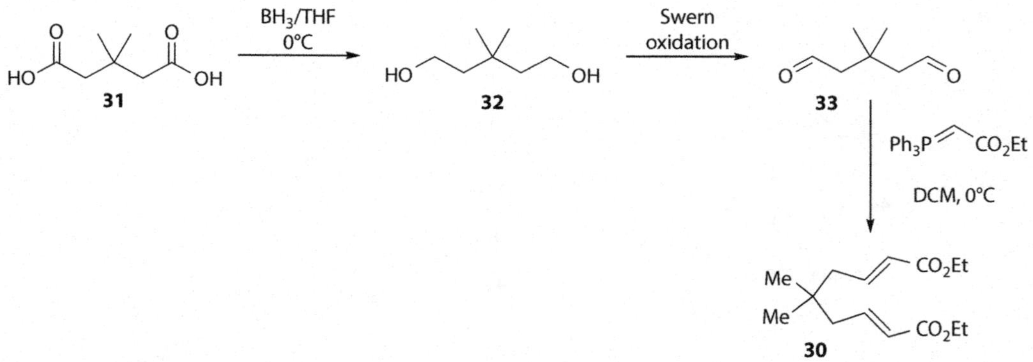

SCHEME 11.8 Synthesis of diester **30**.

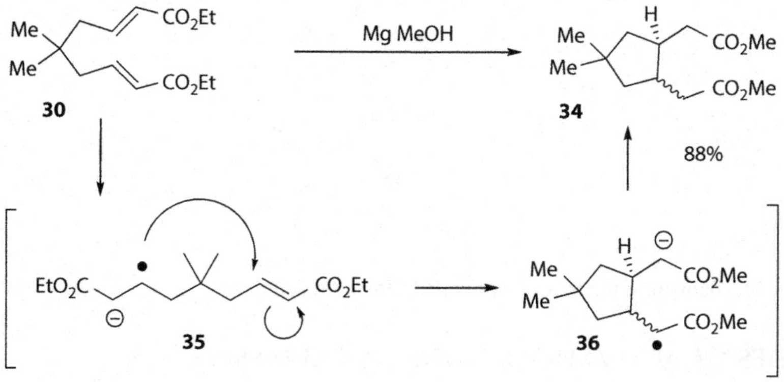

SCHEME 11.9 Radical cyclization.

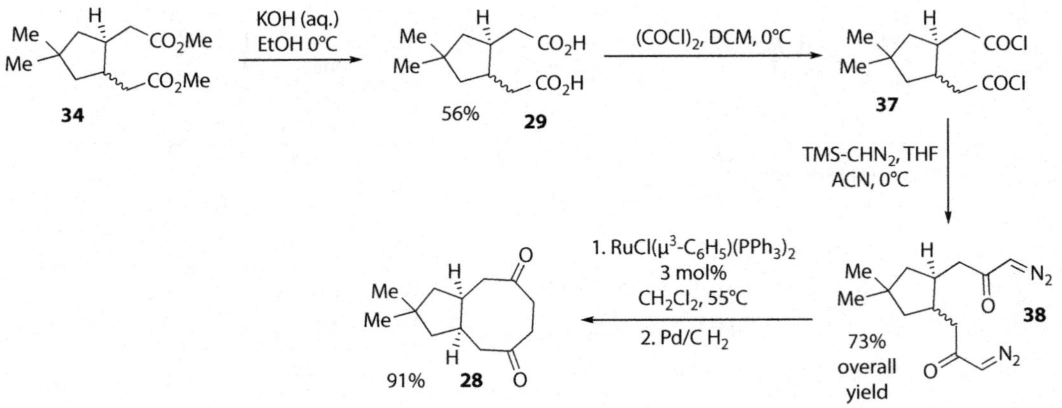

SCHEME 11.10 Synthesis of compound **28**.

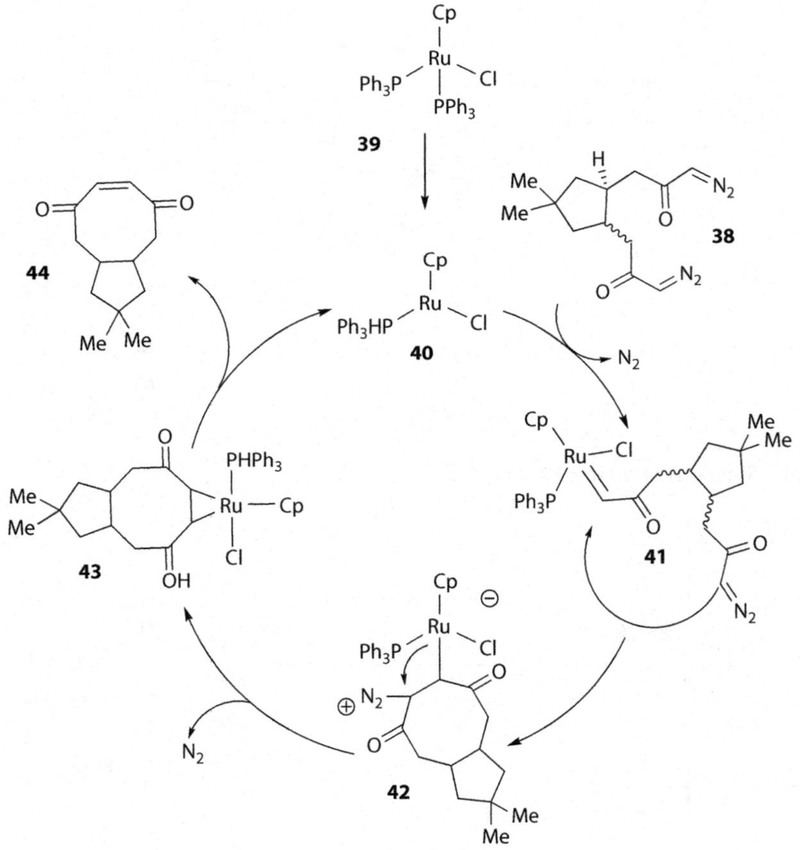

FIGURE 11.5 Mechanism of the Ru-catalyzed ring closing reaction.

11.3 MODES OF ACTIVATION: IMINIUM CATALYSIS[12]

The origins of iminium catalysis can be traced back with the works of MacMillan in 2000. This was clearly inspired by the metal catalysis reactions. As is presented in Figure 11.6a, the use of metal Lewis acid with acyl enals or secondary amines with enals leads to an enhancement of the reactivity

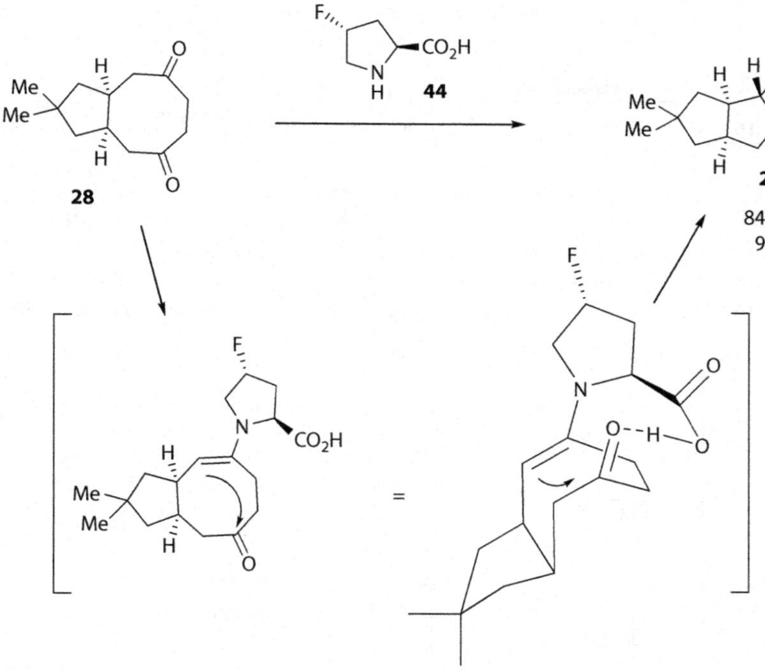

Enantioselective Organocatalysis from Concepts to Applications

SCHEME 11.11 Desymmetrization by intramolecular aldol reaction.

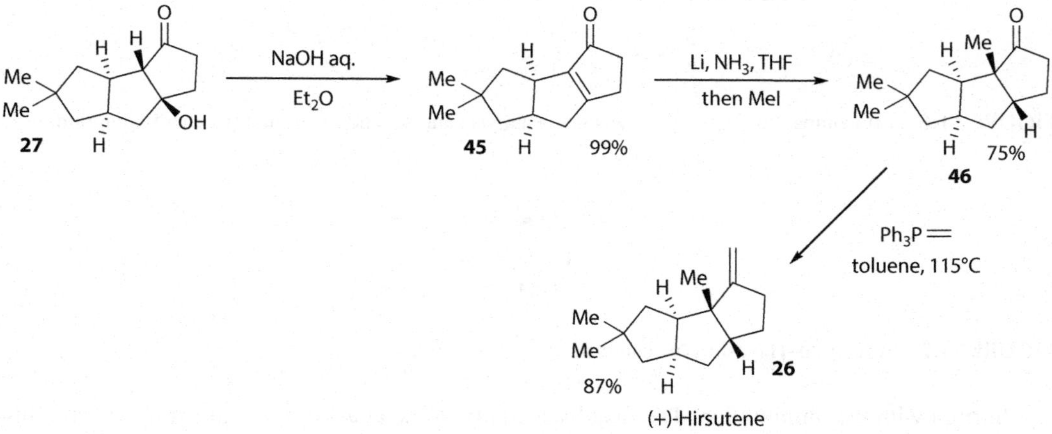

SCHEME 11.12 Final steps of the synthesis of (+)-hirsutene.

as Michael acceptors, decreasing the LUMO energy and decreasing the energy gap of the reaction (as example a Diels–Alder in Figure 11.6b).

Iminium catalysis is now an established strategy for the asymmetric addition of nucleophiles at the β-position of enals. Arguably the most important class of iminium catalysts are the secondary amine catalysts from the Jørgensen–Hayashi catalyst (Figure 11.7).

An example of a typical catalytic cycle for iminium chemistry is shown in Figure 11.8.

The cycle starts with the acid-promoted condensation of the carbonyl **16** with the amine **51** to form the iminium ion **52**, more electrophilic than the starting enal. The equilibrium between the *E*-iminium ion and the *Z*-iminium ion is heavily shifted toward the *E* as it is the more stable one. The nucleophile then attacks the β-position of the iminium with the formation of the anti-enamine in

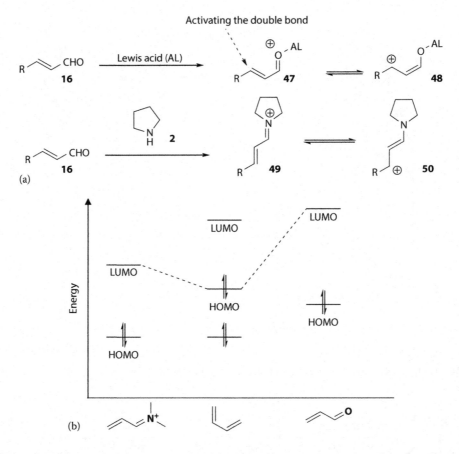

FIGURE 11.6 (a) Comparison between Lewis acid catalysis and secondary amine; (b) energy diagram.

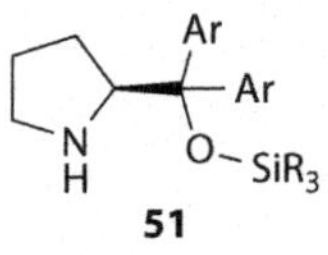

FIGURE 11.7 Jørgensen–Hayashi-type catalyst.

equilibrium with the iminium ion. The facial selectivity of the attack is determined by the chirality of the secondary amine. The bigger the substituents on the pyrrolidine, the better the selectivity will be, but conversely the rate of the reaction will decrease. The hydrolysis of the complex releases the product and the catalyst, which can reenter the catalytic cycle.

Some examples of C–C bond formation based on the iminium activation are the addition of malonates to α,β-unsaturated aldehydes developed by Jørgensen,[13] the addition of fluoromethyl, fluorobis(phenylsulfonyl)methane to enals,[14] or the synthesis of spirocyclic compounds developed by Melchiorre and Rios (Scheme 11.13).[15]

Another important class of catalysts in iminium activation are the MacMillan catalysts, already cited in enamine catalysis (Figure 11.9). The MacMillan's secondary amines can act as excellent iminium catalysts due to the locking conformation of the iminium ion. As in the case of Jørgensen–Hayashi catalysts, MacMillan catalysts take the *E*-configuration of the iminium ion. The bulky group of the catalyst is positioned away of the iminium, and this group and/or R^1 are shielding efficiently the bottom face of the double bond, allowing the nucleophiles attack by the upper face.

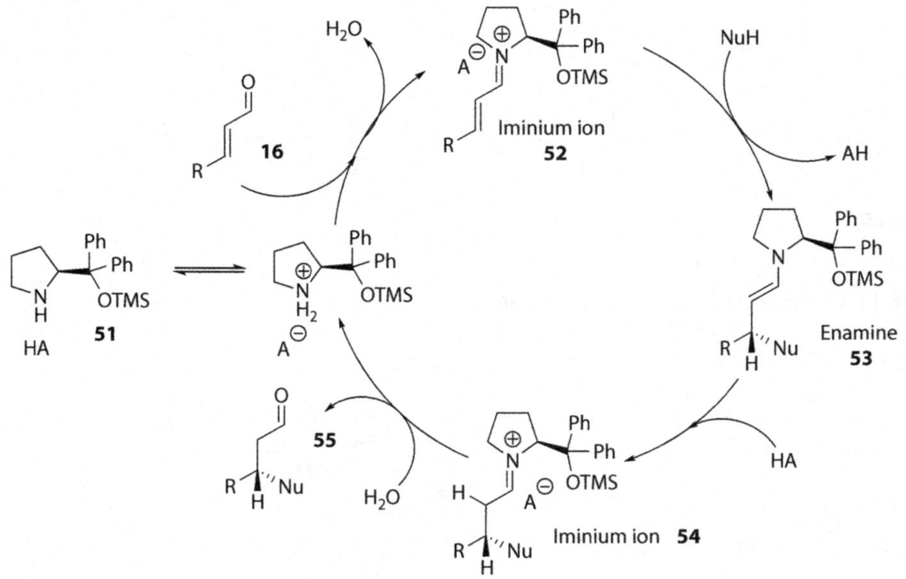

FIGURE 11.8 Catalytic cycle of iminium catalysis.

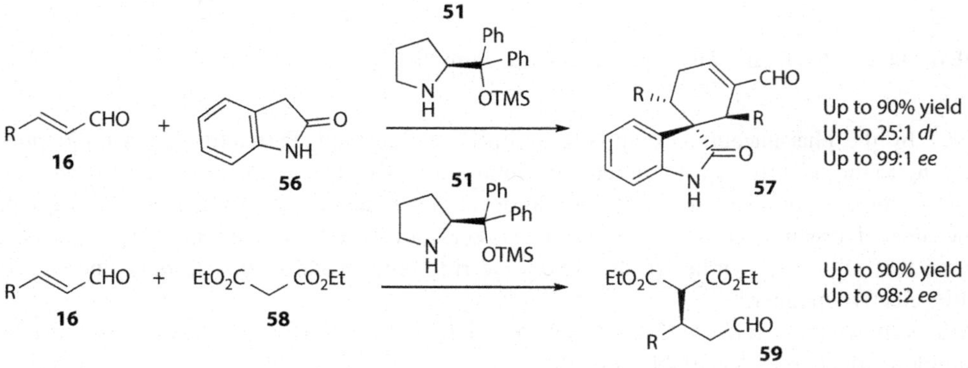

SCHEME 11.13 Examples of organocatalytic nucleophilic addition.

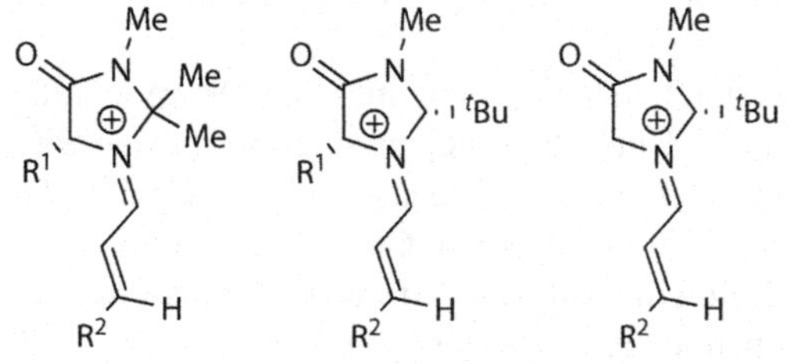

FIGURE 11.9 Iminium anions derived from MacMillan catalysts.

11.3.1 Applications in Total Synthesis: Synthesis of Paroxetine

Paroxetine is a blockbuster drug used for the treatment of major depressive disorders, such as obsessive–compulsive disorder, social anxiety disorder, panic disorder, posttraumatic stress disorder, generalized anxiety disorder, and premenstrual dysphoric disorder. It was first marketed

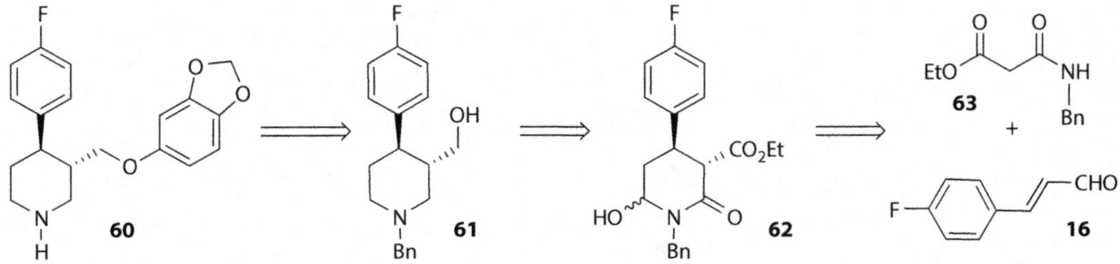

SCHEME 11.14 Proposed retrosynthesis for paroxetine.

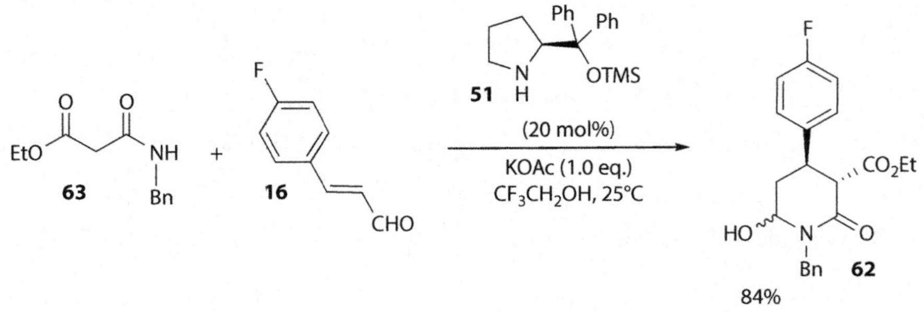

SCHEME 11.15 Synthesis of the piperidine core of paroxetine.

in 1992 by the pharmaceutical company SmithKline Beecham. Paroxetine is an antidepressant belonging to the selective serotonin reuptake inhibitor (SSRI) class. Due to its importance, several research groups have devoted their efforts to develop synthetic routes to achieve this compound. In the area of organocatalysis, Rios and coworkers developed an iminium–hemiacetal cascade reaction that allows to synthesize the piperidine ring core of Paroxetine in only one step in an enantioselective manner.[16]

As it is shown in the retrosynthesis of Scheme 11.14, the reaction of the enal with the corresponding amidomalonate **63** leads to the formation of the piperidine core.

The reaction between the enal **16** and the amidomalonate **63** will take place under secondary amine catalysis using Jørgensen–Hayashi catalyst **51** (20 mol%) in trifluoroethanol, and KOAc as an additive. The desired piperidine derivative **62** was obtained in 84% yield and 90% enantiomeric excess (Scheme 11.15).

As explained before, the stereochemical outcome could be rationalized by the mechanistic proposal outlined in Figure 11.10. Thus, the efficient shielding of the *Re*-face of the chiral iminium intermediate by the bulky aryl groups of **51** leads to stereoselective *Si*-facial nucleophilic conjugate attack at the β-position. Next, intermediate **64** cyclizes spontaneously *via* a favored 6-*exo*-trig ring closure, according to Baldwin's rules, to afford the hemiaminal **62**. It should be noticed that epimerization of the stereochemically labile stereocenter at C3 will establish the thermodynamically more stable (3*S*,4*R*) *trans*-configuration (Figure 11.10).

Once the piperidine core with the desired absolute configuration was obtained, the authors proceed to synthesize the paroxetine: first the piperidine ring is reduced with borane in THF to obtain piperidine **61** in 86% yield, next mesylation of the primary alcohol followed by S_N2 using sesamol, gives the protected paroxetine in 80% yield. Finally, deprotection of the benzyl group by hydrogenation with Pd/C affords paroxetine in 92% yield (Scheme 11.16).

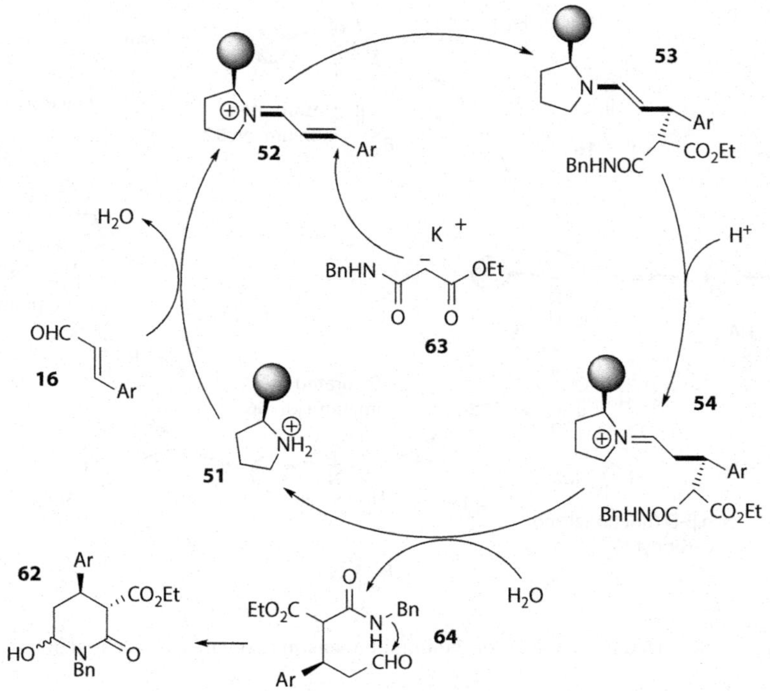

Enantioselective Organocatalysis from Concepts to Applications

FIGURE 11.10 Stereochemical outcome of piperidine's synthesis.

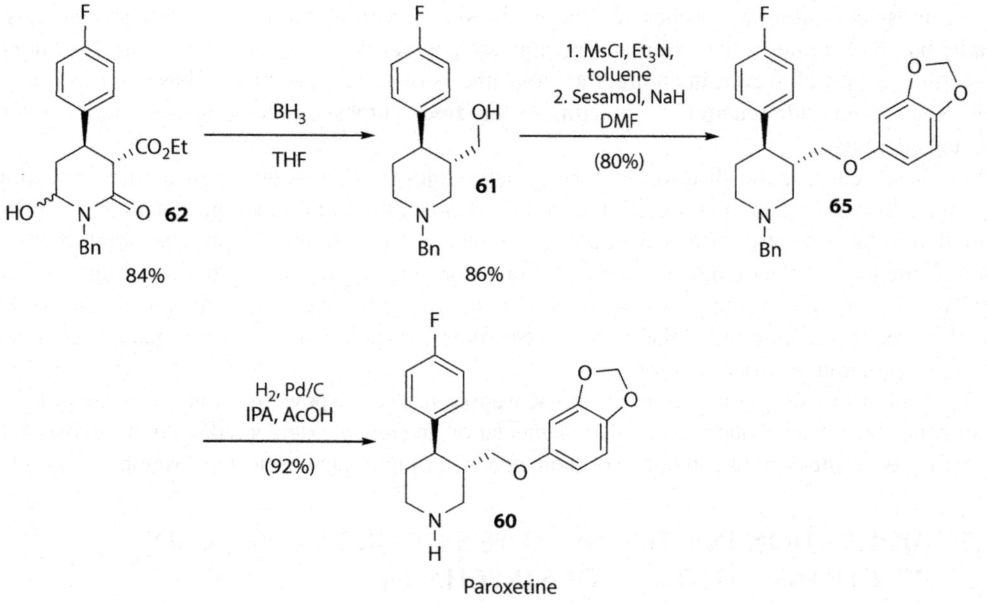

SCHEME 11.16 Final steps for the synthesis of paroxetine.

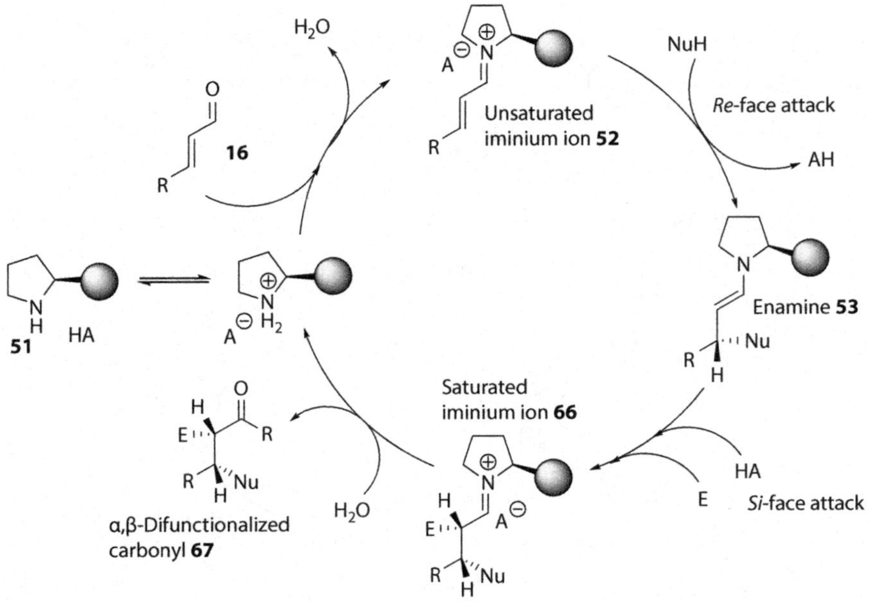

FIGURE 11.11 Generalized mechanism for the iminium–enamine organocascade reactions.

11.4 CASCADE REACTIONS BASED ON IMINIUM/ENAMINE CHEMISTRY

11.4.1 Iminium/Enamine Organocascades[17]

One of the most challenging goals for organic chemists is the synthesis of complex molecules. The high complexity of natural products has always fascinated researchers, encouraging and pushing the limits of organic synthesis methodologies.

The most common approaches for the synthesis of natural products or pharmaceuticals are usually based on a multistep or stop-and-go approach, which requires the use of multiple orthogonal protecting groups. However, in nature, the biosynthesis of natural products is based on several principles such as cascade reactions, protecting group–free synthesis, redox economy, atom economy and step economy.

For these reasons, the discovery of new methodologies that—similar to nature—are able to implement various reaction strategies in a multicomponent domino reaction to achieve multi-bond formation in one operation is a holy grail for organic chemists. In this chapter, we want to draw the readers' attention to the organocascade C–C bond–forming reactions, which play a huge role in the rapid synthesis of highly complex structures. This strategy is atom economical and avoids the necessity of protecting groups and isolation of intermediates. Its goal is to mimic the nature in its highly selective sequential transformations.

The combination of enamine–iminium ion activations has been a powerful strategy for the fast assembly of complex structures due to its easy implementation and reliable stereoprediction that allows enantioselective consecutive formation of two or more bonds in a highly stereoselective fashion (Figure 11.11).

11.5 APPLICATION IN TOTAL SYNTHESIS OF ORGANOCASCADE REACTIONS: SYNTHESIS OF STRYCHNINE

Among the great variety of organic transformations that are amenable to asymmetric organocatalysis, organocascade reactions occupy a preeminent position, due to the structural simplicity of most organocatalysts, the variety of its modes of activation, and the easy prediction of the stereochemical

Enantioselective Organocatalysis from Concepts to Applications 339

outcome of the reaction. This chapter is not intended to be comprehensive, but rather, selected organocatalytic cascade reactions will be outlined in the context of their usefulness in synthetic chemistry.

One of the most remarkable examples of using iminium–enamine cascade reactions for the synthesis of natural products is the synthesis of several complex members of the *Strychnos*, *Aspidosperma*, and *Kopsia* families of alkaloids by D. W. C. MacMillan.[18] Concretely, in this chapter, we will focus on the synthesis of strychnine, the best known member of the *Strychnos* family, which has attracted a huge interest of the synthetic community in the last 50 years.

The most notorious characteristic of strychnine is its pharmacological activity. Strychnine is a neurotoxin that acts as an antagonist of glycine and acetylcholine receptors. It primarily affects the motor nerves in the spinal cord that control muscle contraction. Strychnine is famous for its toxicity and its use as poison for animals and humans.

Strychnine was discovered by Joseph Bienaime and Pierre-Joseph Pelletier in 1818 in the Saint Ignatius' bean. Chemically speaking, strychnine is a terpene indole alkaloid belonging to the *Strychnos* family of *Corynanthe* alkaloids. Their biosynthesis can be traced back to tryptamine and secologanin. The condensation of these two compounds catalyzed by strictosidine synthase followed by a Pictet–Spengler reaction generates the intermediate strictosidine that leads to the final strychnine after several more steps.

In the kingdom of organic synthesis, strychnine has been a common target due to its complexity and relatively low molecular weight; moreover, its pharmacological activities stimulated the interest for the synthesis of derivatives and to study the structure–activity relationships. The first total synthesis of strychnine was reported by Woodward in 1954.

MacMillan and coworkers in 2011 reported the use of an organocascade reaction for the synthesis of common core alkaloids of the *Strychnos*, *Aspidosperma*, and *Kopsia* families. In this work, they synthesized strychnine, aspidospermidine, kopsinine, akuammicine, vincadifformine and kopsanone using this common organocascade approach. The retrosynthesis of strychnine, made by MacMillan, is shown in Figure 11.12. The key steps of the retrosynthesis consist in a Jeffery–Heck cyclization/lactol formation sequence to afford the Wieland–Gumlich aldehyde and a formal organocatalyzed Diels–Alder reaction between propargyl aldehyde and the 2-vinyl indole **74**.

MacMillan and coworkers started the synthesis by preparing 2-vinyl indole. Compound **75** was first protected as PMB by reaction with PMBCl and NaH in DMF. Next, oxidation using SeO_2 in dioxane, followed by hydrolysis and finally Wittig reaction with $(EtO)_2P(O)CH_2SeMe$ **78** using KHDMS as the base at −78°C, furnished the final 2-vinyl indole in good overall yield (63%) (Scheme 11.17).

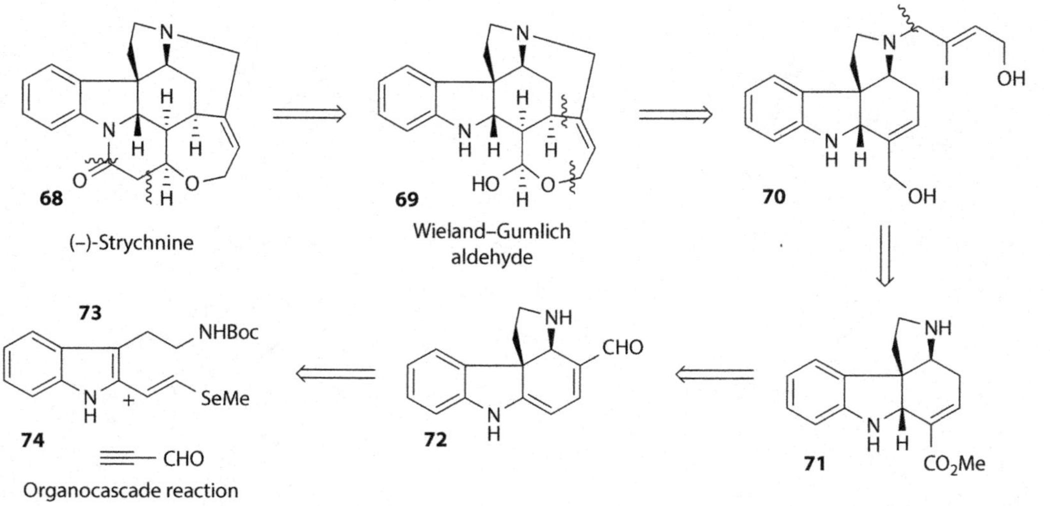

FIGURE 11.12 Retrosynthesis of strychnine proposed by MacMillan.

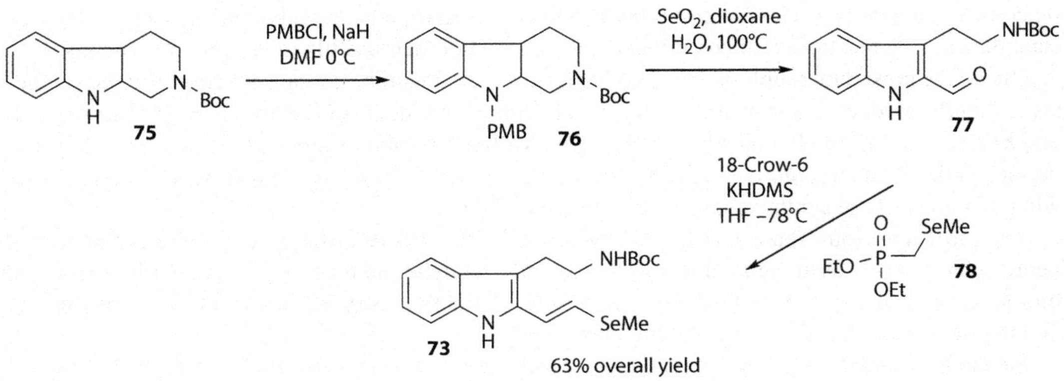

340 Asymmetric Synthesis of Drugs and Natural Products

SCHEME 11.17 Synthesis of compound **73**.

Next, the formal Diels–Alder organocascade reaction took place, followed by an intramolecular organocatalyzed Michael reaction. In this reaction, a modified MacMillan second-generation catalyst **82** was used. In the reaction of the iminium with the propynal, as was explained before, the triple bond would be expected to partition away from the bulky tert-butyl group of the catalyst. Then, the naphthyl group of the catalyst shields the bottom face of the alkyne (Figure 11.13).

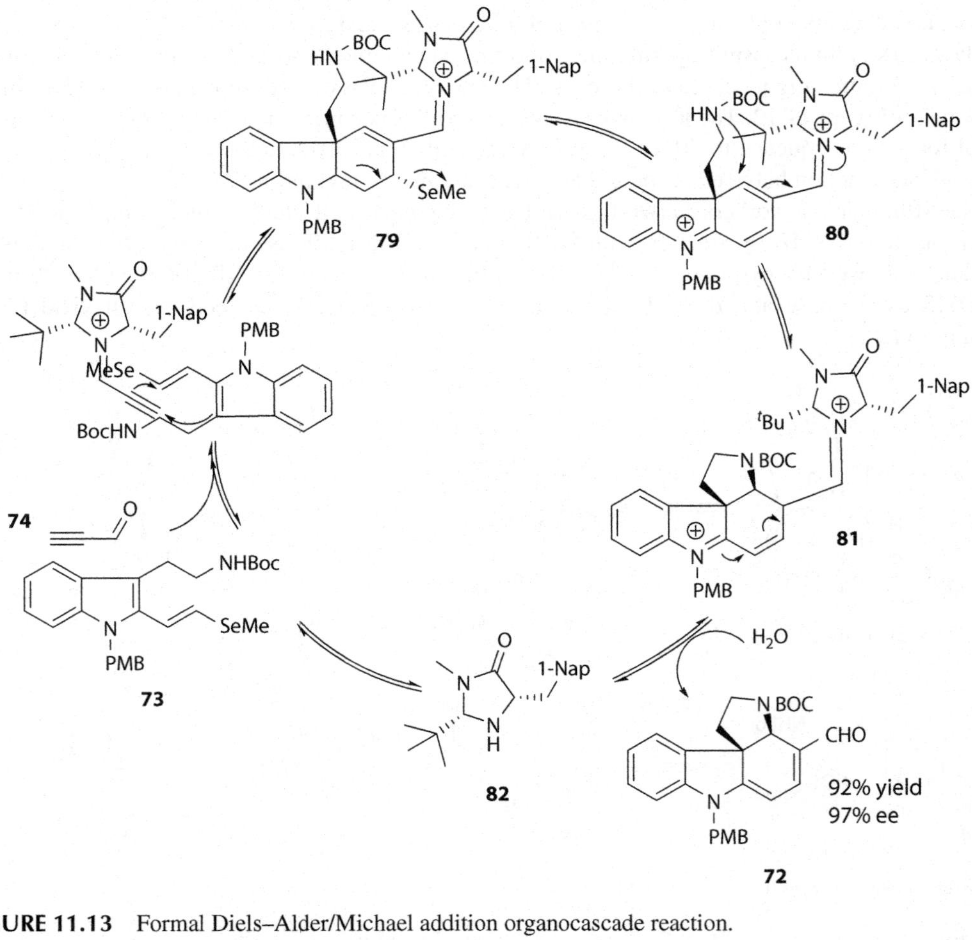

FIGURE 11.13 Formal Diels–Alder/Michael addition organocascade reaction.

Enantioselective Organocatalysis from Concepts to Applications 341

Once the formal Diels–Alder took place, the cycloadduct **79** undergoes an easy β-elimination of the methyl selenide to render the unsaturated iminium ion **80**. Next, the pendant BOC imine reacts with the double bond via a Michael addition to furnish the cis-fused bicyclic ring. After hydrolysis, the tetracyclic ring **72** was obtained with good yield and excellent enantioselectivity (82% yield, 97% ee).

Decarbonylation using Wilkinson's catalyst and treatment with phosgene and methanol introduce the carbomethoxy group at the dienamine position. Next, treatment with DIBAL-H reduces the unsaturated enamine to install the tertiary indoline stereocenter and provides the unsaturated ester. Finally, the treatment with trifluoroacetic acid deprotects the BOC amine and generates compound **85** (Scheme 11.18).

Intermediate **85** was converted into the vinyl iodide **87** using a two-step protocol. First, an allylation with allylbromide takes place, followed by reduction with DIBAL-H to furnish the diol **87** (Scheme 11.19).

Compound **87** is then converted to the protected Wieland–Gumlich aldehyde **90** through a Jeffery–Heck cyclization sequence (Scheme 11.20). First, Pd inserts in the vinyl iodide and subsequent carbopalladation forms the six-membered ring and an alkyl palladium intermediate that undergoes a fast β-elimination to form the enol intermediate. Hemiacetal formation leads to the Wieland–Gumlich intermediate in a 58% yield.

Finally, strychnine was obtained by TFA-mediated removal of the PMB group, followed by addition of malonic acid, acetic anhydride, and sodium acetate at 120°C in 6.4% overall yield (Scheme 11.21).

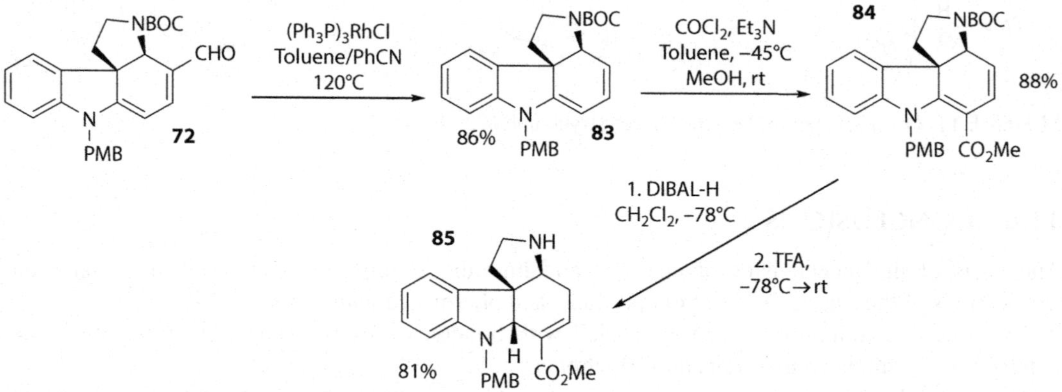

SCHEME 11.18 Synthesis of compound **85**.

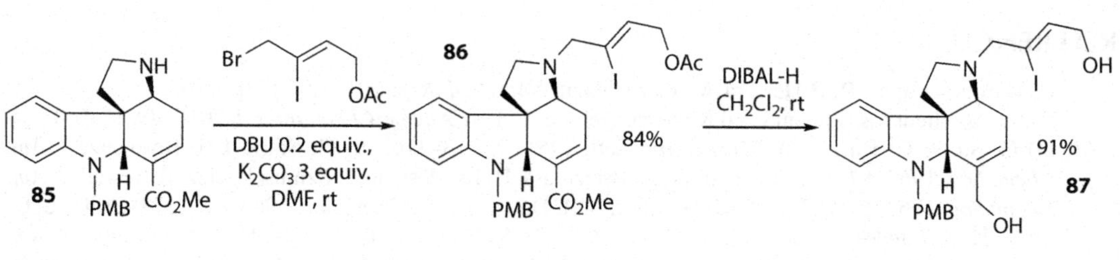

SCHEME 11.19 Synthesis of compound **87**.

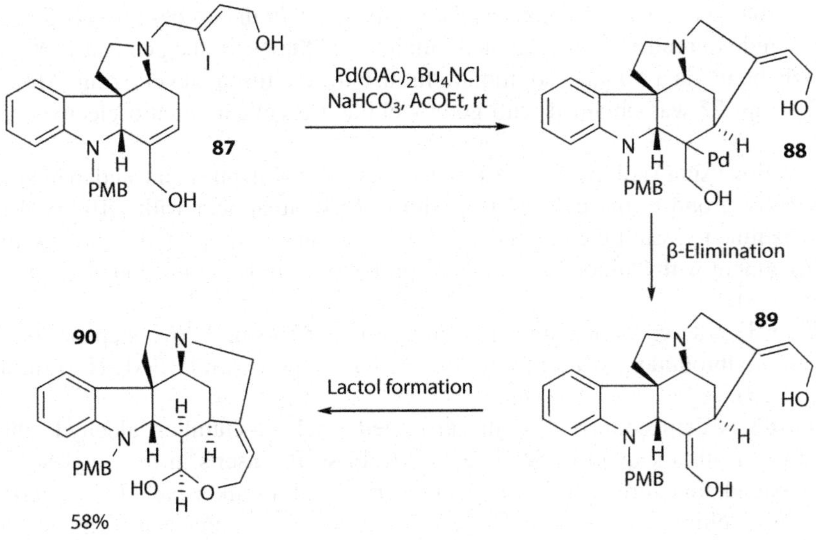

SCHEME 11.20 Synthesis of protected Wieland–Gumlich aldehyde **90**.

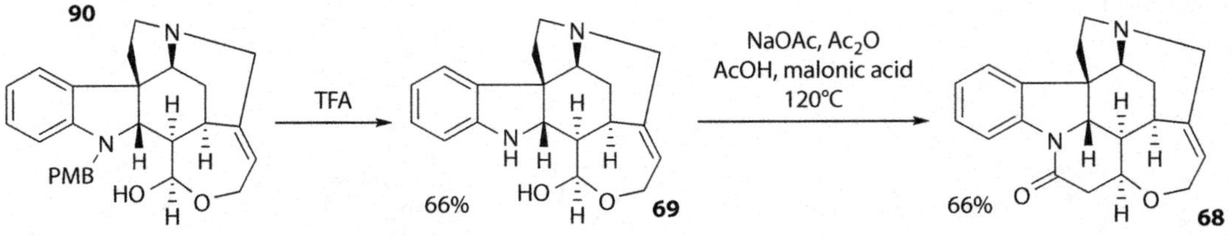

SCHEME 11.21 Last step of the synthesis of strychnine **68**.

11.6 CONCLUSIONS

The basis of the mechanisms of enamine and iminium chemistry illustrates some remarkable applications in the synthesis of natural products and pharmaceuticals.

Enamine and iminium chemistry coupled with organocascade reactions was exploited in the synthesis of hirsutene, paroxetine, and strychnine.

Organocatalysis presents several advantages, in front of its organometallic chemistry couterpartners, like the benign reaction conditions, the high functional group tolerance, and the easy stereoprediction. However, to see a real use of organocatalysis in industry, it is still a long way to go in the area of catalyst loadings, product purification, and catalyst recovery and recycle.

REFERENCES

1. E. Marques-Lopez, R. P. Herrera, M. Christmann, *Nat. Prod. Rep.* 2010, *27*, 1138–1167.
2. For an excellent essay about Emil Knoevenagel see: B. List, *Angew. Chem. Int. Ed.* 2010, *49*, 1730–1734.
3. (a) G. Stork, G. Birnbaum, *Tetrahedron Lett.* 1961, *2*, 313–316; (b) G. Stork, I. J. Borowitz, *J. Am. Chem. Soc.* 1962, *84*, 313; (c) G. Stork, A. Brizzolara, H. Landesman, J. Szmuszkovicz, R. Terrell, *J. Am. Chem. Soc.* 1963, *85*, 207–222; (d) G. Stork, S. R. Dowd, *J. Am. Chem. Soc.* 1963, *85*, 2178–2180; (e) G. Stork, H. K. Landesman, *J. Am. Chem. Soc.* 1956, *78*, 5129–5130; (f) G. Stork, H. K. Landesman, *J. Am. Chem. Soc.* 1956, *78*, 5128–5129; (g) G. Stork, R. Terrell, J. Szmuszkovicz, *J. Am. Chem. Soc.* 1954, *76*, 2029–2030.

4. (a) U. Eder, G. Sauer, R. Wiechert, *Angew. Chem. Int. Ed.* 1971, *10*, 496; (b) Z. G. Hajos, D. R. Parrish, *J. Org. Chem.* 1974, *39*, 1615.
5. B. List, R. A. Lerner, C. F. Barbas III, *J. Am. Chem. Soc.* 2000, *122*, 2395–2396.
6. K. A. Ahrendt, C. J. Borths, D. W. C. MacMillan, *J. Am. Chem. Soc.* 2000, *122*, 4243–4244.
7. S. Mukherjee, J. W. Yang, S. Hoffmann, B. List, *Chem. Rev.* 2007, *107*, 5471–5569.
8. D. Seebach, A. K. Beck, D. M. Badine, M. Limbach, A. Eschenmoser, A. M. Treasurywala, R. Hobi, *Helv. Chim. Acta* 2007, *90*, 425–471.
9. C. Allemann, R. Gordillo, F. R. Clemente, P. H.-Y. Cheong, K. N. Houk, *Acc. Chem. Res.* 2004, *37*, 558–569.
10. J. Franzen, M. Marigo, D. Fielenbach, T. C. Wabnitz, A. Kjrsgaard, K. A. Jørgensen, *J. Am. Chem. Soc.* 2005, *127*, 18296–18304.
11. C. L. Chandler, B. List, *J. Am. Chem. Soc.* 2008, *130*, 6737–6739.
12. A. Erkkilae, I. Majander, P. M. Pihko, *Chem. Rev.* 2007, *72*, 10081–10087.
13. S. Brandau, A. Landa, J. Franzén, M. Marigo, K. A. Jørgensen, *Angew. Chem. Int. Ed.* 2006, *45*, 4305–4309.
14. (a) A.-N. Alba, X. Companyó, A. Moyano, R. Rios, *Chem. Eur. J.* 2009, *15*, 7035–7038; (b) F. Ullah, G.-L. Zhao, L. Deiana, M. Zhu, P. Dziedzic, I. Ibrahem, P. Hammar, J. Sun, A. Córdova, *Chem. Eur. J.* 2009, *15*, 10013–10017; (c) S. Zhang, Y. Zhang, Y. Ji, H. Li, W. Wang, *Chem. Commun.* 2009, *32*, 4886–4888.
15. (a) X. Companyo, A. Zea, A.-N. R. Alba, A. Mazzanti, A. Moyano, R. Rios, *Chem. Commun.* 2010, *46*, 6953–6955; (b) C. Cassani, X. Tian, E. C. Escudero-Adan, P. Melchiorre, *Chem. Commun.* 2011, *47*, 233–235.
16. (a) G. Valero, J. Schimer, I. Cisarova, J. Vesely, A. Moyano, R. Rios, *Tetrahedron Lett.* 2009, *50*, 1943–1946; (b) S. Číhalová, G. Valero, J. Schimer, M. Humpl, M. Dračínský, A. Moyano, R. Rios, J. Vesely, *Tetrahedron* 2011, *67*, 8942–8950.
17. A. Moyano, R. Rios, *Chem. Rev.* 2011, *111*, 4703–4832.
18. S. B. Jones, B. Simmons, A. Mastracchio, D. W. C. MacMillan, *Nature* 2011, *475*, 183–188.

12 Chiral Building Blocks for Drugs Synthesis *via* Biotransformations

Pilar Hoyos, Vittorio Pace, and Andrés R. Alcántara

CONTENTS

12.1 Introduction 346
12.2 Anticancer Drugs 348
    12.2.1 Modification of Natural Products 349
        12.2.1.1 Taxanes 349
        12.2.1.2 Epothilones 363
        12.2.1.3 Cyclopamine Analogues 367
        12.2.1.4 Aureolic Acids 369
12.3 Small Molecules 371
    12.3.1 Pelitrexol 372
    12.3.2 Monastrol 374
    12.3.3 AXL Inhibitors 375
    12.3.4 Crizotinib 377
    12.3.5 Lonafarnib 378
    12.3.6 Odanacatib 380
    12.3.7 Niraparib 381
12.4 Antidiabetic Drugs 383
    12.4.1 Peroxisome Proliferator–Activated Receptor Agonists 383
    12.4.2 Glucagon-Like Peptide-1 Mimetics and Modulators 387
    12.4.3 Dipeptidyl Peptidase-4 (DPP-4) Inhibitors 389
    12.4.4 Sitagliptin 389
    12.4.5 Saxagliptin 390
12.5 Anti-Inflammatory Drugs: Profens 392
    12.5.1 (*S*)-Ibuprofen (Dexibuprofen) 393
    12.5.2 (*S*)-Ketoprofen (Dexketoprofen) 396
    12.5.3 (*S*)-Flurbiprofen 397
    12.5.4 Naproxen 398
    12.5.5 (*S*)-Ketorolac 402
12.6 Drugs for Treatment of Cardiovascular Diseases 404
12.7 Anticholesterol Drugs 404
    12.7.1 Simvastatin 406
    12.7.2 Atorvastatin 408
    12.7.3 Biocatalyzed Synthesis of Ethyl (*R*)-4-Cyano-3-Hydroxybutyrate 408
    12.7.4 Other Biocatalyzed Synthesis of Chiral Building Blocks for Statins Production 412
12.8 Antihypertensive Drugs 413
    12.8.1 β-Blockers 414

12.9	ACE Inhibitors	424
12.10	Calcium Channel Blockers	426
12.11	Conclusion	428
References		429

12.1 INTRODUCTION

There is a growing interest in developing more environmentally acceptable processes in chemical or biotechnological industries. This trend toward what has become known as sustainable technologies can be renamed as either Green Chemistry or White Biotechnology, depending on the emphasis focused either on chemical or biotechnological processes, respectively. The term Green Chemistry was coined by Warner and Anastas[1] and accepted by the "US Green Chemistry Program" from the U.S. Environmental Protection Energy (EPA) in 1993. On the other hand, the European Association for Bioindustries (EuropaBio) defined White Biotechnology as an emerging field within modern biotechnology that serves industry; in fact, the concept White Biotechnology is commonly accepted in the industrial world as the application of biotechnological tools, such as genetically modified organisms (GMO), new enzymes from extremophiles, etc., to produce substances with interest, taking into account the principles of Green Chemistry. Therefore, White Biotechnology can help realize substantial gains for environment, consumers, and industry.

On the other hand, the synthesis of optically pure compounds is increasingly in demand in the pharmaceutical, fine chemicals, and agro-alimentary industries, as it has well established the importance of chirality on the activity and properties of many compounds. If we focus on the pharmaceutical industry, we are talking about a market that is moving an enormous amount of money: in fact, the global pharmaceutical market is predicted to grow to $800 billion by the year 2020.[2] In another recently published study,[3] in 2007, the sales of biotech products were calculated to be around 48 billion Euros worldwide (3.5% of all the chemical sales), active pharmaceutical ingredients (APIs) and cosmetics being the most active areas. These sales were expected to increase up to 135 billion Euros (7.7% of the chemical market) in 2012, and 340 billion Euros (15.4% of global chemical products) by 2017, a fifth of this overall amount corresponding to APIs.

Thus, the development of new processes for obtaining chiral molecules is still an open challenge in organic synthesis, and the production of fine chemicals and drugs by biocatalyzed methodologies is an emerging research field.[4-12] Due to the high chemo-, regio-, and enantioselectivities commonly displayed by biocatalysts, biotransformations have acquired more interest for the production of chiral building blocks, as well as biologically active compounds,[13] offering the development of more environmentally and economically attractive processes. In fact, biocatalysis presents many appealing features in the context of Green Chemistry: gentle reaction conditions (physiological pH and temperature, water as the usual reaction medium, although many green solvents can also be used, as we will mention later) and an environmentally friendly catalyst (an enzyme or a cell) displaying high activities and chemo-, regio-, and stereoselectivities in multifunctional molecules. Additionally, the use of biocatalysts generally circumvents the need for functional group activation, therefore avoiding protection and deprotection steps usually required in traditional organic syntheses. These properties afford processes that are shorter, produce less waste, and are, therefore, both environmentally and economically smarter than conventional routes, so that many of the 12 principles of Green Chemistry are satisfied.[14]

Nevertheless, although enzymes are very active and selective biocatalysts, for industrial purposes, a very common reason to engineer them is to increase their stability under reaction conditions.[15] Actually, the reaction conditions needed for a biocatalytic procedure can differ dramatically from those present in a cell, therefore demanding high temperatures, extremes of pH, high substrate and product concentrations, oxidants, and organic cosolvents. Sometimes an enzyme must tolerate these conditions for only a few minutes or hours, but in a continuous manufacturing process, an enzyme may need to tolerate them for months. There are many ways to increase the robustness of biocatalysts, their immobilization being probably one of the most traditionally studied and used.[16] On the other hand, the use of molecular enzyme engineering techniques, such as directed evolution,[17] has enormously contributed to the preparation of new biocatalysts, which are able to work efficiently in experimental conditions very different from the "natural" ones, in terms of temperature, pH, presence of organic solvents, etc., therefore starting what has been very recently called "the third wave" in biocatalysis.[18]

For sure, another aspect that deserves a close attention is scaling up. Moving one process from the laboratory to industrial scale is not trivial at all and must be perfectly optimized[19] to solve all the specific problems that can arise. Some key aspects to be considered are correct reactor choice,[20] pH control, risks of contamination, logistics, real process feeds, and, last but not least, use of GRAS (Generally Regarded As Safe) solvents.[19] Certainly, regulatory features of APIs' biocatalyzed synthesis have to take into account not only the contribution of the solvent but also the role of the biocatalyst itself. In this sense, it is inappropriate to apply biologic and fermentation guidance, developed for the preparation of recombinant proteins, directly to small molecule API manufacture, where enzymes are used purely as catalysts. A detailed discussion on this topic is out of the scope of this manuscript, so we recommend the recent paper of Wells et al.,[21] which deals with different aspects such as enzyme/biocatalyst source, quality, and specification, processing issues, residues in APIs and strategies for managing potential impurities, and general toxicity and tiered risk assessments.

Anyhow, in fine chemical or pharmaceutical industries, solvents are nowadays used in large quantities compared to both catalyst and product (the ratio solvent/product varies between 100 and 1000). Replacing hazardous chemicals by more environmentally friendly alternatives is currently a matter of intense research, aligned with the Green Chemistry philosophy.[22] The introduction of eco-friendly solvents is a key area,[23–25] because solvents constitute the major source of waste in chemical processes[26]; in this sense, we must consider that solvents

1. Typically account for 80%–90% of the nonaqueous mass of material usage for API manufacture[27,28]
2. Consume about 60% of the overall energy used to produce API
3. On removal are responsible for 50% of the total posttreatment greenhouse gas emissions[29]

Ideally, the setup of chemical solvent-free processes would be highly desirable,[30] because a Green Chemistry axiom stands that "the greenest solvent is no solvent at all." Nevertheless, in many cases solvents are essential to improve mass and heat transfer and may significantly improve reaction rates, selectivities, or the position of chemical equilibria. Likewise, water represents undoubtedly the second best choice after neat conditions, and its ability to act as an efficient solvent in organic reactions is being actively assessed.[31–33] At this point, the differences between reactions carried out "in water" and "on water" have to be stressed; thus, as defined by Sharpless and coworkers,[34] the term "on water" has to be used when insoluble reactants are stirred in aqueous emulsions or suspensions without the addition of any organic cosolvents, while when using the term "in water" we should consider that all the reactants are water soluble. In "on water" conditions, due to hydrophobic effects, the use of water as a solvent not only accelerates reaction rates but also enhances reaction selectivities, even when the reactants are sparingly soluble or insoluble in this medium, so that the old principle *corpora non agunt nisi soluta* (substances do not react unless dissolved) should need to be revised.[32]

Nevertheless, the simple use of water instead of an organic solvent does not automatically improve the environmental impact of the synthetic procedure, because many other parameters, such as atom economy,[35,36] environmental factor,[37–39] final yield, workup, and purification demands, must be considered. Furthermore, wastewater production should not be underestimated in terms of Green Chemistry.

Anyhow, as long as some enzymes, especially lipases, can also perfectly work in organic media[40] and other nonconventional solvents,[41,42] the applicability and versatility of biocatalysis become notoriously amplified in terms of suitability for green solvents.

Representative examples of biocatalytic strategies are to prepare chiral building blocks as key intermediates in the synthesis of pharmaceuticals at the lab and industrial scales. The advantages of using different types of enzymes on each process, such as hydrolases, oxide reductases, transferases, or lyases, will be discussed, as well as the optimization of the biocatalytic reaction conditions. When possible, the therapeutic class of each drug has been extracted from the Anatomical Therapeutic Chemical (ATC) classification, as defined by the World Health Organization (WHO) (http://www.whocc.no/atc_ddd_index); in this classification, the active substances are divided into different groups according to the organ or system on which they act and their therapeutic, pharmacological, and chemical properties. Drugs are classified in groups at five different levels. Thus, drugs are divided into 14 main groups (first level), with pharmacological/therapeutic subgroups (second level). The third and fourth levels are chemical/pharmacological/therapeutic subgroups and the fifth level is the chemical substance. Thus, the second, third, and fourth levels are often used to identify pharmacological subgroups when that is considered more appropriate than therapeutic or chemical subgroups. DB is another code for defining drugs from DrugBank (http://www.drugbank.ca), a comprehensive online database containing extensive biochemical and pharmacological information about drugs, their mechanisms, and their targets.[43]

Noncommunicable diseases (NCDs) is a prominent research area, because 36 million (63%) of the estimated 57 million global deaths in 2008 were due to NCDs[44]. In fact, both populace growth and augmented longevity are leading to a quick rise in the total number of middle-aged and older adults, therefore leading to a parallel increase in the number of deaths caused by NCDs. Inside NCDs, cardiovascular diseases and cancer are the leading pathologies; in fact, it is projected that the annual number of deaths due to cardiovascular disease will increase from 17 million in 2008 to 25 million in 2030, and about 12.7 million cancer cases and 7.6 million cancer deaths are estimated to have occurred in 2008 worldwide,[45] and this last figure is expected to reach 13 million by 2030.[44] As a result of such trends, the total number of annual NCD deaths is projected to reach 55 million by 2030, whereas annual infectious disease deaths are projected to decline over the next 20 years.

12.2 ANTICANCER DRUGS

Cancer is the principal cause of death in economically developed countries and the second leading cause of death in developing countries,[45] where this disease is very rapidly growing as a result of population aging and growth as well as, progressively, the adoption of cancer-associated routine choices such as smoking, physical sedentariness, and "westernized" nourishments.

The majority of cancer patients treated today will receive either pre- or postoperative chemotherapy with any of the 206 antitumor compounds available in the market from 1940 to 2010, as recently compiled by Newman and Cragg.[46] In the time period between 1981 and 2010, 128 new chemical entities (NCE) were approved for cancer treatment, 99 of them (77%) were small molecules, not coming from natural products.[46] This fact clearly shows the increasing policy of pharmaceutical companies, which have neglected in the recent past the development of potential natural drug candidates in favor of combinatorial chemistry and high-throughput synthesis of large compound libraries. The main reason for this lies in the great structural complexity of natural products, which make their synthesis very long, tedious, and difficult to be scaled up.[47] In any case, biocatalysis can help in the preparation of anticancer drugs, both in the modification of natural products and in the synthesis of smaller molecules with antitumor activity.

Biocatalysis is able to provide useful and greener alternatives for the preparation of anticancer drugs, either by modification of natural products or by *de novo* synthesis of small molecules. In most of the examples, hydrolases or oxidoreductases (in any form: wild-type or genetically improved enzymes, whole cells, either free or conveniently immobilized) will be the biocatalysts employed, provided they are the most used enzymes both at the lab[48] and industrial scales,[49] although some examples using transaminases will be also commented. For sure, we do not pretend to be exhaustive because there are many cases described, especially at the lab scale; rather, we would show some attractive examples, trying to highlight the advantages of the biocatalyzed protocol.

12.2.1 Modification of Natural Products

12.2.1.1 Taxanes

Microtubules, filamentous intracellular structures made with a dynamic equilibrium with noncovalently bonded tubulin dimmers, are responsible for several aspects of cell morphology (cytoskeleton formation) and cell movements (they are part of the cilia and flagella). A very important structure generated from microtubules is the mitotic spindle, used by eukaryotic cells to segregate their chromosomes correctly during cell division and allow the transfer of chromosomes of the original cell to the daughter cells. Thus, microtubules represent the best known cancer target, especially for cytotoxic natural products, which disrupt the mentioned dynamic equilibrium, either by binding to tubulin and inhibiting polymerization or by binding to the microtubules and inhibiting depolymerization by stabilizing them.[47]

In this sense, paclitaxel (ATC Code L01CD01, DB Code DB01229, Taxol® BMS, Figure 12.1, (**1**)), a complex polycyclic diterpene belonging to the taxane family of compounds,[50] possesses microtubule-stabilizing activity and is the most important natural product in cancer chemotherapy and one of the most successful cancer drugs ever produced, being widely employed in the treatment of breast, ovarian, and lung carcinomas.

Taxol is one of the most popular anticancer drugs developed in the past 50 years. In 1999, global sales for Taxol produced by Bristol-Myers Squibb (BMS) reached $1.5 billion. Although this company reported a 24% decrease of Taxol sales between 2006 and 2007,[51] this reduction was caused by patent expiry and increased generic competition in Europe, and also by generic entry in Japan. Nevertheless, the total market for Taxol remains well above $1 billion *per* year (www.strategyr.com/Bulk Paclitaxel Market Report.asp) and continues to expand, with new supergeneric versions of Taxol, such as Cell Therapeutics' Xyotax (polyglutamate paclitaxel) and Abraxis Oncology's Abraxane (nanoparticle albumin-bound paclitaxel), increasing their sales, and according to Global

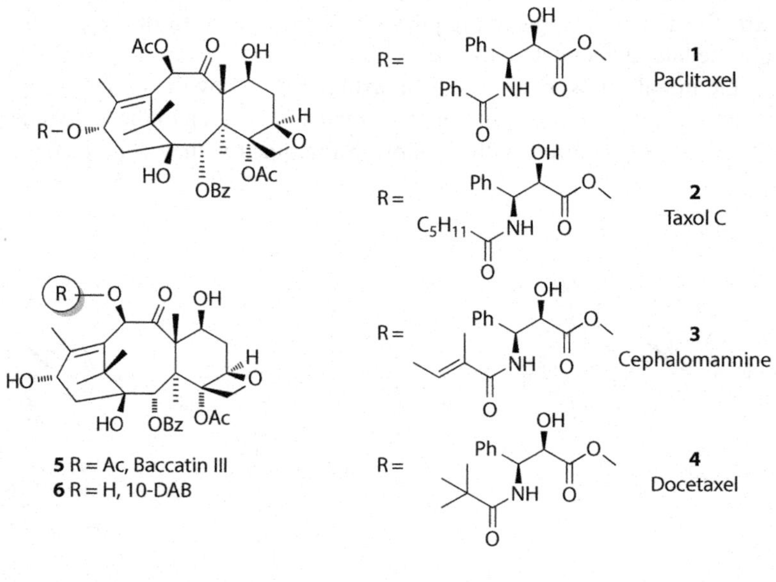

FIGURE 12.1 Structure of some taxanes.

Industry Analysts, Inc., the demand of paclitaxel reached 1040 kg per year in 2012 and is expected to be more than 3000 kg by 2017.[51]

The anticancer activity of paclitaxel was discovered in the 1960s during a large-scale plant-screening program sponsored by the National Cancer Institute (NCI) and was originally isolated from the bark of the yew *Taxus brevifolia*, a relatively uncommon tree occurring most abundantly in the old-growth forests of the Pacific Northwest of the United States, although it was also found in other *Taxus* species. The utility of paclitaxel to treat ovarian cancer was demonstrated in clinical trials in the 1980s, but the continuity of supply was not guaranteed; in fact, yew bark contains only about 0.0004% paclitaxel, and it was obtained in very low (0.07%) yield, so that about 4,000 trees were required to provide 360 g of Taxol for the early clinical trials, and 38,000 trees (which need up to 200 years to mature) were necessary to isolate 25 kg of Taxol to treat 12,000 cancer patients after approval of the use of Taxol for treating advanced ovarian cancer in 1992.

The major differences between the paclitaxel skeleton and the other 200 members of its family are the presence of a side chain at the C-13 position, esterified by an *N*-benzoyl-phenylisoserine group and an oxetanic ring attached to C4–C5 of the cyclohexane ring, being groups necessary for its biological activity.[47,52] The published total syntheses of paclitaxel[53–55] involve about 40 steps with an overall yield of approximately 2%. Some other protocols for its preparation, such as production by vegetable crops,[56] preparation from mushrooms,[57] or direct extraction from the leaves of the *Taxus* species,[52] were not viable to produce paclitaxel at the industrial scale. Thus, different semisynthetic approaches were considered, using baccatin III **5** or 10-deacetylbaccatin III (10-DAB), **6**, as starting material. These compounds, lacking the C-13 side chain and the C-10 acetyl group, were isolated from renewable resources (the twigs and needles from young European yews, *Taxus baccata*) in a higher yielding (1 g/kg), and could be transformed through a relatively simple semisynthetic route into paclitaxel, shown in Scheme 12.1, and also into its more soluble and potent analogue docetaxel (**ATC L01CD02, DB01248**, Taxotere®, Sanofi-Aventis, **4**), which was approved for advanced breast cancer in 1996. The market for docetaxel exceeded $3 billion in 2009 and $1.2 billion in 2010, but the Sanofi-Aventis patent expired in 2010 in Europe, 2012 in Japan, and 2013 in the United States,[51] so that generic versions (from Hospira) are becoming available (http://products.hospira.com/assets/pdfs/Docetaxel_091611.pdf).

In those schemes, naturally obtained products, **5** or **6**, were coupled either with an enantiopure *N*-benzoyl azetidinone (3*R*,4*S*)-**7**[58–60] or with the corresponding (2*R*,3*S*)-*N*-benzoyl-3-phenylisoserine ester ((2*R*,3*S*)-**8**) or derivatives.[61–65] Thus, as compounds **5** and **6** could be obtained in reasonable amounts by extractive methodologies not requiring tree damage, most of the efforts were centered in the creation of the lateral chain at C-13.

In this sense, the initial semisynthesis of paclitaxel published by the Potier group[66] required the esterification of cinnamic acid to baccatin, followed by hydroxylation of the C-13 side chain; it gave very low yields and was not commercially feasible. Subsequently, they demonstrated the synthesis

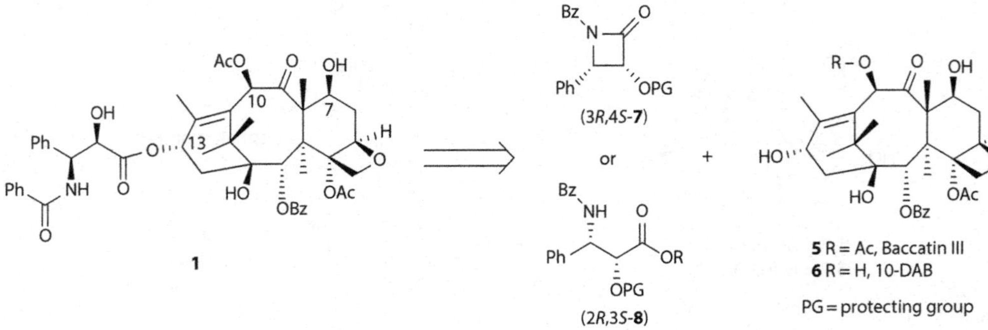

SCHEME 12.1 Semisynthetic strategy for paclitaxel preparation.

of paclitaxel from the esterification of baccatin III with the phenylisoserine side chain protected at the 20-OH position. This process gave only about 40% yield. Subsequent practical and efficient syntheses were developed by Holton[59] and others (e.g., by BMS) using different side chain intermediates and different reaction conditions, and providing much higher yields without any formation of epimers. β-Lactams can serve as excellent precursors of paclitaxel side chains.[60] Methods for the preparation of racemic β-lactam and subsequent conversion to paclitaxel side chains are outlined in Scheme 12.2. *N*-acetyl-β-lactams were prepared by Holton for the purpose of exploring their utility as direct acylating agents.[58]

In contrast, Palomo and coworkers[67] synthesized β-lactam intermediates for preparing suitably protected phenylisoserines. Racemic β-lactam **11** can be prepared in >75% yield by various routes as shown in Scheme 12.2.[60] The Staudinger reaction of acetyl glycolyl chloride with an appropriate imine, in the presence of triethylamine, gives both *cis*-β-lactams **9** in 90% yield, which were separated by a crystallization to furnish the desired (3*R*,4*S*)-**9**. The removal of *p*-methoxyphenyl (PMP) substituent with ceric ammonium nitrate then provides (3*R*,4*S*)-**10** in 92% yield. The acetyl group can be selectively removed by treatment with pyrrolidine, and the resulting alcohol can be protected with a variety of groups to yield (3*R*,4*S*)-**11** in an overall yield of 85%–90%. In an alternate route,

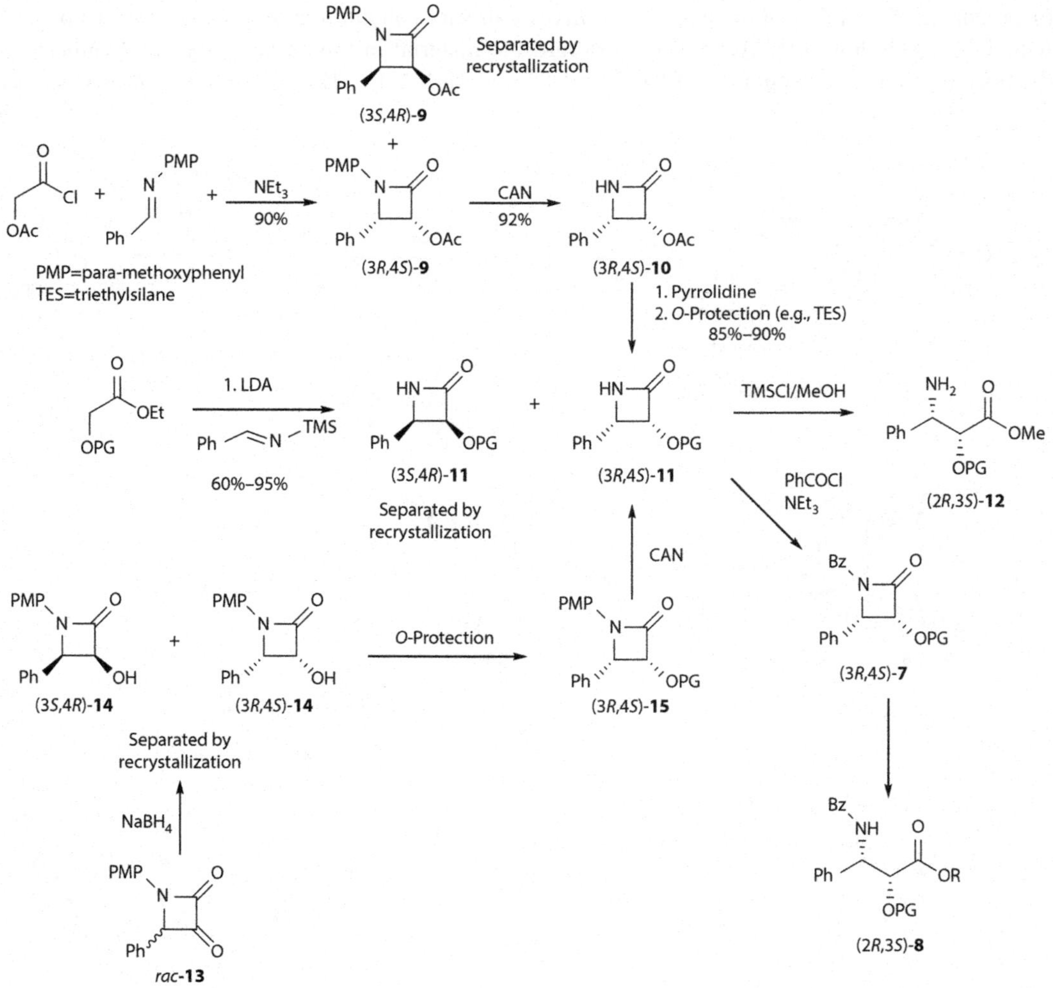

SCHEME 12.2 Chemical synthesis of the enantiopure compounds for building the C-13 lateral chain.

the enolate of glycolate ester, protected with a silyl group, an acetal, or one of several other nonacyl groups, reacts with a trimethylsilylaldimine to provide *rac*-**11** directly in 60%–95% yield depending upon the protecting group used in the reaction. Once again, both enantiomers of *cis*-**11** could be separated via crystallization. In another methodology, Palomo et al.[68] described the reduction of 3-keto azetidin-2-ones such as *rac*-**13** to provide exclusively *cis*-b-lactams **14**. After separation by crystallization, (3R,4S)-**14** can then be O-protected to furnish (3R,4S)-**15**, which upon dearylation gives (3R,4S)-**11**, which undergoes ring opening in the presence of trimethylsilyl chloride/methanol to give amino ester (3R,4S)-**12**, or acylation to give *N*-benzoyl-β-lactam (3R,4S)-**7** in high yield. Treatment of (3R,4S)-**7** with trimethylsilyl chloride/methanol then provides methyl ester (2R,3S)-**8** (R = Me); thus, through these purely chemical processes, these two essential compounds could be prepared, and a semisynthesis of Taxol could be developed,[54,60] as shown in Scheme 12.3.

In any case, those protocols were not developed at a high scale. Thus, to solve this problem, in 1991, NCI signed a Cooperative Research and Development Agreement with Bristol-Myers Squibb (BMS) in which BMS agreed to ensure supply of paclitaxel from yew bark while it developed a semisynthetic route to paclitaxel, which could be easily scalable, so it was mandatory to develop proficient procedures for the preparation of the starting reagents, **5** or **6**, and (3R,4S)-**7** or (2R,3S)-**8**, at the industrial scale.

The first step in this protocol is to develop an efficient methodology for obtaining high amounts of precursors **5** and **6**. In fact, extracts of *Taxus* cultivars contain a complex mixture of taxanes, with **1** in a very low proportion. The most valuable material in this mixture for semisynthesis is the taxane "nucleus" component of baccatin III **5** and 10-DAB **6**. Thus, conversion of mixtures of

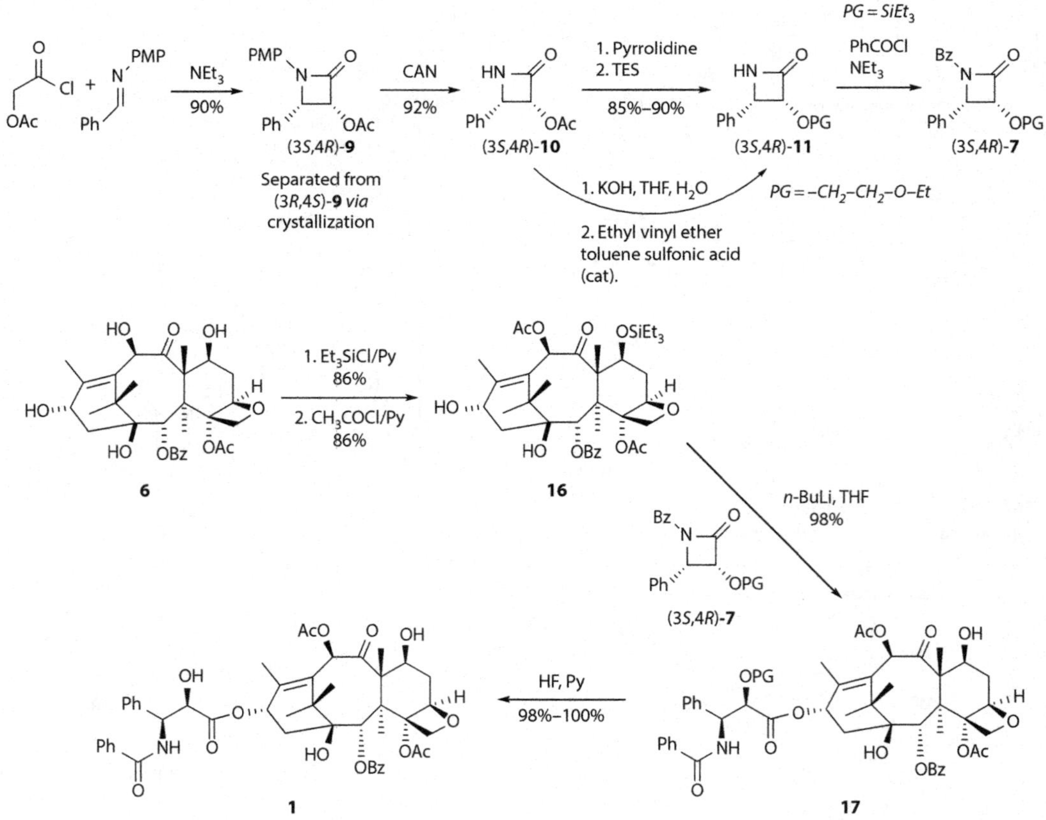

SCHEME 12.3 Semisynthetic strategy for chemical paclitaxel preparation.

taxanes obtained from cultivars to 10-DAB by cleavage of the C-10 acetate and the C-13 paclitaxel side chain was considered a very attractive approach to increase the concentration of **5** and **6** in yew extracts. Cleavage of paclitaxel at C-10 or C-13 was described, however cursing with a nondesired epimerization at C-7.[69]

To solve this problem, the *on water* enzymatic conversion of a complex mixture of taxanes to **6**, by using different approaches, as shown in Scheme 12.4, was described. Thus, using selective enrichment techniques, two strains of *Nocardioides* were isolated from soil samples that contained the novel enzymes C-13 taxolase and C-10 deacetylase.[70,71] The extracellular C-13 taxolase derived from the filtrate of the fermentation broth of *Nocardioides albus* SC 13911 catalyzed the cleavage of the C-13 side chain from paclitaxel and related taxanes such as Taxol C, cephalomannine, 7-β-xylosyltaxol, 7-β-xylosyl-10-deacetyltaxol, and 10-deacetyltaxol. On the other hand, the intracellular C-10 deacetylase derived from fermentation of *Nocardioides luteus* SC 13912 catalyzed the cleavage of the C-10 acetate from paclitaxel, related taxanes, and baccatin III to yield 10-DAB,[71] while the C-7 xylosidase derived from fermentation of *Moraxella* sp. catalyzed the cleavage of the C-7 xylosyl group from various taxanes.[72] Very recently, a novel strain screening protocol for C-7 xylosidase production has been described[73]; thus, the C-7 xylosidase produced was an extracellular inducible enzyme enabling the biotransformation to be carried out directly in microbial suspension cultures. The four strains were identified as *Streptomyces matensi*, *Arthrobacter nicotianae*, *Achromobacter piechaudii*, and *Pseudomonas plecoglossicida*. On the other hand, several chemicals

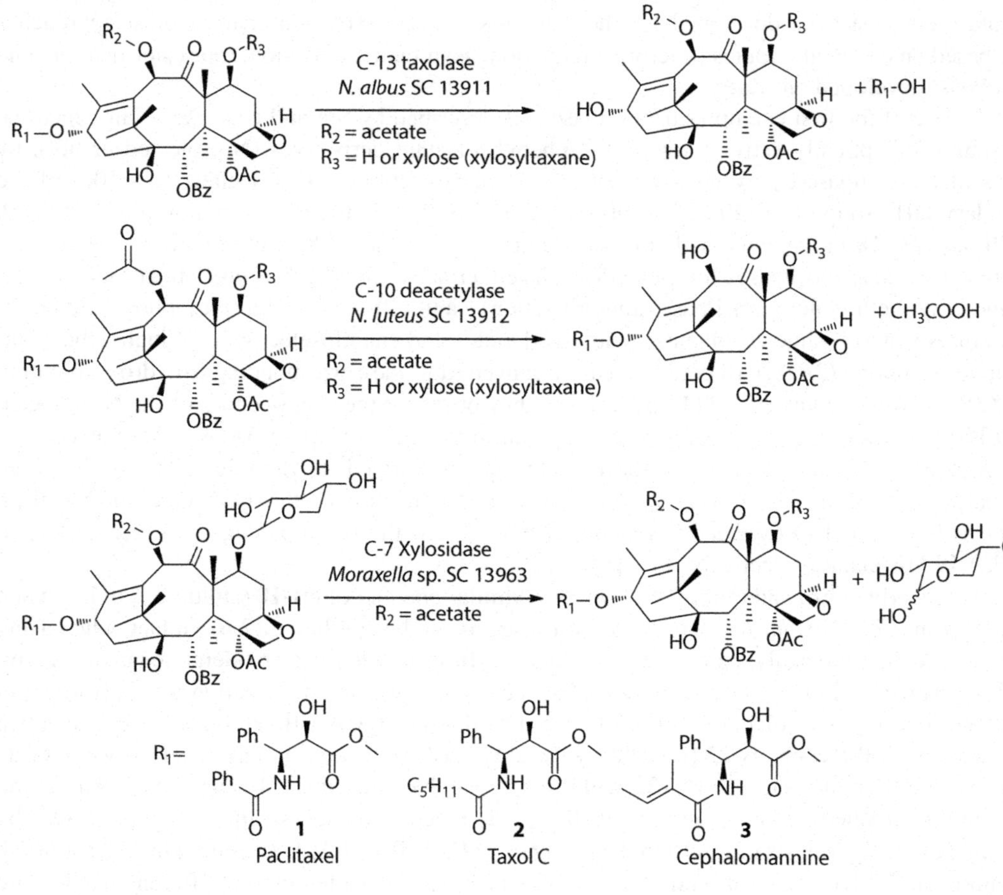

SCHEME 12.4 Enzymatic reaction on the taxane nucleus.

were confirmed as activating the enzyme activity, in which magnesium acetate improved the maximal substrate concentration from 0.1 (maximum obtained with *Moraxella* sp.) to 0.5 g L^{-1} at complete transformation in *S. matensi* suspension cultures. The nonmucous, extracellular activity and high substrate concentration characters of *S. matensi* facilitate both the upstream production of the enzyme and downstream extraction and purification of the enzyme and the product.

Once these processes were optimized at the lab scale, fermentation processes were developed for growth of *N. albus* SC 13911 and *N. luteus* SC 13912 to produce C-13 taxolase and C-10 deacetylase, respectively, in 5000 L batches, for the scaling of the biocatalytic process, which was demonstrated for the conversion of paclitaxel and related taxanes in extracts of *Taxus* cultivars to the single compound 10-DAB using both enzymes. In this way, the concentration of 10-DAB was increased by 5.5 to 24 by treatment with the two enzymes. The bioconversion process was also applied to extracts of the bark of *T. brevifolia* to give a 12-fold increase in 10-DAB concentration. Enhancement of the 10-DAB concentration in yew extracts was useful in increasing the amount and ease of purification of this key precursor for the paclitaxel semisynthetic process using renewable resources.

The other key precursor for paclitaxel semisynthesis, as shown in Scheme 12.1, is the lateral chain at C-13. For its preparation, different biocatalyzed strategies were described for improving the chemical synthesis of the enantiopure azetidinone (3R,4S)-**7** or the preparation of (2R,3S)-*N*-benzoyl-3-phenylisoserine ester ((2R,3S)-**8**), shown in Scheme 12.2. As we said before, these two molecules are the reagents needed for coupling to **5** or **6**, and the chirality of their stereogenic centers is mandatory to furnish the lateral chain. Although many different methods (e.g., asymmetric epoxidation routes, routes involving asymmetric dihydroxylation, strategies using a chiral auxiliary, inverse electron demand Diels–Alder reaction, enol–imine condensation, strategies using asymmetric catalysts) have been developed for the synthesis of the Taxol side chain,[74] most approaches were based on enzymatic routes (microbial reduction, asymmetric acylation, transesterification, and hydrolysis) developed on water.

Thus, one of the first examples using lipase was described by Sih and coworkers[75] in which the preparation of optically active 3-hydroxy-4-phenyl-β-lactam derivatives (N-protected or not), by means of either lipase-catalyzed asymmetric acylation reactions (vinyl acetate, E > 100) of the secondary OH group of *cis*-**20** or hydrolysis (H$_2$O, E > 100) of the ester function at C3 of *cis*-**9**, *cis*-**10**, and *cis*-**18** in aqueous medium, was described (Scheme 12.5). These authors tested some different *Pseudomonas* lipases for the hydrolysis, either native or adsorbed on Celite (a very classical and easy methodology for lipase immobilization)[76] and different media (pure phosphate buffer or mixtures water/organic cosolvents), with good yields and enantioselectivities. Finally, the enantiopure azetidinone (3R,4S)-**18** could be easily converted to the corresponding α-hydroxy-β-amino acid (2R,3S)-**21**. In addition to OAc hydrolysis, they observed the ring opening of the N-protected [C(O)Ph] β-lactam as a competitive reaction when lipase P-30 or lipase AK was used as catalyst with MeOH as the nucleophile and the reaction was performed in ^{t}BuOMe, a "usable" solvent for pharmaceutical industries, as reported.[24,77] The authors mentioned that two penicillinases, from *Escherichia coli* and *Enterobacter cloacae*, also catalyzed the ring-opening reactions leading to (2R,3S)-**19** less enantioselectively (E ≤ 16).

More recently, Forró and Fülöp[78,79] observed a similar opening of *cis*-**10** (amidase activity) when lipase B from *Candida antarctica* (CAL-B) was used as catalyst (Scheme 12.6). In fact, when using lipase PS-IM, the biocatalyzed cleavage of ester *cis*-**10** proceeded with moderate enantioselectivity (E = 24) when H$_2$O (0.5 equiv.) was used as the nucleophile and the reaction was performed in ^{i}Pr$_2$O, not a very green solvent,[24,77] at 50°C. On the other hand, CAL-B-catalyzed enantioselective ring cleavage of β-lactam *cis*-**23** (E > 200) with added H$_2$O (0.5 equiv.) in usable ^{t}BuOMe at 60°C furnished the β-amino acid (2R,3S)-**22** with high ee (>98%) and in good yield (45%). Therefore, these authors applied an enzymatic two-step cascade reaction for the synthesis of (2R,3S)-**22**, by carrying out the hydrolysis of *cis*-**10** in the presence of CAL-B with H$_2$O (1 equiv.) in ^{i}Pr$_2$O at 60°C, obtaining an 100% overall conversion of *cis*-**10**, which was converted in two different enantiopure products, the lactam (3S,4R)-**20** and amino acid (2R,3S)-**22** with ee ≥ 98%.

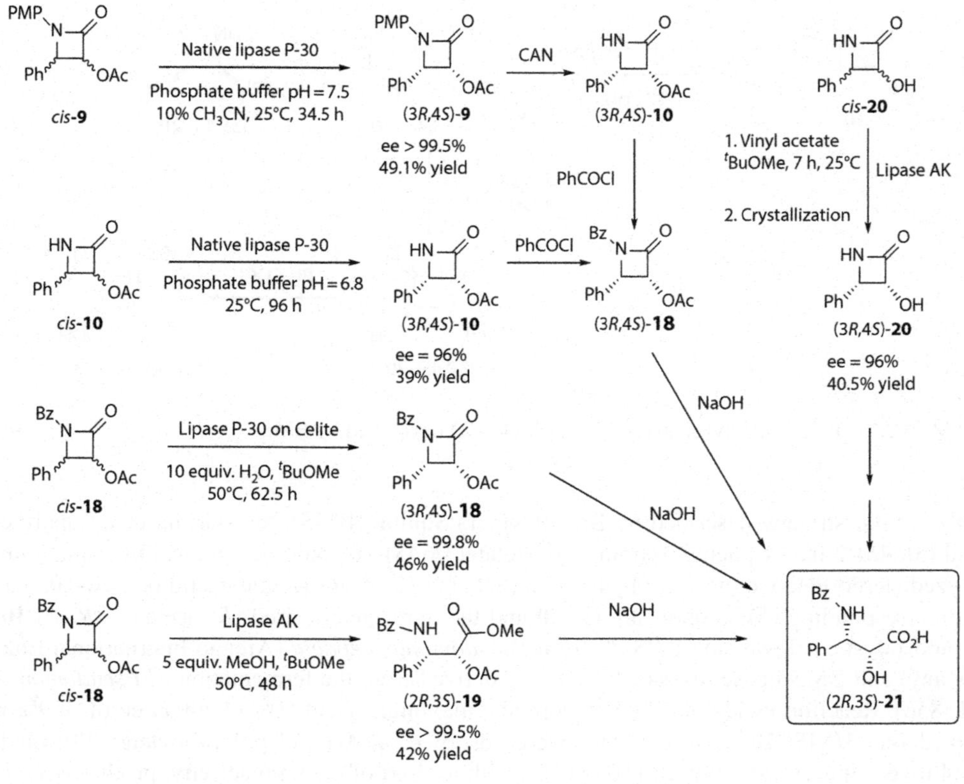

SCHEME 12.5 Lipase-catalyzed synthesis of azetidinones described by Sih et al.[75]

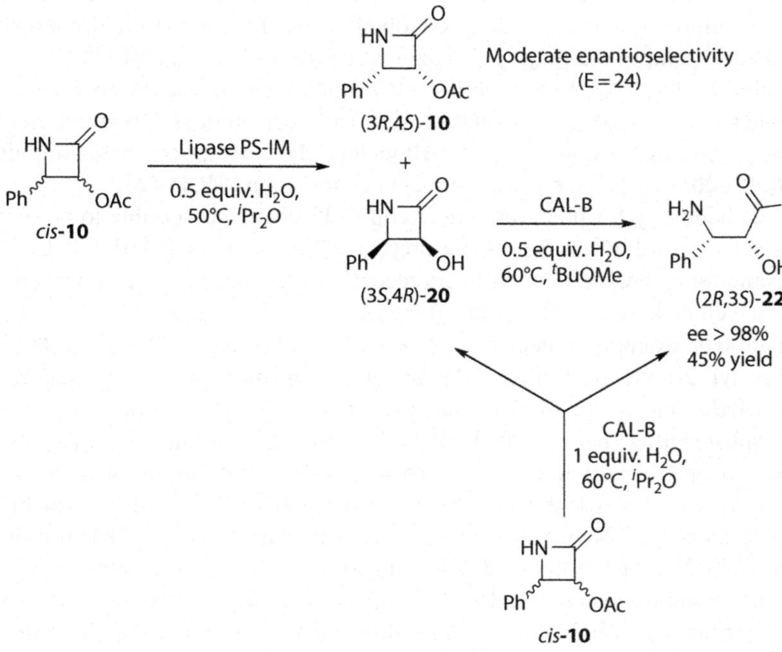

SCHEME 12.6 Lipase opening of *cis*-10 according to Forró and Fülöp.[78]

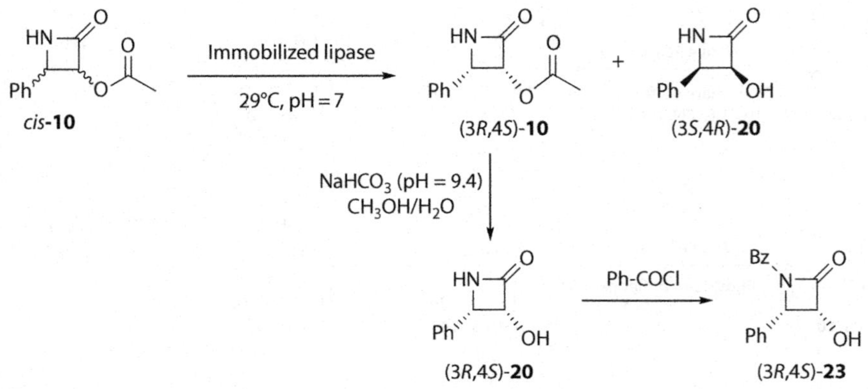

SCHEME 12.7 Lipase-catalyzed synthesis of (3R,4S)-**23** developed by BMS.

In a similar strategy described by Bristol-Myers Squibb (BMS), the starting enantiopure compound (3R,4S)-2-oxo-4-phenylazetidin-3-yl acetate(3R,4S)-**10** could be obtained by using a lipase-catalyzed stereoselective on water hydrolysis (Scheme 12.7) of racemic azetidione cis-**10**, leading to the corresponding (S)-alcohol (3R,4S)-**20** and the unreacted desired (R)-acetate (3R,4S)-**10**.[80,81] The biocatalyst used was lipase PS-30 from *Pseudomonas cepacia* (Amano International Enzyme Company) and BMS lipase (extracellular lipase derived from the fermentation of *Pseudomonas* sp. SC 13856). Reaction yields of >48% (theoretical maximum yield 50%) with an ee of >99% were obtained for (3R,4S)-**10**. Both enzymes were adsorbed on Accurel polypropylene (PP), and the immobilized lipases were reused (10 cycles) without loss of enzyme activity, productivity, or ee of the desired product. The immobilized enzyme can be reused for ten cycles without any loss in activity, productivity, or optical purity of the product. The rate of hydrolysis was determined to be 0.12 g L^{-1}·h^{-1} and remains constant over 10 cycles. At the end of the reaction the temperature is lowered to 5°C and the agitation from 200 to 50 rpm, so that (3R,4S)-**10** precipitates from the reaction mixture, while the immobilized enzyme floats on top of the reactor due to its hydrophobicity and can be easily separated by simply draining. This enzymatic process was susceptible to be scaled up to 250 L (2.5 kg substrate input) using immobilized BMS lipase and lipase PS-30, respectively. From each reaction batch, (3R,4S)-**10** was isolated in 45% yield (theoretical maximum yield 50%) and 99% ee. Finally, a basic hydrolysis of (3R,4S)-**10** yielded the corresponding enantiopure hydroxyl azetidinone (3R,4S)-**20**, which can be easily N-acylated to obtain (3R,4S)-**23**.

In fact, this methodology for preparing the lateral C-13 chain, susceptible to be scaled up, was patented and used by Bristol-Myers Squibb for preparing Taxol at the industrial scale,[82,83] and it has been the main route for its preparation until very recently, when latest plant cell fermentation (PCF) technology was developed, as we will comment later.

In another hydrolytic example, Taneja and coworkers[84] reported the enzymatic synthesis of several 4-substituted (phenyl, 2-furyl, or 2-thienyl) N-protected (4-methoxyphenyl) azetidin-2-one derivatives by lipase (*Arthrobacter* sp., ABL)-catalyzed enantioselective cleavage of the ester at C3 (E > 200) in phosphate buffer (pH 7.0), as shown in Scheme 12.8. The authors tested several lipases, finding that ABL showed the best performance. Thus, the kinetic resolutions of cis-**9**, cis-3-acetoxy-1-(4-methoxyphenyl)-4-(2-thienyl)-2-azetidinone (cis-**26**), and cis-3-acetoxy-1-(4-methoxyphenyl)-4-(2-furyl)-2-azetidinone (cis-**24**) were achieved using ABL with ee ≈ 99%. The use of cosolvents, such as "usable" CH$_3$CN or "undesirable" DMF and DMSO,[24,77] not very green in any case, dramatically improved the enantioselectivity and reduced reaction timings. However, cis-3-acetoxy-1-(4-methoxyphenyl)-4-t-butyl-2-azetidinone (cis-**28**), side chain intermediate for the semisynthesis of docetaxel (Taxotere®; Figure 12.1), could not be directly hydrolyzed by any one of the lipases used, but could be effectively resolved (ee 99%) after its conversion to cis-3-acetoxy-4-t-butyl-2-azetidinone

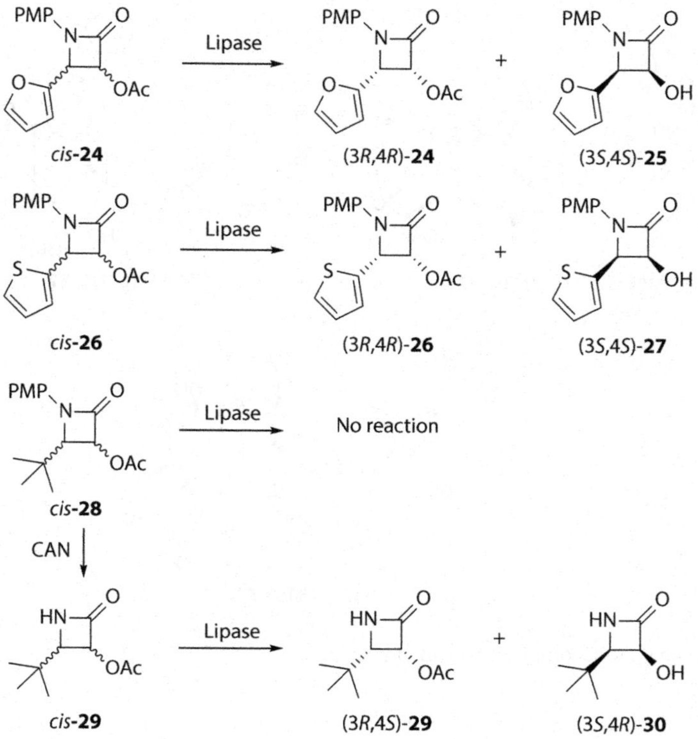

Chiral Building Blocks for Drugs Synthesis via Biotransformations

SCHEME 12.8 Lipase-catalyzed hydrolysis of different azetidinones according to Taneja et al.[84]

(cis-**29**) upon N-dearylation using ceric ammonium nitrate (CAN). There salting cis-3-acetoxy-4-t-butyl-2-azetidinone cis-**29** was now easily amenable to ABL, so it could be hydrolyzed stereoselectively to produce enantiomers (3R,4S)-**29** and (3S,4R)-**30** in 99% ee after 4 h.

As we commented before, docetaxel (Taxotere®, Sanofi-Aventis, **4**), a more water-soluble and potent analogue of paclitaxel, was approved for advanced breast cancer in 1996. Since then, many studies on the structure modifications of paclitaxel have been carried out to design newer analogues, which display improved bioactivity and low toxicity.[52] The phenyl group at the 30-position on the C-13 side chain has been an important and easy target for the synthesis of new compounds. A number of such new generation taxanoids, starting from key building blocks shown in Scheme 12.8, have been developed, such as MAC-321 (TL-00139) **31**,[85] MST-997 (TL-909) **32**,[86] and BMS-275183 **33**[87] (Figure 12.2). This last compound is very promising, and these days it is in phase I studies.[88]

In another synthetic protocol, the target molecules (3R,4S)-**7** (N-benzoylazetidinone) or (2R,3S)-**8** (N-benzoyl-3-phenylisoserine) can be prepared using a biocatalytic reduction, starting from the corresponding 3-oxo-β-lactams rac-**34** or β-oxoesters rac-**35** (Scheme 12.9). When using this kind of biocatalysts, in all the presented examples, water is always the reaction medium, unless specifically mentioned.

These bioreductions require the use of an efficient biocatalyst that can be either isolated enzymes or whole cells.[89,90] The choice between one type of biocatalysts versus the other is sometimes controversial, due to the fact that both of them present different advantages and drawbacks,[91–94] mainly based on the productivity and stereoselectivity of the reduction (generally higher with isolated enzymes[94]), or the absence of a cofactor regeneration system (making whole cells a cheaper option[95]). Overall, whole cells are most frequently used as biocatalysts, while isolated enzymes represent only a quarter of the processes.[10,49]

The first example of a biocatalytic reduction of analogues of rac-**34** with whole baker's yeast cells was described by Kayser et al.,[96] by using a commercially available laboratory strain INVSc1,

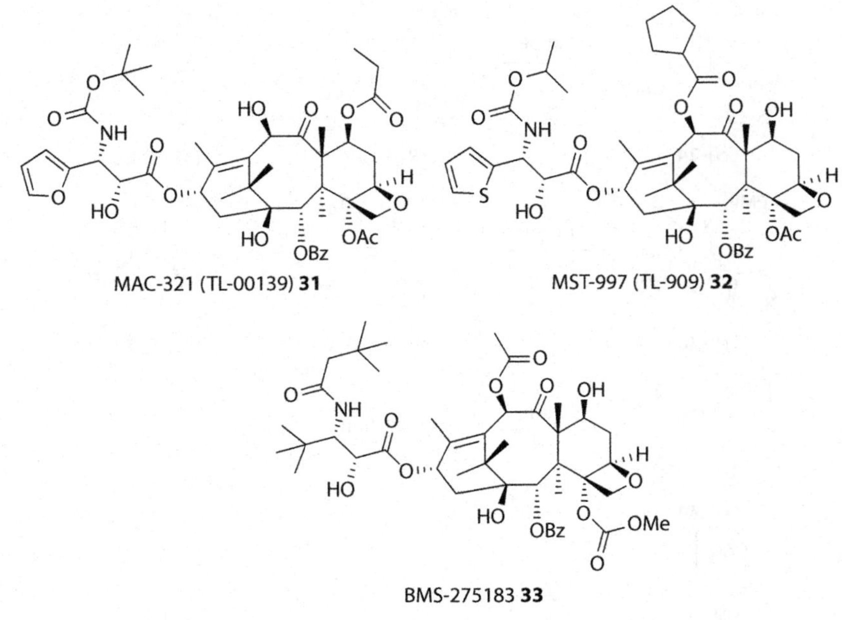

FIGURE 12.2 Some pharmacology-interesting taxanes.

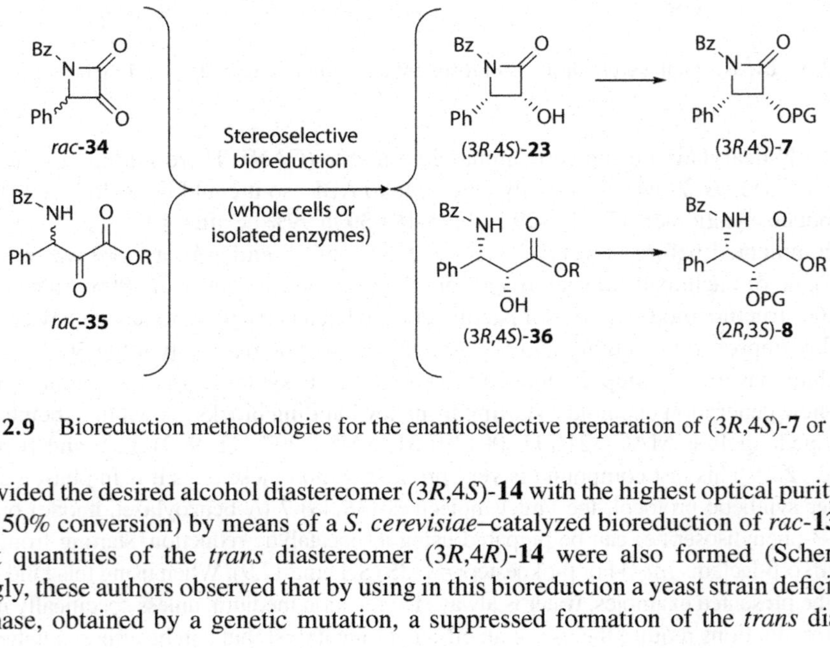

SCHEME 12.9 Bioreduction methodologies for the enantioselective preparation of (3R,4S)-**7** or (2R,3S)-**8**.

which provided the desired alcohol diastereomer (3R,4S)-**14** with the highest optical purity (up to ca. 80% ee at 50% conversion) by means of a *S. cerevisiae*-catalyzed bioreduction of *rac*-**13**, although significant quantities of the *trans* diastereomer (3R,4R)-**14** were also formed (Scheme 12.10). Interestingly, these authors observed that by using in this bioreduction a yeast strain deficient in fatty acid synthase, obtained by a genetic mutation, a suppressed formation of the *trans* diastereomer

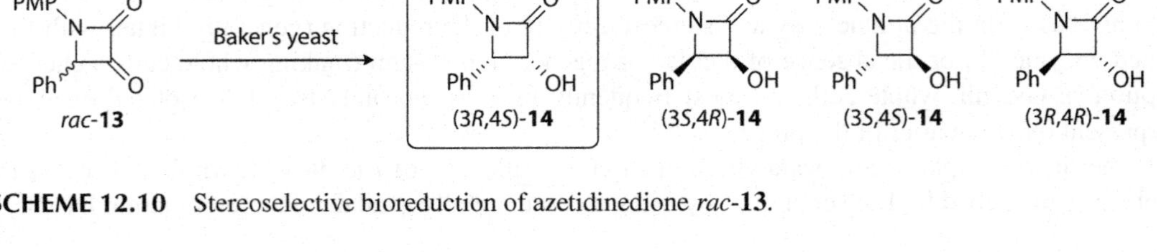

SCHEME 12.10 Stereoselective bioreduction of azetidinedione *rac*-**13**.

Chiral Building Blocks for Drugs Synthesis *via* Biotransformations

was observed, but also with a concomitant diminishing of the optical purity of the desired alcohol (3*R*,4*S*)-**14**. Thus, they concluded that both *cis* diastereomers ((3*R*,4*S*)-**14** and (3*S*,4*R*)-**14**) were products of a single reductase enzyme inside the yeast, with relatively low enantioselectivity that favors (*S*)-**13**. In the presence of fatty acid synthase, however, the apparent enantioselectivity of this enzyme was increased because fatty acid synthase competes effectively for the undesired (*R*)-**13** and converts it to a different diastereomer that can be removed by simple chromatographic separation. They also suggested that a further genetic manipulation of yeast reductases might lead to a whole-cell catalyst with greater enantio- and diastereoselectivities for reductions of such compounds.

In fact, once the yeast genome sequence was available, 49 putative reductases were found,[97] and 19 individual yeast reductases derived from *S. cerevisiae* genome were cloned and expressed in *E. coli* as glutathione (*S*)-transferase(GST) fusion proteins, and their ability as enantioselective reducing agents for different β-keto esters was tested.[98] One of those enzymes, reductase Ara1p, was found to be efficient in the reduction of *rac*-**13**,[99] so that the recombinant *E. coli* strain BL21(DE3)/Ara1p (*E. coli*/Ara1p), also prepared by the same group,[100–102] was tested in the bioreduction of this same substrate (Scheme 12.11). Thus, compared to the commercial catalyst (conversion >98%, 1.6:1 ratio *cis*:*trans*, ee 55% for (3*R*,4*S*-**14**), ee > 99% (3*R*,4*R*)-**14**), when using the recombinant enzyme under shake-flask conditions, a higher stereoselectivity for the desired *cis* diastereomer was found (conversion 81%, 1.4:1 ratio *cis*:*trans*, ee > 99% for (3*R*,4*S*-**14**), ee > 99% (3*R*,4*R*)-**14**), and the process was even more attractive if the recombinant cells were used under fermentation conditions (constant oxygen and glucose supply, control of pH at 7.0), because in this case consumption of *rac*-**13** was complete, and only traces of the *trans* isomer were found, therefore leading to (3*R*,4*S*-**14**) at ee values higher than 98%.

Furthermore, some other racemic N-alkylated or arylated azetidinones were tested as substrates for these yeast reductases.[103] Finally, employing racemic 3-oxo-4-phenyl-β-lactam *rac*-**13** as a probe (Scheme 12.12), 19 individual yeast reductases derived from *S. cerevisiae* genome, cloned and expressed in *E. coli* as glutathione (*S*)-transferase (GST) fusion proteins, were screened for their efficiency and enantioselectivity by Kayser et al.[104] These authors selected four highly selective enzymes, all belonging to the aldoketoreductase (AKR) superfamily, to produce *cis* alcohols with high ee. Two of them (Ybr149w and Ycr107w) were *Re*-face selective and two (Yjr096w and Ydl124w) were *Si*-face selective, the latter ones leading to the desired intermediate (3*R*,4*S*)-**14** at

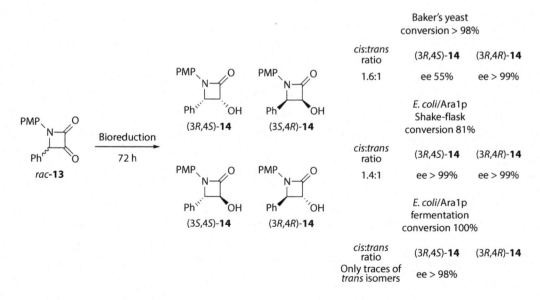

SCHEME 12.11 Bioreduction of *rac*-**13**.

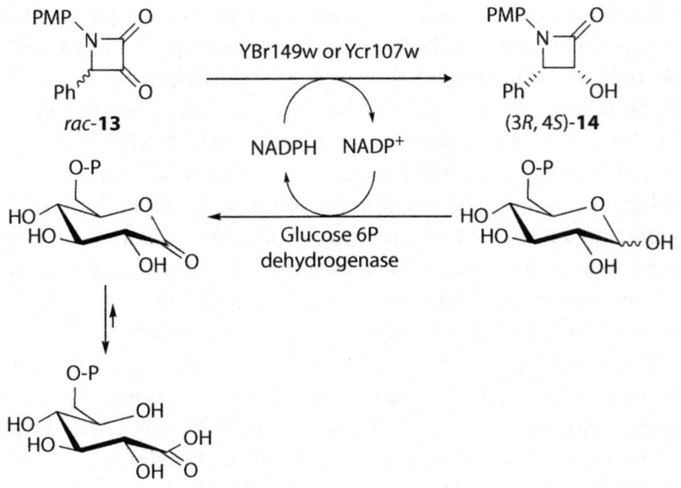

SCHEME 12.12 Stereoselective enzymatic reduction of azetidinedione *rac*-**13**.

high conversion and enantiopurity. In all cases, the bioreduction was carried out at the lab scale (pH = 7–8), and the cofactor (NADPH) was regenerated by a glucose-6-phosphate/glucose-6-phosphate dehydrogenase couple. Finally, these authors also postulated a rational explanation of the *Re* versus *Si* selectivity according to different conformations of a portion of 22 residues of the primary sequence (loop A), quite close to the active site.

If we now consider the other starting substrate for the C-13 lateral chain, (2*R*,3*S*)-*N*-benzoyl-3-phenyl isoserine ester (2*R*,3*S*)-**8**, this compound has also been prepared by means of an enantioselective microbial reduction. Thus, Kearns and Kayser[105] reported in 1994 the reduction of both enantiomers of methyl 3-amino-2-oxo-3-phenylpropanoate **37** by using baker's yeast, leading to moderate yields but good diastereoselectivity (Scheme 12.13). Some years later, Patel et al. described the reduction of racemic ethyl 2-oxo-3-(*N*-benzoylamino)-3-phenyl propionate (*rac*-**29**), using 14 different microorganisms,[106] showing that *Hansenula polymorpha* SC 13865 and *Hansenula fabianii* SC 13894 were the most interesting ones, leading to reaction

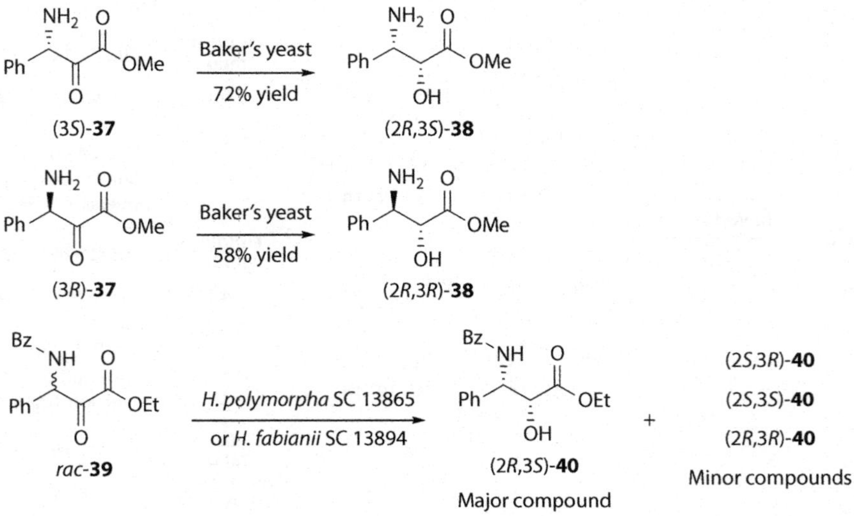

SCHEME 12.13 Whole-cell-catalyzed stereoselective reduction of *rac*-**13**.

Chiral Building Blocks for Drugs Synthesis *via* Biotransformations 361

yields higher than 80%, and ee values higher than 98% for the desired *syn*-(2*R*,3*S*)-**40**. These same authors reported[82,106] the preparative-scale bioreduction of *rac*-**39** by using cell suspensions of both microorganisms in independent experiments. In both batches, a reaction yield of >80% and an ee of >94% were obtained for (2*R*,3*S*)-**40**. A 20% yield of the undesired *anti*-diastereomers was obtained with *H. polymorpha* SC 13865, compared with a 10% yield with *H. fabianii* SC 13894. A 99% ee was obtained with *H. polymorpha* SC 13865 compared with a 94% ee with *H. fabianii* SC 13894. In a single-stage bioreduction process, cells of *H. fabianii* were grown in a 15 L fermenter for 48 h; then the bioreduction process was initiated by the addition of 30 g of substrate and 250 g of glucose and continued for 72 h. A reaction yield of 88% with an ee of 95% was obtained for (2*R*,3*S*)-**40**.

In another approach starting from an open-chain precursor, the enantiocomplementary reduction of an a-chloro-β-oxo ester *rac*-**41** (Scheme 12.14) has been described on a small scale,[107] using two different *Saccharomyces cerevisiae* reductases (expressed as fusion proteins with glutathione *S*-transferase[98]), to yield exclusively those diastereomer precursors for both enantiomers of the *N*-benzoyl phenylisoserine ethyl ester (2*R*,3*S*)-**40** and (2*S*,3*R*)-**40**. Thus, after base-mediated ring closure of the chlorohydrin enantiomers **42**, the corresponding enantiopure epoxides **43** were converted directly to the oxazoline form of the target molecules **44** using a Ritter reaction with benzonitrile, followed by an acidic hydrolysis leading to the ethyl ester form of the Taxol side chain enantiomers.

Very recently, Rimoldi et al.[108] have described the use of nonconventional yeasts, possessing both cell-associated esterase and carbonyl reductase activities, for the two-step, one-pot preparation of the synthon (2*R*,3*S*)-**40** with high yields, furnishing higher yields and a shorter route than previously reported chemoenzymatic protocols. In this approach, shown in Scheme 12.15, 13 different whole cells were tested in the bioconversion of 1-benzamido-3-ethoxy-3-oxo-1-phenylprop-1-en-2-yl ethyl oxalate **45**. As both enzymatic activities are present in the cells, the biotransformation could proceed *via* a previous bioreduction (enoylreductase activity) and a succeeding hydrolysis of the intermediate **46** to render **40**, or alternatively by another pathway starting with a hydrolysis of **45** leading to the oxoester **47**, and a subsequent reduction of this later molecule (ketoreductase activity).

The *syn*- and *anti*-stereoisomers were obtained depending on the yeast used; high diastereoselectivity was sometimes observed. In particular, *Pachysolen tannophilus* CBS 4044 and *Torulopsis*

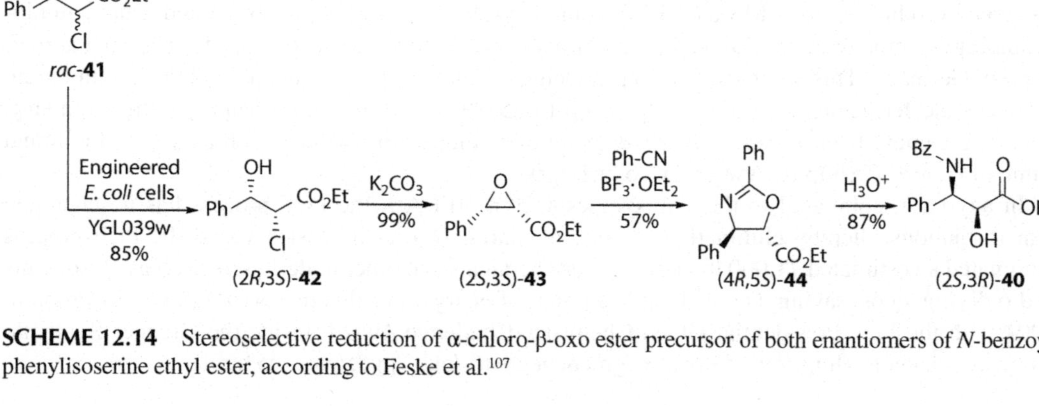

SCHEME 12.14 Stereoselective reduction of α-chloro-β-oxo ester precursor of both enantiomers of *N*-benzoyl phenylisoserine ethyl ester, according to Feske et al.[107]

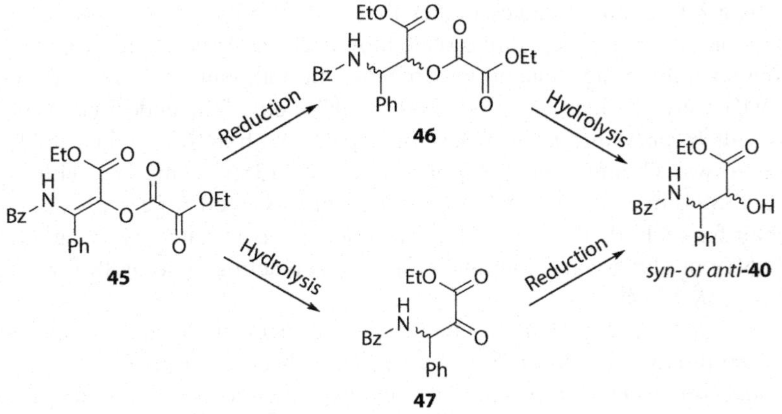

SCHEME 12.15 Bioconversion of 1-benzamido-3-ethoxy-3-oxo-1-phenylprop-1-en-2-yl ethyl oxalate **45** to different stereoisomers of *N*-benzoyl phenylisoserine ethyl ester **40**.

molischiana CBS 837 gave only *anti*-**40** with modest enantioselectivities, while *Lindnera fabianii* gave only *syn*-**40** with notable enantioselectivity. Other yeasts showed poor diastereoselectivity in the formation of *syn*-**40**, but high enantioselectivity. In most cases, the pharmacologically relevant stereoisomer (2*R*,3*S*)-*syn*-**40** was produced with an ee ranging from 59% to 99%, in accordance with Prelog's rule.[109] The best performing yeast in terms of diastereoselectivity was *L. fabianii* (de, diastereomeric excess, 99% *syn*, ee 82% (2*R*,3*S*)-*syn*-**40**), which had been previously used for the stereoselective reduction of **47** into **40**.

Although this semisynthetic process avoided the destruction of many trees, it is still complex and requires 11 chemical transformations and 7 isolations. The semisynthetic process also presents environmental concerns, requiring 13 solvents along with 13 organic reagents and other materials. Thus, although biocatalysis could improve the sustainability of the overall process, as commented up to this point, BMS, the main producers of paclitaxel, decided to develop a more sustainable process using the latest PCF technology. In the cell fermentation stage of the process, calluses (an unorganized, proliferating mass of undifferentiated cells) of a specific *Taxus* cell line are propagated in a wholly aqueous medium in large fermentation tanks under controlled conditions at ambient temperature and pressure.[110] The feedstock for the cell growth consists of renewable nutrients—sugars, amino acids, vitamins, and trace elements—so that paclitaxel can be directly extracted from plant cell cultures and subsequently purified by chromatography and isolated by crystallization.

By replacing leaves and twigs with plant cell cultures, BMS improves the sustainability of the paclitaxel supply, allows year-round harvest, and eliminates solid biomass waste. Currently, all paclitaxel production for BMS uses PCF technology developed by the German and Canadian biotechnology company Phyton Biotech, Inc. (http://www.phytonbiotech.com/) and carried out at their plant in Germany. This company is also producing docetaxel by the same methodology, employing a large-scale fermenter with a capacity of up to 75,000 L. On the other hand, another company, Samyang Genex, from Korea, uses *Taxus* plant cell cultures to produce paclitaxel with the brand name of Genexol® (http://www.genex.co.kr/Eng/).

In any case, compared to the semisynthesis from 10-DAB, the PCF process has no chemical transformations, thereby eliminating six intermediates. During its first 5 years, the PCF process eliminated an estimated 71,000 lb of hazardous chemicals and other materials, as well as 10 solvents and 6 drying steps, saving a considerable amount of energy. For this reason, BMS was awarded in 2004 with the U.S. Presidential Green Chemistry Challenge Award, inside the Greener Synthetic Pathways category (http://www.epa.gov/greenchemistry/pubs/pgcc/presgcc.html).

Chiral Building Blocks for Drugs Synthesis *via* Biotransformations

12.2.1.2 Epothilones

The clinical success of paclitaxel has stimulated research into compounds with similar modes of activity in an effort to emulate its antineoplastic efficacy while minimizing its less desirable aspects, which include no water solubility, difficult synthesis, and emerging resistance.

The epothilones are a novel class of natural product cytotoxic compounds derived from the fermentation of the myxobacterium *Sorangium cellulosum* that are nontaxane microtubule-stabilizing compounds that trigger apoptosis.[111]

Epothilones A and B (**DB03010**, also known as patupilone) (Figure 12.3, **48** and **49**) have demonstrated broad-spectrum antitumor activity *in vitro*, including tumors with paclitaxel resistance,[112] although their *in vivo* activities are reduced due to their limited metabolic stabilities and unfavorable pharmacokinetics.[113] The role of epothilones as a potential paclitaxel successor has initiated interest in their synthesis.[112] Correspondingly, epothilone analogues have been prepared just to optimize the water solubility, *in vivo* metabolic stability, and antitumor efficacy of this class of antineoplastic agents. In this sense, ixabepilone (**ATC L01DC04, DB04845**, Ixempra®, BMS-247550, **50**), a semisynthetic aza-analogue of epothilone B,[114,115] is the first and only representative of the epothilone family that has been approved (in October 2007) to date by the U.S. Food and Drug Administration (FDA).

Thus, a fermentative protocol has been developed for the production of epothilone B, the most potent of naturally occurring epothilones, by a continuous feed of sodium propionate during the fermentation of *S. cellulosum* cells (Scheme 12.16).[7] The inclusion of XAD-16 resin during fermentation to adsorb epothilone B and to carry out volume reduction made the recovery of product

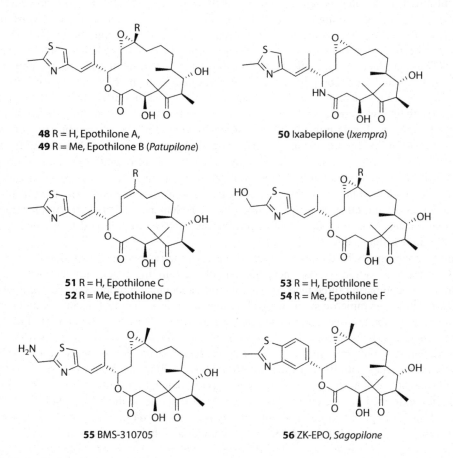

FIGURE 12.3 Some relevant epothilones.

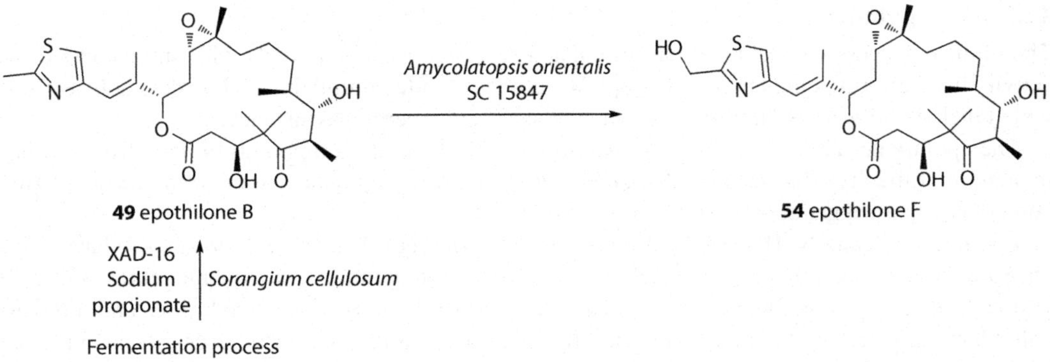

SCHEME 12.16 Preparation of epothilone F from epothilone B.

very simple. In addition, a high level of free epothilone B, which is inhibitory to the growth of the producing culture, was avoided by supplying XAD-16 resin during the fermentation process, so that epothilone B could be obtained at a semi-industrial scale. In a further step, a microbial hydroxylation on water was developed for conversion of epothilone B to epothilone F by *Amycolatopsis orientalis* SC 15847 (Scheme 12.16). A bioconversion yield of 37%–47% was obtained when the process was scaled up to 100–250 L with an intermittent feed of epothilone B6. The reducing power NAD(P)H required for hydroxylation was generated internally during growth by the carbon source glucose that was used. Recently, the epothilone B hydroxylase along with the ferredoxin gene has been cloned and expressed in *Streptomyces rimosus* from *A. orientalis* SC 15847 and variants thereof. This cloned enzyme has been used in the hydroxylation of epothilone B to epothilone F to obtain even higher yields (80%) of product.[116–118] Finally, epothilone F is used as starting material for the preparation of BMS-310705, an antitumor compound that has been evaluated in phase I clinical trials.[119]

In contrast with taxanes, the epothilones can be obtained and modified efficiently by total synthesis due to their less complex structure, so that different protocols describing different synthetic pathways have been already published.[120–125] Some enzymatic protocols for obtaining enantiopure chiral synthons for epothilone preparation have already been described, such as those described in Scheme 12.17. As can be seen, by lipase-catalyzed acylations in organic solvents, different compounds useful for the preparation of 2-methyl thiazole-containing structures (lateral chain of many epothilones) could be obtained. In fact, by using lipase from *Pseudomonas* AK, the kinetic resolution of alcohols **57** and **60** could be efficiently obtained (Scheme 12.17). At room temperature, the reaction was faster for alcohol **60**,[126] but increasing the temperature up to 40°C, good results could be obtained in less than a day.[127]

In another strategy, Scheid et al.[128] reported the asymmetric synthesis and kinetic resolution of a series of acyloins (a-hydroxy ketones) **63**, suitable as building blocks for the northern half of epothilones, by comparing three methods to obtain nonracemic compounds at the eventual epothilone C15-position: asymmetric synthesis with Evans' auxiliary, chemical resolution and enzymatic resolution, finding the last one the most suitable at least for the lab scale. Out of a set of nine lipases and esterases tested, lipases from *Burkholderia cepacia*, *Pseudomonas* sp., lipase B from *Candida antarctica*, and recombinant esterases from *Streptomyces diastatochromogenes* exhibited the highest enantioselectivities with E values ranging from 60 to >200 in the enantioselective hydrolysis of *rac*-**63** in an aqueous biphasic system composed of phosphate buffer/toluene. Very interestingly, pig liver esterase exhibited inverse enantiopreference. After that, the methodology using CAL-B lipase was scaled up to 0.5 g, obtaining yields of 179 mg (42%) for *S*-**64** product (>99% ee) and 186 mg (48%) for the remaining substrate (*R*-**63**, 46% ee), under a global conversion of 31% (E > 200).

On the other hand, sagopilone (ZK-EPO, Bayer, **56**) is a promising molecule,[129,130] actually in phase II trials, showing efficacy in first- and second-line settings of ovarian cancer, prostate

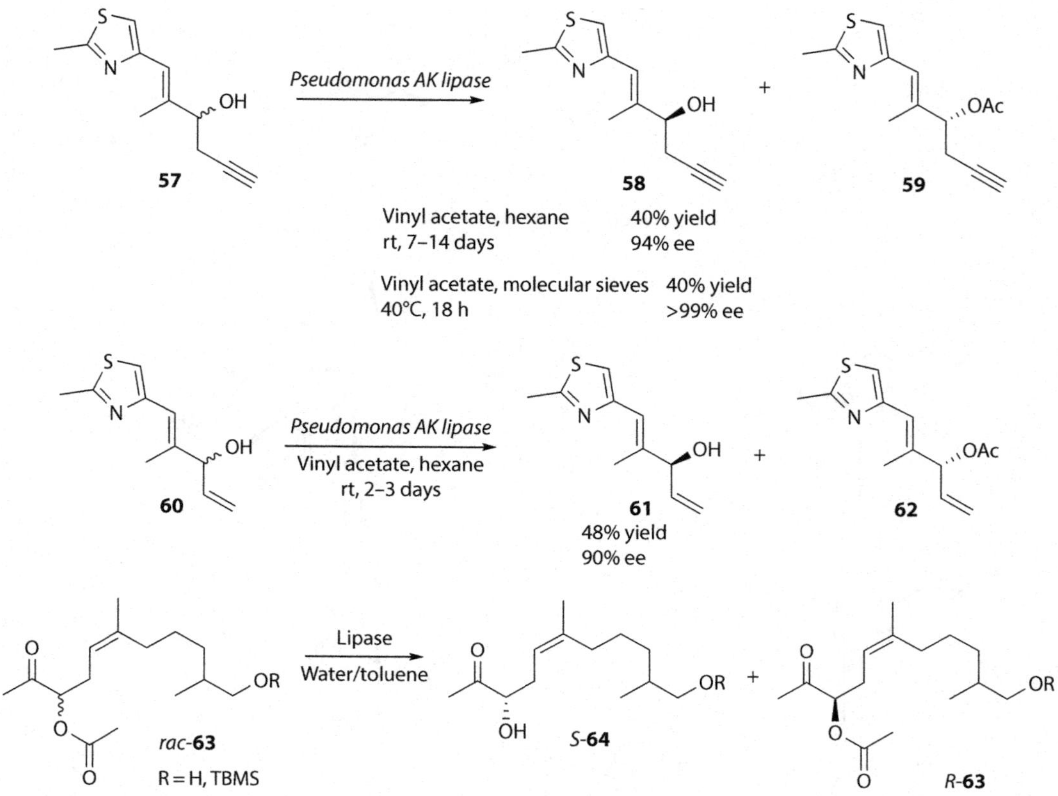

SCHEME 12.17 Kinetic resolutions of chiral building blocks for epothilone synthesis.

cancer, melanoma, and glioblastoma.[131] In this sense, the original research synthesis of **56**,[132] initially designed for the flexible introduction of structural modifications at the 16-membered lactone scaffold, was already optimized to produce about 50 g of material required for preclinical investigations, putting special emphasis in obtaining high enantioselectivity in each step. Nevertheless, due to the length and complexity of the synthesis, for scaling-up purposes, the production should start with multi-100-kg material right at the beginning to produce sagopilone in a multi-kg scale. Thus, very recently, a new synthetic methodology (Scheme 12.18) has been described,[131] carefully optimizing reaction sequences and parameters so that intermediates can be carried over several steps without isolation or purification, because in the original synthesis most of the intermediates were oils, which had to be purified at nearly each step in the research synthesis by chromatography (very efficient on a laboratory scale, but highly expensive and time consuming in a production process).

In this scheme, three main fragments, **65**, **68**, and **70**, were synthesized and connected to furnish the whole structure of sagopilone. In the synthesis of the fragment C1–C2–C5–C8–C9–C10–C11–C12, starting from the very cheap 4-chloro-3-nitrobenzoic acid **59**, and by different chemical steps, the β-keto ester **66** was obtained in 90%–95% overall yield as crystalline material. Using this compound, and after checking different chemical and biocatalyzed reduction, the desired chirality of C15 was obtained by means of a microbiological reduction using the yeast strain *Pichia wickerhamii* in water–DMF mixtures, which produced the alcohol **73** in 83% yield in crystalline form with an enantiomeric excess higher than 99%[133] (Scheme 12.19). This enantiopure alcohol **73** was subsequently transformed to **65** by different chemical steps.

Regarding the synthesis of the fragment C1–C6, the correspondent ketone **70** could be derived either by starting from two compounds commercially available, nitrile **72** or *R*(−)-pantolactone **71**.

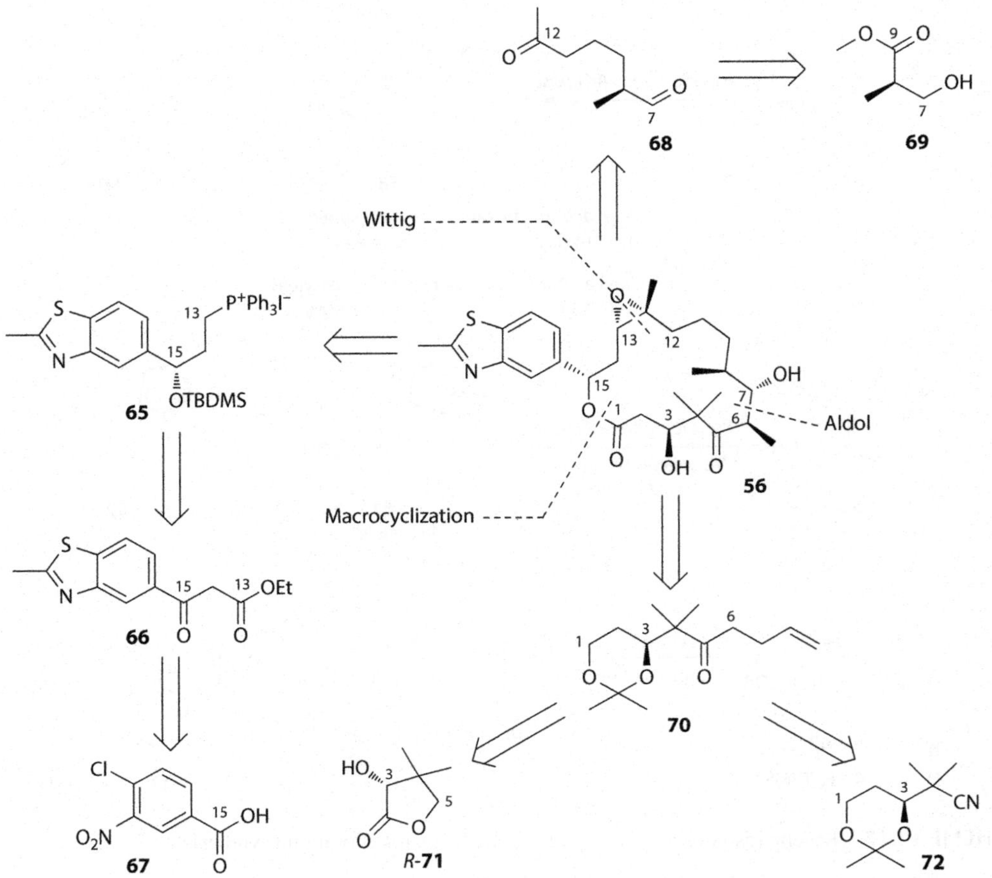

SCHEME 12.18 Retrosynthetic scheme for the preparation of sagopilone **56**, as described by Klar and Platzek.[131]

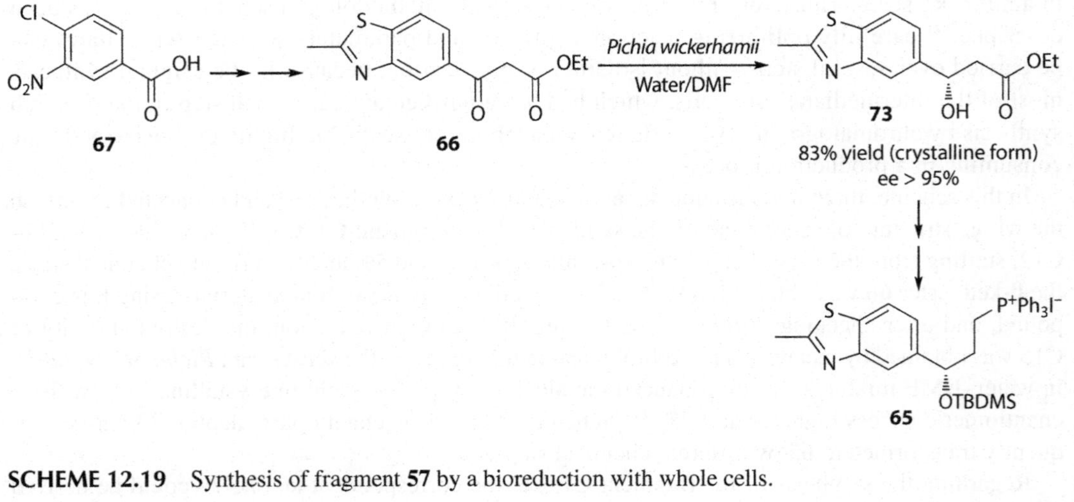

SCHEME 12.19 Synthesis of fragment **57** by a bioreduction with whole cells.

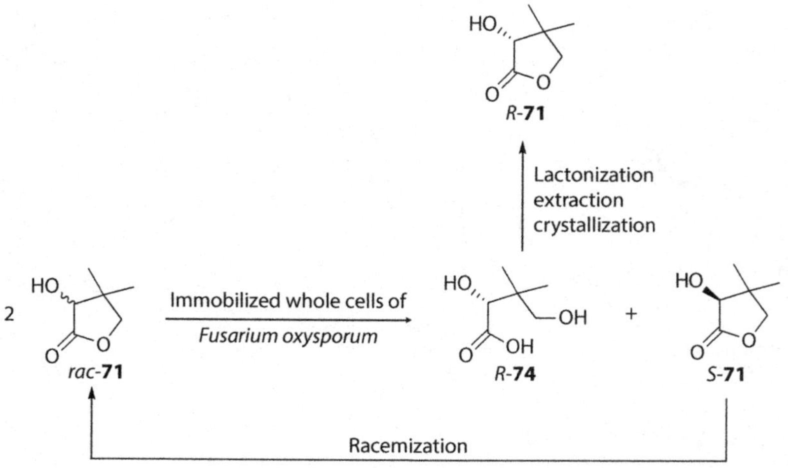

SCHEME 12.20 Biocatalyzed kinetic resolution of pantolactone **71**.

The development synthesis was performed starting from **72**, because of its easier scalability, but it is worth mentioning that the industrial preparation of R(−)-pantolactone **71** can also be carried out by a biocatalyzed protocol. In fact, Daiichi Fine Chemicals Co. Ltd is employing the lactonase activity of whole cells of *Fusariumoxy sporum*[134,135] (immobilized calcium alginate beads and used in a fixed bed reactor) for catalyzing the kinetic resolution of **rac-71**, under very mild reaction conditions (aqueous medium, pH = 6.8–7.2, T = 30°C), as shown in Scheme 12.20. In this industrial process, the immobilized cells retain more than 90% of their initial activity even after 180 days of continuous use, and at the end of the reaction (S)-**71** is extracted and reracemized to **rac-71**, which is recycled into the reactor. Finally, the R-pantoic acid (R-**74**) is chemically lactonized to (R)-**71** and extracted. In this way, conversion up to 95% and ee values of 97% were obtained (capacity 3500 t a^{-1}). Interestingly, by using enantiocomplementary cells of *Brevibacterium protophormia*, it was possible to obtain (S)-**71** in similar yields and purity values.[49]

Using this biocatalytic procedure, several steps that are necessary in the chemical resolution process can be avoided, so that a greener industrial protocol could be implemented, for which Daiichi Fine Chemicals was granted with The Chemical Society of Japan Award for Technological Development for 2000. For sure, many other biocatalytic approaches have been described for the preparation of R-pantolactone, based on other enzymes such as hydroxynitrile lyases[136] or other hydrolases,[137,138] and very recently whole cells of *E. coli* JM109 (DE3) expressing *Fusarium proliferatum* lactonase gene have shown to be an improved biocatalyst for this process.[139]

12.2.1.3 Cyclopamine Analogues

The hedgehog (Hh) signaling pathway regulates cell growth and differentiation during embryonic tissue patterning and plays a pivotal role in tissue homeostasis during embryonic development but is generally silenced in adults.[140] Nevertheless, this pathway is reactivated during tissue repair and regeneration, so that along the last decade there has been increasing evidence that the Hh pathway plays an important role in carcinogenesis,[141] and this knowledge has provided an attractive platform for the development of novel anticancer agents.[142] Among them, cyclopamine (**75**, Figure 12.4), a natural product isolated in the late 1960s from *Veratrum californicum*,[143,144] represents a significant pharmacological tool to validate the Hh pathway in cancer, provided that cyclopamine is able to inhibit tumor growth in several murine models of pancreatic, medulloblastoma, prostate, small cell lung, and digestive tract cancers.[145] However, as the clinical development of cyclopamine is hindered by its limited solubility, acid sensitivity, and weak potency compared to other reported small molecule Hh antagonists, some more active analogues such as **76** (IPI-269609) or **77** have been synthesized.[145]

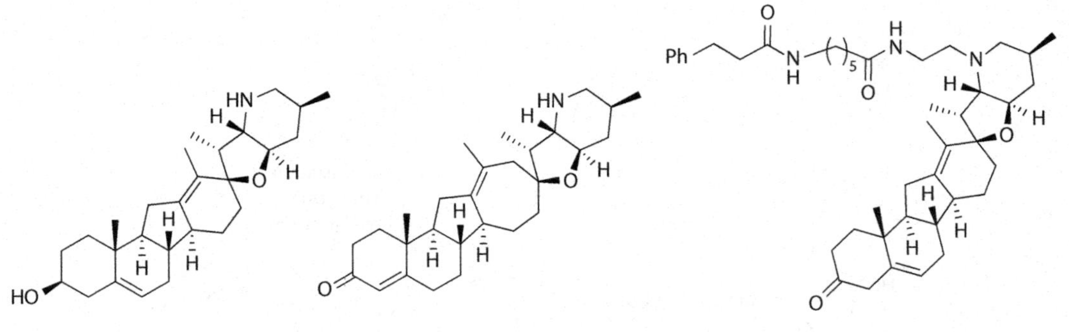

Cyclopamine **75** | IPI-269609 **76** | **77**

FIGURE 12.4 Cyclopamine and analogues.

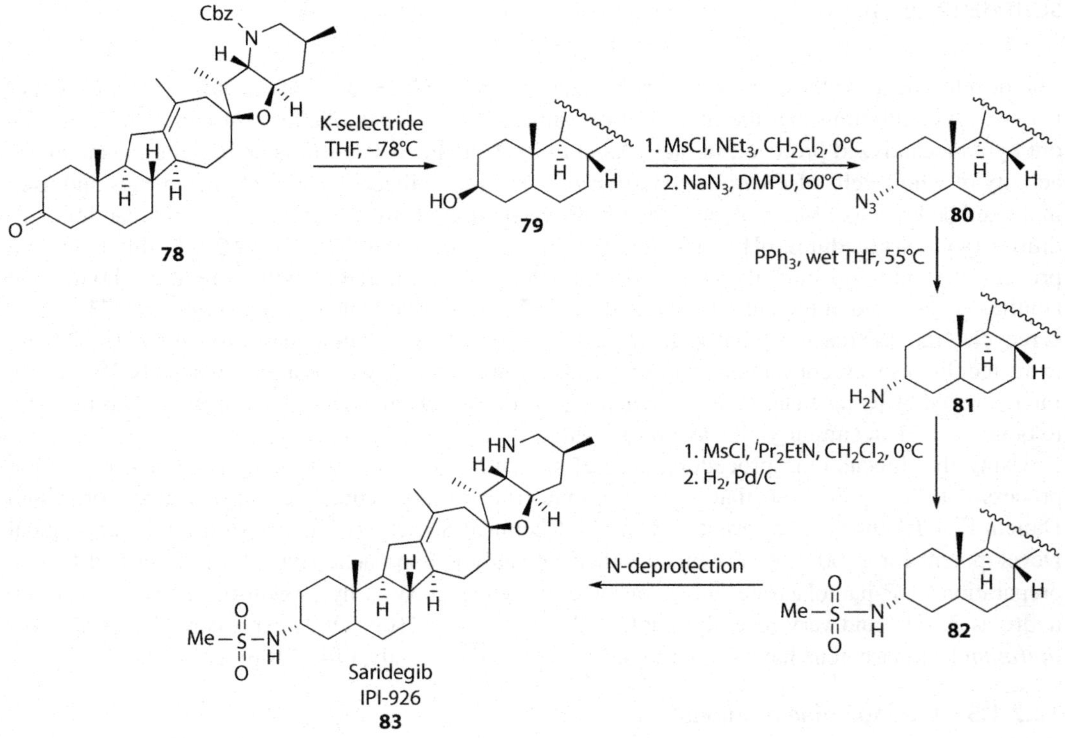

SCHEME 12.21 Chemical synthesis of saridegib **83**, according to Tremblay et al.[145]

In these sense, saridegib, compound IPI-926 **83** (Scheme 12.21), has been shown to be a cyclopamine analogue with substantially improved pharmaceutical properties and potency and a favorable pharmacokinetic profile, leading to a complete tumor regression in a Hh-dependent medulloblastoma allograft model after daily oral administration of 40 mg/kg.[145] Its synthesis starting from N-protected ketone **78** requires several steps and the use of not very sustainable solvents such as THF or CH_2Cl_2 to furnish the intermediate enantiopure amine **81**, which could be finally converted to IPI-926 in three steps (Scheme 12.21).

Very recently, an enzymatic transamination has been disclosed to directly transform **78** into **81**, therefore shortening the synthetic scheme and avoiding the use of some organic solvents by

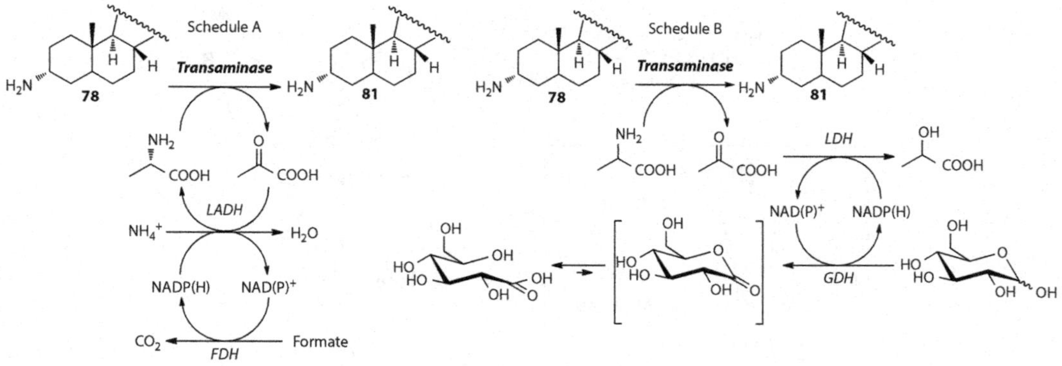

Chiral Building Blocks for Drugs Synthesis *via* Biotransformations 369

SCHEME 12.22 Two enzymatic transamination methodologies to convert **78** into **81**.

replacement by an "on water" methodology.[146] Thus, different commercial transaminases (ATA-113 and ATA-117 from Codexis, *Vibrio fluvialis* omegatransaminase, glutamate pyruvate transaminase, and broad-range transaminase from Fluka) were examined for its efficiency and stereoselectivity, following two strategies as depicted in Scheme 12.22. In fact, as transaminases catalyze an equilibrium reaction, it is necessary to drive this to completion by coupling the main reaction with other enzymatic steps,[147] using secondary substrates and secondary enzymes in a one-pot multienzymatic system. For this purpose, L-alanine dehydrogenase (LADH)/formate dehydrogenase (FDH) promoted transamination (Scheme 12.22, Schedule A) or lactate dehydrogenase (LDH)/glucose dehydrogenase (GDH) promoted transamination (Scheme 12.22, Schedule B) were tested. Best results (up to nearly 50% yield, ee values around 90%–95%) were obtained using some buffer pH = 7.5, adding small amounts of cosolvents (10% MeOH) to improve substrate solubility, running reactions at 37°C for 7 days.

12.2.1.4 Aureolic Acids

The aureolic acids, mithramycin (or plicamycin, **ATC L01DC02**, **DB06810**), chromomycin A3, olivomycin, chromocyclomycin, UCH9, and durhamycin A, constitute a family of polyglycosidic aromatic polyketides bearing a tricyclic core.[148] They are all antineoplastic antibiotics against gram-positive bacteria and also stop the proliferation of tumor cells by inhibiting replication and transcription processes during macromolecular biosynthesis by interacting, in the presence of Mg^{2+}, with GC-rich nucleotide sequences located in the minor groove of DNA. They have been shown to prevent resistance to other antitumor agents by a number of mechanisms, including downregulation of proteins, such as MDR1.[149] Mithramycin (MTM, **84**, Figure 12.5), the most representative member of the family, was approved as an anticancer drug in 1970 and used originally for the treatment of several types of cancer.

MTM is now limited to the control of hypercalcemia in patients with malignant diseases, but new interesting applications are promoting its use; for example, it can be combined with betulinic acid, leading to a novel antiangiogenic therapy for pancreatic cancer.[150] It is known that the acylation of some OH groups of aureolic acids displays an important role in the antitumoral activity of these compounds,[151] so very recently González-Sabin et al.[152,153] described the lipase-catalyzed regioselective acylation of MTM, and some genetically modified analogues, mithramycin SK (MTM-SK, **85**) and mithramycin SDK (MTM-SDK, **86**), by using different lipases and acyl donors.

As can be seen, MTM possess different OH groups that could be susceptible to become acylated, both in the sugar moiety and in the aglycon. These authors tested many acyl donors and alkoxycarbonylating agents (activated esters, anhydrides, and carbonates) and different lipases, being CAL-A (lipase A from *Candida antarctica*) and CAL-B (lipase B from the same origin) very selective in several cases. In fact, CAL-B exclusively acylated the 4′-hydroxyl group with vinyl decanoate and vinyl benzoate, with conversion values of over 90% (Scheme 12.23). In addition, CAL-A gave rise to the corresponding monoacylated derivative on the 3B hydroxyl group using vinyl crotonate,

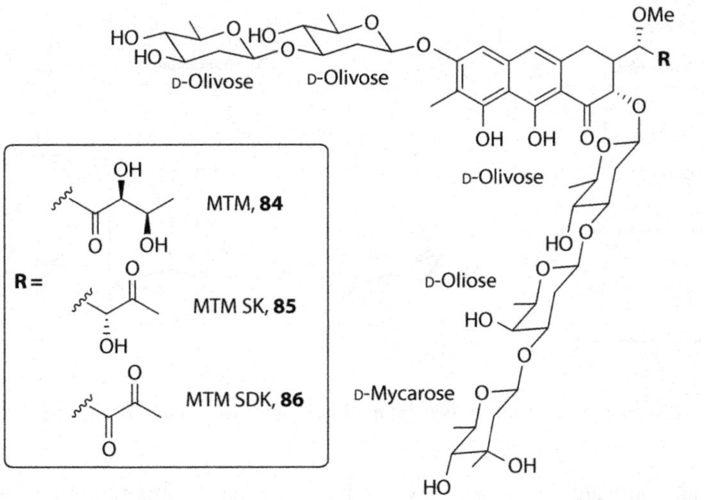

370   Asymmetric Synthesis of Drugs and Natural Products

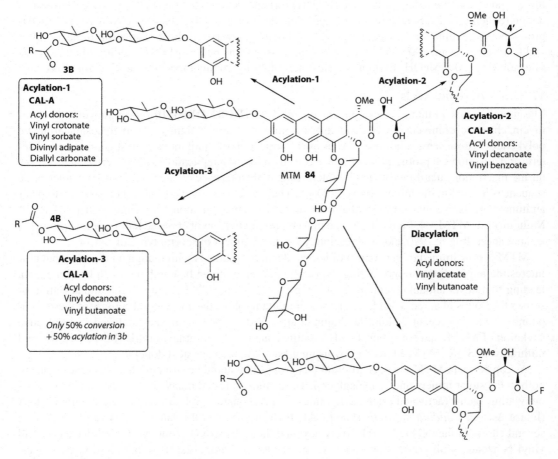

FIGURE 12.5 Structure of aureolic acids.

SCHEME 12.23 Lipase-catalyzed acylation of MTM **84**.

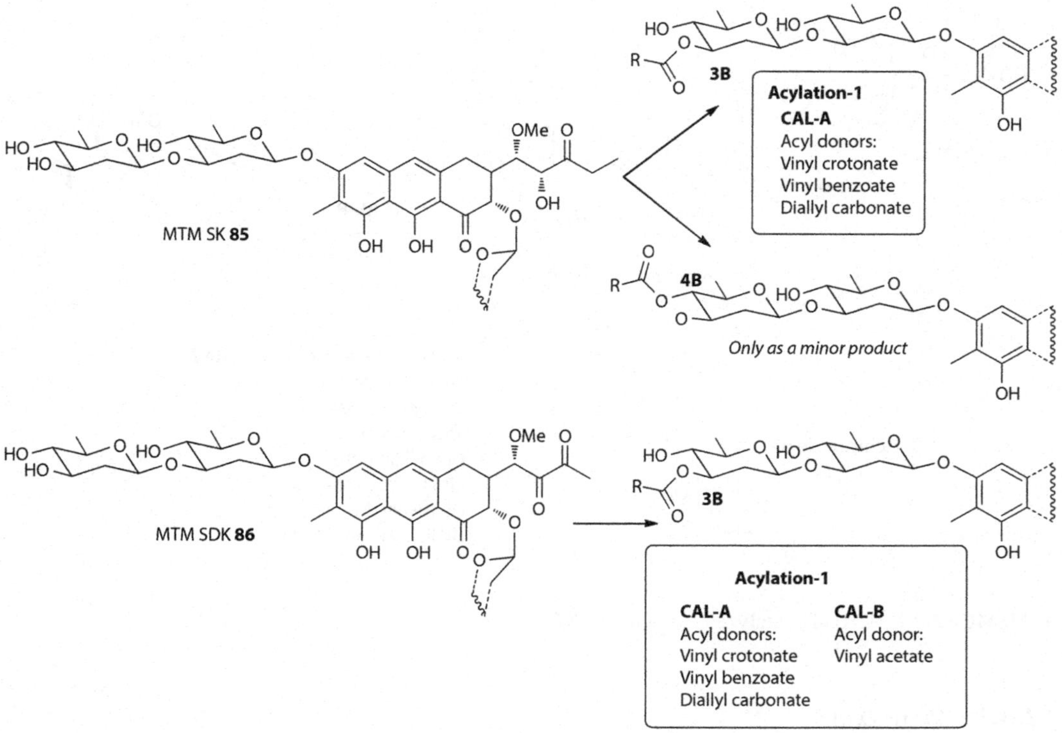

SCHEME 12.24 Lipase-catalyzed acylation of MTM SK **85** and MTM SDK **86**.

vinyl sorbate, divinyl adipate, and diallyl carbonate as acylating agents (conversion values were also higher than 90%). The use of other acyl donors yielded mixtures of products. As a general rule, however, CAL-B favorably produces acylation of 4′ and 3B hydroxyl groups, while CAL-A acylates the 3B and 4B hydroxyl groups. Accordingly, depending on the acyl donor being tested, it could be possible to isolate not only monoacylated products but also the corresponding diacylated derivatives in excellent yields. Similar results were obtained in the acylation of MTM SK **85** and MTM SDK **86** (Scheme 12.24), where the acylation of 3B led to the major products with good conversions.

Some of these acylated structures, concretely 3B-Ac-MTM, 3B-Ac-MTM SK, and 3B-Ac-MTM SDK, showed improved cytotoxicity against different tumor cell lines,[152] so that this enzymatic acylation is able to render more active structures for cancer treatment.

In a similar manner, this same research group has reported the acylation of chromomycin A_3 (**87**, Scheme 12.25), another member of the aureolic acid family.[154] CAL-B lipase was the only biocatalyst active on this substrate, leading to the acylation in 4′ position of the aglycon side chain. The so-obtained 4′-acyl derivatives showed antitumor activity at the micromolar or lower-level concentration. Particularly, chromomycin A_3 4′-vinyladipate showed three to five times higher activity against four tumor cell lines assayed as compared to parental **87**.

12.3 SMALL MOLECULES

As we commented in the "Introduction," most of the pharmaceutical companies are nowadays focusing on small molecules as anticancer drugs, because of their easier synthesis compared to natural products. In this area, biocatalysis has proven to be a very useful tool as we will comment now through different examples in which the key step for generating chirality is a biotransformation performed on water.

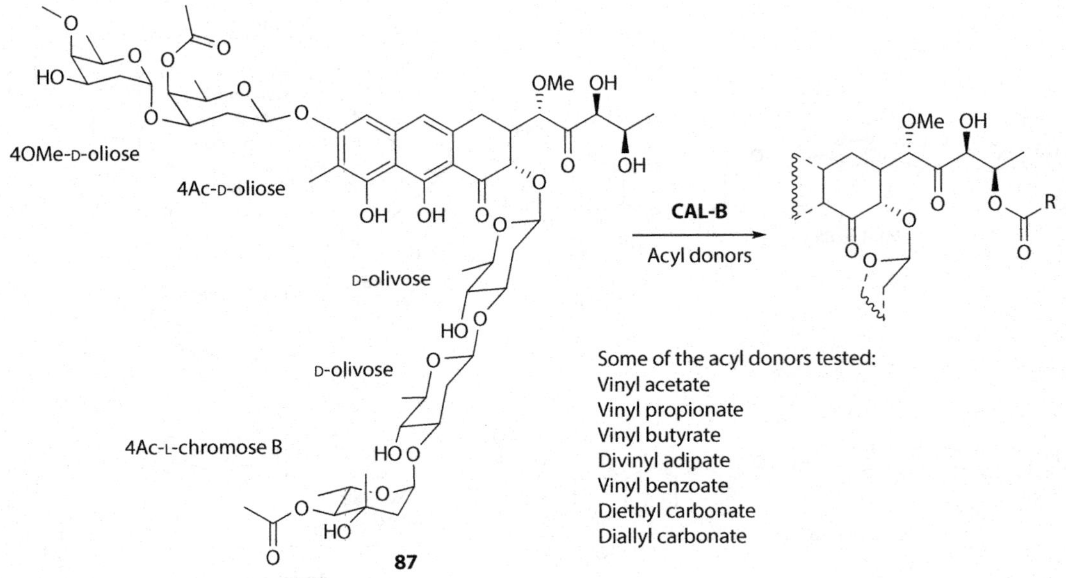

SCHEME 12.25 CAL-B-catalyzed acylation of **87**.

12.3.1 Pelitrexol

Pelitrexol **91** (Scheme 12.26) is a water-soluble antifolate with antiproliferative activity. This compound inhibits activity of glycinamide ribonucleotide formyltransferase (GARFT), the first folate-dependent enzyme of the *de novo* purine synthesis pathway essential for cell proliferation. Enzyme inhibition reduces the purine nucleotide pool required for DNA replication and RNA transcription. As a result, this agent causes cell cycle arrest in S-phase and ultimately inhibits tumor cell proliferation.[155] In the preparation of pelitrexol, the first-generation process[156] required 20 linear chemical transformations with an overall yield of only 2% starting from 2-methyl thiophene (Scheme 12.26). In this synthetic scheme, the key steps include the insertion of Evans' chiral auxiliary 4-benzyl-2-oxazolidinone in **88** after 11 steps of linear chain extension, a further chiral

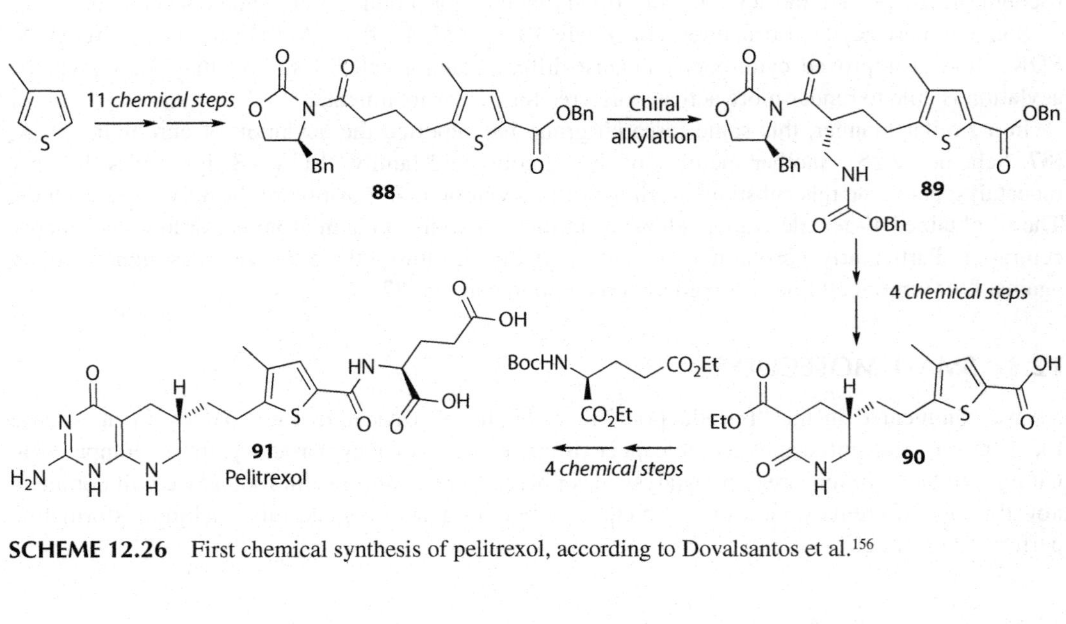

SCHEME 12.26 First chemical synthesis of pelitrexol, according to Dovalsantos et al.[156]

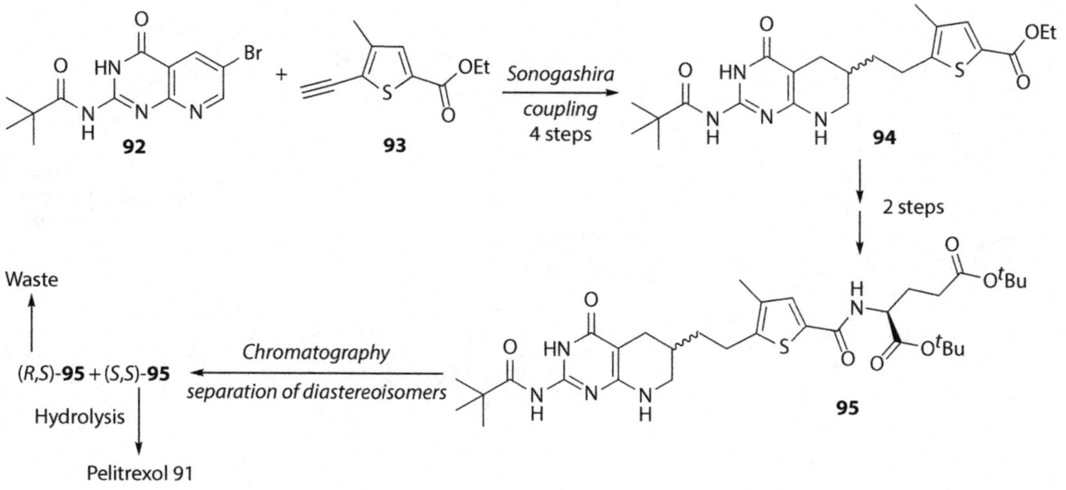

Chiral Building Blocks for Drugs Synthesis *via* Biotransformations 373

SCHEME 12.27 Chemical synthesis of pelitrexol, using a chromatographic separation of diastereomers.[156]

alkylation to introduce an amino methylene side chain **89**, intramolecular cyclization **90**, and finally the building of the dihydropyridopyrimidine moiety by coupling with guanidine to provide the end molecule upon linking with the glutamic ester followed by global deprotection.

In a further redesign of the synthesis,[157,158] a retrosynthetic analysis led to a new process combining Sonogashira coupling with a chromatographic resolution of the diastereomeric mixture **95** (Scheme 12.27). In fact, different asymmetric chemical synthesis of **95** had proved to be remarkably problematic, and the design of this methodology is heavily reliant on the chromatographic separation of two diastereomers of **95** in the penultimate step, which suffers from several concerns. One major drawback involved the wasted wrong diastereomer (*R,S*)-**95**, which cannot be recycled, resulting in a maximum of 50% yield. Moreover, the low solubility of both diastereomers in most column-compatible organic solvents involves hundreds of liters of organic solvent per kilogram of material in this resolution, which provided poor separation and a recovery of less than 25%, even after extensive optimization. Therefore, this protocol is violating most of the 12 principles of Green Chemistry,[14] and a more sustainable alternative using biotransformation was developed (Scheme 12.28).

When the kinetic resolution of racemic **94** was tested, as long as the chiral center is six bonds away from the ester group, direct enzymatic hydrolysis using different lipases gave low selectivity,[157] as could be expected. Nevertheless, using the typical strategy of modifying the substrate structure for favoring the enzymatic recognition,[48] and based upon some previous results from same authors,[158] a labile oxalamic ester protecting group was introduced (*rac*-**96**); hence, the kinetic resolution of this compound using CAL-B yielded the desired enantiomer *S*-**97** with excellent resolution efficiency (100 g/L substrate, 95% ee, 45% conversion within 4–5 h) (Scheme 12.28). In this process, the use of 30% DMF (anyhow not a very green cosolvent), which increases both the solubility of the substrate in water and the enantioselectivity of the enzyme, was crucial to this process. It is worth noting that enzymatic hydrolysis takes place at the thiophene ester rather than the oxalamic ester, this latter one being cleaved during the workup, and the resulting intermediate was finally converted to pelitrexol after coupling with the glutamic ester followed by deprotection. Very interestingly, the nondesired enantiomer (*R*)-**96** could be efficiently recycled under oxidation and hydrogenation conditions, so that the amount of waste generated is considerably reduced. To sum up, the overall process can be considered much more sustainable, as long as it required only nine steps, and the overall yield was increased to 10%–15% from the original 2%.

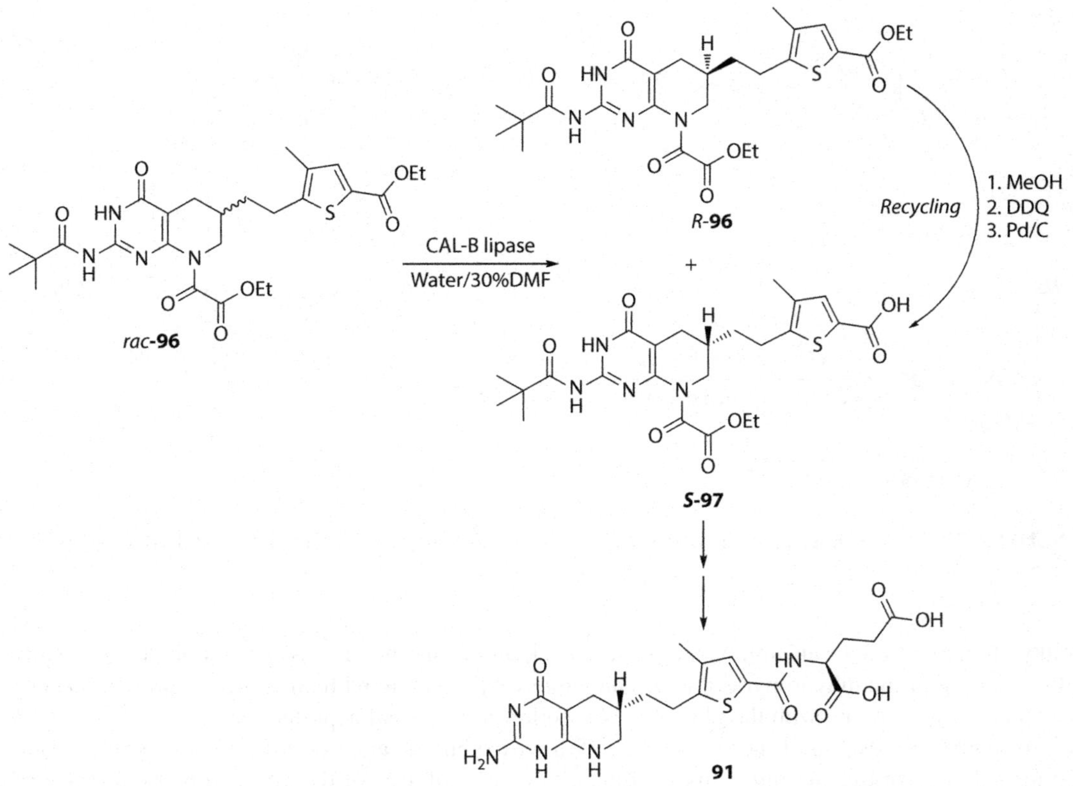

374   Asymmetric Synthesis of Drugs and Natural Products

SCHEME 12.28 Chemical synthesis of pelitrexol, by means of an enzymatic kinetic resolution.[157]

12.3.2 MONASTROL

Monastrol (**DB04331**), **99**, is the first small molecule described possessing inhibitory effect on the mitotic motor Eg5 (kinesin spindle protein, KSP) and represents a promising current lead structure in anticancer research.[159–161] In fact, many KSP inhibitors are being evaluated in clinical trials, and they are considered as very promising anticancer drugs alternative from the existing microtubule targeting agents.[162] Based on monastrol structure, new compounds bearing the common 4-aryl-3,4-dihydropyrimidin-2(1*H*)-thione skeleton are being studied, displaying very potent cell-permeable inhibition of Eg5.[163,164] While both enantiomers of monastrol abolished basal Eg5 ATPase activity, the (*S*)-eutomer shows a 15-fold enhanced activity compared with the corresponding (*R*)-distomer.[160] A chemical procedure for the preparation of racemic monastrol, based on a Yb(OTf)$_3$-catalyzed Biginelli condensation between thiourea, ethyl acetoacetate, and 3-hydroxybenzaldehyde, was described by Dondoni et al.,[165] and these authors also reported the chromatographic separation of both enantiomers of monastrol *via* a previous synthesis of the diastereomers by treating with *N*-3-ribofuranosyl amides. On the other hand, also asymmetric versions of the Biginelli reaction were described by Gong et al.[166] and Zhu and coworkers,[167] using chiral ytterbium catalysts, and some other new asymmetric syntheses have been also published.[168–170]

An enantioselective synthesis of *S*-monastrol has been recently described by Gröger and coworkers,[171] using a lipase-catalyzed resolution of racemic monastrol based on a hydrolytic strategy (Scheme 12.29). In this sense, the direct hydrolysis of racemic monastrol (ester group close to the stereogenic center) was tested with different lipases, showing a very low enantioselectivity; however, if the resolution was carried out starting from phenol esters such as *rac*-**98**, much better results were obtained, when using a biphasic medium water/CH$_2$Cl$_2$, under very mild reaction conditions

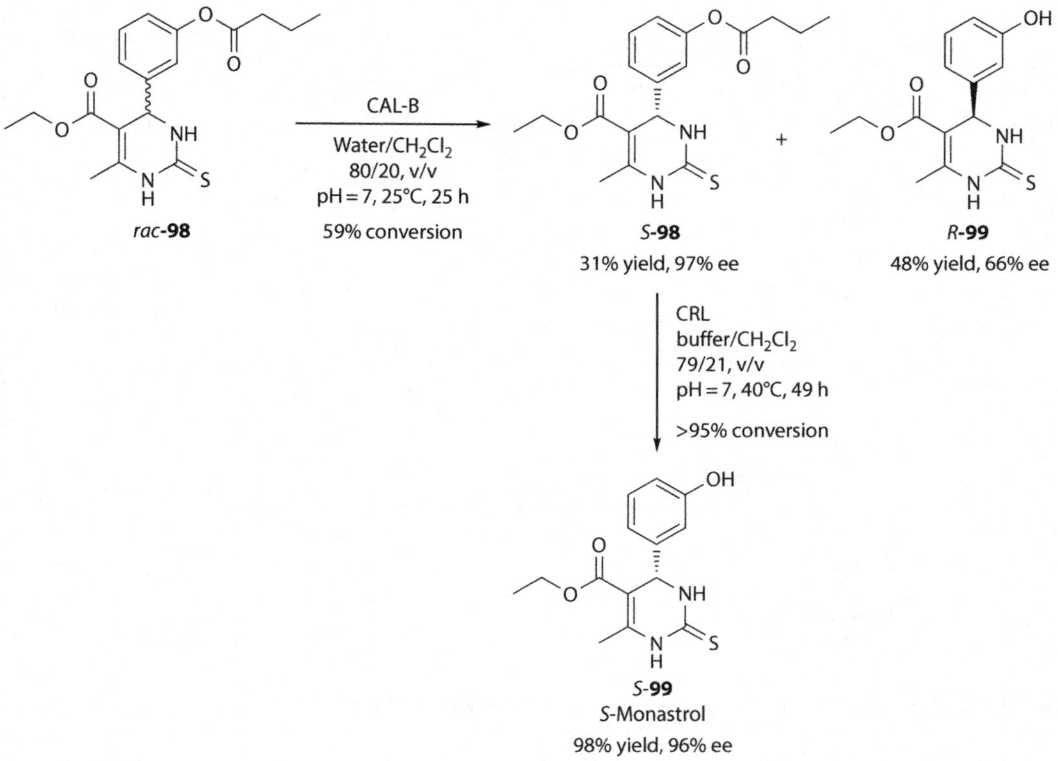

SCHEME 12.29 Enantioselective biocatalytic synthesis of S-monastrol.[171]

(25°C, 25 h). Hence, the use of CAL-B lipase allowed a good kinetic resolution leading to R-monastrol and the corresponding butyric ester (S)-**98**. This later one was subsequently hydrolyzed using CRL (lipase from *Candida rugosa*), which very regioselectively hydrolyzed only the phenol ester, leaving unaltered the ethyl ester attached to the dihydropyrimidine moiety.

Although this process has to be improved according to Green Chemistry postulates, in features such as cosolvent replacement (avoiding the use of highly undesirable CH_2Cl_2, and recycling of the nonconverted enantiomer to reduce waste generation), it is very promising in terms of a future scaling up.

12.3.3 AXL INHIBITORS

AXL is a receptor tyrosine kinase (RTK) involved in the growth, differentiation, survival, and motility of many different cell types.[172] It belongs to the TAM family of RTKs (AXL, MER, TYRO3), which is similar to other RTKs in that they have an intracellular kinase domain and span the cell membrane to an extracellular domain. Nevertheless, the extracellular portion of AXL (and TAM family members) possesses differential properties and structural elements (repetition of units of immunoglobulin and fibronectin type III) similar to cell adhesion molecules, so that it may participate in cell-to-cell contacts not very well understood.[173] In any case, AXL signaling stimulates different cellular responses, which ultimately implies different biological consequences, such as invasion, migration, survival signaling, angiogenesis, resistance to chemotherapeutic and targeted drugs, cell transformation, and proliferation, which eventually represent undesirable traits associated with cancer.[174]

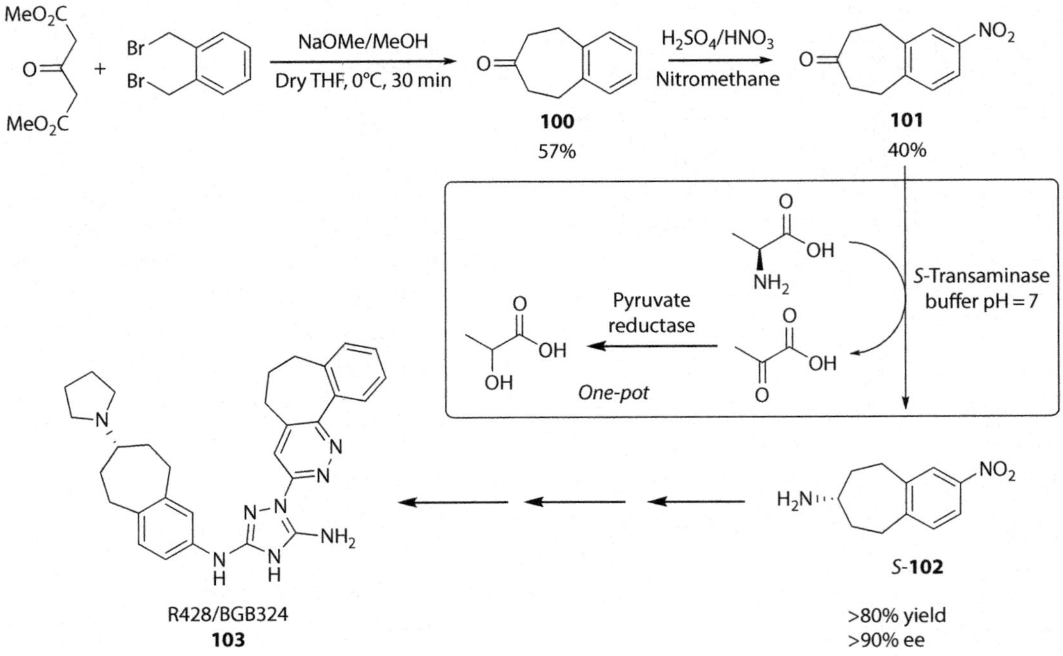

SCHEME 12.30 Rigel's enantioselective biocatalytic synthesis of R428/BGB324 **103**.[178]

Therefore, the search for AXL inhibitors represents a very attractive and modern research field.[175–177] In this area, researchers from Rigel, Inc. have recently described R428 (**103**, Scheme 12.30), a small molecule that blocks the catalytic and procancerous activities of AXL.[175] In fact, R428 inhibits AXL with low nanomolar activity and blocked AXL-dependent events, including Akt phosphorylation, breast cancer cell invasion, and proinflammatory cytokine production, with good pharmacokinetics. After an agreement between Rigel and Norway's BerGenBio, R428 was renamed BGB324.

As can be seen, this compound possesses a stereogenic center with S-absolute configuration; for its enantioselective preparation, this pharmaceutical company has developed a biocatalytic one-pot procedure involving the use of transaminase-reductase, as depicted in Scheme 12.30.[178]

Thus, starting from commercially available dimethyl acetone-1,3-dixarboxylate and o-xylene dibromide, it was possible to synthesize 7-oxo-6,7,8,9-tetrahydro-5H-benzo[7]annulene **100**, which upon nitration yielded 2-nitro-8,9-dihydro-5H-benzo[7]annulene-7(6H)-one **101**. This compound was treated with an excess molar amount of an amino donor molecule (L-alanine) in the presence of a catalytic amount of a transaminase, in this case a (S)-specific transaminase, and a stoichiometric or excess stoichiometric amount of a pyruvate reductase mixture, in order to reduce piruvic side product, thereby driving the reaction into the desired direction, solving in this way one of the major limitations of transaminase use at high concentrations, which is the requirement of shifting the equilibrium toward the desired amine.[179] Several (S)-specific transaminases were used, and the best results were obtained with ATA-103, a (S)-specific transaminase commercialized by Codexis, while the pyruvate reductase mixture used was PRM-102, also from Codexis. The reaction was conducted at room temperature, at a pH of between about 7.5 and about 8.0, and for a period of time of between about 24 h and about 6 days, preferably for about 4 days. Chiral amine (S)-**102** was easily isolated from the reaction mixture (yield >80%, ee > 90%), and this process was scaled up to a gram scale.[178] Finally, (S)-**102** was converted into R428/BGB324 **103** through several steps and an ultimate dialkylation with 1,4-dibromobuthane.

12.3.4 Crizotinib

Crizotinib (R)-**107** (ATC L01XE16) is a potent and selective mesenchymal–epithelial transition factor/anaplastic lymphoma kinase (c-Met/ALK) inhibitor.[180,181] Crizotinib was first approved as XALKORI® (Pfizer) in the United States in August 2011 for the treatment of locally advanced or metastatic non-small cell lung cancer (NSCLC) that is ALK positive, as detected by an FDA-approved test. This drug has also received approval in a number of other countries, including Switzerland, Canada, South Korea, and Japan. Additional applications are under regulatory review in several countries worldwide, and in July 2012 Pfizer announced that the Committee for Medicinal Products for Human Use (CHMP) of the European Medicines Agency (EMA) had adopted a positive opinion recommending that crizotinib be granted conditional marketing authorization in the European Union (EU) for the treatment of adults with previously treated ALK-positive advanced NSCLC.

In order to ensure an uninterrupted supply of **107**, the initial enabling route was used to supply >100 kg while the proposed commercial route was developed. Thus, in a first stage racemic **107** was prepared, but soon after this, route was modified in order to prepare the desired biologically active eutomer (R)-**107**.[182–184] The chiral center was introduced in the first step, by the preparation of chiral alcohol (S)-**105** (Scheme 12.31).[183] This enantiopure compound was then submitted to Mitsunobu reaction conditions to afford the essential building block of the optically pure intermediate (R)-**106**.

Chiral alcohol (S)-**105** was firstly prepared via enzymatic hydrolysis of the corresponding racemic acetate catalyzed by pig liver esterase (PLE, Scheme 12.32),[182] but this resolution approach did not satisfy the yield expectations and new alternative procedures were studied.

In a second stage, the stereoselective bioreduction of 2,6-dichloro-3-fluoroacetophenone **108** to afford the desired chiral alcohol intermediate catalyzed by whole cells or isolated ketoreductases was implemented as the most successful choice.[183,185,186] Among the 188 different microbial strains tested, yeasts UC2387 belonging to *Rhodotorula* species were chosen for the preparation of (S)-**105** at a larger scale.[185] In further studies, better results were obtained by employing genetically modified ketoreductase enzymes. Although naturally occurring ketoreductases from *Lactobacillus* sp. displayed insignificant activity toward the reduction of substituted acetophenone **108**, engineered enzymes derived from a wild-type *Lactobacillus* ketoreductase developed at Codexis[186] showed not only reversed selectivity but also very high activity toward the reduction of **108**, reaching 99%

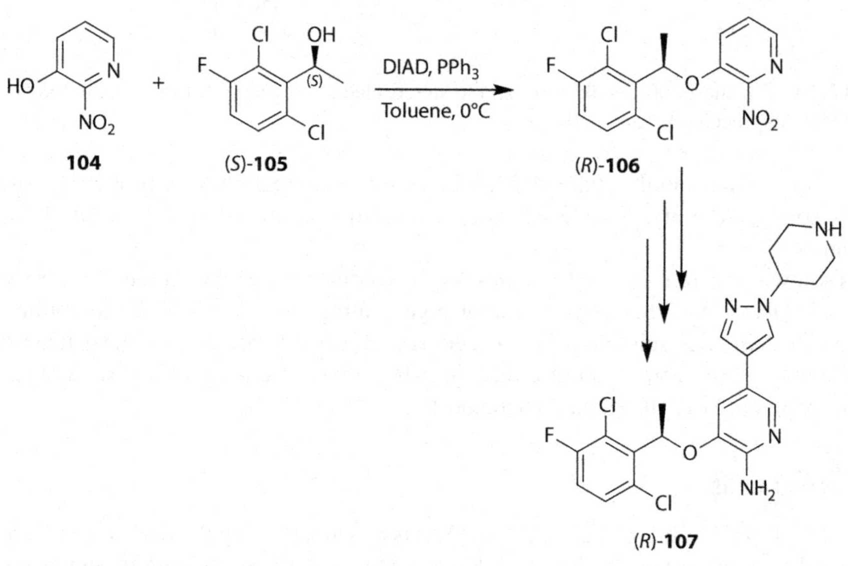

SCHEME 12.31 Synthesis of optically pure crizotinib.[183]

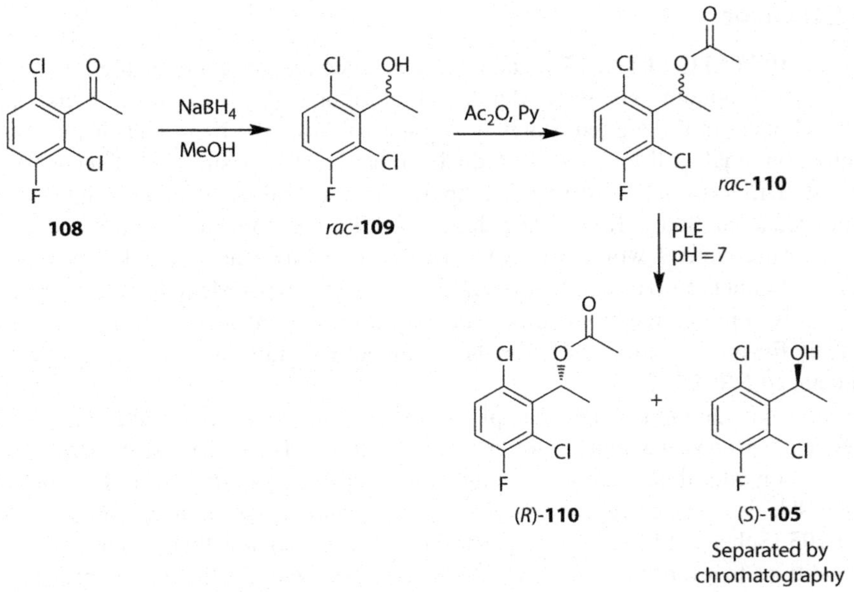

SCHEME 12.32 Preparation of crizotinib (S)-**105** through the kinetic resolution of the corresponding acetate catalyzed by pig liver esterase.

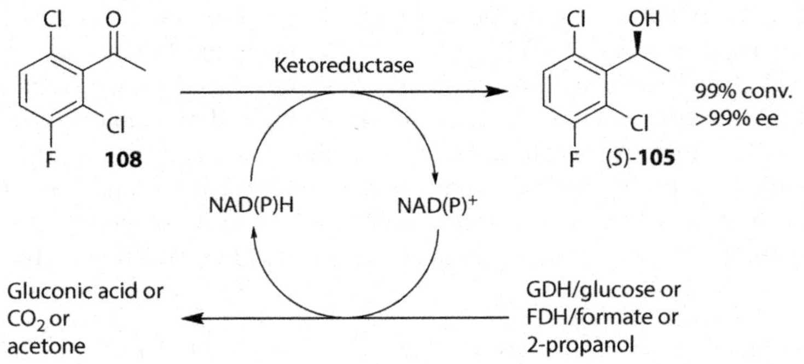

SCHEME 12.33 Synthesis of (S)-**105** through the stereoselective reduction of the substituted acetophenone **108** catalyzed by engineered ketoreductases.

conversion and ee values higher than 99% (Scheme 12.33). Other enzyme properties, such as thermostability, solvent stability, or reduced product inhibition, were also improved in those modified ketoreductases.

As it is known, the reduction of ketones by ketoreductase enzymes requires a cofactor, most commonly NADH or NADPH and a cofactor regenerating system employed in conjunction with the ketoreductase. Several suitable cofactor regeneration systems are proposed, such as glucose and glucose dehydrogenase, formate and formate dehydrogenase, and a secondary alcohol (2-propanol) and secondary alcohol dehydrogenase (Scheme 12.33).[186]

12.3.5 Lonafarnib

Lonafarnib (R)-**115** (U.S. brand name SARASAR) is a synthetic tricyclic derivative of carboxamide with antineoplastic properties. As it is well known, Ras proteins participate in numerous signaling pathways (proliferation, cytoskeletal organization) and play an important role in oncogenesis, as

Chiral Building Blocks for Drugs Synthesis via Biotransformations   379

mutated Ras proteins have been found in a wide range of human cancers.[187] Lonafarnib binds to and inhibits farnesyltransferase, an enzyme involved in the posttranslational modification and activation of Ras proteins, thus terminating the signal for cell proliferation.[178] Although extensive clinical research (phase III) indicates limited activity of lonafarnib in solid tumors, there is recent interest in combinations of farnesyltransferase inhibitors with imatinib or bortezomib in hematological malignancies.[188]

The synthesis of (R)-115 begins with loratadine 111 and the chirality was introduced by the resolution carried out through the formation of a diastereomeric salt with N-Ac-L-phenylalanine and the intermediate 114, followed by recrystallization from EtOH (Scheme 12.34). As a potential alternative to this resolution, enzymatic resolutions of different intermediates were also studied.[189]

The first attempts were proposed through the deacylation of carbamate rac-112 by microbial hydrolysis, but very poor results were obtained. However, the kinetic resolution of the racemic intermediate rac-113 via enzymatic acylation led to excellent conversion and enantiomeric excess. Although this enzymatic process is not carried out on water, it is worth mentioning the selectivity displayed by the biocatalyst. The racemic substrate does not contain a chiral center but exists as a pair of enantiomers due to atropisomerism about the exocyclic double bond.[189] Thus, after an exhaustive screening of reaction conditions (223 commercially available enzymes tested, 12 organic solvents and a wide variety of esters and carbonates as acylating agents tested), the best results were obtained with Toyobo LIP-300 (a lipase from *Pseudomonas aeruginosa*), employing "usable" tert-butyl methyl ether (TBME) as solvent and 2,2,2-trifluoroethyl isobutyrate as acylating donor. After 24 h, 42% conversion and excellent ee values were reached (Scheme 12.35).

In addition, the isobutyramide product (+)-116 allowed its easy separation from the unreacted starting material by simple acid extraction and the undesired (−)-113 could be recovered, racemized thermally by refluxing in di(ethyleneglycol)dibutyl ether, and resubmitted to enzymatic acylation. A 65% overall yield was obtained with ee 98% after three rounds of enzymatic resolution, which was completed at a 50 kg scale.

Furthermore, the kinetic resolution of rac-114 was also examined, obtaining a very selective N-acylation catalyzed by lipase Toyobo LIP-300. However, as the unwanted enantiomer could not be recovered and racemized, this process was not integrated into the final synthesis of lonafarnib.[189]

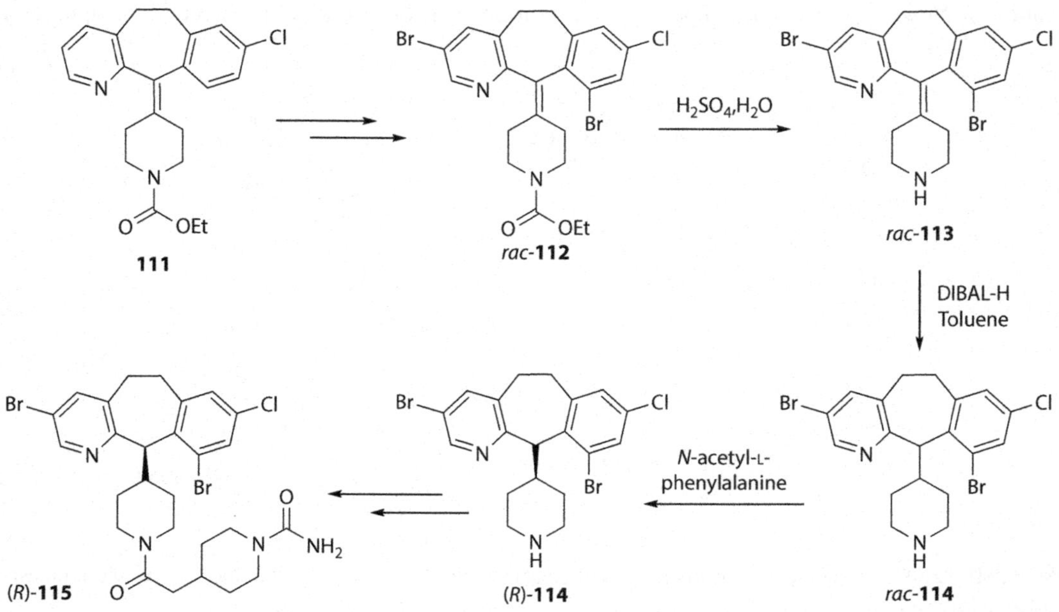

SCHEME 12.34 Synthetic route of lonafarnib (R)-115.

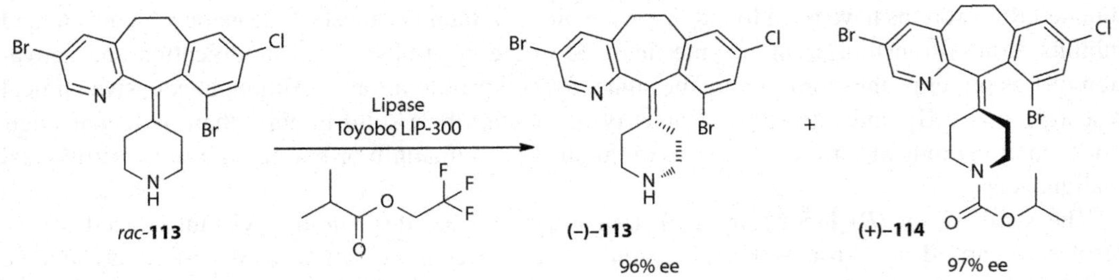

SCHEME 12.35 Enzymatic kinetic resolution of *rac*-113.[189]

12.3.6 ODANACATIB

Cathepsin K is a lysosomal cysteine protease involved in osteoclast-mediated bone resorption. Inhibition of cathepsin K represents a potentially attractive therapeutic approach for treating diseases characterized by excessive bone resorption, such as osteoporosis.[190] Anyway, cathepsin K is also involved in tumor progression mainly in breast and prostate cancer, both of which have a propensity to metastasize to skeleton; in both cancers, levels of cathepsin K have been shown to be significantly increased in bone tumors, as compared with primary tumors and soft tissue metastases from the same patients.[191,192]

Odanacatib (**118**, Scheme 12.36) is an orally bioavailable, potent, and selective cathepsin K inhibitor[193] currently being evaluated in clinical trials. The total synthesis of odanacatib requires an enzyme-mediated dynamic kinetic ethanolysis of azlactone **116** to the desired intermediate, an (*S*)-γ-fluoroleucine ethyl ester **117**.

First described enzymatic DKRs employed Novozyme 435 (commercially accessible CAL-B lipase adsorbed on ion-exchange resin Lewatit) in a batch process (79% yield, 78% ee of the desired product, enzyme/substrate ratio of 1:1 by weight) for ring opening of **116**,[194] although significant improvements both in yield and selectivity (90% yield, 86% ee, enzyme-to-substrate ratio of 1:20) were described by switching to a continuous plug flow column reactor setup.[195] However, deactivation of Novozyme 435 under operational conditions was observed, so that recently Truppo and

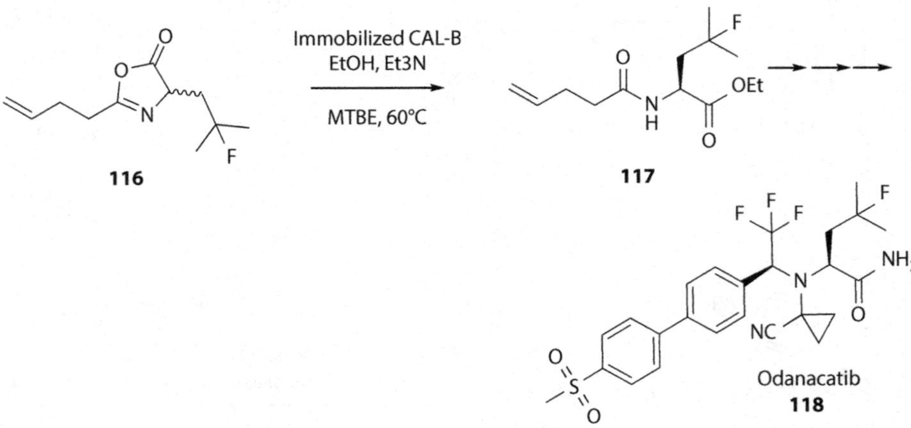

SCHEME 12.36 Dynamic kinetic ethanolysis of an azlactone **116** to render an intermediate for the synthesis of odanacatib **118**.

Hughes have reported the development of a new immobilized form of CAL-B with greatly enhanced stability and activity compared to Novozyme 435.[196] In fact, a preparation of CAL-B immobilized on a polymethacrylate SEPABED resin containing octadecyl functional groups exhibited significantly greater activity, allowing the DKR with 95% yield and 88% ee. The greater stability of the new immobilized CAL-B preparation also provided for a significant reduction in enzyme-to-substrate loading operating at an enzyme-to-substrate ratio of <1:100. Economically speaking, considering the reusability of the new immobilized CAL-B preparation along with cost to manufacture the immobilized enzyme, the new process is 99.9% less expensive than the process utilizing Novozyme 435, with a 3-fold lower E-factor.[37] This process has been demonstrated several times at a 100 kg scale (>90% yield and 88% ee).

12.3.7 Niraparib

Poly(ADP-ribose)polymerase (PARP) has a key role in the repair of single-strand breaks in DNA, and its overexpression in certain cancer types has been implicated in cancer cell proliferation and tumor progression.[197] Hence, PARP inhibitors are very promising drug candidates, and some of them are now in clinical trials.[198] Very recently, Wallace et al.[199] reported a kilogram-scale synthesis (11 total steps, 11% overall yield) of one of these compounds, niraparib p-toluenesulfonate monohydrate (**122**, Scheme 12.37), an orally active PARP inhibitor useful in murine xenograft cancer models.[200] However, this protocol must be improved for allowing larger API deliveries; in this sense, a primary objective is the developing of an efficient asymmetric synthesis of the piperidine moiety of niraparib, because in the previous synthesis a chromatographic resolution was recognized to be a significant bottleneck.[199]

Thus, scientists from Merck[201] described two different biocatalytic methodologies, both of them based on the use of DKRs using transaminases for generating the desired chirality, as depicted in Scheme 12.37. In the first one, shown in Scheme 12.38, the aldehyde **123** (previously synthesized through different chemical steps[201]) was the starting substrate for an enzymatic transamination catalyzed by ATA-301 (commercial enzyme from Codexis), which afforded the lactam **119** through an intermediate amine **124**.

Using a buffer with pH 10.5 was the optimum balance between the rate of the desired transaminase reaction and that of competing ester hydrolysis of **123**. Sufficient enzyme stability was maintained at 45°C to fully consume **123** and enable a satisfactory rate of cyclization of amine intermediate **124**

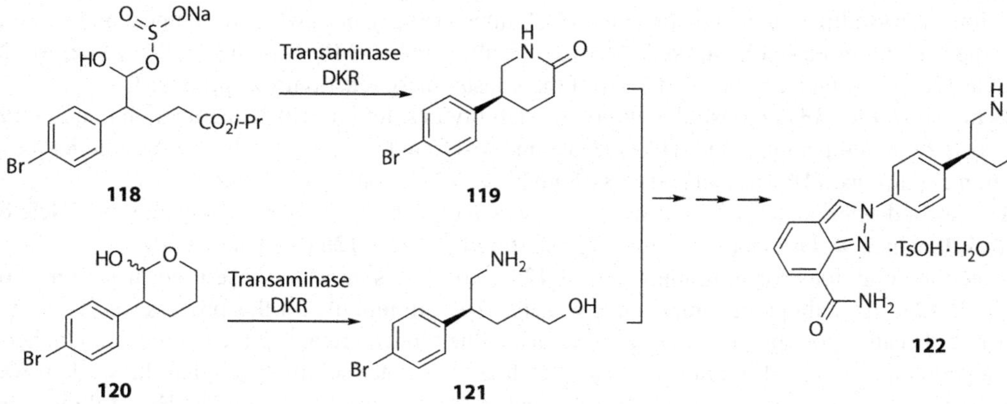

SCHEME 12.37 Two enzymatic strategies for the asymmetric synthesis of niraparib **122**.

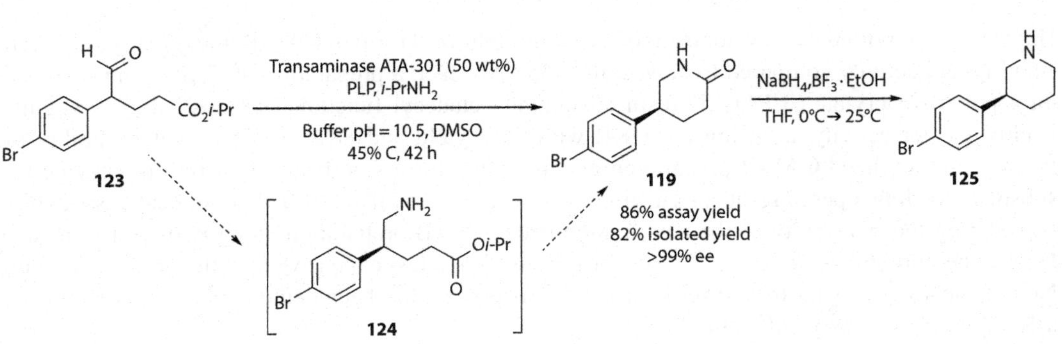

SCHEME 12.38 Dynamic kinetic resolution of starting aldehyde **123** via transamination for the asymmetric synthesis of niraparib intermediate **125**.

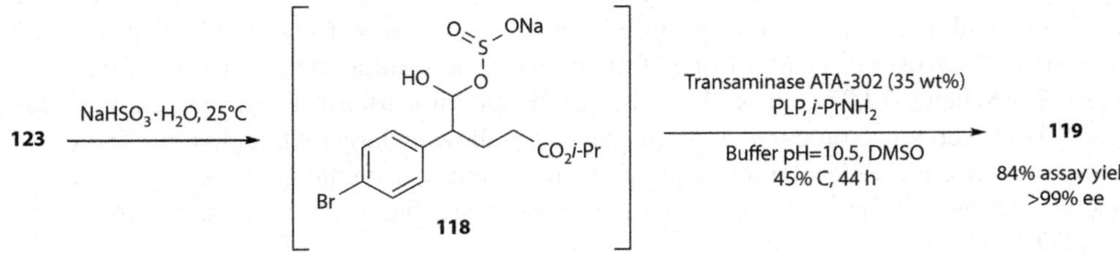

SCHEME 12.39 Dynamic kinetic resolution (DKR) of bisulfite adduct **118** via transamination.

(typically less than 2% of **124** remained at the end of reaction). An acceptable reaction profile could be achieved at a reduced enzyme loading of 50 wt%, albeit at the expense of extended reaction time. Under these optimized conditions, lactam **119** was formed in 86% assay yield and >99% ee starting from 35 g of isopropyl ester **118** (Scheme 12.38). Isolation of the product was accomplished with good recovery by aqueous workup followed by crystallization from IPAc. Finally, **119** could be transformed in the corresponding piperidine-containing compound **125** by a borane reduction, and a subsequent treatment with Boc$_2$O afforded the required chiral moiety of niraparib **122**.

In order to provide better reaction conditions, mainly because aldehyde **123** is an oil, impeding its isolation by crystallization and displaying only limited stability both when stored neat and in solution, the use of a more stable masked aldehyde, bisulfite adduct **118** (Scheme 12.39), which would liberate **123** *in situ* under basic conditions of the transaminase reaction, was proposed.

In fact, bisulfite **118** demonstrated improved stability in solution, with a degradation rate of <1% per day at room temperature, and ATA-302 enzyme was found to be a better biocatalyst; under these reaction conditions, **119** was isolated in 84% yield (>99% ee) on a 100 g scale.

The second alternative proposed by Merck was based in the transaminase-catalyzed DKR of lactol **120**, which was previously synthesized starting from acid **126** (Scheme 12.40).

Reaction conditions for transamination of **120** were very similar to those used in the DKR of aldehyde **123**; the main procedural change was that the optimum pH for this process was 8.5 rather than 10.5, because possibly at lower pH there is a higher concentration of the open-chain aldehyde available in equilibrium. The amino alcohol **121** had high water solubility, so that the workup was done by precipitating the spent enzyme at the end of the reaction using aqueous HCl, followed by removal by filtration. The acidic filtrate was then adjusted to pH 10 and treated directly with Boc$_2$O to obtain a crude readily extracted product. Subsequent chemical steps were conducted to finally afford Boc-piperidine **127** in 94% assay yield (Scheme 12.40).

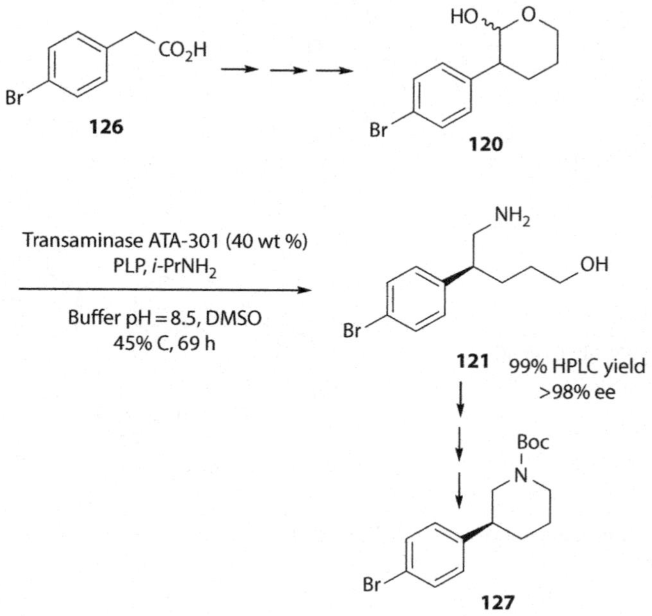

SCHEME 12.40 DKR of lactol **120** via transamination.

12.4 ANTIDIABETIC DRUGS

12.4.1 PEROXISOME PROLIFERATOR–ACTIVATED RECEPTOR AGONISTS

Peroxisome proliferator–activated receptors (PPARs) are ligand-dependent transcription factors. The three mammalian PPARs (α, β/δ, and γ)[202] are crucial regulators of fatty acids and lipoprotein metabolism, glucose homeostasis, cellular proliferation/differentiation, and the immune response. PPARs are key targets in the treatment of metabolic disorders such as insulin resistance and type 2 diabetes mellitus (T2DM); furthermore, PPARs are also involved in chronic inflammatory diseases such as atherosclerosis, arthritis, chronic pulmonary inflammation, pancreatitis, inflammatory bowel disease, psoriasis, blood pressure regulation, neuroinflammation, nerve-cell protection, inflammatory pain reduction, and hypothalamic control of metabolism.[203] Thus, although PPAR agonists have enormous therapeutic potential, current PPAR-based therapies present several side effects, caused by undesired activation of PPARs in nontarget cells, so that there is a growing interest in the development of cell-specific PPAR agonists.

In this sense, AstraZeneca synthesized AZD 4619 (Scheme 12.41, *S*-**129**), an α agonist, by means of an enzymatic dynamic kinetic resolution (DKR) of the corresponding racemic thioester, using an organic base to promote the racemization,[204] as shown in Scheme 12.41.

The thioester *rac*-**128** was resolved with *Pseudomonas cepacia* lipase in the presence of a tert-amine base, trioctylamine. The required *S* acid is stable, and residual *R*-thioester was racemized by deprotonation and reprotonation and catalyzed by the organic base, which cannot make a similar undesired racemization step of *S*-**129**, because the α protons of the carboxylate product are not acidic enough to be deprotonated by tert-amine bases. Although this process was scaled to grams, AZD 4619 was discontinued because of hepatotoxicity problems detected in phase I.[205]

Glitazars (Figure 12.6) are dual PPAR α/γ agonists that improve the lipid profile and exert an antidiabetic action—similar to a combination of a fibrate and a thiazolidinedione, so that they are considered as "two drugs in one."[206] Ragaglitazar (Figure 12.6, **130**) was discontinued by Novo Nordisk and Dr. Reddy's laboratories because of its adverse effects after detecting

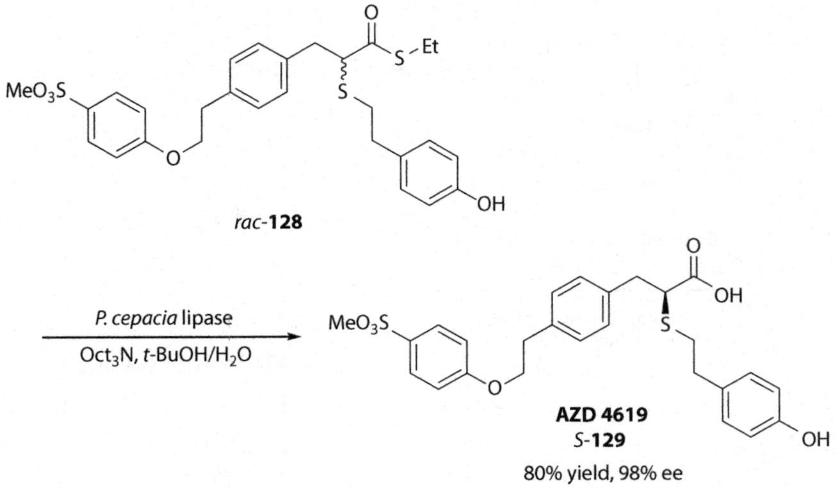

SCHEME 12.41 DKR process to synthesize AZD 46919.

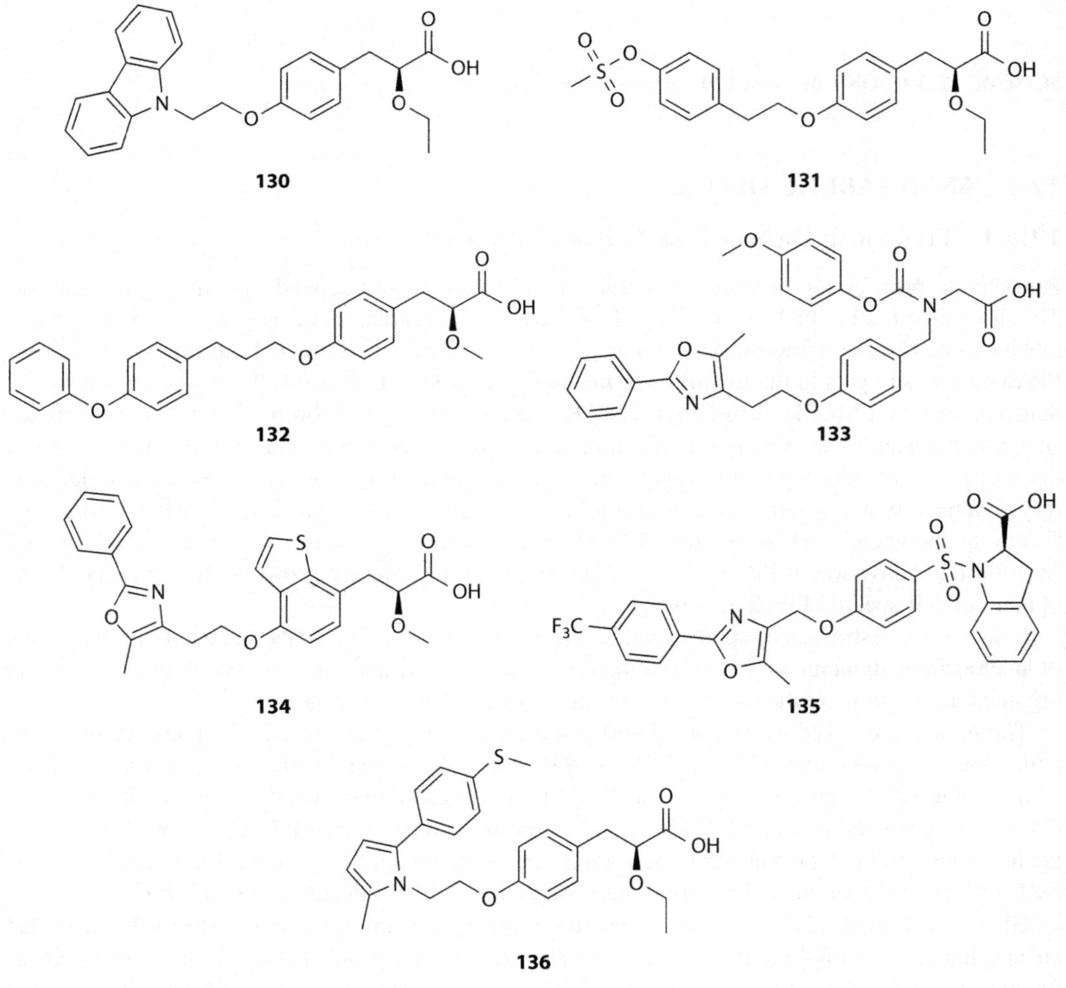

FIGURE 12.6 Some glitazars.

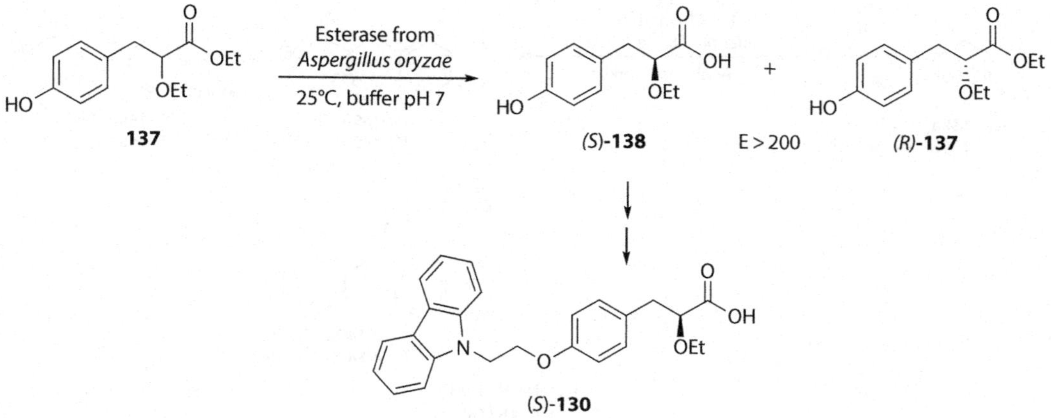

Chiral Building Blocks for Drugs Synthesis *via* Biotransformations   385

SCHEME 12.42 Enzymatic kinetic resolution of *rac*-**137** by an enantioselective hydrolysis.

urinary bladder tumor in mice. After different clinical trials, in May 2006, the two glitazars most advanced in development at that time, muraglitazar (**133**, Pargluva, developed by Bristol-Myers Squibb) and tesaglitazar (**131**, Galida, AstraZeneca), were discontinued. In fact, **133** was associated with an increased incidence of heart failure, while **130** was associated with decreased glomerular filtration.[207] Some other interesting glitazars are aleglitazar **134**, from F. Hoffmann-La Roche AG,[206] which has been discontinued in July 2013 after phase III trials, and cevoglitazar **135**,[208] which is still in phase I.

Although the previous glitazars were discontinued, its synthesis is very interesting form a biocatalytic point of view. Thus, in the synthesis of (*S*)-**130** (Scheme 12.42), the key intermediate (*S*)-2-ethoxy-3-(4-hydroxyphenyl)propanoic acid ((*S*)-**138**) was obtained through a very nice enantioselective hydrolysis of the racemic ethyl ester **137**, catalyzed by an esterase from *Aspergillus oryzae*. Finally, this process was run on a 44 kg pilot scale, producing enantiopure (*S*)-**130** in 43%–48% yields with ee values between 98.8% and 99.6%.[209]

Very recently, Brenna et al. reported two different biocatalytic approaches to obtain enantiopure precursors for the preparation of glitazars. In a first strategy, the preparation of (*S*)-**137a** or (*S*)-**137b** was initiated by a chymotrypsin-catalyzed hydrolysis of racemic **139a** or **139b**, and subsequent workup to finally obtain (*S*)-**137a** or (*S*)-**137b** with moderate overall yields[210] (Scheme 12.43).

Due to the lower yield obtained, this research group decided to change to a reductase-catalyzed strategy, depicted in Scheme 12.44, leading to the corresponding alcohols (*S*)-**137** in good yield of 78% and an excellent ee of 99%, using baker's yeast and an *in situ* substrate feeding product removal (SFPR) technique.[211] Nevertheless, it suffers from (1) an extremely low productivity (0.39 g L^{-1} d^{-1}), (2) a nonquantitative conversion, (3) a complex purification process based on the chemoselective oxidation of the by-product, an undesired allylic alcohol, and (4) a useless and counterproductive reduction of the carbonyl group (by alcohol dehydrogenases present in baker's yeast) since the final target is an ester (such as (*S*)-**137**). Therefore, Brenna and coworkers improved the methodology by using a genetically engineered enoate reductase (Old Yellow Enzyme, OYE) from *Saccharomyces cerevisiae* expressed in *E. coli*, and glucose dehydrogenase for cofactor regeneration, to improve the productivity in the preparation of ethyl-(*S*)-2-ethoxy-3-(*p*-methoxyphenyl)propanoate (EEHP) (*S*)-**140a** up to 55.6 g L^{-1} d^{-1} (74% yield, 98% ee), at a gram scale.[212]

Using this same cloned OYE, a similar strategy (Scheme 12.45) has been recently developed by the same group[213] for the preparation of enantiopure methyl (*S*)-2-bromobutyrate (97% ee) starting from the corresponding Z-a,β-unsaturated ester. By using baker's yeast, it was possible to prepare the corresponding (*S*)-2-bromobutyric acid with good ee (97%) starting from the same substrate. These enantiopure

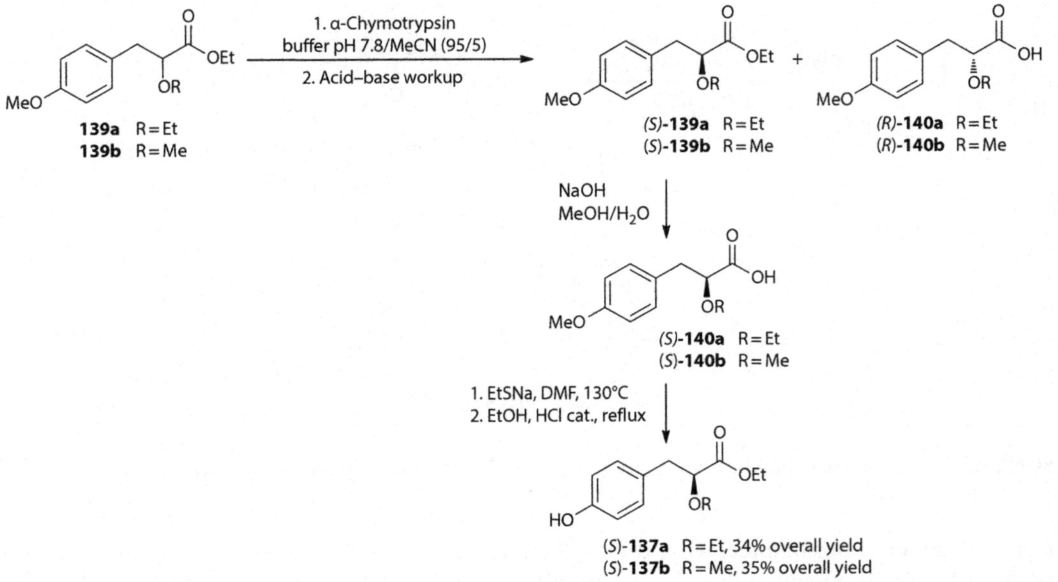

SCHEME 12.43 Enzymatic kinetic resolution of *rac*-**139a** and *rac*-**139b** by an enantioselective hydrolysis.

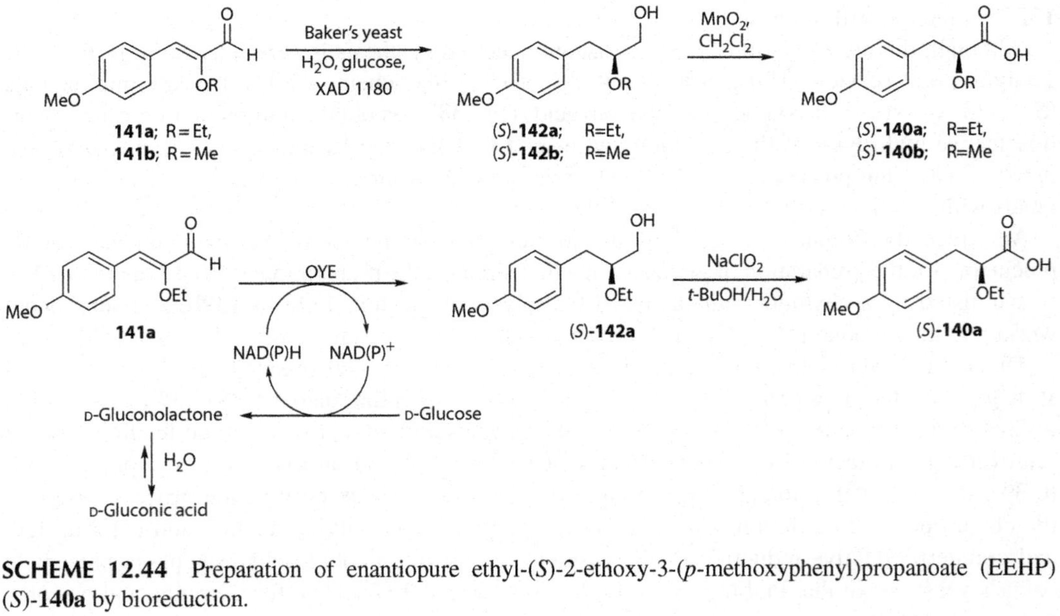

SCHEME 12.44 Preparation of enantiopure ethyl-(*S*)-2-ethoxy-3-(*p*-methoxyphenyl)propanoate (EEHP) (*S*)-**140a** by bioreduction.

molecules can be useful as chiral building blocks in the preparation of compounds such as **146**[214] or **147**,[215] a new type of promising peroxisome proliferator–activated receptor γ modulator for the treatment of T2DM, a very active research area.[216]

Very recently, in June 2013, the Indian company Zydus Cadila has presented saroglitazar (Lipaglyn™, Figure 12.6, **138**, the first glitazar to be approved in the world), accepted for launch in India by the Drug Controller General of India (DCGI) for the treatment of diabetic dyslipidemia or hypertriglyceridemia in patients with T2DM not controlled by stains alone.[217]

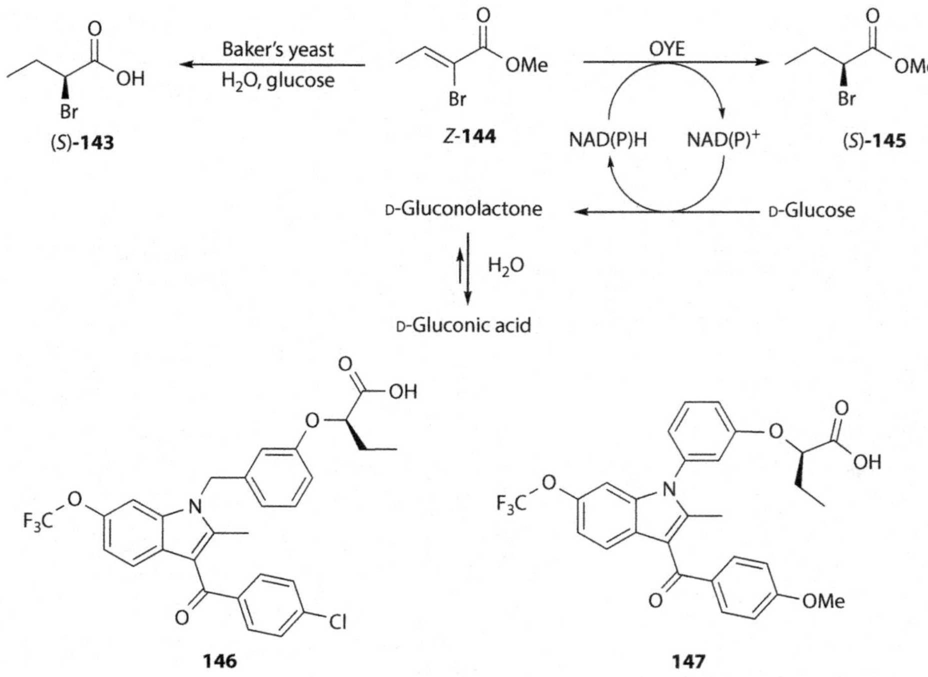

Chiral Building Blocks for Drugs Synthesis *via* Biotransformations 387

SCHEME 12.45 Preparation of enantiopure methyl (*S*)-2-bromobutyrate **145** and (*S*)-2-bromobutyric acid **143** by bioreduction.

12.4.2 GLUCAGON-LIKE PEPTIDE-1 MIMETICS AND MODULATORS

The main role of pancreatic β cells, to synthesize and secrete insulin, is somewhat modulated by a group of heterotrimeric G proteins, which are the immediate downstream targets of diverse G protein–coupled receptors (GPCRs). In this sense, different GPCRs expressed by pancreatic β cells regulate insulin secretion and therefore are potential therapeutic targets for treating T2DM.[218,219] One of them is the receptor for the glucagon-like peptide-1 (GLP-1R), which binds to and is activated by glucagon-like peptide-1 (GLP-1), a 30-amino-acid residue peptide, originated from preproglucagon, synthesized in the L-cells in the distal ileum, in the pancreas, and in the brain. GLP-1 is a member of the incretin hormone family, a term that refers to the observation that orally administered glucose results in a larger increase in plasma insulin levels and insulin-dependent decrease in blood glucose concentration when compared to the same amount of glucose given intravenously.[220] Then, GLP-1 mimetics are interesting targets in the treatment of T2DM.[221]

Thus, (*S*)-2-amino-3-(6-*o*-tolylpyridin-3-yl)propanoic acid (*S*)-**148**, Scheme 12.46, is a key intermediate needed for the synthesis of GLP-1 mimetics or GLP-1R modulators: its synthesis has been described using three different enzymatic procedures[222]; in the first one, depicted in Scheme 12.46, (*S*)-**148** was prepared (gram scale) in 68% solution yield and 54% isolated yield (100% ee), starting from racemic **148** using a recombinant (*R*)-amino acid oxidase from *Trigonopsis variabilis*, cloned and overexpressed in *E. coli* and then immobilized on Celite, and an (*S*)-amino acid dehydrogenase from *Sporosarcina ureae*. The cofactor NADH required for the reductive amination reaction was regenerated using formate and formate dehydrogenase.

In a second strategy (Scheme 12.47), (*S*)-**148** could be prepared in 73% isolated yield with 99.9% ee from racemic amino acid 1 using the same initial enzyme, (*R*)-amino acid oxidase from *T. variabilis* expressed in *E. coli*, but now in combination with an (*S*)-aminotransferase (purified from a soil organism identified as *Burkholderia* sp., also cloned and expressed in *E. coli*), and using (*S*)-aspartate as amino donor. This procedure had the advantage that both enzymes could be added at the

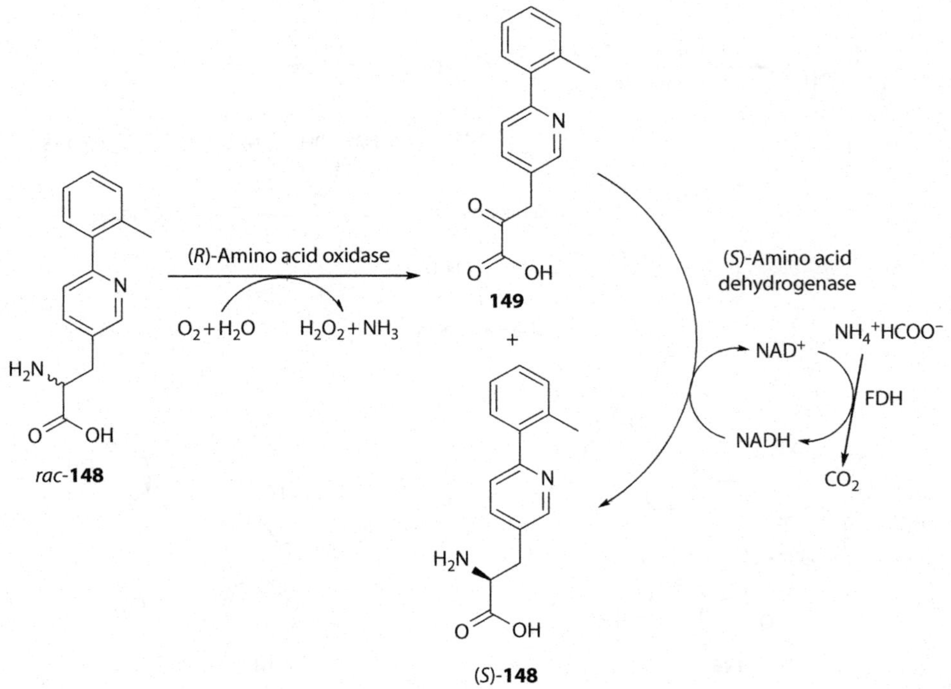

SCHEME 12.46 Synthesis of (*S*)-**148** using a (*R*)-amino acid oxidase and an (*S*)-amino acid dehydrogenase, according to Chen et al.[222]

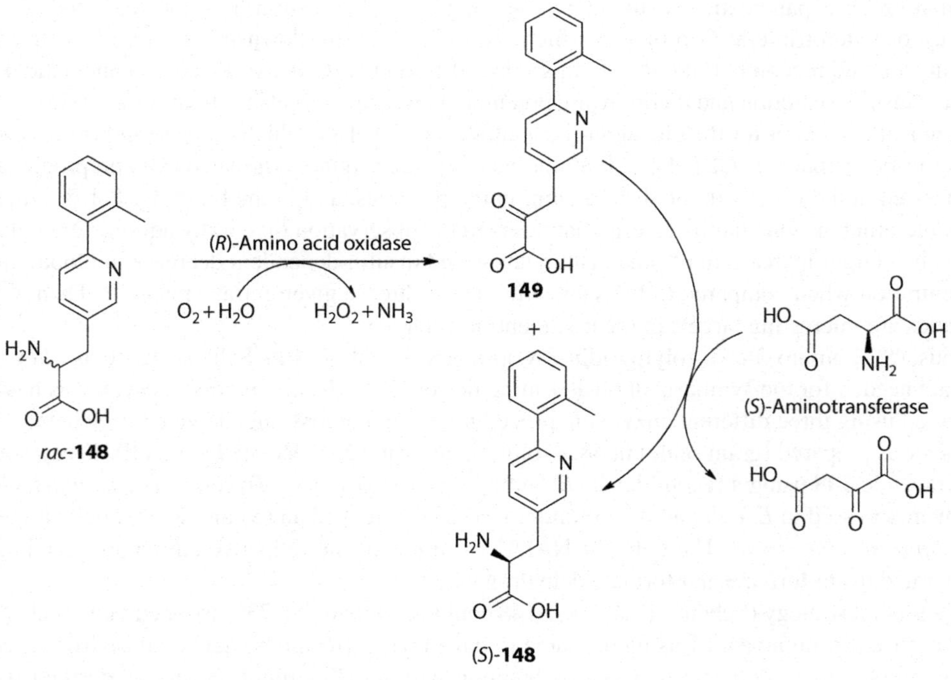

SCHEME 12.47 Synthesis of (*S*)-**148** using a (*R*)-amino acid oxidase and an (*S*)-aminotransferase, according to Chen et al.[222]

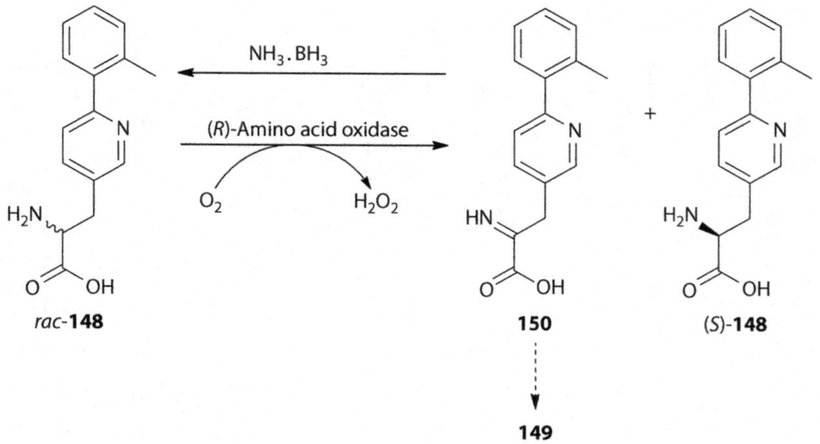

Chiral Building Blocks for Drugs Synthesis *via* Biotransformations 389

SCHEME 12.48 Synthesis of (*S*)-**148** using a (*R*)-amino acid oxidase combined with a chemical racemization protocol, according to Chen et al.[222]

start of the reaction in a one-pot system, and several batches containing 9.11 g (15 g of the monosulfate monohydrate, corrected for potency) of *rac*-**148** were run in a 2 L reactor at 30°C for 22 h for the purpose of developing a procedure for a further scale-up of the bioconversion, to produce 85% yield (73% after crystallization), ee 99.9%. Hence, the reaction was scaled up with 607 g (1 kg of the monosulfate monohydrate corrected for potency) of *rac*-**148** to give a 66% isolated yield of (*S*)-**148** as the monosulfate monohydrate, ee 99.9%.

Finally, a chemoenzymatic dynamic resolution of *rac*-**148** by (*R*)-selective oxidation was developed using Celite-immobilized (*R*)-amino acid oxidase, in combination with chemical imine reduction using borane-ammonia complex (Scheme 12.48). Before the enzyme-bound imine **150** hydrolyzes to the keto acid **149**, borane-ammonia reduces it to regenerate *rac*-**148** in a dynamic resolution process. Through this strategy, a maximum yield of 76%–79% was obtained at pH 6.0–7.0, with ee values reaching >99.9% at pH 6–8 using 10 equiv. of borane-ammonia complex.

12.4.3 Dipeptidyl Peptidase-4 (DPP-4) Inhibitors

Dipeptidyl peptidase-4 (DPP-4), also known as adenosine deaminase complexing protein 2 or CD26 (cluster of differentiation 26), is a protein playing a pivotal role in glucose metabolism, because it is responsible for the degradation of incretins such as GLP-1. Thus, DPP-4 inhibitors are a novel pharmacological class of glucose-lowering agents that open up new perspectives for T2DM treatment because of their unique mechanism of action.[223–225] Furthermore, recently, the cardioprotective effects of these compounds have been described.[226,227]

12.4.4 Sitagliptin

Sitagliptin (Scheme 12.49, **153**) (sold under the trade name Januvia by Merck Sharp & Dohme Corp.) is an oral antihyperglycemic (antidiabetic drug) acting as a noncovalent inhibitor of (DPP-4).[228] Sitagliptin can be used either alone or combined with metformin or thiazolidinedione, another oral antihyperglycemic agent in the treatment of T2DM.[229] The chemical synthesis of sitagliptin[230] involves asymmetric hydrogenation of an enamine using a rhodium-based chiral catalyst (Rh[Josiphos]), at high pressure; nevertheless, this process is not very stereoselective, and the final product is contaminated with rhodium, so that different additional purification steps are required.

Recently, an enzymatic process has substantially improved the efficiency of sitagliptin manufacturing[231,232]; in fact, using an engineered transaminase (developed at Codexis), a biocatalyst with

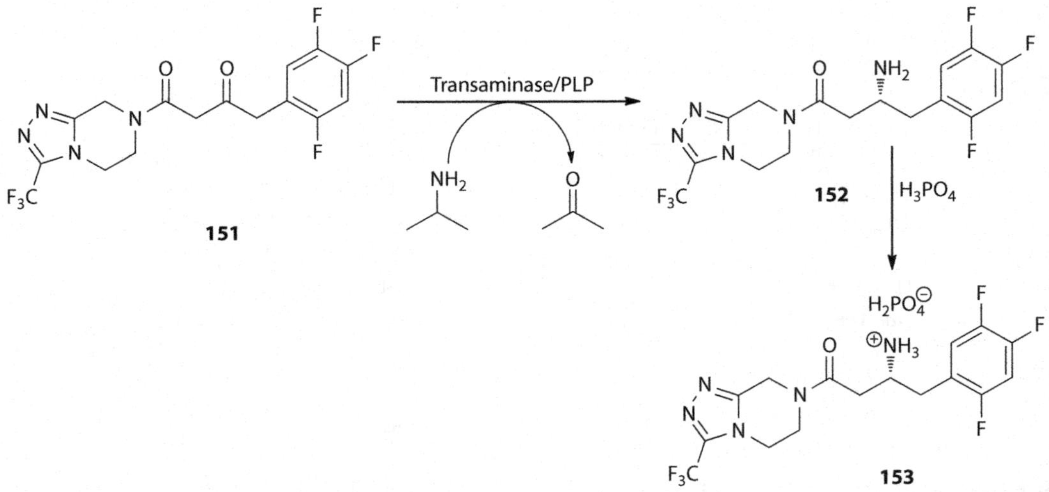

SCHEME 12.49 Enzymatic preparation of sitagliptin **153**.

broad applicability toward the synthesis of chiral amines, an improved protocol has been implemented. Under optimal conditions, the best variant converted 200 g/L prositagliptin ketone **151** (Scheme 12.49) to sitagliptin with a 92% yield and an enantiomeric excess higher than 99%, by using 6 g/L enzyme in 50% DMSO. The biocatalytic process provides sitagliptin with a 10%–13% increase in overall yield compared to the chemical process, a 53% increase in productivity (kg/L per day), a 19% reduction in total waste, the elimination of all heavy metals, and a reduction in total manufacturing cost. Furthermore, the enzymatic reaction is run in multipurpose vessels, so that specialized high-pressure hydrogenation equipment is no longer needed. Full details of this process, which obtained the Presidential Green Chemistry Challenge Award (Greener Reaction Conditions Award) from the U.S. Environmental Protection Agency in 2010 (http://www.epa.gov/greenchemistry/pubs/pgcc/past.html), can be found in a recent publication from Moore et al.[233]

12.4.5 Saxagliptin

Saxagliptin (Onglyza®, **157**) is another inhibitor of dipeptidyl peptidase-4 (DPP-4) developed by Bristol-Myers Squibb.[234] This compound inhibits DDP-4 by covalent bonding to the catalytic serine present in DDP-4 active site.[228] Its synthesis[235] requires (S)-N-Boc-3-hydroxyadamantylglycine **156** as a key chiral intermediate. For its preparation, a process for the conversion of the keto acid **154** to the corresponding amino acid **155** using (S)-amino acid dehydrogenases was developed, as depicted in Scheme 12.50. A modified form of a recombinant phenylalanine dehydrogenase cloned from *Thermoactinomyces intermedius* and expressed in *Pichia pastoris* as well as in *E. coli* was used for this process development and scale-up. NAD+ produced during the reaction was recycled to NADH using formate dehydrogenase cloned and overexpressed in *E. coli*. The modified phenylalanine dehydrogenase contains two amino acid changes at the C-terminus and a 12-amino-acid extension of the C-terminus.[236] Production of multikilogram batches was originally carried out with extracts of *P. pastoris* expressing the modified phenylalanine dehydrogenase from *T. intermedius* and endogenous formate dehydrogenase. The reductive amination process was further scaled up using a preparation of the two enzymes, formate dehydrogenase and phenylalanine dehydrogenase, expressed in single recombinant *E. coli*. The amino acid **155** was directly protected as its Boc derivative **156** without isolation to afford the intermediate. Yields before isolation were close to

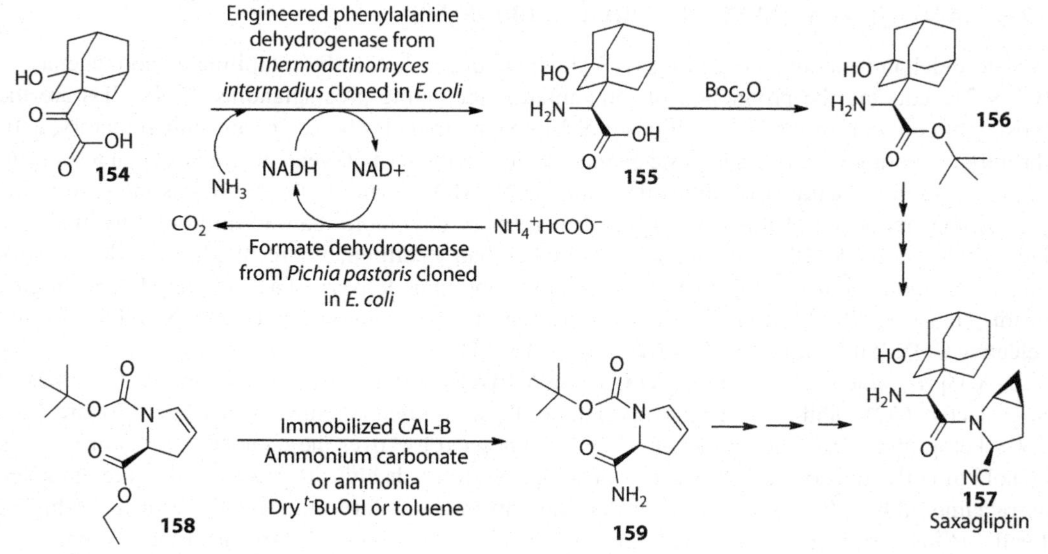

Chiral Building Blocks for Drugs Synthesis via Biotransformations 391

SCHEME 12.50 Enzymatic preparation of two intermediates in the synthesis of saxagliptin.[157]

98% with 100% ee. This process has now been used to prepare several hundred kilograms of **156** to support the development and manufacturing of saxagliptin.

Also, (5S)-5-aminocarbonyl-4,5-dihydro-1H-pyrrole-1-carboxylic acid,1-(1,1-dimethylethyl)-ester **159** is required in the synthetic scheme for obtaining saxagliptin. Direct chemical ammonolysis was hindered by the requirement for aggressive reaction conditions, which resulted in unacceptable levels of amide racemization and side-product formation, whereas milder two-step hydrolysis condensation protocols using coupling agents such as 4-(4,6-dimethoxy-1,3,5-triazin-2-yl)-4-methylmorpholinium chloride (DMT-MM) were compromised by reduced overall yields.[237] To address this issue, a biocatalytic procedure has been developed, based upon the CAL-B-mediated ammonolysis of (5S)-4,5-dihydro-1H-pyrrole-1,5-dicarboxylic acid, 1-(1,1-dimethylethyl)-5-ethyl ester **158** with ammonium carbamate to furnish **159** without racemization and with low levels of side-product formation.[238] Experiments utilized process stream ester feed, which consisted of ~22% w/v (0.91 M) of the ester in toluene. Since the latter precluded the use of free ammonia due to its low solubility in toluene, solid ammonium carbamate was employed. Reactions were performed using a mixture of neat process feed, ammonium carbamate (71 g/L, 2 mol equiv. of ammonia), and biocatalyst (25 g/L) and shaken at 400 rpm, 50°C. Under these conditions, CAL-B provided racemization-free amide with yields of 69%, together with 21% of side products (by HPLC). The inclusion of drying agents such as calcium chloride gave significant improvement (79% amide and 13% side products), as well as the use of soda lime and Ascarite, respectively, at 200 g/L in the reaction headspace (increase in amide yield to 84% and 95%), this presumably by way of adsorption of carbon dioxide liberated from the decomposition of ammonium carbamate. A further increase in yield to 98% was attained via the combined use of 100 g/L of calcium chloride and 200 g/L of Ascarite. A prep-scale reaction with the process ester feed was used. So, in the optimized process, **158** (220 g/L) was reacted with 90 g/L (1.25 mol equiv.) of ammonium carbamate, 33 g/L (15% w/w of ester input) of CAL-B, 110 g/L of calcium chloride, and 216 g/L of Ascarite (in the headspace) and run at 50°C for 3 days. Complete conversion of ester was achieved, with the formation of 96% (182 g/L) of **159** and 4% of side products; finally, after workup, 98% potency amide of >99.9% ee was isolated in 81% yield.[238]

12.5 ANTI-INFLAMMATORY DRUGS: PROFENS

Nonsteroidal anti-inflammatory drugs (NSAIDs) are drugs that act by inhibiting cyclooxygenase-2 (COX-2), reducing the production of inflammation-mediating prostaglandins.[239] NSAIDs are the most commonly prescribed medications globally owing to their efficacy as anti-inflammatory, antithrombotic, antipyretic, and analgesic agents.[240] During the past 30 years, there has been a substantial increase in the number of clinically available NSAIDs, owing to the enormous increase in the prescribing frequency of this category of drugs. Every day, more than 30 million individuals are being exposed to NSAIDs, with annual sales higher than $6 billion; as a result, there is a lot of focus on the development of suitable dosage forms for the optimum delivery of a drug candidate belonging to this category. NSAIDs may be classified as nonselective cyclooxygenase (COX) inhibitors and selective COX-2 inhibitors (coxibs, ATC Code M01AH).

2-Arylpropionic acids (**profens, ATC Code M01AE**) are one of the most important families of nonselective COX inhibitors: the most popular of these compounds, such as ibuprofen (Figure 12.7, **160**), ketoprofen (**161**), or flurbiprofen (**162**) catalogued as "over-the-counter" (OTC) drugs (can be sold directly to a consumer without a prescription from a healthcare professional), are marketed as racemic, although it is well known that the anti-inflammatory effect of these drugs is higher for the (*S*)-enantiomer.[241] Anyhow, (*R*)-profens suffer *in vivo* interconversion to their *S* antipodes by means of an endogenous alpha-methylacyl-CoA racemase,[242–245] which has been recently proposed as a new anticancer target.[246,247] Profens marketed as only the *S*-eutomer are naproxen (**164**) and the enantiopure versions of ibuprofen and ketoprofen, named dexketoprofen and dexibuprofen, respectively.

Nevertheless, there is a great industrial interest in the production of enantiomerically pure (*S*)-2-arylpropionic acids in order to avoid nondesired effects. Anyhow, the (*R*)-enantiomer of some of these molecules presents other properties; in fact, (*R*)-enantiomers of ibuprofen, naproxen, and flurbiprofen are potent "substrate-selective inhibitors" of endocannabinoid oxygenation, so that they can maintain endocannabinoid tone in models of neuropathic pain.[248] Besides, (*R*)-flurbiprofen has shown anticancer effects,[249] what increases the interest on developing effective methodologies for the preparation of both pure enantiomers.

However, although the potential utility of biocatalysis is highly recognized in the pharmaceutical industry, the implementation of enzymatic strategies for the preparation of enantiomerically pure 2-arylpropionic acids at the industrial scale is still an open challenge. As in the case of the synthesis of other chiral molecules, despite the progress in asymmetric synthesis, the resolution of racemates

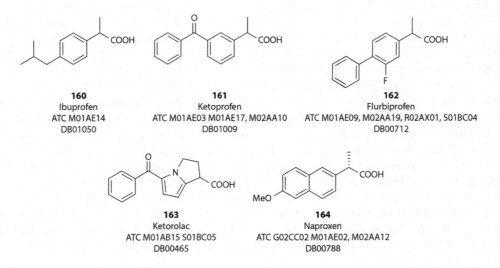

FIGURE 12.7 Some anti-inflammatory profen drugs.

Chiral Building Blocks for Drugs Synthesis *via* Biotransformations   393

is still the strategy of choice in the preparation of chiral profens. Thus, biocatalysis is presented as an attractive and greener alternative, and many efforts have been conducted for the development of innovative and economically competitive enzymatic routes for the synthesis of chiral profens.[250,251]

The use of hydrolases (particularly lipases) for catalyzing either the enantioselective hydrolysis of esters or the stereoselective esterification of acids, is probably the most common strategy to afford chiral profens,[252–258] but other enzymatic methodologies have been also successfully employed.

12.5.1 (S)-Ibuprofen (Dexibuprofen)

Although ibuprofen **160** (ATC Code M01AE14, DrugBank Code DB01050, Figure 12.7) is still widely marketed as a racemic mixture, the consumption of its pure active enantiomer (*S*)-ibuprofen (also known as dexibuprofen) is rapidly increasing. One of the first enzymatic strategies successfully implemented for the production of dexibuprofen employed the lipase from *Candida rugosa* (CRL) immobilized in the pores of a hollow fiber membrane, operating on a batch mode.[49] The use of enzymatic membrane reactors (EMR) has been extensively reported[259,260] and widely employed in the lipase-catalyzed resolution of profens.[261–264] The lipase catalyzed the enantioselective hydrolysis of *rac*-ibuprofen methoxyethylester **161**, showing enantiopreference toward the *S*-enantiomer (Scheme 12.51). The membrane allowed the use of a biphasic system in which reaction and separation occur simultaneously: the hydrophobic substrate is delivered in the organic phase and chiral ibuprofen is extracted by the aqueous phase into the lumen of the hollow fibers. Another membrane module adjusted to a higher pH is combined to separate the product. This system allows the resolution of ibuprofen in a multi-kg scale, and optically pure products are obtained with 96% ee. This EMR technology has also been applied by other research groups to develop an efficient and clean resolution of ibuprofen, employing CRL as catalyst on the enantioselective hydrolysis of ibuprofen ester,[262] or lipase from *Pseudomonas* sp. in the stereoselective esterification or transesterification of the acid substrate.[261]

Many examples can be found in literature about lipase-catalyzed enantioselective esterification of profens for the preparation of optically pure enantiomers, although esterifications are limited by the lower activity of the biocatalysts, as well as by the need to shift the equilibrium to the product side, for example, by removing the water formed in the process. For instance, López-Belmonte et al.[265] studied different variables, including water activity, influencing the esterification of racemic ibuprofen and other profens using lipase from *Rhizomucor miehei*. Recently, several commercial preparations from *Candida rugosa* have been evaluated in the same reaction,[266] but all tested lipases preferentially catalyzed the undesired esterification of the *S*-enantiomer of ibuprofen. On the contrary, lipase B from *Candida antarctica* is able to esterify the *R*-enantiomer.[267] Therefore, based on the complementary selectivities of different lipases, Miyako et al. reported a

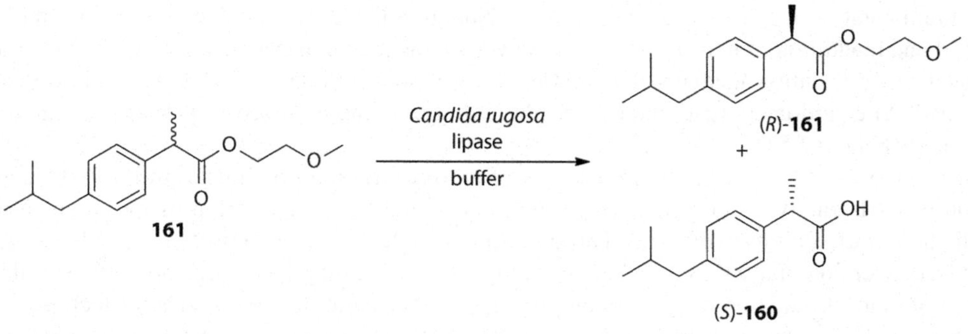

SCHEME 12.51 *Candida rugosa* lipase–catalyzed enantioselective hydrolysis of ibuprofen methoxyethylester.

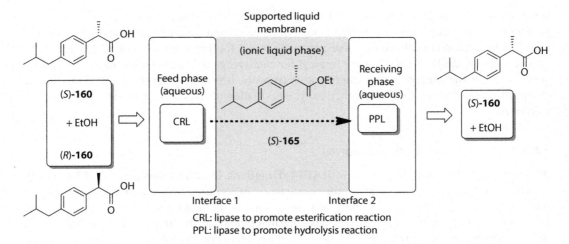

SCHEME 12.52 Lipase-facilitated enantioselective transport of (S)-ibuprofen through a supported liquid membrane.

lipase-catalyzed kinetic resolution of **160** through the use of a supported liquid membrane based on ionic liquids, which allowed a lipase-facilitated selective permeation of (S)-**160** (Scheme 12.52).[268] The first compartment contained *rac*-**160** and *Candida rugosa* lipase in a buffer–ionic liquid mixture, which catalyzed the enantioselective esterification of (S)-**160** to generate the ethyl ester (S)-**165**. The ethyl ester permeated across the membrane toward the second compartment, where it was hydrolyzed by porcine pancreatic lipase (PPL) in a buffer, affording optically pure (S)-ibuprofen (Scheme 12.52).

Carrier-free immobilization techniques have also been applied in the production of chiral ibuprofen. Margolin and coworkers carried out the selective hydrolysis of racemic ibuprofen methyl ester catalyzed by cross-linked crystals from *Candida rugosa* lipase (CRL-CLEC), yielding dexibuprofen in 87% yield and 93% ee.[269] Alternatively, Yu et al. described the enantioselective esterification of ibuprofen using cross-linked enzyme aggregates (CLEAs) of *C. rugosa* lipase.[270] After optimizing the aggregation agent type, lipase and glutaraldehyde concentration, and pH, the carrier-free immobilized lipase showed a 1.8-fold increase in enantioselectivity compared to free enzyme.

Kinetic resolution processes suffer the inherent limitation of 50% maximum conversion. Among the different strategies developed to overcome this problem, DKR is gaining more and more attention in the preparation of bioactive compounds.[271] This methodology implies the coupling of the enzymatic-catalyzed kinetic resolution with the *in situ* racemization of the unreacted enantiomer. Bäckvall and group developed the DKR of primary alcohols with an unfunctionalized stereogenic center in the β-position (Scheme 12.53).[272] The racemization of the alcohol **166** took place through ruthenium-catalyzed dehydrogenation of the alcohol, followed by enolization of the aldehyde formed and readdition of hydrogen to the aldehyde by the action of the metal catalyst. This process coupled to a resolution mediated by *Burkholderia cepacia* (Amano PS-D I) lipase afforded the esterified (S)-enantiomers that could be easily hydrolyzed and oxidized to provide optically pure profens (Scheme 12.53).

Fazlena et al. described the DKR process of ethoxyethyl ibuprofen ester through the combination of CRL-catalyzed enantioselective hydrolysis and the racemization of the remnant ester by the action of sodium hydroxide. Through this methodology, dexibuprofen could be achieved in 86% conversion and 98% ee.[273] Very recently, Uzir and group have developed the simulation of a pilot plant production of (S)-**160** employing an enzymatic membrane reactor technique and considering the synthesis of the racemic ester, the DKR of the ester prepared, and the purification of the product.[263]

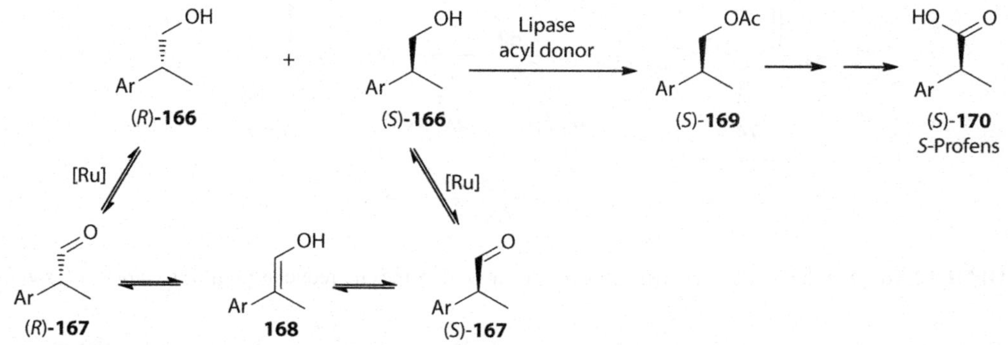

SCHEME 12.53 DKR of primary alcohols **166** as potential methodology for the synthesis of chiral profens.

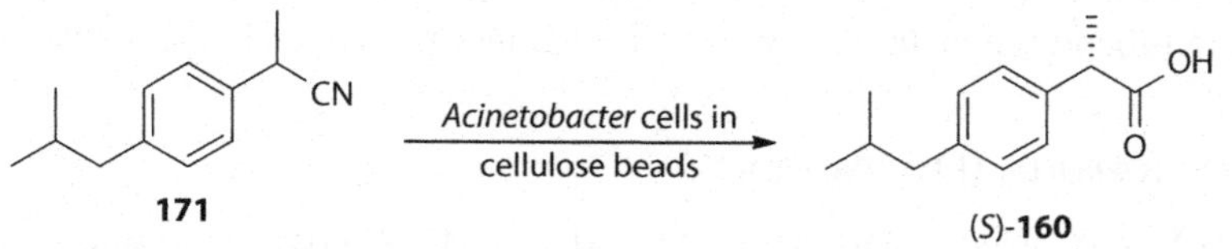

SCHEME 12.54 Stereoselective hydrolysis of ibuprofen nitrile **171**.

On the other hand, not only lipases have been employed in the resolution of racemic profens but also other types of enzymes such as nitrilases, amidases, oxidases, ene-reductases, or alcohol dehydrogenases. The synthesis of enantiomerically pure profens from the corresponding nitrile or profenamide has been carried out through kinetic resolution catalyzed by nitrilases or amidases, respectively.[274] (*S*)-**160** has been achieved through this methodology employing immobilized cells from *Acinetobacter* in cellulose porous beads with pore sizes of 10–30 μm (Scheme 12.54).[275] The beads could be reused up to 12 times maintaining the enantiopurity (99% ee), but the conversion decreases from 48% to 32%.

Galletti et al. described the DKR of 2-arylpropanals (**172**) catalyzed by alcohol dehydrogenases to afford chiral (2*S*)-arylpropanols (**173**), intermediates of the synthesis of enantiomerically pure profens (Scheme 12.55). Best results were obtained by the use of NADH-dependent horse liver alcohol dehydrogenase (HLADH). Ethanol was employed for cofactor recycling and the enzymatic resolution was coupled to the *in situ* chemical base–catalyzed racemization of the unreacted (*R*)-aldehyde, which was mediated through the formation of the achiral enolic form.[276] This strategy allowed the synthesis of (2*S*)-ibuprofenol in 93% yield and 99% ee.

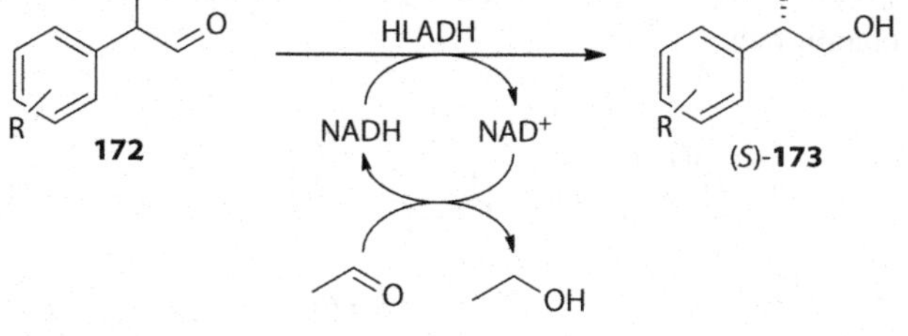

SCHEME 12.55 Synthesis of chiral intermediates for the preparation of profens through the DKR of 2-arylpropanals catalyzed by horse liver alcohol dehydrogenase (HLADH).

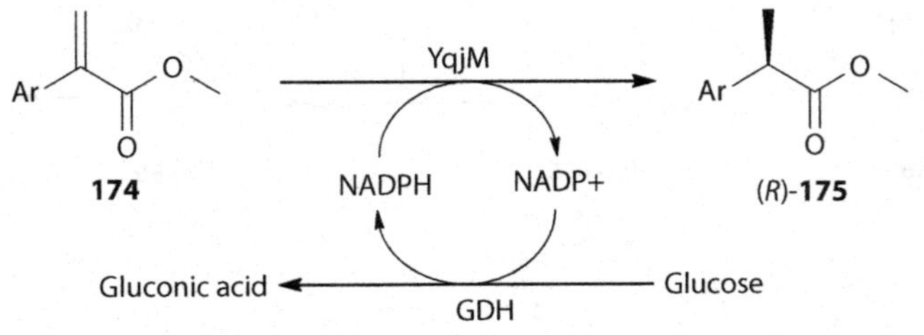

SCHEME 12.56 Synthesis of (R)-profen derivatives catalyzed by the ene-reductase YqjM from *Bacillus subtilis*.

Recently, Pietruszka and Schölzel have presented an innovative methodology for the preparation of optically pure (R)-profen derivatives using the ene-reductase YqjM from *Bacillus subtilis* (Scheme 12.56).[277] Enantioselectivity was excellent in all cases, but the conversion was decreased with increasing the size of the aromatic moiety. After optimizing the process, a semipreparative scale was conducted, using glucose dehydrogenase as cofactor regenerating system. Following this methodology, (R)-ibuprofen methyl ester was afforded in 40% conversion and 99% ee.

12.5.2 (S)-Ketoprofen (Dexketoprofen)

Ketoprofen (**161**, ATC Codes M01AE03 M01AE17, M02AA10, DrugBank Code DB01009, Figure 12.7) is one of the most useful NSAIDs, and thus the interest on synthesizing enantiomerically pure (S)-ketoprofen (dexketoprofen) has received more attention in the last decade. As in the case of preparation of other profens, kinetic resolution processes are the most employed methodologies to afford enantiomerically pure products. CRL has shown (S)-enantiopreference toward the hydrolysis of ketoprofen ester, although not very high enantiomeric excesses have been achieved; thus, different strategies have been applied to improve the enantiopurity and the yield of the process.[257,278–283]

Hence, (S)-**161** was achieved in 42% conversion and 95% ee through the CRL-catalyzed hydrolysis of ketoprofen 2-chloroethyl ester **176** under an extremely acidic condition, pH 2.5, and in the presence of Tween 80 (Scheme 12.57).[284]

The effect of surfactants over this process was investigated, finding out that Tween 80 influenced positively the activity and the enantioselectivity.[279] The enzyme was adsorbed onto a cation ion-exchange resin (SP-Sephadex C-50), integrating two processes for the partial purification and for the immobilization of the crude lipase.[257] Xi and Xu carried out the immobilization of CRL by adsorption on silica gel, which was used to construct a packed bed reactor for the continuous production of dexketoprofen.[281] Maximum enantiomeric excess, conversion of 30%, and productivity of 1.5 mg g^{-1} biocatalyst h^{-1} were afforded.

Enzyme-catalyzed enantioselective esterification of racemic ketoprofen has also successfully applied in the production of optically pure dexketoprofen. CAL-B catalyzes the enantioselective esterification of the (R)-enantiomer, allowing the recovery of unreacted chiral (S)-ketoprofen.[285] This strategy was developed by D'Antona et al. in a large-scale preparation of dexketoprofen. The use of methanol in dichloropropane allowed large scale separation, giving the desired S-ketoprofen with 96% ee as unreacted enantiomer.[286]

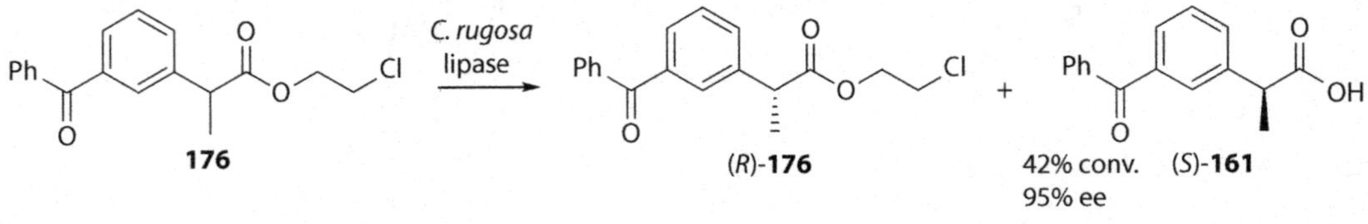

SCHEME 12.57 *Candida rugosa* lipase–catalyzed kinetic resolution of ketoprofen 2-chloroethyl ester.

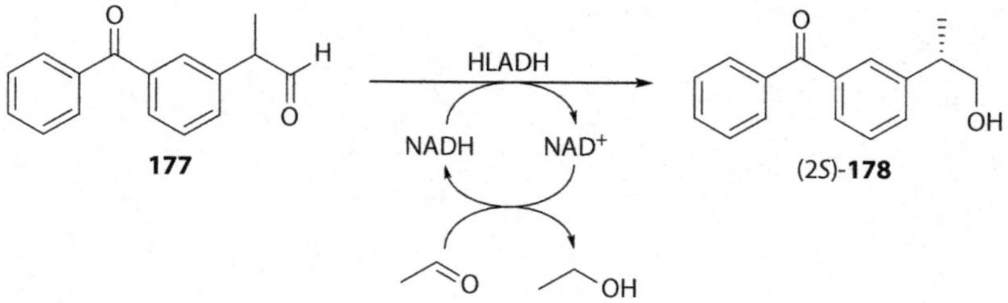

Chiral Building Blocks for Drugs Synthesis via Biotransformations

SCHEME 12.58 Stereo- and chemoselective reduction of ketoprofenal **177** to prepare optically pure ketoprofenol (2S)-**178**, intermediate in the synthesis of dexketoprofen.

Apart from lipases, other enzymes have been employed in the preparation of optically pure **161**, such as esterases, which have been widely used as biocatalysts in the kinetic resolution of ketoprofen esters.[287–289] It is worth mentioning the synthesis of enantiomerically pure (2S)-ketoprofenol ((2S)-**178**) as an intermediate for the preparation of chiral (S)-**161** through a DKR process of the corresponding racemic aldehyde, employing HLADH.[276] This enzyme showed not only high stereoselectivity but also very high chemoselectivity in the reduction of ketoprofenal, as only (2S)-**178** was obtained, without detecting traces of the possible ketone reduction by-products (Scheme 12.58).

12.5.3 (S)-FLURBIPROFEN

As it was commented earlier, both enantiomers of flurbiprofen (**162**, ATC Codes M01AE09, M02AA19, R02AX01, S01BC04, DrugBank Code DB00712, Figure 12.7) are valuable for pharmaceutical applications: the (S)-enantiomer presents anti-inflammatory properties and the (R)-enantiomer has shown anticancer effects *in vivo* and *in vitro*.[239,290]

Different biocatalytic processes have been described in the last years for the preparation of optically pure flurbiprofen through resolution methodologies, mainly through the selective hydrolysis of flurbiprofen esters.[291,292]

Molinari and group carried out the resolution of *rac*-**162** catalyzed by lyophilized mycelia of molds, which present very interesting economic and technological benefits, as avoiding costly and time-consuming purifications.[293] Among all catalysts tested, best results were obtained using dry mycelia of *Aspergillus oryzae* MIM, which catalyzed the enantioselective esterification of (R)-**162** (Scheme 12.59). Recovery of the unreacted (S)-**162** and its separation from the ester were easily accomplished by removing the biomass by centrifugation and extraction of the acid from the organic solvent.

The same group has recently described the lipase-catalyzed kinetic resolution of **162** performed in a flow-chemistry reactor, reducing the reaction time from 6 h to 15–60 min compared to the batch method. Immobilized CAL-B, which had previously shown enantiopreference toward the esterification of the (R)-enantiomer,[294,295] was employed as biocatalyst. The scalability of the process up to 80 mM concentration was also evaluated and an in-line purification step was also implemented.[296]

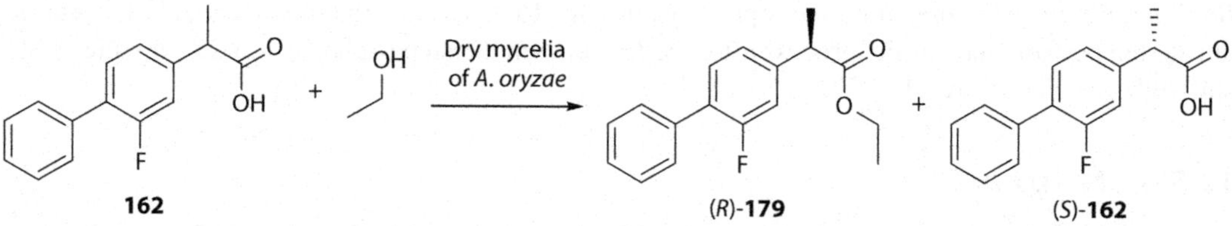

SCHEME 12.59 Kinetic resolution of flurbiprofen catalyzed by dry mycelia of *A. oryzae*.

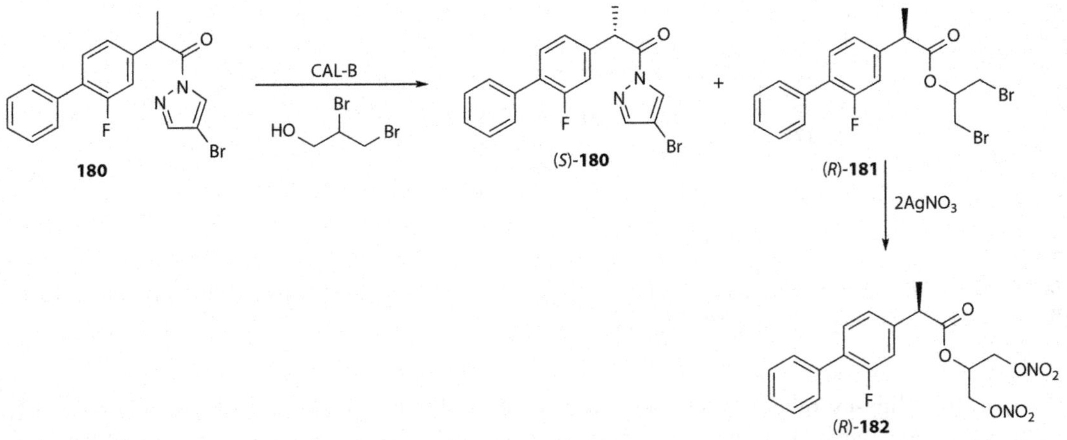

SCHEME 12.60 Kinetic resolution of racemic flurbiprofenil bromopyrazolide **180**.

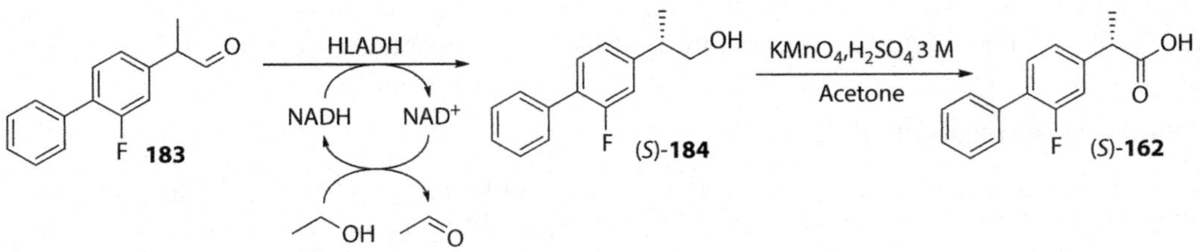

SCHEME 12.61 Synthesis of optically pure (S)-**162** through a bioreduction.

It is worth mentioning the lipase-catalyzed resolution of racemic flurbiprofen azolides developed by Ciou et al.[297] (R,S)-N-profenyl-1,2,4-triazoles have been recently described as novel excellent substrates instead of their corresponding ester analogues, for preparing optically pure profens through a lipase-catalyzed kinetic resolution process.[298,299] This group employed CAL-B as biocatalyst of the enantioselective alcoholysis of (R,S)-flurbiprofenyl-4-bromopyrazolide **180** and 2,3-dibromo-1-propanol (Scheme 12.60). As it has been shown that the preparation of prodrugs allows the development of safer profens, the chiral (R)-flurbiprofenyl-2,3-dibromo-1-propyl ester (R)-**181** obtained after the alcoholysis was then employed for the synthesis of optically pure (R)-flurbiprofenyl 2,3-bisnitrooxypropyl ester prodrug (R)-**182** (Scheme 12.60).

As it was described for ibuprofen, chiral (S)-**184** was afforded by Galletti et al. through the reduction of the correspondent 2-arylpropanal **183** catalyzed by HLADH, combined with the chemical base–catalyzed racemization of the remnant substrate in a DKR process.[276] This group demonstrated that the oxidation of the primary alcohol toward the formation of enantiomerically pure (S)-**162** could be performed maintaining the enantiopurity of the product (Scheme 12.61).

Pietruszka and Schölzel applied the strategy based on the use of the ene-reductase YqjM from *Bacillus subtilis* to the preparation of optically pure (R)-**162**, ensuring that no racemization occurred during the performance of the process after the first enzymatic step (Scheme 12.62), on which 68% and 99% ee had been reached.[277]

12.5.4 NAPROXEN

Naproxen (**164**, ATC Codes G02CC02 M01AE02, M02AA12, DB00788, Figure 12.7) was the first member of the profen family of NSAIDs that was originally marketed (1976, Syntex) as S-eutomer.[300]

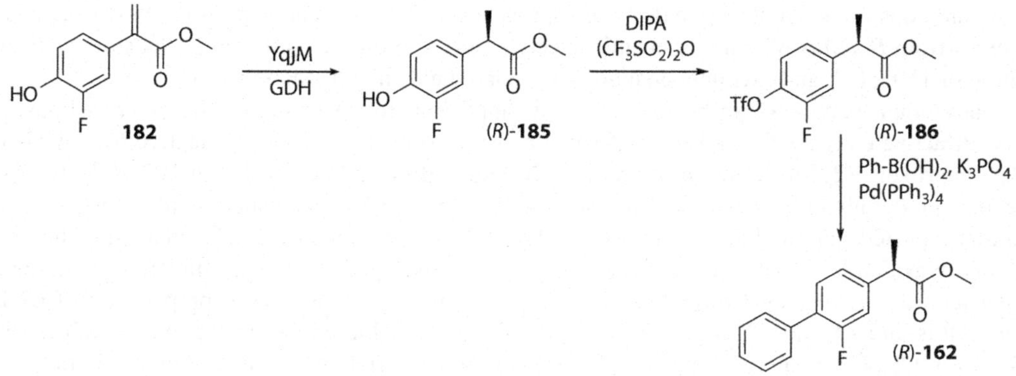

SCHEME 12.62 Chemoenzymatic synthesis of (R)-**162**.

The original process, producing 500 kg of material, was described by Harrison et al. in 1970,[301] and the racemic acid was resolved using diastereoisomer formation with cinchonidine; the same methodology based on the chiral pool was used in the first naproxen large-scale manufacturing process, in place in 1972–1975, while in the second large-scale manufacturing process (1976–1993) cinchonidine was substituted by N-alkylglucamine.[300] In those processes, one-third of the total naproxen production cost was associated with the preparation of racemic acid, while two-thirds of the production cost (primarily labor) was in the resolution–racemization, so that new strategies were developed by direct chemical asymmetric synthesis in order to solve this drawback. Thus, while the Zambon process[302] used a ketalization with (R,R)-dimethyltartrate to introduce the chirality in the synthetic schedule, Ohta et al. described a strategy based on a stereoselective hydrogenation of a precursor naphthacrylic acid using a ruthenium (S)-BINAP catalyst.[303] Other asymmetric methodologies are the hydroformylation of 2-methoxy-6-vinylnaphthalene using a rhodium catalyst complexed with BINAPHOS[304] or the direct synthesis starting from enantiopure (S)-ethyl lactate developed by Syntex in 1982.[300]

Yet, several biocatalytic approaches have been proposed for obtaining enantiopure naproxen. Thus, hydrolase-catalyzed resolutions were the first ones described (Scheme 12.63), starting in the 1970s when Gist-Brocades, in a patent, disclosed the preparation of (S)-(+)-naproxen through enantioselective hydrolysis of alkyl esters, using a biocatalyst from *Bacillus thai* and other microorganisms.[305] The resolution process was performed at low substrate concentration and required longer reaction time (5–6 days). Bianchi et al., by using a 500 mL column bioreactor filled with immobilized CRL, reported the production of 1.8 kg of optically pure (S)-(+)-naproxen after 1200 h of continuous operation with a slight loss of enzymatic activity.[306]

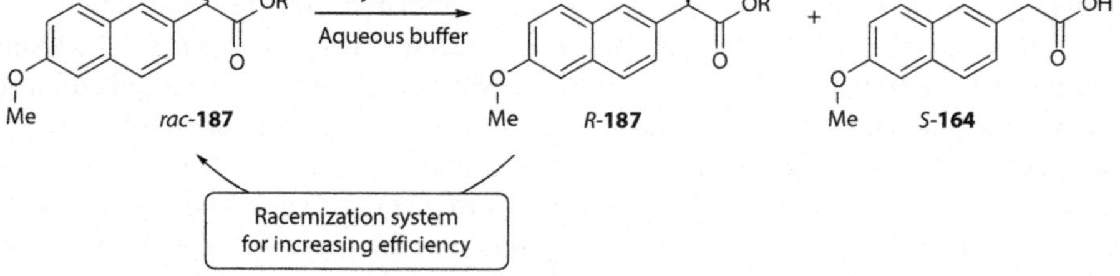

SCHEME 12.63 Hydrolase-catalyzed resolution of racemic alkyl 2-(6-methoxynaphthalen-2-yl)propanoate (*rac*-**187**).

As an industrial example for hydrolytic approaches, in 1994, researchers from DSM isolated an esterase from *Bacillus NP*, highly enantioselective toward profens,[307] allowing a substrate concentration of 150 g L^{-1}, and excellent enantiopurities and yields in 5 h reaction time.

Many other examples can be found for this approach; for instance, *Carica papaya* lipase, a very attractive enzyme,[308] has been described as very useful for the stereoselective hydrolysis of (*R,S*)-naproxen 2,2,2-trifluoroethyl ester and thioester, affording an E value of 122.[309–311] For this reaction model also lipase from *Carica pentagona* Heilborn has been successfully applied.[312] In another reported sample, Steenkamp and Brady,[313] after a previous screening, selected eight commercially available enzymes for optimization of enantioselectivity through statistically designed experiments, finding that ChiroCLEC-CR (a cross-linked enzyme crystal preparation of CRL) from Altus and ESL001-01 from Diversa provided acceptable enantiomeric excess values, but only ChiroCLEC-CR met the specification set for an industrial approach: enantiomeric ratio (E) should preferably be >100, which means that it could yield (*S*)-naproxen with an enantiomeric excess of more than 98%, a substrate conversion in excess of 40% of the racemate. These same authors described the enantioselective resolution of *rac*-naproxen ester with formaldehyde stabilized carboxylesterase NP, yielding the *S*-eutomer and *R*-naproxen methyl ester (*R*-**187**) with a higher efficiency than previously reported with other enzymes,[314] at 150 g/L ester in 0.1 M sodium phosphate buffer at pH 8.75 at 45°C in the presence of 1% Tween 80, and pH maintenance with either 2.5 M NaOH or NH$_4$OH. Up to 46.9% conversion was achieved in only 5 h (13.3 g L^{-1} h^{-1}) with an ee of 99.0% and an E of 576, which is more than double that we previously attained with CLECs of CRL.[313] Furthermore, a recycling of the *R*-naproxen methyl ester *via* racemization with DBU improved the commercial viability of the process; in fact, DBU at catalytic concentrations was found to have significant influence neither on the biocatalytic resolution reaction rate nor on the quality of the product during substrate recycling experiments. With the implementation of this racemization and recycling, yield was increased to >90%. Thus, overall yield for the whole process starting with *rac*-naproxen and esterification thereof to the racemic NME, conversion, racemization, and purification was ≈77%. Final product purity was >99%, with residual surfactant, protein, and formaldehyde levels in the ppm range.[314] In another recent example, Liu et al. have reported a (*S*)-enantioselective esterase from *Bacillus subtilis* ECU0554, named BsENP01, which was cloned and overexpressed in a heterologous host *E. coli* BL21.[315] This enzyme could catalyze the selective hydrolysis of the (*S*)-enantiomer of racemic naproxen methyl ester, giving optically pure (*S*)-naproxen with 98% enantiomeric excess. A mechanic-grinding approach to substrate dispersion was also reported, an alternative to using surfactants such as Tween 80, with improved performance of the hydrolysis. Batch production of (S)-naproxen was repeatedly carried out in a solid–water biphasic system at a 2 L scale, achieving an average total yield of about 85% after ten runs with complete recycling of (*R*)-substrate.[315]

Not only isolated enzymes but also whole cells have been employed for the hydrolysis of racemic naproxen esters; hence, *Trichosporon* sp. (TSL) whole cells selectively hydrolyze racemic alkyl esters, showing very high enantioselectivity (ee > 99%, E = 500) at a substrate concentration of 100 g L^{-1}. Temperature in the range of 28°C–32°C and pH between 7 and 8 were found to be optimum for the best time–space yield and enantioselectivity, and racemization using sodium metal or sodium methoxide for enriched (*R*)-(−)-ester was also optimized to afford a final yield higher than 98%.[316]

Some examples of enantioselective esterification of racemic naproxen can be also found in literature, although as we mentioned earlier, due to the water produced in the esterification reaction, at high conversion grade, the favored hydrolysis of the enantioform preferentially recognized by the lipase occurs, with consequent final damage of the enantiomeric excess of both ester (*R*-form) and unreacted acid (*S*-form). To solve this problem, Morrone et al.[317] described the use of dimethyl carbonate alcohol donor in the esterification of naproxen in the presence of immobilized lipase B from *Candida antarctica* (Novozyme 435). The conjugation of hydrolysis of dimethyl carbonate and esterification of acid created irreversible operative conditions, depicted in Scheme 12.64, therefore allowing the recovery of *S*-naproxen in high ee (>98%).

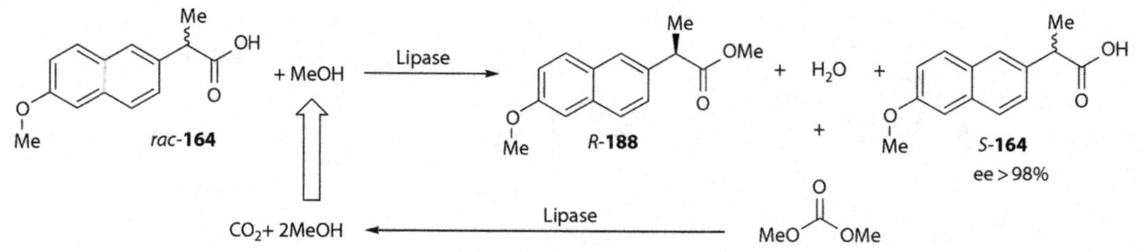

Chiral Building Blocks for Drugs Synthesis via Biotransformations 401

SCHEME 12.64 Irreversible enantioselective esterification of racemic naproxen as described by Morrone et al.[317]

For avoiding the generation of water, transesterification is an interesting alternative; as previously commented, the use of azolides ((R,S)-N-profenyl-1,2,4-triazoles) for preparing optically pure profens through a lipase-catalyzed kinetic resolution process is an attractive strategy.[298,299] Thus, for CAL-B-catalyzed alcoholysis of (R,S)-naproxenyl 1,2,4-triazolide at the optimal conditions (anhydrous MTBE as the solvent, and methanol as the acyl acceptor at 45°C), both the enzyme enantioselectivity and specific activity for the fast-reacting (R)-azolide were greatly improved in comparison with the results obtained using (R,S)-naproxenyl 2,2,2-trifluoroethyl ester as substrate.[299]

Focusing now on redox bioprocesses, enzymatic kinetic resolutions have also been reported by oxidases through stereoselective biooxidations; thus, whole cells from *Gluconobacter oxydans* showed high enantioselectivity in the oxidation of 2-arylpropanols,[318] while Miyamoto et al. reported that whole cells of *Brevibacterium* sp. have excellent enantioselectivity (E = 117) in the oxidation of 2-naphtylpropanol.[319] Unfortunately, the corresponding precursor of naproxen was resolved with poor selectivity (E = 3), so that protein engineering could be an option for further optimization of the biocatalyst.

Changing to bioreductions, in 2010, Friest et al. reported a DKR of racemic naproxenal **189** using a thermophile dehydrogenase from the hyperthermophile *Sulfolobus solfataricus* (SsADH),[320] affording the synthesis of (S)-naproxenol S-**190** with excellent enantiopurities on a gram scale. The higher solubility of the aldehydes **189** at elevated temperatures allowed the reduction of the amount of water-soluble cosolvent to 5% EtOH, which also served as auxiliary substrate for cofactor regeneration. For working up, cooling down at room temperature allowed product precipitation and separation by filtration; in this way, the very stable SsADH could be recycled and reused for up to 5 consecutive cycles (Scheme 12.65).

Finally, another alternative for profen synthesis is the enzymatic decarboxylation of precursor prochiral malonic acid esters **191**, which desymmetrization would lead to profens in 100% theoretical yield without any required racemization reaction (Scheme 12.66). Arylmalonate decarboxylase (AMDase) from *Bordetella bronchiseptica*, a suitable enzyme for this reaction, was first described

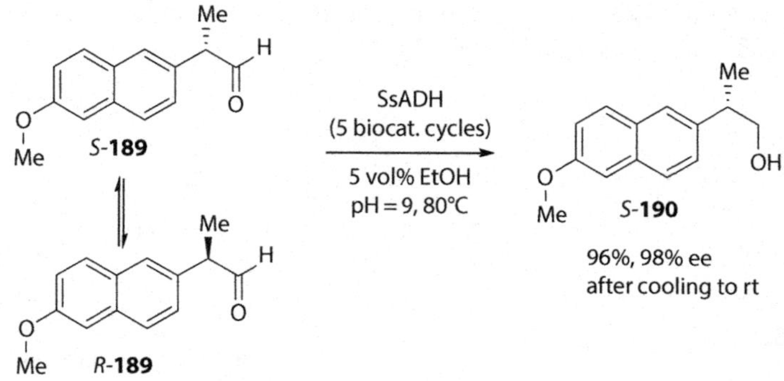

SCHEME 12.65 DKR of naproxenal **189** by bioreduction.

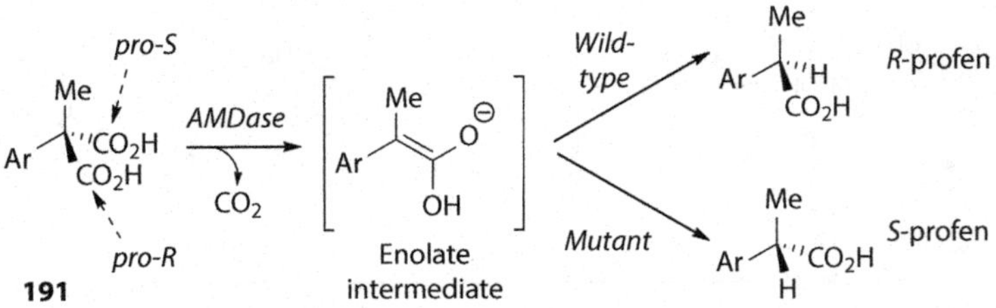

SCHEME 12.66 Desymmetrization of 2-aryl-2-methyl malonate esters **191**.

in the early 1990s[321]; the activity and enantioselectivity are excellent in the synthesis of naproxen but are much lower with the precursors of ketoprofen and ibuprofen, the latter not being converted at all. Unfortunately, wild-type AMDase produces the undesired (*R*)-profens[321]; for this reason, this enzyme has been genetically modified[322] and its variant G74C was shown to display also a racemizing activity toward profens, therefore being able to produce *S*-profens. Finally, using three rounds of structure-guided directed evolution, the catalytic activity of the (*S*)-selective arylmalonate decarboxylase variant G74C/C188S could be increased up to 920-fold,[323] and the best variant had a 220-fold improved activity in the production of (*S*)-naproxen with excellent enantioselectivity (>99% ee).

12.5.5 (*S*)-Ketorolac

Ketorolac (*rac*-5-benzoyl-2,3-dihydro-1*H*-pyrrolizine-1-carboxylic acid, **163**, ATC Codes ATC M01AB15 S01BC05; DrugBank Code DB00465, Figure 12.8) is a chiral NSAID marketed as the racemic mixture, in tromethamine salt form (Figure 12.8). The anti-inflammatory activity of the levorotatory isomer of the drug is twice that of dextrorotatory isomer.[324,325] Ketorolac is administered as tromethamine salt **192** orally, intramuscularly, intravenously, and as a topical ophthalmic solution.[240]

The classical synthetic approaches to racemic ketorolac, as summarized by Muchowski,[326] involve different methodologies; hence, in 1978, Muchoswki and Kluge, from Syntex, firstly described a flexible route leading to a ketorolac precursor molecule through an initial Hantzsch synthesis of a conveniently functionalized pyrrole, and a subsequent cyclization to yield the bicyclic scaffold, reporting 18%–20% overall yield.[327] The first large-scale synthesis, described by Franco et al.,[328] started directly from pyrrole, and the cyclization to furnish the bicyclic structure proceeded *via* a base-catalyzed intramolecular displacement of methanesulfinate ion at position 5 of a 2-benzoyl pyrrole by the sodium enolates of properly disposed substituted malonate esters. The third-generation synthesis, depicted in Scheme 12.67, also starting from pyrrole,[329] obtained the bicyclic structure *via* an initial 2-alkylation of **193** with dihalide **194** and a subsequent intramolecular nucleophilic substitution of the resulting **195** using phase transfer catalysis. An acylation and ulterior hydrolysis afforded racemic **163**, with an overall 45% yield.

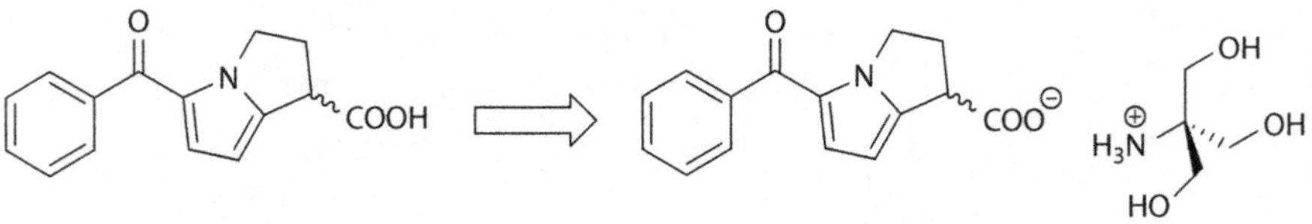

FIGURE 12.8 Ketorolac and ketorolac tromethamine.

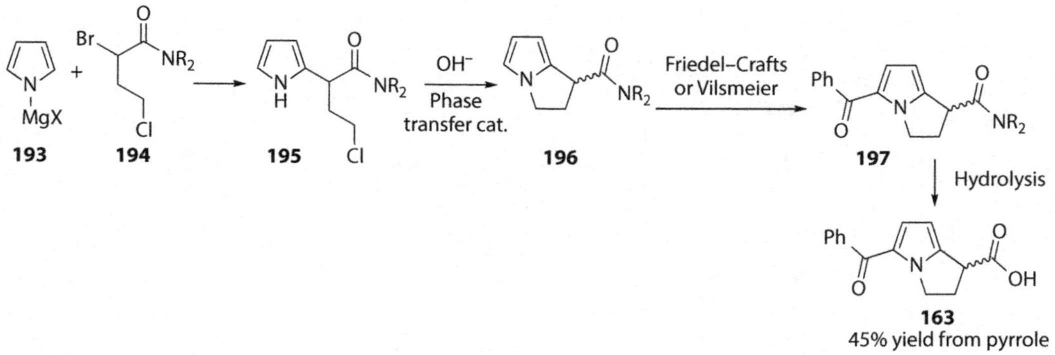

SCHEME 12.67 The third-generation synthesis of racemic ketorolac **163**.[329]

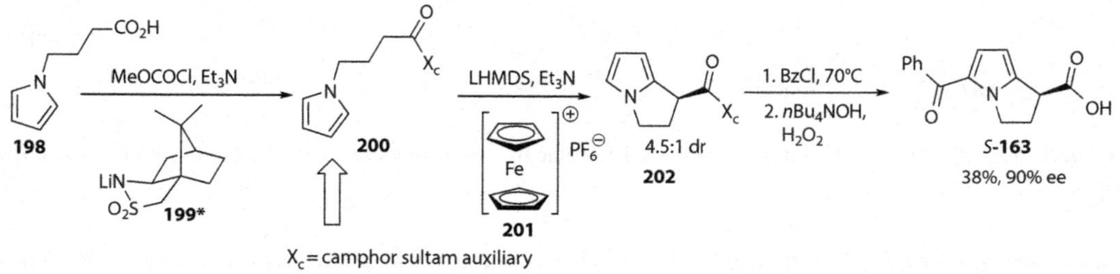

SCHEME 12.68 First asymmetric synthesis of S-ketorolac.

Although the efficiency of this drug is very high (the usual dosage is 10 mg), it has some adverse effects such as gastrointestinal bleeding, renal impairment, and platelet inhibition with altered hemostasis.[240] Therefore, one way to minimize these effects is to prescribe at the lowest dosage necessary; if the eutomer could be used, the dosage might be halved. Thus, it is important to resolve such compounds. In this sense, the first asymmetric synthesis yielding the S-eutomer of ketorolac was described in 2005 by Baran et al.[330] and is shown in Scheme 12.68. The synthesis commenced by installing the appropriate Oppolzer's sultam **199** as a chiral auxiliary on pyrrole acid **198** to furnish **200**. The stereoselective cyclization was obtained when ferrocenium hexafluorophosphate **201** was used as the oxidant, providing the annulated product **202**, where no other attempted oxidants were successful. As **202** was extremely unstable, it was immediately benzoylated and hydrolyzed using tetrabutylammonium hydroxide and hydrogen peroxide, giving S-**163** in good yield and enantiopurity. Highlights of this route include the avoidance of protecting groups, conservation of oxidation state, and the stereochemical induction observed in the key coupling reaction.[331]

Not many biocatalytic examples can be found in literature; in fact, in 1987, Fulling and Sih described the enantioselective hydrolysis of racemic ketorolac esters **203** using lipases and proteases.[332] Among some 15 lipases examined, several of these were found to be highly enantioselective, but most of them preferentially cleaved the (+)-R ester. The only exception was the porcine pancreatic lipase (PPL), which exhibited low enantiospecificity (E = 2). Using proteases, the results were much better; more concretely, the basic hydrolysis of racemic ketorolac ethyl ester XX-30 using protease from *Streptomyces griseus* reported a nice DKR (through the *in situ* racemization of the ester) affording S-ketorolac with 92% isolated yield and 94% ee (Scheme 12.69).

In another example, racemic ketorolac was resolved by esterification with various alcohols in organic solvents using *C. antarctica* lipase B,[333] as shown in Scheme 12.70. Among them, the resolving efficiency was very high in n-octanol. To check solvent effects, racemic **163** reacted with n-octanol in chlorinated solvents such as dichloromethane and 1,2-dichloromethane and each

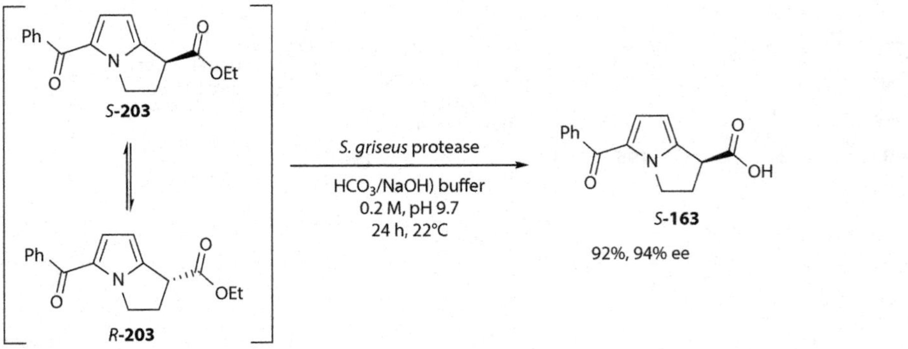

SCHEME 12.69 Dynamic-kinetic resolution of racemic ketorolac ethyl esters **203** using *S. griseus* protease.[332]

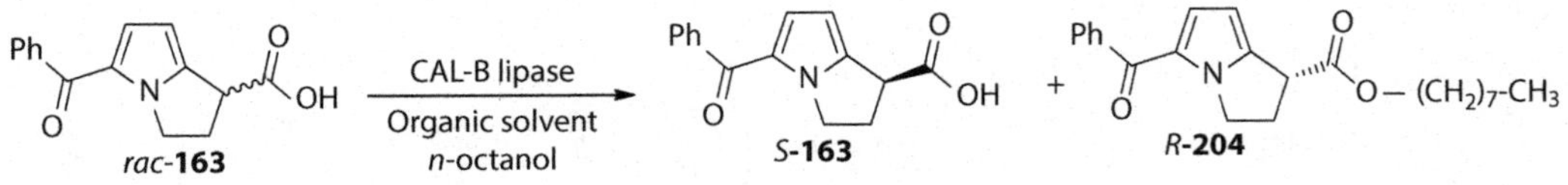

SCHEME 12.70 Kinetic resolution of racemic ketorolac by enantioselective esterification using *C. antarctica* lipase B.[333]

enantiomer was resolved, with yield around 50% and up to 99% ee. To get pure (*S*)-acid, the temperature was acceptable from 34°C to 60°C.

12.6 DRUGS FOR TREATMENT OF CARDIOVASCULAR DISEASES

The largest proportion of NCD deaths was caused by cardiovascular diseases (48%), and it has been estimated that the annual number of deaths due to these pathologies (any disease that affects the cardiovascular system, principally cardiac disease, vascular diseases of the brain and kidney, and peripheral arterial disease) will increase from 17 million in 2008 to 25 million in 2030.[44] There are different causes leading to these maladies, although atherosclerosis and hypertension are the most usual. Furthermore, there are different physiological and morphological changes related to aging that can impede normal cardiovascular functions, therefore leading to increased risk of suffering cardiovascular malfunctions.[334] Finally, behavioral risk factors (tobacco use, physical inactivity, unhealthy diet, or the harmful use of alcohol) are estimated to be responsible for about 80% of coronary heart disease and cerebrovascular disease.[335] These behavioral risk factors are mainly associated with four key metabolic and/or physiological changes—raised blood pressure, increased weight leading to obesity, hyperglycemia, and hyperlipidemia. In Section 12.4, we have presented some examples of biocatalyzed synthesis of drugs for treating hyperglycemia, and now we will present some examples showing how the use of biocatalysis can help in the preparation of different drugs for the treatment of cardiovascular diseases.

12.7 ANTICHOLESTEROL DRUGS

Raised cholesterol increases the risks of heart disease and stroke. Globally, a third of ischemic heart disease is attributable to high cholesterol, and according to the WHO (http://www.who.int/gho/ncd/risk_factors/cholesterol_text/en/index.html), raised cholesterol is estimated to cause 2.6 million deaths (4.5% of total) and 29.7 million disability-adjusted life years.

In this sense, inhibitors of 3-hydroxy-3-methylglutaryl coenzyme A (HMG-CoA) reductase, commonly known as statins (Figure 12.9), are broadly prescribed drugs in the pharmacological

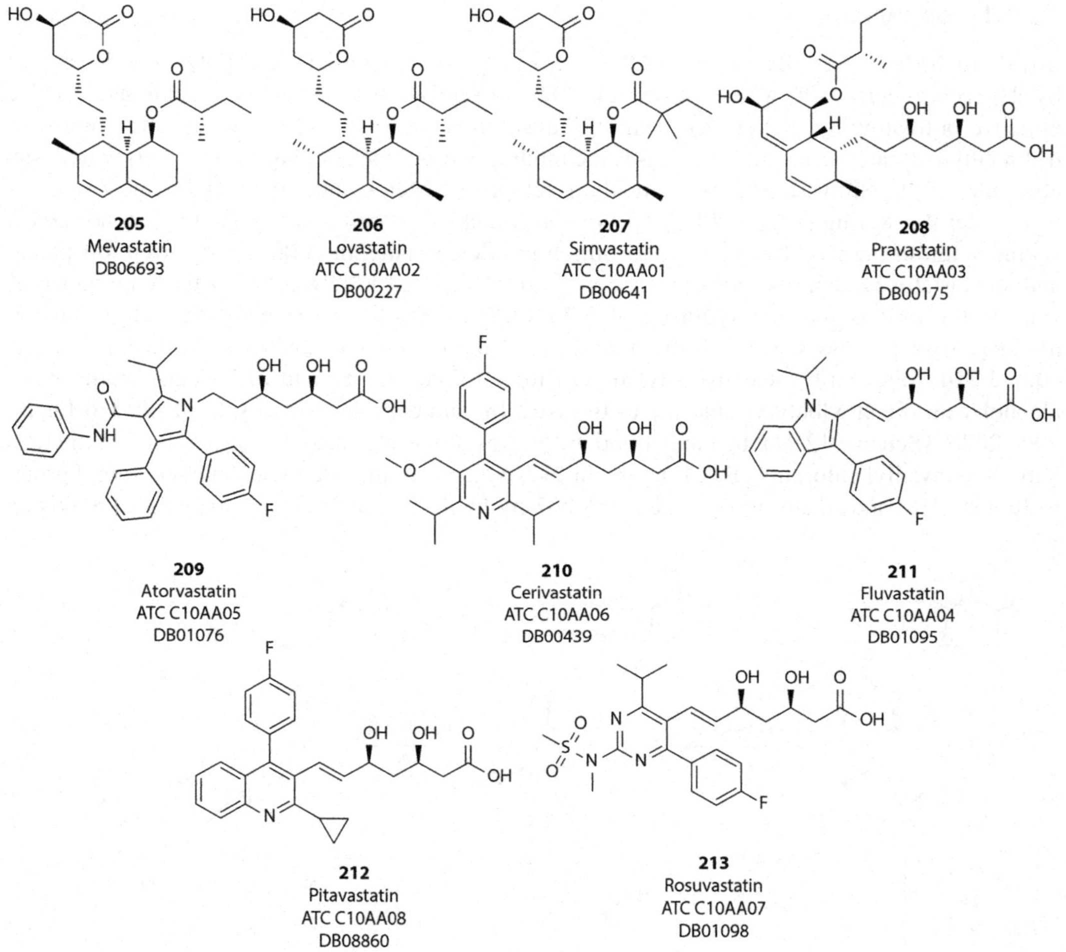

Chiral Building Blocks for Drugs Synthesis *via* Biotransformations

205 Mevastatin DB06693

206 Lovastatin ATC C10AA02 DB00227

207 Simvastatin ATC C10AA01 DB00641

208 Pravastatin ATC C10AA03 DB00175

209 Atorvastatin ATC C10AA05 DB01076

210 Cerivastatin ATC C10AA06 DB00439

211 Fluvastatin ATC C10AA04 DB01095

212 Pitavastatin ATC C10AA08 DB08860

213 Rosuvastatin ATC C10AA07 DB01098

FIGURE 12.9 Some statins, inhibitors of 3-hydroxy-3-methylglutaryl coenzyme A (HMG-CoA) reductase.

treatment of hypercholesterolemia and dyslipidemia.[336] These drugs act by reversibly and competitively inhibiting the rate-limiting step of the mevalonate pathway, so that they block the *de novo* synthesis of cholesterol and isoprenoid by-products. Although in recent years, statins have shown to possess very different biological activities not related to their cholesterol-lowering effect, such as anti-Alzheimer[337] or anticancer,[338,339] they are mainly used for reducing high cholesterol levels.

Since the discovery of the first statins from natural sources, mevastatin (Figure 12.9, **205**), also named compactin, from the fungi *Penicillium citrinum*[340] and *Penicillium brevicompactum*,[341] lovastatin (Figure 12.9, **206**), Mevinolin, found in *Aspergillus terreus*[342] and food such as oyster mushrooms[343] or red yeast rice,[344] simvastatin (Figure 12.9, **207**), Mevacor, also isolated from *Aspergillus terreus*,[345] and pravastatin (Figure 12.9, **208**), initially known as CS-514, originally identified in the bacterium *Nocardia autotrophica*,[346] synthetic and more potent compounds (Figure 12.5, **209**–**213**), also known as superstatins, were introduced in the drug market.[347,348] As can be observed, the common structure of these compounds is formed by a central core of different heterocyclic aromatic rings containing nitrogen and a lateral chain derived from (3*R*,5*R*)-3,5-dihydroxyheptanoic acid. The absolute configuration of both stereogenic centers is crucial for the activity of these compounds; hence, for its preparation, biocatalysis is a very useful tool, as we will show in the following examples.

12.7.1 SIMVASTATIN

Lovastatin (Mevacor; **206**, Figure 12.9) is a naturally occurring fungal polyketide produced by *Aspergillus terreus*,[349] while simvastatin **207** is a semisynthetic analogue of **206** and is more effective in treating hypercholesterolemia.[345] Substitution of the α-methylbutyrate side chain with α-dimethylbutyrate significantly increases the inhibitory properties of **206**, while lowering undesirable side effects. Simvastatin was originally developed by Merck under the brand name Zocor® as a cholesterol-lowering drug; in 2005, Zocor® was Merck's best-selling drug and the second-largest selling statin in the world with about $5 billion in sales. Finally, in 2006, Zocor® went off patent and simvastatin became the most-prescribed statin in the world.[350] Because of the importance of simvastatin, various multistep syntheses of **207** starting from **206** have been described, for example, a widely used process starts with the hydrolysis of the C8 ester in **206** to yield the triol monacolin J **214**, followed by selective silylation of the C13 alcohol to yield **215**, esterification of C8 alcohol with dimethylbutyryl chloride to furnish **216**, and deprotection of C13 alcohol to finally yield **207**[351] (Scheme 12.71). In another option,[352] lovastatin was treated with *n*-butylamine and tert-butyldimethylsilyl chloride (TBSCl) to obtain **217**, which was alkylated with another methyl group to furnish **218**, and finally transformed into **207** by hydrolysis and lactonization. Both multistep

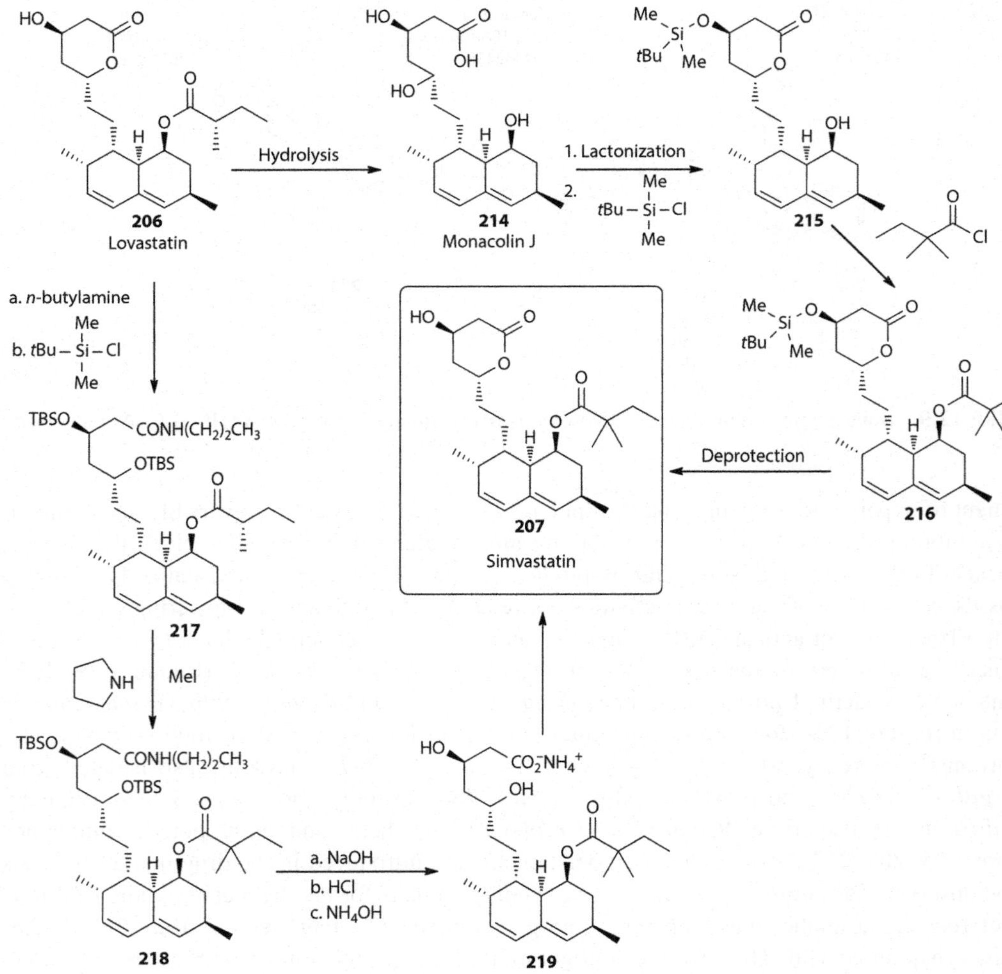

SCHEME 12.71 Chemical transformations of lovastatin **206** into simvastatin **207**.

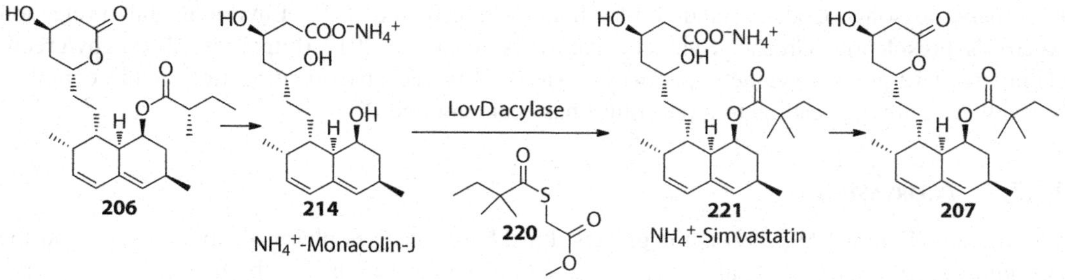

SCHEME 12.72 Biocatalyzed transformations of lovastatin **206** into simvastatin **207**.

processes shown in Scheme 12.79 were laborious, thus contributing to simvastatin being nearly five times more expensive than lovastatin.[353]

Some enzymatic transformations using lipases and esterases were investigated as alternatives to chemical hydrolysis leading to monacolin.[354,355] However, the requirement of regioselective esterification of the C8 alcohol requires protection of other reactive alcohol groups in 3, and generally leads to lowered overall yield. Therefore, a specific reagent that is able to selectively acylate C8 of 3 is important for the efficient synthesis of 2 and additional statin analogues.

In this sense, Tang and coworkers[350] described an acyltransferase (LovD) able to catalyze the last step of lovastatin biosynthesis, as shown in Scheme 12.72, by transferring α-2-methylbutyryl acyl group from dimethylbutyryl-*S*-methylmercaptopropionate (DMB-SMMP, **220**) regioselectively to the C8 hydroxyl of monacolin J **214**, the immediate biosynthetic precursor of lovastatin. The reaction proceeds via a ping-pong mechanism and is inhibited by monacolin J at moderate substrate concentrations. LovD displayed broad substrate specificity toward the acyl carrier, the acyl substrate, and the decalin core of the acyl acceptor. This same group developed a one-step, whole-cell biocatalytic process for the synthesis of simvastatin from monacolin J by using an *E. coli* strain overexpressing LovD, leading to >99% conversion of monacolin J to simvastatin without the use of any chemical protection steps.[353] The process was scaled up for gram-scale synthesis of simvastatin, also showing that simvastatin synthesized via this method could be readily purified from the fermentation broth with >90% recovery and >98% purity as determined by high-performance liquid chromatography.

Codexis improved the enzyme as well as the process chemistry to enable a large-scale simvastatin manufacturing process, by carrying out nine iterations of *in vitro* evolution, creating 216 libraries and screening 61,779 variants to develop a LovD variant with improved activity, in-process stability, and tolerance to product inhibition. The approximately 1000-fold improved enzyme and the new process pushed the reaction to completion at high substrate loading and minimized the amounts of acyl donor and of solvents for extraction and product separation. This process possesses many advantageous characteristics from a Green Chemistry point of view:

- Catalyst is produced efficiently from renewable feedstocks.
- Reduced use of toxic and hazardous substances like *tert*-butyl dimethyl silane chloride, methyliodide, and *n*-butyl lithium.
- Improved energy efficiency as the reaction is run at ambient temperature and at near atmospheric pressure.
- Reduction in solvent use due to the aqueous nature of the reaction conditions.
- The only by-product (methyl 3-mercaptopropionic acid) is recycled.
- The major waste streams generated are biodegraded in biotreatment facilities.
- Codexis' process can produce simvastatin with yields of 97%, significant when compared to <70% with other manufacturing routes.

For these reasons, Codexis and Prof. Tang obtained the U.S. Environmental Protection Agency's Presidential Greener Synthetic Pathways Award in 2012 (http://www2.epa.gov/green-chemistry/2012-greener-synthetic-pathways-award). Very recently, identification of the complete biosynthetic pathway leading to monacolin J has been reported.[356]

12.7.2 Atorvastatin

Atorvastatin (Figure 12.9, **209**) is the greatest blockbuster drug in pharmaceutical history and the best known representative of superstatins. It was first synthesized in 1985 by Bruce Roth of Parke-Davis Warner-Lambert Company (now Pfizer), which commercialized it under the name of Lipitor. Since it was approved in 1996, sales exceed US$125 billion, and the drug has topped the list of best-selling branded pharmaceuticals in the world for nearly a decade. When Pfizer's patent on Lipitor expired in the United States by the end of 2011 and in Europe in mid-2012, generic atorvastatin from other companies became available.

The chemical synthesis of atorvastatin originally described by researchers at Warner-Lambert Company[357] is shown in Scheme 12.73. The starting chiral building block was ethyl (*R*)-4-cyano-3-hydroxybutyrate (Scheme 12.73, **222**), also known as "hydroxynitrile" (**HN**), and the second stereogenic center of **224** was obtained by diastereomeric induction, using cryogenic borohydride reduction of a boronate derivative of the 5-hydroxy-3-keto intermediate **223** derived from **HN**.

12.7.3 Biocatalyzed Synthesis of Ethyl (*R*)-4-Cyano-3-Hydroxybutyrate

For preparing starting **HN**, different methodologies were described, as shown in Scheme 12.74. As shown, first synthetic protocols involved kinetic resolutions using microbes, or the opening of (*S*)-3-hydroxy butyrolactone *S*-**233** (produced from chiral pool raw materials, lactose or malic acid) to yield the corresponding ethyl (*S*)-4-bromo-3-hydroxybutanoate (*S*-BHBE, *S*-**232**).[358] The synthesis of the corresponding *S*-**233** has been described using lipase-catalyzed resolutions, as depicted in Scheme 12.75. Thus, it can be synthesized by enzymatic hydrolysis

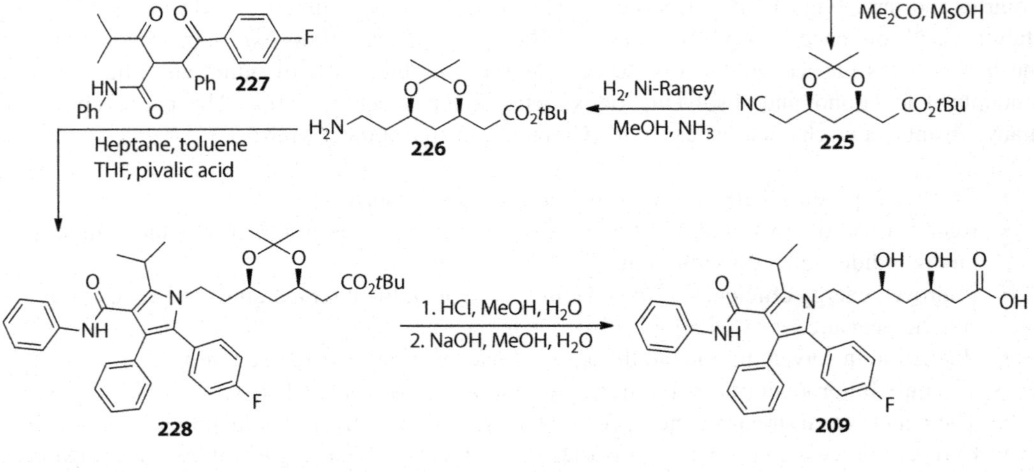

SCHEME 12.73 Chemical synthesis of atorvastatin **209**.

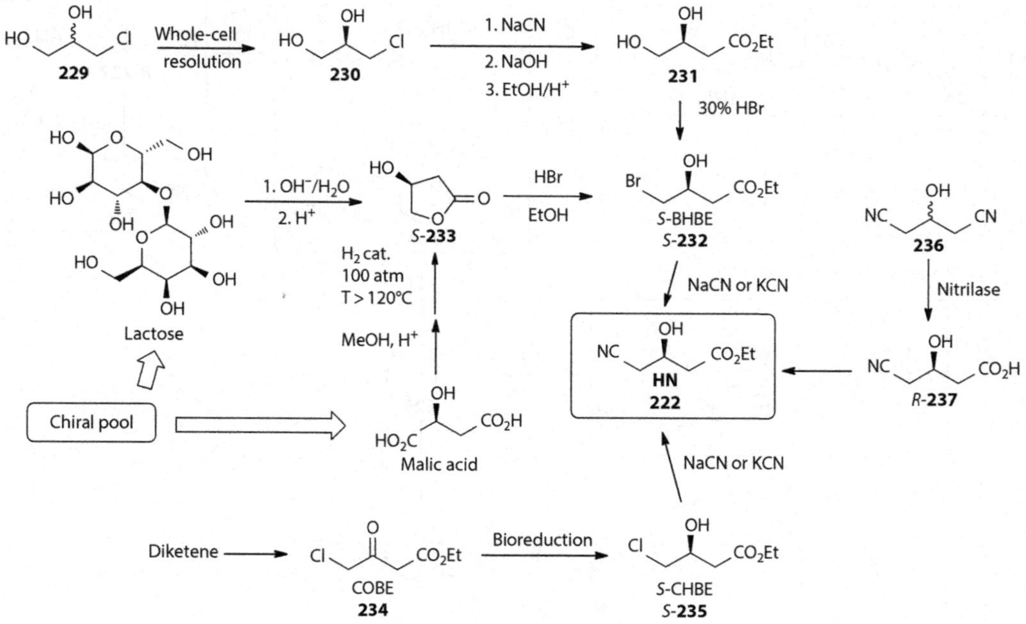

SCHEME 12.74 Different methodologies for the preparation of HN **222**.

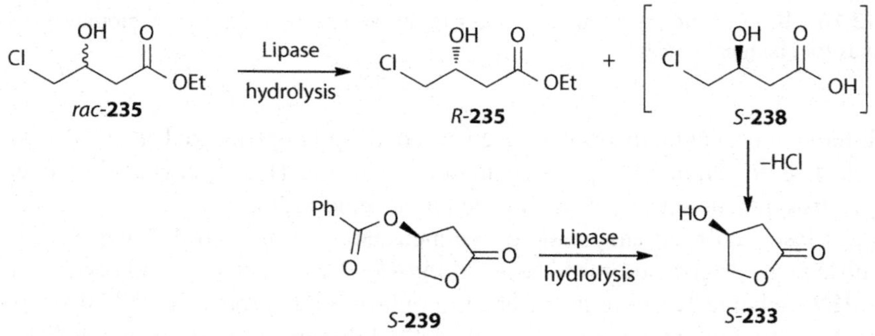

SCHEME 12.75 Different methodologies for the preparation of (S)-3-hydroxy butyrolactone **233**.

of the racemic ethyl 4-chloro-3-hydroxybutanoate (CHBE, rac-**235**) in aqueous phase.[359] The lipase stereospecifically hydrolyzed only the (S)-enantiomer, however, the resulting acid S-**238** is unstable, and it readily loses one HCl molecule to give the corresponding lactone of high enantiopurity (>99% ee). However, the enantiopurity of the lactone rapidly decreased when the process is operated at yields of more than 40%. The hydrolysis of the enantiopure benzoic ester of (S)-hydroxybutyrolactone S-**239** has also been described using CRL immobilized on amberlite XAD-7 as polymeric support, with an ee of 99%.[360] This enzymatic hydrolysis was observed to be nonstereoselective in nature, because the enzymatic hydrolysis of the racemic benzoic ester yielded the racemic lactone, so that a chiral pool precursor (L-malic acid) for this process was necessary. Anyhow, this method has been scaled up to a ton scale, with an overall yield of over 80% and a reaction time of 14 h.[361] Very recently, a platform pathway for the production of 3-hydroxyacids was described as an alternative biosynthetic route to generate the enantiopure lactone as shown in Scheme 12.75.[362]

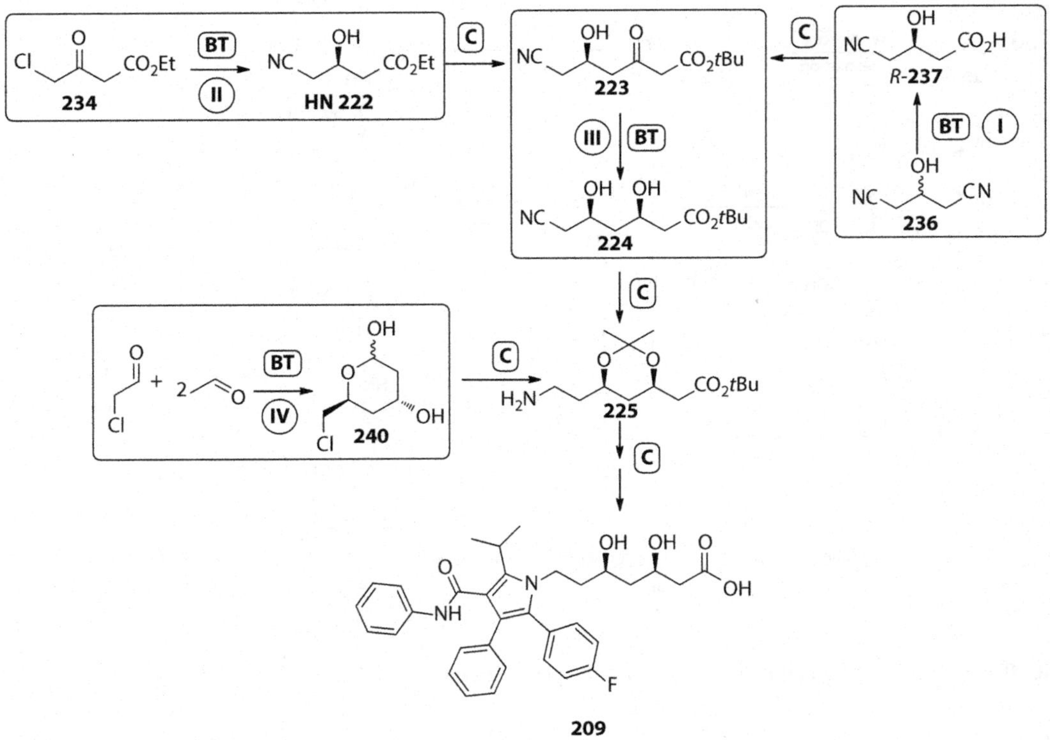

SCHEME 12.76 Biocatalytic retrosynthetic routes to atorvastatin. BT represents biotransformations step, while C stands for chemical processes.

Thus, different biocatalytic routes have been proposed and implemented at an industrial scale for the stereoselective preparation of lateral chain of atorvastatin. This process can be focused from a biocatalytic retrosynthetic scheme,[363] as depicted in Scheme 12.76.

Hence, nitrilase-catalyzed enzymatic desymmetrization of prochiral 3-hydroxyglutaronitrile **236** and subsequent esterification of the resulting (*R*)-3-hydroxy-4-cyanobutyric acid (*R*-**237**) can lead to **HN** (Scheme 12.76, route I). The use of enzymatic protocols for hydrolyzing nitriles is a green alternative compared to chemical methodologies,[364] because of the harsh reaction conditions required, demanding either strong mineral acids (e.g., hydrochloric or phosphoric acid) or bases (e.g., potassium or sodium hydroxide) and relatively high reaction temperatures. Moreover, chemical procedures sometimes give low yields due to unwanted by-products formation and generate concentrated contaminating waste salt streams (e.g., 6 mol L^{-1}) when the acid or base is neutralized prior to disposal.[365] Thus, researchers at Diversa described a wild-type nitrilase enzyme that catalyzed desymmetrization of **236** at high substrate concentration (3 M) at a lab-scale reaction, with an ee of 88%.[366] A mutant nitrilase, obtained by directed evolution using gene site saturation mutagenesis (GSSM), and showing Ala190His single mutation, resulted in an excellent biocatalyst; hence, after a 15 h reaction at 20°C, **236** was isolated in 96% yield, with an excellent ee of 98.5% and a volumetric productivity of 619 g L^{-1} d^{-1}.[367] Subsequently, Dow Chirotech, a subsidiary of Dow Chemical Company, has developed the Diversa nitrilase further into a biocatalysis process[368] and using the Pfenex expression system (a *Pseudomonas fluorescens*–based host expression system) to overproduce the enzyme. In this way, optimal reaction conditions for desymmetrization of **236** were as follows: 3 M (330 g L^{-1}), pH 7.5, 27°C, under 16 h reaction time. A conversion of 100% and 99% product ee were obtained, and the so-obtained *R*-**237** was subsequently esterified to give **HN**. Overall, a highly efficient three-stage synthesis of **HN** starting from low-cost epichlorohydrin (required to produce **236**) was achieved with an overall yield of 23%, 98% ee, and 97% purity.[368]

Later routes have involved asymmetric reduction of ethyl 4-chloroacetoacetate (COBE, **234**), produced from diketene, to furnish ethyl (*S*)-4-chloro-3-hydroxybutanoate (*S*-CHBE, *S*-**235**), using either chemical or biocatalytic reductions (Scheme 12.74); finally, the corresponding halohydrin (*S*-**232** or *S*-**235**) could be converted to **HN** by treatment with cyanide. In this sense, the enzymatic asymmetric reduction of 4-bromo-3-oxobutyrate esters has hardly been investigated compared to the corresponding chlorine analogue, because of the lower reactivity and enantioselectivity of enzymes toward brominated compounds, although (*S*)-4-bromo-3-hydroxybutanoate esters would be better substrates for the ulterior cyanide treatment; anyhow, some examples can be found in literature, starting from methyl 4-bromo-3-oxobutyrate (BAM), using *E. coli* engineered cells containing a mutant β-keto ester reductase (KER-L54Q) from *Penicillium citrinum* and a cofactor regeneration enzyme such as glucose dehydrogenase (GDH) or *Leifsonia* sp. alcohol dehydrogenase (LSADH).[369,370]

Regarding chlorine-containing oxoesters, the seminal paper of Patel et al. using glucose-, acetate-, or glycerol-grown cell (10% w/v) suspensions of *Geotrichum candidum* SC 5469[371] to produce *S*-**235** in reaction yield of 95% and optical purity of 96%, starting from 10 mg mL^{-1} of **234**, showed how bioreduction could be an interesting alternative to asymmetrical chemical reduction. Furthermore, the optical purity of *S*-**235** was increased to >99% by heat treatment of cell suspensions (55°C for 30 min) prior to conducting bioreduction at 28°C.

Ye et al.[372] have reviewed a list of different yeasts able to reduce **234** to furnish *S*-**235**, such as *Candida etchellsii*,[373] *Candida parapsilosis*,[374] *Candida magnoliae*,[375] *Saccharomycopsis lipolytica*,[373] or *Candida macedoniensis*,[376] but in many cases the stereoselectivity values obtained were not very high. Also fungi such as *Aureobasidium pullulans* CGMCC 1244,[377] *Cylindrocarpon sclerotigenum* IFO 31855,[378] *Penicillium oxalicum* IFO 5748,[374] *Botrytis allii* IFO 9430,[374] or *Pichia stipitis* CBS 6054[379] can produce *S*-**235** with a higher enantiomeric excess compared with yeasts. This same group, through genome database mining of this yeast *Pichia stipitis*, found two carbonyl reductases (PsCRI and PsCRII) leading to *S*-**235** with >99% enantiomeric excess, which were subsequently characterized, cloned, and expressed in *E. coli*.[372] Very recently, Cai et al.[380] described a substrate-coupled biocatalytic process based on the reactions catalyzed by an NADPH-dependent sorbose reductase (SOU1) from *Candida albicans* in which **234** was reduced to *S*-**235**, while NADPH was regenerated by the same enzyme via oxidation of sugar alcohols (sorbitol, mannitol, or xylitol). Optimization of COBE and sorbitol proportions yielded 2340 mM of *S*-**235** starting from 2500 mM **234**, with an enantiomeric excess of 99%. This substrate-coupled system maintained a stable pH and a robust intracellular NADPH circulation, so that pH adjustment and addition of extra coenzymes were unnecessary, therefore making this system very attractive.

Nevertheless, because of the great overall demand of **HN** required for atorvastatin synthesis (estimated to be in excess of 100 mT),[358] it is highly desirable to reduce the wastes and hazards involved in its manufacture, while reducing its cost and maintaining or, preferably, improving its quality. This has been successfully carried out on a multiton scale by Codexis (Scheme 12.77, route II) by means of a three-enzyme two-step process, which is depicted in detail in Scheme 12.78.

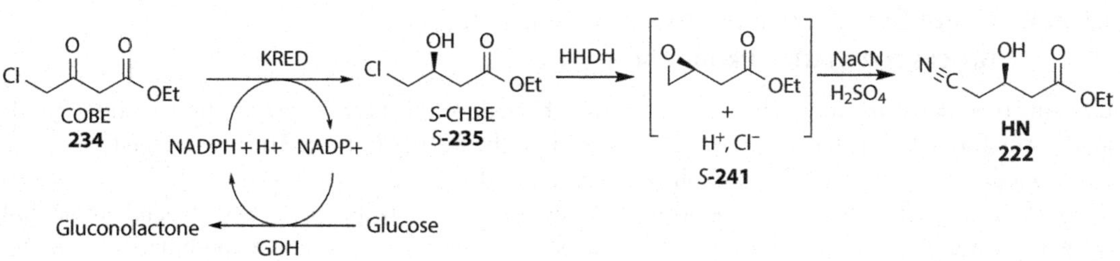

SCHEME 12.77 Codexis synthesis of HN.

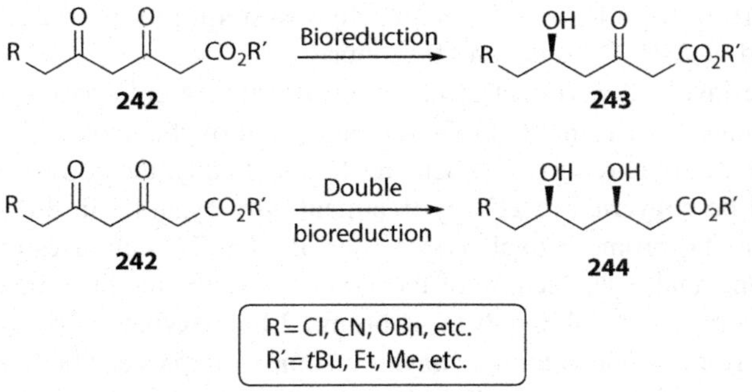

SCHEME 12.78 Bioreductions to produce chiral building blocks for statins.

Hence, the first step involves the biocatalytic reduction of **234**, using a ketoreductase (KRED) in combination with glucose and an NADP-dependent glucose dehydrogenase (GDH) for cofactor regeneration, leading to *S*-**235** in 96% isolated yield and >99.5% ee. In the second step, a halohydrin dehalogenase (HHDH), an enzyme capable of catalyzing the elimination of halides from vicinal haloalcohols, resulting in epoxide ring formation,[381] was employed to catalyze a nucleophilic substitution of chloride by cyanide, using HCN at neutral pH and ambient temperature. The efficiency and greenness of this protocol (Codexis was awarded the U.S. Environmental Protection Agency's Presidential Green Chemistry Challenge Award in 2006 for this work[382]) is based on the fact that all previous manufacturing routes to **HN** shown in Scheme 12.74 involved, as the final step, a standard but troublesome S_N2 substitution of halide with cyanide ion in alkaline solution (pH = 10) at high temperatures (80°C), being this reaction substituted in the Codexis protocol. In fact, in the S_N2 chlorine substitution, both *S*-**235** and **HN** are base-sensitive molecules, and extensive by-product formation is observed, leading to high E values.[358] Moreover, the product is a high-boiling oil, and a troublesome high-vacuum fractional distillation is required to recover **HN**, resulting in further yield losses and waste, and clearly contravening the first and sixth principles of Green Chemistry.[14] Thus, conducting the cyanation reaction under milder conditions at neutral pH, by employing the enzyme HHDH, is the key step for increasing the greenness of the overall process. Awkwardly, both wild-type KRED and GDH displayed very low activities, so that huge enzyme amounts were required to obtain an economically feasible reaction rate, originating troublesome emulsions, which hampered the subsequent downstream processing; additionally, severe product inhibition and poor stability under operating conditions. To enable a practical large-scale process, the three enzymes were optimized by *in vitro* enzyme evolution using gene shuffling technologies according to predefined criteria and process parameters, resulting in an overall process in which the volumetric productivity per mass catalyst load of the cyanation process was improved ~2500-fold, comprising a 14-fold reduction in reaction time, a 7-fold increase in substrate loading, a 25-fold reduction in enzyme use, and a 50% improvement in isolated yield.[358]

12.7.4 OTHER BIOCATALYZED SYNTHESIS OF CHIRAL BUILDING BLOCKS FOR STATINS PRODUCTION

By using bioreductions, some other strategies have been developed for the preparation of chiral building blocks for statin synthesis. Thus, as depicted in Scheme 12.76, route III, monoreduction of the corresponding 6-substituted-3,5-dioxohexanoates **242** (R = Cl, R' = *t*Bu, Scheme 12.78) to furnish **243** (similar to **223**, Scheme 12.73) has been described[383,384] by using NADP(H)-dependent alcohol dehydrogenase of *Lactobacillus brevis*. This enzyme was overexpressed in a recombinant *E. coli* and the cell extracts were then employed for carrying out the biocatalytic reactions on a gram scale, to give the corresponding hydroxyl ketoester **243** (R = Cl, R' = *t*Bu) in >99.5% ee and isolated yield of

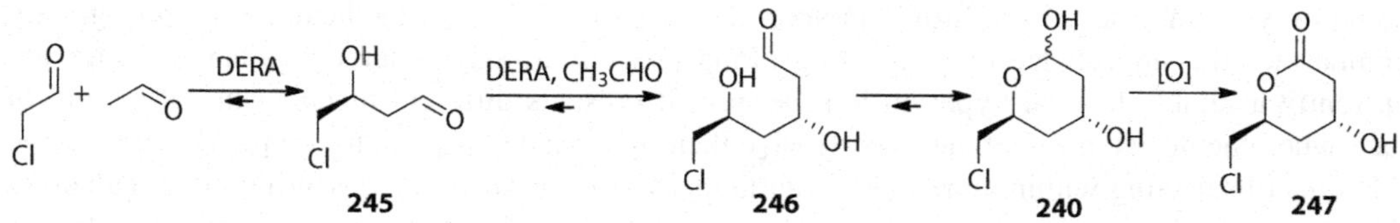

SCHEME 12.79 Aldolase-catalyzed synthesis of chiral building blocks for statins.

72%, at 24 h. Alcohol dehydrogenase itself recycles its cofactor by a substrate-coupled methodology, by oxidation of 2-propanol to acetone. This process was scaled up to 100 g scale[385] by using a fed-batch reactor, with the conversion of more than 90% attained in a total reaction time of 24 h.

More interestingly, the double reduction to obtain structures such **244** would overcome the requirement of the cryogenic borohydride reduction of the boronate derivative of the 5-hydroxy-3-oxo intermediate **223** (Scheme 12.73). Thus, whole cells of *L. brevis*, which contains two different types of alcohol dehydrogenase, are able to convert **242** (R = Cl, R′ = *t*Bu) into the dihydroxy ester **244** (99% ee in a total yield of 47.5% with reaction time being 22 h[386]). The cofactor NADP(H) was regenerated by the usual glucose metabolism of the cell. The double bioreduction has been also described using isolated enzymes, from *Acinetobacter* species; in fact, Patel et al. originally described the bioreduction of **242** (R = OBn, R′ = Et) using both whole cells and cell extracts from *Acinetobacter calcoaceticus*,[387] and some years later, they also cloned and overexpressed[388] the diketoreductase responsible for the double reduction, which was efficiently carried out with the engineered enzyme.[389] Similarly, a diketoreductase from *Acinetobacter baylyi* ATCC 33305 was cloned and heterogeneously expressed in *E. coli* by Wu et al.,[390] showing an excellent biocatalytic performance at substrate concentration around 100 g L^{-1}.[391] Interestingly, the 3D structure of this enzyme has been recently reported, and the catalytic mechanism has been revealed.[392–394]

Aldolases can also be used in the preparation of chiral building blocks for statin synthesis. This would correspond to route IV in Scheme 12.76. In fact, Gijsen and Wong[395,396] firstly described the use of 2-deoxy-D-ribose 5-phosphate aldolase (DERA) in the preparation of intermediate **240**, in a reaction mixture consisting of 133 mg of chloroacetaldehyde and 264 mg of acetaldehyde in a total reaction volume of 20 mL (Scheme 12.79). The atorvastatin intermediate lactone **247** can be easily formed by oxidation of lactol **240**. However, aldolase showed low affinity to chloroacetaldehyde and was promptly inactivated at required aldehyde concentrations, so that a huge amount of aldolase was required. Furthermore, very long reaction time of 6 days was required because of reversible nature of aldol reactions, making this process unpractical for scaling up.

Subsequent studies by Liu et al.[397] described a mutant aldolase, leading to an increased yield of **247** to 43%, in comparison to 25% for the wild-type aldolase, although the other reaction drawbacks were not overpassed. The process was markedly improved and scaled up by Greenberg et al.[398] of Diversa Corporation, by genetically modifying DERA by means of high-throughput screenings of environmental DNA libraries, focusing on chloroacetaldehyde resistance and higher productivity; in a second step, the process was further improved by using a fed-batch bioreactor, to avoid significant substrate inhibition. Thus, the final synthesis of **247** was conducted on a 100 g scale in a total reaction time of 3 h with an ee of >99.9% and a 10-fold reduction in catalyst load over previous method.[399] Very recently, the use of whole-cell systems is being evaluated for this process,[400,401] as well as new strategies for improving DERA by genetic engineering.[402]

12.8 ANTIHYPERTENSIVE DRUGS

Hypertension, or elevated blood pressure, is one of the most common risk factors for coronary artery disease, heart failure, stroke, and renal failure. Approximately 50 million Americans have a systolic/diastolic blood pressure above 140/90 mm Hg (the onset of hypertension) and most commonly during the fourth, fifth, and sixth decades of life.[403] The medical community greatly emphasizes the importance of controlling blood pressure. Today, a large number of drugs are currently available to treat hypertension, because it presents different mechanisms of action. In fact, when the decision to initiate hypertensive therapy is made, physicians are presented with the dilemma of selecting within more than 80 antihypertensive products, representing more than eight different drug classes: diuretics, sympatholytic drugs (centrally acting drugs, ganglionic blocker drugs, adrenergic neuron blocking drugs, β-adrenergic blocking drugs, α-adrenergic blocking drugs, and mixed α/β-adrenergic blocking drugs), vasodilators (arterial or arterial and venous), calcium channel blockers, angiotensin-converting enzyme inhibitors, and angiotensin receptor antagonists.[404] In the following section, we will discuss some examples of biocatalyzed synthesis of some of these compounds.

12.8.1 β-Blockers

In the years since Ahlquist's original classification of adrenoreceptors into α and β types,[405] according to their responses to different catecholamine-type agonists (norepinephrine **249**, epinephrine **248**, and isoprenaline **250**; Figure 12.10), additional small molecule agonists and antagonists have been used to allow further subclassification of α- and β-receptors into α_1 and α_2 subtypes of α-receptors and the β_1, β_2, and β_3 subtypes of β-adrenoceptors. The powerful tools of molecular biology have been used to clone, sequence, and identify even more subtypes of alpha receptors for a total of six. Currently, three types of α_1-adrenoceptors, called α_{1A}, α_{1B}, and α_{1D}, are known. (There is no α_{1C}, because identification of a supposed α_{1C} was found to be incorrect.) Currently, three subtypes of α_2, known as α_{2A}, α_{2B}, and α_{2C}, also are known.[406] At this time, however, only the α_1-, α_2-, β_1-, and

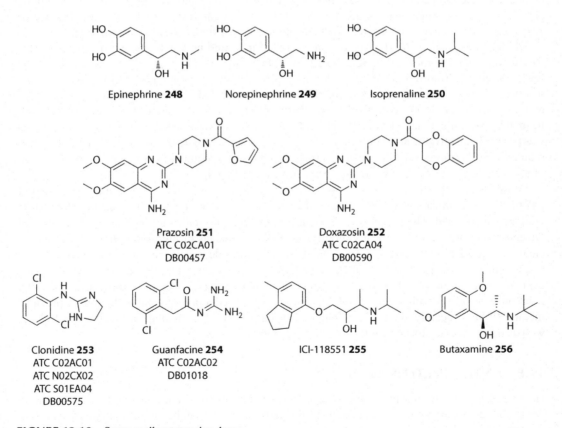

FIGURE 12.10 Some antihypertensive drugs.

Chiral Building Blocks for Drugs Synthesis *via* Biotransformations 415

β$_2$-receptor subtypes are sufficiently well differentiated by their small molecule binding characteristics to be clinically significant in pharmacotherapeutics.

Therefore, clinical utility of receptor-selective drugs becomes obvious when one considers the adrenoreceptor subtypes and effector responses of only a few organs and tissues innervated by the sympathetic nervous system. Focusing on hypertension, it is well known that the predominant response to adrenergic stimulation of smooth muscles of the peripheral vasculature is constriction, causing a rise in blood pressure. Because this response is mediated through α$_1$-receptors, an **α$_1$-antagonist** would be expected to cause relaxation of the blood vessels and a drop in blood pressure with clear implications for treating hypertension; drugs such as prazosin **251** or doxazosin **252** belong to this class (ATC C02CA). On the other hand, **α$_2$-agonists** such as clonidine or guanfacine (ATC C02AC) are also useful for treating hypertension. In addition, a smaller number of β2-receptors act on vascular smooth muscles for mediating arterial dilation, particularly to skeletal muscles, so that a few antihypertensives such as butaxamine or ICI-118551 (Figure 12.10) act through stimulation of these β2-receptors (**β2-antagonists**). Finally, adrenergic stimulation of the heart causes an increase in rate and force of contraction, which is mediated primarily by β1-receptors, so that **β1-antagonists** would slow the heart rate and decrease the force of contraction, and will be useful for treating hypertension, as well as angina, and certain cardiac arrhythmias. Inside this last kind of compounds, β-antagonists, many examples of biocatalyzed synthesis have been reported.

In fact, these β-blockers impede the binding of norepinephrine **249** or epinephrine **248** (Figure 12.10) and therefore causing inhibition of normal sympathetic effects. That is the reason why they are called sympatholytic drugs.[406] The first goal in the development of these agents was to achieve selectivity for β-receptors with respect to α-receptors. Isoprenaline **250** (Figure 12.10) was chosen as the lead compound of common structures possessing a 1-aryl-2-alkylaminoethanol motif **257**: in fact, although **250** is an agonist, it was initially selected as lead molecule because it was active only at β-receptors.[407] Thus, drugs such as dichloroisoprenaline **258** or pronethanol **259** were designed, although their use was restricted because of its carcinogenicity.

Because of the *R*-chirality of catecholamines, the preparation of *R*-enantiomers of 1-aryl-2-alkylaminoethanols was an attractive task. For instance, it is well known that only the *R*-isomer of nifenalol **261** (Figure 12.11) displays biological activity[408]; on the other hand, both enantiomers of sotalol **260** have similar class I antiarrhythmic effects, while the *R*-(−)-enantiomer is responsible for virtually all of the beta-blocking activity and potassium-channel-blocking (class III) properties.[409] The *S*-(+)-enantiomer has class III properties similar to those of *R*-(−)-sotalol, but its affinity for β-adrenergic receptors is 30–60 times lower than the affinity of the *R*-isomer.[410]

Kapoor et al.[411] described a kinetic resolution for obtaining building blocks for the synthesis of this kind of molecules; in fact, they used 4-nitrophenacyl-bromide **263**, a nonoptically active molecule to generate *rac*-2-bromo-1-(4-nitrophenyl) ethanol **264** as shown in Scheme 12.80. The racemate was then transesterified enzymatically using a lipase, and a simultaneous process was developed. Excellent ee values were obtained, although long reaction times were required. Native *Pseudomonas cepacia* lipase (PCL) gave higher enantiomeric ratio and shorter reaction time compared to lipases from porcine pancreas, *Candida rugosa*, *Chromobacterium viscosum*, and *Rhizomucor miehei*.[412] Immobilization of PCL on diatomite allowed a higher optical purity and better yields compared to those obtained with PCL immobilized on ceramic particles (PS-C), immobilized *Mucor meihei* lipase (Lipozyme) or its native form.[413]

As described in Scheme 12.80, enantiopure *S*-(+)-sotalol *S*-**260** and *R*-(−)-nifenalol *R*-**261** were prepared from the enantiomers separated enzymatically from the compound **264**: the *S*-bromoester **266** was used to form *S*-**260**, while the *R*-bromoalcohol **265** gave *R*-**261**. This route thus reduces the well-known drawback of a kinetic resolution, always limited by the fact that even in the ideal case (one of the enantiomers is completely converted into the product while the other remains unchanged, therefore reaching a maximum 50% conversion and an ee value higher than 99%), there is always a 50% of remnant substrate which is not converted. This one can be considered waste, so that the process cannot be considered very sustainable, unless a recycling strategy is implemented, or, as described here, both the converted and the nonconverted enantiomer could be useful.

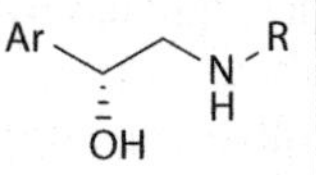

257

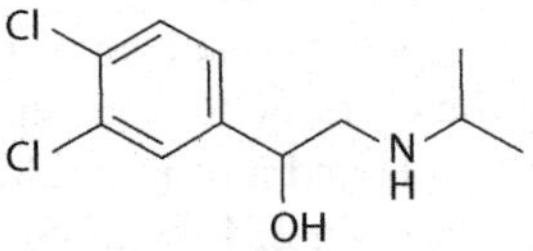

Dichloroisoprenaline
258

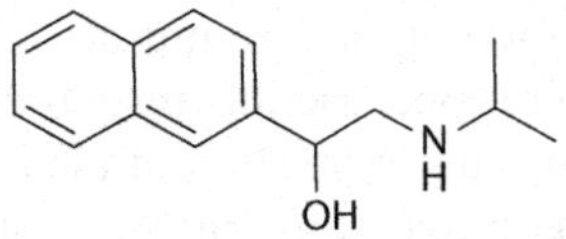

Pronethalol
259

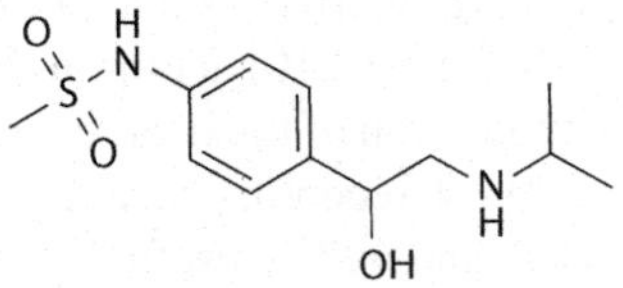

Sotalol
260
ATC C07AA07
ATC C07AA57
DB00489

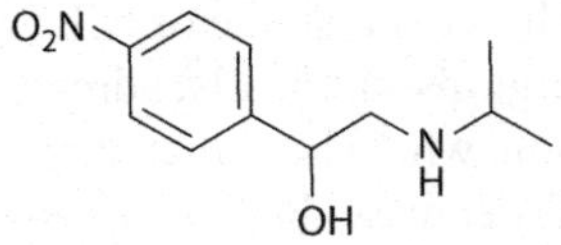

Nifenalol
261

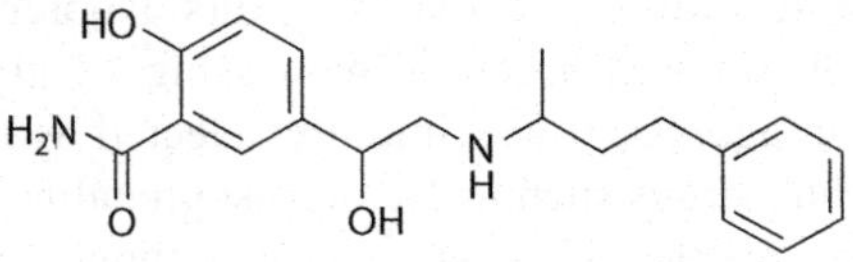

Labetalol
262
ATC C07AG01
DB00598

FIGURE 12.11 Some β-blockers possessing 1-aryl-2-alkylaminoethanol structure.

In a classical example using a bioreduction employing whole cells, Yang et al. reported the reduction of the aromatic ketone **267** to the corresponding (*R*)-alcohol **265**, which can be directly transformed into *R*-nifenalol *R*-**261** (Scheme 12.81).[414]

Another example of chemoenzymatic synthesis of *R*-nifenalol was described in 1997 by Furstoss and coworkers[415] by using a combined chemoenzymatic approach that allowed the preparation of *R*-**261** in an enantioconvergent manner, through a consecutive use of (1) an enantioselective hydrolysis of the epoxide moiety obtained by using either a whole-cell suspension or a soluble enzymatic extract of the fungus *Aspergillus niger* and (2) an acid-catalyzed hydrolysis of the remaining epoxide, with a 58% overall yield following a four-step strategy including a resolution step, as shown in Scheme 12.82.

A crucial step in the preparation of adrenergic β-blockers was the discovery of the effect of the insertion of an oxymethylene bridge (–OCH_2–) into the arylethanolamine structure of pronethalol **259** to afford propranolol **270** (Table 12.1), the first aryloxypropanolamine designed and the first clinically successful β-blocker.[404] In fact, these types of β-blockers have been used in the management of many sympathetic nervous system–associated cardiovascular disorders[416] such as hypertension,[417,418] angina pectoris,[419] cardiac arrhythmia,[420] and other disorders such as depression, loss of

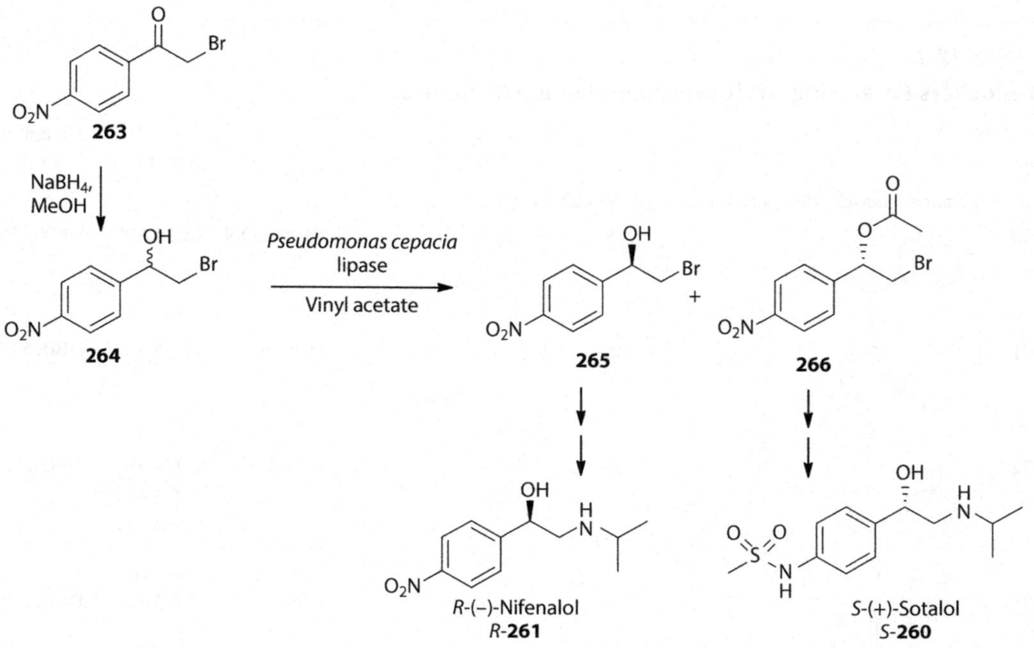

Chiral Building Blocks for Drugs Synthesis *via* Biotransformations 417

SCHEME 12.80 Chemoenzymatic process of *S*-(+)-sotalol and *R*-(−)-nifenalol.

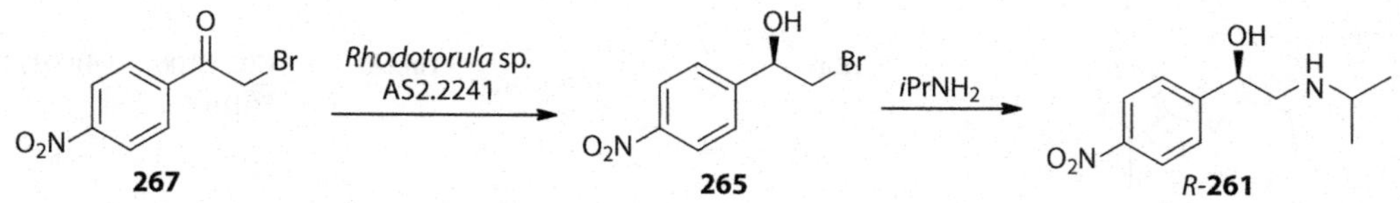

SCHEME 12.81 Synthesis of *R*-nifenalol by a chemoenzymatic approach including a bioreduction.

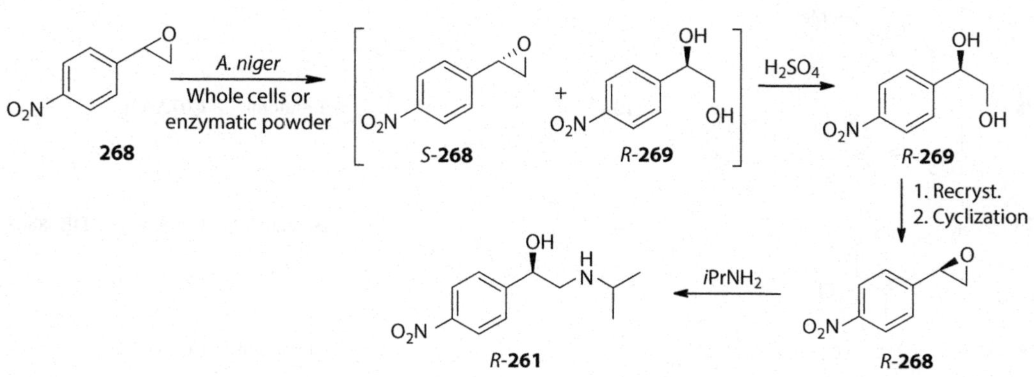

SCHEME 12.82 Synthesis of *R*-nifenalol by a chemoenzymatic approach including a biocatalyzed epoxide opening.

TABLE 12.1
β-Blockers Possessing Aryloxypropanolamine Structure

Ar-	R-	Name	ATC Code	DrugBank Code
First generation: Nonselective, antagonists of β1- and β2-receptors				
270 (1-naphthyl)	iPr–	Propranolol	C07AA05	DB00571
271 (2-allyloxyphenyl)	iPr–	Oxprenolol	C07AA02	DB01580
272 (4-indolyl)	iPr–	Pindolol	C07AA03 C07AA14 C07AA17	DB00960
273 (2-cyclopentylphenyl)	tBu–	Penbutolol	C07AA23	DB01359
274 (2,3-dihydroxy-5,6,7,8-tetrahydronaphthyl)	tBu–	Nadolol	C07AA12	DB01359
275 (4-morpholino-1,2,5-thiadiazol-3-yl)	tBu–	Timolol	C07AA06 S01ED01	DB00373
276 (2-allylphenyl)	iPr–	Alprenolol	C07AA01	DB00866
277 (2-methyl-4-indolyl)	iPr–	Mepindolol	C07AA14	
278 (thiochroman-8-yl)	tBu–	Tertatolol	C07AA16	
279 (4-methyl-2-chlorophenyl)	tBu–	Bupranolol	C07AA19	DB08808
280 (2,5-dichloro-... phenyl)	tBu–	Cloranolol	C07AA27	

(Continued)

TABLE 12.1 (*Continued*)
β-Blockers Possessing Aryloxypropanolamine Structure

Ar-	R-	Name	ATC Code	DrugBank Code
Second generation: Selective antagonists of β1-receptors				
281 (Me-C(=O)-NH-C6H4-)	*i*Pr–	Practolol	C07AB01	DB 01297
282 (Me-O-CH2CH2-C6H4-)	*i*Pr–	Metoprolol	C07AB02	DB00264
283 (H2N-C(=O)-CH2-C6H4-)	*i*Pr–	Atenolol	C07AB03	DB00335
284 (Me-CH2-C(=O)-NH-C6H3(Me)(C(=O)Me)-)	*i*Pr–	Acebutolol	C07AB04	DB01193
285 (3-methylphenyl-)	(3,4-dimethoxyphenylpropyl)	Bevantolol	C07AB06	DB001295
286 (Me-CH(Me)-O-CH2CH2-O-CH2-C6H4-)	*i*Pr–	Bisoprolol	C07AB07	DB00612
287 (Me-O-C(=O)-CH2CH2-C6H4-)	*i*Pr–	Esmolol	C07AB09	DB00187
288 (cyclohexyl-NH-C(=O)-NH-C6H4-)	*t*Bu–	Talinolol	C07AB12	
Third generation: Nonselective antagonists, with additional activities				
289 (5-methyl-3,4-dihydroquinolin-2(1H)-one-)	*t*Bu–	Carteolol	C07AA15 S01ED05	DB00521
290 (4-methylcarbazol-HN-)	(2-methoxyphenoxy-propyl)	Carvedilol	C07AG02	DB01136
291 (2-cyanophenyl-)	(Me2C-CH2-indol-3-yl, NH)	Bucindolol		

(*Continued*)

TABLE 12.1 (*Continued*)
β-Blockers Possessing Aryloxypropanolamine Structure

Ar-	R-	Name	ATC Code	DrugBank Code
Third generation: Selective antagonists of β1-receptors, with additional activities				
292 [cyclopropylmethoxy-ethyl-phenyl]	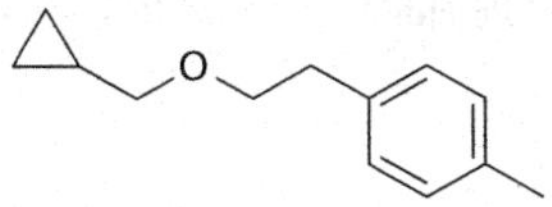 *i*Pr–	Betaxolol	C07AB05 S01ED02	DB00195
293 [diethylureido-acetyl-methylphenyl]	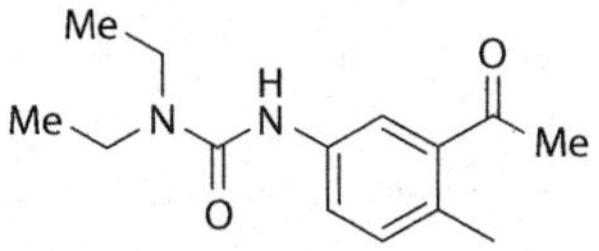 *t*Bu–	Celiprolol	C07AB08	
294 [cyano-methylphenyl]	[–NH–CH₂CH₂–C(O)–CH₂–C₆H₄–OH]	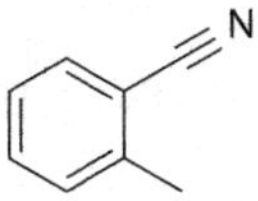 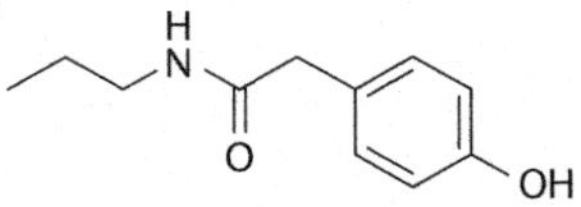 Epanolol	C07AB10	
295 [methylphenyl] 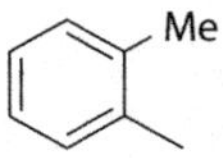 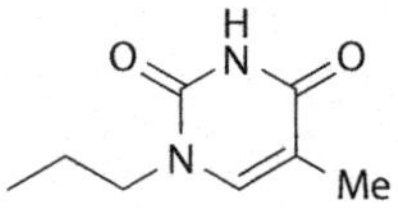	[pyrimidinedione-propyl-Me]	Primidolol		
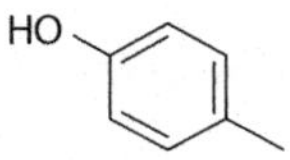 296 HO–[methylphenyl]	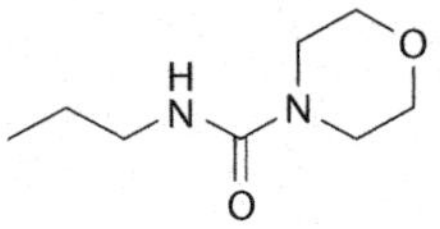 [–NH–CH₂CH₂–NH–C(O)–morpholine]	Xamoterol		

appetite, asthma, migraine, glaucoma, and so on. All these β-blockers possessing the aryloxypropanolamine moiety have at least one stereogenic center in their structure, so that they have two enantiomers,[421] the (*S*)-enantiomer of β-blockers being more potent antagonists than the corresponding (*R*)-enantiomer.[422] It is not surprising that the eutomer of aryloxypropanolamines is the *S*-enantiomer, while for arylethanolamines it is the *R*-enantiomer; this can be easily understood by superimposing both structures as shown in Figure 12.12; as can be seen, the critical functional groups occupy the

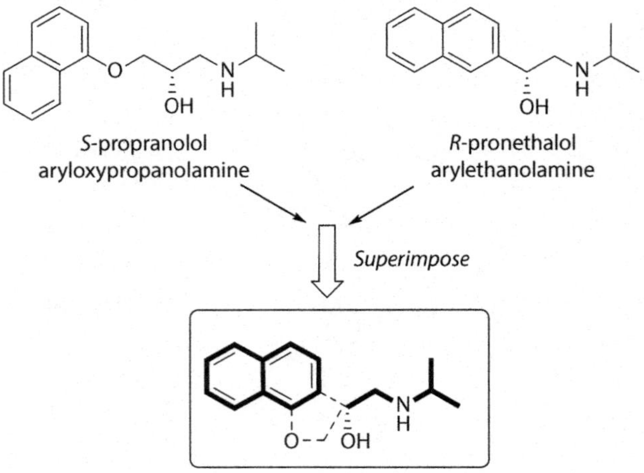

S-propranolol
aryloxypropanolamine

R-pronethalol
arylethanolamine

Superimpose

FIGURE 12.12 Superimposition of *S*-aryloxypropanolamines and *R*-arylethanolamines.

Chiral Building Blocks for Drugs Synthesis *via* Biotransformations

same approximate regions in space, as indicated by the bold lines in the superimposed drawings, while the dotted lines are those parts that do not overlap but are not necessary to receptor binding.[404]

Thus, different biocatalytic approaches have been developed for the preparation of this (S)-enantiopure β-blockers, such as chemoenzymatic kinetic resolution catalysis of a chiral synthon leading to the drug, chemoenzymatic DKR (catalyzed in both cases by hydrolases) and reductive protocols.[423,424] More concretely, many hydrolases (mainly lipases) have been described in the preparation of enantiopure chiral building blocks (halohydrins, diols, hydroxyesters, azidoalcohols, or nitriles) for the ulterior synthesis of β-blockers, as reviewed by Zelaszczyk and Kiec-Kononowicz.[423] This was motivated by the fact that *N,O*-diacetyl-propranolol failed to be a hydrolyzable substrate for proteases, lipases, and esterases,[423] but when the acyl moiety was changed by removing the methyl from the carbonyl group (resulting in *rac*-*N,O*-bis(methoxycarbonyl) derivative of propranolol (*rac*-**297**, Scheme 12.83) and exposed to various esterases (pig liver, cholesterol, and porcine pancreatic), the racemic compound was readily hydrolyzed to yield *R*-*N*-methoxycarbonylpropranolol *R*-**298**. The best percent of conversion (40%) and enantiomeric excess (ee = 83%) was obtained with the use of porcine pancreatic esterase. The enzymatic hydrolysis of racemic *O*-butyryl propanol *rac*-**299** was also described by Ávila et al.[425]; these authors performed a screening of different lipases (from *C. rugosa*, *C. antarctica*, *P. fluorescens*, *M. miehei*, and *R. niveus*, respectively), immobilized on Octyl-Sepharose, and by means of an experimental design concluded that lipase from *P. fluorescens*, using tetrahydrofuran as solvent, and 15 mM buffer phosphate concentration at 37°C, led to obtain *S*-isomer with 87% ee; if the R-isomer is the required target, lipase from *C. rugosa*, acetone, 15°C, acetone, buffer phosphate 25 mM and 15°C were the best conditions.

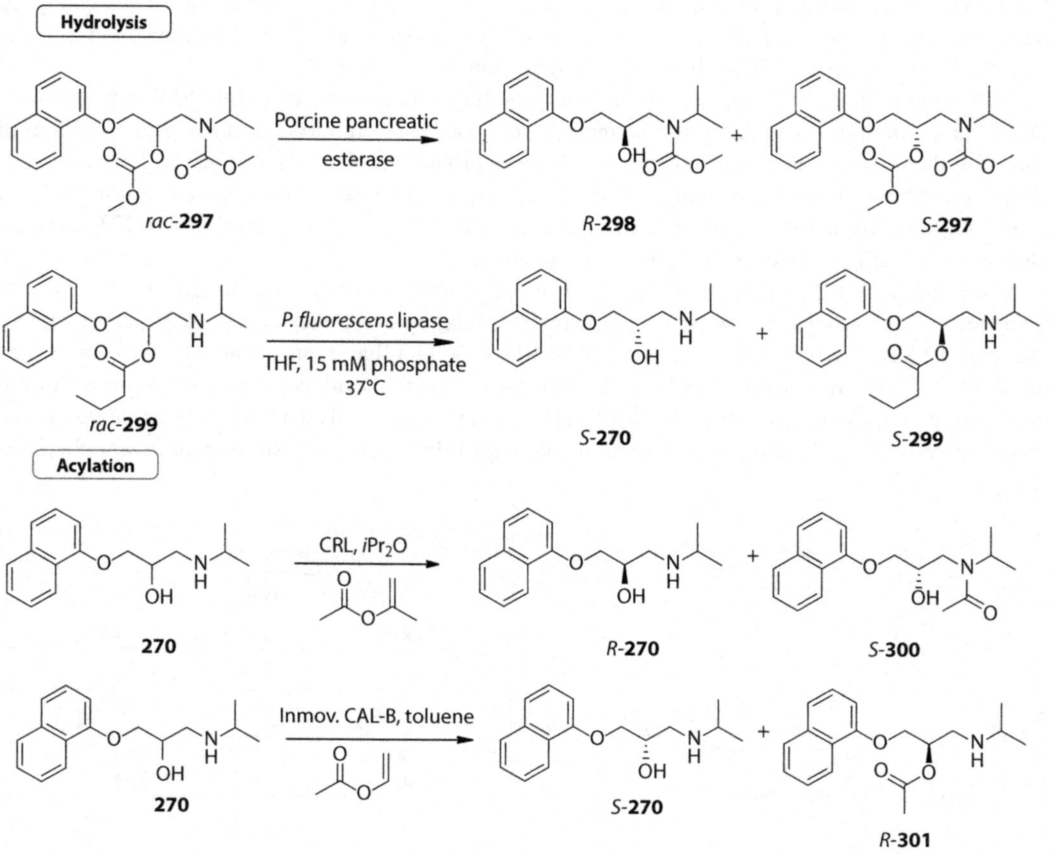

SCHEME 12.83 Some kinetic resolutions of propranolol.

Kinetic resolution of propranolol was also achieved by a lipase-catalyzed N-acetylation of the secondary amino group,[426] as shown in Scheme 12.83. Among several enzymes used, *Candida rugosa* lipase (CRL) was found to be the most reactive and enantioselective for catalyzing the N-acetylation of propranolol with isopropenyl acetate in isopropylether. The outcome of this process showed only moderate ee values of the obtained inactive (R)-propranolol (67%) and slightly better for (S)-amide **300** (73.4%). More recently, Barbosa et al. have described the use of lipase B from *Candida antarctica* immobilized on Eupergit in the resolution of propranolol by O-acetylation with vinyl acetate in toluene,[427] thus rendering the desired S-eutomer (S-**270**) with a moderate 30% yield and 96% ee.

As can be seen, only moderate resolutions have been described using propranolol as substrate, so that another very common strategy has been the preparation of different enantiopure building blocks, which could be easily converted to the desired S-aryloxypropanolamine through a further step. In this sense, this research area was opened by the pioneering paper of Bevinakatti and Banerji[428] on the resolution of racemic halohydrins either by hydrolysis of its ester *rac*-**302** or by direct acylation, as shown in Scheme 12.84, to produce structures such as R-**302** or R-**303**, which, upon treatment with the corresponding amine, would afford the desired enantiopure aryloxypropanolamine S-**304**. In this sense, for preparing S-propranolol S-**270** (Ar = 1-naphthyl, R_1 = *i*Pr), a purified lipase from *Pseudomonas cepacia* (Lipase PS from Amano) was found to show an excellent selectivity toward the S-isomer; thus, at a conversion rate of around 50%, the transesterification reaction of acetate **302** (Ar = 1-naphthyl, R = Me, X = Cl) using *n*-butanol in diisopropyl ether (DIPE) produced chlorohydrin S-**303** and acetate R-**302** in >95% ee (E = >100). While hydrolysis worked faster than transesterification, the ease of workup and isolated yields were in favor of the latter. Hence, in the acylation of *rac*-**303**, LPSA also showed excellent selectivity toward the S-isomer using vinyl acetate (VA) or acetic anhydride as acylating agent. Another notable factor for all the LPSA-catalyzed reactions was that the initial rate of reaction drastically dropped down after 40% conversion, coming practically to stop around 50% after consuming the entire S-isomer.[428]

This same protocol was applied for preparing both enantiomers of atenolol **283**[429] and practolol **281**,[430] and simultaneously Ader and Schneider published three reports concerning the problem of choosing the most suitable biocatalyst,[24] synthetic intermediates for β-blockers,[25] and the influence of the substitution pattern on both activity and enantioselectivity of the selected enzyme.[26] As a result, biologically active S-isomers of propranolol **270**, alprenolol **276**, penbutolol **273**, atenolol **283**, practolol **281**, and oxprenolol **271** were obtained.

Many other chiral synthons can be used as substrates for enzymatic kinetic resolutions, as reviewed by Zelaszczyk and Kiec-Kononowicz[423]: diols, hydroxyesters, azidoalcohols, or nitriles (Scheme 12.85). Anyhow, KRs are always limited by the fact that the maximum conversion is limited to 50%. To overcome this drawback, DKR is an excellent alternative. As the previous technologies, this process exploits many chemical catalysts and a biocatalyst in the sequential reactions, therefore combining the enzymatic kinetic resolution of the racemic substrate with *in situ* chemical

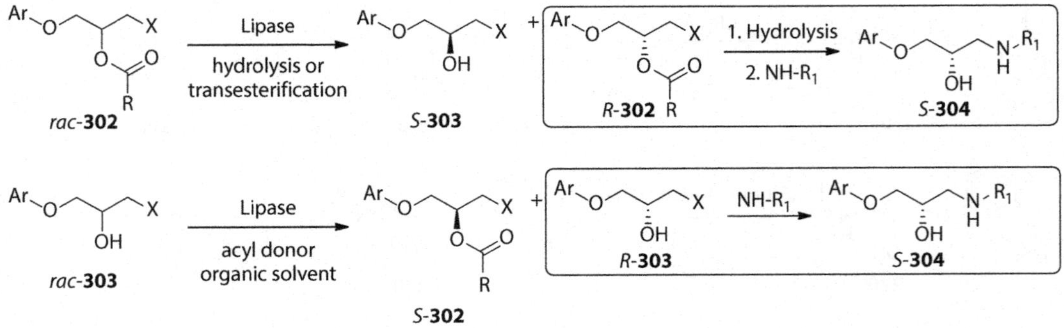

SCHEME 12.84 Two different resolutions of halohydrins for preparing S-aryloxypropanolamines.

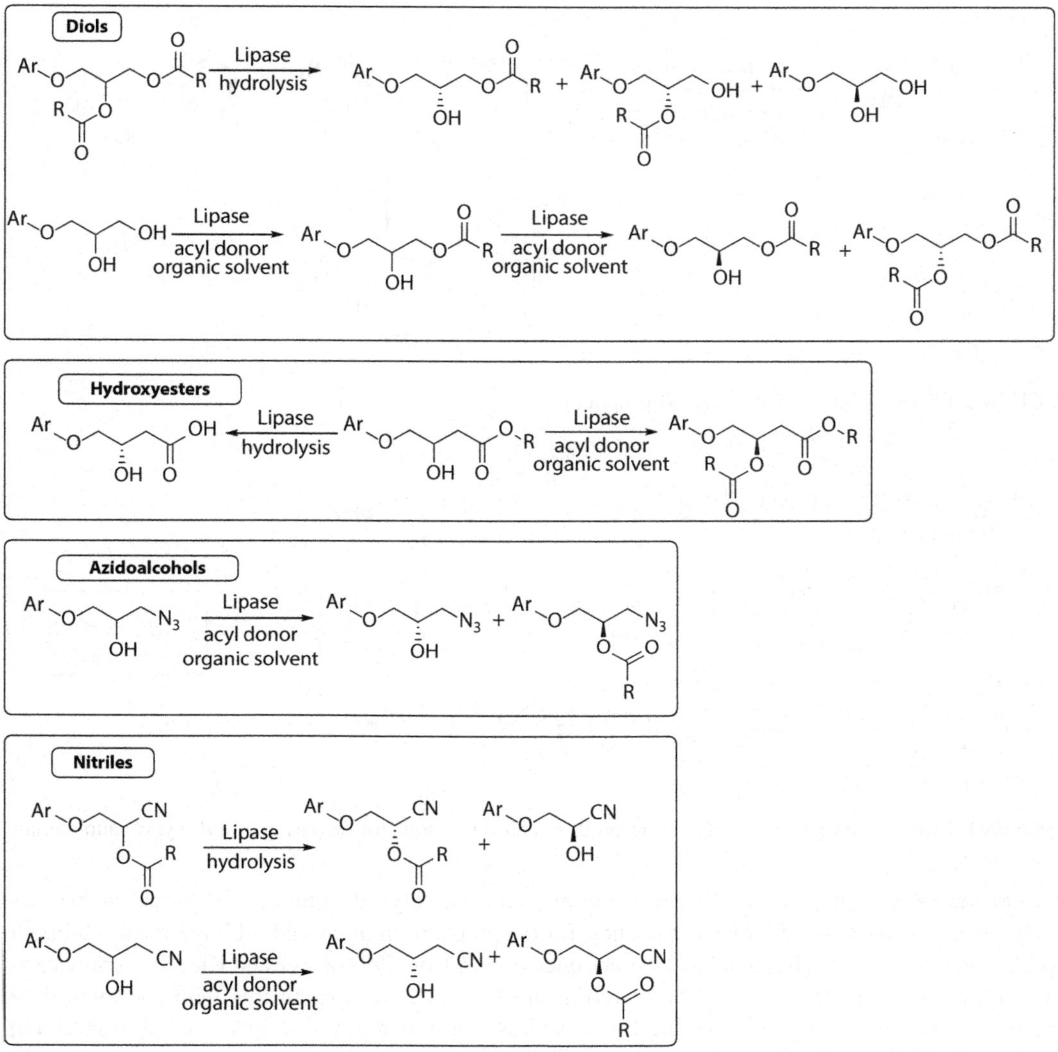

SCHEME 12.85 Kinetic resolutions of diols, hydroxyesters, azidoalcohols, and nitriles to produce chiral synthons for preparing S-aryloxypropanolamines.

racemization of the undesired enantiomers.[271] During the process, the racemic substrate is split enzymatically to form the desired enantiomer wherein the undesired enantiomers are racemized chemically to give the racemic mixture, then resolved again by the enzymatic resolution. For instance, (S)-propranolol **270**, archetypical reference of 1-alkylamino-3-aryloxypropanols, could have been prepared by using this technology,[431] as depicted in Scheme 12.85. The one-pot DKR of racemic azidoalcohol **305** was conducted by employing Novozyme 435 (CAL-B lipase adsorbed on the ion-exchange resin Lewatit) and p-chlorophenyl acetate, and commercial Shvo's ruthenium catalyst as the racemization agent. This step produced azidoacetate S-**305** in 71% yield and 86% ee. Hydrolysis of the acetate with lithium hydroxide in methanol at room temperature followed by azide reduction and *in situ* reductive alkylation using Adam's catalyst in the presence of acetone at the same temperature produced the (S)-propranolol in almost enantiomerically pure form (Scheme 12.86).

Some other examples of DKRs for the preparation of enantiomerically pure chiral building blocks for β-blocker structures and many other bioactive compounds can be found in our recent review.[271]

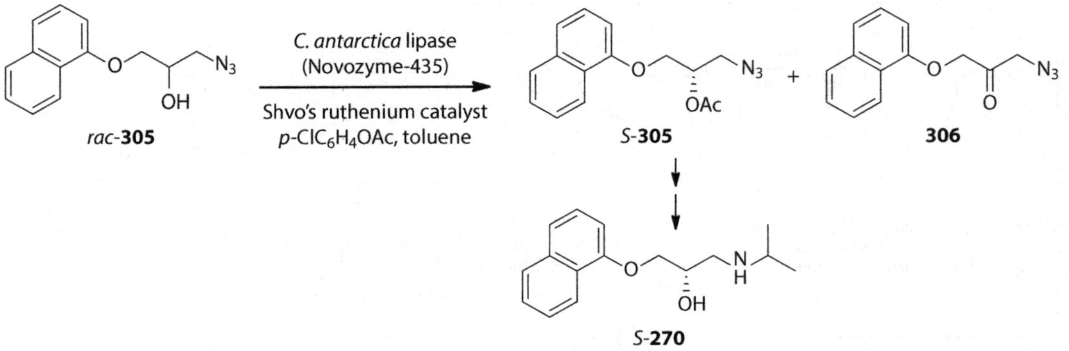

SCHEME 12.86 DKR for obtaining (S)-propranolol.

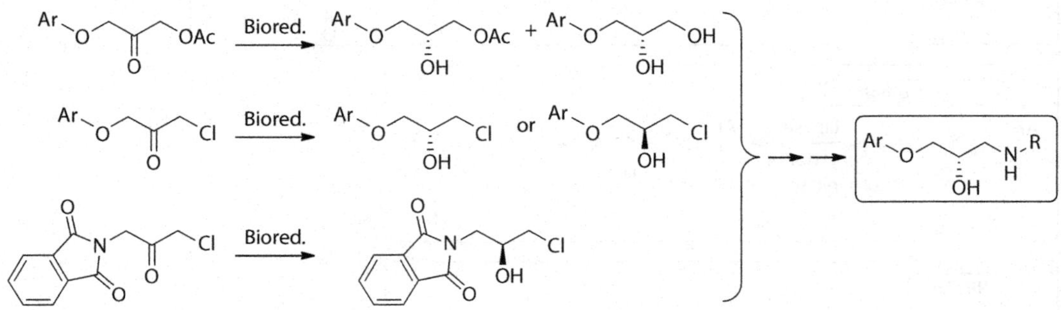

SCHEME 12.87 Bioreduction strategies to produce chiral synthons for preparing S-aryloxypropanolamines.

On the other hand, many different examples of biocatalyzed reduction of prochiral ketones, either using whole cells or isolated enzymes, for the preparation of chiral building blocks leading to β-blocker structures can be found in the mentioned review from Zelaszczyk and Kiec-Kononowicz,[423] as outlined in Scheme 12.87. Our research group has been working in this field for some time, mainly using whole cells,[432–435] and in this sense it is worth mentioning the work of Sinisterra et al. in the bioreduction of prochiral 1-chloro-3-(phthalimidyl)-propan-2-one, an interesting chiral building block that can be used to conventionally prepare the (S)-adrenergic β-blockers.[434–436] Finally, although traditionally the use of whole cells is more common, examples of bioreduction of similar prochiral ketones using isolated enzymes are becoming more frequent.[36,37]

12.9 ACE INHIBITORS

Certain antihypertensive drugs may target the renin angiotensin pathway, a set of chemical reactions that ultimately boosts blood pressure when it gets low. Though the process is complicated, the main steps involve the conversion of angiotensin I to angiotensin II by angiotensin-converting enzyme (ACE) in the lungs. Thus, ACE inhibitors are capable to decrease blood pressure.

Captopril, 1-[(2S)-3-mercapto-2-methylpropionyl]-L-proline **307** (Scheme 12.88), prevents the conversion of angiotensin I to angiotensin II by the inhibition of ACE. The potency of captopril as an inhibitor of ACE depends critically on the configuration of the mercaptoalkanoyl moiety; the compound with the (S)-configuration is about 100 times more active than its corresponding R-isomer.[437] The synthesis of the S-side chain of captopril by the lipase-catalyzed enantioselective hydrolysis of racemic 3-acetylthio-2-methyl propanoic acid **308** (Scheme 12.88) to yield desired S-**308** as well as R-**309** and acetic acid has been described.[438] In fact, lipase from *Rhizopus oryzae* ATCC 24563

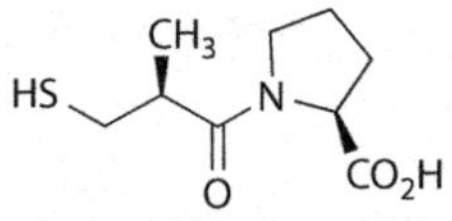

307 Captopril

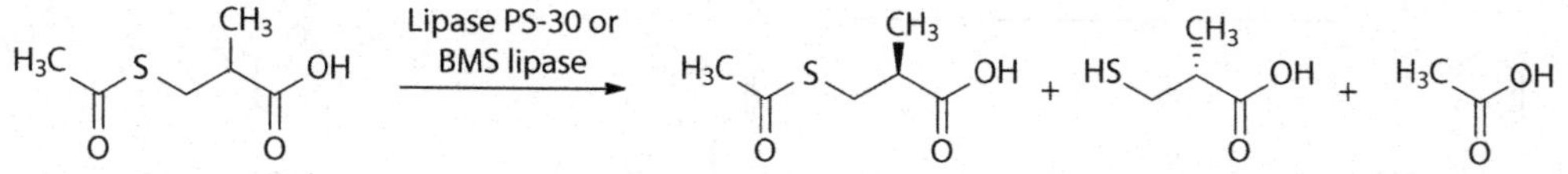

(a) *rac*-**308** *S*-**308** *R*-**309**

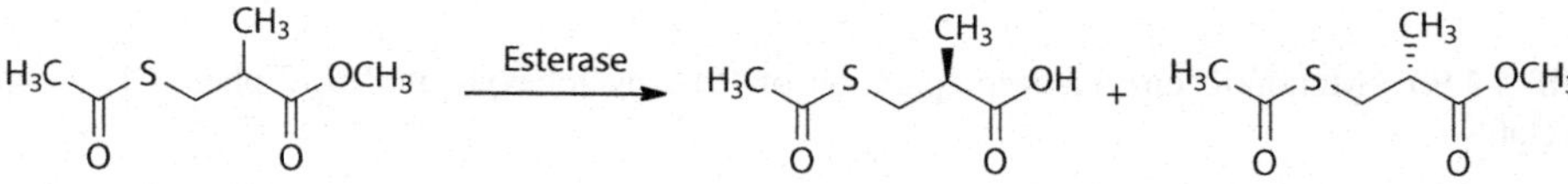

(b) *rac*-**310** *S*-**308** *R*-**310**

SCHEME 12.88 Captopril: (a) enantioselective hydrolysis of racemic 3-acetylthio-2-methyl propanoic acid by lipase; (b) enantioselective hydrolysis of racemic 3-mercapto-2-methylpropionic acid methyl ester by esterase.

(heat-dried cells) and lipase PS-30 from *P. cepacia* in toluene catalyzed the hydrolysis of the thioester bond of the undesired enantiomer to yield the desired *S*-**308** in >28% yield (theoretical max is 50%) and ee of >95% using each enzyme. In an alternate process, *S*-**308** was prepared by enantioselective hydrolysis of racemic 3-mercapto-2-methylpropionic acid, methyl ester **310** by using an esterase from *Pseudomonas* sp. MRC strain. A reaction yield of 49% (max yield is 50%) and an ee of 99.9% were obtained for *S*-**308**.[439]

Following an alternative protocol for the synthesis of the lateral chain of **307** and zofenopril **313** (Scheme 12.89), an enantioselective esterification of racemic 3-benzoilthio-2-methyl propanoic acid *rac*-**311** was described by means of lipases in toluene to yield *R*-**312**, thus leaving behind the desired enantiopure acid *S*-**311**.[440] Using lipase PS-30, methanol, and toluene, *S*-**311** was obtained (37% yield, 97% ee). When this lipase was immobilized by adsorption on polypropylene Accurel (PP), better results were obtained (45% yield, 97.5% ee of *S*-**311**), allowing up to 23 repetitive cycles without any significant loss of productivity.

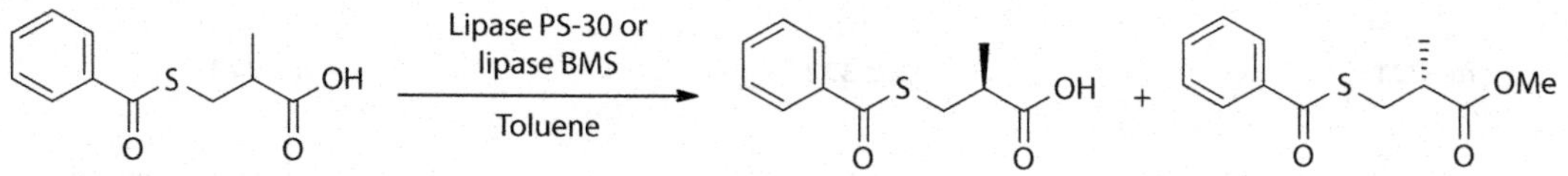

rac-**311** *S*-**311** *R*-**312**

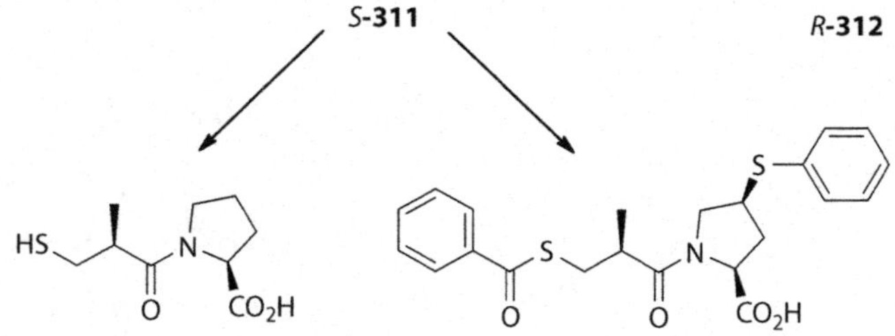

307 **313** Zofenopril

SCHEME 12.89 Preparation of captopril **307** and zofenopril **313**.

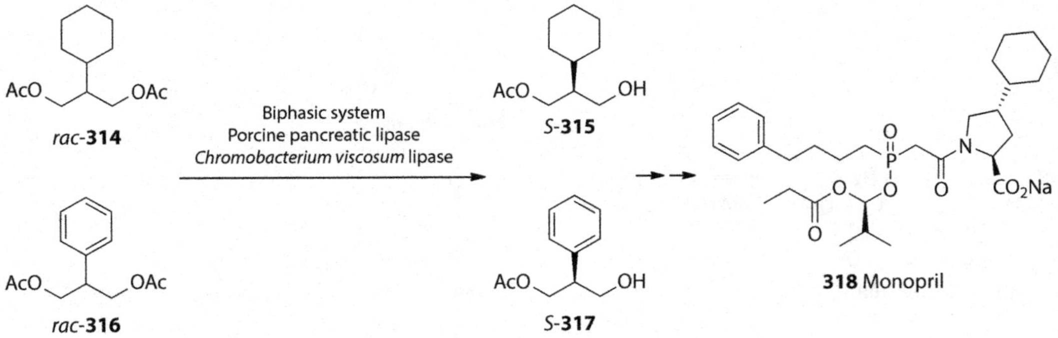

SCHEME 12.90 Monopril: enzymatic preparation of (S)-2-cyclohexyl- and (S)-2-phenyl 1,3-propanediol monoacetates.

Enantiopure monoacetates S-**315** and S-**317** (Scheme 12.90) are key chiral intermediates for the chemoenzymatic synthesis of Monopril **318**, another antihypertensive drug that acts as an ACE inhibitor.[441] The asymmetric hydrolysis of racemic diesters by PPL and *Chromobacterium viscosum* lipase was described.[439] Thus, in a biphasic system using 10% toluene, reaction yields of >65% with ee's of 99% were obtained for S-**315** using each enzyme. S-**317** was obtained in 90% reaction yield with 99.8% ee using *C. viscosum* lipase under similar conditions.

12.10 CALCIUM CHANNEL BLOCKERS

Calcium channel blockers (CCBs) are drugs that inhibit transport of calcium across cell membranes. They primarily act on the smooth muscle layer of arteries, causing relaxation of smooth muscles, increased arterial diameter, and lowered blood pressure. Certain drugs in this class, such as diltiazem, also reduce heart muscle contractility.[442]

Compounds belonging to the class of 4-aryl-1,4-dihydropyridines, such as nifedipine (**320**, Scheme 12.91), nitrendipine, nimodipine, etc., are the most studied calcium channel modulators and

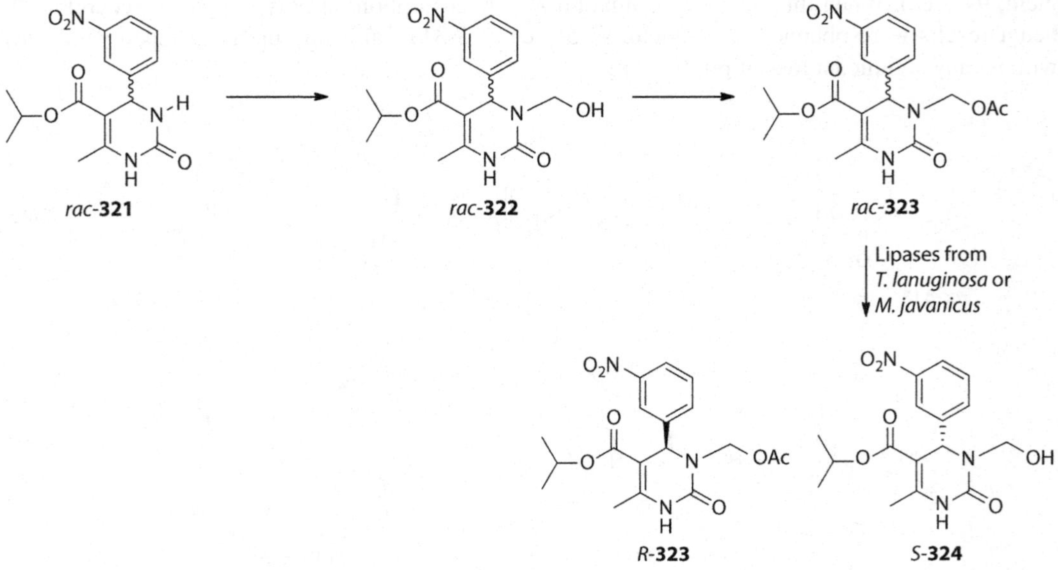

SCHEME 12.91 Kinetic resolution of *N*-3-acetoxymethylated 4-aryl-3,4-dihydropyrimidinone.

Chiral Building Blocks for Drugs Synthesis via Biotransformations

have become almost indispensable for the treatment of cardiovascular diseases such as hypertension and angina since their introduction into clinical medicine in the 1970s.[443]

Although compounds of this class have symmetrical scaffold, a lot of desymmetrization studies have been carried out to generate chiral dihydropyridines, keeping in mind that chirality plays an important role in biological activities.[444–448]

4-Aryl-3,4-dihydropyrimidinones R and S-**319**, which have similar structure as nifedipine **320**, also exhibited calcium channel modulating activity, but in contrast to the latter compounds they are inherently asymmetric (Figure 12.13). The calcium channel modulation activity of 4-aryl-3,4-dihydropyrimidinones is dependent on the absolute configuration at the stereogenic center C-4. The orientation (R-/S-configuration) of C-4 aryl group in 4-aryl-3,4-dihydropyrimidinones acts as a molecular switch between calcium channel blocking (antagonist) and activating activities (agonist).[449] For example, only R-**319** carries the therapeutically desired calcium channel blocking activity, and not the S-**319**.[450]

The use of biocatalyzed protocols is really advantageous from the Green Chemistry point of view, because very mild conditions can be used to produce only (or mainly) the desired enantiomer. For instance, Schnell et al.[451,452] described the excellent resolution of rac-**323** using lipases from *Mucor javanicus* (Amano M) and *Thermomyces lanuginosus* (Amano CE) in mixed aqueous organic solvents in the presence of enhancer, polyethylene glycol methyl ether (PEG, M. Wt. 5,000) and dextran (M. Wt. 260,000), respectively (Scheme 12.91). The resolution of rac-**321** has been affected using N-3-acetoxymethyl group as a handle for lipase-catalyzed deacetylation of N-3-acetoxymethylated 4-aryl-3,4-dihydropyrimidinone rac-**323**, which was synthesized by the reaction of corresponding dihydropyrimidinone with formaldehyde, followed by chemical acetylation of the resulted rac-**322**.

Similarly, Prasad et al.[453] have developed a biocatalytic resolution of these types of dihydropyrimidinones using acetoxyl group on the phenyl ring and Novozyme 435 as biocatalyst with good yields and ee.

Diltiazem (Scheme 12.92, **328**) is a nondihydropyridine (non-DHP) member of CCBs used in the treatment of hypertension, angina pectoris, and some types of arrhythmia.[454] Mitsubishi Tanabe Pharma Corporation have commercialized a lipase-catalyzed protocol for carrying out the resolution of trans-**325**, key step for diltiazem synthesis, employing a lipase secreted by *Serratia marcescens* Sr41 8000.[455–458] Thus, the enzymatic hydrolysis is "cleaning" the desired (2S,3R)-**325** from the by-product (2R,3S)-**326**, which spontaneously decarboxylates to generate aldehyde **329**. In the Tanabe process, shown in Scheme 12.92, a membrane reactor is used to carry out hydrolysis, separation, and crystallization of (2S,3R)-**325**. Thus, toluene dissolves the starting substrate in the crystallizer module and takes it to the membrane containing the immobilized lipase. When hydrolysis is going on, nondesired acid (2R,3S)-**326** passes through the membrane to the aqueous phase, gets decarboxylated, and is neutralized with sodium bisulfite to produce **330**, thus avoiding any lipase deactivation promoted by free aldehyde. Thus, (2S,3R)-**325** stays in the toluene phase and it is taken back to the crystallizer, where it becomes accumulated. After eight cycles, fresh lipase has to be recharged in the reactor to compensate the progressive decrease of enzymatic activity. Final yield is higher than 43% with 100% optical purity.

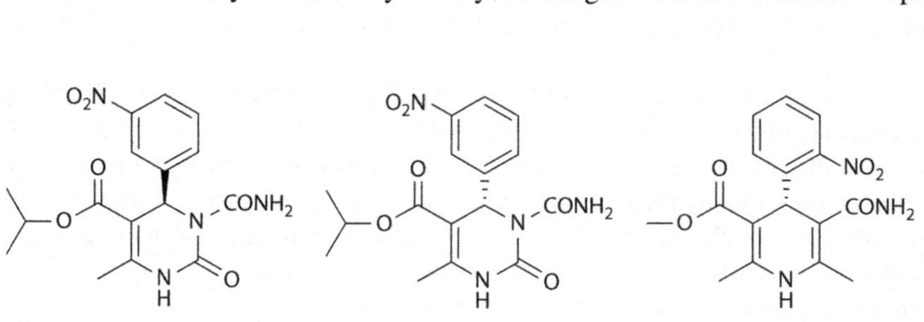

FIGURE 12.13 Nifedipine and analogues.

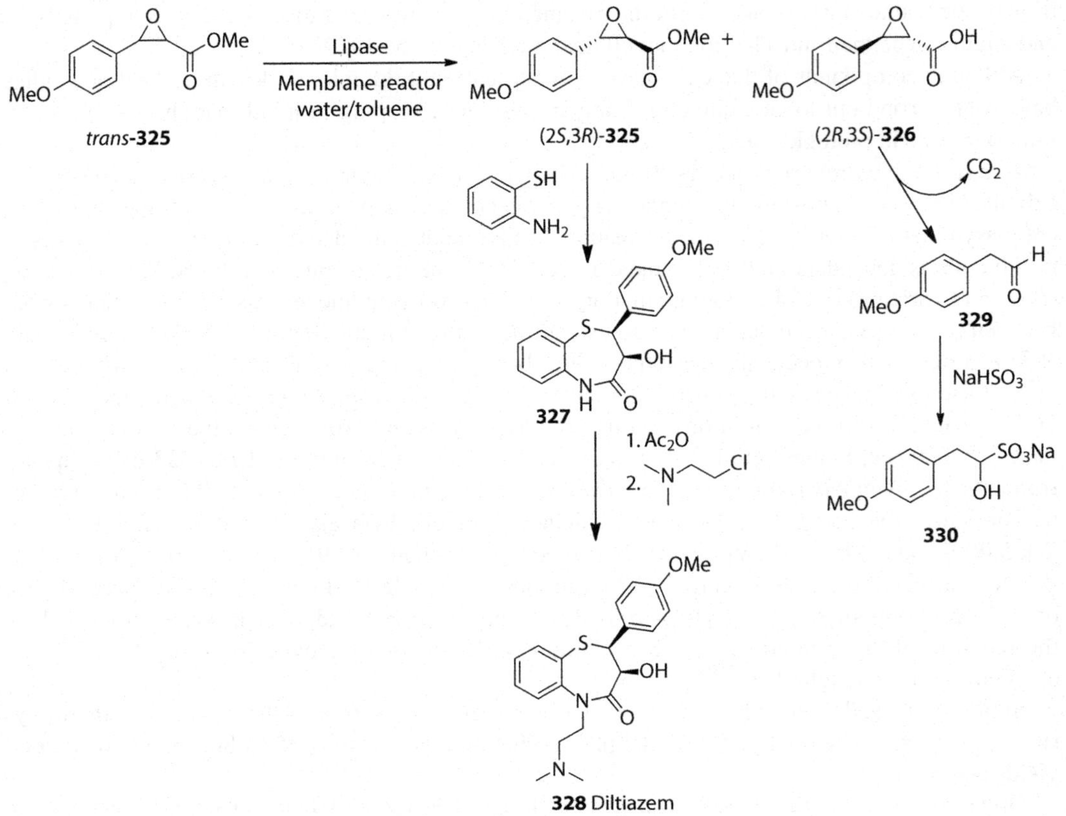

SCHEME 12.92 Tanabe synthesis of diltiazem **328**.

12.11 CONCLUSION

It is far from our intention to try to be exhaustive in this paper, but, as can be easily inferred from the examples so far presented, biocatalysis represents a greener alternative to classical organic synthesis in the preparation of chiral building blocks leading to many drugs, especially when chemo-, regio-, and stereoselectivities are the desired target. Putting those protocols into practice on water even expands the sustainability of the process, because of the inherent environmental advantages associated with its use, mainly at the industrial scale. Additionally, for those biocatalytic processes in which nonaqueous media are required, the implementation of green solvents is a growing trend[23,41]; among them, 2-methyltetrahydrofuran is becoming a very attractive alternative.[459,460]

Thus, a growing increase in the use of biocatalysis and biotransformations for the preparation of drugs and fine chemicals is mainly promoted by the enhancement of biocatalysts' performance through chemical modification and genetic engineering, as shown in some examples presented. This asseveration is based on many recent references in some of the most prestigious scientific journals in which this tendency is clearly revealed.[5,9,11,233,461–465] Inside the area of pharmaceutical industry, the significance of biocatalysis implementation is certainly magnified,[10,12,21,462–469] because pharma industry generates a huge amount of waste materials[470] that can be reduced by the employment of such a mild technology as biocatalysis.

REFERENCES

1. Anastas, P. T.; Warner, J. C. *Green Chemistry: Theory and Practice*. Oxford University Press: Oxford, U.K., 1998.
2. Kristan, K.; Stojan, J.; Adamski, J.; Rizner, T. L. Rational design of novel mutants of fungal 17 beta-hydroxy steroid dehydrogenase. *J. Biotechnol.* 2007, *129*, 123–130.
3. Festel, G. Drivers and barriers for industrial biotechnology. *Int. Sugar J.* 2011, *113*, 19–23.
4. Hoyos, P.; Quezada, M. A.; Sinisterra, J. V.; Alcántara, A. R. Optimised dynamic kinetic resolution of benzoin by a chemoenzymatic approach in 2-MeTHF. *J. Mol. Catal. B: Enzym.* 2011, *72*, 20–24.
5. Nestl, B. M.; Nebel, B. A.; Hauer, B. Recent progress in industrial biocatalysis. *Curr. Opin. Chem. Biol.* 2011, *15*, 187–193.
6. Patel, R. N. Biocatalysis: Synthesis of key intermediates for development of pharmaceuticals. *ACS Catal.* 2011, *1*, 1056–1074.
7. Patel, R. N. Biocatalytic routes to chiral intermediates for development of drugs. In *Biocatalysis for Green Chemistry and Chemical Process Development*, Tao, J. and Kazlauskas, R., Eds. John Wiley & Sons, Inc.: Hoboken, NJ, 2011; pp. 89–149.
8. Sanchez, S.; Demain, A. L. Enzymes and bioconversions of industrial, pharmaceutical, and biotechnological significance. *Org. Process Res. Dev.* 2011, *15*, 224–230.
9. Wenda, S.; Illner, S.; Mell, A.; Kragl, U. Industrial biotechnology—The future of green chemistry? *Green Chem.* 2011, *13*, 3007–3047.
10. Muñoz Solano, D.; Hoyos, P.; Hernáiz, M. J.; Alcántara, A. R.; Sánchez-Montero, J. M. Industrial biotransformations in the synthesis of building blocks leading to enantiopure drugs. *Bioresour. Technol.* 2012, *115*, 196–207.
11. Clouthier, C. M.; Pelletier, J. N. Expanding the organic toolbox: A guide to integrating biocatalysis in synthesis. *Chem. Soc. Rev.* 2012, *41*, 1585–1605.
12. Watson, W. J. W. How do the fine chemical, pharmaceutical, and related industries approach green chemistry and sustainability? *Green Chem.* 2012, *14*, 251–259.
13. de Carvalho, C. C. C. R. Enzymatic and whole cell catalysis: Finding new strategies for old processes. *Biotechnol. Adv.* 2011, *29*, 75–83.
14. Li, C.-J.; Anastas, P. T. Green chemistry: Present and future. *Chem. Soc. Rev.* 2012, *41*, 1413–1414.
15. Tao, J.; Kazlaukas, R. J., *Biocatalysis for Green Chemistry and Chemical Process Development*. John Wiley & Sons, Ltd.: Hoboken, NJ, 2011.
16. Hanefeld, U.; Gardossi, L.; Magner, E. Understanding enzyme immobilisation. *Chem. Soc. Rev.* 2009, *38*, 453–468.
17. Behrens, G. A.; Hummel, A.; Padhi, S. K.; Schätzle, S.; Bornscheuer, U. T. Discovery and protein engineering of biocatalysts for organic synthesis. *Adv. Synth. Catal.* 2011, *353*, 2191–2215.
18. Bornscheuer, U. T.; Huisman, G. W.; Kazlauskas, R. J.; Lutz, S.; Moore, J. C.; Robins, K. Engineering the third wave of biocatalysis. *Nature* 2012, *485*, 185–194.
19. Tufvesson, P.; Fu, W.; Jensen, J. S.; Woodley, J. M. Process considerations for the scale-up and implementation of biocatalysis. *Food Bioprod. Process.* 2010, *88*, 3–11.
20. Palomares, L. A.; Ramírez, O. T. Bioreactor scale-up. In *Encyclopedia of Industrial Biotechnology*, Flickinger, M. C., Ed. John Wiley & Sons, Inc.: Oxford, U.K., 2009.
21. Wells, A. S.; Finch, G. L.; Michels, P. C.; Wong, J. W. Use of enzymes in the manufacture of active pharmaceutical ingredients—A science and safety-based approach to ensure patient safety and drug quality. *Org. Process Res. Dev.* 2012, *16*, 1986–1993.
22. Anastas, P.; Eghbali, N. Green chemistry: Principles and practice. *Chem. Soc. Rev.* 2010, *39*, 301–312.
23. Jessop, P. G. Searching for green solvents. *Green Chem.* 2011, *13*, 1391–1398.
24. Henderson, R. K.; Jimenez-Gonzalez, C.; Constable, D. J. C.; Alston, S. R.; Inglis, G. G. A.; Fisher, G.; Sherwood, J.; Binks, S. P.; Curzons, A. D. Expanding GSK's solvent selection guide—Embedding sustainability into solvent selection starting at medicinal chemistry. *Green Chem.* 2011, *13*, 854–862.
25. Moity, L.; Durand, M.; Benazzouz, A.; Pierlot, C.; Molinier, V.; Aubry, J.-M. Panorama of sustainable solvents using the COSMO-RS approach. *Green Chem.* 2012, *14*, 1132–1145.
26. Constable, D. J. C.; Curzons, A. D.; Cunningham, V. L. Metrics to 'green' chemistry—Which are the best? *Green Chem.* 2002, *4*, 521–527.
27. Constable, D. J. C.; Jimenez-Gonzalez, C.; Henderson, R. K. Perspective on solvent use in the pharmaceutical industry. *Org. Process Res. Dev.* 2007, *11*, 133–137.

28. Jiménez-González, C.; Poechlauer, P.; Broxterman, Q. B.; Yang, B.-S.; am Ende, D.; Baird, J.; Bertsch, C.; Hannah, R. E.; Dell'Orco, P.; Noorman, H.; Yee, S.; Reintjens, R.; Wells, A.; Massonneau, V.; Manley, J. Key green engineering research areas for sustainable manufacturing: A perspective from pharmaceutical and fine chemicals manufacturers. *Org. Process Res. Dev.* 2011, *15*, 900–911.
29. Jiménez-González, C.; Curzons, A. D.; Constable, D. J. C.; Cunningham, V. L. Expanding GSK's solvent selection guide—Application of life cycle assessment to enhance solvent selections. *Clean Technol. Environ. Policy* 2005, *7*, 42–50.
30. Tanaka, K.; Kaupp, G. *Solvent-Free Organic Synthesis*, 2nd edn. Wiley VCH GmbH & Co, KGaA: Weinheim, Germany, 2009.
31. Li, C.-J.; Chen, L. Organic chemistry in water. *Chem. Soc. Rev.* 2006, *35*, 68–82.
32. Chanda, A.; Fokin, V. V. Organic synthesis "on water". *Chem. Rev.* 2009, *109*, 725–748.
33. Simon, M.-O.; Li, C.-J. Green chemistry oriented organic synthesis in water. *Chem. Soc. Rev.* 2012, *41*, 1415–1427.
34. Narayan, S.; Muldoon, J.; Finn, M. G.; Fokin, V. V.; Kolb, H. C.; Sharpless, K. B. "On water": Unique reactivity of organic compounds in aqueous suspension. *Angew. Chem. Int. Ed.* 2005, *44*, 3275–3279.
35. Trost, B. M. Atom economy—A challenge for organic synthesis: Homogeneous catalysis leads the way. *Angew. Chem. Int. Ed.* 1995, *34*, 259–281.
36. Trost, B. The atom economy—A search for synthetic efficiency. *Science* 1991, *254*, 1471–1477.
37. Sheldon, R. A. Atom utilisation, E factors and the catalytic solution. *C. R. Acad. Sci. II C.* 2000, *3*, 541–551.
38. Sheldon, R. A. The E factor: Fifteen years on. *Green Chem.* 2007, *9*, 1273–1283.
39. Sheldon, R. A. E factors, green chemistry and catalysis: An odyssey. *Chem. Commun.* 2008, 3352–3365.
40. Adlercreutz, P. Fundamentals of biocatalysis in neat organic solvents. In *Organic Synthesis with Enzymes in Non-Aqueous Media*, Carrea, G.; Riva, S., Eds. Wiley-VCH Verlag GmbH & Co. KGaA: Weinheim, Germany, 2008; pp. 3–24.
41. Hernáiz, M.; Alcántara, A. R.; García, J. I.; Sinisterra, J. V. Applied biotransformations in green solvents. *Chem. Eur. J.* 2010, *16*, 9422–9437.
42. de Maria, P. D.; Maugeri, Z. Ionic liquids in biotransformations: From proof-of-concept to emerging deep-eutectic-solvents. *Curr. Opin. Chem. Biol.* 2011, *15*, 220–225.
43. Law, V.; Knox, C.; Djoumbou, Y.; Jewison, T.; Guo, A. C.; Liu, Y.; Maciejewski, A.; Arndt, D.; Wilson, M.; Neveu, V.; Tang, A.; Gabriel, G.; Ly, C.; Adamjee, S.; Dame, Z. T.; Han, B.; Zhou, Y.; Wishart, D. S. DrugBank 4.0: Shedding new light on drug metabolism. *Nucleic Acids Res.* 2013, *42*, D1091–D1097.
44. World Health Statistics. Global Health Observatory (GHO). World Health Organization: Geneva, Switzerland, 2012.
45. Jemal, A.; Bray, F.; Center, M. M.; Ferlay, J.; Ward, E.; Forman, D. Global cancer statistics. *CA Cancer J. Clin.* 2011, *61*, 69–90.
46. Newman, D. J.; Cragg, G. M. Natural products as sources of new drugs over the 30 years from 1981 to 2010. *J. Nat. Prod.* 2012, *75*, 311–335.
47. Avendaño, C.; Menéndez, J. C. *Medicinal Chemistry of Anticancer Drugs*. Elsevier: Amsterdam, the Netherlands, 2008.
48. Faber, K. *Biotransformations in Organic Chemistry: A Textbook*, 6th edn. Springer-Verlag: Berlin, Germany, 2011.
49. Liese, A.; Seelbach, K.; Wandrey, C. *Industrial Biotransformations*, 2nd edn. John Wiley & Sons, Inc. Verlag GmbH & Co. kGaA: Weinheim, Germany, 2006.
50. Wang, Y. F.; Shi, Q. W.; Dong, M.; Kiyota, H.; Gu, Y. C.; Cong, B. Natural taxanes: Developments since 1828. *Chem. Rev.* 2011, *111*, 7652–7709.
51. Chang, S. H.; Ho, C. K.; Chen, F. H. H. Propagation and bioreactor technology of medicinal plants: Case studies on paclitaxel, 10-deacetylbaccatin III, and camptothecin. In *Medicinal Plants—Recent Advances in Research and Development*, Tsay, H. S.; Shyur, L. F.; Agrawal, D. C.; Wu, Y. C.; Wang, S. Y., Eds. Springer: Singapore, 2016; pp. 257–272.
52. Kingston, D. G. I. Taxol, a molecule for all seasons. *Chem. Commun.* 2001, 867–880.
53. Nicolaou, K. C.; Yang, Z.; Liu, J. J.; Ueno, H.; Nantermet, P. G.; Guy, R. K.; Claiborne, C. F.; Renaud, J.; Couladouros, E. A.; Paulvannan, K.; Sorensen, E. J. Total synthesis of Taxol. *Nature* 1994, *367*, 630–634.
54. Holton, R. A.; Kim, H. B.; Somoza, C.; Liang, F.; Biediger, R. J.; Boatman, P. D.; Shindo, M.; Smith, C. C.; Kim, S. C.; Nadizadeh, H.; Suzuki, Y.; Tao, C. L.; Vu, P.; Tang, S. H.; Zhang, P. S.; Murthi, K. K.; Gentile, L. N.; Liu, J. H. First total synthesis of taxol 2. Completion of the C-ring and D-ring. *J. Am. Chem. Soc.* 1994, *116*, 1599–1600.

55. Holton, R. A.; Somoza, C.; Kim, H. B.; Liang, F.; Biediger, R. J.; Boatman, P. D.; Shindo, M.; Smith, C. C.; Kim, S. C.; Nadizadeh, H.; Suzuki, Y.; Tao, C. L.; Vu, P.; Tang, S. H.; Zhang, P. S.; Murthi, K. K.; Gentile, L. N.; Liu, J. H. First total synthesis of taxol 1. Functionalization of the B-ring. *J. Am. Chem. Soc.* 1994, *116*, 1597–1598.
56. Gibson, D. M.; Ketchum, R. E. B.; Vance, N. C.; Christen, A. A. Initiation and growth of cell-lines of *Taxus brevifolia* (pacific yew). *Plant Cell Rep.* 1993, *12*, 479–482.
57. Stierle, A.; Strobel, G.; Stierle, D. Taxol and taxane production by *Taxomyces andreanae*, an endophytic fungus of pacific yew. *Science* 1993, *260*, 214–216.
58. Holton, R. A. Method for preparation of taxol using an oxazinone. U.S. Patent 5,175,315, 1991.
59. Holton, R. A. Method for preparation of taxol using ß-lactam. US005175315A, 1992.
60. Holton, R. A. Semi-synthesis of taxane derivatives using metal alkoxides and oxazinones. EP 0568203, 1993.
61. Denis, J. N.; Greene, A. E.; Guenard, D.; Guerittevoegelein, F.; Mangatal, L.; Potier, P. A. Highly efficient, practical approach to natural taxol. *J. Am. Chem. Soc.* 1988, *110*, 5917–5919.
62. Colin, M.; Guenard, D.; Gueritte-Voegelein, F.; Potier, P. Taxol derivatives their preparation and pharmaceutical compositions containing them. U.S. Patent 4814470, March 21, 1989.
63. Colin, M.; Guenard, D.; Gueritte-Voegelein, F.; Potier, P. Process for the preparation of taxol and 10-deacetyltaxol. U.S. Patent 4857653, August 15, 1989.
64. Colin, M.; Guenard, D.; Gueritte-Voegelein, F.; Potier, P. Process for preparing derivatives of baccatine III and of 10-deacetyl baccatine III. U.S. Patent 4924012, May 8, 1990.
65. Denis, J.-N.; Greene, A. E.; Guenard, D.; Gueritte-Voegelein, F. Process for preparing taxol, U.S. Patent 4,924,011, 1990.
66. Guéritte-Voegelein, F.; Sénilh, V.; David, B.; Guénard, D.; Potier, P. Chemical studies of 10-deacetyl baccatin III: Hemisynthesis of taxol derivatives. *Tetrahedron* 1986, *42*, 4451–4460.
67. Palomo, C.; Arrieta, A.; Cossio, F. P.; Aizpurua, J. M.; Mielgo, A.; Aurrekoetxea, N. Highly stereoselective synthesis of alpha-hydroxy beta-amino acids through beta-lactams. Application to the synthesis of the taxol and bestatin side-chains and related systems. *Tetrahedron Lett.* 1990, *31*, 6429–6432.
68. Palomo, C.; Aizpurua, J. M.; Concepción López, M.; Aurrekoetxea, N.; Oiarbide, M. Addition of α-bromoesters to azetidine-2,3-diones promoted by zinc-trimethylchlorosilane: A general synthesis of 3-trimethyisilyloxyazetidin-2-ones and α-alkylidene β-lactams. *Tetrahedron Lett.* 1990, *31*, 6425–6428.
69. Patel, R. N. Tour de paclitaxel: Biocatalysis for semisynthesis. *Annu. Rev. Microbiol.* 1998, *52*, 361–395.
70. Nanduri, V. B.; Hanson, R. L.; Laporte, T. L.; Ko, R. Y.; Patel, R. N.; Szarka, L. J. Fermentation and isolation of c10-deacetylase for the production of 10-deacetylbaccatin-III from baccatin-III. *Biotechnol. Bioeng.* 1995, *48*, 547–550.
71. Hanson, R. L.; Wasylyk, J. M.; Nanduri, V. B.; Cazzulino, D. L.; Patel, R. N.; Szarka, L. J. Site-specific enzymatic-hydrolysis of taxanes at c-10 and c-13. *J. Biol. Chem.* 1994, *269*, 22145–22149.
72. Hanson, R. L.; Howell, J. M.; Brzozowski, D. B.; Sullivan, S. A.; Patel, R. N.; Szarka, L. J. Enzymic hydrolysis of 7-xylosyltaxanes by xylosidase from *Moraxella* sp. *Biotechnol. Appl. Biochem.* 1997, *26*, 153–158.
73. Wang, X. H.; Zhang, C.; Yang, L. L.; Li, S.; Zhang, Y.; Gomes-Laranjo, J. Screening and identification of microbial strains that secrete an extracellular C-7 xylosidase of taxanes. *World J. Microbiol. Biotechnol.* 2011, *27*, 627–635.
74. Borah, J. C.; Boruwa, J.; Barua, N. C. Synthesis of the c-13 side-chain of taxol. *Curr. Org. Synth.* 2007, *4*, 175–199.
75. Brieva, R.; Crich, J. Z.; Sih, C. J. Chemoenzymatic synthesis of the C-13 side-chain of taxol—Optically-active 3-hydroxy-4-phenyl beta-lactam derivatives. *J. Org. Chem.* 1993, *58*, 1068–1075.
76. Pace, V.; Sinisterra, J. V.; Alcántara, A. R. Celite-supported reagents in organic synthesis: An overview. *Curr. Org. Chem.* 2010, *14*, 2384–2408.
77. Alfonsi, K.; Colberg, J.; Dunn, P. J.; Fevig, T.; Jennings, S.; Johnson, T. A.; Kleine, H. P.; Knight, C.; Nagy, M. A.; Perry, D. A.; Stefaniak, M. Green chemistry tools to influence a medicinal chemistry and research chemistry based organisation. *Green Chem.* 2008, *10*, 31–36.
78. Forro, E.; Fulop, F. New enzymatic two-step cascade reaction for the preparation of a key intermediate for the taxol side-chain. *Eur. J. Org. Chem.* 2010, *2010*, 3074–3079.
79. Forro, E.; Fulop, F. A. New enzymatic strategy for the preparation of (2*R*,3*S*)-3-phenylisoserine: A key intermediate for the Taxol side chain. *Tetrahedron: Asymmetry* 2010, *21*, 637–639.
80. Patel, R. N.; Banerjee, A.; Ko, R. Y.; Howell, J. M.; Li, W. S.; Comezoglu, F. T.; Partyka, R. A.; Szarka, L. Enzymatic preparation of (3*R-cis*)-3-(acetyloxy)-4-phenyl-2-azetidinone—A taxol side-chain synthon. *Biotechnol. Appl. Biochem.* 1994, *20*, 23–33.

81. Patel, R. N.; Szarka, L. J.; Partyka, R. A. Enzymatic processes for resolution of enantiomeric mixtures of compounds useful as intermediates in the preparation of taxanes. EP0552041, January 15, 1993.
82. Patel, R. N.; Banerjee, A.; Mcnamee, C. G.; Thottathil, J. K.; Szarka, L. J. Enzymatic reduction method for the preparation of compounds useful for preparing taxanes. US5686298, 1995.
83. Li, W.-s.; Thottathil, J. K. Reduction and resolution methods for the preparation of compounds useful as intermediates for preparing taxanes. US5602272, 1997.
84. Anand, N.; Kapoor, M.; Ahmad, K.; Koul, S.; Parshad, R.; Manhas, K. S.; Sharma, R. L.; Qazi, G. N.; Taneja, S. C. *Arthrobacter* sp.: A lipase of choice for the kinetic resolution of racemic arylazetidinone precursors of taxanoid side chains. *Tetrahedron: Asymmetry* 2007, *18*, 1059–1069.
85. Sampath, D.; Discafani, C. M.; Loganzo, F.; Beyer, C.; Liu, H.; Tan, X.; Musto, S.; Annable, T.; Gallagher, P.; Rios, C.; Greenberger, L. M. MAC-321, a novel taxane with greater efficacy than paclitaxel and docetaxel in vitro and in vivo. *Mol. Cancer Ther.* 2003, *2*, 873–884.
86. Sampath, D.; Greenberger, L. M.; Beyer, C.; Hari, M.; Liu, H.; Baxter, M.; Yang, S.; Rios, C.; Discafani, C. Preclinical pharmacologic evaluation of MST-997, an orally active taxane with superior in vitro and in vivo efficacy in paclitaxel- and docetaxel-resistant tumor models. *Clin. Cancer Res.* 2006, *12*, 3459–3469.
87. Mastalerz, H.; Cook, D.; Fairchild, C. R.; Hansel, S.; Johnson, W.; Kadow, J. F.; Long, B. H.; Rose, W. C.; Tarrant, J.; Wu, M. J.; May, Q. F.; Zhang, G. F.; Zoeckler, M.; Vyas, D. M. The discovery of BMS-275183: An orally efficacious novel taxane. *Bioorg. Med. Chem.* 2003, *11*, 4315–4323.
88. Heath, E.; LoRusso, P.; Ramalingam, S.; Awada, A.; Egorin, M.; Besse-Hamer, T.; Cardoso, F.; Valdivieso, M.; Has, T.; Alland, L.; Zhou, X.; Belani, C. A phase 1 study of BMS-275183, a novel oral analogue of paclitaxel given on a daily schedule to patients with advanced malignancies. *Invest. New Drugs* 2011, *29*, 1426–1431.
89. Parachin, N. S.; Carlquist, M.; Gorwa-Grauslund, M. F. Bioreduction. In *Encyclopedia of Industrial Biotechnology: Bioprocess, Bioseparation, and Cell Technology*, Flickinger, M. C., Ed. John Wiley & Sons, Inc.: Oxford, U.K., 2010.
90. Carballeira, J. D.; Fernandez-Lucas, J.; Quezada, M. A.; Hernaiz, M. J.; Alcantara, A. R.; Simeó, Y.; Sinisterra, J. V. Biotransformations. In *Encyclopedia of Microbiology*, Moselio, S., Ed. Academic Press: Oxford, U.K., 2009; pp. 212–251.
91. Sheldon, R. A.; Arends, I.; Hanefeld, U. (eds.) Catalytic reductions. In *Green Chemistry and Catalysis*. Wiley-VCH Verlag GmbH & Co. KGaA: Weinheim, Germany, 2007; pp. 91–131.
92. Gröger, H.; Borchert, S.; Krauber, M.; Hummel, W. Enzyme-catalyzed asymmetric reduction of ketones. In *Encyclopedia of Industrial Biotechnology: Bioprocess, Bioseparation, and Cell Technology*, Flickinger, M. C., Ed. John Wiley & Sons, Inc.: Oxford, U.K., 2010.
93. Hiroaki, Y. Carbonyl reductase. In *Encyclopedia of Industrial Biotechnology: Bioprocess, Bioseparation, and Cell Technology*, Flickinger, M. C., Ed. John Wiley & Sons, Inc.: Oxford, U.K., 2010.
94. Hollmann, F.; Arends, I. W. C. E.; Holtmann, D. Enzymatic reductions for the chemist. *Green Chem.* 2011, *13*, 2285–2314.
95. Carballeira, J. D.; Quezada, M. A.; Hoyos, P.; Simeo, Y.; Hernaiz, M. J.; Alcantara, A. R.; Sinisterra, J. V. Microbial cells as catalysts for stereoselective red-ox reactions. *Biotechnol. Adv.* 2009, *27*, 686–714.
96. Kayser, M. M.; Mihovilovic, M. D.; Kearns, J.; Feicht, A.; Stewart, J. D. Baker's yeast-mediated reductions of alpha-keto esters and an alpha-keto-beta-lactam. Two routes to the paclitaxel side chain. *J. Org. Chem.* 1999, *64*, 6603–6608.
97. Kirst, H.; Yeh, W. K. *Enzyme Technologies for Pharmaceutical and Biotechnological Applications*. Taylor & Francis: Boca Raton, FL, 2001.
98. Kaluzna, I. A.; Feske, B. D.; Wittayanan, W.; Ghiviriga, I.; Stewart, J. D. Stereoselective, biocatalytic reductions of alpha-chloro-beta-keto esters. *J. Org. Chem.* 2005, *70*, 342–345.
99. Yang, Y.; Drolet, M.; Kayser, M. M. The dynamic kinetic resolution of 3-oxo-4-phenyl-beta-lactam by recombinant *E. coli* overexpressing yeast reductase Ara1p. *Tetrahedron: Asymmetry* 2005, *16*, 2748–2753.
100. Rodriguez, S.; Kayser, M. M.; Stewart, J. D. Highly stereoselective reagents for beta-keto ester reductions by genetic engineering of baker's yeast. *J. Am. Chem. Soc.* 2001, *123*, 1547–1555.
101. Rodriguez, S.; Schroeder, K. T.; Kayser, M. M.; Stewart, J. D. Asymmetric synthesis of beta-hydroxy esters and alpha-alkyl-beta-hydroxy esters by recombinant *Escherichia coli* expressing enzymes from baker's yeast. *J. Org. Chem.* 2000, *65*, 2586–2587.
102. Rodriguez, S.; Kayser, M.; Stewart, J. D. Improving the stereoselectivity of bakers' yeast reductions by genetic engineering. *Org. Lett.* 1999, *1*, 1153–1155.
103. Yang, Y.; Kayser, M. M.; Rochon, F. D.; Rodriguez, S.; Stewart, J. D. Assessing substrate acceptance and enantioselectivity of yeast reductases in reactions with substituted alpha-keto beta-lactams. *J. Mol. Catal. B: Enzym.* 2005, *32*, 167–174.

104. Kayser, M. M.; Drolet, M.; Stewart, J. D. Application of newly available bio-reducing agents to the synthesis of chiral hydroxy-beta-lactams: Model for aldose reductase selectivity. *Tetrahedron: Asymmetry* 2005, *16*, 4004–4009.
105. Kearns, J.; Kayser, M. M. Application of yeast-catalyzed reductions to synthesis of (2*R*,3*S*)-phenylisoserine. *Tetrahedron Lett.* 1994, *35*, 2845–2848.
106. Patel, R. N.; Banerjee, A.; Howell, J. M.; McNamee, C. G.; Brozozowski, D.; Mirfakhrae, D.; Nanduri, V.; Thottathil, J. K.; Szarka, L. J. Microbial synthesis of (2*R*,3*S*)-(−)-*N*-benzoyl-3-phenyl isoserine ethyl ester-a taxol side-chain synthon. *Tetrahedron: Asymmetry* 1993, *4*, 2069–2084.
107. Feske, B. D.; Kaluzna, I. A.; Stewart, J. D. Enantiodivergent, biocatalytic routes to both taxol side chain antipodes. *J. Org. Chem.* 2005, *70*, 9654–9657.
108. Rimoldi, I.; Pellizzoni, M.; Facchetti, G.; Molinari, F.; Zerla, D.; Gandolfi, R. Chemo- and biocatalytic strategies to obtain phenylisoserine, a lateral chain of Taxol by asymmetric reduction. *Tetrahedron: Asymmetry* 2011, *22*, 2110–2116.
109. Prelog, V. Specification of the stereospecificity of some oxido-reductases by diamond lattice sections. *Pure Appl. Chem.* 1964, *9*, 119–130.
110. Exposito, O.; Bonfill, M.; Moyano, E.; Onrubia, M.; Mirjalili, M. H.; Cusido, R. M.; Palazon, J. Biotechnological production of taxol and related taxoids: Current state and prospects. *Anticancer Agents Med Chem.* 2009, *9*, 109–121.
111. Gerth, K.; Pradella, S.; Perlova, O.; Beyer, S.; Muller, R. Myxobacteria: Proficient producers of novel natural products with various biological activities—Past and future biotechnological aspects with the focus on the genus *Sorangium*. *J. Biotechnol.* 2003, *106*, 233–253.
112. Goodin, S.; Kane, M. P.; Rubin, E. H. Epothilones: Mechanism of action and biologic activity. *J. Clin. Oncol.* 2004, *22*, 2015–2025.
113. Cortes, J.; Baselga, J. Targeting the microtubules in breast cancer beyond taxanes: The epothilones. *Oncologist* 2007, *12*, 271–280.
114. Borzilleri, R. M.; Zheng, X. P.; Schmidt, R. J.; Johnson, J. A.; Kim, S. H.; DiMarco, J. D.; Fairchild, C. R.; Gougoutas, J. Z.; Lee, F. Y. F.; Long, B. H.; Vite, G. D. A novel application of a Pd(0)-catalyzed nucleophilic substitution reaction to the regio- and stereoselective synthesis of lactam analogues of the epothilone natural products. *J. Am. Chem. Soc.* 2000, *122*, 8890–8897.
115. Souza, A. C. S.; de Fatima, A.; da Silveira, R. B.; Justo, G. Z. Seek and destroy: The use of natural compounds for targeting the molecular roots of cancer. *Curr. Drug Targets* 2011, *13*, 1072–1082.
116. Basch, J. D.; Chiang, S.-j. D.; Liu, S.-w.; Nayeem, A.; Sun, Y.; You, L. Compositions and methods for hydroxylating epothilones. U.S. Patent 6884608, April 26, 2005.
117. Basch, J.; Chiang, S. J. Cloning and expression of a cytochrome P450 hydroxylase gene from *Amycolatopsis orientalis*: Hydroxylation of epothilone B for the production of epothilone F. *J. Ind. Microbiol. Biotechnol.* 2007, *34*, 171–176.
118. Nayeem, A.; Chiang, S. J.; Liu, S. W.; Sun, Y. H.; You, L.; Basch, J. Engineering enzymes for improved catalytic efficiency: A computational study of site mutagenesis in epothilone-B hydroxylase. *Protein Eng. Des. Sel.* 2009, *22*, 257–266.
119. Sessa, C.; Perotti, A.; Llado, A.; Cresta, S.; Capri, G.; Voi, M.; Marsoni, S.; Corradino, I.; Gianni, L. Phase I clinical study of the novel epothilone B analogue BMS-310705 given on a weekly schedule. *Ann. Oncol.* 2007, *18*, 1548–1553.
120. Balog, A.; Meng, D. F.; Kamenecka, T.; Bertinato, P.; Su, D. S.; Sorensen, E. J.; Danishefsky, S. Total synthesis of (−)-epothilone A. *Angew. Chem. Int. Ed. Engl.* 1996, *35*, 2801–2803.
121. Schinzer, D.; Limberg, A.; Bauer, A.; Bohm, O. M.; Cordes, M. Total synthesis of (−)-epothilone A. *Angew. Chem. Int. Ed. Engl.* 1997, *36*, 523–524.
122. Nicolaou, K. C.; Winssinger, N.; Pastor, J.; Ninkovic, S.; Sarabia, F.; He, Y.; Vourloumis, D.; Yang, Z.; Li, T.; Giannakakou, P.; Hamel, E. Synthesis of epothilones A and B in solid and solution phase (vol 387, pg 268, 1997). *Nature* 1997, *390*, 100.
123. Nicolaou, K. C.; Winssinger, N.; Pastor, J.; Ninkovic, S.; Sarabia, F.; He, Y.; Vourloumis, D.; Yang, Z.; Li, T.; Giannakakou, P.; Hamel, E. Synthesis of epothilones A and B in solid and solution phase. *Nature* 1997, *387*, 268–272.
124. Nicolaou, K. C.; Ninkovic, S.; Sarabia, F.; Vourloumis, D.; He, Y.; Vallberg, H.; Finlay, M. R. V.; Yang, Z. Total syntheses of epothilones A and B via a macrolactonization-based strategy. *J. Am. Chem. Soc.* 1997, *119*, 7974–7991.
125. Zhu, B.; Panek, J. S. Total synthesis of epothilone A. *Org. Lett.* 2000, *2*, 2575–2578.
126. Zhu, B.; Panek, J. S. Methodology based on chiral silanes in the synthesis of polypropionate-derived natural products—Total synthesis of epothilone A. *Eur. J. Org. Chem.* 2001, *2001*, 1701–1714.

127. Watkins, E. B.; Chittiboyina, A. G.; Avery, M. A. Recent developments in the syntheses of the epothilones and related analogues. *Eur. J. Org. Chem.* 2006, *18*, 4071–4084.
128. Scheid, G.; Ruijter, E.; Konarzycka-Bessler, M.; Bornscheuer, U. T.; Wessjohann, L. A. Synthesis and resolution of a key building block for epothilones: A comparison of asymmetric synthesis, chemical and enzymatic resolution. *Tetrahedron: Asymmetry* 2004, *15*, 2861–2869.
129. Klar, U.; Hoffmann, J.; Giurescu, M. Sagopilone (ZK-EPO): From a natural product to a fully synthetic clinical development candidate. *Expert Opin. Investig. Drugs* 2008, *17*, 1735–1748.
130. Galmarini, C. M. Sagopilone, a microtubule stabilizer for the potential treatment of cancer. *Curr. Opin. Investig. Drugs* 2009, *10*, 1359–1371.
131. Klar, U.; Platzek, J. Asymmetric total synthesis of the epothilone sagopilone—From research to development. *Synlett* 2012, *23*, 1291–1299.
132. Klar, U.; Buchmann, B.; Schwede, W.; Skuballa, W.; Hoffinann, J.; Lichtner, R. B. Total synthesis and antitumor activity of ZK-EPO: The first fully synthetic epothilone in clinical development. *Angew. Chem. Int. Ed. Engl.* 2006, *45*, 7942–7948.
133. Platzek, J.; Zorn, L.; Buchmann, B.; Skuballa, W.; Petrov, O. Optically active, heteroaromatic ß-hydroxy esters, processes for their preparation from ß-keto esters and processes for the preparation of these ß-keto esters. WO/2005/064006, July 14, 2005.
134. Kataoka, M.; Shimizu, K.; Sakamoto, K.; Yamada, H.; Shimizu, S. Optical resolution of racemic pantolactone with a novel fungal enzyme, lactonohydrolase. *Appl. Microbiol. Biotechnol.* 1995, *43*, 974–977.
135. Shimizu, S.; Kataoka, M. Optical resolution of pantolactone by a novel fungal enzyme, lactonolaydrolase. In *Enzyme Engineering XIII*, Dordick, J. S.; Russell, A. J., Eds. New York Academy Sciences: New York, 1996; Vol. 799, pp. 650–658.
136. Pscheidt, B.; Liu, Z. B.; Gaisberger, R.; Avi, M.; Skranc, W.; Gruber, K.; Griengl, H.; Glieder, A. Efficient biocatalytic synthesis of (R)-pantolactone. *Adv. Synth. Catal.* 2008, *350*, 1943–1948.
137. Haughton, L.; Williams, J. M. J.; Zimmermann, J. A. Enzymatic kinetic resolution of pantolactone: Relevance to chiral auxiliary chemistry. *Tetrahedron: Asymmetry* 2000, *11*, 1697–1701.
138. Kesseler, M.; Friedrich, T.; Hoffken, H. W.; Hauer, B. Development of a novel biocatalyst for the resolution of rac-pantolactone. *Adv. Synth. Catal.* 2002, *344*, 1103–1110.
139. Chen, B.; Fan, L. Q.; Xu, J. H.; Zhao, J.; Zhang, X.; Ouyang, L. M. Biocatalytic properties of a recombinant *Fusarium proliferatum* lactonase with significantly enhanced production by optimal expression in *Escherichia coli*. *Appl. Biochem. Biotechnol.* 2010, *162*, 744–756.
140. Altaba, A. R.; Sanchez, P.; Dahmane, N. Gli and hedgehog in cancer: Tumours, embryos and stem cells. *Nat. Rev. Cancer* 2002, *2*, 361–372.
141. Sahebjam, S.; Siu, L. L.; Razak, A. A. The utility of hedgehog signaling pathway inhibition for cancer. *Oncologist* 2012, *17*, 1090–1099.
142. Yun, J. I.; Kim, H. R.; Park, H.; Kim, S. K.; Lee, J. Small molecule inhibitors of the hedgehog signaling pathway for the treatment of cancer. *Arch. Pharm. Res.* 2012, *35*, 1317–1333.
143. Keeler, R. F. Teratogenic compounds of veratrum californicum (durand). 4. First isolation of veratramine and alkaloid Q and a reliable method for isolation of cyclopamine. *Phytochemistry* 1968, *7*, 303–306.
144. Keeler, R. F. Teratogenic compounds of veratrum californicum (durand). 6. Structure of cyclopamine. *Phytochemistry* 1969, *8*, 223–225.
145. Tremblay, M. R.; Lescarbeau, A.; Grogan, M. J.; Tan, E.; Lin, G.; Austad, B. C.; Yu, L. C.; Behnke, M. L.; Nair, S. J.; Hagel, M.; White, K.; Conley, J.; Manna, J. D.; Alvarez-Diez, T. M.; Hoyt, J.; Woodward, C. N.; Sydor, J. R.; Pink, M.; MacDougall, J.; Campbell, M. J.; Cushing, J.; Ferguson, J.; Curtis, M. S.; McGovern, K.; Read, M. A.; Palombella, V. J.; Adams, J.; Castro, A. C. Discovery of a potent and orally active hedgehog pathway antagonist (IPI-926). *J. Med. Chem.* 2009, *52*, 4400–4418.
146. Austad, B.; Bahadoor, A.; Belani, J. D.; Janardanannair, S.; Johannes, C. W.; Keany, G. F.; Lo, C. K.; Wallerstein, S. L. Enzymatic transamination of cyclopamine analogs. WO/2011/017551, February 10, 2011.
147. Truppo, M.; Janey, J. M.; Grau, B.; Morley, K.; Pollack, S.; Hughes, G.; Davies, I. Asymmetric, biocatalytic labeled compound synthesis using transaminases. *Catal. Sci. Tec.* 2012, *2*, 1556–1559.
148. Lombo, F.; Menendez, N.; Salas, J. A.; Mendez, C. The aureolic acid family of antitumor compounds: Structure, mode of action, biosynthesis, and novel derivatives. *Appl. Microbiol. Biotechnol.* 2006, *73*, 1–14.
149. Mir, M. A.; Majee, S.; Das, S.; Dasgupta, D. Association of chromatin with anticancer antibiotics, mithramycin and chromomycin A(3). *Bioorg. Med. Chem.* 2003, *11*, 2791–2801.
150. Gao, Y.; Jia, Z. L.; Kong, X. Y.; Li, Q.; Chang, D. Z.; Wei, D. Y.; Le, X. D.; Huang, S. D.; Huang, S. Y.; Wang, L. W.; Xie, K. P. Combining betulinic acid and mithramycin A effectively suppresses pancreatic cancer by inhibiting proliferation, invasion, and angiogenesis. *Cancer Res.* 2011, *71*, 5182–5193.

151. Garcia, B.; Gonzalez-Sabin, J.; Menendez, N.; Brana, A. F.; Elena Nunez, L.; Moris, F.; Salas, J. A.; Mendez, C. The chromomycin CmmA acetyltransferase: A membrane-bound enzyme as a tool for increasing structural diversity of the antitumour mithramycin. *Microb. Biotechnol.* 2011, *4*, 226–238.
152. Gonzalez-Sabin, J.; Nunez, L. E.; Brana, A. F.; Mendez, C.; Salas, J. A.; Gotor, V.; Moris, F. Regioselective enzymatic acylation of aureolic acids to obtain novel analogues with improved antitumor activity. *Adv. Synth. Catal.* 2012, *354*, 1500–1508.
153. Núñez, L. E.; García-Fernández, B.; Pérez, M.; Fernández, A.; Menéndez, N.; González-Sabin, J.; Morís-Varas, F.; Méndez, C.; Salas, J. A. Aureolic acid derivatives, the method for preparation thereof and the uses thereof. US 2012/0270823 A1, 2012.
154. Gonzalez-Sabin, J.; Nunez, L. E.; Menendez, N.; Brana, A. F.; Mendez, C.; Salas, J. A.; Gotor, V.; Moris, F. Lipase-catalyzed preparation of chromomycin A(3) analogues and biological evaluation for anticancer activity. *Bioorg. Med. Chem. Lett.* 2012, *22*, 4310–4313.
155. Robert, F.; Garrett, C.; Dinwoodie, W. R.; Sullivan, D. M.; Bishop, M.; Amantea, M.; Zhang, M.; Reich, S. D. Results of 2 phase I studies of intravenous (IV) pelitrexol (AG2037), a glycinamide ribonucleotide formyltransferase (GARFT) inhibitor, in patients (pts) with solid tumors. *J. Clin. Oncol.* 2004, *22*, 3075.
156. Dovalsantos, E. Z.; Flahive, E. J.; Halden, B. J.; Mitchell, M. B.; Notz, W. R. L.; Tian, Q.; O'neil-sla Wecki, S. A. Convergent synthesis of a garft inhibitor containing a methyl substitute thiophene core and a tetrahydropyrido[2,3-*d*]pyrimidine ring system and intermediates therefor. WO/2004/113337, December 29, 2004.
157. Hu, S. H.; Kelly, S.; Lee, S.; Tao, J.; Flahive, E. Efficient chemoenzymatic synthesis of pelitrexol via enzymic differentiation of a remote stereocenter. *Org. Lett.* 2006, *8*, 1653–1655.
158. Hu, S. H.; Tat, D.; Martinez, C. A.; Yazbeck, D. R.; Tao, J. H. An efficient and practical chemoenzymatic preparation of optically active secondary amines. *Org. Lett.* 2005, *7*, 4329–4331.
159. Mayer, T. U.; Kapoor, T. M.; Haggarty, S. J.; King, R. W.; Schreiber, S. L.; Mitchison, T. J. Small molecule inhibitor of mitotic spindle bipolarity identified in a phenotype-based screen. *Science* 1999, *286*, 971–974.
160. Maliga, Z.; Kapoor, T. M.; Mitchison, T. J. Evidence that monastrol is an allosteric inhibitor of the mitotic kinesin Eg5. *Chem. Biol.* 2002, *9*, 989–996.
161. Sarli, V.; Giannis, A. Targeting the kinesin spindle protein: Basic principles and clinical implications. *Clin. Cancer Res.* 2008, *14*, 7583–7587.
162. Rath, O.; Kozielski, F. Kinesins and cancer. *Nat. Rev. Cancer* 2012, *12*, 527–539.
163. Wan, J. P.; Pan, Y. Recent advance in the pharmacology of dihydropyrimidinone. *Mini-Rev. Med. Chem.* 2012, *12*, 337–349.
164. Gartner, M.; Sunder-Plassmann, N.; Seiler, J.; Utz, M.; Vernos, I.; Surrey, T.; Giannis, A. Development and biological evaluation of potent and specific inhibitors of mitotic kinesin Eg5. *ChemBioChem* 2005, *6*, 1173–1177.
165. Dondoni, A.; Massi, A.; Sabbatini, S. Improved synthesis and preparative scale resolution of racemic monastrol. *Tetrahedron Lett.* 2002, *43*, 5913–5916.
166. Gong, L. Z.; Chen, X. H.; Xu, X. Y. Asymmetric organocatalytic biginelli reactions: A new approach to quickly access optically active 3,4-dihydropyrimidin-2-(1*H*)-ones. *Chem. Eur. J.* 2007, *13*, 8920–8926.
167. Huang, Y. J.; Yang, F. Y.; Zhu, C. J. Highly enantioseletive biginelli reaction using a new chiral ytterbium catalyst: Asymmetric synthesis of dihydropyrimidines. *J. Am. Chem. Soc.* 2005, *127*, 16386–16387.
168. Xu, F. X.; Huang, D.; Lin, X. F.; Wang, Y. G. Highly enantioselective Biginelli reaction catalyzed by SPINOL-phosphoric acids. *Org. Biomol. Chem.* 2012, *10*, 4467–4470.
169. Sohn, J. H.; Choi, H. M.; Lee, S.; Joung, S.; Lee, H. Y. Probing the mode of asymmetric induction of biginelli reaction using proline ester salts. *Eur. J. Org. Chem.* 2009, *23*, 3858–3862.
170. Panda, S. S.; Khanna, P.; Khanna, L. Biginelli reaction: A green perspective. *Curr. Org. Chem.* 2012, *16*, 507–520.
171. Blasco, M. A.; Thumann, S.; Wittmann, J.; Giannis, A.; Groger, H. Enantioselective biocatalytic synthesis of (*S*)-monastrol. *Bioorg. Med. Chem. Lett.* 2010, *20*, 4679–4682.
172. Linger, R. M.; Keating, A. K.; Earp, H. S.; Graham, D. K. Taking aim at Mer and Axl receptor tyrosine kinases as novel therapeutic targets in solid tumors. *Expert Opin. Ther. Targ.* 2010, *14*, 1073–1090.
173. Rescigno, J.; Mansukhani, A.; Basilico, C. A putative receptor tyrosine kinase with unique structural topology. *Oncogene* 1991, *6*, 1909–1913.
174. Holland, S. J.; Powell, M. J.; Franci, C.; Chan, E. W.; Friera, A. M.; Atchison, R. E.; McLaughlin, A.; Swift, S. E.; Pali, E. S.; Yam, G.; Wong, S.; Lasaga, J.; Shen, M. R.; Yu, S.; Xu, W. D.; Hitoshi, Y.; Bogenberger, J.; Nor, J. E.; Payan, D. G.; Lorens, J. B. Multiple roles for the receptor tyrosine kinase Axl in tumor formation. *Cancer Res.* 2005, *65*, 9294–9303.

175. Holland, S. J.; Pan, A.; Franci, C.; Hu, Y. M.; Chang, B.; Li, W. Q.; Duan, M.; Torneros, A.; Yu, J. X.; Heckrodt, T. J.; Zhang, J.; Ding, P. Y.; Apatira, A.; Chua, J.; Brandt, R.; Pine, P.; Goff, D.; Singh, R.; Payan, D. G.; Hitoshi, Y. R428, a selective small molecule inhibitor of Axl kinase, blocks tumor spread and prolongs survival in models of metastatic breast cancer. *Cancer Res.* 2010, *70*, 1544–1554.

176. Holland, S. J.; Hu, Y. M.; Chang, B.; Pan, A.; Franci, C.; Li, W. Q.; Duan, M.; Bagos, A.; Torneros, A.; McLaughlin, J.; Zhang, J.; Yu, J. X.; Ding, P.; Heckrodt, T. J.; Litvak, J.; Stauffer, E.; Clemens, G. R.; Daniel-Issakani, S.; Pine, P.; Goff, D.; Singh, R.; Payan, D. G.; Hitoshi, Y. Novel small molecule inhibitors of the AxI receptor tyrosine kinase block tumor growth. *Mol. Cancer Ther.* 2007, *6*, 3505s.

177. Mollard, A.; Warner, S. L.; Call, L. T.; Wade, M. L.; Bearss, J. J.; Verma, A.; Sharma, S.; Vankayalapati, H.; Bearss, D. J. Design, synthesis, and biological evaluation of a series of novel AXL kinase inhibitors. *ACS Med. Chem. Lett.* 2011, *2*, 907–912.

178. Hitoshi, Y.; Holland, S.; Payan, D. G. Axl inhibitors for use in combination therapy for preventing, treating or managing metastatic cancer. U.S. Patent 0196511, August 5, 2010.

179. Truppo, M. D.; Rozzell, J. D.; Turner, N. J. Efficient production of enantiomerically pure chiral amines at concentrations of 50 g/L using transaminases. *Org. Process Res. Dev.* 2010, *14*, 234–237.

180. Zou, H. Y.; Li, Q.; Lee, J. H.; Arango, M. E.; McDonnell, S. R.; Yamazaki, S.; Koudriakova, T. B.; Alton, G.; Cui, J. J.; Kung, P.-P.; Nambu, M. D.; Los, G.; Bender, S. L.; Mroczkowski, B.; Christensen, J. G. An orally available small-molecule inhibitor of c-met, PF-2341066, exhibits cytoreductive antitumor efficacy through antiproliferative and antiangiogenic mechanisms. *Cancer Res.* 2007, *67*, 4408–4417.

181. Hallberg, B.; Palmer, R. H. Crizotinib—Latest champion in the cancer wars? *New Engl. J. Med.* 2010, *363*, 1760–1762.

182. Martinez, C. A.; Keller, E.; Meijer, R.; Metselaar, G.; Kruithof, G.; Moore, C.; Kung, P.-P. Biotransformation-mediated synthesis of (1S)-1-(2,6-dichloro-3-fluorophenyl)ethanol in enantiomerically pure form. *Tetrahedron: Asymmetry* 2010, *21*, 2408–2412.

183. de Koning, P. D.; McAndrew, D.; Moore, R.; Moses, I. B.; Boyles, D. C.; Kissick, K.; Stanchina, C. L.; Cuthbertson, T.; Kamatani, A.; Rahman, L.; Rodriguez, R.; Urbina, A.; Sandoval, A.; Rose, P. R. Fit-for-purpose development of the enabling route to Crizotinib (PF-02341066). *Org. Process Res. Dev.* 2011, *15*, 1018–1026.

184. Cui, J. J.; Funk, L. A.; Jia, L.; Kung, P.-p.; Meng, J. J.; Nambu, M. D.; Pairish, M. A.; Shen, H.; Tran-dube, M. Enantiomerically pure aminoheteroaryl compounds as protein kinase inhibitors. U.S. Patent 0046991, March 2, 2006.

185. Kung, P.-p.; Martinez, C. A.; Tao, J. Enantioselective biotransformation for preparation of protein tyrosine kinase inhibitor intermediates. WO/2006/021885, March 2, 2006.

186. Liang, J.; Mundorff, E.; Ching, C.; Gruber, J. M.; Krebber, A.; Huisman, G. W. Ketoreductase polypeptides for the reduction of acetophenones. WO/2009/036404, March 19, 2009.

187. Prior, I. A.; Lewis, P. D.; Mattos, C. A comprehensive survey of Ras mutations in cancer. *Cancer Res.* 2012, *72*, 2457–2467.

188. Wong, N. S.; Morse, M. A. Lonafarnib for cancer and progeria. *Expert Opin. Investig. Drug.* 2012, *21*, 1043–1055.

189. Morgan, B.; Zaks, A.; Dodds, D. R.; Liu, J. C.; Jain, R.; Megati, S.; Njoroge, F. G.; Girijavallabhan, V. M. Enzymatic kinetic resolution of piperidine atropisomers: Synthesis of a key intermediate of the farnesyl protein transferase inhibitor, SCH66336. *J. Org. Chem.* 2000, *65*, 5451–5459.

190. Boonen, S.; Rosenberg, E.; Claessens, F.; Vanderschueren, D.; Papapoulos, S. Inhibition of cathepsin K for treatment of osteoporosis. *Curr. Osteoporos. Rep.* 2012, *10*, 73–79.

191. Podgorski, I. Future of anticathepsin K drugs: Dual therapy for skeletal disease and atherosclerosis? *Future Med. Chem.* 2009, *1*, 21–34.

192. Desmarais, S.; Masse, F.; Percival, M. D. Pharmacological inhibitors to identify roles of cathepsin K in cell-based studies: A comparison of available tools. *Biol. Chem.* 2009, *390*, 941–948.

193. Gauthier, J. Y.; Chauret, N.; Cromlish, W.; Desmarais, S.; Duong, L. T.; Falgueyret, J.-P.; Kimmel, D. B.; Lamontagne, S.; Leger, S.; LeRiche, T.; Li, C. S.; Masse, F.; McKay, D. J.; Nicoll-Griffith, D. A.; Oballa, R. A.; Palmer, J. T.; Percival, M. D.; Riendeau, D.; Robichaud, J.; Rodan, G. A.; Rodan, S. B.; Seto, C.; Therien, M.; Truong, V.-L.; Venuti, M. C.; Wesolowski, G.; Young, R. N.; Zamboni, R.; Black, W. C. The discovery of odanacatib (MK-0822), a selective inhibitor of cathepsin K. *Bioorg. Med. Chem. Lett.* 2008, *18*, 923–928.

194. Limanto, J.; Shafiee, A.; Devine, P. N.; Upadhyay, V.; Desmond, R. A.; Foster, B. R.; Gauthier, D. R.; Reamer, R. A.; Volante, R. P. An efficient chemoenzymatic approach to (S)-gamma-fluoroleucine ethyl ester. *J. Org. Chem.* 2005, *70*, 2372–2375.

195. Truppo, M. D.; Pollard, D. J.; Moore, J. C.; Devine, P. N. Production of (S)-gamma-fluoroleucine ethyl ester by enzyme mediated dynamic kinetic resolution: Comparison of batch and fed batch stirred tank processes to a packed bed column reactor. *Chem. Eng. Sci.* 2008, *63*, 122–130.
196. Truppo, M. D.; Hughes, G. Development of an improved immobilized CAL-B for the enzymatic resolution of a key intermediate to Odanacatib. *Org. Process Res. Dev.* 2011, *15*, 1033–1035.
197. Jagtap, P.; Szabo, C. Poly(ADP-ribose) polymerase and the therapeutic effects of its inhibitors. *Nat. Rev. Drug Discov.* 2005, *4*, 421–440.
198. Kummar, S.; Chen, A.; Parchment, R.; Kinders, R.; Ji, J.; Tomaszewski, J.; Doroshow, J. Advances in using PARP inhibitors to treat cancer. *BMC Med.* 2012, *10*, 25.
199. Wallace, D. J.; Baxter, C. A.; Brands, K. J. M.; Bremeyer, N.; Brewer, S. E.; Desmond, R.; Emerson, K. M.; Foley, J.; Fernandez, P.; Hu, W.; Keen, S. P.; Mullens, P.; Muzzio, D.; Sajonz, P.; Tan, L.; Wilson, R. D.; Zhou, G.; Zhou, G. Development of a fit-for-purpose large-scale synthesis of an oral PARP inhibitor. *Org. Process Res. Dev.* 2011, *15*, 831–840.
200. Jones, P.; Altamura, S.; Boueres, J.; Ferrigno, F.; Fonsi, M.; Giomini, C.; Lamartina, S.; Monteagudo, E.; Ontoria, J. M.; Orsale, M. V.; Palumbi, M. C.; Pesci, S.; Roscilli, G.; Scarpelli, R.; Schultz-Fademrecht, C.; Toniatti, C.; Rowley, M. Discovery of 2-{4-[(3S)-Piperidin-3-yl]phenyl}-2H-indazole-7-carboxamide (MK-4827): A novel oral poly(ADP-ribose)polymerase (PARP) inhibitor efficacious in BRCA-1 and -2 mutant tumors. *J. Med. Chem.* 2009, *52*, 7170–7185.
201. Chung, C. K.; Bulger, P. G.; Kosjek, B.; Belyk, K. M.; Rivera, N.; Scott, M. E.; Humphrey, G. R.; Limanto, J.; Bachert, D. C.; Emerson, K. M. Process development of C–N cross-coupling and enantioselective biocatalytic reactions for the asymmetric synthesis of Niraparib. *Org. Process Res. Dev.* 2014, *18*, 215–227.
202. Nevin, D. K.; Lloyd, D. G.; Fayne, D. Rational targeting of peroxisome proliferating activated receptor subtypes. *Curr. Med. Chem.* 2011, *18*, 5598–5623.
203. Menendez-Gutierrez, M. P.; Roszer, T.; Ricote, M. Biology and therapeutic applications of peroxisome proliferator-activated receptors. *Curr. Top. Med. Chem.* 2012, *12*, 548–584.
204. Brown, D.; Gilday, J. P.; Hopes, P. A.; Moseley, J. D.; Snape, E. W.; Wells, A.; Hoppes, P. A. Preparing enantiomerically enriched arylalkylthiopropionic acid derivatives, useful to treat lipid disorders, comprises hydrolyzing optionally heterosubstituted methanesulfonyloxy derivatives with an enzyme e.g. *Mucor miehei* lipase. WO2006064213, 2006.
205. Thulin, P.; Rafter, I.; Stockling, K.; Tomkiewicz, C.; Norjavaara, E.; Aggerbeck, M.; Hellmold, H.; Ehrenborg, E.; Andersson, U.; Cotgreave, I.; Glinghammar, B. PPARα regulates the hepatotoxic biomarker alanine aminotransferase (ALT1) gene expression in human hepatocytes. *Toxicol. Appl. Pharmacol.* 2008, *231*, 1–9.
206. Wilding, J. P. H. PPAR agonists for the treatment of cardiovascular disease in patients with diabetes. *Diabetes Obes. Metab.* 2012, *14*, 973–982.
207. Conlon, D. Goodbye glitazars? *Br. J. Diab. Vasc. Dis.* 2006, *6*, 135–137.
208. Chen, H.; Dardik, B.; Qiu, L.; Ren, X. L.; Caplan, S. L.; Burkey, B.; Boettcher, B. R.; Gromada, J. Cevoglitazar, a novel peroxisome proliferator-activated receptor-alpha/gamma dual agonist, potently reduces food intake and body weight in obese mice and cynomolgus monkeys. *Endocrinology* 2010, *151*, 3115–3124.
209. Deussen, H. J.; Zundel, M.; Valdois, M.; Lehmann, S. V.; Weil, V.; Hjort, C. M.; Ostergaard, P. R.; Marcussen, E.; Ebdrup, S. Process development on the enantioselective enzymatic hydrolysis of S-ethyl 2-ethoxy-3-(4-hydroxyphenyl)propanoate. *Org. Process Res. Dev.* 2003, *7*, 82–88.
210. Brenna, E.; Fuganti, C.; Gatti, F. G.; Parmeggiani, F. Enzyme-mediated synthesis of EEHP and EMHP, useful pharmaceutical intermediates of PPAR agonists. *Tetrahedron: Asymmetry* 2009, *20*, 2594–2599.
211. Brenna, E.; Fuganti, C.; Gatti, F. G.; Parmeggiani, F. New stereospecific synthesis of Tesaglitazar and Navaglitazar precursors. *Tetrahedron: Asymmetry* 2009, *20*, 2694–2698.
212. Bechtold, M.; Brenna, E.; Femmer, C.; Gatti, F. G.; Panke, S.; Parmeggiani, F.; Sacchetti, A. Biotechnological development of a practical synthesis of ethyl (S)-2-ethoxy-3-(p-methoxyphenyl)propanoate (EEHP): Over 100-fold productivity increase from yeast whole cells to recombinant isolated enzymes. *Org. Process Res. Dev.* 2012, *16*, 269–276.
213. Brenna, E.; Gatti, F. G.; Manfredi, A.; Monti, D.; Parmeggiani, F. Enoate reductase-mediated preparation of methyl (S)-2-bromobutanoate, a useful key intermediate for the synthesis of chiral active pharmaceutical ingredients. *Org. Process Res. Dev.* 2012, *16*, 262–268.
214. Liu, W. G.; Liu, K.; Wood, H. B.; McCann, M. E.; Doebber, T. W.; Chang, C. H.; Akiyama, T. E.; Einstein, M.; Berger, J. P.; Meinke, P. T. Discovery of a peroxisome proliferator activated receptor gamma (PPAR gamma) modulator with balanced PPAR alpha activity for the treatment of type 2 diabetes and dyslipidemia. *J. Med. Chem.* 2009, *52*, 4443–4453.

215. Acton, J. J.; Akiyama, T. E.; Chang, C. H.; Colwell, L.; Debenham, S.; Doebber, T.; Einstein, M.; Liu, K.; McCann, M. E.; Moller, D. E.; Muise, E. S.; Tan, Y. G.; Thompson, J. R.; Wong, K. K.; Wu, M.; Xu, L. B.; Meinke, P. T.; Berger, J. P.; Wood, H. B. Discovery of (2R)-2-(3-{3-(4-methoxyphenyl) carbonyl-2-methyl-6-(trifluoromethoxy)-1H-indol-1-yl}phenoxy)butanoic acid (MK-0533): A novel selective peroxisome proliferator-activated receptor gamma modulator for the treatment of type 2 diabetes mellitus with a reduced potential to increase plasma and extracellular fluid volume. *J. Med. Chem.* 2009, *52*, 3846–3854.

216. Pirat, C.; Farce, A.; Lebegue, N.; Renault, N.; Furman, C.; Millet, R.; Yous, S.; Speca, S.; Berthelot, P.; Desreumaux, P.; Chavatte, P. Targeting peroxisome proliferator-activated receptors (PPARs): Development of modulators. *J. Med. Chem.* 2012, *55*, 4027–4061.

217. Agrawal, R. The first approved agent in the Glitazar's class: Saroglitazar. *Curr. Drug Targets* 2014, *15*, 151–155.

218. Ahren, B. Islet G protein-coupled receptors as potential targets for treatment of type 2 diabetes. *Nat. Rev. Drug Discov.* 2009, *8*, 369–385.

219. Lovshin, J. A.; Drucker, D. J. Incretin-based therapies for type 2 diabetes mellitus. *Nat. Rev. Endocrinol.* 2009, *5*, 262–269.

220. Rondas, D.; D'Hertog, W.; Overbergh, L.; Mathieu, C. Glucagon-like peptide-1: Modulator of beta-cell dysfunction and death. *Diabetes Obes. Metab.* 2013, *15*, 185–192.

221. Garber, A. J. Novel GLP-1 receptor agonists for diabetes. *Expert Opin. Investig. Drugs* 2012, *21*, 45–57.

222. Chen, Y. J.; Goldberg, S. L.; Hanson, R. L.; Parker, W. L.; Gill, I.; Tully, T. P.; Montana, M. A.; Goswami, A.; Patel, R. N. Enzymatic preparation of an (*S*)-amino acid from a racemic amino acid. *Org. Process Res. Dev.* 2011, *15*, 241–248.

223. Scheen, A. J. DPP-4 inhibitors in the management of type 2 diabetes: A critical review of head-to-head trials. *Diabetes Metab.* 2012, *38*, 89–101.

224. Deacon, C. F.; Holst, J. J. Dipeptidyl peptidase-4 inhibitors for the treatment of type 2 diabetes: Comparison, efficacy and safety. *Expert Opin. Pharmacother.* 2013, *14*, 2047–2058.

225. Mize, D. L. E.; Salehi, M. The place of GLP-1-based therapy in diabetes management: Differences between DPP-4 inhibitors and GLP-1 receptor agonists. *Curr. Diab. Rep.* 2013, *13*, 307–318.

226. Dai, Y.; Dai, D.; Mercanti, F.; Ding, Z.; Wang, X.; Mehta, J. L. Dipeptidyl peptidase-4 inhibitors in cardioprotection: A romising therapeutic approach. *Acta Diabetol.* 2013, *50*, 827–835.

227. Wang, X. M.; Yang, Y. J.; Wu, Y. J. The emerging role of dipeptidyl peptidase-4 inhibitors in cardiovascular protection: Current position and perspectives. *Cardiovasc. Drugs Ther.* 2013, *27*, 297–307.

228. Aroda, V. R.; Henry, R. R.; Han, J.; Huang, W. Y.; DeYoung, M. B.; Darsow, T.; Hoogwerf, B. J. Efficacy of GLP-1 receptor agonists and DPP-4 inhibitors: Meta-analysis and systematic review. *Clin. Ther.* 2012, *34*, 1247–1258.

229. Kim, D.; Wang, L. P.; Beconi, M.; Eiermann, G. J.; Fisher, M. H.; He, H. B.; Hickey, G. J.; Kowalchick, J. E.; Leiting, B.; Lyons, K.; Marsilio, F.; McCann, M. E.; Patel, R. A.; Petrov, A.; Scapin, G.; Patel, S. B.; Roy, R. S.; Wu, J. K.; Wyvratt, M. J.; Zhang, B. B.; Zhu, L.; Thornberry, N. A.; Weber, A. E. (2*R*)-4-oxo-4-[3-(trifluoromethyl)-5,6-dihydro[1,2,4]triazolo[4,3-a]pyrazin-7(8*H*)-yl]-1-(2,4,5-trifluorophenyl) butan-2-amine: A potent, orally active dipeptidyl peptidase IV inhibitor for the treatment of type 2 diabetes. *J. Med. Chem.* 2005, *48*, 141–151.

230. Hansen, K. B.; Yi, H.; Xu, F.; Rivera, N.; Clausen, A.; Kubryk, M.; Krska, S.; Rosner, T.; Simmons, B.; Balsells, J.; Ikemoto, N.; Sun, Y.; Spindler, F.; Malan, C.; Grabowski, E. J. J.; Armstrong, J. D. Highly efficient asymmetric synthesis of sitagliptin. *J. Am. Chem. Soc.* 2009, *131*, 8798–8804.

231. Savile, C. K.; Janey, J. M.; Mundorff, E. C.; Moore, J. C.; Tam, S.; Jarvis, W. R.; Colbeck, J. C.; Krebber, A.; Fleitz, F. J.; Brands, J.; Devine, P. N.; Huisman, G. W.; Hughes, G. J. Biocatalytic asymmetric synthesis of chiral amines from ketones applied to sitagliptin manufacture. *Science* 2010, *329*, 305–309.

232. Savile, C.; Gruber, J. M.; Mundoroff, E.; Huisman, G. W.; Collier, S. J. Ketoreductase polypeptides for the production of a 3-aryl-3-hydroxypropanamine from a 3-aryl-3-ketopropanamine. Patent WO/2010/025238, April 3, 2010.

233. Moore, J. C.; Savile, C. K.; Pannuri, S.; Kosjek, B.; Janey, J. M. Industrially relevant enzymatic reductions. In *Comprehensive Chirality*, Carreira, E. C.; Yamamoto, H., Eds. Elsevier: Amsterdam, the Netherlands, 2012; pp. 318–341.

234. Kania, D. S.; Gonzalvo, J. D.; Weber, Z. A. Saxagliptin: A clinical review in the treatment of type 2 diabetes mellitus. *Clin. Ther.* 2011, *33*, 1005–1022.

235. Savage, S. A.; Jones, G. S.; Kolotuchin, S.; Ramrattan, S. A.; Vu, T.; Waltermire, R. E. Preparation of saxagliptin, a novel DPP-IV inhibitor. *Org. Process Res. Dev.* 2009, *13*, 1169–1176.

236. Hanson, R. L.; Goldberg, S. L.; Brzozowski, D. B.; Tully, T. P.; Cazzulino, D.; Parker, W. L.; Lyngberg, O. K.; Vu, T. C.; Wong, M. K.; Patel, R. N. Preparation of an amino acid intermediate for the dipeptidyl peptidase IV inhibitor, saxagliptin, using a modified phenylalanine dehydrogenase. *Adv. Synth. Catal.* 2007, *349*, 1369–1378.
237. Kunishima, M.; Kawachi, C.; Hioki, K.; Terao, R.; Tani, S. Formation of carboxamides by direct condensation of carboxylic acids and amines in alcohols using a new alcohol- and water-soluble condensing agent: DMT-MM. *Tetrahedron* 2001, *57*, 1551–1558.
238. Gill, I.; Patel, R. Biocatalytic ammonolysis of (5*S*)-4,5-dihydro-1*H*-pyrrole-1,5-dicarboxylic acid, 1-(1,1-dimethylethyl)-5-ethyl ester: Preparation of an intermediate to the dipeptidyl peptidase IV inhibitor Saxagliptin. *Bioorg. Med. Chem. Lett.* 2006, *16*, 705–709.
239. Masferrer, J. L.; Zweifel, B. S.; Colburn, S. M.; Ornberg, R. L.; Salvemini, D.; Isakson, P.; Seibert, K. The role of cyclooxygenase-2 in inflammation. *Am. J. Ther.* 1995, *2*, 607–610.
240. Sinha, V. R.; Kumar, R. V.; Singh, G. Ketorolac tromethamine formulations: An overview. *Expert Opin. Drug Deliv.* 2009, *6*, 961–975.
241. Landoni, M. F.; Soraci, A. Pharmacology of chiral compounds: 2-arylpropionic acid derivatives. *Curr. Drug Metab.* 2001, *2*, 37–51.
242. Chen, C. S.; Shieh, W. R.; Lu, P. H.; Harriman, S.; Chen, C. Y. Metabolic stereoisomeric inversion of ibuprofen in mammals. *Biochim. Biophys. Acta* 1991, *1078*, 411–417.
243. Reichel, C.; Brugger, R.; Bang, H.; Geisslinger, G.; Brune, K. Molecular cloning and expression of a 2-arylpropionyl-coenzyme A epimerase: A key enzyme in the inversion metabolism of ibuprofen. *Mol. Pharmacol.* 1997, *51*, 576–582.
244. Lloyd, M. D.; Darley, D. J.; Wierzbicki, A. S.; Threadgill, M. D. alpha-Methylacyl-CoA racemase—An 'obscure' metabolic enzyme takes centre stage. *FEBS J.* 2008, *275*, 1089–1102.
245. Qu, X.; Allan, A.; Chui, G.; Hutchings, T. J.; Jiao, P.; Johnson, L.; Leung, W. Y.; Li, P. K.; Steel, G. R.; Thompson, A. S.; Threadgill, M. D.; Woodman, T. J.; Lloyd, M. D. Hydrolysis of ibuprofenoyl-CoA and other 2-APA-CoA esters by human acyl-CoA thioesterases-1 and-2 and their possible role in the chiral inversion of profens. *Biochem. Pharmacol.* 2013, *86*, 1621–1625.
246. Lloyd, M. D.; Yevglevskis, M.; Lee, G. L.; Wood, P. J.; Threadgill, M. D.; Woodman, T. J. alpha-Methylacyl-CoA racemase (AMACR): Metabolic enzyme, drug metabolizer and cancer marker P504S. *Prog. Lipid Res.* 2013, *52*, 220–230.
247. Woodman, T. J.; Wood, P. J.; Thompson, A. S.; Hutchings, T. J.; Steel, G. R.; Jiao, P.; Threadgill, M. D.; Lloyd, M. D. Chiral inversion of 2-arylpropionyl-CoA esters by human alpha-methylacyl-CoA racemase 1A (P504S)—A potential mechanism for the anti-cancer effects of ibuprofen. *Chem. Commun.* 2011, *47*, 7332–7334.
248. Duggan, K. C.; Hermanson, D. J.; Musee, J.; Prusakiewicz, J. J.; Scheib, J. L.; Carter, B. D.; Banerjee, S.; Oates, J. A.; Marnett, L. J. (R)-Profens are substrate-selective inhibitors of endocannabinoid oxygenation by COX-2. *Nat. Chem. Biol.* 2011, *7*, 803–809.
249. Wynne, S.; Djakiew, D. NSAID inhibition of prostate cancer cell migration is mediated by Nag-1 induction via the p38 MAPK-p75[NTR] pathway *Mol. Cancer Res.* 2010, *8*, 1656–1664.
250. Alcántara, A. R.; Sánchez-Montero, J. M.; Sinisterra, J. V. Chemoenzymatic preparation of enantiomerically pure *S*(+)-2-arylpropionic acids with anti-inflammatory activity. In *Stereoselective Biocatalysis*, Patel, R. N., Ed. Dekker: New York, 2000; pp. 659–702.
251. Kourist, R.; Dominguez de Maria, P.; Miyamoto, K. Biocatalytic strategies for the asymmetric synthesis of profens—Recent trends and developments. *Green Chem.* 2011, *13*, 2607–2618.
252. Csuk, R. Chiral switches: Problems, strategies, opportunities, and experiences. In *Biocatalysis in the Pharmaceutical and Biotechnology Industries*, Patel, R. N., Ed. CRC Press: Boca Raton, FL, 2007; pp. 699–716.
253. Hernáiz, M. J.; Sánchez-Montero, J. M.; Sinisterra, J. V. Hydrolysis of (*R,S*)2-aryl propionic esters by pure lipase B from *Candida cylindracea*. *J. Mol. Catal. A: Chem.* 1995, *96*, 317–327.
254. Chang, C.-S.; Tsai, S.-W.; Lin, C.-N. Enzymatic resolution of (*RS*)-2-arylpropionic acid thioesters by *Candida rugosa* lipase-catalyzed thiotransesterification or hydrolysis in organic solvents. *Tetrahedron: Asymmetry* 1998, *9*, 2799–2807.
255. Zhang, J.; Hou, Z.; Yao, C.; Yu, Y. Purification and properties of lipase from a *Bacillus* strain for catalytic resolution of (*R*)-Naproxen. *J. Mol. Catal. B: Enzym.* 2002, *18*, 205–210.
256. Qin, B.; Liang, P.; Jia, X.; Zhang, X.; Mu, M.; Wang, X.-Y.; Ma, G.-Z.; Jin, D.-N.; You, S. Directed evolution of *Candida antarctica* lipase B for kinetic resolution of profen esters. *Catal. Commun.* 2013, *38*, 1–5.
257. Liu, Y.-Y.; Xu, J.-H.; Wu, H.-Y.; Shen, D. Integration of purification with immobilization of *Candida rugosa* lipase for kinetic resolution of racemic ketoprofen. *J. Biotechnol.* 2004, *110*, 209–217.

258. Marszałł, M. P.; Siódmiak, T. Immobilization of *Candida rugosa* lipase onto magnetic beads for kinetic resolution of (R,S)-ibuprofen. *Catal. Commun.* 2012, *24*, 80–84.
259. Bouwer, S. T.; Cuperus, F. P.; Derksen, J. T. P. The performance of enzyme-membrane reactors with immobilized lipase. *Enzyme Microb. Technol.* 1997, *21*, 291–296.
260. Balcão, V. M.; Paiva, A. L.; Xavier Malcata, F. Bioreactors with immobilized lipases: State of the art. *Enzyme Microb. Technol.* 1996, *18*, 392–416.
261. Ceynowa, J.; Rauchfleisz, M. High enantioselective resolution of racemic 2-arylpropionic acids in an enzyme membrane reactor. *J. Mol. Catal. B: Enzym.* 2003, *23*, 43–51.
262. Long, W. S.; Kamaruddin, A.; Bhatia, S. Chiral resolution of racemic ibuprofen ester in an enzymatic membrane reactor. *J. Membr. Sci.* 2005, *247*, 185–200.
263. Lau, S. Y.; Gonawan, F. N.; Bhatia, S.; Kamaruddin, A. H.; Uzir, M. H. Conceptual design and simulation of a plant for the production of high purity (S)-ibuprofen acid using innovative enzymatic membrane technology. *Chem. Eng. J.* 2011, *166*, 726–737.
264. Sakaki, K.; Giorno, L.; Drioli, E. Lipase-catalyzed optical resolution of racemic naproxen in biphasic enzyme membrane reactors. *J. Membr. Sci.* 2001, *184*, 27–38.
265. Lopez-Belmonte, M. T.; Alcántara, A. R.; Sinisterra, J. V. Enantioselective esterification of 2-arylpropionic acids catalyzed by immobilized *Rhizomucor miehei* lipase. *J. Org. Chem.* 1997, *62*, 1831–1840.
266. Siodmiak, T.; Ruminski, J. K.; Marszall, M. P. Application of lipases from *Candida Rugosa* in the enantioselective esterification of (R,S)-ibuprofen. *Curr. Org. Chem.* 2012, *16*, 972–977.
267. Arroyo, M.; Sinisterra, J. V. High enantioselective esterification of 2-arylpropionic acids catalyzed by immobilized lipase from *Candida antarctica*—A mechanistic approach. *J. Org. Chem.* 1994, *59*, 4410–4417.
268. Miyako, E.; Maruyama, T.; Kamiya, N.; Goto, M. Enzyme-facilitated enantioselective transport of (S)-ibuprofen through a supported liquid membrane based on ionic liquids. *Chem. Commun.* 2003, 2926–2927.
269. Lalonde, J. J.; Govardhan, C.; Khalaf, N.; Martinez, A. G.; Visuri, K.; Margolin, A. L. Cross-linked crystals of *Candida rugosa* lipase: Highly efficient catalysts for the resolution of chiral esters. *J. Am. Chem. Soc.* 1995, *117*, 6845–6852.
270. Yu, H. W.; Chen, H.; Wang, X.; Yang, Y. Y.; Ching, C. B. Cross-linked enzyme aggregates (CLEAs) with controlled particles: Application to *Candida rugosa* lipase. *J. Mol. Catal. B: Enzym.* 2006, *43*, 124–127.
271. Hoyos, P.; Pace, V.; Alcántara, A. R. Dynamic kinetic resolution via hydrolase-metal combo catalysis in stereoselective synthesis of bioactive compounds. *Adv. Synth. Catal.* 2012, *354*, 2585–2611.
272. Strübing, D.; Krumlinde, P.; Piera, J.; Bäckvall, J.-E. Dynamic kinetic resolution of primary alcohols with an unfunctionalized stereogenic center in the β-position. *Adv. Synth. Catal.* 2007, *349*, 1577–1581.
273. Fazlena, H.; Kamaruddin, A. H.; Zulkali, M. M. D. Dynamic kinetic resolution: Alternative approach in optimizing S-ibuprofen production. *Bioprocess Biosyst. Eng.* 2006, *28*, 227–233.
274. Wang, M.-X.; Lu, G.; Ji, G.-J.; Huang, Z.-T.; Meth-Cohn, O.; Colby, J. Enantioselective biotransformations of racemic α-substituted phenylacetonitriles and phenylacetamides using *Rhodococcus* sp. AJ270. *Tetrahedron: Asymmetry* 2000, *11*, 1123–1135.
275. Takagi, M.; Shirokaze, J.-I.; Oishi, K.; Otsubo, K.; Yamamoto, K.; Yoshida, N.; Fujimatsu, I. Production of S-(+)-ibuprofen with high optical purity from a nitrile compound by cells immobilized on cellulose porous beads. *J. Ferment. Bioeng.* 1994, *78*, 191–193.
276. Galletti, P.; Emer, E.; Gucciardo, G.; Quintavalla, A.; Pori, M.; Giacomini, D. Chemoenzymatic synthesis of (2S)-2-arylpropanols through a dynamic kinetic resolution of 2-arylpropanals with alcohol dehydrogenases. *Org. Biomol. Chem.* 2010, *8*, 4117–4123.
277. Pietruszka, J.; Schölzel, M. Ene reductase-catalysed synthesis of (R)-profen derivatives. *Adv. Synth. Catal.* 2012, *354*, 751–756.
278. Kim, M. G.; Lee, E. G.; Chung, B. H. Improved enantioselectivity of *Candida rugosa* lipase towards ketoprofen ethyl ester by a simple two-step treatment. *Process Biochem.* 2000, *35*, 977–982.
279. Liu, Y.-Y.; Xu, J.-H.; Hu, Y. Enhancing effect of Tween-80 on lipase performance in enantioselective hydrolysis of ketoprofen ester. *J. Mol. Catal. B: Enzym.* 2000, *10*, 523–529.
280. Chamorro, S.; Alcántara, A. R.; de la Casa, R. M.; Sinisterra, J. V.; Sánchez-Montero, J. M. Small water amounts increase the catalytic behaviour of polar organic solvents pre-treated *Candida rugosa* lipase. *J. Mol. Catal. B: Enzym.* 2001, *11*, 939–947.
281. Xi, W.-W.; Xu, J.-H. Preparation of enantiopure (S)-ketoprofen by immobilized *Candida rugosa* lipase in packed bed reactor. *Process Biochem.* 2005, *40*, 2161–2166.

282. Ong, A. L.; Kamaruddin, A. H.; Bhatia, S. Current technologies for the production of (S)-ketoprofen: Process perspective. *Process Biochem.* 2005, *40*, 3526–3535.
283. Long, Z.-D.; Xu, J.-H.; Zhao, L.-L.; Pan, J.; Yang, S.; Hua, L. Overexpression of *Serratia marcescens* lipase in *Escherichia coli* for efficient bioresolution of racemic ketoprofen. *J. Mol. Catal. B: Enzym.* 2007, *47*, 105–110.
284. Wu, H.-Y.; Xu, J.-H.; Liu, Y.-Y. A practical enzymatic method for preparation of (S)-ketoprofen with a crude *Candida rugosa* lipase. *Synth. Commun.* 2001, *31*, 3491–3496.
285. De Crescenzo, G.; Ducret, A.; Trani, M.; Lortie, R. Enantioselective esterification of racemic ketoprofen in non-aqueous solvent under reduced pressure. *J. Mol. Catal. B: Enzym.* 2000, *9*, 49–56.
286. D'Antona, N.; Lombardi, P.; Nicolosi, G.; Salvo, G. Large scale preparation of enantiopure S-ketoprofen by biocatalysed kinetic resolution. *Process Biochem.* 2002, *38*, 373–377.
287. Shen, D.; Xu, J.-H.; Wu, H.-Y.; Liu, Y.-Y. Significantly improved esterase activity of *Trichosporon brassicae* cells for ketoprofen resolution by 2-propanol treatment. *J. Mol. Catal. B: Enzym.* 2002, *18*, 219–224.
288. Kim, G.-J.; Choi, G.-S.; Kim, J.-Y.; Lee, J.-B.; Jo, D.-H.; Ryu, Y.-W. Screening, production and properties of a stereospecific esterase from *Pseudomonas* sp. S34 with high selectivity to (S)-ketoprofen ethyl ester. *J. Mol. Catal. B: Enzym.* 2002, *17*, 29–38.
289. Kim, J.-Y.; Choi, G.-S.; Kim, Y.-J.; Ryu, Y.-W.; Kim, G.-J. A new isolate *Bacillus stearothermophilus* JY144 expressing a novel esterase with high enantioselectivity to (R)-ketoprofen ethyl ester: Strain selection and gene cloning. *J. Mol. Catal. B: Enzym.* 2002, *18*, 133–145.
290. Jin, H. F.; Wang, Z. P.; Liu, L. L.; Gao, L. C.; Sun, L.; Li, X. H.; Zhao, H. X.; Pan, Y. L.; Shi, H.; Liu, N.; Hong, L.; Liang, J.; Wu, Q.; Yang, Z. P.; Wu, K. C.; Fan, D. M. R-flurbiprofen reverses multidrug resistance, proliferation and metastasis in gastric cancer cells by p75(NTR) induction. *Mol. Pharm.* 2010, *7*, 156–168.
291. Shin, G.-S.; Lee, K.-W.; Lee, Y.-H. Fed-batch production of (S)-flurbiprofen in lipase-catalyzed dispersed aqueous phase reaction system induced by succinyl β-cyclodextrin and its extractive purification. *J. Mol. Catal. B: Enzym.* 2005, *37*, 109–111.
292. Bae, H.-A.; Lee, K.-W.; Lee, Y.-H. Enantioselective properties of extracellular lipase from *Serratia marcescens* ES-2 for kinetic resolution of (S)-flurbiprofen. *J. Mol. Catal. B: Enzym.* 2006, *40*, 24–29.
293. Gandolfi, R.; Converti, A.; Pirozzi, D.; Molinari, F. Efficient and selective microbial esterification with dry mycelium of *Rhizopus oryzae*. *J. Biotechnol.* 2001, *92*, 21–26.
294. Ghanem, A. Direct enantioselective HPLC monitoring of lipase-catalyzed kinetic resolution of flurbiprofen. *Chirality* 2010, *22*, 597–603.
295. Duan, G.; Ching, C. B. Preparative scale enantioseparation of flurbiprofen by lipase-catalysed reaction. *Biochem. Eng. J.* 1998, *2*, 237–245.
296. Tamborini, L.; Romano, D.; Pinto, A.; Bertolani, A.; Molinari, F.; Conti, P. An efficient method for the lipase-catalysed resolution and in-line purification of racemic flurbiprofen in a continuous-flow reactor. *J. Mol. Catal. B: Enzym.* 2012, *84*, 78–82.
297. Ciou, J.-F.; Wang, P.-Y.; Wu, A.-C.; Tsai, S.-W. Lipase-catalyzed alcoholytic resolution of (R,S)-flurbiprofenyl azolides for preparation of (R)-NO-flurbiprofen ester prodrugs. *Process Biochem.* 2011, *46*, 960–965.
298. Wang, P.-Y.; Chen, Y.-J.; Wu, A.-C.; Lin, Y.-S.; Kao, M.-F.; Chen, J.-R.; Ciou, J.-F.; Tsai, S.-W. (R,S)-azolides as novel substrates for lipase-catalyzed hydrolytic resolution in organic solvents. *Adv. Synth. Catal.* 2009, *351*, 2333–2341.
299. Wu, A.-C.; Wang, P.-Y.; Lin, Y.-S.; Kao, M.-F.; Chen, J.-R.; Ciou, J.-F.; Tsai, S.-W. Improvements of enzyme activity and enantioselectivity in lipase-catalyzed alcoholysis of (R,S)-azolides. *J. Mol. Catal. B: Enzym.* 2010, *62*, 235–241.
300. Harrington, P. J.; Lodewijk, E. Twenty years of naproxen technology. *Org. Process Res. Dev.* 1997, *1*, 72–76.
301. Harrison, I. T.; Lewis, B.; Nelson, P.; Rooks, W.; Roszkows, A.; Tomoloni, A.; Fried, J. H. Nonsteroidal antiinflammatory agents. 1. 6-substituted 2-naphthylacetic acids. *J. Med. Chem.* 1970, *13*, 203–205.
302. Giordano, C.; Castaldi, G.; Cavicchioli, S.; Villa, M. A. stereoconvergent strategy for the synthesis of enantiomerically pure (R)-(−) and (S)-(+)-2-(6-methoxy-2-naphthyl)-propanoic acid (Naproxen). *Tetrahedron* 1989, *45*, 4243–4252.
303. Ohta, T.; Takaya, H.; Kitamura, M.; Nagai, K.; Noyori, R. Asymmetric hydrogenation of unsaturated carboxylic-acids catalyzed by BINAP-ruthenium(II) complexes. *J. Org. Chem.* 1987, *52*, 3174–3176.
304. Hiyama, T.; Wakasa, N.; Kusumoto, T. Hydroesterification of 6-methoxy-2-naphthylethene. *Synlett* 1991, *1991*, 569–570.

305. Bertola, M. A.; Quax, W. J.; Robertson, B. W.; Marx, A. F.; Vanderlake, C. J.; Koger, H. S.; Phillips, G. T.; Watts, P. D.; Van Der Laken, C. J. Stereoselective prepn. of 2-arylpropionic acids—from ester(s) using a microorganism or derived substance such as an enzyme. EP233656-A1; EP233656-A; AU8667072-A; NO8700040-A; FI8700035-A; PT84060-A; ZA8700086-A; JP63045234-A; CN8700031-A; DK8700049-A; US4886750-A; US5037751-A; IL81127-A; NO9104029-A; EP233656-B1; DE3780184-G; FI90565-B; NO174350-B; ES2051727-T3; JP8242853-A; JP2729044-B2; CA1340382-C; KR9616868-B1; DK172862-B; CA1341324-C; IE82714-B, EP233656-A1; EP233656-A, August 26, 1987.
306. Battistel, E.; Bianchi, D.; Cesti, P.; Pina, C. Enzymatic resolution of (S)-(+)-naproxen in a continuous reactor. *Biotechnol. Bioeng.* 1991, *38*, 659–664.
307. Quax, W. J.; Broekhuizen, C. P. Development of a new *Bacillus* carboxyl esterase for use in the resolution of chiral drugs. *Appl. Microbiol. Biotechnol.* 1994, *41*, 425–431.
308. de María, P. D.; Sinisterra, J. V.; Tsai, S. W.; Alcántara, A. R. *Carica papaya* lipase (CPL): An emerging and versatile biocatalyst. *Biotechnol. Adv.* 2006, *24*, 493–499.
309. Ng, I. S.; Tsai, S. W. Investigation of lipases from various *Carica papaya* varieties for hydrolysis of olive oil and kinetic resolution of (R,S)-profen 2,2,2-trifluoroethyl thioesters. *Process Biochem.* 2006, *41*, 540–546.
310. Ng, I. S.; Tsai, S. W. Partially purified *Carica papaya* lipase: A versatile biocatalyst for the hydrolytic resolution of (R,S)-2-arylpropionic thioesters in water-saturated organic solvents. *Biotechnol. Bioeng.* 2005, *91*, 106–113.
311. Chen, C. C.; Tsai, S. W. *Carica papaya* lipase: A novel biocatalyst for the enantioselective hydrolysis of (R,S)-naproxen 2,2,2-trifluoroethyl ester. *Enzyme Microb. Technol.* 2005, *36*, 127–132.
312. Chen, C. C.; Tsai, S. W.; Villeneuve, P. Enantioselective hydrolysis of (R,S)-naproxen 2,2,2-trifluoroethyl ester in water-saturated solvents via lipases from *Carica pentagona* Heilborn and *Carica papaya*. *J. Mol. Catal. B: Enzym.* 2005, *34*, 51–57.
313. Steenkamp, L.; Brady, D. Screening of commercial enzymes for the enantio selective hydrolysis of *R,S*-naproxen ester. *Enzyme Microb. Technol.* 2003, *32*, 472–477.
314. Steenkamp, L.; Brady, D. Optimisation of stabilised carboxylesterase NP for enantioselective hydrolysis of naproxen methyl ester. *Process Biochem.* 2008, *43*, 1419–1426.
315. Liu, X. A.; Xu, J. H.; Pan, J. A.; Zhao, J. Efficient production of (S)-Naproxen with (R)-substrate recycling using an overexpressed carboxylesterase BsE-NP01. *Appl. Biochem. Biotechnol.* 2010, *162*, 1574–1584.
316. Koul, S.; Parshad, R.; Taneja, S. C.; Qazi, G. N. Enzymatic resolution of naproxen. *Tetrahedron: Asymmetry* 2003, *14*, 2459–2465.
317. Morrone, R.; D'Antona, N.; Lambusta, D.; Nicolosi, G. Biocatalyzed irreversible esterification in the preparation of S-naproxen. *J. Mol. Catal. B: Enzym.* 2010, *65*, 49–51.
318. Gao, K. L.; Wei, D. Z. Asymmetric oxidation by *Gluconobacter oxydans*. *Appl. Microbiol. Biotechnol.* 2006, *70*, 135–139.
319. Miyamoto, K.; Fujimori, K.; Hirano, J. I.; Ohta, H. Microbial kinetic resolution of 2-substituted-1-propanol. *Biocatal. Biotransformation* 2009, *27*, 66–70.
320. Friest, J. A.; Maezato, Y.; Broussy, S.; Blum, P.; Berkowitz, D. B. Use of a robust dehydrogenase from an archael hyperthermophile in asymmetric catalysis-dynamic reductive kinetic resolution entry into (S)-profens. *J. Am. Chem. Soc.* 2010, *132*, 5930–5931.
321. Miyamoto, K.; Ohta, H. Purification and properties of a novel arylmalonate decarboxylase from *Alcaligenes bronchisepticus* KU-1201. *Eur. J. Biochem.* 1992, *210*, 475–481.
322. Kourist, R.; Miyauchi, Y.; Uemura, D.; Miyamoto, K. Engineering the promiscuous racemase activity of an arylmalonate decarboxylase. *Chem. Eur. J.* 2011, *17*, 557–563.
323. Miyauchi, Y.; Kourist, R.; Uemura, D.; Miyamoto, K. Dramatically improved catalytic activity of an artificial (S)-selective arylmalonate decarboxylase by structure-guided directed evolution. *Chem. Commun.* 2011, *47*, 7503–7505.
324. Guzman, A.; Yuste, F.; Toscano, R. A.; Young, J. M.; Vanhorn, A. R.; Muchowski, J. M. Absolute configuration of (−)-5-benzoyl-1,2-dihydro-3H-pyrrolo [1,2-a] pyrrole-1-carboxylic acid, the active enantiomer of ketorolac. *J. Med. Chem.* 1986, *29*, 589–591.
325. Mroszczak, E.; Combs, D.; Chaplin, M.; Tsina, I.; Tarnowski, T.; Rocha, C.; Tam, Y.; Boyd, A.; Young, J.; Depass, L. Chiral kinetics and dynamics of ketorolac. *J. Clin. Pharmacol.* 1996, *36*, 521–539.
326. Muchoswki, J. M. The development of ketorolac: Impact on pyrrole chemistry and on pain therapy. In *Advances in Medicinal Chemistry*, Maryanoff, B. E.; Maryanoff, C. A., Eds. Jai Press Inc.: Greenwich, CT, 1992; Vol. 1, pp. 109–135.
327. Kluge, A. F.; Muchowski, J. M. 5-Aroyl-1,2-dihydro-3H-pyrrolo[1,2-a]pyrrole-1-carboxylic acid derivatives and process for the production thereof. U.S. Patent 4089969 A, 1978.

328. Franco, F.; Greenhouse, R.; Muchowski, J. M. Novel syntheses of 5-aroyl-1,2-dihydro-3H-pyrrolo [1,2]-a pyrrole-1-carboxylic acids. *J. Org. Chem.* 1982, *47*, 1682–1688.
329. Harrington, P. J.; Khatri, H. N.; Schloemer, G. C. Preparation of ketorolac. U.S. Patent 6197976, 2001.
330. Baran, P. S.; Richter, J. M.; Lin, D. W. Direct coupling of pyrroles with carbonyl compounds: Short enantioselective synthesis of (S)-Ketorolac. *Angew. Chem. Int. Ed.* 2005, *44*, 609–612.
331. Richter, J. M.; Whitefield, B. W.; Maimone, T. J.; Lin, D. W.; Castroviejo, M. P.; Baran, P. S. Scope and mechanism of direct indole and pyrrole couplings adjacent to carbonyl compounds: Total synthesis of acremoauxin A and oxazinin 3. *J. Am. Chem. Soc.* 2007, *129*, 12857–12869.
332. Fulling, G.; Sih, C. J. Enzymatic 2nd-order asymmetric hydrolysis of ketorolac esters—In situ racemization. *J. Am. Chem. Soc.* 1987, *109*, 2845–2846.
333. Kim, Y. H.; Cheong, C. S.; Lee, S. H.; Kim, K. S. Enzymatic kinetic resolution of ketorolac. *Tetrahedron: Asymmetry* 2001, *12*, 1865–1869.
334. Dantas, A. P.; Jiménez-Altayó, F.; Vila, E. Vascular aging: Facts and factors. *Front. Physiol.* 2012, *3*, 325.
335. Lloyd-Jones, D. M. Cardiovascular risk prediction basic concepts, current status, and future directions. *Circulation* 2010, *121*, 1768–1777.
336. Stein, E. A. The power of statins: Aggressive lipid lowering. *Clin. Cardiol.* 2003, *26*, 25–31.
337. Silva, T.; Teixeira, J.; Remiao, F.; Borges, F. Alzheimer's disease, cholesterol, and statins: The junctions of important metabolic pathways. *Angew. Chem. Int. Ed.* 2013, *52*, 1110–1121.
338. Caporaso, N. E. Statins and cancer-related mortality—Let's work together. *New Engl. J. Med.* 2012, *367*, 1848–1850.
339. Osmak, M. Statins and cancer: Current and future prospects. *Cancer Lett.* 2012, *324*, 1–12.
340. Endo, A.; Kuroda, M.; Tsujita, Y. ML-236A, ML-236B, and ML-236C, new inhibitors of cholesterogenesis produced by *Penicillium citrinum*. *J. Antibiot.* 1976, *29*, 1346–1348.
341. Brown, A. G.; Smale, T. C.; King, T. J.; Hasenkamp, R.; Thompson, R. H. Crystal and molecular-structure of compactin, a new antifungal metabolite from *Penicillium brevicompactum*. *J. Chem. Soc., Perkin Trans. 1* 1976, 1165–1173.
342. Moore, R. N.; Bigam, G.; Chan, J. K.; Hogg, A. M.; Nakashima, T. T.; Vederas, J. C. Biosynthesis of the hypocholesterolemic agent mevinolin by *Aspergillus terreus*. Determination of the origin of carbon, hydrogen, and oxygen-atoms by 13C NMR and mass spectrometry. *J. Am. Chem. Soc.* 1985, *107*, 3694–3701.
343. Gunde-Cimerman, N.; Cimerman, A. Pleurotus fruiting bodies contain the inhibitor of 3-hydroxy-3-methylglutaryl-coenzyme A reductase—Lovastatin. *Exp. Mycol.* 1995, *19*, 1–6.
344. Liu, J.; Zhang, J.; Shi, Y.; Grimsgaard, S.; Alraek, T.; Fonnebo, V. Chinese red yeast rice (*Monascus purpureus*) for primary hyperlipidemia: A meta-analysis of randomized controlled trials. *Chin. Med.* 2006, *1*, 4.
345. Mol, M.; Erkelens, D. W.; Leuven, J. A. G.; Schouten, J. A.; Stalenhoef, A. F. H. Simvastatin (MK-733)—A potent cholesterol-synthesis inhibitor in heterozygous familial hypercholesterolemia. *Atherosclerosis* 1988, *69*, 131–137.
346. Yoshino, G.; Kazumi, T.; Kasama, T.; Iwatani, I.; Iwai, M.; Inui, A.; Otsuki, M.; Baba, S. Effect of CS-514, an inhibitor of 3-hydroxy-3-methylglutaryl coenzyme A reductase, on lipoprotein and apolipoprotein in plasma of hypercholesterolemic diabetics. *Diabetes Res. Clin. Pr.* 1986, *2*, 179–181.
347. Li, J. J. *Triumph of the Heart: The Story of Statins*. Oxford University Press: New York, 2009.
348. Casar, Z. Historic overview and recent advances in the synthesis of super-statins. *Curr. Org. Chem.* 2010, *14*, 816–845.
349. Barrios-Gonzalez, J.; Miranda, R. U. Biotechnological production and applications of statins. *Appl. Microbiol. Biotechnol.* 2010, *85*, 869–883.
350. Xie, X. K.; Watanabe, K.; Wojcicki, W. A.; Wang, C. C. C.; Tang, Y. Biosynthesis of lovastatin analogs with a broadly specific acyltransferase. *Chem. Biol.* 2006, *13*, 1161–1169.
351. Hoffman, W. F.; Alberts, A. W.; Anderson, P. S.; Chen, J. S.; Smith, R. L.; Willard, A. K. 3-Hydroxy-3-methylglutaryl-coenzyme a reductase inhibitors. 4. Side-chain ester derivatives of mevinolin. *J. Med. Chem.* 1986, *29*, 849–852.
352. Askin, D.; Verhoeven, T. R.; Liu, T. M. H.; Shinkai, I. Synthesis of synvinolin: Extremely high conversion alkylation of an ester enolate. *J. Org. Chem.* 1991, *56*, 4929–4932.
353. Xie, X. K.; Tang, Y. Efficient synthesis of simvastatin by use of whole-cell biocatalysis. *Appl. Environ. Microbiol.* 2007, *73*, 2054–2060.
354. Schimmel, T. G.; Borneman, W. S.; Conder, M. J. Purification and characterization of a lovastatin esterase from *Clonostachys compactiuscula*. *Appl. Environ. Microbiol.* 1997, *63*, 1307–1311.

355. Chen, L. C.; Lai, Y. K.; Wu, S. C.; Lin, C. C.; Guo, J. H. Production by *Clonostachys compactiuscula* of a lovastatin esterase that converts lovastatin to monacolin. *J. Enzyme Microb. Technol.* 2006, *39*, 1051–1059.
356. Xu, W.; Chooi, Y. H.; Choi, J. W.; Li, S.; Vederas, J. C.; Da Silva, N. A.; Tang, Y. LovG: The thioesterase required for dihydromonacolin L release and lovastatin nonaketide synthase turnover in lovastatin siosynthesis. *Angew. Chem. Int. Ed.* 2013, *52*, 6472–6475.
357. Butler, D. E.; Le, T. V.; Millar, A.; Nanninga, T. N. Process for the synthesis of (5*R*)-1,1-dimethylethyl-6-cyano-5-hydroxy-3-oxo-hexanoate. U.S. Patent 5155251, October 13, 1992.
358. Ma, S. K.; Gruber, J.; Davis, C.; Newman, L.; Gray, D.; Wang, A.; Grate, J.; Huisman, G. W.; Sheldon, R. A. A green-by-design biocatalytic process for atorvastatin intermediate. *Green Chem.* 2010, *12*, 81–86.
359. Chung, S.; Hwang, Y. Stereoselective hydrolysis of racemic ethyl 4-chloro-3-hydroxybutyrate by a lipase. *Biocatal. Biotransformation* 2008, *26*, 327–330.
360. Lee, S. H.; Park, O. J.; Uh, H. S. A chemoenzymatic approach to the synthesis of enantiomerically pure (*S*)-3-hydroxy-gamma-butyrolactone. *Appl. Microbiol. Biotechnol.* 2008, *79*, 355–362.
361. Patel, J. M. Biocatalytic synthesis of atorvastatin intermediates. *J. Mol. Catal. B: Enzym.* 2009, *61*, 123–128.
362. Martin, C. H.; Dhamankar, H.; Tseng, H. C.; Sheppard, M. J.; Reisch, C. R.; Prather, K. L. J. A platform pathway for production of 3-hydroxyacids provides a biosynthetic route to 3-hydroxy-gamma-butyrolactone. *Nat. Commun.* 2013, *4*, 1414.
363. Turner, N. J.; O'Reilly, E. Biocatalytic retrosynthesis. *Nat. Chem. Biol.* 2013, *9*, 285–288.
364. Brady, D. Biocatalytic hydrolysis of nitriles. In *Handbook of Green Chemistry, Volume 3: Biocatalysis*, Crabtree, R. H., Ed. Wiley-VCH Verlag GmbH & Co. KGaA, Weinheim, 2009.
365. Faber, K. Biotransformations of non-natural compounds: State of the art and future development, *Pure Appl. Chem.* 1997, *69*, 1613–1632.
366. DeSantis, G.; Zhu, Z. L.; Greenberg, W. A.; Wong, K. V.; Chaplin, J.; Hanson, S. R.; Farwell, B.; Nicholson, L. W.; Rand, C. L.; Weiner, D. P.; Robertson, D. E.; Burk, M. J. An enzyme library approach to biocatalysis: Development of nitrilases for enantioselective production of carboxylic acid derivatives. *J. Am. Chem. Soc.* 2002, *124*, 9024–9025.
367. DeSantis, G.; Wong, K.; Farwell, B.; Chatman, K.; Zhu, Z. L.; Tomlinson, G.; Huang, H. J.; Tan, X. Q.; Bibbs, L.; Chen, P.; Kretz, K.; Burk, M. J. Creation of a productive, highly enantioselective nitrilase through gene site saturation mutagenesis (GSSM). *J. Am. Chem. Soc.* 2003, *125*, 11476–11477.
368. Bergeron, S.; Chaplin, D. A.; Edwards, J. H.; Ellis, B. S. W.; Hill, C. L.; Holt-Tiffin, K.; Knight, J. R.; Mahoney, T.; Osborne, A. P.; Ruecroft, G. Nitrilase-catalysed desymmetrisation of 3-hydroxyglutaronitrile: Preparation of a statin side-chain intermediate. *Org. Process Res. Dev.* 2006, *10*, 661–665.
369. Asako, H.; Shimizu, M.; Itoh, N. Biocatalytic production of (*S*)-4-bromo-3-hydroxybutyrate and structurally related chemicals and their applications. *Appl. Microbiol. Biotechnol.* 2009, *84*, 397–405.
370. Asako, H.; Shimizu, M.; Makino, Y.; Itoh, N. Biocatalytic reduction system for the production of chiral methyl (*R*)/(*S*)-4-bromo-3-hydroxybutyrate. *Tetrahedron Lett.* 2010, *51*, 2664–2666.
371. Patel, R. N.; McNamee, C. G.; Banerjee, A.; Howell, J. M.; Robison, R. S.; Szarka, L. J. Stereoselective reduction of beta-keto-esters by *Geotrichum candidum*. *Enzyme Microb. Technol.* 1992, *14*, 731–738.
372. Ye, Q.; Ouyang, P. K.; Ying, H. J. A review-biosynthesis of optically pure ethyl (*S*)-4-chloro-3-hydroxybutanoate ester: Recent advances and future perspectives. *Appl. Microbiol. Biotechnol.* 2011, *89*, 513–522.
373. Yasohara, Y.; Kizaki, N.; Hasegawa, J.; Takahashi, S.; Wada, M.; Kataoka, M.; Shimizu, S. Synthesis of optically active ethyl 4-chloro-3-hydroxybutanoate by microbial reduction. *Appl. Microbiol. Biotechnol.* 1999, *51*, 847–851.
374. Kita, K.; Kataoka, M.; Shimizu, S. Diversity of 4-chloroacetoacetate ethyl ester-reducing enzymes in yeasts and their application to chiral alcohol synthesis. *J. Biosci. Bioeng.* 1999, *88*, 591–598.
375. Wada, M.; Kataoka, M.; Kawabata, H.; Yasohara, Y.; Kizaki, N.; Hasegawa, J.; Shimizu, S. Purification and characterization of NADPH-dependent carbonyl reductase, involved in stereoselective reduction of ethyl 4-chloro-3-oxobutanoate, from *Candida magnoliae*. *Biosci. Biotechnol. Biochem.* 1998, *62*, 280–285.
376. Kataoka, M.; Doi, Y.; Sim, T. S.; Shimizu, S.; Yamada, H. A novel NADPH-dependent carbonyl reductase of *Candida macedoniensis*. Purification and characterization. *Arch. Biochem. Biophys.* 1992, *294*, 469–474.
377. He, J. Y.; Sun, Z. H.; Ruan, W. Q.; Xu, Y. Biocatalytic synthesis of ethyl (*S*)-4-chloro-3-hydroxy-butanoate in an aqueous-organic solvent biphasic system using *Aureobasidium pullulans* CGMCC 1244. *Process Biochem.* 2006, *41*, 244–249.
378. Saratani, Y.; Uheda, E.; Yamamoto, H.; Nishimura, A.; Yoshizako, F. Stereoselective reduction of ethyl 4-chloro-3-oxobutanoate by fungi. *Biosci. Biotechnol. Biochem.* 2001, *65*, 1676–1679.

379. Ye, Q.; Yan, M.; Xu, L.; Cao, H.; Li, Z.; Chen, Y.; Li, S.; Ying, H. A novel carbonyl reductase from *Pichia stipitis* for the production of ethyl (*S*)-4-chloro-3-hydroxybutanoate. *Biotechnol. Lett.* 2009, *31*, 537–542.
380. Cai, P.; An, M. D.; Xu, L.; Xu, S.; Hao, N.; Li, Y.; Guo, K.; Yan, M. Development of a substrate-coupled biocatalytic process driven by an NADPH-dependent sorbose reductase from *Candida albicans* for the asymmetric reduction of ethyl 4-chloro-3-oxobutanoate. *Biotechnol. Lett.* 2012, *34*, 2223–2227.
381. Schallmey, M.; Floor, R. J.; Hauer, B.; Breuer, M.; Jekel, P. A.; Wijma, H. J.; Dijkstra, B. W.; Janssen, D. B. Biocatalytic and structural properties of a highly engineered halohydrin dehalogenase. *ChemBioChem* 2013, *14*, 870–881.
382. Ritter, S. K. Going green keeps getting easier. *Chem. Eng. News* 2006, *84*, 24–27.
383. Wolberg, M.; Hummel, W.; Muller, M. Biocatalytic reduction of beta,delta-diketo esters: A highly stereoselective approach to all four stereoisomers of a chlorinated beta,delta-dihydroxy hexanoate. *Chem. Eur. J.* 2001, *7*, 4562–4571.
384. Wolberg, M.; Hummel, W.; Wandrey, C.; Muller, M. Highly regio- and enantioselective reduction of 3,5-dioxocarboxylates. *Angew. Chem. Int. Ed.* 2000, *39*, 4306–4308.
385. Wolberg, M.; Villela, M.; Bode, S.; Geilenkirchen, P.; Feldmann, R.; Liese, A.; Hummel, W.; Muller, M. Chemoenzymatic synthesis of the chiral side-chain of statins: Application of an alcohol dehydrogenase catalysed ketone reduction on a large scale. *Bioprocess Biosyst. Eng.* 2008, *31*, 183–191.
386. Pfruender, H.; Amidjojo, M.; Hang, F.; Weuster-Botz, D. Production of *Lactobacillus kefir* cells for asymmetric synthesis of a 3,5-dihydroxycarboxylate. *Appl. Microbiol. Biotechnol.* 2005, *67*, 619–622.
387. Patel, R. N.; Banerjee, A.; McNamee, C. G.; Brzozowski, D.; Hanson, R. L.; Szarka, L. J. Enantioselective microbial reduction of 3,5-dioxo-6-(benzyloxy) hexanoic acid, ethyl-ester. *Enzyme Microb. Technol.* 1993, *15*, 1014–1021.
388. Guo, Z.; Chen, Y.; Goswami, A.; Hanson, R. L.; Patel, R. N. Synthesis of ethyl and *t*-butyl (3*R*,5*S*)-dihydroxy-6-benzyloxy hexanoates via diastereo- and enantioselective microbial reduction. *Tetrahedron: Asymmetry* 2006, *17*, 1589–1602.
389. Goldberg, S.; Guo, Z.; Chen, S.; Goswami, A.; Patel, R. N. Synthesis of ethyl-(3*R*,5*S*)-dihydroxy-6-benzyloxyhexanoates via diastereo- and enantioselective microbial reduction: Cloning and expression of ketoreductase III from *Acinetobacter* sp. SC 13874. *Enzyme Microb. Technol.* 2008, *43*, 544–549.
390. Wu, X.; Liu, N.; He, Y.; Chen, Y. Cloning, expression, and characterization of a novel diketoreductase from *Acinetobacter baylyi*. *Acta Biochim. Biophys. Sin.* 2009, *41*, 163–170.
391. Chen, Y.; Chen, C.; Wu, X. Dicarbonyl reduction by single enzyme for the preparation of chiral diols. *Chem. Soc. Rev.* 2012, *41*, 1742–1753.
392. Lu, M.; Huang, Y.; White, M. A.; Wu, X.; Liu, N.; Cheng, X.; Chen, Y. Dual catalysis mode for the dicarbonyl reduction catalyzed by diketoreductase. *Chem. Commun.* 2012, *48*, 11352–11354.
393. Huang, Y.; Lu, Z.; Ma, M.; Liu, N.; Chen, Y. Functional roles of tryptophan residues in diketoreductase from *Acinetobacter baylyi*. *BMB Rep.* 2012, *45*, 452–457.
394. Huang, Y.; Lu, Z.; Liu, N.; Chen, Y. Identification of important residues in diketoreductase from *Acinetobacter baylyi* by molecular modeling and site-directed mutagenesis. *Biochimie* 2012, *94*, 471–478.
395. Gijsen, H. J. M.; Wong, C. H. Sequential 3-substrate and 4-substrate aldol reactions catalyzed by aldolases. *J. Am. Chem. Soc.* 1995, *117*, 7585–7591.
396. Gijsen, H. J. M.; Wong, C. H. Unprecedented asymmetric aldol reactions with 3 aldehyde substrates catalyzed by 2-deoxyribose-5-phosphate aldolase. *J. Am. Chem. Soc.* 1994, *116*, 8422–8423.
397. Liu, J. J.; Hsu, C. C.; Wong, C. H. Sequential aldol condensation catalyzed by DERA mutant Ser238Asp and a formal total synthesis of atorvastatin. *Tetrahedron Lett.* 2004, *45*, 2439–2441.
398. Greenberg, W. A.; Varvak, A.; Hanson, S. R.; Wong, K.; Huang, H. J.; Chen, P.; Burk, M. J. Development of an efficient, scalable, aldolase-catalyzed process for enantioselective synthesis of statin intermediates. *Proc. Natl. Acad. Sci. USA* 2004, *101*, 5788–5793.
399. Jennewein, S.; Schurmann, M.; Wolberg, M.; Hilker, I.; Luiten, R.; Wubbolts, M.; Mink, D. Directed evolution of an industrial biocatalyst: 2-deoxy-D-ribose 5-phosphate aldolase. *Biotechnol. J.* 2006, *1*, 537–548.
400. Oslaj, M.; Cluzeau, J.; Orkic, D.; Kopitar, G.; Mrak, P.; Casar, Z. A highly productive, whole-cell DERA chemoenzymatic process for production of key lactonized side-chain intermediates in statin synthesis. *PLoS One* 2013, *8*, e62250.
401. Rucigaj, A.; Krajnc, M. Optimization of a crude deoxyribose-5-phosphate aldolase lyzate-catalyzed process in synthesis of statin intermediates. *Org. Process Res. Dev.* 2013, *17*, 854–862.
402. You, Z. Y.; Liu, Z. Q.; Zheng, Y. G.; Shen, Y. C. Characterization and application of a newly synthesized 2-deoxyribose-5-phosphate aldolase. *J. Ind. Microbiol. Biotechnol.* 2013, *40*, 29–39.

403. Mancia, G.; Fagard, R.; Narkiewicz, K.; Redon, J.; Zanchetti, A.; Bohm, M.; Christiaens, T.; Cifkova, R.; De Backer, G.; Dominiczak, A.; Galderisi, M.; Grobbee, D. E.; Jaarsma, T.; Kirchhof, P.; Kjeldsen, S. E.; Laurent, S.; Manolis, A. J.; Nilsson, P. M.; Ruilope, L. M.; Schmieder, R. E.; Sirnes, P. A.; Sleight, P.; Viigimaa, M.; Waeber, B.; Zannad, F.; Task Force, M. ESH/ESC guidelines for the management of arterial hypertension. *Eur. Heart J.* 2013, *34*, 2159–2219.
404. Lemke, T. L.; Williams, D. A. *Foye's Principles of Medicinal Chemistry*. Wolters Kluwer Health: Philadelphia, PA, 2012.
405. Ahlquist, R. P. A study of the adrenotropic receptors. *Am. J. Physiol.* 1948, *153*, 586–600.
406. Harrison, J. K.; Pearson, W. R.; Lynch, K. R. Molecular characterization of alpha-1-adrenoceptors and alpha-2-adrenoceptors. *Trends Pharmacol. Sci.* 1991, *12*, 62–67.
407. Patrick, G. L. *An Introduction to Medicinal Chemistry*. OUP: Oxford, U.K., 2013.
408. Murmann, W.; Rumore, G.; Gamba, A. Pharmacological properties of 1-(4′-nitrophenyl)-2-isopropylamino-ethanol (INPEA), a new beta-adrenergic receptor antagonist. V. Effects of the optical isomers D(minus) and L(plus) INPEA on heart rate, oxygen consumption and body temperature and on the cardiac and metabolic effects of adrenaline and noradrenaline in urethane-anesthetized rats. *Boll. Chim. Farm.* 1967, *106*, 251–268.
409. Blay, G.; Hernandez-Olmos, V.; Pedro, J. R. Synthesis of (S)-(+)-sotalol and (R)-(−)-isoproterenol via a catalytic enantioselective Henry reaction. *Tetrahedron: Asymmetry* 2010, *21*, 578–581.
410. Beyer, T.; Brachmann, J.; Kubler, W. Comparative effects of D-sotalol and L-sotalol on the atrioventricular node of the rabbit heart. *J. Cardiovasc. Pharmacol.* 1993, *22*, 240–246.
411. Kapoor, M.; Anand, N.; Ahmad, K.; Koul, S.; Chimni, S. S.; Taneja, S. C.; Qazi, G. N. Synthesis of β-adrenergic blockers (R)-(−)-nifenalol and (S)-(+)-sotalol via a highly efficient resolution of a bromohydrin precursor. *Tetrahedron: Asymmetry* 2005, *16*, 717–725.
412. Wünsche, K.; Schwaneberg, U.; Bornscheuer, U. T.; Meyer, H. H. Chemoenzymatic route to β-blockers via 3-hydroxy esters. *Tetrahedron: Asymmetry* 1996, *7*, 2017–2022.
413. Kamal, A.; Khanna, G. B. R.; Krishnaji, T.; Tekumalla, V.; Ramu, R. New chemoenzymatic pathway for β-adrenergic blocking agents. *Tetrahedron: Asymmetry* 2005, *16*, 1485–1494.
414. Yang, W.; Xu, J.-H.; Xie, Y.; Xu, Y.; Zhao, G.; Lin, G.-Q. Asymmetric reduction of ketones by employing *Rhodotorula* sp. AS2.2241 and synthesis of the beta-blocker (R)-nifenalol. *Tetrahedron: Asymmetry* 2006, *17*, 1769–1774.
415. Pedragosa Moreau, S.; Morisseau, C.; Baratti, J.; Zylber, J.; Archelas, A.; Furstoss, R. Microbiological transformations. 37. An enantioconvergent synthesis of the beta-blocker (R)-Nifenalol® using a combined chemoenzymatic approach. *Tetrahedron* 1997, *53*, 9707–9714.
416. Conolly, M. E.; Kersting, F.; Dollery, C. T. Clinical pharmacology of beta-adrenoceptor-blocking drugs. *Prog. Cardiovasc. Dis.* 1976, *19*, 203–234.
417. Frishman, W. H.; Alwarshetty, M. Beta-adrenergic blockers in systemic hypertension—Pharmacokinetic considerations related to the current guidelines. *Clin. Pharmacokinet.* 2002, *41*, 505–516.
418. Sica, D. A.; Black, H. R. Pharmacologic considerations in the positioning of beta-blockers in antihypertensive therapy. *Curr. Hypertens. Rep.* 2008, *10*, 330–335.
419. Shu, D. F.; Dong, B. R.; Lin, X. F.; Wu, T. X.; Liu, G. J. Long-term beta blockers for stable angina: Systematic review and meta-analysis. *Eur. J. Prev. Cardiol.* 2012, *19*, 330–341.
420. Aidietis, A.; Laucevicius, A.; Marinskis, G. Hypertension and cardiac arrhythmias. *Curr. Pharm. Design* 2007, *13*, 2545–2555.
421. Siebert, C. D.; Hansicke, A.; Nageo, T. Stereochemical comparison of nebivolol with other beta-blockers. *Chirality* 2008, *20*, 103–109.
422. Agustian, J.; Kamaruddin, A. H.; Bhatia, S. Single enantiomeric beta-blockers—The existing technologies. *Process Biochem.* 2010, *45*, 1587–1604.
423. Zelaszczyk, D.; Kiec-Kononowicz, K. Biocatalytic approaches to optically active beta-blockers. *Curr. Med. Chem.* 2007, *14*, 53–65.
424. Campo, C.; Llama, E. F.; Bermudez, J. L.; Sinisterra, J. V. Methodologies for the stereoselective synthesis of adrenergic beta-blockers: An overview. *Biocatal. Biotransformation* 2001, *19*, 163–180.
425. Avila, R.; Ruiz, R.; Amaro-Gonzalez, D.; Diaz, O.; Gonzalez, J. A.; Nunez, A. J. Increased racemate resolution of propranolol esters by lipase immobilized catalysis. *Lat. Am. Appl. Res.* 2005, *35*, 307–311.
426. Chiou, T. W.; Chang, C. C.; Lai, C. T.; Tai, D. F. Kinetic resolution of propranolol by a lipase-catalyzed N-acetylation. *Bioorg. Med. Chem. Lett.* 1997, *7*, 433–436.
427. Barbosa, O.; Ariza, C.; Ortiz, C.; Torres, R. Kinetic resolution of (R/S)-propranolol (1-isopropylamino-3-(1-naphtoxy)-2-propanolol) catalyzed by immobilized preparations of *Candida antarctica* lipase B (CAL-B). *N. Biotechnol.* 2010, *27*, 844–850.

428. Bevinakatti, H. S.; Banerji, A. A. Practical chemoenzymatic synthesis of both enantiomers of propranolol. *J. Org. Chem.* 1991, *56*, 5372–5375.
429. Bevinakatti, H. S.; Banerji, A. A. Lipase catalysis in organic-solvents—Application to the synthesis of (*R*)-atenolol and (*S*)-atenolol. *J. Org. Chem.* 1992, *57*, 6003–6005.
430. Thakkar, N. V.; Banerji, A. A.; Bevinakatti, H. S. F. Chemoenzymatic synthesis of *S*(−) practolol. *Biotechnol. Lett.* 1995, *17*, 217–218.
431. Pàmies, O.; Bäckvall, J. E. Dynamic kinetic resolution of beta-azido alcohols. An efficient route to chiral aziridines and beta-amino alcohols. *J. Org. Chem.* 2001, *66*, 4022–4025.
432. Martinez, F.; Del Campo, C.; Sinisterra, J. V.; Llama, E. F. Preparation of halohydrin beta-blocker precursors using yeast-catalysed reduction. *Tetrahedron: Asymmetry* 2000, *11*, 4651–4660.
433. Lagos, F. M.; Del Campo, C.; Llama, E. F.; Sinisterra, J. V. New yeast strains for enantioselective production of halohydrin precursor of (*S*)-propranolol. *Enzyme Microb. Technol.* 2002, *30*, 895–901.
434. Lagos, F. M.; Carballeira, J. D.; Bermudez, J. L.; Alvarez, E.; Sinisterra, J. V. Highly stereoselective reduction of haloketones using three new yeasts: Application to the synthesis of (*S*)-adrenergic beta-blockers related to propranolol. *Tetrahedron: Asymmetry* 2004, *15*, 763–770.
435. Martinez-Lagos, F.; Sinisterra, J. V. Enantioselective production of halohydrin precursor of propranolol catalysed by immobilized yeasts. *J. Mol. Catal. B: Enzym.* 2005, *36*, 1–7.
436. Carballeira, J. D.; Alvarez, E.; Campillo, M.; Pardo, L.; Sinisterra, J. V. Diplogelasinospora grovesii IMI 171018, a new whole cell biocatalyst for the stereoselective reduction of ketones. *Tetrahedron: Asymmetry* 2004, *15*, 951–962.
437. Ondetti, M. A.; Cushman, D. W. Inhibition of the renin-angiotensin system—A new approach to the therapy of hypertension. *J. Med. Chem.* 1981, *24*, 355–361.
438. Patel, R. N.; Howell, J. M.; McNamee, C. G.; Fortney, K. F.; Szarka, L. J. Stereoselective enzymatic-hydrolysis of alpha-(acetylthio)methyl benzenepropanoic acid and 3-acetylthio-2-methylpropanoic acid. *Biotechnol. Appl. Biochem.* 1992, *16*, 34–47.
439. Patel, R. N. Synthesis of chiral pharmaceutical intermediates by biocatalysis. *Coord. Chem. Rev.* 2008, *252*, 659–701.
440. Patel, R. N.; Howell, J. M.; Banerjee, A.; Fortney, K. F.; Szarka, L. J. Stereoselective enzymatic esterification of 3-benzoylthio-2-methylpropanoic acid. *Appl. Microbiol. Biotechnol.* 1991, *36*, 29–34.
441. Lednev, O. A.; Zaslavskaia, R. M.; Buniatian, N. D.; Sergeev, S. V. The use of monopril in elderly patients with arterial hypertension and coronary heart disease. *Klin. Med. (Mosk).* 2009, *87*, 48–49.
442. Triggle, D. J.; Janis, R. A. Calcium-channel ligands. *Annu. Rev. Pharmacol.* 1987, *27*, 347–369.
443. Toal, C. B.; Meredith, P. A.; Elliott, H. L. Long-acting dihydropyridine calcium-channel blockers and sympathetic nervous system activity in hypertension: A literature review comparing amlodipine and nifedipine GITS. *Blood Press.* 2012, *21*, 3–10.
444. Sobolev, A.; Franssen, M. C. R.; Makarova, N.; Duburs, G.; de Groot, A. *Candida antarctica* lipase-catalyzed hydrolysis of 4-substituted bis(ethoxycarbonylmethyl) 1,4-dihydropyridine-3,5-dicarboxylates as the key step in the synthesis of optically active dihydropyridines. *Tetrahedron: Asymmetry* 2000, *11*, 4559–4569.
445. Sobolev, A.; Franssen, M. C. R.; Vigante, B.; Cekavicus, B.; Makarova, N.; Duburs, G.; de Groot, A. An efficient chemoenzymatic approach to enantiomerically pure 4-2-(difluoromethoxy)phenyl substituted 1,4-dihydropyridine-3,5-dicarboxylates. *Tetrahedron: Asymmetry* 2001, *12*, 3251–3256.
446. Sobolev, A.; Franssen, M. C. R.; Vigante, B.; Cekavicus, B.; Zhalubovskis, R.; Kooijman, H.; Spek, A. L.; Duburs, G.; de Groot, A. Effect of acyl chain length and branching on the enantioselectivity of *Candida rugosa* lipase in the kinetic resolution of 4-(2-difluoromethoxyphenyl)-substituted 1,4-dihydropyridine 3,5-diesters. *J. Org. Chem.* 2002, *67*, 401–410.
447. Sobolev, A.; Franssen, M. C. R.; Duburs, G.; De Groot, A. Chemoenzymatic synthesis of enantiopure 1,4-dihydropyridine derivatives. *Biocatal. Biotransformation* 2004, *22*, 231–252.
448. deCastro, M. S.; Salazar, L.; Sinisterra, J. V. Influence of the N-MOM group in the enantioselective lipase catalyzed methanolysis of racemic 1,4-dihydropyridine dicarboxylates. *Tetrahedron: Asymmetry* 1997, *8*, 857–858.
449. Rovnyak, G. C.; Kimball, S. D.; Beyer, B.; Cucinotta, G.; DiMarco, J. D.; Gougoutas, J.; Hedberg, A.; Malley, M.; McCarthy, J. P. Calcium entry blockers and activators: Conformational and structural determinants of dihydropyrimidine calcium channel modulators. *J. Med. Chem.* 1995, *38*, 119–129.
450. Atwal, K. S.; Swanson, B. N.; Unger, S. E.; Floyd, D. M.; Moreland, S.; Hedberg, A.; O'Reilly, B. C. Dihydropyrimidine calcium channel blockers. 3. 3-Carbamoyl-4-aryl-1,2,3,4-tetrahydro-6-methyl-5-pyrimidinecarboxylic acid esters as orally effective antihypertensive agents. *J. Med. Chem.* 1991, *34*, 806–811.

451. Schnell, B.; Krenn, W.; Faber, K.; Kappe, C. O. Synthesis and reactions of Biginelli-compounds. Part 23. Chemoenzymatic syntheses of enantiomerically pure 4-aryl-3,4-dihydropyrimidin-2(1H)-ones. *J. Chem. Soc., Perkin Trans. 1* 2000, 4382–4389.
452. Schnell, B.; Strauss, U. T.; Verdino, P.; Faber, K.; Kappe, C. O. Synthesis of enantiomerically pure 4-aryl-3,4-dihydro-pyrimidin-2(1H)-ones via enzymatic resolution: Preparation of the antihypertensive agent (R)-SQ 32926. *Tetrahedron: Asymmetry* 2000, *11*, 1449–1453.
453. Prasad, A. K.; Mukherjee, C.; Singh, S. K.; Brahma, R.; Singh, R.; Saxena, R. K.; Olsen, C. E.; Parmar, V. S. Novel selective biocatalytic deacetylation studies on dihydropyrimidinones. *J. Mol. Catal. B: Enzym.* 2006, *40*, 93–100.
454. Corelli, F.; Manetti, F.; Tafi, A.; Campiani, G.; Nacci, V.; Botta, M. Diltiazem-like calcium entry blockers: A hypothesis of the receptor-binding site based on a comparative molecular field analysis model. *J. Med. Chem.* 1997, *40*, 125–131.
455. Matsumae, H.; Furui, M.; Shibatani, T. Lipase-catalyzed asymmetric hydrolysis of 3-phenylglycidic acid ester, the key intermediate in the synthesis of diltiazem hydrochloride. *J. Ferment. Bioeng.* 1993, *75*, 93–98.
456. Matsumae, H.; Furui, M.; Shibatani, T.; Tosa, T. Production of optically-active 3-phenylglycidic acid ester by the lipase from serratia-marcescens on a hollow-fiber membrane reactor. *J. Ferment. Bioeng.* 1994, *78*, 59–63.
457. Matsumae, H.; Shibatani, T. Purification and characterization of the lipase from *Serratia marcescens* SR41-8000 responsible for asymmetric hydrolysis of 3-phenylglycidic acid-esters. *J. Ferment. Bioeng.* 1994, *77*, 152–158.
458. Shibatani, T.; Omori, K.; Akatsuka, H.; Kawai, E.; Matsumae, H. Enzymatic resolution of diltiazem intermediate by *Serratia marcescens* lipase: Molecular mechanism of lipase secretion and its industrial application. *J. Mol. Catal. B: Enzym.* 2000, *10*, 141–149.
459. Pace, V.; Hoyos, P.; Castoldi, L.; Domínguez de María, P.; Alcántara, A. R. 2-Methyltetrahydrofuran (2-MeTHF): A biomass-derived solvent with broad application in organic chemistry. *ChemSusChem* 2012, *5*, 1369–1379.
460. Simeó, Y.; Sinisterra, J. V.; Alcántara, A. R. Regioselective enzymatic acylation of pharmacologically interesting nucleosides in 2-methyltetrahydrofuran, a greener substitute for THF. *Green Chem.* 2009, *11*, 855–862.
461. Liese, A.; Hilterhaus, L. Evaluation of immobilized enzymes for industrial applications. *Chem. Soc. Rev.* 2013, *42*, 6236–6249.
462. DiCosimo, R.; McAuliffe, J.; Poulose, A. J.; Bohlmann, G. Industrial use of immobilized enzymes. *Chem. Soc. Rev.* 2013, *42*, 6437–6474.
463. Wells, A. Industrial applications of biocatalysis: An overview. In *Comprehensive Chirality*, Carreira, E. C.; Yamamoto, H., Eds. Elsevier: Amsterdam, the Netherlands, 2012; pp. 253–287.
464. Sheldon, R. A. Industrial applications of asymmetric synthesis using cross-linked enzyme aggregates. In *Comprehensive Chirality*, Carreira, E. C.; Yamamoto, H., Eds. Elsevier: Amsterdam, the Netherlands, 2012; pp. 353–366.
465. Reetz, M. T. Biocatalysis in organic chemistry and biotechnology: Past, present, and future. *J. Am. Chem. Soc.* 2013, *135*, 12480–12496.
466. Huisman, G. W.; Collier, S. J. On the development of new biocatalytic processes for practical pharmaceutical synthesis. *Curr. Opin. Chem. Biol.* 2013, *17*, 284–292.
467. Dunn, P. J. Pharmaceutical green chemistry process changes—How long does it take to obtain regulatory approval? *Green Chem.* 2013, *15*, 3099–3104.
468. Bryan, M. C.; Dillon, B.; Hamann, L. G.; Hughes, G. J.; Kopach, M. E.; Peterson, E. A.; Pourashraf, M.; Raheem, I.; Richardson, P.; Richter, D.; Sneddon, H. F. Sustainable practices in medicinal chemistry: Current state and future directions. *J. Med. Chem.* 2013, *56*, 6007–6021.
469. Meyer, H.-P.; Eichhorn, E.; Hanlon, S.; Lutz, S.; Schurmann, M.; Wohlgemuth, R.; Coppolecchia, R. The use of enzymes in organic synthesis and the life sciences: Perspectives from the Swiss Industrial Biocatalysis Consortium (SIBC). *Catal. Sci. Technol.* 2012, *3*, 29–40.
470. Jimenez-Gonzalez, C.; Constable, D. J. C.; Ponder, C. S. Evaluating the "Greenness" of chemical processes and products in the pharmaceutical industry—A green metrics primer. *Chem. Soc. Rev.* 2012, *41*, 1485–1498.

13 Chiral Medicines

Apurba Bhattacharya and Rakeshwar Bandichhor

CONTENTS

13.1 Introduction 449
13.2 Sitagliptin **1** 449
    13.2.1 Medicinal Chemistry Route 449
13.3 Aprepitant **2** 454
    13.3.1 Medicinal Chemistry Route 454
13.4 Simvastatin **3** 454
    13.4.1 Medicinal Chemistry Route 454
13.5 Paroxetine **4** 454
    13.5.1 Medicinal Chemistry Route 454
13.6 Levetiracetam **5** 454
    13.6.1 Medicinal Chemistry Route 455
13.7 Pregabalin **6** 456
    13.7.1 Medicinal Chemistry Route 456
13.8 Sertraline **7** 456
    13.8.1 Medicinal Chemistry Route 456
13.9 Conclusions 457
References 459

13.1 INTRODUCTION

Organic molecules of biological origin, for example, DNA, proteins, enzymes, carbohydrates, hormones, and other endogenously functional materials, are chiral ones.

In the pharmaceutical industry, nearly 60% of the drugs currently in use are chiral in nature. Although asymmetric center–containing molecules and their enantiomers have the same chemical structure, they exhibit entirely different biological activities. There are 15 top medicines and out of those 8 are small to medium-size organic molecules. Apart from aripiprazole (Abilify), all (Sovaldi, Crestor, Nexium, Lyrica, Spiriva, Januvia, and Copaxone) are chiral ones as shown in Table 13.1.

In the following sections, the medicinal chemistry routes of 7 chiral medicines (sitagliptin **1**, aprepitant **2**, simvastatin **3**, paroxetine **4**, levetiracetam **5**, pregabalin **6**, and sertraline **7**) as shown in Figure 13.1 are discussed.

13.2 SITAGLIPTIN 1

Sitagliptin phosphate, widely known as Januvia, was developed by Merck and approved by the U.S. FDA for the treatment of type 2 diabetes mellitus in late 2006.[1]

13.2.1 MEDICINAL CHEMISTRY ROUTE

One of the early medicinal chemistry routes for sitagliptin **1** was demonstrated by coupling of β-amino acid **8** with fused heterocycle **9**.[2] The corresponding β-amino acid **8** was synthesized by homologation of the α-amino acid **10**.[3] Alkylation of **11** with the bromide derivative gave **12**,

TABLE 13.1
Top 15 Medicines That Did More Than Billion USD Business till 2014

Sl. No.	Category	API/Antibody	Brand Name	Innovator	Business ($)
1.	For arthritis	Antibody	Humira	AbbVie	11.8 bn/2014
2.	Antidiabetes	Insulin analog	Lantus	Sanofi	10.3 bn/2014
3.	For hepatitis C	Sofosbuvir	Sovaldi	Gilead	9.4 bn/2014
4.	For mental illness	Aripiprazole	Abilify	Otsuka	9.3 bn/2014
5.	For arthritis	Fusion protein	Enbrel	Amgen	8.7 bn/2014
6.	For high cholesterol	Rosuvastatin	Crestor	AstraZeneca	8.5 bn/2014
7.	For arthritis	Antibody	Remicade	JNJ	8.1 bn/2014
8.	Anti–acid reflux	Esomeprazole	Nexium	AstraZeneca	7.7 bn/2014
9.	Anticancer	Chimeric antibody	Rituxan	Roche	6.6 bn/2014
10.	Anticancer	Antibody	Avastin	Roche	6.1 bn/2014
11.	Nerve disorder	Pregabalin	Lyrica	Pfizer	6.0 bn/2014
12.	Anticancer	Antibody	Herceptin	Roche	5.6 bn/2014

(Continued)

Chiral Medicines

TABLE 13.1 (*Continued*)
Top 15 Medicines That Did More Than Billion USD Business till 2014

Sl. No.	Category	API/Antibody	Brand Name	Innovator	Business ($)
13.	COPD	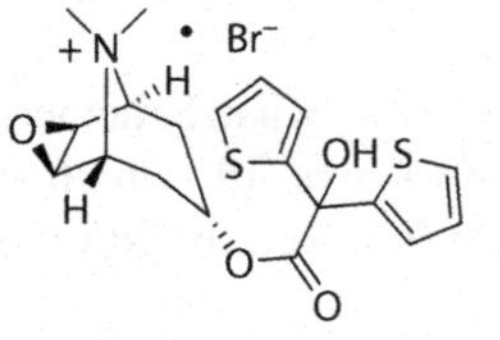 Tiotropium bromide	Spiriva	BI	5.5 bn/2014
14.	Antidiabetic	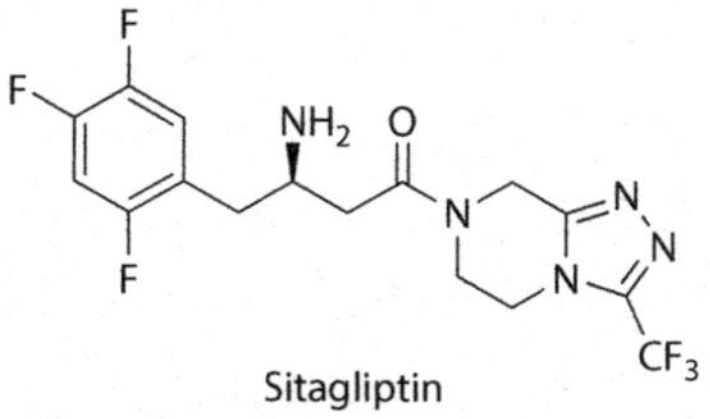 Sitagliptin	Januvia	Merck	5.0 bn/2014
15.	Multiple sclerosis	Random copolymer of amino acid: glatiramer	Copaxone	Teva	4.8 bn/2014

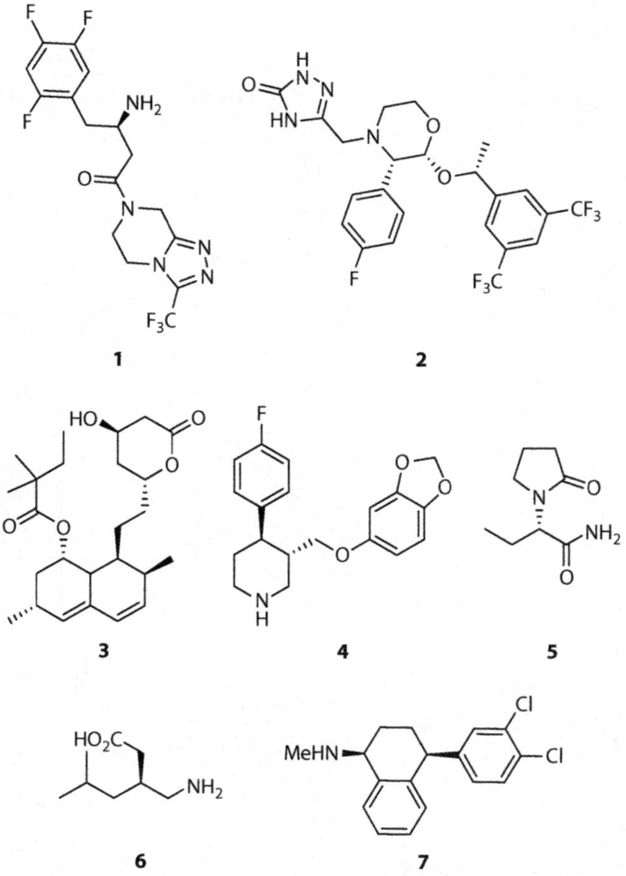

FIGURE 13.1 Structure of APIs.

452   Asymmetric Synthesis of Drugs and Natural Products

which was treated with hydrochloric acid followed by di-*tert*-butyl dicarbonate to give ester **13**. Hydrolysis afforded the α-amino acid **10**, which was treated with isobutyl chloroformate followed by diazomethane to give diazo ketone **14**. Subsequent rearrangement of diazo ketone **14** to ester **15** followed by hydrolysis gave the desired β-amino acid **8**. Alternatively, the 2,4,5-trifluophenyl-substituted β-amino acid **8** could also be prepared in one step by sonication of diazo ketone **14** in the presence of silver benzoate (Scheme 13.1).[4]

Synthesis of the piperazine derivative **16** was achieved as shown in Scheme 13.2. Hydrazinopyrazine **18**, prepared from chloropyrazine **17**,[5] was acylated to give **19**, which was then condensed in polyphosphoric acid (PPA) to give triazolopyrazine **20** followed by Pd/C-catalyzed hydrogenation affording desired heterocycle **16**.

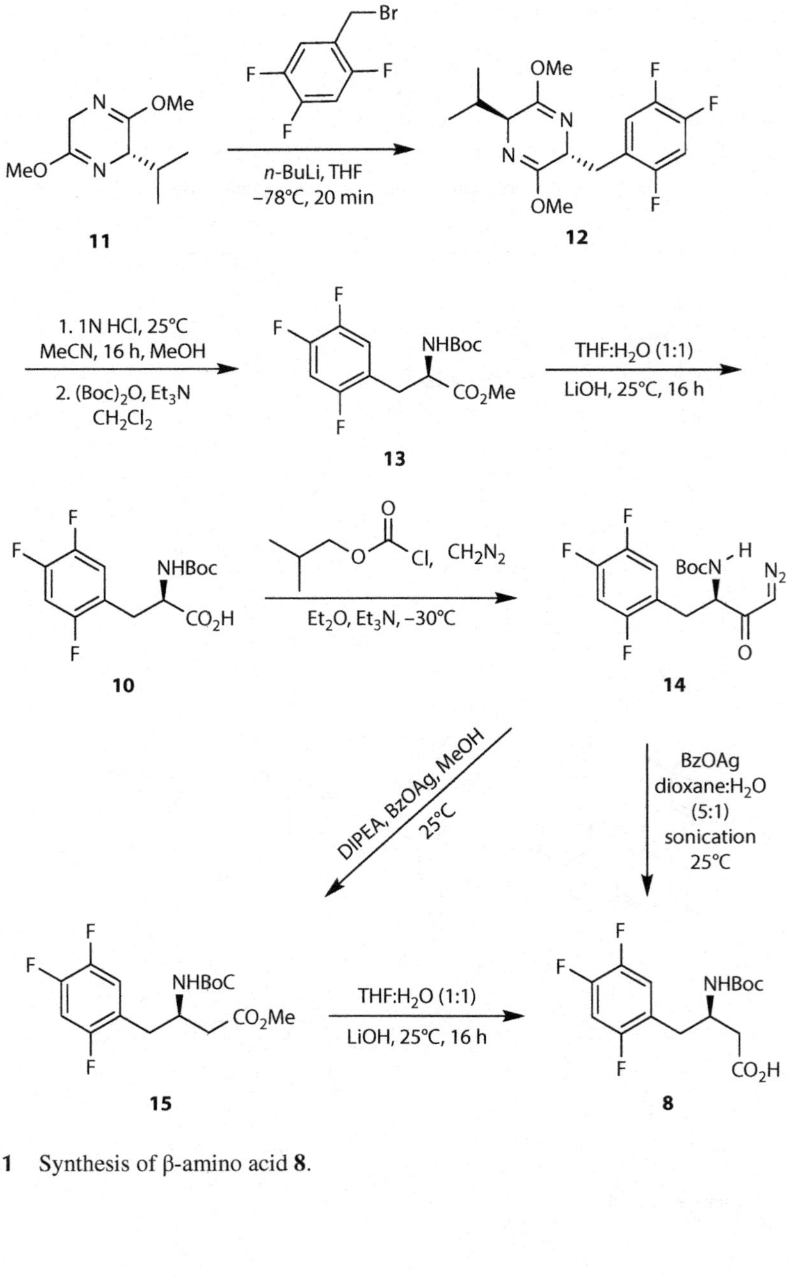

SCHEME 13.1 Synthesis of β-amino acid **8**.

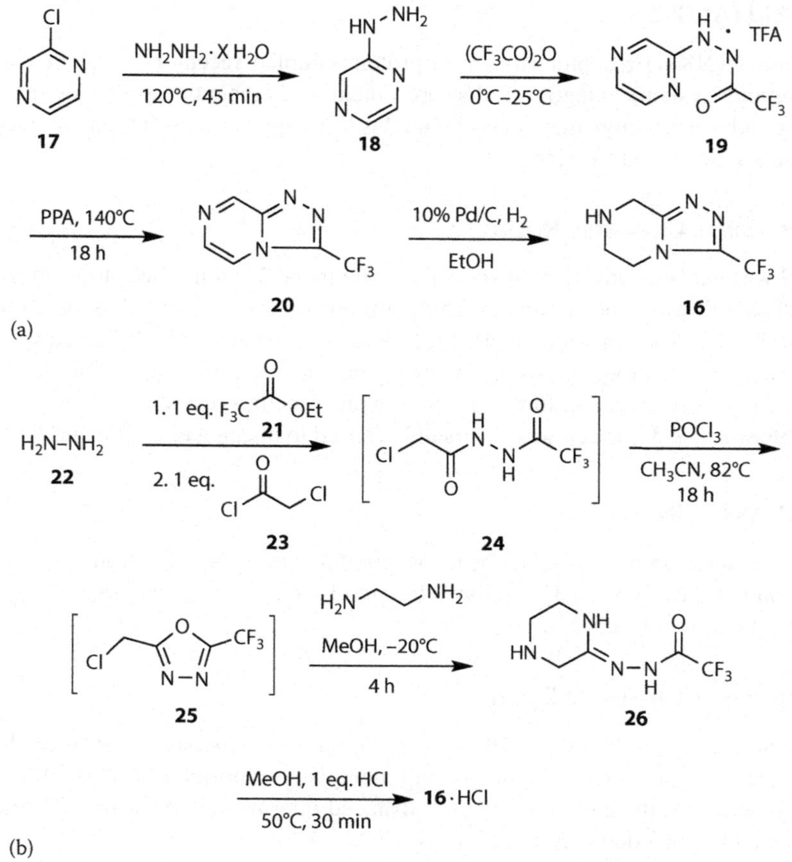

SCHEME 13.2 (a) Synthesis of **16** starting from chloropyrazine **17**. (b) Synthesis of **16** HCl salt starting from hydrazine **22**.

Alternatively, the reaction of ethyl trifluoroacetate **21** with hydrazine **22** afforded the corresponding hydrazide, which was condensed with chloroacetyl chloride **23** to yield bishydrazide **24**. Cyclization of compound **24** by means of phosphorous oxychloride (POCl$_3$) in acetonitrile afforded oxadiazole **25**, which was treated with ethylenediamine **26** to provide **27**. Cyclization of hydrazide **27** yielded HCl salt of **16**.

Coupling of the β-amino acid **8** with triazolopiperazines **16** followed by Boc deprotection of the amine provided the HCl salt of desired compound **1** (Scheme 13.3).[6]

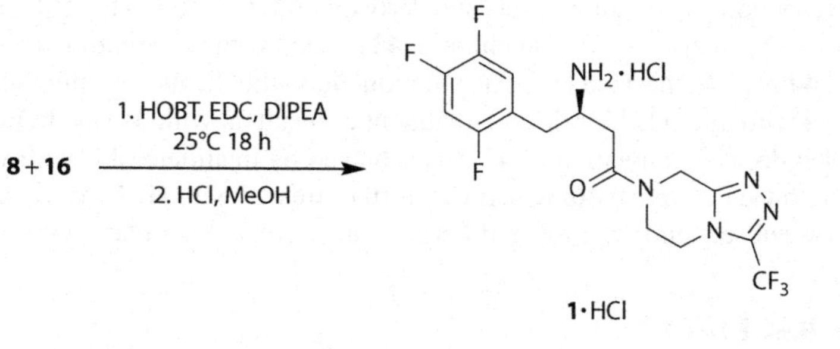

SCHEME 13.3 Synthesis of HCl salt of sitagliptin **1**.

13.3 APREPITANT 2

The neurokinin-1 (NK-1) receptor (it is a G-protein coupled receptor traced in the central and peripheral nervous system) antagonists that are potential therapeutic agents for the treatment of chemotherapy-induced mainly emesis, aprepitant **2** is FDA approved (2003) one of the potent orally active antagonists developed by Merck.[7]

13.3.1 Medicinal Chemistry Route

Synthesis of **2** was marked with the reduction of morpholinone **27** with L-Selectride affording intermediate lactol **28**, which was condensed with 3,5-bis(trifluoromethyl)benzoyl chloride **29** to give rise to acyl acetal **30**. Further reaction with dimethyl titanocene offered ether **31**. Pd/C-catalyzed hydrogenation of double bond and hydrogenolysis of N-benzyl group offered 8:1 mixture of diastereomers and the major isomer **32** was isolated, and further reaction with N-methoxycarbonyl-2-chloroacetamidrazone **33** afforded intermediate **34**, which was eventually cyclized to obtain API **2** (Scheme 13.4).[8]

13.4 SIMVASTATIN 3

Simvastatin[9] is a semisynthetic product of lovastatin that can be isolated from *Aspergillus terreus*. Orally active simvastatin (Zocor) **3**, developed by Merck, is a prodrug that targets a specific enzyme responsible for lowering the cholesterol level.[10]

13.4.1 Medicinal Chemistry Route

As shown in Scheme 13.5, lovastatin **70** on deacylation with aqueous LiOH offered free diol **36**, which was selectively protected with *tert*-butyldimethylsilyl chloride (TBSCl), affording monosilylated derivative **37**. At the end, acylation of advanced intermediate **37** with 2,2-dimethyl butyryl chloride followed by TBS deprotection gave rise to API **3**.

13.5 PAROXETINE 4

Paroxetine **4** under the trade names Seroxat and Paxil was developed by GlaxoSmithKline. It is found to be effective in first-line therapy for generalized anxiety disorder.[11]

13.5.1 Medicinal Chemistry Route

In the medicinal chemistry route, the synthesis started with the reaction of methyl ester **38** and 4-fluorophenylmagnesium bromide **39** that gave rise to a mixture of 1-methyl-4-(4-fluorophenyl)piperidine-3-carboxylic acid methyl ester **40a/40b** as shown in Scheme 13.6. Base-mediated hydrolysis/equilibration of *cis/trans*-4-(4-fluorophenyl)piperidine-1,3-dicarboxylic acid dimethyl esters **40a/40b** afforded the corresponding *trans* carboxylic acid derivative **41**, which was treated with thionyl chloride to obtain acid chloride derivative **42**. Acid chloride **42** reacted with (−)-menthol to afford menthol ester **43**. The carbinol **44** was synthesized by means of reduction with the use of lithium aluminum hydride. The reaction of **44** with thionyl chloride and treatment of resultant with 3,4-methylenedioxyphenoxide (sesamol) afforded **45**. Demethylation of **45** was achieved by treatment of it with vinylchloroformate to afford venylurethane derivative **46**, which was further treated with HCl gas to yield chlorourethane derivative that was subsequently hydrolyzed in refluxing methanol to obtain API **4**.[12]

13.6 LEVETIRACETAM 5

Levetiracetam **5** is used for the management of various epileptic disorders since more than two decades and emerged as one of the most prescribed antiepileptic medicines.[13]

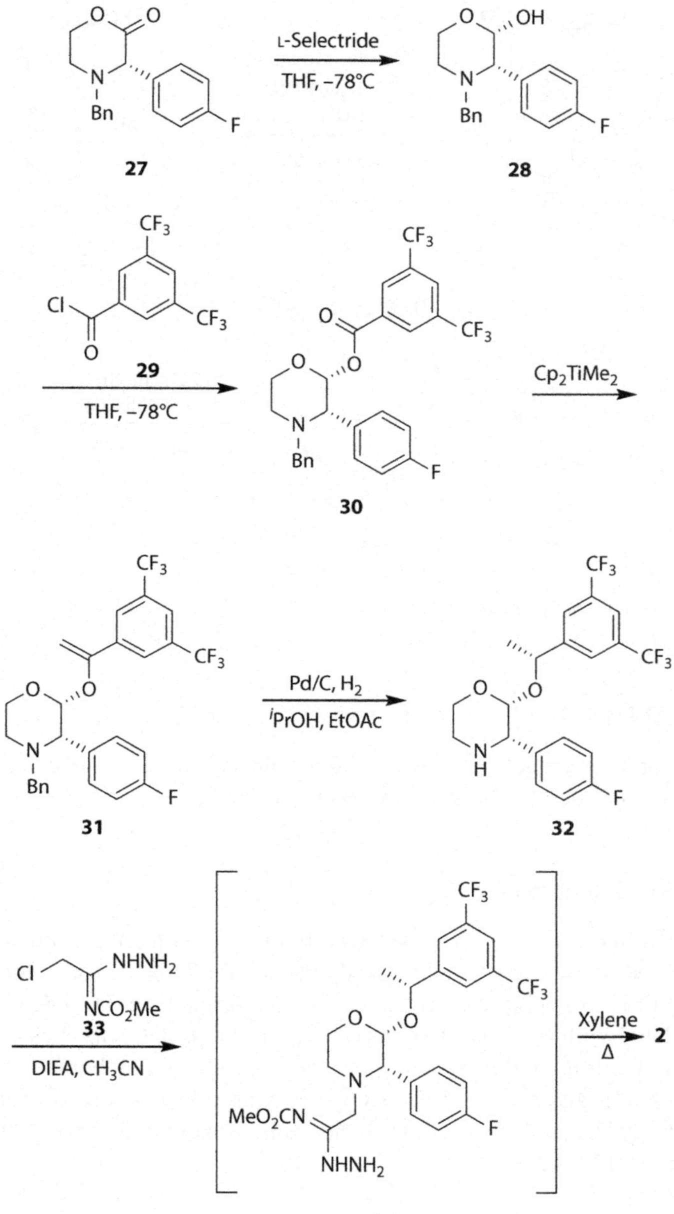

Chiral Medicines

SCHEME 13.4 Medicinal chemistry route for **2**.

13.6.1 Medicinal Chemistry Route

As shown in Scheme 13.7, medicinal chemistry synthesis of **5** involves protection–deprotection strategy (**49**).[14] In this case, synthesis starts with benzoyl protection of (*S*)-aminobutanoic acid **47** to afford corresponding *N*-benzoyl-protected (*S*)-aminobutyric acid **48**. After *N*-benzoyl deprotection and subsequent amidation, (*S*)-aminobutyramide **48** was obtained. Chemoselective butyrolactam ring formation, using 4-chlorobutyryl chloride and amide **48**, was performed to obtain levetiracetam **5**.

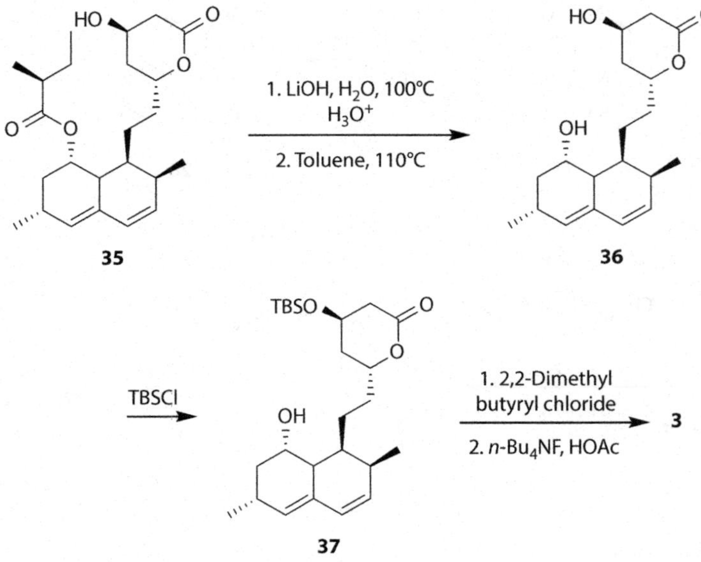

SCHEME 13.5 Medicinal chemistry route for **3**.

13.7 PREGABALIN 6

Pregabalin is used for the management of neuropathic pain and epilepsy—one of the fastest-growing drugs in the market with sales reaching nearly $6 billion in 2014.[14]

13.7.1 MEDICINAL CHEMISTRY ROUTE

Synthesis of pregabalin **6** was developed by Pfizer. In this particular approach, 4-methylpentanoic acid **50** was converted to the corresponding acid chloride in the presence of thionyl chloride and further reaction with Evans' chiral auxiliary to afford intermediate **51**. Thereafter, the intermediate **51** was alkylated with bromoacetate derivative to obtain the benzyl ester intermediate **52**. Oxidative removal of chiral auxiliary, bisulfate work-up, followed by borane reduction led to the access of corresponding alcohol **53**. The alcohol **53** was subsequently converted to the corresponding azide **54**. Hydrogenolysis and simultaneous hydrogenation of intermediate **54** afforded **6** as shown in Scheme 13.8 (**49**).[15]

13.8 SERTRALINE 7

Sertraline hydrochloride **7**, a selective inhibitor of serotonin reuptake, is used for the management of depression and other anxiety-related disorders (**52–54**).[16–18]

13.8.1 MEDICINAL CHEMISTRY ROUTE

In medicinal chemistry route (Scheme 13.9) (**53**),[17] 3,4-dichlorobenzophenone **55** on reaction with diethyl succinate **56** affords intermediate **57**. Intermediate **57** was further subjected to HBr–glacial acetic acid to obtain intermediate **58**. Thereafter, hydrogenation of **58** afforded intermediate **59**. Subsequently, acid chloride formation followed by Friedel–Crafts acylation afforded

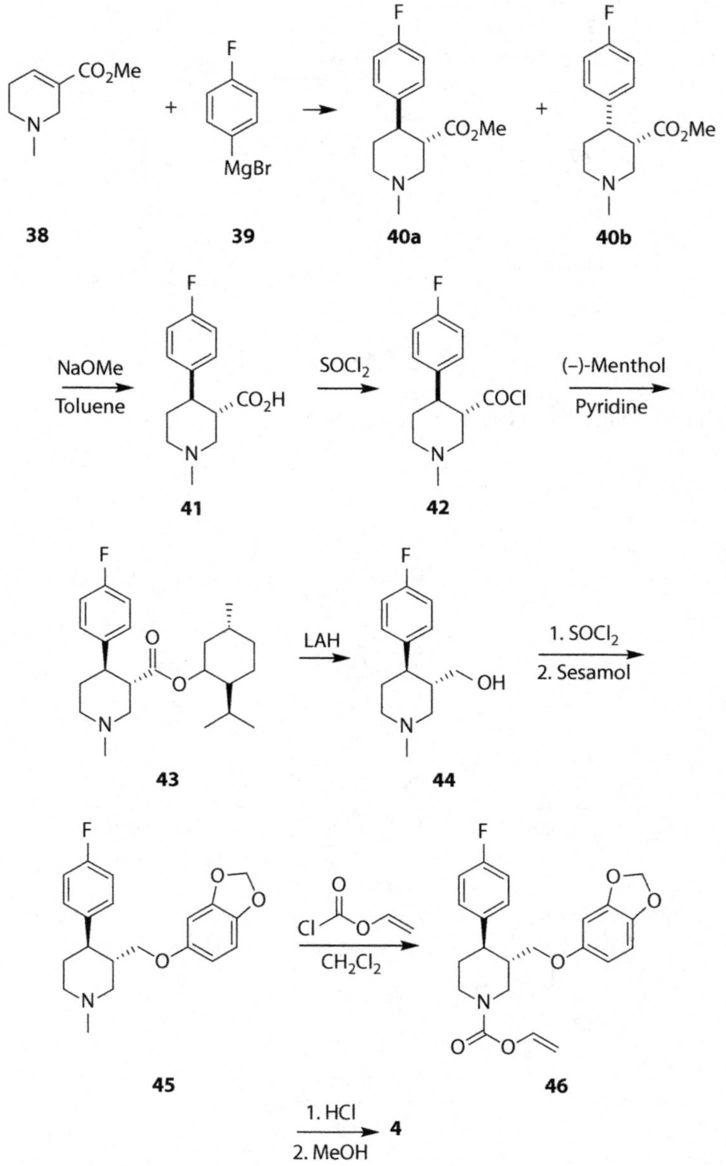

Chiral Medicines

SCHEME 13.6 Medicinal chemistry route for **4**.

intermediate **60**. Titanium tetrachloride–mediated dehydration of **60** offered Schiff base **61** followed by sodium borohydride–mediated reduction, which afforded intermediate **62**. This intermediate was dissolved in EtOAc and partitioned with 20% NaOH to obtain *cis*-racemate **63**, which was further resolved employing (D)-mandelic acid to afford **7**.

13.9 CONCLUSIONS

Atom economy and the concept of process mass intensity must be introduced during the early stages of synthetic route design to achieve sustainability in processes and products. A holistic consideration is mandatory for developing a greener route right from commercial viability

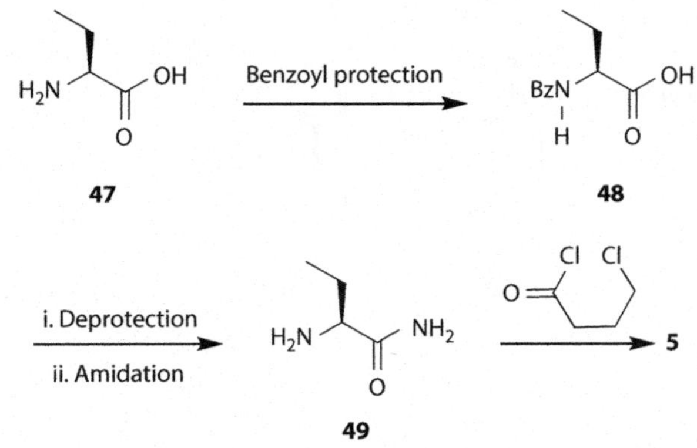

SCHEME 13.7 One of the medicinal chemistry routes for **5**.

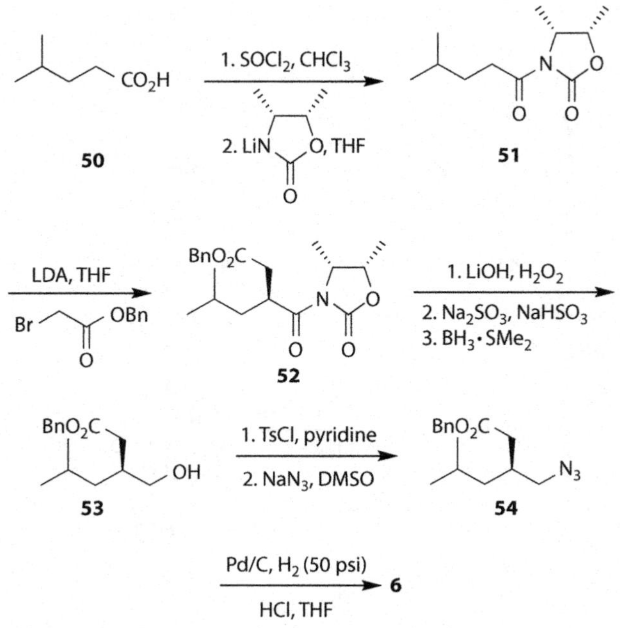

SCHEME 13.8 Medicinal chemistry route for **6**.

to intellectual property generation. The discovery approaches that have been discussed in this chapter are initial synthetic efforts. Moreover, these routes have been discovered 5–15 years back. A lot of advances have been made in the area of organic chemistry that led to the development of second- and third-generation synthesis (deliberately not discussed) of most of the medicines that are in use. Academic researcher may try to design fourth- and fifth-generation synthesis of medicines by considering 12 green chemistry principles that would signify successful incorporation of green principle in the education system. As a matter of fact, asymmetric synthesis, essential to the development of chiral medicines, is one of the enablers to improve atom economy and reduce the process mass intensity.

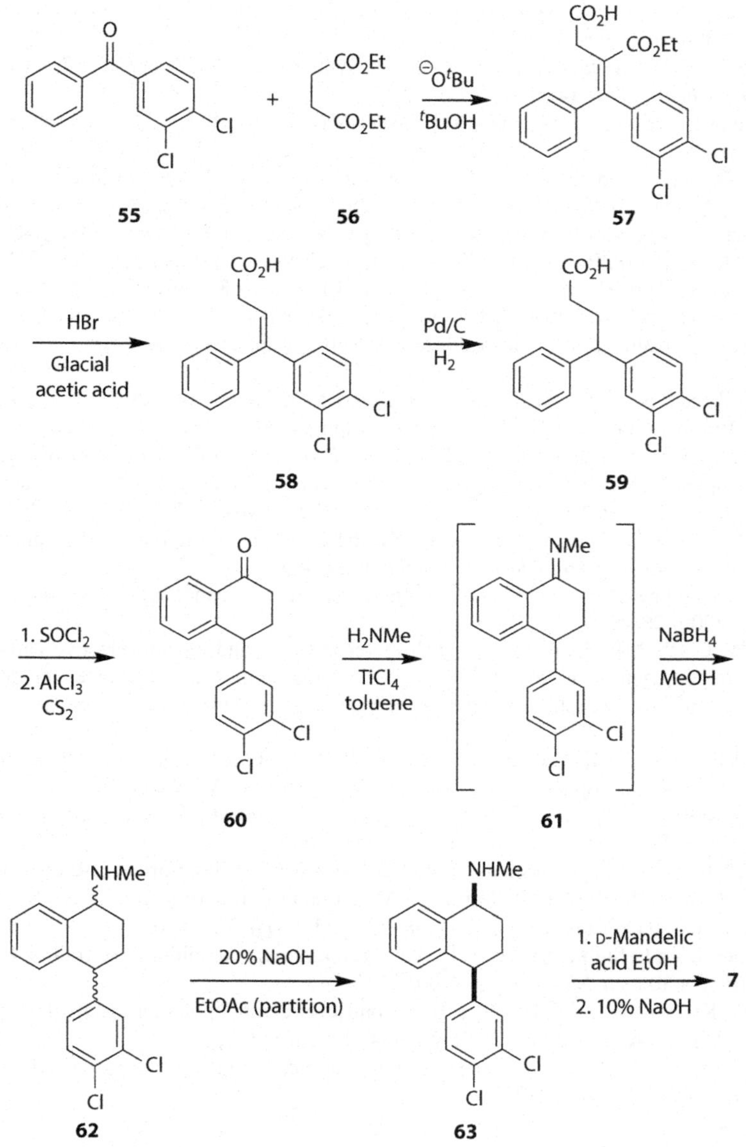

Chiral Medicines

SCHEME 13.9 Medicinal chemistry route for **7**.

REFERENCES

1. Drucker D, Easley C, Kirkpatrick P: Sitagliptin. *Nat Rev Drug Discov* (2007) **6**: 109–110.
2. (a) Holst JJ: Glucagon-like peptide 1 (GLP-1) a newly discovered GI hormone. *Gastroenterology* (1994) **107**: 1048–1055; (b) Drucker DJ: Glucagon-like peptides. *Diabetes* (1998) **47**: 159–169; (c) Deacon CF, Holst JJ, Carr RD: Glucagon-like peptide 1: A basis for new approaches to the management of diabetes. *Drugs Today* (1999) **35**: 159–170; (d) Livingston JN, Schoen WR: Glucagon and glucagon-like peptide-1. *Annu Rep Med Chem* (1999) **34**: 189–198.
3. (a) Evans DA, Britten TC, Ellman JA, Dorow RL: The asymmetric synthesis of α-amino acids. Electrophilic azidation of chiral imide enolates, a practical approach to the synthesis of (*R*)- and (*S*)-α-azido carboxylic acids. *J Am Chem Soc* (1990) **112**: 4011–4030; (b) Deng C, Groth U, Schoellkopf U: Enantioselective syntheses of (*R*)-amino acids using L-valine as chiral agent. *Angw Chem Int Ed Engl* (1981) **20**: 798–799.

4. Muller A, Vogt C, Sewald N: Synthesis of Fmoc-β-homoamino acids by ultrasound-promoted Wolff rearrangement. *Synthesis* (1988) **6**: 837–841.
5. Huynh-Dinh T, Sarfati RS, Gouyette C, Igolen J: Synthesis of *C*-nucleosides. 17. *s*-triazolo[4,3-α]pyrazines. *J Org Chem* (1979) **44**: 1028–1035.
6. Kim D, Wang L, Beconi M, Eiermann GJ, Fisher MH, He K, Hickey GJ, Kowalchick JE, Leiting B, Lyons K, Marsilio F, McCann ME, Patel RA, Petrov A, Scapin G, Patel SB, Roy R, Wu J, Wyvratt MJ, Zhang BB, Zhu L, Thornberry NA, Weber AE: (2*R*)-4-Oxo-4-[3-(trifluoromethyl)-5,6-dihydro[1,2,4]triazolo[4,3-*a*]pyrazin 7(8*H*)-yl]-1-(2,4,5-trifluorophenyl)butan-2-amine: A potent, orally active dipeptidyl peptidase IV inhibitor for the treatment of type 2 diabetes. *J Med Chem* (2005) **48**: 141–151.
7. Brands KMJ, Payack JF, Rosen JD, Nelson TD, Candelario A, Huffman MA, Zhao MM, Li J, Craig B, Song ZJ, Tschaen DM, Hansen K, Devine PN, Pye PJ, Rossen K, Dormer PG, Reamer RA, Welch CJ, Mathre DJ, Tsou NN, McNamara JM, Reider PJ: Efficient synthesis of NK$_1$ receptor antagonist aprepitant using a crystallization induced diastereoselective transformation. *J Am Chem Soc* (2003) **125**: 2129–2135.
8. Hale JJ, Mills SG, MacCoss M, Finke PE, Cascieri MA, Sadowski S, Ber E, Chicchi GG, Kurtz M, Metzger J, Eiremann G, Tsou NN, Tattersall D, Rupniak MJ, Williams AR, Rycroft W, Hargreaves R, MacIntyre DE: Structural optimization affording 2-(*R*)-(1-(*R*)-3,5-bis(trifluoromethyl)phenylethoxy)-3-(*S*)-(4-fluoro)phenyl-4-(3-oxo-1,2,4-triazol-5-yl)methylmorpholine, a potent, orally active, long acting morpholine acetal Human NK-1 receptor antagonist. *J Med Chem* (1998) **41**: 4607–4614.
9. Askin D, Verhoeven TR, Liu TMH, Shinkai I: Synthesis of synvinolin: Extremely high conversion alkylation of an ester enolate. *J Org Chem* (1991) **56**: 4929–4932.
10. Pollard DJ, Woodley JM: Biocatalysis for pharmaceutical intermediates: The future is now. *Trends Biotechnol* (2006) **25**: 66–73.
11. Johnson TA, Curtis MD, Beak P: Highly diastereoselective and enantioselective carbon-carbon bond formations in conjugate additions of lithiated N-Boc allylamines to nitroalkenes: enantioselective synthesis of 3,4- and 3,4,5-substituted piperidines including (−)-paroxetine. *J Am Chem Soc* (2001) **123**: 1004–1005.
12. SmithKline Beecham PLC, Crowe D, Jones DA, Ward N: Process for the preparation of 1-methyl-3-carbomethoxy-4-(4′-fluorophenyl)-piperidine. WO-2001017966 A1(2001).
13. Hauser WA: *Overview: Epidemiology, Pathology, and Genetics in Epilepsy*. A comprehensive textbook. Lippincott-Raven Publishers, Philadelphia, PA (1997), pp. 11–13.
14. Stebbins S: http://247wallst.com/special-report/2016/04/26/top-selling-drugs-in-the-world/2/.
15. Yuen P-W, Kanter GD, Taylor CP, Vartanian MG: Enantioselective synthesis of PD144723: A potent stereospecific anticonvulsant. *Bioorg Med Chem Lett* (1994) **4**: 823–826.
16. MacQueen G, Born L, Steiner M: The selective serotonin reuptake inhibitor sertraline: Its profile and use in psychiatric disorders. *CNS Drug Rev* (2001) **7**: 1–24.
17. Welch WM, Kraska AR, Sarges R, Koe BK: Nontricyclic antidepressant agents derived from *cis*- and *trans*-1-amino-4- aryltetralins. *J Med Chem* (1984) **27**: 1508–1515.
18. McRae AL, Brady KT: Review of sertraline and its clinical application in psychiatric disorders. *Expert Opin Pharmacother* (2001) **2**: 883–892.

14 Drug Delivery Systems

Ahindra Nag

CONTENTS

14.1 Introduction 461
14.2 Classification of Drugs 462
14.3 Prodrug 462
14.4 Mode of Action of Drug 464
14.5 Sites of Drug Action 464
14.6 Goal of Drug Delivery 465
    14.6.1 Physiochemical Properties 465
    14.6.2 Biological Properties 465
    14.6.3 Absorption 465
    14.6.4 Distribution 466
    14.6.5 Metabolism 466
    14.6.6 Elimination 467
14.7 Safety Considerations 467
14.8 Drug Distribution 468
    14.8.1 Oral Administration 468
    14.8.2 Nasal Administration 468
    14.8.3 Rectal Administration 469
    14.8.4 Transdermal Delivery Systems 470
    14.8.5 Targeted Drug Delivery System 470
        14.8.5.1 Nanotechnology 471
        14.8.5.2 Liposomes 472
        14.8.5.3 Resealed Erythrocytes 472
        14.8.5.4 Injection 473
        14.8.5.5 Biological Drugs 473
Further Reading 474

14.1 INTRODUCTION

Drugs are compounds that cause physiological change in the body, and the dose level of a drug determines whether it will act as medicine or poison. The beneficial dose level of a drug at a lower dose versus at a higher dose of harmful effect is a measure of the therapeutic index of the drug. A **brand name** drug is a medicine that's discovered, developed, and marketed by a pharmaceutical company. At this point the drug has two names: a generic name, which is the drug's common scientific name, and a brand name to make it stand out in the marketplace. **Generic drugs** have the same active ingredients as brand name drugs already approved by the Food and Drug Administration (FDA). The choice of drug depends on the lowest toxicity and the widest therapeutic range of the available agents. Generally, chirality greatly influences a drug's pharmaceutical properties, and nearly two-thirds of drugs on today's market are chiral drugs.

14.2 CLASSIFICATION OF DRUGS

Traditionally, drugs were obtained through extraction from medicinal plants but also by organic synthesis. To prepare a successful product, a judicious choice of drug is essential. Drugs are classified on the basis of their origin.*

1. **Drugs from natural origin**, which are extracted from plants or herbs such as vincristine, digoxin, Taxol, reserpine, ephedrine, etc.
2. **Drugs from animal sources**, such as gonadotropins, insulin, hormones, and enzymes.
3. **Drugs derived from microbial origin**, such as penicillin, streptomycin, and tetracycline.
4. **Drugs derived from minerals**, such as medicated springs; hot water containing iron, calcium, magnesium, sodium, potassium, etc.
5. **Drugs derived from chemical synthesis** in which drugs are chemically pure. The World Health Organization keeps a list of essential medicines, such as
    a. Antipyretics: reducing fever (pyrexia/pyresis)
    b. Analgesics: reducing pain (painkillers)
    c. Antimalarial drugs: treating malaria
    d. Antibiotics: inhibiting germ growth
    e. Antiseptics: preventing germ growth near burns, cuts, and wounds
    f. Mood stabilizers: lithium and valpromide
    g. Hormone replacements: Premarin
    h. Oral contraceptives: Enovid, "biphasic" pill, and "triphasic" pill
    i. Stimulants: methylphenidate and amphetamine
    j. Tranquilizers: meprobamate, chlorpromazine, reserpine, chlordiazepoxide, diazepam, and alprazolam
6. **Drugs derived by biotechnology and genetic engineering**, such as different types of cancer, diabetes mellitus, infectious diseases (e.g., AIDS virus/HIV), as well as cardiovascular, neurological, respiratory, and autoimmune diseases, among others. The pharmaceutical industry has used different technologies to obtain new and promising active ingredients, as exemplified by the fermentation technique, recombinant DNA technique, and hybridoma technique, especially for different types of vaccines.
7. **Drugs derived from radioactive substances** that are used in the field of nuclear medicine as radioactive tracers in medical imaging and in therapy for many diseases (e.g., brachytherapy).

14.3 PRODRUG

It relates to biologically inert derivatives of drug molecules that undergo an enzymatic and chemical conversion in vivo to release the pharmacologically active parent drug, which was first introduced in 1958 by Adrien Albert, a leading authority in the development of medicinal chemistry in Australia. It is estimated that about 10% of the drugs approved worldwide can be classified as prodrugs. The basic applications of the prodrug strategy include the ability to improve oral absorption and aqueous solubility, lipophilicity, and active transport, as well as achieve site-selective delivery. As for example, esters are susceptible to metabolic hydrolysis in vivo by esterase, a metabolic enzyme, and they are used deliberately to mask polar functional groups such as carboxylic acid, alcohol, and phenol, so in the blood supply the ester is hydrolyzed to release the active drug as prodrug strategy (Figure 14.1).

---

* *Source*: U.S. Federal Food, Drug, and Cosmetic Act, SEC. 210., (g)(1)(B). Accessed August 17, 2008.

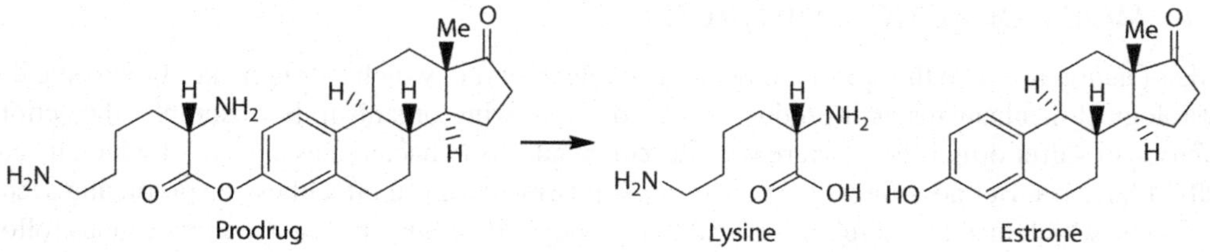

Drug Delivery Systems

FIGURE 14.1 Metabolic hydrolysis of lysine ester of estrone as prodrug to produce estrone to improve water solubility and nontoxic compound (lysine).

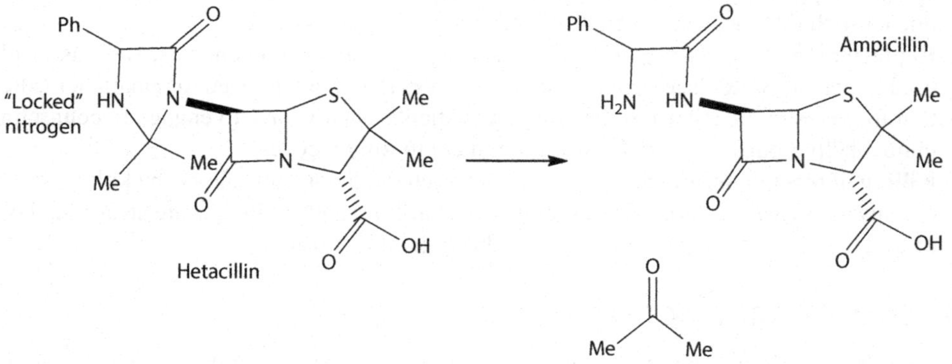

FIGURE 14.2 Chemical stability of ampicillin by prodrug.

Prodrugs alter toxicity and elimination and improve water solubility and chemical stability. In Hetacillin (a prodrug) nitrogen is locked up for decomposing ampicillin in aqueous solution (Figure 14.2).

The activity of drug can be prolonged by the application of prodrugs that are susceptible to pH, or chemical degradation can be effective in targeting drugs or increasing stability in solution prior to injection. Reducing polarity of prodrugs can be increased by introducing N-methyl group or activated by light, which is the basis for photodynamic theory. There are two classes of prodrugs: (1) **carrier-linked prodrugs** and (2) **bioprecursor prodrugs**.

1. **Carrier-linked prodrugs**: In the carrier-linked prodrugs, the drug is temporarily linked to a carrier that should be nonimmunogenic, easy to synthesize at a low cost, stable under administration, and can undergo biodegradation to nonactive metabolites. Examples of carrier-linked prodrugs include sulfapyridine (5-amino-salicylic acid), indomethacin (paracetamol), L-DOPA (entacapone), gabapentin (pregabalin), cytarabine (5-fluoro-uracil), dexamethasone triamcinolone (5-fluorouracil), etc.
2. **Bioprecursor prodrugs**: These prodrugs do not contain a carrier but result from molecular modifications of the active compound itself. The bioprecursor prodrug is transformed metabolically or chemically by hydration (e.g., lactones, such as some statins), oxidation (e.g., dexpanthenol, nabumetone), or reduction (e.g., sulindac, platinum(IV) complexes).

Two major aspects of drug delivery systems are *spatial placement*, which relates to targeting the drug to a specific organ or tissue, and *temporal delivery systems*, which refers to controlling the rate of drug delivery to the target organ or tissue. Drug delivery technologies modify drug release profile, absorption, distribution, and elimination for the benefit of improving product efficacy and safety, as well as patient convenience and compliance.

14.4 MODE OF ACTION OF DRUG

Mode of action refers to the specific biochemical interaction in which a drug makes biochemical and physiological mechanisms as it produces a response in living organism. For example, the action of antibiotic or sulfur drug is to interfere with the cell synthesis in bacteria, as a result of which it causes death of bacteria. But no drug will produce single pharmacological effect besides producing secondary effects, which may be useful or harmful to the body. The mode of drug action may be as follows:

1. **Stimulation and depression**: Drugs act by stimulating or depressing normal physiological functions. Stimulation produces activity but depression is the opposite of it.
2. **Irritation**: Inflammation or painful reaction to cell-lining damage or pain induced by some allergic response due to exposure of some allergens. Certain drugs, such as senna or castor oil, act such that they cause irritation in the body.
3. **Replacement**: Drugs serve as replacement of essential body chemicals, such as insulin used in diabetics. Replacement therapy is not short term but will need to remain in medication for years in order for it to be effective. Patients usually have to engage in counseling, mutual help groups, or other forms of treatment to fully recover.
4. **Killing of foreign organisms**: Drugs can act as chemotherapeutic agents by killing foreign organisms such as bacteria, viruses, or worms. Vaccines and antibiotics are an artificial way of tricking the immune system to prevent bacterial infections.

14.5 SITES OF DRUG ACTION

The site of drug action may be of three types: (1) enzyme inhibition, (2) drug–receptor interaction, and (3) nonspecific interactions.

1. **Enzyme inhibition**: An enzyme inhibitor is a molecule that binds to an enzyme within the cell and decreases its activity. Many drugs are enzyme inhibitors, such as antibiotics, acetylcholinesterase agents, monoamine oxidase inhibitors, and diuretics.
2. **Drug–receptor interaction**: A drug's ability to affect a given receptor is related to the drug's affinity, and its activity is determined by its chemical structure. Molecules (e.g., drugs, hormones, neurotransmitters) that bind to a receptor are called ligands. The binding can be specific and reversible. The pharmacological effect is also determined by the duration of time that the drug–receptor complex persists (residence time). The lifetime of the drug–receptor complex is affected by conformational changes of the complex that control the rate of drug association and dissociation from the target. There are **agonists-activated receptors** that produce the desired response, and they can differ in both affinity and efficacy of receptors. Higher-efficacy agonists are full agonists because they elicit maximal effects. Examples of drugs are phenylephrine, isoproterenol, and benzodiazepines. Generally, reversible antagonists dissociate from their receptor, whereas irreversible antagonists form a stable, permanent chemical bond with their receptor. **Antagonists-activated receptors** prevent receptor activation by decreasing cellular function, and thus antagonists are essentially zero-efficacy drugs. They may be two types: **competitive antagonism**, where binding of the antagonist to the receptor prevents binding of the agonist to the receptor, or **noncompetitive antagonism**, where the agonist and antagonist can be bound simultaneously, but antagonist binding reduces or prevents the action of the agonist.
3. **Nonspecific interactions**: Generally, nonspecific interactions are not directed against a particular agent but rather have a general effect. Many drugs interact nonspecifically with a wide variety of tissue components. Localization of a drug in a particular organ, subcellular particle, or macromolecule does not necessarily imply a site of action there. Drugs may act outside of the cell membrane by chemical interactions, such as neutralization of stomach acids by antacids.

Drug Delivery Systems

14.6 GOAL OF DRUG DELIVERY

The goal of drug delivery system approaches formulations, technologies, and systems for transporting a pharmaceutical compound in the body that is required to achieve safety for its desired therapeutic effect. It may involve scientific site targeting within the body or it might involve facilitating systemic pharmacokinetics, which is typically concerned with both quantity and duration of drug presence.

The undesired adverse effect of a drug is dependent on the physical and chemical properties of the drug at the site of action, which in turn depends upon the dosage form and the extent of absorption and metabolic conversion of the drug (biological properties) at the site of action. Drug selections based on the physicochemical and biological properties of the drug are as follows.

14.6.1 Physiochemical Properties

The choice of drugs depends on solubility, crystallinity, molecular weight <400, polarity, and melting point <200. Water-soluble molecules and ions below 50 Da enter the cell through aqueous-filled channels, whereas those of larger size are restricted unless a specialized transport size exists for them. Aqueous solubility of the drug influences its dissolution rate and the absorptive power of the drug. Generally, water-soluble drugs will reside in the blood and fat-soluble drugs will reside in the cell membrane, adipose tissue, and other fat-rich areas. According to pH-partition hypothesis, basic drugs present in the blood readily enter into acidic tissues and intercellular fluids (pH 7.4), whereas acidic drugs concentrate in the relatively more alkaline fluids. Acidic drugs commonly bind to albumin, while basic drugs often bind to α_1-acid glycoprotein and lipoprotein. Many endogenous substances, steroids, vitamins, and metal ions are bound to globulins. Partition coefficient log P (octanol–water) between 1.0 and 4 must be considered during drug selections. Drug stability is an important factor because a drug in the solid state undergoes degradation at a much slower rate than a drug in solution. A drug that is unstable in the gastrointestinal tract can be improved by using a sustained-release drug delivery system. Protein-binding characteristics of a drug can play an important role in the therapeutic effect regardless of the dosage form because the distribution of a drug in the extravascular space is governed by the equilibrium process of dissociation of the drug from the protein, and the drug protein can serve therefore as a reservoir in the vascular space for sustained drug release to extravascular tissues. Drugs that are charged bind to serum proteins and protein-bound drugs form macromolecular complexes that cannot cross the biological membranes and remain confined in the bloodstream.

14.6.2 Biological Properties

Among the various properties of chemical compounds, biological properties play a crucial role because biological activity depends critically on the fulfillment of the ADME criteria (absorption, distribution, metabolism, and excretion). To be an effective drug, a compound not only must be active against a target but also possess the appropriate ADME properties necessary to make it suitable for use as a drug. Some drugs interact with the food material, such as tetracycline may bind to the calcium ions of milk, which inhibit absorption of drug.

14.6.3 Absorption

For a definite drug delivery system, the rate, extent, and uniformity of absorption of a drug are significant considerations. A rapid release of absorption of the drug relative to its release is essential to making the system successful.

14.6.4 Distribution

It is a process whereby an absorbed drug moves away from the site of absorption to other areas of the body. Distribution rate and extent depend on (1) the chemical structure of drug and drug solubility, (2) rate of blood flow, (3) ease of transport through the membrane, (4) drug binding to plasma and tissue proteins irreversibly, and (5) elimination process.

The volume of distribution (VD), also known as apparent volume of distribution, is a pharmacological term used to quantify the distribution of a medication between plasma and the rest of the body after oral or parenteral dosing. It is defined as the volume in which the amount of drug would need to be uniformly distributed to produce the observed blood concentration:

$$V_D = \frac{\text{Total amount of drug in the body at equilibrium}}{\text{Plasma drug concentration}}$$

Drugs that bind strongly to plasma protein have lower volume of distribution than drugs that do not bind to plasma protein (↑ Protein binding = ↓ V_D). For drugs that obey a one-compartment model, the apparent volume of distribution is

$$V = \frac{\text{Dose}}{C_0}$$

where C_0 is the initial drug concentration.

The one-compartment model assumption is that there is a rapid equilibrium in drug concentrations throughout the body; however, it does not mean the concentration is same throughout the body.

If a drug has a large molecular weight or binds extensively to plasma proteins, it is too large to move out through the endothelial slit junctions of the capillaries, thus restricted within vascular compartments, for example, warfarin. When two drugs are given, each with a high affinity for plasma protein, they compete for the available binding sites. Cholestyramine, used to lower cholesterol, can bind to warfarin, preventing thrombosis and thromboembolism, and also bind to the thyroid drug levothyroxine sodium.

The following conditions will affect the distribution of drug flow (1) decrease blood flow, (2) decrease lipid–water partition coefficient, (3) decrease size of the organ, and (4) decrease plasma protein binding.

14.6.5 Metabolism

Drug metabolism is a process which may result in pharmacologically active, inactive, or toxic metabolites, which involves the enzymatic conversion of therapeutically important chemical species to a new molecule inside the human body. It includes the following factors: (1) termination of drug action because only water-soluble substances undergo excretion, whereas lipid-soluble substances are passively reabsorbed from renal or extrarenal excretory sites into the blood by virtue of their lipophilicity; (2) activation of prodrug; (3) bioactivation and toxication; and (4) carcinogenesis and teratogenesis.

Drug metabolism is the sum total of all the enzyme-catalyzed reactions that alter the physicochemical properties of foreign chemicals (drug/xenobiotics) from those that favor absorption across biological membranes (lipophilicity) to those favoring elimination in the urine or bile (hydrophilicity). Drug metabolism can be categorized in two ways: (1) Phase I reaction, which involves hydrolysis, oxidation, and reduction, and (2) Phase II reaction, which involves conjugation.

Drug Delivery Systems

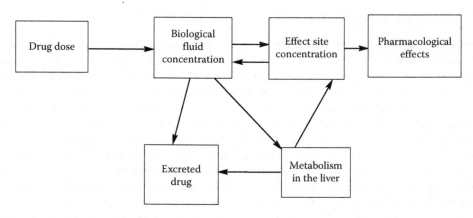

FIGURE 14.3 ADME criteria of drug.

14.6.6 ELIMINATION

It is the ability of organs like the kidney or liver to clear the drug from the bloodstream. The volume of fluid is completely cleared of any drug quantitatively by its biological half-life period $t_{1/2}$. This half-life of a drug is related to its apparent volume of distribution (V) and its systematic clearance:

$$t_{1/2} = 0.693 \ V/Cl = 0.693 \ AUC/dose$$

Clearance (Cl) is related to AUC dose where AUC is defined as the area under the curve in a plot of concentration of the drug in the blood plasma against time. Typically, the area is computed starting at the time the drug is administered and ending when the concentration in plasma is negligible.

Generally, clearance varies with the body weight and also with the degree of protein binding. Elimination by the kidney is major and metabolism is minor. It is a significant site for biotransformation activities for both Phase I and Phase II metabolism. The renal cortex and outer and inner medulla exhibit different profiles of drug metabolism. Most metabolizing enzymes are localized mainly in the proximal tubules, although various enzymes are distributed in all segments of the nephron. Metabolism in the liver is important as (1) Phase 1 and II reactions (2) change a lipid-soluble molecule to a more water-soluble molecule to excrete from the kidney, (3) with the possibility of active metabolites with same or different properties as the parent molecule. All people respond to a similar dose in a same manner as it depends on the following factors: (1) usually immunologically sensitized responses and (2) pharmacodynamic factors contribute to different physiological responses to the same drug concentration; and (3) pharmacokinetic factors contribute to different concentrations of the drug at the target area (Figure 14.3).

14.7 SAFETY CONSIDERATIONS

The most widely used measure of the margin of safety of a drug is its therapeutic index, TI:

$$TI = \frac{TD_{50}}{ED_{50}}$$

where
  TD_{50} is the median toxic dose
  ED_{50} is the median effective dose

The value of TI varies from as little as unity, where ED also produces toxic symptoms, to several thousands. For very "potent" drugs, whose therapeutic concentration range is narrow, the value of TI is small. The larger the value of TI, the safer the drug.

14.8 DRUG DISTRIBUTION

Generally, drug delivery is a concept heavily integrated with dosage form and route of administration, that is, through noninvasive peroral (through the mouth), topical (skin), transmucosal (nasal, buccal/sublingual, vaginal, ocular, and rectal), and inhalation routes. But vaccine- and gene-based drugs may not be delivered using these routes because they might be susceptible to enzymatic degradation or cannot be absorbed into the systemic circulation efficiently due to molecular size and charge issues to be therapeutically effective, and therefore they have to be delivered by injection or a nanoneedle array. Nasal and pulmonary routes of drug delivery are gaining increasing importance because these routes provide promising alternatives to parenteral drug delivery, particularly for peptide and protein therapeutics. Similar developments with other compounds have produced a plethora of new devices, concepts, and techniques that have together been termed controlled release technology (CRT), such as nasal and buccal aerosol sprays, drug-impregnated lozenges, encapsulated cells, oral soft gels, etc. Hence, drug delivery system depends on drug release profile, absorption, distribution, and elimination for the benefit of improving product efficacy and safety as well as patient convenience.

Many medications such as peptide-, protein-, vaccine-, antibody-, and gene-based drugs, in general, may not be delivered using these routes because they might be susceptible to enzymatic degradation or cannot be absorbed into the systemic circulation efficiently due to molecular size and charge issues to be therapeutically effective. For this reason, these drugs have to be delivered by injection or a nanoneedle array.

Drug delivery into the body is commonly achieved in the following ways: (1) oral (through swallowing), (2) nasal, (3) rectal, (4) transdermal (through skin), and (5) using a targeted drug delivery system: (a) nanotechnology, (b) liposomes, (c) injection.

14.8.1 Oral Administration

Among all these administrations, oral administration (by swallowing through mouth) is very well accepted to patients. Drugs can be taken as pills, capsules, or solutions, which is a more preferred form. The advantages of oral delivery are as follows:

1. Controls total dose as low value and reduces dosing frequency
2. Understands better the patients' acceptance and compliance, that is, vomiting or uneasiness
3. Reduces gastroenterology side effects and less fluctuation at plasma drug level
4. Allows drugs to be distributed more uniformly
5. Is easy to produce and is of low cost

The disadvantages of the process are (1) stability problem of drug, (2) dose dumping, and (3) reduction of potential for accurate dose adjustment. The design of oral drug is done by two ways:

1. Continuous release system where drug is released continuously over an extended period of time
2. Pulsatile release system, which is characterized by a time period of no release, followed by a rapid and complete drug release

14.8.2 Nasal Administration

Nasal drug administration has been used as an alternative route for drug administration in recent years. About 2% of the overall drug delivery is administered through the nasal route. The drug is absorbed through the cell linings of the respiratory tract and ultimately moves into the blood supply. The various delivery systems are (1) nasal spray, (2) nasal drops, (3) aerosol spray, (4) metered dose nebulizer, (5) saturated cotton pledget, (6) insufflators, (7) microspheres, and (8) mucosal atomizer device.

Drug Delivery Systems

Nasal sprays such as salbutamol, ipratropium, and montelukast and a large number of inhalational anesthetic agents are being used commonly. Recently, antimigraine drugs, available by the trade names of Imitrex (sumatriptan), Zomig (zolmitriptan), Migranal (dihydroergotamine), and the OTC nasal spray Sinol-M, are also administered through the nasal route. Some of the drugs are absorbed through inhalation or smoking, such as cocaine, marijuana, heroin, nicotine, etc. Peptide drug is used for the treatment of diabetes as for both nasal and oral administration. Recently, the upper part of the nasal cavity, as high as the cribriform plate, has been proposed for drug delivery to the brain.

The advantages of this process are as follows:

1. Rapid absorption of the process.
2. It is parenteral drug administration, which means drug can be administered by non-oral means as well. It is generally interpreted as relating to injecting directly into the body, bypassing the skin and mucous membrane.
3. Ease of self-administration such as painless, noninvasive, and needle-free administration mode.
4. Lower dose and less side effects.
5. Quicker onset of pharmacological activity. It treats macular edema in central retinal vein occlusion i.e intravitreal Bevacizumab (Avastin), and drug antidotes such as hydroxocobalamin (antidote to cyanide poisoning) are being developed as intranasal medications.
6. Deposition of an active compound in the nasal cavity results in its degradation through "first-pass" metabolism.

The disadvantages of the process are as follows:

1. Once administered, removal of the therapeutic agent from the site of absorption is difficult.
2. Environmental conditions such as infections, smaller absorption, smaller surface area of nasal cavity, and intersubject variability can lead to inconsistent absorption.
3. Short time span is available for absorption due to rapid clearance.
4. There is a risk of local side effects and irreversible damage of the cilia on the nasal mucosa as constituent added to the dosage form because the histological toxicity of absorption enhancers used in the nasal drug delivery system is not clearly established.

14.8.3 Rectal Administration

When the drug is destroyed in the GI tract or expelled through vomiting or when the patient is unconscious or incapable of swallowing oral formulations, the drug is administered through the rectum. Many oral forms of medications can be crushed and suspended in water to be given via a rectal catheter. The process has the following advantages:

1. It has a faster onset and shorter duration than the oral route.
2. It induces less nausea and prevents loss due to vomiting.
3. It has higher bioavailability, that is, the drug will reach the circulatory system with significantly less alteration and in greater concentrations.
4. There is less chance of anxiety or bacterial infections.
5. The rectal route enables a rapid, safe, and lower-cost alternative to administration of medication.

The major disadvantages of the process are as follows:

1. Patients may suffer Rectal irritation as the process is not preferred by many and is found inconvenient.
2. Rectal absorption of most drugs is frequently erratic and unpredictable as some suppositories "leak" or are expelled after insertion.

14.8.4 TRANSDERMAL DELIVERY SYSTEMS

It is a route of administration that employs a skin portal to the systemic circulation at a predetermined rate and maintains clinically effective concentration over a prolonged period of time. Examples of this system include transdermal patches for medicine delivery and transdermal implants for aesthetic purposes. The drug must traverse three layers: the stratum corneum (the toughest barrier), the epidermis, and the dermis. The advantages of this process are as follows:

1. Easy to apply and monitor, that is, maintenance of constant drug level in the systemic circulation within the therapeutic window.
2. Reduction of intra- and interpatient variability.
3. First-pass metabolism in the liver and GI tract can be avoided.
4. Extension of duration of drug action by a single administration thereby reducing the frequency of dosing.
5. Termination of drugs in emergencies simply by removing the patches.

The disadvantages of the process are as follows:

1. Poor diffusion of large molecules.
2. Tolerance-inducing compounds are not good choices for this mode of administration.
3. Skin irritation.
4. The system is limited to potent and very potent drug molecules, typically those requiring a daily dose on the order of 10 mg or less.
5. The pharmacokinetic and pharmacodynamic characteristics of the drug must be such that the relatively sustained and slow input provided by the transdermal delivery systems produces the desirable effect.
6. The process is more expensive than oral drug administration.

14.8.5 TARGETED DRUG DELIVERY SYSTEM

This system, sometimes called **smart drug delivery**, is a method of drug delivery system where medicament is selectively targeted or delivered only to its site of action or absorption and not to the nontargeted organs or tissues or cells. In traditional drug delivery systems, only a small portion of medication reaches the affected organs, as in chemotherapy, 99% of the drugs do not reach the tumor site. A targeted drug delivery system has the following objectives:

1. Treatment or prevention of diseases
2. Increase in pharmaceutical drug instability in conventional dosage form solubility
3. High membrane binding
4. Large volume distribution
5. Better therapeutic index
6. Improving efficacy and reducing side effects

The disadvantages of the process are as follows:

1. Expensive
2. Yield comparatively very less
3. Stability issues important for chemical, physical, and biological
4. Technical skill required

The processes of a targeted drug delivery system are as follows.

14.8.5.1 Nanotechnology

This is known as colloidal drug delivery system or nanomedicine. A nanoparticle is a particle that contains dispersed drug within a diameter of 200–500 nm, which is similar to that of most biological molecules and structures. Hence, nanomaterials can be useful both in vivo and in vitro, that is, they can be easily administered through intravenous injection. The advantages of this process are as follows:

1. Smaller size and higher surface area
2. Prevention of drug from biological degradation
3. Low dose required
4. Uniform delivery of drug
5. Effective targeting
6. Patient compliance
7. Can be administered via different routes
8. Ability to incorporate hydrophilic and hydrophobic drug molecules
9. Low-cost research compared to that for discovery of new chemical entities
10. Product life extension
11. Minimizing drug usage significantly reduces the effective cost of the drug, which would give financial benefit to the patients
12. Reduces toxicity, for example, application of platinum-bearing nanoparticles to cancer tumors reduces the toxic state of platinum in which it can kill cancer cells

The disadvantages of nanoparticles are as follows:

1. Physical handling of nanoparticles is difficult in liquid and dry forms.
2. Small size and large surface area can lead to particle aggregation.
3. Limited drug loading.
4. Toxic metabolites (when some substances are metabolized, the resulting metabolites are toxic to the body and can cause fatal results) may form.

The materials used for the preparation of nanoparticles should be sterile, nontoxic, and biodegradable. There are two types of nanoparticles:

1. **Nanospheres**, which are matrix-type structures in which the drug is dispersed
2. **Nanocapsules**, which are membrane wall structure with a polymeric system containing drug

Characterization of nanoparticles can be done by the following:

1. Particle size and specific surface area
2. Density
3. Molecular weight
4. Structure and crystallinity
5. Surface charge, surface hydrophobicity, and electronic mobility
6. Nanoparticles yield
7. In vitro release
8. Drug entrapment efficiency, which is defined as

$$\text{Drug entrapment efficiency}\% = \frac{\text{Mass of drug in nanoparticles}}{\text{Mass of drug in formulation}}$$

The behavior of nanoparticles in vivo is the same as exhibited by other colloidal systems.

14.8.5.2 Liposomes

Liposomes are concentrated microscopic spheres made of bilayered vesicles, predominately of phospholipids, and can vary in size from 0.025 to 2.5 μm. Complementary activation by liposomes becomes a clinical problems as they directly encounter large amount of C proteins. Liposomes encapsulation technology refers to entrapment of drugs in capsules, which is a recent drug delivery technology, to administer drugs to a target cell, such as tumor cell, or tissue with or without expression of target recognition molecules on lipid membrane.

The advantages of liposome drug delivery are as follows:

1. Liposomes are biocompatible, flexible, and nonimmunogenic for systematic and nonsystematic administration. They are suitable to administer via various routes.
2. Liposome stability is increased via encapsulation, hence suitable for controlled release of hydrophilic, hydrophobic, and amphipathic drugs.
3. Liposomes increase the efficacy and therapeutic index of drug for antiviral, antituberculosis, antifungal, antimicrobial, vaccine, and gene therapy.
4. Liposomes reduce site avoidance effect and also help to reduce the exposure of sensitive tissues to toxic drugs. Liposomal anthracyclines have achieved highly efficient drug encapsulation, resulting in significant anticancer activity with reduced cardiotoxicity.
5. Liposome constructs are being developed for the delivery of other drugs.

The disadvantages of liposomes are as follows:

1. Short half-life
2. Low solubility
3. Leakage and fusion of encapsulated drug; sometimes phospholipid undergoes oxidation and hydrolysis reactions
4. High production cost

Liposomes are characterized by physical (specially sizing of liposomes), chemical, and biological parameters. Loading of drugs into liposomes has proved to be a measure of their utility. If there is a poor loading, there is a great loss of the active drug and the use of liposomes as the pharmaceutical vehicle becomes uneconomical.

The method for preparing liposomes is as follows:

1. First of all, liposomes are made soluble in organic solvents, specially in ether or ethyl acetate, then filtered, and the solvent evaporated in a rotary evaporator.
2. They are then dispersed in aqueous medium.
3. After analysis of the purified product liposomes are loaded by **passive or active loading techniques**.

Anti-HER2 immunoliposomes, in which mAb fragments are conjugated to liposomes, have been developed with either Fab′ or scFv fragments linked to long-circulating liposomes. Anti-HER2 immunoliposomes loaded with doxorubicin displayed potent and selective anticancer activity against HER2-overexpressing tumors. Immunoliposomes also appear to be nonimmunogenic and capable of long circulation even with repeated administration.

14.8.5.3 Resealed Erythrocytes

This is one of the methods of targeted delivery system. When erythrocytes are suspended in a hypotonic medium, they swell to about one and a half times their normal size. As a result, the membrane ruptures resulting in the formation of pores with diameters of 200–500 Å. The pores allow equilibration of the intracellular and extracellular solutions. If the ionic strength of the medium then is adjusted to its tonicity and the cells are incubated at 37°, the pores will close and cause the erythrocytes to "reseal." Using this technique with a drug present in the extracellular solution, it is possible to entrap up to 40% of the drug inside the resealed erythrocyte and use this system for targeted delivery via intravenous injection.

This system has limited application as the ability of resealed erythrocytes to deliver drug to the liver or spleen can be viewed as a disadvantage in that other organs and tissues are inaccessible. Another disadvantage is metal toxicity, where the site of drug action is in the reticuloendothelial system.

14.8.5.4 Injection

One of the common methods by which drugs can be introduced into the body is through **intramuscular**, **subcutaneous**, or **intrathecal** injection.

For **intramuscular injection**, gluteal muscle is used where the patient lies in a prone position, the muscle is palpated, and the needle is inserted at a 90° angle to the distance required to reach the center of the muscle. The needle is then removed quickly and the area is massaged gently with a disinfectant sponge to aid in the dispersion of the solution. This method is useful for those types of medicines that are irritating to the subcutaneous tissue, and a large amount of fluid can be injected into the muscle tissue where absorption through a muscle is faster than through subcutaneous tissue. A disadvantage of this process is the danger of damaging nerves and blood vessels.

The **subcutaneous** injection sites are the following: the outer area of the upper arm, abdomen, from the rib margin to the iliac crest avoiding a 2 in. circle around the navel (this has the fastest rate of absorption among the sites), the front of the thigh, midway to the outer side, 4 in. below the top of the thigh to 4 in. above the knee (this has a slower rate of absorption than the upper arm), the upper area of the buttock, just behind the hip bone (this is also the slowest rate of absorption among the sites). Subcutaneous injection involves the use of sterile equipment and supplies. These include a syringe, a needle, medication, a swab, and a disinfectant to clean the skin. Medication is administered slowly, about 10 s/mL. This route has the advantage to almost complete absorption, provided the patient's circulation is good; therefore, an accurate measure of the amount of drug absorbed is possible. Medicines administered in this manner are not affected by gastric disturbances, nor is their administration dependent upon the consciousness or rationality of the patient. The chief disadvantage of this method is that by introducing a needle through the skin one of the body's barriers against infection is broken. It is therefore important that aseptic technique be used for all needle injections.

Intrathecal administration is an injection of drugs into the spinal canal, or into the subarachnoid space to avoid the blood–brain barrier. The process is useful in spinal anesthesia, chemotherapy, pain management, and certain infections, particularly postneurosurgical applications. Currently, only four medicines are licensed for intrathecal chemotherapy, which are methotrexate, cytarabine (Ara-C), hydrocortisone, and, rarely, thiotepa. Intrathecal chemotherapy is a treatment in which anticancer drugs are injected into the fluid-filled space between the thin layers of tissue that cover the brain and spinal cord.

14.8.5.5 Biological Drugs

Recently, there has been interest on biological drugs based on peptide- and protein-based therapeutics produced using recombinant technology in mammalian, bacterial, or yeast cells. Nutropin, a marketable drug from Genentech/Alkermes, is an intramuscular injectable slow-release formulation based on poly(lactic-co-glycolic acid) (PLGA) microparticles for the delivery of recombinant human growth hormone. Regranex, a carboxymethyl cellulose–based gel with immobilized platelet-derived growth factor, is prescribed for wound healing as part of the treatment of diabetic ulcers. Conjugation of polymers to the proteins of bioconjugation (N-hydroxysuccinimide derivative of PEG) is a one-step conjugation to the peptidic amine groups on, lysine or chain terminus and resulting conjugates are administered via injection. But still it is under research for its fine products en route from laboratory to clinical testing, stage of acceptance to the market (inhaled insulin), and proper scientific reactions in vivo. For this reason, predicting the future of commercial formulations for the delivery of biological drugs is challenging.

FURTHER READING

Chandrasekaran, S. K. and Hilman, R., Heterogeneous model of drug release from polymeric matrix, *J. Pharma. Sci.* 69, 1311, 1980.

Chein, Y. W., *Novel Drug Delivery Systems*, Marcel Dekker, New York, 1982.

Delgado, J. N. and Remers, W. A., *Text Book of Organic Medicinal and Pharmaceutical Chemistry*, Lippincott Williams & Wilkins, Philadelphia, PA, 2007.

Kar, A., *Medicinal Chemistry*, New Age International(P) Limited Publishers, New Delhi, India, 2008.

Kenakin, T., *Pharmacological Analysis of Drug receptor Interactions*, Raven Press, New York, 1993.

Lee, V. H. K., Robison, J. R., and Lee, V. H. I., *Controlled Drug Delivery*, Marcel Dekker, New York, 1987.

Nadendla, R. R., *Principles of Organic Medicinal Chemistry*, New Age International(P) Limited Publishers, New Delhi, India, 2010.

Nag, A. and Dey, B., *Computer-Aided Drug Design and Delivery Systems*, McGraw Hill Publishers, New York, 2010.

Patrick, G. L., *An Introduction to Medicinal Chemistry*, Oxford University Press Inc., New York, 2009.

Thomas, G., *Chemistry for Pharmacy and Life Sciences*, Prentice-Hall, New York, 1996.

Problems to Be Solved

1. Classify the following objects as to whether they are chiral or achiral.
   (a) Screw, (b) plain spoon, (c) fork, (d) cup, (e) foot, (f) ear, (g) shoe, (h) spiral staircase.
2. Imides such as phthalimides are sufficiently acidic and dissolve in alkali metal hydroxide solution to form salts.
3. Some of the molecules listed here have chiral carbons; some do not. Write two-dimensional formulas for the enantiomers of those molecules that do have chiral carbons.
   a. 1-Chloropropane
   b. Chlorobromoiodomethane
   c. 2-Methyl-1-chloropropane
   d. 2-Methyl-2-chloropropane
   e. 2-Bromobutane
   f. 1-Chloropentane
   g. 2-Chloropentane
   h. Chloropentane
4. Which one of the objects listed in Problem 1 possesses a plane of symmetry?
   Write the two-dimensional formula and designate a plane of symmetry of all the achiral molecules in Problem 2.
5. The amount of enol present at equilibrium is much larger in acetylacetone than in ethyl acetoacetate.
6. An optically pure sample of S-(+)-2-butanol shows a specific rotation of +13.52° ($\alpha_D^{25} = +13.52°$). What relative molar proportions of S-(+)-2-butanol would give a specific rotation α_D^{25} equal to +6.76°?
7. Identify heterotopic atoms or group in each of the following. Indicate whether the group is enantiotopic or diastereotopic.

8. Which of the following are *meso* forms?

9. Which of the following are prochiral compounds?
   a. Methanol
   b. Propan-1-ol
   c. Propan-2-ol
   d. 2-Methylpropan-2-ol
10. Assign *R* and *S* designations to each of the following compounds.

11. Designate configuration (*R* or *S*) of the following compounds (each chiral center).

12. All the stereoisomers of 2,3-dibromopentane are represented by the following Fischer projections:

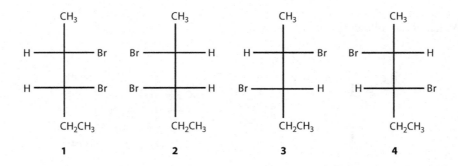

Problems to Be Solved

    a. What is the stereoisomeric relation between compounds **2** and **3**?
    b. Between **1** and **4**?
    c. Between **2** and **4**?
    d. Would compounds **1** and **2** have the same boiling point?
    e. Would compounds **1** and **3**?

13. In each of the following cases, is the compound shown resolvable?

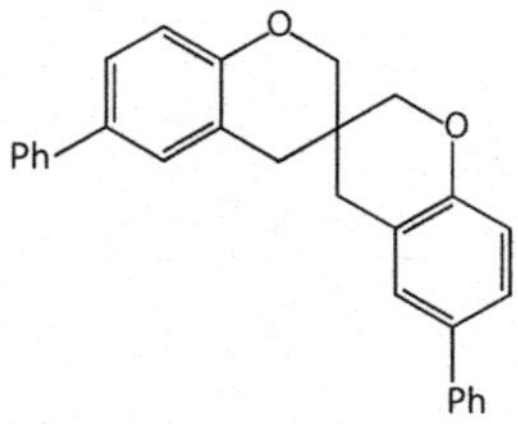

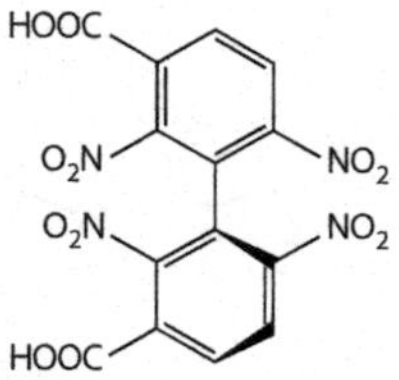

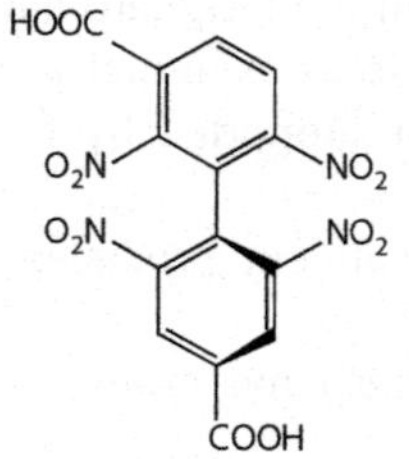

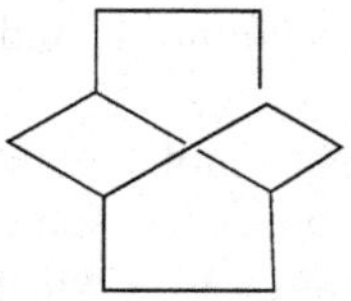

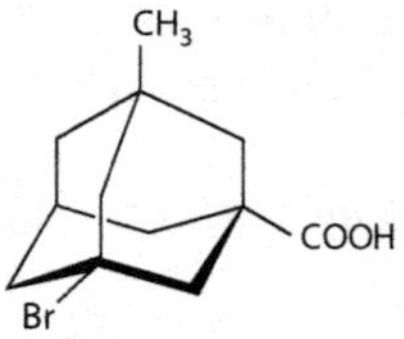

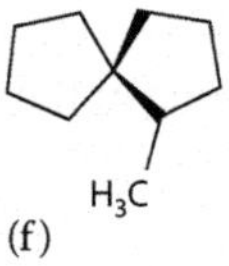

14. In each case, answer the two questions: (1) How many asymmetric carbon atoms does the molecule contain? (2) How many isomers are predicted?

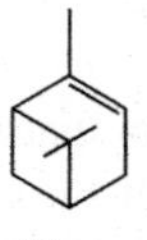

a-Pinene

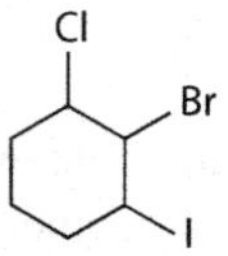

15. The following shows the structure of a prochiral compound, with the enantiotopic ligands.

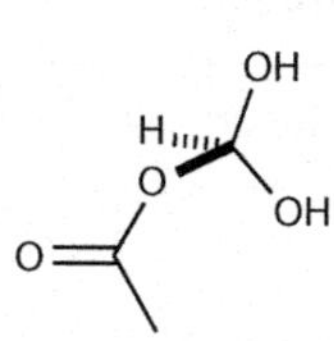

Which one of the enantiotopic ligands is the pro-*R* ligand?

16. What are the characteristics of stereoselective synthesis?
    a. Synthesis of a pair of diastereomers greater in yield than the rest
    b. Synthesis of one diastereomer more in yield than the rest
    c. Both (a) and (b)
    d. None
17. Explain Prelog's rule. (*Hint*: Nucleophilic addition to chiral glyoxylate or α-keto esters)

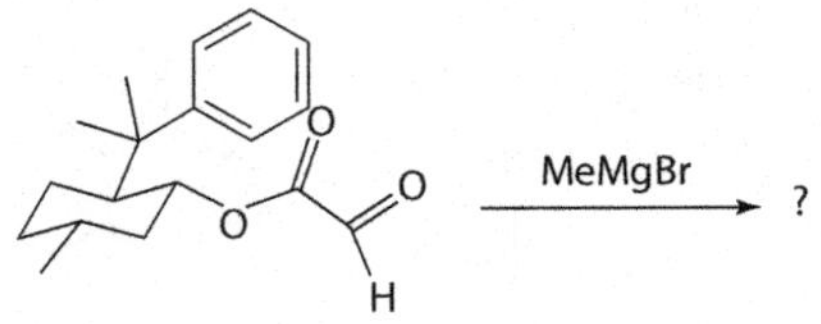

[Chemical scheme: chiral glyoxylate ester + MeMgBr → ?]

Explain the sense of asymmetric induction in the Grignard addition to the following chiral glyoxylate ester. How was the absolute sense of asymmetric induction determined in this case? Is the sense of asymmetric induction supported by Prelog's rule? (Jawdosiuk and Umiński, 1982)

18. Why *cis*-4-bromocyclohexanol on alkali treatment produces *cis*- and *trans*-cyclohexanol whereas epioxidation from *trans*-isomer?
19. What is true about Cram's rule when the carbonyl group is adjacent to one asymmetric carbon?
    a. –C = O is positional between large and medium groups.
    b. –C = O is positional between large and small groups.
    c. –C = O is positional between medium and small groups.
    d. None.
20. Explain stereochemical outcome of the following reaction using Cram's rule (based on Donald and Elhafez, 1952).

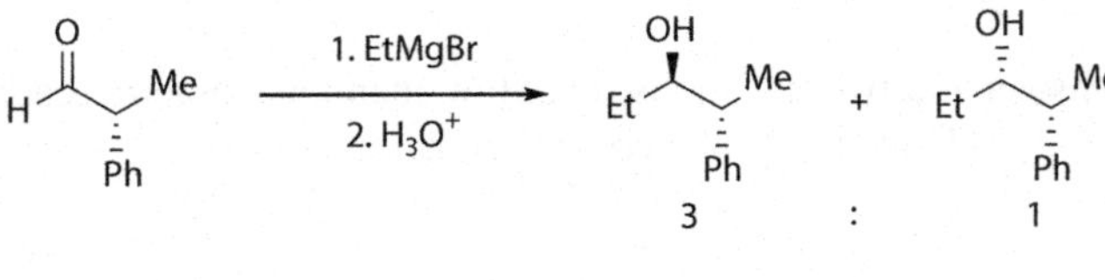

[Chemical scheme: PhCH(Me)–CHO with 1. EtMgBr, 2. H₃O⁺ → two diastereomeric alcohols in 3:1 ratio]

21. Name the main drawbacks of Cram's rule in comparison with the Felkin–Ahn rule.
22. Predict major **A** and minor **B** diastereomers of the following reaction (based on Yamamoto and Yamada, 1987).

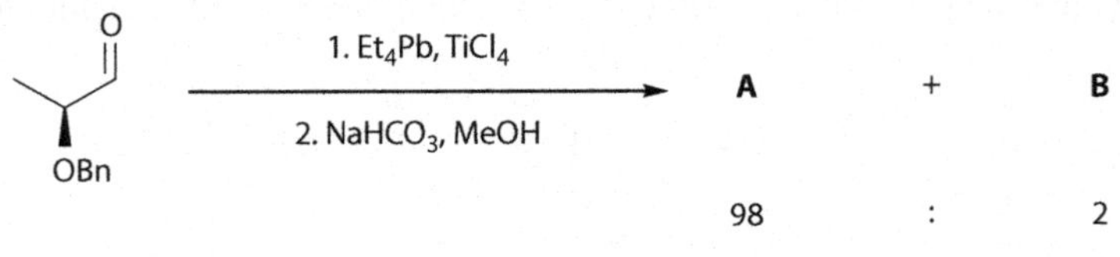

[Chemical scheme: α-OBn ketone with 1. Et₄Pb, TiCl₄; 2. NaHCO₃, MeOH → **A** + **B**, 98:2]

23. Draw the possible transition state that explains the stereochemical outcome of the following reaction (based on Manfred, 1984).

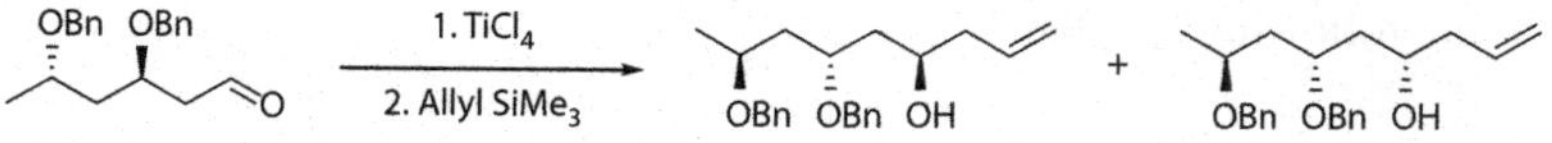

Problems to Be Solved

24. Draw the products of the following reactions, giving a detailed mechanism that justifies the stereochemistry of major diastereomers.

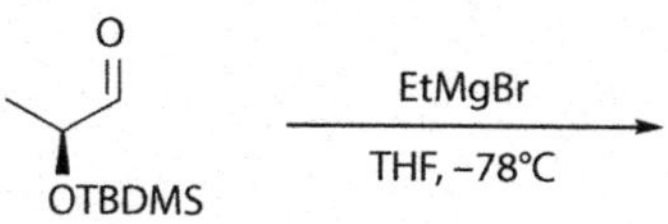

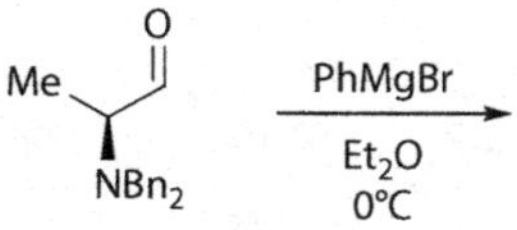

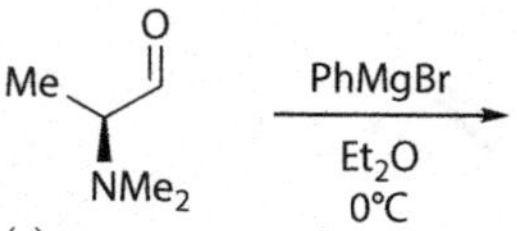

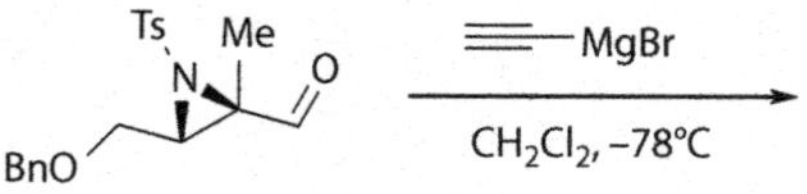

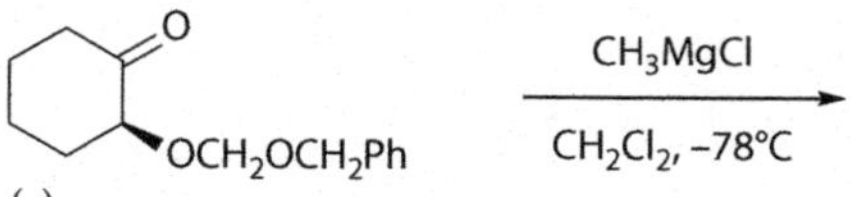

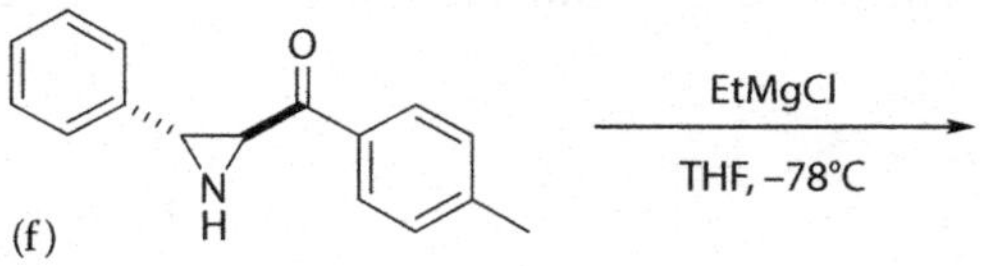

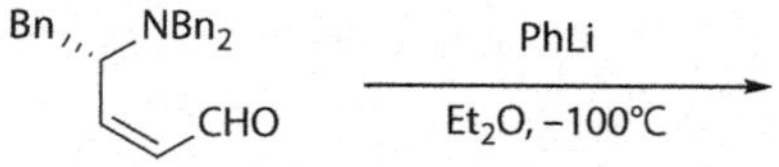

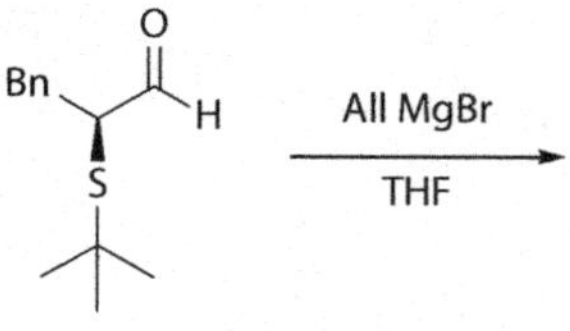

25. The following is a prochiral diol.

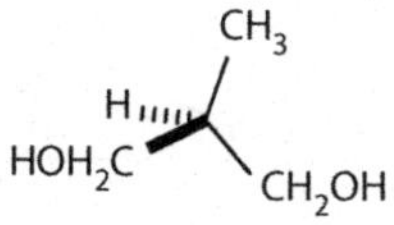

If the hydroxy group on the pro-*S* ligand of this compound is acetylated, what the absolute configuration of the generated monoester will be?

26. The (*R*)-alcohol is created as a result of selective attack on the (*Re/Si*) face.

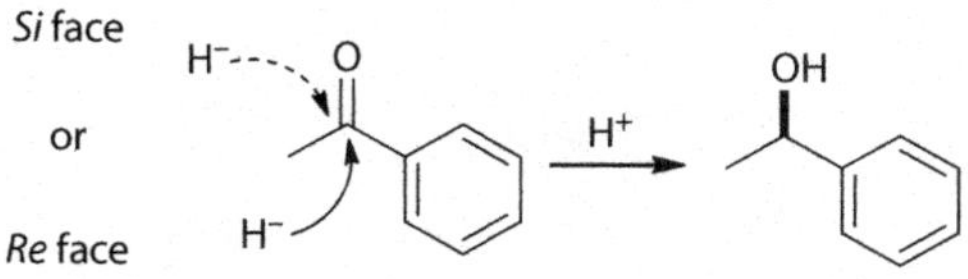

27. Show how to convert 3-methylbutan-2-one to (*R*)-2-hydroxy-3-methylbutane.

28. Complete the following scheme.

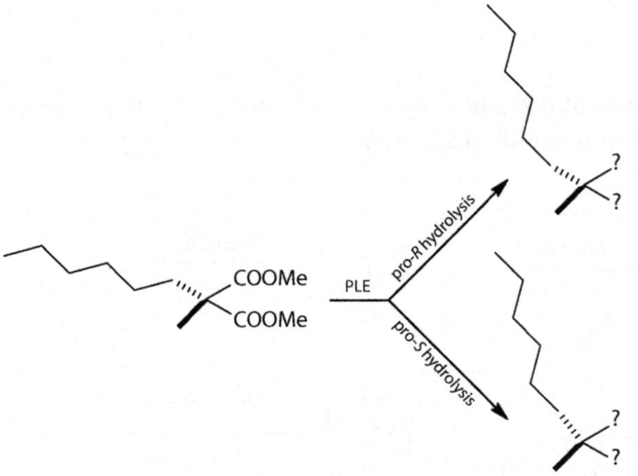

29. The following shows the structure of a triglyceride.

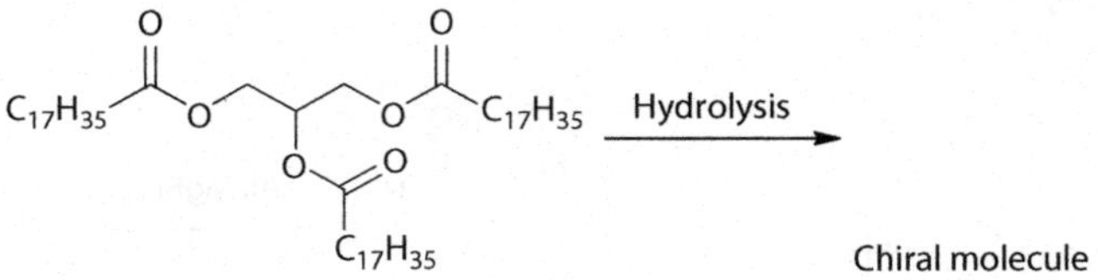

The compound can be converted into a chiral molecule if only one of the three acyl groups is hydrolyzed. Hydrolysis of which –OCOC$_{17}$H$_{35}$ group makes it chiral.

30. Complete the following scheme.

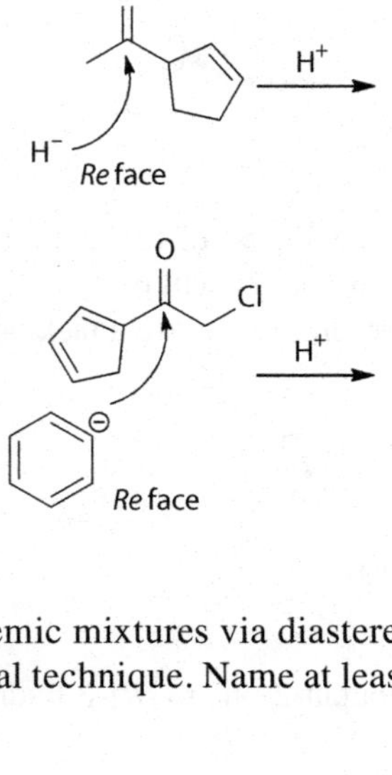

31. To date, the resolution of racemic mixtures via diastereomeric salt formation has been the most commonly used industrial technique. Name at least three criteria for a good resolving agent.

Problems to Be Solved

32. D-*p*-Hydroxyphenylglycine, a building block for semisynthetic penicillins and cephalosporins, is prepared by classical resolution utilizing (+)-3-bromocamphor-8-sulfonic acid ammonium salt (BCSA) as follows:

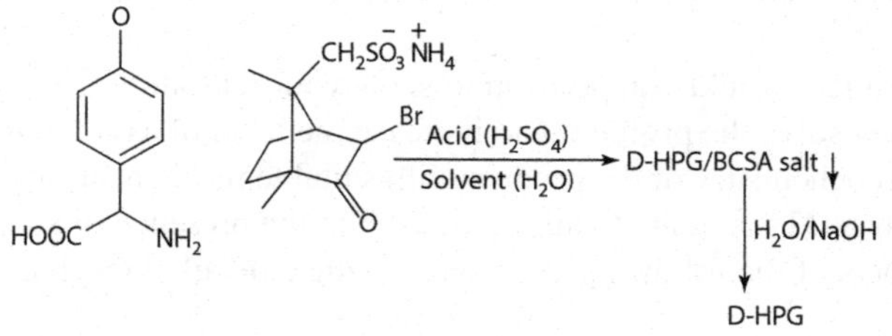

The maximum theoretical yield in this process is 50% (actual yield is ≅30%). How would you propose to solve this problem and increase the yield in this process? (Bhattcharya et al., 1994)

33. a. How would you experimentally distinguish whether a racemic modification exists as racemic mixture (conglomerate) or racemic compound?
    b. The hypotensive activity of α-methyldopa has been shown to reside only in the *S*-isomer.

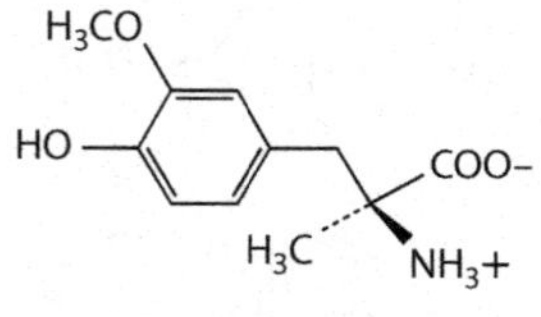

S-Methyl DOPA

Describe the Merck process for the production of *S*-methyl DOPA. (Eliel and Wilen, 1996)

34. Joel Hawkins and Greg Fu (MIT) resolved the following dihydroazepine analog via the dibenzoyl-L-tartrate salt. (Hawkins and Fu, 1986)

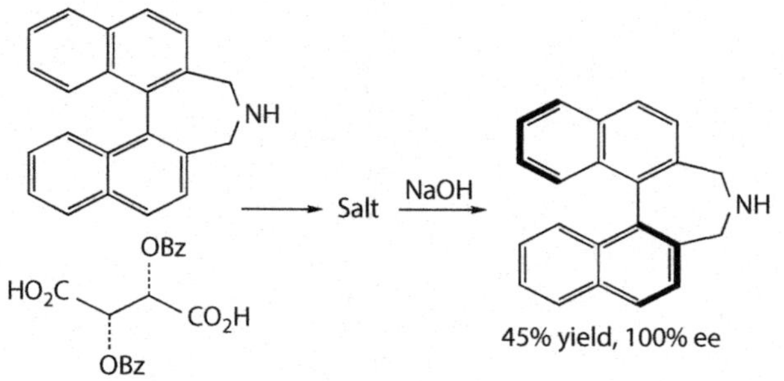

45% yield, 100% ee

This base was used to study the asymmetric carbon–carbon bond formation in the Michael addition on crotonate esters.

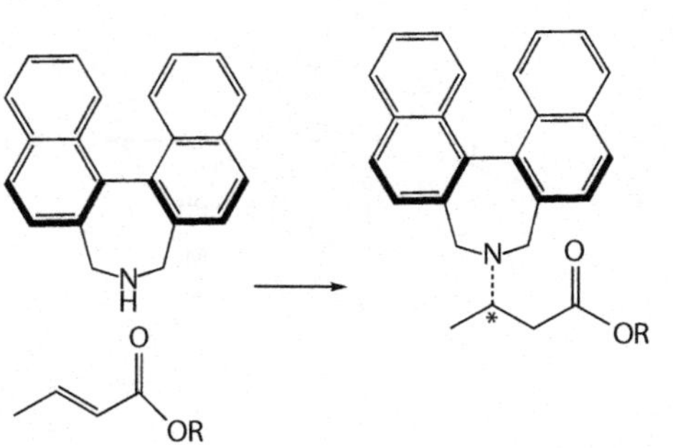

a. Is it necessary to resolve the chiral azepine base to ascertain whether there is any asymmetric induction in the process? Explain.
b. It was found that during the addition of the free amine to methyl crotonate (utilizing HPLC analysis), the ratio of isomers started at 4.1:1 and ultimately leveled off at 1:1. Why?
c. What experiment would you perform to show your reasoning?
d. How did they solve the problem and increase the level of asymmetric induction to 98%?

35. Explain the stereochemistry of the product with suitable mechanism when *threo* diastereomer of PhCHMeCHMeOTs is treated with acetic acid in the presence of sodium acetate.

36. Draw the products of the following reactions, giving a detailed mechanism that justifies their stereochemistry.

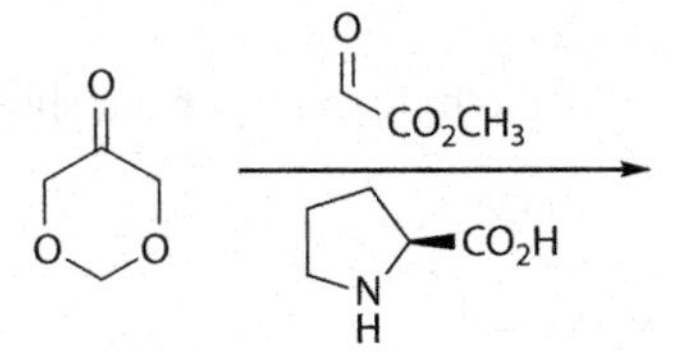

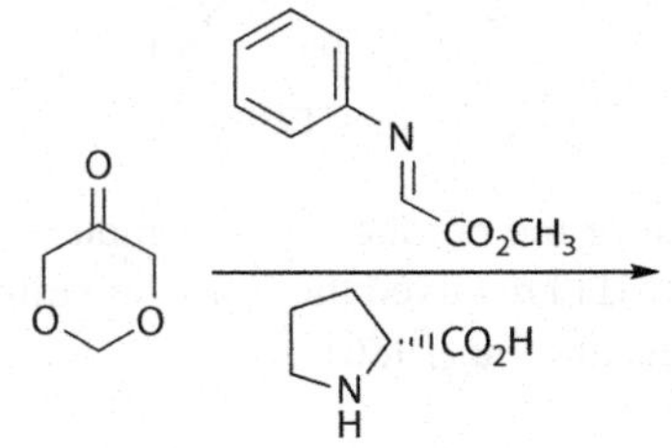

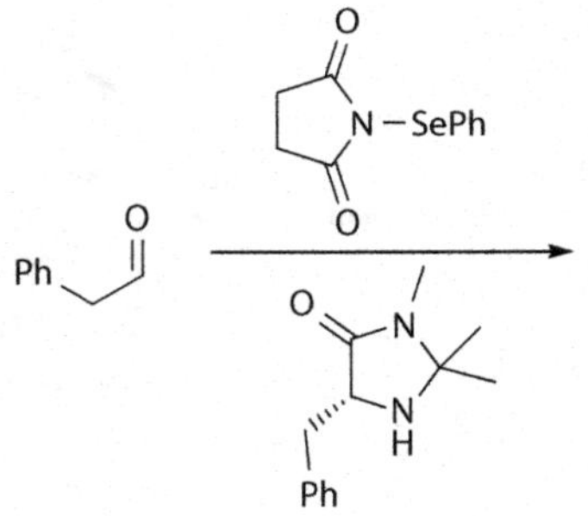

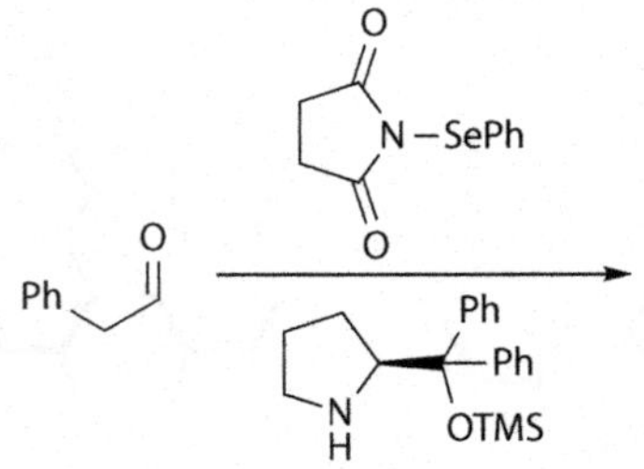

37. Draw the products of the following reactions and justify the final stereochemistry.

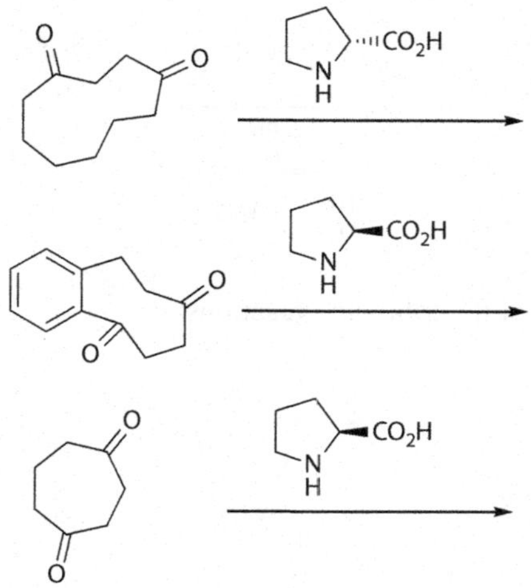

38. Draw the transition state and justify the stereochemistry of the product.

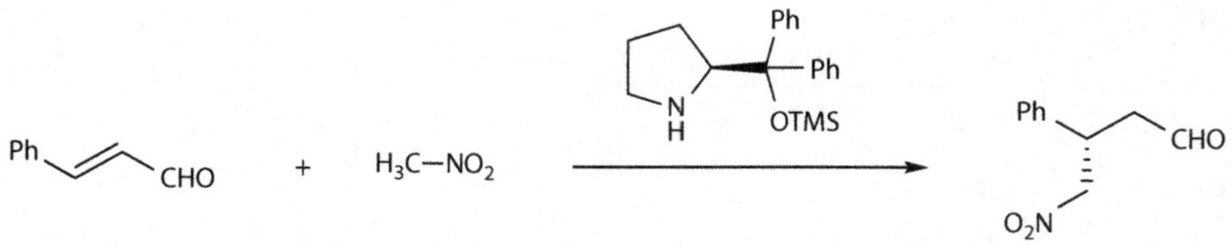

39. Explain why enals (*E*) and (*Z*) render the same stereoisomer of the final product.

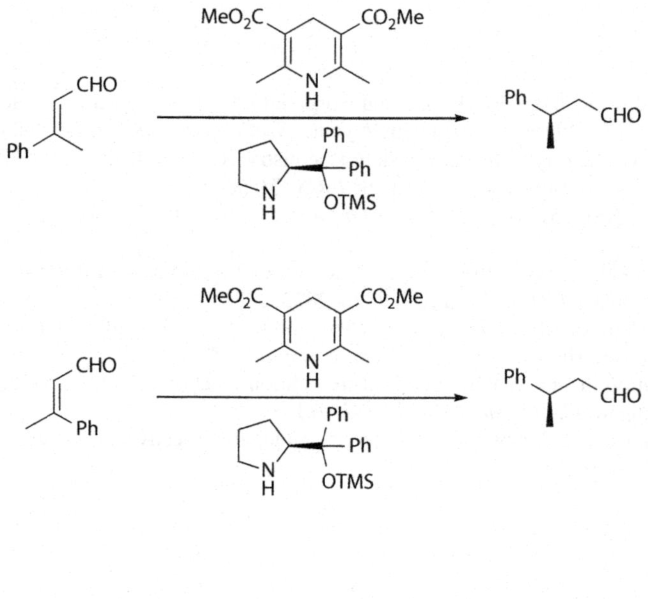

40. Propose a mechanism for the following reaction and justify the stereochemical outcome for carbons **a** and **b**.

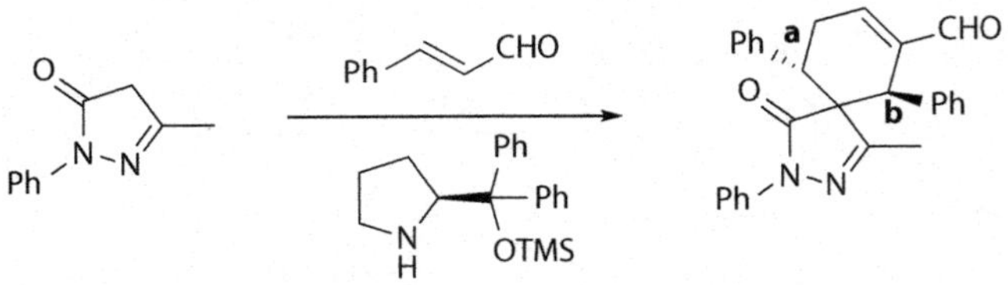

41. Discuss the mechanism of the following reactions:

   (a) [structure with N(CH$_3$)$_2$ ylide] $\xrightarrow{\Delta}$ methylcyclopentene (97%) + methylenecyclopentane (3%)

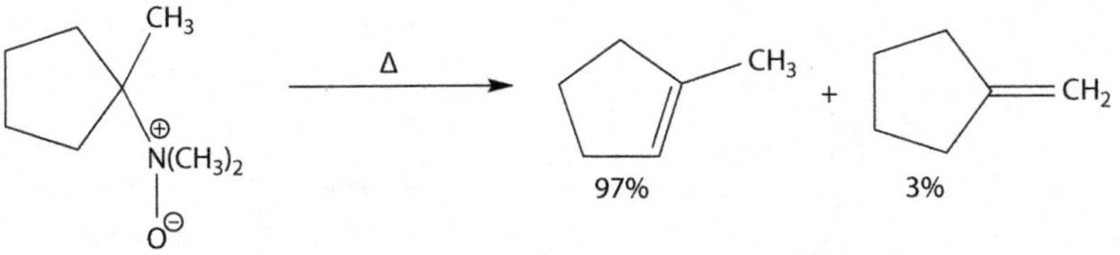

   (b) $R-C\equiv C-R \xrightarrow{B_2H_6}$?? $\xrightarrow{CH_3CH_2COOH}$??
   $\xrightarrow{H_2O_2/\overset{\ominus}{O}H}$??

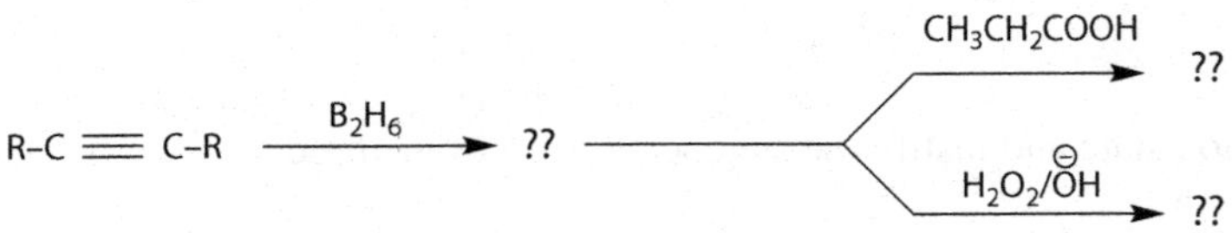

   (c) norbornene $\xrightarrow{KMnO_4, H_2O, NaOH}$?

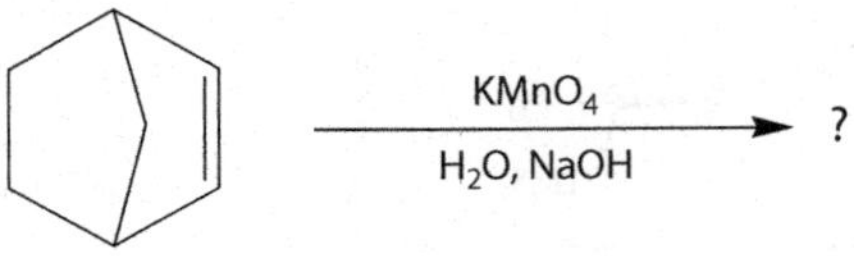

   (d) 1,2-dimethylcyclopentene $\xrightarrow[\text{Ether}]{OsO_4}$? $\xrightarrow{\text{Aq. NaHSO}_3}$?

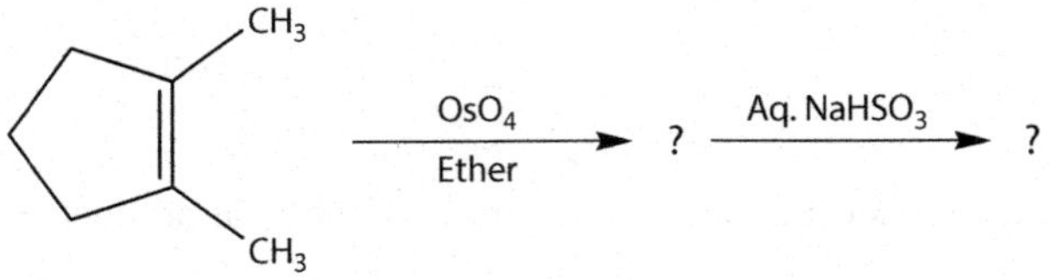

   (e) *Cis* and *trans*-2-chloro cyclohexanol behave differently toward alkali.

REFERENCES

Bhattcharya, A., Araullo-McadamsHoechs, C., and Corporatio, C.; Crystallization induced asymmetric transformation: Synthesis of D-p-hydroxyphenylglycine; *Synth. Commun.*, 24, 2449–2459, 1994.

Donald, J.C. and Abd Elhafez, F.A.; Studies in stereochemistry. X, The rule of "Steric Control of Asymmetric Induction" in the syntheses of acyclic system; *JACS*, 74, 5832–5835, 1952.

Eliel, E.L. and Wilen, S.H.; *Stereochemistry of Organic Compounds*, Hoboken, NJ: John Wiley & Sons, pp. 392–393.

Hawkins, J.M. and Fu, G.C.; Asymmetric Michael reactions of 3,5-dihydro-4H-dinaphth[2,1-c:1′,2′-e]azepine with methyl crotonate; *J. Org. Chem.*, 51, 2820–2822, 1986.

Jawdosiuk, M. and Umiński, M.; β-elimination of the isonitrile group in alkylation reactions of C–H acids activated by the isonitrile function; *JCS Chem. Commun.*, 17, 988, 1982.

Manfred, T.R.; Chelation or non-chelation control in addition reactions of chiral α- and β- alkoxy carbonyl Compounds; *Angewandte Chime*, 23(8), 556–569, 1984.

Yamamoto, Y. and Yamada, J.; A new type of stable, storable, and selective alkylating reagent, tetraalkyllead; *JACS*, 109, 4395–4396, 1987.

Index

A

Absolute configuration, 9–10
ACC, see 1-Aminocyclopropane-1-carboxylic acid
Acetobacter pasteurianus DSM 8937, 254
Acetylcholinesterase (AChE), 196–198, 221–222
Achromobacter piechaudii, 353
Acinetobacter baylyi ATCC 33305, 413
Acinetobacter calcoaceticus, 413
Active pharmaceutical ingredients (APIs), 346
Acyclic chiral ketones
    anti-Felkin–Ahn product, 63–64
    cyclic ketones, 68
    nitrogen in α-position, 66–67
    oxygen in α-position, 63–66
    sulfur atom, 67
AD, *see* Alzheimer's disease
Agonists-activated receptors, 464
Alcaligenes sp. lipase (QL), 232
Aldehyde
    achiral organometallic reagents, 43–44
    α-nitrogen substituent
        aziridine carbaldehydes, 51, 53–54
        chelating conditions, 51–52
        diastereoisomeric *anti/syn* β-amino alcohols, 47, 50
        examples, 57, 59
        Garner's aldehyde **79**, 55, 58
        N-Boc-protected prolinal **75**, 55, 57
        nonchelating conditions, 50–51
        N-protected group, 51–52
        N-tosyl-protected 2,3-disubstituted aziridine-2-carbaldehydes **67**, 53–54
        N-tritylprolinal **77**, 55, 57
        organocerium compounds, 56, 59
        stereocontrolled 1,2-additions, 54–56
        3-substituted-4-oxoazeti-dien-2-carboxaldehydes **72**, 54–56
    α-oxygen substituent, 44–45
    α-position, 57, 59–63
    β-C-glycoside aldehyde **46**, 47, 49
    dimer of acrolein **42**, 45, 47
    examples, 45–46
    high diastereoselective addition, 47, 50
    2,3-isopropylideneglyceraldehyde (**49**), 47, 49
    6-Membered Cyclic Aldehydes **53**, 47, 50
    organolithium/Grignard reagents, 45–46
    syn-2,3-dialkoxy groups, 47–48
Aldol reaction, 326
α-chymotrypsin, 219
α-glucosidase, 198–201
α-substituted butenedioic esters, 253
Alzheimer's disease (AD), 196–198
Amano PS lipase, 225
AMDase, *see* Arylmalonate decarboxylase
1-Aminocyclopropane-1-carboxylic acid (ACC), 225
Amycolatopsis mediterranei, 230
Amycolatopsis orientalis SC 15847, 364

Anatomical Therapeutic Chemical (ATC) classification, 348
Angiotensin-converting enzyme (ACE) inhibitors
    captopril, 424–425
    enantiopure monoacetates, 426
    Monopril, 426
    zofenopril, 425
Anomerization, 2
Anomers, 2, 4
Antagonists-activated receptors, 464
Anticancer drugs
    modification of natural products, 348
        aureolic acids, 369–371
        cyclopamine analogues, 367–369
        epothilones, 363–367
        taxanes, 349–362
    NCE, 348
    small molecules, *de novo* synthesis, 348, 371
        AXL inhibitors, 375–376
        crizotinib, 377–378
        lonafarnib, 378–380
        monastrol, 374–375
        niraparib, 381–383
        odanacatib, 380–381
        pelitrexol, 372–374
Anticholesterol drug
    atorvastatin, 405, 408
    ethyl (*R*)-4-cyano-3-hydroxybutyrate
        biocatalytic retrosynthetic routes, 410
        bioreductions, 411–412
        enantiopure benzoic ester hydrolysis, 409
        ethyl 4-chloroacetoacetate, 411
        ethyl (*S*)-4-chloro-3-hydroxybutanoate, 411
        glucose dehydrogenase, 411–412
        HN, Codexis synthesis, 411
        HN preparation, 408–409
        KRED, 412
        L-malic acid, 409
        nitrilase-catalyzed enzymatic desymmetrization, 410
        prochiral 3-hydroxyglutaronitrile, 410
        (*R*)-3-hydroxy-4-cyanobutyric acid, 410
        racemic ethyl 4-chloro-3-hydroxybutanoate, 409
        (*S*)-4-bromo-3-hydroxybutanoate esters, 411
        (*S*)-3-hydroxy butyrolactone preparation, 408–409
        yeasts, 411
    isoprenoid by-products, 405
    lovastatin, 405
    Mevacor, 405
    mevastatin, 405
    Mevinolin, 405
    pravastatin, 405
    simvastatin, 406–408
    statins
        Acinetobacter species, 413
        alcohol dehydrogenase, 413
        aldolase-catalyzed synthesis, 412–413

2-deoxy-D-ribose 5-phosphate aldolase
(DERA), 413
HMG-CoA reductase inhibitors, 404–405
Antidiabetic drugs
DPP-4 inhibitors, 389
glucagon-like peptide-1 mimetics and modulators, 387–389
PPARs, 383–387
saxagliptin, 390–391
sitagliptin, 389–390
Anti-elimination, 6
Anti-HER2 immunoliposomes, 472
Antihypertensive drugs
β-blockers
adrenoreceptor subtypes, 415
α_1-adrenoceptors, 414
α_1-antagonist, 415
α_2-agonist, 415
aromatic ketone, 416–417
1-aryl-2-alkylaminoethanol structure, 415–416
aryloxypropanolamine structure, 416, 418–420
azidoacetate, 423
β_2-antagonists, 415
catecholamine-type agonists, 414
chemoenzymatic DKR, 421
chemoenzymatic process, 415, 417
chiral synthons, 422, 424
CRL, 422
diols, hydroxyesters, azidoalcohols, and nitriles, 422–423
DKR, 423–424
halohydrins, S-aryloxypropanolamines, 422
nifenalol, 415–417
O-acetylation, 422
prochiral 1-chloro-3-(phthalimidyl)-propan-2-one, 424
propranolol, kinetic resolutions, 421–424
(R)-alcohol, 416–417
R-arylethanolamines, 420
R-chirality of catecholamines, 415
receptor-selective drugs, 415
S-aryloxypropanolamines, 420, 423–424
sotalol, 415–417
sympatholytic drugs, 415
ulterior synthesis, 421
hypertension, 413
Anti-inflammatory profen drugs
2-arylpropionic acids, 392
naproxen
AMDase, 401–402
2-aryl-2-methyl malonate ester desymmetrization, 401–402
azolides ((R,S)-N-profenyl-1,2,4-triazoles), 401
BsENP01, 400
Candida antarctica, 400
Carica papaya lipase, 400
Carica pentagona Heilborn, 400
ChiroCLEC-CR, 400
cinchonidine, 399
DKR, 401
enantioselective esterification, 400
first member, profen family, 398
hydrolase-catalyzed resolutions, 399
hyperthermophile SsADH, 401
irreversible enantioselective esterification, 400–401
2-methoxy-6-vinylnaphthalene hydroformylation, 399
naphthacrylic acid, 399
racemic alkyl 2-(6-methoxynaphthalen-2-yl) propanoate, 399
redox bioprocesses, 401
S-eutomer, 399
(S)-selective arylmalonate decarboxylase variant, 402
Zambon process, 399
NSAIDs, 392
(S)-flurbiprofen
Aspergillus oryzae MIM, 397
chemoenzymatic synthesis, 398–399
enantioselective esterification, 397
ene-reductase YqjM from *Bacillus subtilis*, 398
lipase-catalyzed kinetic resolution, 397–398
optically pure (S)-**162**, 398
racemic flurbiprofenil bromopyrazolide, 398
(S)-ibuprofen (dexibuprofen)
2-arylpropanals, 395
Candida rugosa lipase, 393–394
carrier-free immobilization techniques, 394
DKR process, 394–395
EMR, 393
ene-reductase YqjM from *Bacillus subtilis*, 396
kinetic resolution processes, 394
lipase-facilitated enantioselective transport, 394
rac-ibuprofen methoxyethylester, 393
(R)-profen derivatives, 396
ruthenium-catalyzed dehydrogenation of alcohol, 394
stereoselective hydrolysis, 395
(S)-ketoprofen (dexketoprofen), 396–397
(S)-ketorolac
acylation and ulterior hydrolysis, 402
enantioselective esterification, *C. antarctica* lipase B, 403–404
first asymmetric synthesis, 403
ketorolac and ketorolac tromethamine, 402
Oppolzer's sultam, 403
S-eutomer, 403
stereoselective cyclization, 403
Streptomyces griseus protease, 403–404
third-generation synthesis, 402–403
Antimicrobial agents, 191–193
Antirenin, 201–202
APIs, *see* Active pharmaceutical ingredients
Aprepitant, 454–455
Aromatic dioxygenases, 259–261
Arthrobacter nicotianae, 353
Artificial/semisynthetic biocatalysts
regio- and enantioselective preparation
chiral amine synthesis, 183
chiral Me-imidacloprid, 184
DKR process, 183
enzyme–ruthenium organometallic combo DKR, 184
KR process, 183
norsertraline, 184
racemic aryl amines, DKR, 185
racemic mixture, 182

Index

regioselective artificial hydrolase-catalyzed C–C bond formation
  CAL-B-catalyzed decarboxylative aldol reaction, 186
  CAL-B-catalyzed decarboxylative Knoevenagel reaction, 186
  catalytic promiscuity, 185
  Michael addition reaction, 186–187
  non-natural substrate-promoted catalytic promiscuity, 185
  PPL, 185
  Suzuki–Miyaura reaction, 186–187
regioselective preparation
  monodeprotected nucleosides, 180, 182
  peracetylated carbohydrates, 178–181
Artocarpus nobilis, 195–196
Arylmalonate decarboxylase (AMDase), 401–402
2-Arylpropionic acids, 392
Aspergillus alliaceus, 287
Aspergillus fumigatus, 11
Aspergillus glaucus, 276
Aspergillus niger, 416
Aspergillus niger lipase, 232
Aspergillus oryzae protease (protease A), 216
Aspergillus terreus, 405
AstraZeneca, 383
Asymmetric intramolecular cyclization reactions
  acyclic bromoalkenes, unactivated primary C(sp³)–H bonds, 307–308
  allylamides, 305–306
  2-azabicyclo[3.1.0]hexane derivatives, 306
  2,2′-biphenol linkage, 309
  (−)-blennolide A, 308
  C(sp³)–H alkenylation approach, 308
  chiral chroman, 310
  chiral 2,3-*cis*-dimethyldihydrobenzofuran, 308–309
  chiral diphosphine, 305
  cyclopropane-fused chiral lactams synthesis, 306–307
  cyclopropane-fused γ-lactam products, 306–307
  Dieckmann cyclization, 309
  η³-allyl palladium intermediates, 305
  five-membered nitrogen heterocycles, 307
  iodinated monomeric tetrahydroxanthenone synthon, 309–310
  (±)-kainic acid, 305
  lactones and lactams, 304
  mono- and bicyclic α-alkylidene-γ-lactams, 307
  palladium-catalyzed C3–C4 ring closure reactions, 304, 306
  polyketide–terpenoids, 308
  trans pyrrolidone, 304–305
  pyrrolizidine alkaloid analogue, 307–308
  reductive Heck cyclization reaction, 308–309
  secalonic acid E, 308–310
  syn-5-*exo* or *anti*-5-*exo* isomers, 305
  unsaturated phosphono-allylacetamide, 305
  Wacker-type cyclization, 309–311
Atorvastatin, 405, 408
Atropisomerism, 5, 271
Atropodiastereoselective synthesis
  calphostin, 276–277
  diazonamide A, 286
  lignin, 276–277

Atropoenantioselective synthesis
  1-aryl-5,6,7,8-tetrahydroisoquinolines, 287, 289
  (+)-isokotanin A, 287–288
  planar-chiral [(arene)Cr(CO)₃] complex, 279
Atroposelective ring opening reaction, lactones, 287
Aureobasidium pullulans SC 13849, 247
Aureolic acids
  antineoplastic antibiotics, 369
  CAL-B-catalyzed acylation, 371–372
  chromocyclomycin, 369
  chromomycin A3, 369
  durhamycin A, 369
  MTM, lipase-catalyzed acylation, 369–371
  olivomycin, 369
  polyglycosidic aromatic polyketides, 369
  structure, 369–370
  UCH9, 369
Axial bonds, 5
Axial isomers, 271
Axially chiral biaryls
  asymmetric synthesis, aromatic ring
    asymmetric [2+2+2] cycloaddition, 287, 289
    atropoenantioselective synthesis, 1-aryl-5,6,7, 8-tetrahydroisoquinolines, 287, 289
    bis(chromium carbene) species, 290
    double benzannulation, 290
  atroposelective synthesis, 274–275
  axially chiral but conformationally unstable biaryl compounds conversion
    additional substituent in *ortho* posit, 285
    bridge formation, 285–286
    cleavage of a bridge, 286–289
  intramolecular coupling
    with chiral leaving groups, 279–280
    with chiral *ortho* substituents, 277–280
    with chiral tethers, 274–277
    with element of planar chirality, 279, 281
  natural compounds
    dimeric sesquiterpenes, 273
    examples, 272–273
AXL inhibitors, 375–376
Azetidinedione *rac*-13
  bioreduction, 358
  stereoselective enzymatic reduction, 359–360

B

Bacillus megaterium, 256
Baeyer–Villiger monooxygenases (BVMO), 259
Barleria prionitis, 194
β-amino acid, 452
β-hydride elimination, 30
β-ketodiester, 222–223
β-ketoesters, 246–247
β-sheets, 133–138, 140–142, 156
β-turns, 133–135, 138, 141, 143, 156
Biaryl compounds
  asymmetric axially chiral biaryl synthesis, aromatic ring
    asymmetric [2+2+2] cycloaddition, 287, 289
    atropoenantioselective synthesis, 1-aryl-5,6,7, 8-tetrahydroisoquinolines, 287, 289
    bis(chromium carbene) species, 290
    double benzannulation, 290

axially chiral biaryls
  atroposelective synthesis, 274–275
  intramolecular coupling with chiral leaving groups, 279–280
  intramolecular coupling with chiral *ortho* substituents, 277–280
  intramolecular coupling with chiral tethers, 274–277
  intramolecular coupling with element of planar chirality, 279, 281
axially chiral but conformationally unstable biaryl compounds conversion
  additional substituent in *ortho* posit, 285
  bridge formation, 285–286
  cleavage of a bridge, 286–289
bridged biaryls, 273
chirality, 273
natural axially chiral biaryl compounds
  dimeric sesquiterpenes, 273
  examples, 272–273
nonbridged biaryls, 274
oxidative homocoupling, chiral additives, 281–282
prostereogenic aryl compounds, desymmetrization, 284–285
redox-neutral cross-coupling, 282–283
stable but achiral biaryls and conformationally unstable but chiral biaryls, 284
Biochemical separation, 9, 11–12
Biphenyl dioxygenase (BPDO), 259–261
Boat conformation, 5
Brevibacterium protophormia, 367
Bulky ketones, 242
Burkholderia cepacia (BCL), 206, 208, 394
Butaxamine/ICI-118551, 414–415
BVMO, *see* Baeyer-Villiger monooxygenases

C

Caesalpinia bonduc, 196
Cahn–Ingold–Prelog (CIP) system, 206
Calcium channel blockers (CCBs)
  4-aryl-3,4-dihydropyrimidinone, 426–427
  biocatalyzed protocols, 427
  diltiazem, 427–428
  nifedipine and analogues, 427
  Novozyme 435, 427
  Tanabe synthesis, 427–428
Candida albicans, 411
Candida antarctica, 393
Candida antarctica lipase B (CAL-B), 186, 214, 223–224, 403–404
Candida rugosa lipase (CRL), 215, 393–394, 422
Captopril, 424–425
Carboxamidates
  heteroleptic complexes, 121, 123–125
  homoleptic complexes
    allylic α-cyano-α-diazoacetates, 118
    axial carbene ligand, 114, 116
    (2)-*cis* configuration, 113–114
    cyclopropan(en)ations, 120
    double intramolecular cyclopropanation, 120
    geometrical isomers, 113, 116
    intramolecular C–H insertion, 114, 116
    intramolecular cyclopropenation, 121–122
    matched and mismatched configurations, 114, 117
    2-methallyl diazoacetate, 119
    methyl diazoacetate, 120–121
    methyl phenyldiazoacetate, 120–121
    N-allyl diazoacetamide, 119–120
    PETT analogue **93**, 116, 118
    $Rh_2(4S\text{-MEAZ})_4$, 116–117
    $Rh_2(S\text{-PTPI})_4$ and $Rh_2(S\text{-BPTPI})_4$ catalysts, 121–122
    structures, 113–115
    transition state, 119
    vinyl diazolactone **94**, 118
Carvone, 16
CCBs, *see* Calcium channel blockers
CDA, *see* Chiral derivatizing agent
Chair conformation, 5
Chemical separation, 9–11
Chiral, 1
Chiral alcohols, 242
Chiral building blocks, biotransformations
  ACE inhibitors
    captopril, 424–425
    enantiopure monoacetates, 426
    Monopril, 426
    zofenopril, 425
  advantages, 348
  anticancer drugs
    de novo synthesis, small molecules, 348, 371–383
    modification of natural products, 348–372
    new chemical entities, 348
  anticholesterol drug
    atorvastatin, 405, 408
    ethyl (*R*)-4-cyano-3-hydroxybutyrate, 408–412
    isoprenoid by-products, 405
    lovastatin, 405
    Mevacor, 405
    mevastatin, 405
    Mevinolin, 405
    pravastatin, 405
    simvastatin, 406–408
    statins, 404–405, 412–413
  antidiabetic drugs
    DPP-4 inhibitors, 389
    glucagon-like peptide-1 mimetics and modulators, 387–389
    PPARs, 383–387
    saxagliptin, 390–391
    sitagliptin, 389–390
  antihypertensive drugs
    β-blockers, 414–424
    hypertension, 413
  anti-inflammatory profen drugs
    2-arylpropionic acids, 392
    naproxen, 398–402
    NSAIDs, 392
    (*S*)-flurbiprofen, 397–398
    (*S*)-ibuprofen (dexibuprofen), 393–396
    (*S*)-ketoprofen (dexketoprofen), 396–397
    (*S*)-ketorolac, 402–404
  APIs, 346–347
  ATC classification, 348
  biocatalysts, 346–347
  cardiovascular disease treatment drugs, 404
  CCBs, 426–428

Index

ecofriendly solvents, 347
Green Chemistry, 346
"in water" and "on water" conditions, 347
lipases, 348
molecular enzyme engineering techniques, 347
Chiral carbonyl compounds
  achiral reagents, 43
  acyclic chiral ketones
    anti-Felkin–Ahn product, 63–64
    cyclic ketones, 68
    nitrogen in α-position, 66–67
    oxygen in α-position, 63–66
    sulfur atom, 67
  aldehyde
    achiral organometallic reagents, 43–44
    α-nitrogen substituent, 47, 50–57
    α-oxygen substituent, 44–45
    α-position, 57–63
    β-C-glycoside aldehyde **46**, 47, 49
    dimer of acrolein **42**, 45, 47
    examples, 45–46
    high diastereoselective addition, 47, 50
    2,3-isopropylideneglyceraldehyde (**49**), 47, 49
    6-Membered Cyclic Aldehydes **53**, 47, 50
    organolithium/Grignard reagents, 45–46
    syn-2,3-dialkoxy groups, 47–48
  α-hydroxy acids **2**, 30
  1,2-asymmetric carbonyl addition
    Ahn–Eisenstein presumption, 38–39
    axial and equatorial alcohols, 34
    Cieplak model, 35, 37–38
    Cram-chelation rule, 31–33
    EFOE, 38, 40
    1,2-endo,endo-disubstituted-7-norbornanones, 36–37
    Felkin–Ahn rule, 33–35
    Grignard reagents and alkyl lithiums, 31–32
    nucleophilic additions, 32–33
    PDAS, 38, 40
    perpendicular orientation of R_L, 33
    stereoselective reactions, 33–34
  1,3-asymmetric carbonyl addition
    chelation-control models, 42–43
    Evans model, 41–42
    Jacques model, 41–42
    open-chain transition state, 41, 43
    transition states, 40
Chiral derivatizing agent (CDA), 13–14
Chiral medicines, 449–451
  APIs structure, 449, 451
  aprepitant, 454–455
  levetiracetam, 454–455, 458
  paroxetine, 454, 457
  pregabalin, 456, 458
  sertraline, 456–457, 459
  simvastatin, 454, 456
  sitagliptin, 449, 452–453
Chiral N-protected amino acid ligands
  cyclopropane carboxylic acids, 110–113
  heteroleptic complexes, 109–110
  homoleptic complexes
    Adly and Ghanem complexes, 106–107
    α-alkyl substituents, 94–95
    α-diazopropionates, 95, 97
    α-nitro-α-diazo-p-methoxyacetophenone, 94, 96

α-PMP-ketone group, 95–96
asymmetric synthesis, 97–98
β-lactam derivatives, 97–98
chiral crown cavity, 94
2-chlorophenyl aryldiazoacetate derivative, 100–101
1,1-cyclopropane diesters, 101, 103
diastereo- and enantioselectivity levels, 101, 103
diazo ketones, 100
diazooxindole, 94–95
dimethyl α-diaz obenzylphosphonate, 106–107
dimethyl 1,2-diphenylcyclopropylphosphonate, 98–99
2,4-dimethyl-3-pentyl-α-alkyl-α-diazoacetates, 96–97
electron-deficient alkenes, 100–101
Fox's proposal, 94
intramolecular cyclopropanation, 104–105
lower symmetry, 105–106
molecular structure, 97, 99
nitrile-substituted cyclopropanes, 98, 100
square-shaped cavity, 107–108
structures, 92–93
transformation, 101–102
transition states, 103–104
trifluoromethyl-substituted cyclopropanes, 98, 100
2D Heteronuclear NOESY studies, 94
x-ray crystal structure, 105–106
Chiral resolution
  biochemical separation, 9, 11–12
  chemical separation, 9–11
  chromatographic method, 9, 12–14
  crystal picking, 9–10
  kinetic resolution, 9
  methods, 9
Chromatographic method, 9, 12–14
Chromobacterium viscosum lipase, 426
Chronic lymphocytic leukemia (CLL) patients, 194
Cieplak model, 35, 37–38, 40
Circular dichroism (CD) spectrum, 140–141
Clostridium spp., 251
Comamonas sp. NCIMB 9872 CPMO, 259
Compactin, *see* Mevastatin
Competitive antagonism, 464
Conformational formulas, 5
Conformational isomerism, 271
Conformations, 5–8
Conglomerates, 10
Cram-chelation rule, 31–33
Crizotinib, 377–378
Crystal picking, 9
Cunninghamella echinulata, 192
Curvularia lunata, 192
Curvularia lunata CECT 2130 fungi, 250
Cyclic alcohols, 230
Cyclohexanone monooxygenase (CHMO), 259
Cyclopamine analogues, 367–369
Cytochrome P450 monooxygenases (CYPs)
  Acinetobacter sp. NCIB 9871 CHMO, 259
  BVMO, 259
  E. coli BL21 (DE3)/pMM4 CHMO, 258–259
  examples, 255–256
  3-hydroxyisobutyric acid, 256–257
  mutant enzymes BM-3, 256
  prochiral olefins, 257–258
  reactions, 255
  simvastatin, 257

D

Daiichi Fine Chemicals Co. Ltd, 367
Danaus chrysippus (African monarch), 225
Datura stramonium, 245
Density functional theory (DFT) calculations, 80–81
2-Deoxyribose-5-phosphate aldolase (DERA), 170
Dexibuprofen, *see* Ibuprofen
Dexketoprofen, *see* Ketoprofen
Dextrorotatory, 1
Diastereomeric excess (De), 2
Diastereomers, 2
Diastereoselectivity, 2
Dichloroisoprenaline, 415–416
Dimethylformamide (DMF), 147–148
Dipeptidyl peptidase-4 (DPP-4) inhibitors, 389
Dirhodium(II)
    axial ligands, 124–126
    carboxamidates
        heteroleptic complexes, 121, 123–125
        homoleptic complexes, 113–122
    chiral N-protected amino acid ligands
        cyclopropane carboxylic acids, 110–113
        heteroleptic complexes, 109–110
        homoleptic complexes, 92–108
    conformational mobility, 83–84
    electronic modifications, 82
    history
        alkene approach, 80–81
        lantern structure, 76
        mechanism, 77–79
        model examples, 76–77
        modes, 80–81
        types, 79–80
    prolinate-based ligands
        alkynyldiazoacetates, 86–87
        allyl vinyldiazoacetate, 91
        asymmetric cyclopropanations, 87–88, 91
        Davies complexes, 84–85
        enantioselective cyclopropenation, 89–90
        heteroaryldiazoacetates, 85, 87
        late transition state, 89–90
        p-bromophenyldiazoacetate, 89
        pyrroles/furans, 87–88
        $Rh_2(S\text{-}DOSP)_4$, 84–86
        second-generation, 89, 91
        Su prolinate-based complexes, 91–92
        tamoxifen, 84, 86
    steric modifications, 83
DMF, *see* Dimethylformamide
Dolabrifera dolabrifera, 230
Dötz benzannulation, 290
Doxazosin, 414–415
Doxifluridine, chemo-enzymatic synthesis, 180, 182
DrugBank, 348
Drug delivery systems
    absorption, 465
    biological properties, 465
    distribution, 466
    elimination, 467
    inhalation route, 468
    metabolism, 466
    nasal administration, 468–469
    noninvasive peroral route, 468
    oral administration, 468
    physiochemical properties, 465
    rectal administration, 469
    spatial placement, 463
    targeted drug delivery system, 470–474
    temporal delivery systems, 463
    topical route, 468
    transdermal delivery systems, 470
    transmucosal route, 468
Drug–receptor interaction, 464
Drugs
    brand name, 461
    chirality, 461
    classification, 462
    dose level, 461
    FDA approval, 461
    generic name, 461
    mode of action, 464
    prodrug
        ampicillin, chemical stability of, 463
        applications, 462
        carrier-linked prodrugs, 463
        Hetacillin, 463
    safety considerations, 467
    scientific name, 461
    sites of action, 464
Drypetes gossweileri, 199
Dynamic kinetic resolution (DKR)
    aldehyde, 381–382
    bisulfite, 382
    lactol, 382–383
Dynamic stereochemistry, 7

E

Eclipsed conformation, 5
Electron-donating group (EDG), 87–88
Electrophilicity, 79–80
Enamine chemistry
    cascade reactions, 338
    enamine catalysis
        Brønsted acid cocatalyst AH, 327
        catalytic cycle efficiency, 327
        chiral 2–substituted pyrrolidine, 327
        enantioselective α-functionalization, 326
        enolizable aldehydes and ketones, 326
        general mechanism, 327
        Hajos–Parrish aldol reaction, 328
        Houk–List transition state, 328–329
        intermolecular Mannich reaction, 329
        MacMillan catalyst, 329–330
        proline, 327–328
        transition state models, electrophilic attack, 327–328
    hirsutene synthesis
        desymmetrization, 330, 333
        diester synthesis, 330–331
        diketone synthesis, 330, 332
        (+)-hirsutene, 330, 333
        intramolecular aldol reaction, 330, 333
        List's retrosynthesis, 330–331
        organometallic/cycloaddition approaches, 330
        radical cyclization, 330–331
        Ru-catalyzed ring closing reaction mechanism, 330, 332
    by Stork, 325–326

Enantiomeric enrichment, *see* Chiral resolution
Enantiomeric excess, 2
Enantiomeric ratio (E), 2
Enantiomers, 1
   L-(+) ascorbic acid, 15
   chirality, 16
   D-Glucose, 15
   one-pot operations, 16–17
   propranolol, 16
   single-enantiomer products, 14–15
   stereopharmacology, 14
   synthesis, 17
Enantioselective enzymatic desymmetrization (EED)
   additives, 218
   double-step kinetics, 211
   ee$_P$ parameter, 211–212
   effect of temperature, 217–218
   free energy difference, 209–210
   optimal pH, 217
   P and Q products, 210
   selectivity, 215–217
   single-step kinetics, 210–211
   stability and reactivity, 212–215
   yield and optical purity, 212
Enantioselectivity, 325
Endo compounds, 7
Ene-reductases
   α-substituted butenedioic esters, 253
   carbon–carbon double bond, 251
   flavoproteins, 251
   nonracemic α-methyl dihydrocinnamaldehyde derivatives, 251–252
   (*R*)-3-hydroxy-2-methylpropanoate, 253–254
   saturated nitriles, 253
   unsaturated aldehydes, 252
Enzymatic asymmetric synthesis
   EED
      additives, 218
      double-step kinetics, 211
      ee$_P$ parameter, 211–212
      effect of temperature, 217–218
      free energy difference, 209–210
      optimal pH, 217
      P and Q products, 210
      selectivity, 215–217
      single-step kinetics, 210–211
      stability and reactivity, 212–215
      yield and optical purity, 212
   hydrolases, 218–219
   KR methods, 206–207
   prochirality *meso* compounds
      achiral center of, 206
      BCL, 206–208
      carbonyl group sides, differentiation, 209
      centrosymmetric substrates, 209–210
      pro-R ester group, 207–208
      $R_1R_2C=X$ molecule, 207, 209
      $R_1R_2CX_2$ molecule, 206, 208
   symmetrical compounds
      α-oxidase, 260–261
      aromatic dioxygenases, 259–261
      CYPs, 255–259
      dehydrogenases, 237–250, 254–255
      ene-reductases, 251–254
      esterases and proteases, 219–223
      lipases, 222–237
      peroxidases, 260–264
Enzymatic membrane reactors (EMR), 393
Enzyme inhibition, 464
Enzymes
   advantages in biotransformations, 177
   lipases, 177
   regioselectivity, 177
   stability, activity, and selectivity, 178
   stereoselectivity, 177
Epilobium angustifolium, 200–201
Epimerization, 2
Epimers, 2–3
Epinephrine, 414–415
Epothilones
   acyloins, 364–365
   enantioselective hydrolysis of *rac*-**63**, 364
   epothilone B, 363–364
   epothilone F, 363–364
   fragment **57** synthesis, 365–366
   ixabepilone, 363
   2-methyl thiazole-containing structures, 364
   myxobacterium *Sorangium cellulosum*, 363
   pantolactone, biocatalyzed kinetic resolution, 367
   patupilone, 363
   sagopilone **56**, 365–366
   total synthesis, 364–365
Equatorial bonds, 5
Escherichia coli (*E. coli*), 387, 390
Ethacrynic acid (**19**), 194
Ethyl (*R*)-4-cyano-3-hydroxybutyrate
   biocatalytic retrosynthetic routes, 410
   bioreductions, 411–412
   (*S*)-4-bromo-3-hydroxybutanoate esters, 411
   enantiopure benzoic ester hydrolysis, 409
   ethyl 4-chloroacetoacetate, 411
   GDH, 411–412
   HN, Codexis synthesis, 411
   HN preparation, 408–409
   (*S*)-3-hydroxy butyrolactone preparation, 408–409
   (*R*)-3-hydroxy-4-cyanobutyric acid, 410
   KRED, 412
   L-malic acid, 409
   nitrilase-catalyzed enzymatic desymmetrization, 410
   prochiral 3-hydroxyglutaronitrile, 410
   racemic ethyl 4-chloro-3-hydroxybutanoate, 409
   S-CHBE, 411
   yeasts, 411
Ethyl (*S*)-4-chloro-3-hydroxybutanoate (*S*-CHBE), 411
1-Etoxyvinyl esters, 224
European Association for Bioindustries (EuropaBio), 346
Exo compounds, 7
Exterior Frontier Orbital Extension (EFOE), 38, 40

F

Felkin–Ahn model, 33–34, 44, 56, 63, 69
Fischer projection, 9
Flagpole interaction, 5
Flurbiprofen, 392
   Aspergillus oryzae MIM, 397
   chemoenzymatic synthesis, 398–399
   enantioselective esterification, 397

ene-reductase YqjM from *Bacillus subtilis*, 398
lipase-catalyzed kinetic resolution, 397–398
optically pure (S)-**162**, 398
racemic flurbiprofenil bromopyrazolide, 398
Food and Drug Administration (FDA), 461
Formate dehydrogenase (FDH), 369
D-Fructose-6-phosphate aldolase (FSA), 170–171
Fusariumoxy sporum, 367
Fusarium proliferatum lactonase gene, 367
Fusarium solani, 223

G

Galactose oxidase (GAO), 171
Gas chromatography (GC), 13
Gauche, 5
Generic drugs, 461
Gene site saturation mutagenesis (GSSM), 410
Geotrichum candidum, 238, 240
Global pharmaceutical market, 346
Glucagon-like peptide-1 mimetics and modulators
30-amino-acid residue peptide, 387
chemoenzymatic dynamic resolution, 389
(S)-2-amino-3-(6-o-tolylpyridin-3-yl)propanoic acid (S)-**148**, 387–389
T2DM treatment, 387
Glucose dehydrogenase (GDH), 369, 411–412
Glutathione *S*-transferase (GST), 194–196, 359
Glycinamide ribonucleotide formyltransferase (GARFT), 372
Green Chemistry
biocatalysis features, 346
EPA, 346
philosophy, 347
US Green Chemistry Program, 346

H

Hajos–Parrish–Eder–Sauer–Wiechert reaction, 326
Hansenula fabianii SC 13894, 360–361
Hansenula polymorpha SC 13865, 360–361
Heterochiral, 9
Hh-dependent medulloblastoma allograft model, 368
High-performance liquid chromatography (HPLC), 13–14
Homochiral, 9
Horseradish peroxidase (HRP), 146
3-Hydroxy-3-methylglutaryl coenzyme A (HMG-CoA) reductase inhibitors, 404–405
(R)-3-Hydroxy-2-methylpropanoate, 253–254

I

Ibuprofen, 392
2-arylpropanals, 395
Candida rugosa lipase, 393–394
carrier-free immobilization techniques, 394
DKR process, 394–395
EMR, 393
ene-reductase YqjM from *Bacillus subtilis*, 396
kinetic resolution processes, 394
lipase-facilitated enantioselective transport, 394
(R)-profen derivatives, 396
rac-ibuprofen methoxyethylester, 393
ruthenium-catalyzed dehydrogenation of alcohol, 394
stereoselective hydrolysis, 395
ICD, *see* Induced circular dichroism

Iminium chemistry, 326
cascade reactions, 338
iminium catalysis
anions derived from MacMillan catalysts, 334–335
catalytic cycle, 333–335
energy diagram, 333–334
Jørgensen–Hayashi catalyst, 333–334
Lewis acid catalysis *vs.* secondary amine, 332, 334
organocatalytic nucleophilic addition, 334–335
paroxetine synthesis
benzyl group deprotection, 336–337
depressive disorder treatments, 335
piperidine core, 336
retrosynthesis, 336
SSRI class, 336
stereochemical outcome, piperidine's synthesis, 336–337
Iminium/enamine organocascades
generalized mechanism, 338
strychnine synthesis
compound **85** synthesis, 341
compound **87** synthesis, 341
Corynanthe alkaloids, 339
decarbonylation, Wilkinson's catalyst, 341
formal Diels–Alder organocascade reaction, 340
modified MacMillan second-generation catalyst, 340
neurotoxin, 339
Pictet–Spengler reaction, 339
protected Wieland–Gumlich aldehyde, 341–342
retrosynthesis by MacMillan, 339
*Strych*nos family, 339
TFA-mediated removal, 341–342
2-vinyl indole, 339–340
Imino- and azasugars, 230–231
Indian tobacco, 232–233
Induced circular dichroism (ICD), 149
Intramolecular coupling reaction, axially chiral biaryls
with chiral leaving groups, 279–280
with chiral *ortho* substituents
aryl Grignard reagents, 277–278
aryl halides, 278
atropisomers P-configured, 278
calphostin, 279
chiral oxazoline moiety, 277
gossypol, 279–280
nucleophilic aromatic substitution, 277–278
S_N2Ar reaction of oxazolinylbenzene derivative, 277–278
(M)-steganone, 278
Ullmann homocoupling, 279
with chiral tethers
aryl halides, 274
biaryl diacids, Ullmann coupling, 274–275
(M)-binol, 274
calphostin, 276–277
diethers, 276
intramolecular aromatic nucleophilic substitution, 274
lignin, 276–277
oxidative coupling reaction, 276
with element of planar chirality, 279, 281
Intramuscular injection, 473
Intrathecal administration, 473

Index

(+)-Isokotanin A-bicoumarin, 287
Isoprenaline, 414–415
Isopropenyl acetate (IPA), 224
Isozymes, 194

J

Jeffery–Heck cyclization sequence, 341

K

Ketoprofen, 396–397
Ketoreductase (KRED), 377–378, 412
Ketorolac
    acylation and ulterior hydrolysis, 402
    enantioselective esterification, *C. antarctica* lipase B, 403–404
    first asymmetric synthesis, 403
    ketorolac and ketorolac tromethamine, 402
    Oppolzer's sultam, 403
    S-eutomer, 403
    stereoselective cyclization, 403
    Streptomyces griseus protease, 403–404
    third-generation synthesis, 402–403
Kinetic resolution (KR) methods, 9, 206–207
Klebsiella pneumoniae IFO 3319, 247

L

Lactate dehydrogenase (LDH), 369
Lactobacillus brevis, 413
Lactobacillus kefiri, 238
Lactobacillus ketoreductase, 377
L-alanine dehydrogenase (LADH), 369
Leustroducsin B, 227
Levetiracetam, 454–455, 458
Levorotatory, 1
Limonene, 16
List–Houk model, 327–329
(−)-Lobeline, 232–233
Lonafarnib, 378–380
Lovastatin, 405–407

M

Medicinal chemistry route
    aprepitant, 454–455
    levetiracetam, 455, 458
    paroxetine, 454, 457
    pregabalin, 456, 458
    sertraline, 456–457, 459
    simvastatin, 454, 456
    sitagliptin, 449, 452–453
Meso compound, 3–4
Meso diols, 230
Metal-free catalysis, 325
Metal ion complexes, 325
Methyl diazoacetate, 80–81
Methyl vinyldiazoacetate, 80–81
Mevacor, 405
Mevastatin, 405
Mevinolin, 405
Microbacterium campoquemadoensis (MB5614), 247
Minimum inhibitory concentration (MIC), 192–193

Mithramycin (MTM)
    lipase-catalyzed acylation, 369–371
    structure, 369–370
Mitsunobu stereoinversion, 183
Monastrol, 374–375
Monopril, 426
Mucor hiemalis, 247
Mucor javanicus, 238

N

Nanocapsules, 471
Nanospheres, 471
Naphthalene dioxygenase (NDO), 259–261
Naproxen, 392
    AMDase, 401–402
    2-aryl-2-methyl malonate ester desymmetrization, 401–402
    azolides ((*R*,*S*)-*N*-profenyl-1,2,4-triazoles), 401
    BsENP01, 400
    Candida antarctica, 400
    Carica papaya lipase, 400
    Carica pentagona Heilborn, 400
    ChiroCLEC-CR, 400
    cinchonidine, 399
    DKR, 401
    enantioselective esterification, 400
    first member, profen family, 398
    hydrolase-catalyzed resolutions, 399
    hyperthermophile SsADH, 401
    irreversible enantioselective esterification, 400–401
    2-methoxy-6-vinylnaphthalene hydroformylation, 399
    naphthacrylic acid, 399
    racemic alkyl 2-(6-methoxynaphthalen-2-yl) propanoate, 399
    redox bioprocesses, 401
    S-eutomer, 399
    (*S*)-selective arylmalonate decarboxylase variant, 402
    Zambon process, 399
Nasal administration, 468–469
Nauclea latifolia, 194–195
New chemical entities (NCE), 348
Nicotinamide adenine dinucleotide (NADH), 237
Nifenalol, 415–416
Niraparib
    asymmetric synthesis, 381–382
    ATA-301, 381
    DKR
        aldehyde, 381–382
        bisulfite, 382
        lactol, 382–383
    PARP inhibitors, 381
    transaminases, 381
N-methyl-2-pyrrolidone (NMP), 147–148
Nocardia autotrophica, 257
Noncommunicable diseases (NCDs), 348
Noncompetitive antagonism, 464
Nonspecific interactions, 464
Norepinephrine, 414–415

O

Odanacatib, 380–381
Old Yellow Enzyme (OYE), 385
Onglyza®, *see* Saxagliptin

Optically active polyanilines (PAn)
　anilines
　　chirality-organized structures, 145–146
　　chiral nanotubes, 146
　　chiroptical properties, 144
　　2,3-dichloro-5,6-dicyanobenzoquinone, 145
　　emeraldine salts, 143–144
　　HRP, 146
　　2-methoxyaniline-5-sulfonic acid, 144–145
　　PEOA, 145
　　sulfonated fibers, 146–147
　　synthesis, 146–147
　application, 156
　chiral groups
　　chiral ether substituents, 149
　　covalent attachment, 149
　　neutral emeraldine base, 148
　　phenylenediamine derivative, 152–154
　　π-conjugated moieties, 149–151
　　syn-conformation, 154
　emeraldine bases
　　chiral complexation, 154–156
　　chiral CSA–anion, 147–148
　　chiroptical properties, 148
　　organic solvents, 147
Oral administration, 468
Organocascade reactions, 338–342
Organocatalysis
　enamine chemistry
　　cascade reactions, 338
　　enamine catalysis, 326–330
　　hirsutene synthesis, 330–333
　　by Stork, 325–326
　Hajos–Parrish–Eder–Sauer–Wiechert reaction, 326
　iminium chemistry, 326
　　cascade reactions, 338
　　iminium catalysis, 332–335
　　paroxetine synthesis, 335–337
　renaissance, 326
Ortho-bromobenzoic acid, 286

P

Paclitaxel
　Abraxane, 349
　β-lactam, 351
　Bristol-Myers Squibb (BMS), 349, 352
　C-7 xylosidase, 353
　C-10 acetate, 353
　C-10 deacetylase, 353–354
　C-13 lateral chain, 351
　C-13 paclitaxel side chain, 353
　C-13 taxolase, 353–354
　docetaxel, 350, 357
　esterification of baccatin III, 350–351
　nucleus, enzymatic reaction, 353
　PCF technology, 362
　plant cell cultures, 362
　semisynthetic strategy, 350, 352
　S. matensi, 353–354
　structure, 349
　structure modifications, 357
　sustainability, 362
Taxus brevifolia, 350
total syntheses, 350
Xyotax, 349
Palladium(II), 155–156
Palladium-catalyzed asymmetric allylic alkylations
　allyl enol carbonates, 297
　α-2-propenyl benzyl motifs, 301–302
　aryl propionic acid NSAIDs, 301, 303
　(–)-aspewentin, 300
　with benzylic nucleophiles, 301, 303
　β-keto esters, 297–299
　bicyclic ketone, 299
　C–C bond-forming process, 297
　(–)-cyanthiwigin F, 299
　cycloalkenones, 301, 303–304
　cyclohexadione, 299
　cyclohexanone-derived natural products, 304
　decarboxylative asymmetric allylic alkylation, 297–298
　enantioenriched α-arylalkanoic acid, 301, 303
　enantioselective double allylic alkylation, 299–300
　general mechanism, 297–298
　hard nucleophiles, 301
　(+)-harveynone, 304
　(–)-harveynone, 304
　hydrogen bond activation, allylic ethers, 300–302
　Lewis acids, 300
　meso-dibenzoates, 301
　nonsteroidal anti-inflammatory drug (NSAID) analogue, 301
　nucleophile, 300
　π-allylpalladium intermediate, 297
　pyrrolidine, 300
　(2R,5R)-2,5-diallyl-2,5-dimethylcyclohexane-1,4-dione, 299
　soft carbon and heteroatom nucleophiles, 300
　softened hard pronucleophiles, 301–302
　vinyl triflate, 299
Palladium-catalyzed carbonylation reactions
　acylpalladium species, 311
　alkenyl-palladium(II) intermediate, 312–313
　alkoxycarbonylation reactions, 313–314
　aminocarbonylation reactions, 311
　androst-16-ene-17,17'-dicarboxamides, 311–313
　(–)-blennolide C, 318
　(+)-boronolide
　　δ-lactone, 313, 315
　　retrosynthetic approach, 313–314
　　synthesis, 313–314
　chroman, 319
　(+)-deacetylboronolide, 313–314
　9-demethylneopeltolide
　　epimer, 317
　　macrolactone, 317
　　Mitsunobu reaction, 316
　　retrosynthetic analysis, 316
　　synthesis *via* alkoxycarbonylative macrolactonization, 316–317
　17-iodosteroids, 312–313
　macrocyclic structural motifs, 315
　(+)-monocerin
　　cis-fused tetrahydrofuran, 314
　　intramolecular alkoxycarbonylation, 314–316
　　retrosynthetic analysis, 314–315
　　tandem dihydroxylation-SN2 cyclization sequence, 314

Index

natural α,β-unsaturated δ-lactones, 313
(−)-neopallavicinin, 318
organic halides, 311–312
oxypalladation mechanism, 317
palladium-catalyzed domino Wacker/
　　methoxycarbonylation, 318–319
palladium(II) complex, 311
(+)-pallambin B, 318
steroidal dicarboxamides, 311–312
Palladium-catalyzed reactions, organic synthesis
　alkynes, amines, and alkenes functionalization, 295
　asymmetric allylic alkylations, 297–304
　asymmetric intramolecular cyclization reactions,
　　304–311
　carbonylation, 311–319
　C–C, C–H, and C–heteroatom bonds, 293
　citations, WEB OF SCIENCE, 293–294
　coupling reactions
　　catalytic cycle, 295
　　designations, 293–295
　　general mechanisms, 295–296
　　triphenylphosphine, 296
　diversity, 295
　economic and ecological advantages, 293
　ligand properties, 296
　phosphine-based ligands, 296–297
　retrosynthetic approach, 297
　σ-donor and π-acceptor properties, 296
　stereo-inductive effects, 296
　steric effects, 296
　Tolman's cone angle values, 296–297
　2010 Chemistry Nobel Prize laureates, 293–294
Pancreatic β cells, 387
Paroxetine, 219, 221, 454, 457
Pelitrexol
　cell proliferation, 372
　chemical synthesis
　　diastereomers, chromatographic separation of, 373
　　enzymatic kinetic resolution, 373–374
　　first-generation process, 372
　　retrosynthetic analysis, 373
　GARFT, 372
Penicillium glaucum, 11
Peptides
　organic molecular scaffold
　　β-sheet structure, 138–139
　　cyclic peptide, 135–137
　　epindolidione scaffold **5**, 135
　　intramolecular interaction, 133–134
　　rigid aromatic spacer, 135–137
　　urea unit, 137–138
　organometallic ferrocene
　　1′-aminoferrocene-1-carboxylic acid (Fc-ac),
　　　142–143
　　circular dichroism, 140–141
　　dipeptide chain, 140–141
　　ferrocene **16**, 140
　　hydrogen-bonding ability, 138–139
　　inter-ring spacing, 138
　　1,n′-diaminoferrocene (Fc-aa), 142–143
　　P- and *M*-helical conformations, 140
　　P-helical chirality, 141–142
　PAn's, 134–135
　secondary structures, 133–134

Peroxisome proliferator–activated receptors (PPARs)
　AZD 4619, 383–384
　chymotrypsin-catalyzed hydrolysis, 385
　enzymatic kinetic resolution
　　ethyl-(*S*)-2-ethoxy-3-(*p*-methoxyphenyl)propanoate
　　　(EEHP) (*S*), 385–386
　　rac-**137**, 385
　　rac-**139a** and *rac*-**139b**, 385–386
　glitazars, 383–385
　metabolic disorder treatment, 383
　methyl (*S*)-2-bromobutyrate and (*S*)-2-bromobutyric
　　acid, 385, 387
　racemic thioester, 383
　saroglitazar, 386
　SFPR technique, 385
Pharmacokinetic differences, 15
Phenylethylthiazoylthiourea (PETT) analogue **93**,
　116, 118
Pichia pastoris, 390
Pichia wickerhamii, 365
Pig liver esterase (PLE), 213
Pig pancreas lipase (PPL), 185
π-plane-divided accessible space (PDAS), 38, 40
p-methoxyphenyl group (PMP), 171
Podophyllum peltatum, 234
Poly(ADP-ribose)polymerase (PARP) inhibitors, 381
Polyanilines (PAn's), 134–135
Poly(*o*-ethoxyaniline) (PEOA), 145
Polyhydroxylated pyrrolidines, 230–231
Polyketones, 247–248
Poly(*o*-toluidine) (POT), 154–155
Porcine pancreatic lipase (PPL), 225
Pravastatin, 405
Prazosin, 414–415
Preferential crystallization, 10
Pregabalin, 456, 458
Prolinate-based ligands
　alkynyldiazoacetates, 86–87
　allyl vinyldiazoacetate, 91
　asymmetric cyclopropanations, 87–88, 91
　Davies complexes, 84–85
　enantioselective cyclopropenation, 89–90
　heteroaryldiazoacetates, 85, 87
　late transition state, 89–90
　p-bromophenyldiazoacetate, 89
　pyrroles/furans, 87–88
　Rh$_2$(*S*-DOSP)$_4$, 84–86
　second-generation, 89, 91
　Su prolinate-based complexes, 91–92
　tamoxifen, 84, 86
Pronethanol, 415–416
Propranolol
　DKR, 423–424
　kinetic resolutions, 421–422
　N-acetylation, 422
　O-acetylation with vinyl acetate, 422
　(*S*)-propranolol, 423
Pseudomonas aeruginosa, 223
Pseudomonas cepacia, 225, 234–235, 237
Pseudomonas cepacia lipase (PCL), 218
Pseudomonas fluorescens lipase (PFL),
　224, 228
Pseudomonas plecoglossicida, 353
Pseudomonas sp. lipase (PSL), 225

R

Racemic temperature (T$_{rac}$), 217–218
Ragaglitazar, 383–384
Receptor tyrosine kinase (RTK), 375
Rectal administration, 469
Redox switching, 154
Regio- and enantioselective preparation
    chiral amine synthesis, 183
    chiral Me-imidacloprid, 184
    DKR process, 183
    enzyme–ruthenium organometallic combo DKR, 184
    KR process, 183
    norsertraline, 184
    racemic aryl amines, DKR, 185
    racemic mixture, 182
Regioselective artificial hydrolase-catalyzed C–C bond formation
    CAL-B-catalyzed decarboxylative aldol reaction, 186
    CAL-B-catalyzed decarboxylative Knoevenagel reaction, 186
    catalytic promiscuity, 185
    Michael addition reaction, 186–187
    non-natural substrate-promoted catalytic promiscuity, 185
    PPL, 185
    Suzuki–Miyaura reaction, 186–187
Regioselective preparation
    monodeprotected nucleosides, 180, 182
    peracetylated carbohydrates
        β-O-napthylmethyl-lactosamine peracetate, 180–181
        glycoconjugates, 178
        immobilization effect, hydrolysis, 178–179
        monosaccharides, 180
        peracetylated glucal, 180
        regioselective biocatalytic process, 178–179
        stereoselective glycosylation, 178
        thiol–disulfide exchange ligation, 180
Renin–angiotensin system (RAS), 201–202
Rhizomucor miehei, 393
Rhizomucor miehei lipase (RML), 232
Rhodotorula mucilaginosa (CBS 2378), 209
Rhodotorula rubra, 240
Rhodotorula species, 377
Ring-closing metathesis (RCM), 171–172
Roche ester, 253–254

S

Saccharomyces cerevisiae, 212, 358–359
Saridegib, 368
Saxagliptin, 390–391
Seebach–Eschenmoser model, 327–328
Selective serotonin reuptake inhibitor (SSRI), 336
Sertraline, 456–457, 459
Simvastatin, 257, 406–408, 454, 456
Sitagliptin, 389–390, 449, 452–453
Smart drug delivery, *see* Targeted drug delivery system
Sorangium cellulosum, 363–364
Sotalol, 415–416
Sphaeranthus indicus, 192
Sphingomonas sp. cells, 255
Sporosarcina ureae, 387
Stereogenic center, 1
Stereomutation, 6–7
Stereopharmacology, 14
Stereoselective reaction, 7–8
Stereospecific reaction, 7–8
Steric acceleration, 5
Steric assistance, 5
Steric model, enamine, 327–328
Streptomyces matensi (*S. matensi*), 353
Subcutaneous injection, 473
Substrate feeding product removal (SFPR) technique, 385
Sulfolobus solfataricus (SsADH), 401
Supercritical fluid chromatography (SFC), 13
Swertia corymbosa, 199–200
Symmetrical compounds
    α-oxidase, 260–261
    aromatic dioxygenases, 259–261
    CYPs
        Acinetobacter sp. NCIB 9871 CHMO, 259
        BVMO, 259
        E. coli BL21 (DE3)/pMM4 CHMO, 258–259
        examples, 255–256
        3-hydroxyisobutyric acid, 256–257
        mutant enzymes BM-3, 256
        prochiral olefins, 257–258
        reactions, 255
        simvastatin, 257
    dehydrogenases
        alcohol dehydrogenases, 237–238
        aliphatic and aromatic ketonitriles, 250
        baker's yeast, 238–239
        β-bromoalcohols, 240
        bulky ketones, 242
        carrot root, 240
        cofactor recycling, 237–238
        desymmetrization, 242, 244
        3,5-diketo-6-(benzyloxy)hexanoic acid ethyl ester, 248, 249
        EED, 247–249
        examples, 242–243
        Geotrichum candidum, 240
        isolated enzymes, 242
        ketoesters, 245–247
        Klebsiella pneumoniae IFO 3319, 247
        meso and prochiral diols, 254–255
        microorganisms, 240–241
        polyketones, 247
        stereochemical outcome, 240
        trifluoromethyl-substituted alkyl aryl ketones, 240
        tropinone reductase-I, 245
    ene-reductases
        α-substituted butenedioic esters, 253
        carbon–carbon double bond, 251
        flavoproteins, 251
        nonracemic α-methyl dihydrocinnamaldehyde derivatives, 251–252
        (*R*)-3-hydroxy-2-methylpropanoate, 253–254
        saturated nitriles, 253
        unsaturated aldehydes, 252
    esterases and proteases, 219–223
    lipases
        ACC, 225
        active site, 223–224

Index

acyl group transfer, 222–223
aquayamicin, 233–234
chemoenzymatic processes, 234, 236
cyclic alcohols, 230–231
five-membered heterocycles, 230–231
hydrophobic helical lid, 223
(–)-lobeline, 232–234
meso cis-cyclohexane-1,3-dicarboxylic acid diesters, 234–235, 237
meso 5-t-butyldimethylsilyloxy-2-cyclopentene-1,4-diol, 231–232
N-protected serinol derivatives, 225
pentane-1,5-diol derivatives, 228–229
PFL, 224
piperidine ring, 232–233
primary meso diols, 233, 235
prochiral 3-substituted glutarates, 228–229
propane-1,3-diol derivatives, 225–227
rifamycin S, 230
2-substituted glycerine derivatives, 225, 228
tetracyclic meso diacetates, 234, 236
peroxidases, 260–264
Sympatholytic drugs, 415
Syn-elimination, 6–7

T

Targeted drug delivery system
  biological drugs, 473–474
  disadvantages, 470
  injection, 473
  liposomes, 472
  nanotechnology, 471
  objectives, 470
  resealed erythrocytes, 472–473
Tautomerism, 4–5
Taxanes
  1-benzamido-3-ethoxy-3-oxo-1-phenylprop-1-en-2-yl ethyl oxalate bioconversion, 361–362
  bioreduction
    azetidinedione rac-13, 358
    enantioselective preparation, 357–358
    glucose-6-phosphate/glucose-6-phosphate dehydrogenase couple, 360
    rac-13, 359
    rac-41, 361
  diastereoselectivity, 362
  E. coli, 359
  enantioselective microbial reduction, 360
  fatty acid synthase, 359
  GST fusion proteins, 359
  lipase
    azetidinones, 354–357
    BMS lipase, 356
    CAL-B, 354
    Celite, 354
    cis-10 opening, 354–355
    hydrolysis, 356–357
    3-hydroxy-4-phenyl-β-lactam derivatives, 354
    immobilized lipases, 356
    lipase PS-30, 356
    ring-opening reactions, 354
    (3R,4S)-23, 356
    oxoester, 361–362
  paclitaxel
    Abraxane, 349
    β-lactam, 351
    Bristol-Myers Squibb (BMS), 349, 352
    C-7 xylosidase, 353
    C-10 acetate, 353
    C-10 deacetylase, 353–354
    C-13 lateral chain, 351
    C-13 paclitaxel side chain, 353
    C-13 taxolase, 353–354
    docetaxel, 350, 357
    esterification of baccatin III, 350–351
    nucleus, enzymatic reaction, 353
    PCF technology, 362
    plant cell cultures, 362
    semisynthetic strategy, 350, 352
    S. matensi, 353–354
    structure, 349
    structure modifications, 357
    sustainability, 362
    Taxus brevifolia, 350
    total syntheses, 350
    Xyotax, 349
  pharmacology-interesting taxanes, 357–358
  semisynthetic process, 362
  stereoselective reduction
    α-chloro-β-oxo ester precursor, 361
    azetidinedione rac-13, 359–360
    whole-cell-catalyzed stereoselective reduction, rac-13, 360
  syn- and anti-stereoisomers, 361–362
  yeast genome sequence, 359
Taxol®, see Paclitaxel
Thermoactinomyces intermedius, 390
Thermoanaerobium brockii (ADH-TB), 242
Thin-layer chromatography (TLC), 13
Toluene dioxygenase (TDO), 259–261
Toyobo LIP-300, 379
Transdermal delivery systems, 470
Triage, 10
Trigonopsis variabilis, 387
3,4,5-Trihydroxypiperidines
  asymmetric syntheses, 171–173
  bio-catalyzed syntheses, 170–171
  chiral pool, syntheses
    allylic alcohol 42, 168–169
    bis-epoxides, 166–168
    5-bromo-5-deoxy-D-arabinonolactone 15, 165–166
    bromopyranose sugar intermediates, 168
    chemical synthesis, 170
    iminosugar 1, 163–164
    pyrrolidine and piperidine, 170
    ring expansion rearrangement, 165–166
    D-Threo 48, 169–170
    D-xylose derivatives, 164–165
  configurations, 163–164
Trimethylsilylethyl (TMSE), 89
Triphenyl phosphine (TPP), 172

U

U.S. Environmental Protection Energy (EPA), 346

V

Valence tautomerism, 4–5
Vancomycin
   antibiotic heptapeptide, 271
   asymmetric Suzuki cross-coupling, 283
   structure, 271–272
Veratrum californicum, 367
Vinyl acetate (VA), 224

W

White Biotechnology; *see also* Green Chemistry
   biotechnological tools, 346
   EuropaBio, 346
   substantial gains, 346
Wilson's disease, 15

X

XALKORI®, *see* Crizotinib

Z

Zimmerman–Traxler cyclic transition, 328, 329
Zocor®, *see* Simvastatin
Zofenopril, 425
Zymomonas mobilis (NCR), 251